S. G. Krantz

S0-ACJ-875

Stanisław Łojasiewicz

Introduction to Complex Analytic Geometry

Translated from the Polish
by Maciej Klimek

1991   Birkhäuser Verlag
Basel · Boston · Berlin

Author's address:

Dr. Stanisław Łojasiewicz
Jagiellonian University
Department of Mathematics
ul. Reymonta 4
PL-30-059 Cracow (Poland)

Originally published as:
Wstęp do geometrii analitycznej zespolonej
© PWN – Państwowe Wydawnictwo Naukowe, Warszawa, 1988

Deutsche Bibliothek Cataloging-in-Publication Data

Łojasiewicz, Stanisław:
Introduction to complex analytic geometry / Stanisław Łojasiewicz. Transl. from the Polish by Maciej Klimek. – Basel; Boston; Berlin: Birkhäuser, 1991
Einheitssacht.: Wstęp do geometrii analityczmei zespolonej ‹engl.›
ISBN 3-7643-1935-6 (Basel ...) Gb.
ISBN 0-8176-1935-6 (Boston) Gb.

This work is subject to copyright. All rights are reserved, whether the whole or part of the material is concerned, specifically those of translation, reprinting, re-use of illustrations, broadcasting, reproduction by photocopying machine or similar means, and storage in data banks. Under § 54 of the German Copyright Law, where copies are made for other than private use a fee is payable to «Verwertungsgesellschaft Wort», Munich.

© 1991 for the English edition: Birkhäuser Verlag Basel
Printed from the translator's camera-ready manuscript on acid-free paper in Germany
ISBN 3-7643-1935-6
ISBN 0-8176-1935-6

*Without optimism one cannot
prove a theorem.*

Aldo Andreotti

PREFACE

TO THE POLISH EDITION

The subject of this book is analytic geometry understood as the geometry of analytic sets (or, more generally, analytic spaces), i.e., sets described locally by systems of analytic equations ([1]). Except for the last chapter, mostly local problems are investigated and, throughout the book, only the complex case is studied. From the purely geometric point of view, the real case is more natural and more general. But it displays fewer regularities and – by and large – the corresponding theory is more difficult. The complex structure is richer. Hence one can expect deeper results. Indeed, some phenomena, such as analyticity of the set of singular points (see IV. 24) or analyticity of proper images (Remmert's theorem, see V. 5.1), do not have counterparts in the real case. More than anything else, the beauty of the interplay between the geometric and algebraic phenomena constitutes the main attraction of the "complex" theory ([2]).

This book should be regarded as an introduction. It does not pretend to reflect the entire theory. Its aim is to familiarize the reader with the basic range of problems, using means as elementary as possible. They belong to complex analysis, commutative algebra, and set topology (the methods of algebraic topology have not been employed), and are gathered in the first three preliminary chapters. The author's intention was to provide the reader with access to complete proofs without the need to rely on so called "well-known"

([1]) More or less, until the 50's, the name "analytic geometry" had been associated with an undergraduate level course dealing with the use of Cartesian coordinates in the study of linear and quadratic objects.

([2]) To an even greater degree this is true in the case of algebraic geometry (i.e., in simple terms, the geometry of sets defined by systems of polynomial equations). There, the geometric and algebraic aspects are interconnected in almost every problem.

facts. An elementary acquaintance with topology, algebra, and analysis (including the notion of a manifold) is sufficient as far as the understanding of this book is concerned. All the necessary properties and theorems have been gathered in the preliminary chapters - either with proofs or with references to standard and elementary textbooks.

The first chapter of the book is devoted to a study of the rings $\mathcal{O}_a$ of holomorphic functions. The notions of analytic sets and germs are introduced in the second chapter. Its aim is to present elementary properties of these objects, also in connection with ideals of the rings $\mathcal{O}_a$. The case of principal germs (§5) and one–dimensional germs (Puiseux theorem, §6) are treated separately.

The main step towards understanding of the local structure of analytic sets is Rückert's descriptive lemma proved in Chapter III. Among its consequences is the important Hilbert Nullstellensatz (§4).

In the fourth chapter, a study of local structure (normal triples, §1) is followed by an exposition of the basic properties of analytic sets. The latter includes theorems on the set of singular points, irreducibility, and decomposition into irreducible branches (§2). The role played by the ring $\mathcal{O}_A$ of an analytic germ is shown (§4). Then, the Remmert–Stein theorem on removable singularities is proved (§6). The last part of the chapter deals with analytically constructible sets (§7).

The fifth chapter is concerned with holomorphic mappings between analytic sets. To begin with, the theorem on multiplicities and Rouché's theorem are proved for mappings between manifolds of the same dimension (§2). Next – theorems concerning dimension of fibres are shown in the general case (§3). After the introduction of (reduced) analytic spaces (via atlases) and extending properties of analytic sets (onto these spaces), Remmert's proper and open mapping theorems are proved (§§5 and 6). Finally, the Andreotti–Stoll theorem on the structure of finite mappings is presented (§7).

Normal spaces and the normalization theorem are the main topic of the sixth chapter. A necessary tool here is the Cartan–Oka theorem (§1), which is proved without introducing the full notion of an analytic sheaf.

The seventh chapter is connected with ideas of the well-known paper "GAGA" by Serre [38] about the phenomenon of the "necessary" algebraicity of analytic objects in projective spaces. Serre's theorem (which requires the use of the theory of analytic sheaves and cohomologies) is not presented though. After an elementary discussion of the manifold structure on the projective and Grassmann spaces (§§2 and 4), blowings-up (§5) are introduced. The projective and Grassmann spaces are also investigated later on in §16 as algebraic manifolds. Next, Chow's theorems on algebraicity of analytic sets (§6) as well as Rudin–Sadullaev criteria of algebraicity in $\mathbb{C}^n$ are proved (§7). Then constructible sets are introduced and Chevalley's theorem is shown.

The algebraic version of Rückert's lemma (§9) is followed by the Hilbert Nullstellensatz (§10), the theorem about the degree of algebraic sets (§11), and Bézout's theorem (§13). One part of the chapter is devoted to a study of meromorphic and rational functions. After a series of characterizations of such functions, the Siegel–Thimm theorem, the Hurwitz theorem, and a theorem of Zariski on constructible graphs are proved. The last theorem is used in the proof of Serre's algebraic graph theorem (§16). Then, Zariski's theorem on analytic normality is shown (§16). Subsequently, algebraic spaces ("variétés algébriques" in the sense of Serre) are introduced and Chow's theorem for such spaces is proved. Hironaka's example of a non-projective algebraic variety is given (§17). The purpose of the last part of the chapter is the exposition of Chow's characterization of biholomorphic mappings of Grassmann manifolds (§20). Its proof is based on the theorem on biholomorphic mappings of factorial sets (§18) and on the Andreotti–Salmon theorem saying that Grassmann's cone is factorial (§19).

While working on this material, I have used many sources, particularly the books by Bochner and Martin [11], Hervé [24], Narasimhan [33], and Whitney [44]. Among them, the last one is the closest in character to this book.

I am grateful to my colleagues for numerous valuable comments. Above all, I would like to thank Kamil Rusek and Tadeusz Winiarski. Their enthusiasm, encouragement, and stimulation certainly played an important role in the writing of this book. Also, I would like to express my gratitude to Prof. Stanisław Balcerzyk and Prof. Krzysztof Maurin, as well as to Sławomir Cynk, Zbigniew Hajto, Marek Jarnicki, Grzegorz Jasiński, Krzysztof Kurdyka, Wiesław Pawłucki and Piotr Tworzewski from Cracow, Adam Parusiński from Gdańsk, Piotr Pragacz from Toruń, Fabbrizio Catanese and Arturo Vaz Ferreira from Pisa, Hans–Jörg Reiffen from Osnabrück, Jean-Jaques Risler from Paris, and M. Umemura from Nagoya.

I wish to thank Maria Rataj for her dedicated work in the preparation of the typescript.

Also, I am most grateful to Jaś Ciaptak-Gąsienica and Hania Fronek–Gąsienica for creating for me a very good environment for the writing of this book.

Krzeptówki, September 1983   Stanisław Łojasiewicz.

PREFACE
TO THE ENGLISH EDITION

The text of the English edition has been slightly extended. Among other changes, Section 7 of Chapter V (on finite holomorphic mappings) has been modified and extended by the addition of the Grauert–Remmert formula (the primitive element theorem) with some of its consequences. Furthermore, the Grauert–Remmert theorem on uniform convergence (with a simple proof following from an observation due to S. Cynk), preceded by Cartan's closedness theorem, has been included in Section 1 of Chapter VI. Some facts related to complete intersections have been added in Chapter VI (on normalization); in particular: Schumacher's, Tsikh's, and Oka–Abhyankar's theorems, as well as Serre's normality criterion. Naturally, this has required a series of modifications in the preceding chapters.

I would like to convey here my sincere thanks to Professors Edward Bierstone, Sean Dineen, and Stephen G. Krantz for reading the text and for their valuable remarks.

I wish to thank very warmly the staff of Birkhäuser Verlag for their special effort which they put into this English edition.

Finally, many thanks should go to Dr Maciej Klimek who has undertaken the translation of the Polish edition with the new supplements.

Kraków, 21 March 1991   Stanisław Łojasiewicz.

Górole, górole, góralska muzyka,

Cały świat obyńdzies, nima takiej nika!

CONTENTS

PRELIMINARIES

COMPLEX ANALYTIC GEOMETRY

PRELIMINARIES

CHAPTER A

ALGEBRA

§1. Rings, fields, modules, ideals, vector spaces

1. By a *ring* we will mean a commutative ring with identity. We will be assuming that all *subrings* contain the identity, all *ring homomorphisms* preserve the identity and all modules satisfy the condition $1x = x$. By a *field* we will mean a commutative field. In any non-zero ring and in any field, 1 is not equal to 0 ([1]) . If $\mathbf{K}$ is a field, then *vector spaces* over $\mathbf{K}$ are defined as modules over $\mathbf{K}$ and *linear mappings* between them are just homomorphisms of modules. The dimension of a vector space X over $\mathbf{K}$ is denoted as $\dim X$ or $\dim_{\mathbf{K}} X$.

By $\mathbf{C}$, $\mathbf{R}$, $\mathbf{Q}$, $\mathbf{Z}$, $\mathbf{N}$ we will denote the fields of complex, real, and rational numbers, the ring of integers, and the set of natural numbers, respectively.

Every commutative group is a module over $\mathbf{Z}$. Its subgroups coincide with submodules over $\mathbf{Z}$ (and its quotient groups with quotient modules). Homomorphisms of commutative groups coincide with homomorphisms of modules over $\mathbf{Z}$.

The symbol $\approx$ will denote isomorphism (for rings, fields, modules over the same ring).

The equivalence class of an element x of a ring (module) in a quotient ring (module) will usually be denoted by $\bar{x}$ (provided this does not cause any confusion).

Any ring (field) containing a ring (field) A as a subring (subfield) is called an *extension of the ring (field)* A. By an *A–homomorphism* (*A-isomorphism*)

([1]) A module over the zero ring contains only the zero element.

we mean any homomorphism (isomorphism) $h : B \longrightarrow C$ of extensions B and C of the ring A, such that $h(x) = x$ for $x \in A$. If there is an A–isomorphism between B and C, we say that B and C are A–isomorphic and we write $B \overset{A}{\approx} C$.

The *Cartesian product* $A \times B$ of rings A and B furnished with the operations

$$(x, y) + (x', y') = (x + x', y + y')$$

and

$$(x, y)(x', y') = (xx', yy')$$

is a ring. The same is true for Cartesian products of finite families of rings.

2. Since the intersection of an arbitrary family of subrings (subfields) of a ring (field) A is a subring (subfield), for any subset $E \subset A$ there exists a *subring (subfield) generated* by E, i.e., the smallest subring (subfield) containing E.

3. Let $f : A \longrightarrow B$ be a ring homomorphism. Then the inverse image of an ideal of B is an ideal of A. Moreover, if f is an epimorphism, then the image of an ideal in A is an ideal in B. If I, J are ideals of the rings A, B, respectively, and $f(I) \subset J$, the induced homomorphism $A/I \ni \bar{u} \longrightarrow \overline{f(u)} \in B/I$ is well-defined. Its kernel is the set $f^{-1}(J)/I$. In particular, if $I \subset \ker f$, then we have the induced homomorphism $A/I \ni \bar{u} \longrightarrow f(u) \in B$. If f is an epimorphism, then the induced homomorphism $A/\ker f \longrightarrow B$ is an isomorphism, i.e., $A/\ker f \approx B$.

4. Let A be a ring. Obviously, A is a module over A and the ideals of the ring A are the same as the submodules of the module A.

Let M be a module over A. If $G, H \subset M$, then $G + H$ denotes the set $\{x + y : x \in G, y \in H\}$. If $E \subset A$ and $G \subset M$, then EG denotes the set $\left\{ \sum t_i x_i : t_i \in E, x_i \in G \right\}$ $(^2)$. If E is an ideal or G is a submodule, then EG is a submodule. In particular, for $E, F \subset A$, we have $E + F = \{t + u : t \in E, u \in F\}$ and $EF = \left\{ \sum t_i u_i : t_i \in E, u_i \in F \right\}$. If E is an ideal, then so is EF. (The above set operations are associative and commutative; we have $\sum_1^k G_i = \left\{ \sum_1^k x_i : x_i \in G_i \right\}$ for $G_i \subset M$, and $E_1, \ldots, E_k = \left\{ \sum_\nu t_{1\nu} \ldots t_{k\nu} : t_{i\nu} \in E_i \right\}$ for $E_i \subset A$.)

$(^2)$ In particular, we have $sG = \{sx : x \in G\}$ for $s \in A$, and $Ez = \{tz : t \in E\}$ for $z \in M$.

We will often use the following implication:

$$G \subset N \subset G + Z \Longrightarrow N \subset G + (Z \cap N),$$

where G and Z are subsets and N is a submodule of a module M.

We will say that $M = N + L$ is a direct sum of submodules L and N if $L \cap N = 0$, i.e., $L \times N \ni (x,y) \longrightarrow x + y \in M$ is an isomorphism.

Note also that if $f : M \longrightarrow L$ is a homomorphism of modules over A, then $f(G + H) = f(G) + f(H)$ and $Ef(G) = f(EG)$. In particular, $E(M/N) = (EM+N)/N$ when N is a submodule of the module M (see n° 10 below). If $f : A \longrightarrow B$ is a ring homomorphism, then $f(E+F) = f(E)+f(F)$ and $f(EF) = f(E)f(F)$.

5. If I and J are ideals (of a ring A), then so are $I + J$, $I \cap J$, and IJ (in particular, $I^n = \underbrace{I \ldots I}_{n \text{ times}}$ for $n > 0$ is an ideal; we assume that $I^0 = A$). If I is an ideal, then so is its *radical* rad $I = \{t \in A : t^k \in I$ for some $k\}$. Clearly, rad $I \supset I$. Also, rad (rad I) = rad I and rad $(IJ) = $ rad $(I \cap J) = ($rad $I) \cap$ (rad J). Thus rad $(I^n) = $ rad I. We also define $I : J = \{t \in A : tJ \subset I\}$. We have $(I : J_1) : J_2 = I : (J_1 J_2)$. Finally, for $a \in A$, we have the ideal $I : a = I : Aa = \{t \in A : ta \in I\}$. (See [1], Chapter I, §1; or [4], Chapter 1.)

6. We say that an element $t \in A$ is a *zero divisor* in M if $tx = 0$ for some $x \in M \setminus 0$. (In particular, if $M = A$, then (t is a zero divisor)$\Longleftrightarrow$(t is a zero divisor in A)). Obviously, the invertible elements of the ring A are not zero divisors in M.

An element $t \in A$ is not a zero divisor in M precisely when the mapping $M \ni x \longrightarrow tx \in M$ is a monomorphism. Therefore, if the elements t, u are not zero divisors in M, neither is tu. Hence, if an ideal I consists only of zero divisors in M, then so does its radical rad I.

7. Since the intersection of an arbitrary family of submodules of a module M is a submodule of M, for any set $E \subset M$ there exists the *submodule generated by* E, i.e., the smallest submodule containing E. It coincides with AE (³) . The module M is said to be *finitely generated* or *finite over A* if it is generated by a finite set. A family $\{x_\iota\}$ of elements of the module M is said to be a *system of generators* of the module M if the set of all x_ι generates M. We also say that x_ι *generate* the module M or that they are its *generators*.

(³) The empty set generates the zero submodule 0.

Thus $x_1, \ldots, x_n$ is a sequence of generators of M if and only if $M = \sum_1^n Ax_i$, that is, if every element of the module M is a *(linear) combination* [4] *of the elements* $x_1, \ldots, x_n$ with coefficients in A. We denote by $g(M)$ or $g_A(M)$ the smallest number of generators of the module M. (We put $g(M) = \infty$ if M is not finitely generated.) Of course, $g(M) = 0 \iff M = 0$. If A is a field, then $g(M) = \dim M$ and

$$(M \text{ is finite over } A) \iff (M \text{ is finite dimensional}).$$

In particular, *the ideal generated by the set* $E \subset A$ is the set AE. An ideal that has a finite number of generators is said to be *finitely generated*. The elements $x_1, \ldots, x_n \in A$ generate the ideal $I = \sum_1^n Ax_i$. Then $I^k = \sum_{|p|=k} Ax^p$, and so $\{x^p\}_{|p|=k}$ is a system of generators for the ideal I^k (where $x = (x_1, \ldots, x_n)$, $p = (p_1, \ldots, p_n) \in \mathbf{N}^n$, $|p| = p_1 + \ldots + p_n$, $x^p = x_1^{p_1} \ldots x_n^{p_n}$). Any ideal generated by a single element (i.e., any ideal of the form Ac) is called a *principal ideal*.

If C is a subring of the ring A, then every module M over A is naturally a module over C. In particular, A is a module over C and so is every ideal of the ring A. Notice that if A is finite over C and M is finite over A, then M is finite over C.

Observe also that if $h : B \longrightarrow C$ is homomorphism of some extensions B and C of the ring A, then

$$(h \text{ is an } A-\text{homomorphism}) \iff (h \text{ is a homomorphism of modules over } A).$$

If N is a submodule of the module M generated by $y_1, \ldots, y_n$ and $x_1, \ldots, x_r \in M$, then

$$(\bar{x}_1, \ldots, \bar{x}_r \text{ generate } M/N) \iff (x_1, \ldots, x_r, y_1, \ldots, y_n \text{ generate } M).$$

Hence $g(M) \le g(N) + g(M/N)$.

8. Let $f : M \longrightarrow M'$ be a homomorphism of modules over A. Then the image (resp., inverse image) of a submodule of M (resp., M') is a submodule of M' (resp., M). Let $N \subset M$ be a submodule. If $N \subset \ker f$, the induced homomorphism $M/N \ni \bar{x} \longrightarrow f(x) \in M'$ is well-defined. If f is an epimorphism, then the induced homomorphism $M/\ker f \longrightarrow M'$ is an isomorphism and $M/\ker f \approx M'$. Let $N' \subset M'$ be a submodule. If $f(N) \subset N'$,

---

[4] In what follows, we will drop the word "linear".

then the induced homomorphism $M/N \ni \bar{x} \longrightarrow \overline{f(x)} \in M'/N'$ is well-defined, the set $f^{-1}(N')/N$ is its kernel, and $\left(f(M) + N'\right)/N'$ is its range. In particular, if $M \subset M'$ and $N \subset N'$, then we have a natural homomorphism $M/N \longrightarrow M'/N'$ (induced by $M \hookrightarrow M'$) with the kernel $(M \cap N')/N$ and the range $(M + N')/N'$. Therefore, if $L, N \subset M$ are submodules, the natural homomorphism $L/L \cap N \longrightarrow (L + N)/N$ is an isomorphism, and $L/L \cap N \approx (L+N)/N$. For any submodules $L \subset N \subset M$, the natural homomorphism $M/L \longrightarrow M/N$ is an epimorphism with the kernel N/L, and hence $(M/L)/(N/L) \approx M/L$. If the mapping is an isomorphism, then $L = N$.

Let $g : M' \longrightarrow M$ be another homomorphism. If we have

$$g \circ f = \mathrm{id}_M \quad (^5) ,$$

then g is an epimorphism, f is a monomorphism, and

$$M' = \ker g + \mathrm{im} \; f$$

is a direct sum $(^6)$.

If $x_1, \ldots, x_n$ generate M, then $f(x_1), \ldots, f(x_k)$ generate $f(M)$. Therefore $g\left(f(M)\right) \leq g(M)$. If $h : A \longrightarrow B$ is a ring epimorphism and an ideal I in the ring A is generated by a set E, then the ideal $f(I)$ in the ring B is generated by the set $f(E)$.

9. If I is an ideal such that $IM = 0$, then M is also a module over A/I with the same addition and with the multiplication well-defined by $\bar{u}x = ux$, where $u \in A$ and $x \in M$; it is called an *induced* module. In this case we have the same $(^7)$ submodules and the same systems of generators. Therefore $g_{A/I}(L) = g_A(L)$ when $L \subset M$ is a submodule. We have here the same homomorphisms into any module N over A, such that $IN = 0$. Hence M and N can be isomorphic (over A and A/I) only simultaneously. Note also that an element $u \in A$ is a zero divisor in M precisely when $\bar{u}$ is a zero divisor in M.

$(^5)$ We will denote by id_M the identity mapping of the set M, i.e., the mapping $M \ni x \longrightarrow x \in M$.

$(^6)$ Because $g\left(y - f(g(y))\right) = 0$ for $y \in M'$.

$(^7)$ When regarded as subsets. Each of them, as a submodule over A/I, is induced by itself considered as a module over A. The same applies to quotients of submodules: if $L \subset N \subset M$ are submodules, then the quotient N/L of submodules over A/I is induced by the quotient N/L of submodules over A.

In particular, for an arbitrary ideal I, the module M/IM is also [8] a module over A/I. Then $\bar{u}\bar{x} = ux$ for $u \in A$ and $x \in M$.

Notice that the module $(A/I)^n$ over A/I is induced for the module $(A/I)^n$ over A.

10. Let N be a submodule of the module M. We have the natural epimorphism $\kappa : M \ni x \longrightarrow \bar{x} \in M/N$ with the kernel N. The image of a submodule $L \subset M$ is $(L + N)/N$. The mapping $L \longrightarrow L/N$ of the set of submodules of the module M that contain N onto the set of submodules of the module M/N preserves inclusion. (Its inverse is given by $L' \longrightarrow \kappa^{-1}(L')$ [9].)

In particular, let I be an ideal in the ring A. In this case, the natural mapping $\kappa : A \ni u \longrightarrow \bar{u} \in A/I$ is simultaneously an epimorphism of rings and modules (over A). The image of an ideal J of the ring A is the ideal $(J + I)/I$ of the ring A/I. We have the bijection given by

$$(*) \qquad\qquad J \longrightarrow J/I$$

that preserves inclusion and maps the set of ideals of the ring A which contain I onto the set of ideals of the ring A/I. Its inverse is given by $J' \longrightarrow \kappa^{-1}(J')$.

11. An ideal of a ring A is *proper* (i.e., different from A) precisely when it does not contain the identity or, equivalently, if it does not contain any invertible elements.

An ideal I is said to be *prime* if it is proper and $xy \in I \implies (x \in I$ or $y \in I)$ [10] . In this case, rad $I = I$. If I contains the intersection of the ideals $J_1, \ldots, J_k$, then it must contain one of them. In particular, $I = J_1 \cap \ldots \cap J_k \implies I = J_i$ for some i. If an ideal is contained in the union of a finite family of prime ideals, then it must be contained in one of them. Thus, if J is an ideal and E, F are finite unions of prime ideals, then $(E \not\subset J, F \not\subset J) \implies E \cup F \not\subset J$. Finally, if I is prime and $J \not\subset I$, then $I : J \subset I$. (See [1], Chapter I, §1; or [4], Chapter 1).

If $f : C \longrightarrow A$ is a ring homomorphism, then the inverse image of any prime ideal of A is a prime ideal of C. Moreover, if f is an epimorphism, then

---

[8] This is because $I(M/IM) = 0$.

[9] Indeed, if $L \supset N$ is a submodule of the module M, then $\kappa(L) = L/N$ and $x \in L \implies \bar{x} \in L/N$, and hence $\kappa^{-1}(\kappa(L)) = L$. On the other hand, if L' is a submodule of the module M/N, then $\kappa^{-1}(L') \supset N$ and $\kappa(\kappa^{-1}(L')) = L'$.

[10] Equivalently, $I : a = I$ for $a \in A \setminus I$.

an ideal of the ring A is prime if and only if its inverse image is a prime ideal. In particular, the bijection (∗) (from n° 10) maps the set of prime ideals of the ring A that contain I onto the set of prime ideals of the ring A/I.

A ring A is an *integral domain* if and only if 0 is a prime ideal. The ring A/I is an integral domain if and only if the ideal I is prime.

A maximal element (with respect to inclusion) in the set of all proper ideals of the ring A is called a *maximal ideal* in the ring A. It is always a prime ideal, too. A ring A is a field if and only if the ideal 0 is maximal, i.e., if 0 and A are the only ideals of the ring A. Therefore a homomorphism from a field to a non-zero ring is always a monomorphism. The ring A/I is a field if and only if the ideal I is maximal. (See [1], Chapter I, §1; or [4], Chapter 1.)

12. A ring A is said to be a *ring of principal ideals* if each of its ideals is principal. In such a case, a common divisor d of two elements $x, y \in A$ is their greatest common divisor if and only if $d \in Ax + Ay$. Indeed, $Ax + Ay = Ac$ for some c, thus c is a common divisor of x and y. Now, if d is the greatest common divisor, then c is a divisor of d and hence $d \in Ac$.

13. Assume now that A is an integral domain. An element of the ring A is said to be *irreducible* if it is different from zero, non-invertible, and is not a product of two non-invertible elements $(^{10a})$. An element c of the ring A is said to be *prime* if the ideal Ac is prime and non-trivial or, in other words, if c is a non-zero, non-invertible element such that divisibility of ab by c implies divisibility of a or b by c. Each prime element is irreducible. (See [1a], Chapter IX, §3.) Also the opposite implication is true, provided that A is a unique factorization domain (see §6 below).

14. We say that a ring A is of *characteristic zero* if the mapping $\mathbf{Z} \ni s \longrightarrow s \cdot 1 \in A$ is a monomorphism $(^{11})$. In this case, $\mathbf{Z}$ is a subring of A (after the identification via the monomorphism). If A is a subring of the ring B, then A is of characteristic zero if and only if B is of characteristic zero. A field $\mathbf{K}$ of characteristic zero contains $\mathbf{Q}$ as a subfield (after the identification via the monomorphism $\mathbf{Q} \ni p/q \longrightarrow (p \cdot 1) \cdot (q \cdot 1)^{-1} \in \mathbf{K}$). If A is an integral domain of characteristic zero, then $(s \neq 0,\ x \neq 0) \Longrightarrow sx \neq 0$ for $s \in \mathbf{Z}$ and $x \in A$. Note also that if a ring A contains (as a subring) a field of characteristic zero,

---

$(^{10a})$ An element of A is said to be *reducible* if it is a product of two non-invertible elements.

$(^{11})$ Or, equivalently, if $s \cdot 1 \neq 0$ for all $s \in \mathbf{Z} \setminus \{0\}$.

then, for each proper ideal I, the ring A/I is of characteristic zero. (This is because, in such a case, the elements $s \cdot 1$, where $s \in \mathbf{Z} \setminus \{0\}$, are invertible and hence they do not belong to I.)

15. If A is a subring of a field L, then by *the field of fractions of the ring A in the field L* we mean the subfield generated by A. It is equal to the set of elements of the form xy^{-1} where $x \in A$ and $y \in A \setminus 0$, and is isomorphic to the field of fractions of the ring A via the A–isomorphism $xy^{-1} \longrightarrow x/y$. In particular, when A is a subring of an integral domain B, the field of fractions of A will be identified with the subfield (generated by A) of the field of fractions of the ring B. If A, A' are subrings of the fields L, L', respectively, then every isomorphism between the rings A and A' can be extended to an isomorphism between their fields of fractions in L and L', respectively. Let $f : L \longrightarrow L'$ be a field isomorphism. If K is the field of fractions of the subring A in L, then $f(K)$ is the field of fractions of the subring $f(A)$ in L'. If K is a subfield of the field L and $K' = f(K)$, then $\dim_K L = \dim_{K'} L'$.

15a. Let A be a ring, and let S be the complement of the set of zero divisors in A. There exists a unique (up to an A–isomorphism) extension of the ring A in which all elements of S are invertible and every element of which is of the form xy^{-1}, where $x \in A$, $y \in S$. The proofs of existence and uniqueness as well as the construction of the canonical extension with these properties are exactly the same as in the case of the field of fractions of an integral domain. The only difference is that the "denominators" are taken from S, i.e., one begins the construction with the set $A \times S$ (see [1], Chapter 1, §4; or [4], Chapter 3). The extension is called *the ring of fractions of the ring A*. Obviously, when A is an integral domain, one gets the field of fractions of A (since, in this case, $S = A \setminus 0$).

Let A be a subring of a ring A', and let R, R' be the rings of fractions of the rings A and A', respectively. If all elements of the ring A which are not zero divisors in A are not zero divisors in A', then – after identification via the well-defined monomorphism $R \ni x/y \longrightarrow x/y \in R'$ ([12]) – R is a subring of the ring R'.

Note also that if $R_1, \ldots, R_k$ are the rings of fractions of the rings $A_1, \ldots, A_k$, respectively, then the ring of fractions R of the ring $A_1 \times \ldots \times A_k$

([12]) Then the diagram

$$
\begin{array}{ccc}
A & \hookrightarrow & A' \\
\downarrow & & \downarrow \\
R & \longrightarrow & R'
\end{array}
$$

commutes.

is isomorphic to the ring $R_1 \times \ldots \times R_k$ via the well-defined natural $(A_1 \times \ldots \times A_k)$–isomorphism

$$R_1 \times \ldots \times R_k \ni (x_1/y_1, \ldots, x_k/y_k) \longrightarrow (x_1, \ldots, x_k)/(y_1, \ldots, y_k) \in R .$$

If B is an extension of the ring A, then a *denominator* (in A) *of an element* $b \in B$ is an element $a \in A$ which is not a zero divisor and is such that $ab \in A$. When each element of B has a denominator in A, the ring B can be identified – in a natural way – with a subring of the ring of fractions R of the ring A. The identification is obtained via the (well-defined) monomorphism $B \ni b \longrightarrow ab/a \in R$ (where a is a denominator of the element b). An element $a \in A$ which is not a zero divisor and is such that $aB \subset A$ (i.e., a common denominator for all elements of B) is called *a universal denominator of the extension* B (or – more precisely – of *the ring* B *over the ring* A). Obviously, if B is an integral domain and has a universal denominator over A, then the fields of fractions of both rings coincide.

16. A finite module M over A is said to be *free* if it has a basis, i.e., a sequence of linearly independent generators $x_1, \ldots, x_n$. If that is the case, $M \approx A^n$ because $A^n \ni (t_1, \ldots, t_n) \longrightarrow \sum_1^n t_i x_i \in M$ is an isomorphism. For any n, the module A^n is free and it has the *canonical basis* $e_1 = (1, 0, \ldots, 0), \ldots, e_n = (0, \ldots, 0, 1)$. Hence a module (which is finite over A) is free precisely when it is isomorphic with one of the modules A^n. Clearly, a (finite dimensional) vector space is always a free module.

Obviously, if M is a free module, there exists a unique homomorphism of the module M into another one that takes arbitrarily given values on an arbitrarily given basis of M. It follows that if $f : L \longrightarrow M$ is an epimorphism (of a module L), there exists a monomorphism $g : M \longrightarrow L$ such that $f \circ g = \mathrm{id}_M$.

With every module homomorphism $f : A^n \longrightarrow A^m$ one can associate its matrix C_f. This is the unique $(m \times n)$–matrix $[c_{ij}] \in A_m^n$ (with entries $c_{ij} \in A$) such that

$$(y_1, \ldots, y_m) = f(x_1, \ldots, x_n) \iff y_i = \sum_{j=1}^n c_{ij} x_j \text{ for } i = 1, \ldots, m .$$

If $n = m$, the mapping $f \longrightarrow C_f$ is an isomorphism between the non-commutative ring of all endomorphisms of the module A^n and the non-commutative ring A_n^n of all $(n \times n)$–matrices with entries from A. We define $\det f$ as $\det C_f$.

CRAMER'S THEOREM . *For each matrix $C \in A_n^n$ there exists a matrix $D \in A_n^n$ such that $CD = DC = (\det C)I$, where I is the identity matrix. If $x_1, \ldots, x_n$ are elements of a module over A and $\sum_{j=1}^n c_{ij}x_j = 0$ for $i = 1, \ldots, n$, then $(\det c_{ij})x_s = 0$ for $s = 1, \ldots, n$. If $\det C$ is invertible, then the matrix C is invertible. If $\det f$ is invertible, then f is an automorphism.*

(See [3], Chapter XIII, §4; and 2.2a below.)

Let M be a module over a ring A. We have the following

MATHER-NAKAYAMA LEMMA . *Let C be a subring (not necessarily with identity, e.g. an ideal) of the ring A, and let $\xi \in A$. If M is finite over C*[13], *then $\eta M = 0$ for some $\eta \in \xi^m + C\xi^{m-1} + \ldots + C$.*

Indeed, let $a_1, \ldots, a_m$ be generators of M over C. Then

$$\xi a_i = \sum_{j=1}^m \gamma_{ij} a_j \, ,$$

where $\gamma_{ij} \in C$, and in view of Cramer's theorem $\big(\det(\xi\delta_{ij} - \gamma_{ij})\big)a_s = 0$, $s = 1, \ldots, m$.

16a. Let B be an extension of a ring A. We say that B is *flat* over A if every solution $(x_1, \ldots, x_n) \in B^n$ of any linear equation $\sum_1^n x_i a_i = 0$ with coefficients $a_1, \ldots, a_n \in A^r$ is a linear combination (with coefficients from B) of solutions of this equation that belong to A^n (see [11a], Chapter I, 2.11).

An equivalent reformulation of this definition is as follows. Let an A–homomorphism be any module homomorphism $h : B^n \longrightarrow B^r$ such that $h(A^n) \subset A^r$. Note that A–homomorphisms are exactly the mappings of the form $B^n \ni (x_1, \ldots, x_n) \longrightarrow \sum_1^n x_i a_i \in B^r$, where $a_1, \ldots, a_n \in A^r$. Thus B is flat over A if and only if the kernel of every A–homomorphism $B^n \longrightarrow B^r$ is a submodule generated by a subset of A^n.

If B is flat over A, then the ring of polynomials $B[X]$ is flat over the ring of polynomials $A[X]$. This can be verified easily by expressing linear equations involving polynomials in terms of linear equations involving the coefficients of the polynomials instead (see 2.1 below).

17. Let M be a vector space. *The codimension of a subspace L of M,* i.e., the dimension of the space M/L, is denoted by codim L or $\mathrm{codim}_M L$. Let H and L be subspaces. If $M = L + H$, then codim $L \leq \dim H$. If $L \cap H = 0$, then codim $L \geq \dim H$. Hence, if $M = L + H$ is a direct sum,

---

[13] i.e., $M = \sum Cx_i$ with some $x_1, \ldots, x_k \in M$ (as for rings with identity).

then codim $L = \dim H$. In particular, $\mathrm{codim}L = \dim M - \dim L$, provided that M is finite dimensional. (Since the natural linear mapping $H \longrightarrow M/L$ is surjective and injective, respectively, in the previously considered cases.) Therefore L has finite dimension if and only if $M = L + H$ for a finite dimensional subspace H. If $L \subset H$, then codim $L \geq$ codim H, and hence H has a finite codimension if L does. Moreover, if codim $L =$ codim $H < \infty$, then $L = H$. (For the natural epimorphism $M/L \longrightarrow M/H$ becomes an isomorphism in such a case.)

18. Let M be an n–dimensional vector space over a field $\mathbf{K}$. We say that the subspaces $L_1, \ldots, L_k \subset M$ *intersect transversally* (or that they *are in general position*) if $\bigcap_1^k L_i' \neq \emptyset$ for all systems of affine subspaces $L_1', \ldots, L_k'$ which are parallel to $L_1, \ldots, L_k$, respectively. This happens precisely when

$$(\tau) \qquad \mathrm{codim} \bigcap_1^k L_i = \sum_1^k \mathrm{codim}\ L_i\ .$$

Indeed, let us take the natural linear mapping $\varphi_i : M \longrightarrow M/L_i$. The fact that subspaces L_i intersect transversally means exactly that the mapping

$$(\varphi_1, \ldots, \varphi_n): M \longrightarrow \tilde{M} = (M/L_1) \times \ldots \times (M/L_k)$$

(with the kernel $\bigcap_1^k L_i$) is surjective. Since $\dim \tilde{M} = \sum_1^k \mathrm{codim}L_i$, this is equivalent to (τ). Note that the inequality

$$\mathrm{codim} \bigcap_1^k L_i \leq \sum_1^k \mathrm{codim}\ L_i$$

is always true.

Observe also that the subspaces L_i intersect transversally if and only if in some linear coordinate system ($\varphi : M \longrightarrow K^n$) they are of the form

$$T_i = (x_1, \ldots, x_n): x_\nu = 0 \quad \text{for} \quad \nu \in I_i \quad (^{14})\ ,$$

where $I_1, \ldots, I_n$ are disjoint subsets of $\{1, \ldots, n\}$.

In fact, the condition is sufficient, as we have $\mathrm{codim} \bigcap T_i = \#\bigcup I_i = \sum \#I_i = \sum \mathrm{codim}T_i$. Conversely, if the L_i's intersect transversally, then the sum $\sum L_i^\perp = (\bigcap L_i)^\perp$ $(^{15})$ is direct because $\dim \sum L_i^\perp = \mathrm{codim} \bigcap L_i = \sum \mathrm{codim}L_i = \sum \dim L_i^\perp$. Hence we

---

$(^{14})$ It means that $\varphi(L_i) = T_i$.

$(^{15})$ For any subspace $L \subset M$ of dimension k, $L^\perp = \{\varphi \in M : \varphi_L = 0\}$ is a subspace of dimension $n - k$ of the dual space M^*. See B. 6.4.

can choose a basis $\varphi_1, \ldots, \varphi_n$ of the dual space M^*, such that φ_ν, $\nu \in I_i$, generate $L_i^\perp$ ($i = 1, \ldots, n$) and the I_i's are disjoint. Then $\varphi = (\varphi_1, \ldots, \varphi_n) : M \longrightarrow K^n$ (^{16}) is an isomorphism and $L_i = \{\varphi_\nu = 0 \quad \text{for} \quad \nu \in I_i\} = \varphi^{-1}(T_i)$.

It follows that if $L_1, \ldots, L_k$ intersect transversally, then so do $L_{\alpha_1}, \ldots, L_{\alpha_s}$, where $1 \le \alpha_1 < \ldots < \alpha_s \le k$.

Two subspaces $L, H \subset M$ intersect transversally if and only if $L + H = M$. (Since, in this case, the condition (τ) reduces to the equality $\dim(L + H) = n$.) If, in addition, we know that $\dim L + \dim H = n$, the condition of transversality of the intersection of L and H can be expressed as $L \cap H = 0$.

If $H_1, \ldots, H_n$ are hyperplanes, then they intersect transversally if and only if $\bigcap_1^n H_i = 0$.

In the same way one defines *transversality of the intersection* of a family of affine spaces $L_1, \ldots, L_k \subset M$ by requiring that $\bigcap_1^k L_i \ne \emptyset$ and condition (τ) is fulfilled (^{17}) .

19. Notice that if $\varphi : M \longrightarrow X$ is an epimorphism of vector spaces, then for any isomorphism of the form $\chi = (\varphi, \pi) : M \longrightarrow X \times Y$, where $\pi : M \longrightarrow Y = \ker \varphi$ is a projection, and for the natural projection $p : X \times Y \longrightarrow X$, the diagram

$$
\begin{array}{ccc}
M & \xrightarrow{\chi} & X \times Y \\
{\scriptstyle\varphi}\searrow & & \swarrow{\scriptstyle p} \\
& X &
\end{array}
$$

commutes. Then

$$\chi(\varphi^{-1}(E)) = E \times Y \quad \text{for} \ E \subset X \ .$$

20. Let M and L be vector subspaces over a field $\mathbf{K}$, and let $f : M^k \longrightarrow L$ be a k–linear alternating mapping. If $\alpha = (\alpha_1, \ldots, \alpha_n)$ is a permutation of the set $\{1, \ldots, k\}$, then

$$f(x_{\alpha_1}, \ldots, x_{\alpha_k}) = \varepsilon_\alpha f(x_1, \ldots, x_k) \quad \text{for} \ x_1, \ldots, x_k \in M,$$

where ε_α denotes the sign of the permutation (^{18}) . If $1 \le k \le n$, then we have the identity

$$f\left(\sum_1^n a_{1j}x_j, \ldots, \sum_1^n a_{kj}x_j\right) = \sum_{1 \le \lambda_1 < \ldots < \lambda_k \le n} (\det a_{i\lambda_j})f(x_{\lambda_1}, \ldots, x_{\lambda_k}) \, ,$$

(^{16}) See footnote (^{22}).

(^{17}) The codimension of an affine subspace $L \subset M$ is defined in the same way as before, i.e., $\operatorname{codim} L = n - \dim L$.

(^{18}) That is, $\varepsilon_\alpha = 1$ or $\varepsilon_\alpha = -1$ when the permutation is even or odd, respectively.

for $a_{ij} \in \mathbf{K}$ $(i = 1, \ldots, k \, ; \, j = 1, \ldots, n)$ and $x_1, \ldots, x_n \in M$. (See [3], Chapter XIII, §4; ([19])).

21. Let m be a finite dimensional vector space over a field $\mathbf{K}$. A vector from the space $\bigwedge^k M$ is said to be *simple* if it is of the form $z_1 \wedge \ldots \wedge z_k$, where $z_1, \ldots, z_k \in M$.

A *direction vector of a subspace* $U \subset M$ is an arbitrary non-zero vector from $\bigwedge^m U \subset \bigwedge^m M$, where $m = \dim U$, i.e., a vector of the form $u_1 \wedge \ldots \wedge u_m$, where $u_1, \ldots, u_m$ is a basis for the subspace U. It is determined up to a non-zero factor from $\mathbf{K}$.

Assume now that M is the direct sum of subspaces X and Y.

If $z \in \bigwedge^k X$, $w \in \bigwedge^l Y$, then $(z \neq 0, w \neq 0) \implies z \wedge w \neq 0$. Hence, if $z \in \bigwedge^k X \setminus 0$ and $w', w'' \in \bigwedge^l Y$, then

$$(z \wedge w' = z \wedge w'') \implies w' = w'' \ .$$

We also have the direct sum $\bigwedge^k M = \sum_{i+j=k} \bigwedge^{ij}(X,Y)$, $(i, j \geq 0)$, where $\bigwedge^{ij}(X,Y)$ denotes the subspace generated by the vectors of the form $u \wedge v$, where $u \in \bigwedge^i X$, $v \in \bigwedge^j Y$ ([20]) .

Let $z \in \bigwedge^k X \setminus 0$, $w \in \bigwedge^j Y$. If the vectors w and $z \wedge w$ are simple, so is the vector w. In fact, one can assume that $\dim X = k$. Let U be a subspace with the direction vector $z \wedge w$. Take a basis $w_1, \ldots, w_s$ for the kernel of the projection of U onto X, parallel to Y. Extend it to a basis $u_1 + v_1, \ldots, u_r + v_r, w_1, \ldots, w_s$ for U, where $u_i \in X$, $v_i \in Y$. Then $u_1, \ldots, u_r$ are linearly independent and

$$z \wedge w = a(u_1 + v_1) \wedge \ldots \wedge (u_r + v_r) \wedge w_1 \wedge \ldots \wedge w_s \in \overset{kl}{\bigwedge}(X,Y) \, ,$$

where $a \in \mathbf{K} \setminus 0$. This yields $r = k$, $s = l$, and $z \wedge w = au_1 \wedge \ldots \wedge u_k \wedge w_1 \wedge \ldots \wedge w_l$. Thus $u_1 \wedge \ldots \wedge u_k = bz$, where $b \in \mathbf{K} \setminus 0$, and hence $w = abw_1 \wedge \ldots \wedge w_l$.

If $z \in \bigwedge^k X \setminus 0$, $w \in \bigwedge^k Y \setminus 0$, and $k \geq 2$, then the vector $z + w$ is not simple. For if it were simple, we would have the equality

$$z + w = (u_1 + v_1) \wedge \ldots \wedge (u_k + v_k),$$

([19]) The above identity can be checked in the same way as in the case $n = k$.

([20]) It is enough to observe that if $\{e'_\nu\}$ and $\{e''_\mu\}$ are bases for the spaces X, Y, respectively, then the subspace $\bigwedge^{ij}(X,Y)$ is generated by the elements $e'_{\alpha_1} \wedge \ldots \wedge e'_{\alpha_i} \wedge e''_{\beta_1} \wedge \ldots \wedge e''_{\beta_j}$, $(\alpha_1 < \ldots < \alpha_i, \beta_1 < \ldots < \beta_j)$.

where $u_i \in X$, $v_i \in Y$. Thus, by taking the components of both sides in the subspaces $\bigwedge^{k0}(X,Y)$ and $\bigwedge^{0k}(X,Y)$, we would get $z = u_1 \wedge \ldots \wedge u_k$ and $w = v_1 \wedge \ldots \wedge v_k$. That would imply that the elements $v_1, \ldots, v_k$ are linearly independent. Therefore, for $0 < s < k$, the component of the right hand side belonging to $\bigwedge^{s,k-s}(X,Y)$ would be non-zero, while the corresponding component of the left hand side would vanish.

22. Let M be a vector space over a field $\mathbf{K}$. A subset S of M is said to be a *cone* if $S \neq \emptyset$ and $\lambda S \subset S$ for all $\lambda \in \mathbf{K}$. Then $0 \in S$ and $\lambda S = S$ for all $\lambda \in \mathbf{K} \setminus 0$. Thus a cone in the space M is precisely the union of a non-empty family of straight lines passing through the origin or the set $\{0\}$. The intersection of any family of cones is a cone. Hence, for any set $E \subset M$, there is *the cone generated by the set* E, i.e., the smallest cone in M containing E. If $E \setminus 0 \neq \emptyset$, it coincides with the union of all straight lines passing through 0 that have non-empty intersections with $E \setminus 0$ [21]. The union of a non-empty family of cones is a cone. If $S, T \subset M$ are cones, then $(S \setminus T) \cup 0$ is a cone. Both the image and the inverse image of a cone under a linear mapping are cones. The Cartesian product of cones is a cone.

Let $S \subset M$, $T \subset N$ (where N is a vector space over $\mathbf{K}$) be cones, and let $k \in \mathbf{N}$. The mapping $F: S \longrightarrow T$ is said to be *homogeneous of degree* k if

$$F(tx) = t^k F(x) \quad \text{for} \quad T \in \mathbf{K}, \ x \in S.$$

If, in addition, we assume that $T = N = \mathbf{K}$, then we call F a *homogeneous function of degree* k. The homogeneous mappings of order zero are precisely the constant mappings. For any homogeneous mapping F of positive degree, $F(0) = 0$. Excluding the zero mapping, the degree of a homogeneous mapping is determined uniquely. The restriction of a homogeneous mapping of degree k to a cone is homogeneous of degree k. The diagonal product [22] and the Cartesian product [23] of mappings (from a cone to a cone) is homogeneous of degree k if and only if each of these mappings is homogeneous of degree k. A non-empty composition of homogeneous mappings of degrees $k_1, \ldots, k_r$

---

[21] It is equal to 0 if $E \subset 0$.

[22] By the *diagonal product of the mappings* $f_i: E \longrightarrow F_i$, $i = 1, \ldots, m$, we mean the mapping

$$(f_1, \ldots, f_m): E \ni x \longrightarrow (f_1(x), \ldots, f_m(x)) \in F_1 \times \ldots \times F_m.$$

[23] The *Cartesian product of the mappings* $f_i: E_i \longrightarrow F_i$, $i = 1, \ldots, m$, is defined as the mapping

$$f_1 \times \ldots \times f_m: E_1 \times \ldots \times E_m \ni (x_1, \ldots, x_m) \longrightarrow (f_1(x_1), \ldots, f_m(x_m)) \in F_1 \times \ldots \times F_m.$$

is a homogeneous mapping of degree $k_1 \ldots k_r$. The product of homogeneous functions of degrees $k_1, \ldots, k_r$ is a homogeneous function of degree $k_1 + \ldots + k_r$. Note also that if F is a homogeneous mapping, then $F^{-1}(0) \cup 0$ is a cone ([24]) .

§2. Polynomials

1. Let A be a ring. We denote the ring of polynomials with respect to the variables $X_1, \ldots, X_n$ over A by $A[X_1, \ldots, X_n]$. If the elements of the ring A are identified with the constant polynomials, $A[X_1, \ldots, X_n]$ is an extension of the ring A. Each polynomial $P \in A[X_1, \ldots, X_n]$ can be uniquely expressed in the form

$$P = \sum a_p X^p,$$

where $a_p \in A$, $p = (p_1, \ldots, p_n) \in \mathbf{N}^p$, $X^p = X_1^{p_1} \ldots X_n^{p_n}$, and $a_p = 0$ for all but finite number of indices p. If P is a non-zero polynomial, then its *degree* is equal to $\max\{|p| : a_p \neq 0\}$, where $|p| = p_1 + \ldots + p_n$ ([25]) .

A polynomial of the form $\sum_{|p|=k} a_p X^p$ is called a *form of degree k*. The set of all forms of degree k constitutes a module over A. The product of s forms of degrees $k_1, \ldots, k_s$, respectively, is a form of degree $k_1 + \ldots + k_s$. Every polynomial has a unique *decomposition into forms*. Namely, $P = \sum P_i$, where P_i is a form of degree i and $P_i = 0$ for sufficiently large i ([26]) . The form P_i is said to be the *homogeneous component of degree i of the polynomial P*. If the polynomial P is non-zero, then its degree coincides with $\max\{i : P_i \neq 0\}$. Note that if $P = \sum P_i$, $Q = \sum Q_i$, and $PQ = \sum R_i$ are the decompositions of P, Q, and PQ into forms, then $P_i = \sum_{\nu=0}^{i} P_\nu Q_{i-\nu}$. A polynomial is said to be a *homogeneous polynomial* if it is a form (of any degree). Hence, if $k \geq 0$, a homogeneous polynomial of degree k is just a non-zero form of degree k.

If a form S is a divisor of a polynomial P, then it also divides each of the homogeneous components of the polynomial P. (Indeed, if $P = QS$ and $Q = \sum Q_i$ is the decomposition of Q into forms, then $P = \sum Q_i S$ is also a decomposition into forms).

The ring $A[X_{\alpha_1}, \ldots, X_{\alpha_k}]$, where $1 \leq \alpha_1 < \ldots < \alpha_k \leq n$, can be identified with a subring of the ring $A[X_1, \ldots, X_n]$ consisting of polynomials

---

([24]) The image of a cone under a homogeneous mapping is a cone. The inverse image, on the other hand, is a cone or the empty set.

([25]) One assumes that the degree of the zero polynomial is $-\infty$.

([26]) If $P = \sum a_p X^p$, then $P_i = \sum_{|p|=i} a_p X^p$.

that are independent of $X_{\beta_1}, \ldots, X_{\beta_{n-k}}$. Here $\{\beta_1, \ldots, \beta_{n-k}\} = \{1, \ldots, n\} \setminus \{\alpha_1, \ldots, \alpha_k\}$, and $\beta_1 < \ldots < \beta_{n-k}$. Also the rings

$$A[X_{\alpha_1}, \ldots, X_{\alpha_n}][X_{\beta_1}, \ldots, X_{\beta_{n-k}}] \text{ and } A[X_1, \ldots, X_n]$$

can be identified. The *degree of a polynomial* from $A[X_1, \ldots, X_n]$ *with respect to the indeterminates* $X_{\beta_1}, \ldots, X_{\beta_{n-k}}$ is defined as the degree of the same polynomial regarded as an element of $A[X_{\alpha_1}, \ldots, X_{\alpha_k}][X_{\beta_1}, \ldots, X_{\beta_{n-k}}]$. If A is a subring of a ring B, then $A[X_1, \ldots, X_n]$ is a subring of the ring $B[X_1, \ldots, X_n]$. Every ring homomorphism (monomorphism, epimorphism, isomorphism) $h : A \longrightarrow C$ *induces the homomorphism (monomorphism, epimorphism, isomorphism, respectively) of the rings of polynomials*

$$A[X_1, \ldots, X_n] \ni \sum a_p X^p \longrightarrow \sum h(a_p) X^p \in C[X_1, \ldots, X_n] \ .$$

The induced homomorphism extends the homomorphism h. (See [1a], Chapter VI, §9; [3], Chapter V, §2; [7], Chapter I, §18.)

2. Let C be an extension of the ring A. The *value of a polynomial* $P = \sum a_p X^p \in A[X_1, \ldots, X_n]$ at $c = (c_1, \ldots, c_n) \in C^n$ is defined as $P(c) = \sum a_p c^p$. The mapping $A[X_1, \ldots, X_n] \ni P \longrightarrow P(c) \in C$ is an A–homomorphism. If $d = P(c)$ for some $P \in A[X_1, \ldots, X_n]$, we say that the element d can be expressed as a polynomial with respect to the elements $c_1, \ldots, c_n$ with coefficients from A. Let $A \ni x \longrightarrow x' \in B$ be a ring homomorphism, and let $A[X_1, \ldots, X_n] \ni P \longrightarrow P' \in B[X_1, \ldots, X_n]$ and $A[Y_1, \ldots, Y_m] \ni Q \longrightarrow Q' \in B[Y_1, \ldots, Y_m]$ be the induced homomorphisms. Let $P \in A[X_1, \ldots, X_n]$. Then

$$P(Q_1, \ldots, Q_n)' = P'(Q'_1, \ldots, Q'_n) \text{ for } Q_1, \ldots, Q_n \in A[Y_1, \ldots, Y_m]$$

and, in particular,

$$P(a_1, \ldots, a_n)' = P'(a'_1, \ldots, a'_n) \text{ for } a_1, \ldots, a_n \in A \ .$$

Thus

$$P(Q_1, \ldots, Q_n)(c) = P(Q_1(c), \ldots, Q_n(c))$$

for $Q_1, \ldots, Q_n \in A[Y_1, \ldots, Y_m]$ and $c \in C^m$. (This is because $A[Y_1, \ldots, Y_m] \ni Q \longrightarrow Q(c) \in C$ is an A–homomorphism.) Hence, if an element (of the ring C) can be expressed as a polynomial with respect to $Q_1(c), \ldots, Q_m(c)$, it can be also expressed as a polynomial with respect to $c_1, \ldots, c_m$ (with coefficients in A).

Similarly, for any polynomial $P = \sum a_p X^p \in \mathbf{Z}[X_1, \ldots, X_n]$, we can define its value at an n–tuple $c = (c_1, \ldots, c_n)$ of elements of any ring C by the formula $P(c) = \sum a_p c^p$. Then the mapping $\mathbf{Z}[X_1, \ldots, X_n] \ni P \longrightarrow P(c) \in C$ is a homomorphism. If $h : C \longrightarrow C'$ is a ring homomorphism, then

$$h\big(P(c_1, \ldots, c_n)\big) = P\big(h(c_1), \ldots, h(c_n)\big) \ .$$

In particular, if $Q_1, \ldots, Q_n \in \mathbf{Z}[Y_1, \ldots, Y_m]$ and $b = (b_1, \ldots, b_m)$ is an m–tuple of elements of a ring, then

$$P(Q_1, \ldots, Q_n)(b) = P\big(Q_1(b), \ldots, Q_n(b)\big) \ .$$

For any polynomial $P \in A[X_1, \ldots, X_n]$, the mapping $\tilde{P} : A^n \ni x \longrightarrow P(x) \in A$ is called the *polynomial function* associated with the polynomial. Note that, for $Q_1, \ldots, Q_n \in A[Y_1, \ldots, Y_m]$, we have $P(Q_1, \ldots, Q_m)\tilde{} = \tilde{P} \circ (\tilde{Q}_1, \ldots, \tilde{Q}_n)$. The polynomial functions associated with all the polynomials from $A[X_1, \ldots, X_n]$ constitute a ring which is isomorphic to $A[X_1, \ldots, X_n]$ if A is of characteristic zero. Moreover, $P \longrightarrow \tilde{P}$ is an isomorphism. (We have $\tilde{P} = 0 \Longrightarrow P = 0$. Indeed, if $n = 1$, then $\sum_0^k a_\nu s^\nu = 0$ for $s = 0, \ldots, k$, implies that $a_0 = \ldots = a_k = 0$, because $\det[s^\nu]_{\nu, s=0, \ldots, n} \neq 0$. Now we can apply induction with respect to n $(^{27})$). Hence we will often identify polynomials over a ring of characteristic zero with the corresponding polynomial functions. Note that the polynomial $P(Q_1, \ldots, Q_n) \in A[Y_1, \ldots, Y_m]$, where $Q_1, \ldots, Q_n \in A[Y_1, \ldots, Y_m]$, is identified with $\tilde{P} \circ (\tilde{Q}_1, \ldots, \tilde{Q}_n)$.

2a. The determinant of n-th order regarded as a polynomial of all its "variables" is an element of the ring of polynomials in the variables $X = (X_{ij})_{i,j=1,\ldots,n}$ over $\mathbf{Z}$, i.e.,

$$\det = \det X = \det X_{ij} \in \mathbf{Z}[X] = \mathbf{Z}[X_{ij} \ ; \ i, j = 1, \ldots, n] \ .$$

The value it takes for a matrix $C = [c_{ij}]_{i,j=1,\ldots,n}$ (with entries c_{ij} belonging to any ring) is just the determinant of the matrix, i.e., $\det(C) = \det C = \det c_{ij}$.

In the ring of polynomials $\mathbf{Z}[X, T]$, we have *Cramer's identities*

$$\sum_{j=1}^n X_{ij} \det X = T_i \det X^{(j)}, \ i = 1, \ldots, n \ ,$$

$(^{27})$ In view of the identification of $A[X_1, \ldots, X_{n-1}][X_n]$ with $A[X_1, \ldots, X_n]$, we have $P(c_1, \ldots, c_{n-1})(c_n) = P(c_1, \ldots, c_n)$.

where $T = (T_1, \ldots, T_n)$ and $X^{(j)}$ is the matrix obtained from X by replacing its j-th column by T $(^{28})$. They imply *Cramer's identities* for an $n \times n$–matrix C and an $n \times m$–matrix D (with entries belonging to any ring) :

$$CB = (\det C)D \ ,$$

where $B = (\det B_{ij})_{i=1,\ldots,n; j=1,\ldots,m}$ and B_{ij} is the matrix obtained from C by replacing its i-th column by the j-th column of the matrix D $(^{29})$.

3. Assume that A is an integral domain. Then the ring of polynomials $A[X_1, \ldots, X_n]$ is also an integral domain (see [1a], Chapter VI, §9). In this case, the degree of the product of two non-zero polynomials is equal to the sum of their degrees. Thus the invertible elements in the ring $A[X_1, \ldots, X_n]$ are precisely invertible elements in the ring A. Also, the irreducible elements of degree zero in the ring $A[X_1, \ldots, X_n]$ coincide with the irreducible elements of the ring A. In particular, a polynomial from $A[X_1, \ldots, X_r]$ (where $r \leq n$) is irreducible if and only if it is irreducible in $A[X_1, \ldots, X_n]$.

Note that if the product of non-zero polynomials is homogeneous, then each of the polynomials must be homogeneous. (It is enough to check this for two polynomials, say P and Q. We have the decompositions into forms: $P = \sum_k^{k'} P_i$, $Q = \sum_l^{l'} Q_i$, and $PQ = \sum_{k+l}^{k'+l'} R_i$, where $P_k, P_{k'}, Q_l, Q_{l'}$ are non-zero. Thus $R_{k+l} = P_k Q_l \neq 0$ and $R_{k'+l'} = P_{k'} Q_{l'} \neq 0$, and hence $k + l = k' + l'$. Therefore $k = k'$ and $l = l'$.) Consequently, a non-zero form of degree 1 is irreducible precisely when its coefficients are relatively prime (see §6).

In particular, it follows that the determinant (regarded as a polynomial with respect to all its "variables") is irreducible. Indeed, excluding the trivial case, we may assume that $n \geq 2$. Expanding with respect to the first column, we can see that $\det X_{ij}$, as a polynomial from $A[X_{ij} : j \geq 2][X_{11}, \ldots, X_{n1}]$, is a non-zero form of degree 1 whose coefficients $D_1, \ldots, D_n$ are polynomials with the coefficients ± 1. Moreover, D_i is of degree 0 with respect to the variables $X_{i2}, \ldots, X_{in}$. Therefore every common divisor of the polynomials D_i must be of degree zero, i.e., must be an element of A. Hence it divides the coefficients of these polynomials, which means that it has to be invertible. Consequently, the polynomials D_i are relatively prime.

$(^{28})$ Obviously, they follow from the same identities in every ring of characteristic zero (because of the isomorphism $P \longrightarrow \tilde{P}$). At the same time, by considering the values taken by the polynomials, we obtain the same identities in any ring.

$(^{29})$ Indeed, if $C = [c_{ij}]$ and $D = [d_{ik}]$ then, by considering the values at (C, D_k), where $D_k = (d_{1k}, \ldots, d_{nk})$, we get $\sum_{j=1}^n c_{ij} \det B_{jk} = d_{ik} \det C$.

4. Every non-zero polynomial from $A[X]$ can be uniquely written in the form $a_0 X^n + a_1 X^{n-1} + \ldots + a_n$, where $a_0 \neq 0$. In this case, a_0 is called the *leading coefficient of the polynomial*. If the ring A is an integral domain, then the leading coefficient of the product of non-zero polynomials in $A[X]$ is equal to the product of their leading coefficients. A non-zero polynomial whose leading coefficient is 1 is said to be *monic*. If a polynomial $g \in A[X]$ is monic, then, for each $f \in A[X]$, there exists a unique pair $q, r \in A[X]$ such that $f = qg + r$ and the degree of r is less than that of g. In such a case, g is a divisor of f if and only if $r = 0$. Thus an element $a \in A$ is a root of the polynomial f if and only if f is divisible by $X - a$. If the ring A is an integral domain, then a polynomial $P \in A[X]$ of degree $n \geq 0$ can have at most n roots in A (counted with their multiplicities). If it can be factorized into linear factors, i.e., $P = a(X - \zeta_1) \ldots (X - \zeta_n)$, then we call $\zeta_1, \ldots, \zeta_n$ a *complete sequence of roots* of P (it is determined by P up to order). (See [1a], Chapter VI, §§4 and 6.)

5. For every polynomial $P \in A[X]$ one can define its *derivative* $P' \in A[X]$. (See [3], Chapter V, §8.) If the ring A is of characteristic zero and the degree of P is $k > 0$, then P' is of degree $k - 1$. The partial derivative $\frac{\partial P}{\partial X_s} \in A[X_1, \ldots, X_n]$ of a polynomial $P \in A[X_1, \ldots, X_n]$ is defined as its derivative in $A[X_1, \ldots, X_{s-1}, X_{s+1}, \ldots, X_n][X_s]$. The sum and product rules for derivatives of polynomials are the same as in the case of differentiable functions ([30]).

6. Let C be an extension of the ring A. We say that $\zeta \in C$ is an *algebraic element* over A if it is a root of a non-zero polynomial from $A[X]$. If A is a field, such a polynomial can always be chosen to be monic. An element $\zeta \in C$ which is not algebraic (over A) is called *transcendental* (over A). An element ζ is transcendental if and only if $A[X] \ni P \longrightarrow P(\zeta) \in C$ is a monomorphism.

7. Let C be an extension of the ring A. If $c_1, \ldots, c_k \in C$, then the set $A[c_1, \ldots, c_k] = \{P(c_1, \ldots, c_k) : P \in A[X_1, \ldots, X_n]\}$ is a subring of the ring C generated by the set $A \cup \{c_1\} \cup \ldots \cup \{c_k\}$. It is called the *ring generated over A* (in C) by the elements $c_1, \ldots, c_k$. Notice that $A[c_1, \ldots, c_k] = A[c_1, \ldots, c_{k-1}][c_k]$. If an element $\zeta \in C$ is transcendental (over A), then $A[X] \ni P \longrightarrow P(\zeta) \in A[\zeta]$ is an A–isomorphism (and vice versa).

([30]) When $A = \mathbf{R}$ or $A = \mathbf{C}$, differentiation of polynomials coincides with differentiation of the corresponding polynomial functions (which, in this case, can be identified with the polynomials).

8. Let K be a field, and let L be its extension. The *field generated over K* (in L) *by the elements* $\xi_1, \ldots, \xi_k \in L$ is defined as the subfield $K(\xi_1, \ldots, \xi_k)$ of the field L generated by the set $K \cup \{\xi_1\} \cup \ldots \cup \{\xi_k\}$. We have $K(\xi_1, \ldots, \xi_k) = K(\xi_1, \ldots, \xi_{k-1})(\xi_k)$. If A is a subring of the field L, $\xi_1, \ldots, \xi_k \in L$, and K is the field of fractions of the ring A in the field L, then $K(\xi_1, \ldots, \xi_k)$ is the field of fractions of the ring $A[\xi_1, \ldots, \xi_k]$ in the field L.

9. Let K be a field. The invertible elements in the ring $K[X_1, \ldots, X_n]$ are precisely the constant non-zero polynomials. Therefore a polynomial in $K[X_1, \ldots, X_n]$ is irreducible if and only if its degree is positive and it is not equal to the product of two polynomials of positive degrees. In particular, all polynomials of degree one are irreducible.

If K is a field, then $K[X]$ is a principal ideal domain (see [1a], Chapter VII, §4).

§3. Polynomial mappings

Let **K** be a field of characteristic zero. In this case we can identify the polynomials in $\mathbf{K}[X_1, \ldots, X_n]$ with their polynomial functions (see 2.2).

Any linear isomorphism of an n–dimensional vector (or affine) space over **K** onto the space $\mathbf{K}^n$ is called a *linear* (or *affine*) *coordinate system* on that vector space [31].

1. Let X and Y be finite dimensional vector spaces over **K**. A *polynomial mapping* from the space X into the space Y is defined as follows. If $X = \mathbf{K}^n$ and $Y = \mathbf{K}^m$, such a mapping is the diagonal product of m polynomial functions on $\mathbf{K}^n$. In the general case, it is defined as a polynomial mapping in some (and hence in all) linear coordinate systems on X and Y [32].

When $Y = \mathbf{K}$, the polynomial mappings are called *polynomials on X*. A function $f : X \longrightarrow \mathbf{K}$ is a polynomial on X precisely when it is a polynomial in some (and thus in every) linear coordinate system on X. Its degree (which

---

[31] Recall that if $e_1, \ldots, e_n$ is the canonical basis for $\mathbf{K}^n$ (see 1.16), then we have a natural bijection $\varphi \longrightarrow (\varphi^{-1}(e_1), \ldots, \varphi^{-1}(e_n))$ from the set of linear coordinate systems on X onto the set of all bases of X. In such a case, if $c_1, \ldots, c_n$ is a basis, then $X \ni \sum_1^n t_i c_i \longrightarrow (t_1, \ldots, t_n) \in \mathbf{K}^n$ is the corresponding coordinate system.

[32] Let $\varphi : X \longrightarrow \mathbf{K}^n$ and $\psi : Y \longrightarrow \mathbf{K}^m$ be linear coordinate systems. By the mapping $f : X \longrightarrow Y$ *in the coordinate systems* φ and ψ we mean the mapping $\psi \circ f \circ \varphi^{-1}$. By a function $f : X \longrightarrow \mathbf{K}$ *in the coordinate system* φ we mean the function $f \circ \varphi^{-1}$. In particular, the function f in the coordinate system corresponding to a basis $c_1, \ldots, c_n$ of

is independent of the choice of a coordinate system) is called the *degree of the polynomial* f. The polynomials on X constitute a ring which we will denote by $\mathcal{P}(X)$. (It is isomorphic to the ring of polynomials of n variables over $\mathbf{K}$, where $n = \dim X$.) Clearly, $\mathcal{P}(X)$ is also a vector space over $\mathbf{K}$ and the polynomials of degree $\leq k$ constitute a finite dimensional subspace (for any k).

More generally, we define the *degree of a polynomial mapping* as the maximum of the degrees of its components in arbitrary coordinate systems on X and Y. (The definition is obviously independent of the choice of such systems.) The polynomial mappings of degree ≤ 1 are exactly the affine mappings. The polynomial mappings which are homogeneous of degree 1 are precisely the linear mappings (see n° 2).

The composition of polynomial mappings is a polynomial mapping. The Cartesian and diagonal products of mappings are polynomial if and only if the component mappings are polynomial. If $f : X \longrightarrow Y$ is a polynomial mapping, $U \subset X$ and $V \subset Y$ are subspaces, and $f(U) \subset V$, then the restriction $f|_U : U \longrightarrow V$ is also a polynomial mapping.

Note also that a k–linear mapping $f : X_1 \times \ldots X_k \longrightarrow Y$ (where X_k are finite dimensional vector spaces over $\mathbf{K}$) is polynomial and homogeneous of degree k. (Indeed, if $X_k = \mathbf{K}^m$ and $Y = K$, then $f\big(u, (t_1, \ldots, t_m)\big) = \sum_1^m t_i a_i(u)$, where $a_i : X_1 \times \ldots \times X_{k-1} \longrightarrow K$ are $(k-1)$–linear. Thus it suffices to apply induction with respect to k.)

The definition of *polynomial mappings between affine spaces* (and *polynomials on affine spaces*) is the same (the linear coordinate systems must be replaced by affine ones), and the above properties are true (except those involving homogeneity). Polynomial mappings between vector spaces are the same as polynomial mappings between the same spaces regarded as affine spaces. Affine mappings are polynomial. The polynomials on an affine space X constitute a ring (that is isomorphic to the ring of polynomials of n variables over $\mathbf{K}$, where $n = \dim X$). We will also denote this ring by $\mathcal{P}(X)$.

2. Let X be a finite dimensional vector space over $\mathbf{K}$. A *form of degree k on X* is a polynomial on X which is a homogeneous function of degree k (see 1.22) or, equivalently, which is, in some (and thus in each) linear coordinate

---

the space X (see footnote (^{31}) above) is the function

$$\mathbf{K}^n \ni (t_1, \ldots, t_n) \longrightarrow f(\sum_1^n t_i c_i) \, .$$

system, a form of degree k when regarded as an element of $K[X_1, \ldots, X_n]$. Here $n = \dim X$. (Indeed, if $P \in K[X_1, \ldots, X_n]$ and $P = \sum P_i$ is the decomposition into forms, then the condition $P(tx) = t^k P(x)$ for $t \in \mathbf{K}$ and $x \in \mathbf{K}^n$ means that $\sum t^i P(x) = t^k \sum P_i(x)$ for $t \in \mathbf{K}$ and $x \in \mathbf{K}^n$; hence $P_i = 0$ for $i \neq k$.)

Therefore every polynomial $f \in \mathcal{P}(X)$ of degree $\leq k$ has a unique *decomposition into forms* $f = f_0 + \ldots + f_k$, where f_i is a form of degree i on X ($i = 0, \ldots, k$). The form f_i is called the *homogeneous component of degree i of the polynomial* f.

The forms of degree k on X constitute a finite dimensional vector subspace of the space $\mathcal{P}(X)$ (for any $k \in \mathbf{N}$). Forms of degree 1 coincide with the linear mappings. The product of forms of degrees $k_1, \ldots, k_r$, respectively, is a form of degree $k_1 + \ldots + k_r$.

For every polynomial $f \in \mathcal{P}(X)$ of degree $\leq k$ there exists a unique form F of degree k on $\mathbf{K} \times X$ such that $f(x) = F(1, x)$. Namely, $F(t, x) = \sum_{\nu=0}^{k} t^{k-\nu} f_\nu(x)$, where $f = f_0 + \ldots + f_k$ is the decomposition into forms. If $X = \mathbf{K}^n$ and $f(x) = \sum_{|p| \leq k} a_p x^p$, then $F(t, x) = \sum_{\nu+|p|=k} a_p t^\nu x^p$. The mapping $f \longrightarrow F$ is an isomorphism between the vector space of all polynomials of degree $\leq k$ and the vector space of all forms of degree k on $\mathbf{K} \times X$.

Similarly, for every polynomial $f \in \mathcal{P}(X_1 \times \ldots \times X_r)$ (where X_i are finite dimensional vector spaces over $\mathbf{K}$) there is a polynomial $F(t_1, x_1, \ldots, t_r, x_r)$ on $\mathbf{K} \times X_1 \times \ldots \times \mathbf{K} \times X_r$, with the following properties. It is homogeneous with respect to each pair of variables (t_i, x_i), $i = 1, \ldots, r$, and it satisfies the equality $f(x_1, \ldots, x_r) = F(1, x_1, \ldots, 1, x_r)$. In order to prove this, it is enough to use the expansion $f = \sum f_{k_1, \ldots, k_r}$, where $f_{k_1, \ldots, k_r}$ is a homogeneous polynomial of degree k_i with respect to x_i, $i = 1, \ldots, r$, and $f_{k_1, \ldots, k_r} = 0$, when $\max(k_1, \ldots, k_r) > k$. Here k is the degree of the polynomial f. We then define $F = \sum t_1^{k-k_1} \ldots t_r^{k-k_r} f_{k_1, \ldots, k_r}$ [33] .

3. Note that if every element of a field $\mathbf{K}$ has roots of all orders, then

---

[33] In any linear coordinate systems on the spaces X_i, we have

$$f(x_1, \ldots, x_r) = \sum a_{p_1, \ldots, p_r} x_1^{p_1} \ldots x_r^{p_r} .$$

Thus we can define

$$f_{k_1, \ldots, k_r} = \sum_{|p_1|=k_1, \ldots, |p_r|=k_r} a_{p_1, \ldots, p_r} x_1^{p_1} \ldots x_r^{p_r} .$$

every polynomial on X of degree $r > 0$ is monic of degree r with respect to each of its variables (^{34}) in some coordinate system.

Indeed, let $f = f_0 + \ldots + f_r$ be the decomposition of a polynomial f on an n–dimensional space into forms, and let $f_r \neq 0$. It is enough to find a basis $c_1, \ldots, c_n$ in X such that $f_r(c_i) \neq 0$ $(i = 1, \ldots, n)$. Namely, for such a basis, $f_r(t_i c_i) = t_i^r f_r(c_i) = 1$ with some $t_i \in \mathbf{K} \setminus 0$. Hence, as φ is the coordinate system corresponding to the basis $t_1 c_1, \ldots, t_n c_n$ and $f \circ \varphi^{-1} = \sum a_p x^p$, we have $a_{0,\ldots,r,\ldots,0} = (f_r \circ \varphi^{-1})(0, \ldots, 1, \ldots, 0) = 1$, that is, $f \circ \varphi^{-1} = x_1^r + \ldots x_n^r + \sum_{|p| < r} a_p x^p$.

Now take $c_1 \in X$ such that $f_r(c_1) \neq 0$. Having chosen linearly independent $c_1, \ldots, c_k$ (where $1 \leq k \leq n$), we take $a \in X \setminus \sum_1^k \mathbf{K} c_i$. Then the polynomial $K \ni t \longrightarrow f_r(c_1 + ta)$ is not identically zero. Hence $f_r(c_1 + ua) \neq 0$ for some $u \in \mathbf{K} \setminus 0$, and thus it suffices to set $c_{k+1} = c_1 + ua$ (since $c_{k+1} \notin \sum_1^k \mathbf{K} c_i$).

§4. Symmetric polynomials. Discriminant

Let A be a ring.

1. A polynomial $P \in A[X_1, \ldots, X_n]$ is said to be *symmetric* if for each permutation $(\alpha_1, \ldots, \alpha_n)$ we have $P(X_{\alpha_1}, \ldots X_{\alpha_n}) = P$. In the ring $A[T, X_1, \ldots, X_n]$ we have the polynomial identity $(T - X_1) \ldots (T - X_n) = T^n + \sigma_1 T^{n-1} + \ldots + \sigma_n$, where

$$\sigma_i = (-1)^i \sum_{\nu_1 < \ldots < \nu_i} X_{\nu_1} \ldots X_{\nu_i}, \quad i = 1, \ldots, n,$$

are symmetric polynomials from $A[X_1, \ldots, X_n]$. These are called the *basic symmetric polynomials*. Therefore, if $\xi_1, \ldots, \xi_n \in A$, the following polynomial identity holds in $A[X]$

$$(X - \xi_1) \ldots (X - \xi_n) = X^n + \sigma_1(\xi_1, \ldots, \xi_n) X^{n-1} + \ldots + \sigma_n(\xi_1, \ldots, \xi_n).$$

This means that the coefficients of a monic polynomial which can be factorized into linear factors are equal to the values of the basic symmetric polynomials at the sequence of its roots.

(^{34}) That is, for each i it has the form $x_i^r + a_{i1} x_i^{r-1} + \ldots + a_{ir}$, where $a_{i\nu}$ are polynomials in the variables $x_1, \ldots, x_{i-1}, x_{i+1}, \ldots, x_n$.

THEOREM ON SYMMETRIC POLYNOMIALS. *If a polynomial P that belongs to $A[X_1, \ldots, X_n]$ is symmetric, then there is a unique polynomial $Q \in A[Y_1, \ldots, Y_n]$ such that $P = Q(\sigma_1, \ldots, \sigma_n)$.* (See [3], Chapter V, §9.)

2. Consider the symmetric polynomials

$$s_\nu = X_1^\nu + \ldots + X_n^\nu \in \mathbf{Z}[X_1, \ldots, X_n],$$

where $\nu = 1, 2, \ldots$.

Each of the basic symmetric polynomials $\sigma_i \in \mathbf{Z}[X_1, \ldots, X_n]$ can be expressed as a polynomial in $s_1, \ldots, s_n$ with rational coefficients, i.e., $\sigma_i \in \mathbf{Q}[s_1, \ldots, s_n]$. This follows recursively from *Newton's identities*:

$$s_k + \sigma_1 s_{k-1} + \ldots + \sigma_{k-1} s_1 + k\sigma_k = 0, \quad k = 1, \ldots, n .$$

Indeed, by putting $\sigma_0 = 1$ and $s_0 = n$ we get

$$(\sigma_0 + \ldots + \sigma_n)(s_0 + \ldots + s_n) = (1 - X_1) \ldots (1 - X_n) \sum_{r=1}^{n} (1 + X_r + \ldots + X_r^n)$$

$$= \sum_{r=1}^{n} (1 - X_r^{n+1}) \prod_{\nu \neq r} (1 - X_\nu) = n + (n-1)\sigma_1 + \ldots + \sigma_{n-1} + B ,$$

where B is a sum of monomials of degree $> n$. This is because

$$\sum_{r-1}^{n} \prod_{\nu \neq r} (T - X_\nu) = \frac{\partial}{\partial T} \left((T - X_1) \ldots (T - X_n) \right) .$$

Thus (for $k = 1, \ldots, n$) we must have $\sum_{\nu=0}^{k} \sigma_\nu s_{k-\nu} = (n - k)\sigma_k$, i.e., the k-th of Newton's identities.

Alternate proof (by F.Catanese): Fix $k \geq 1$. The identity is true when $n = k$; it is enough to substitute $T = X_i$ in the equation $(T - X_1) \ldots (T - X_n) = T^n + \sigma_1 T^{n-1} + \ldots + \sigma_n$ and to add up all the equations for $i = 1, \ldots, n$. Now assume that $n > k$ and the identity holds for $n - 1$ variables. Put $P = s_k + \sigma_1 s_{k-1} + \ldots + k\sigma_k$. Let $\sigma'_i, s'_i \in \mathbf{Z}[X_1, \ldots, X_{n-1}]$ be polynomials defined in a fashion similar to σ_i, s_i, respectively. We have $P' = s_k + \sigma'_1 s'_{k-1} + \ldots + k\sigma'_k = 0$. Take the homomorphism

$$\chi : \mathbf{Z}[X_1, \ldots, X_n] \ni P \longrightarrow P(X_1, \ldots, X_{n-1}, 0) \in \mathbf{Z}[X_1, \ldots, X_{n-1}]$$

and note that $\chi(\sigma_i) = \sigma'_i$, $\chi(s_i) = s'_i$, and $\ker \chi = \{\sum a_p X^p : p_n = 0 \Longrightarrow a_p = 0\}$. Thus $\chi(P) = P' = 0$, i.e., $P \in \ker \chi$. On the other hand, P is symmetric, hence $p_i = 0 \Longrightarrow a_p =$

0 $(i = 1, \ldots, n)$, where $P = \sum a_p X^p$. Thus $a_p \neq 0 \implies |p| \geq n$. Consequently $P = 0$, as the degree of P is less than or equal to $k < n$.

Note also that Newton's identities remain true for an arbitrary $k \geq 1$, provided we put $\sigma_k = 0$ for $k > n$. This is the case because, if $k > n$, it suffices to substitute $X_{n+1} = \ldots = X_k = 0$ in the k–th Newton identity in $\mathbf{Z}[X_1, \ldots, X_k]$.

Therefore, *if a ring A contains a field of characteristic zero, the theorem on symmetric polynomials is true when $\sigma_1, \ldots, \sigma_n$ are replaced by the polynomials $s_1, \ldots, s_n$* [35] .

3. Since the polynomial $\prod_{i<j}(X_i - X_j)^2 \in \mathbf{Z}[X_1, \ldots, X_n]$ is symmetric, the theorem on symmetric polynomials implies the existence of a unique polynomial $D_n \in \mathbf{Z}[Y_1, \ldots, Y_n]$ such that

$$\prod_{i<j}(X_i - X_j)^2 = D_n(\sigma_1, \ldots, \sigma_n) .$$

The discriminant of a monic polynomial $P = X^n + a_n X^{n-1} + \ldots + a_n \in A[X]$ is defined to be the element $D_n(a_1, \ldots, a_n) \in A$ [36] . If the polynomial P can be factorized in $A[X]$ into linear factors, i.e., $P = (X - \xi_1) \ldots (X - \xi_n)$, where $\xi_\nu \in A$, then (since $a_i = \sigma_i(\xi_1, \ldots, \xi_n)$) we have

$$D_n(a_1, \ldots, a_n) = \prod_{i<j}(\xi_i - \xi_j)^2 = (-1)^{\binom{n}{2}} \prod_{1}^{n} P'(\xi_i) .$$

If δ is the discriminant of a monic polynomial $P \in A[X]$, then

$$GP + HP' = \delta$$

for some $G, H \in A[X]$ (see [3], Chapter V, §10).

Let $h : A \longrightarrow B$ be a ring homomorphism, and let $H : A[X] \longrightarrow B[X]$ denote the induced homomorphism. If δ is the discriminant of a monic polynomial $P \in A[X]$, then $h(\delta)$ is the discriminant of the monic polynomial $H(P) \in B[X]$.

---

[35] The uniqueness is a direct consequence of the following easy-to-verify statement: if the Jacobian of the polynomials $P_1, \ldots, P_n \in A[X_1, \ldots, X_n]$ is not a zero divisor, then the polynomials are algebraically independent.

[36] We assume that $D_0 = 1$. Thus the discriminant of the polynomial 1 is equal to 1.

§5. Extensions of fields

Let K be a field, and let L be a field extension of K. Then L can be regarded as a vector space over K and its dimension $\dim_K L$ is called *the dimension of the field L over the field K* ([37]).

1. If $f, g \in K[X]$ and g divides f in $L[X]$, then the same is true in $K[X]$. Indeed, we may assume that g is monic. Then $f = qg + r$, where $q, r \in K[X]$ and the degree of r is smaller than the degree of g. It follows from our hypothesis that $r = 0$ (see 2.4).

The greatest common divisor of two polynomials from $K[X]$ is also their greatest common divisor in $L[X]$. This is so because the greatest common divisor of polynomials $f, g \in K[X]$ has the form $pf + qg$ (where $p, q \in K[X]$), since $K[X]$ is a principal ideal domain (see 1.12 and 2.9).

2. Let $\zeta \in L$ be an algebraic element over K. There is a unique monic polynomial in $K[X]$ of minimal degree, such that ζ is its root. It is called *the minimal polynomial of ζ over K*. It is irreducible in $K[X]$ and divides each polynomial from $K[X]$ for which ζ is a root.

Indeed, the set $I = \{f \in K[X] : f(\zeta) = 0\}$ is a non-zero ideal in the ring $K[X]$. Its monic generator g is the unique monic polynomial in I of minimal degree. It must be irreducible, because I is prime.

An irreducible monic polynomial in $K[X]$ is a minimal polynomial for each of its roots in L. For if it were not minimal for a root ζ, it would be divisible by a polynomial of a smaller degree, namely, by the minimal polynomial of ζ.

Thus, if K is of characteristic zero and a polynomial $g \in K[X]$ is irreducible, then each of its roots $\zeta \in L$ is *simple*, i.e., $(X - \zeta)^2$ is not a divisor of $g \in L[X]$. Otherwise, ζ would be a root of the derivative $g' \in K[X]$, i.e., a root of a non-zero polynomial whose degree is smaller than that of g.

3. If the elements $\zeta_1, \ldots, \zeta_k \in L$ are algebraic over K, then $K(\zeta_1, \ldots \ldots, \zeta_k) = K[\zeta_1, \ldots, \zeta_k]$ (see [1a], Chapter X, §1).

We say that a field L is a *finite extension of the field K* if $\dim_K L < \infty$. This happens precisely when L is generated over K by a finite number of algebraic elements. Then every element of the field L is algebraic over K. (See [2], Chapter I, §§2.2, 2.3; or [3], Chapter VII, §1.)

([37]) It is often denoted by $[L : K]$ and called the *degree of the field L over the field K*.

If $\zeta \in L$ is an algebraic element over K, then $1, \zeta, \ldots, \zeta^{n-1}$ is a basis for $K(\zeta)$ regarded as a vector space over K, where $n = \dim_K K(\zeta)$ is the degree of the minimal polynomial for the element ζ. This n is called the *degree of the algebraic element ζ over K*. (See [1a], Chapter X, §§1 and 2.)

Let K' be a field with an extension L', and let $\eta \in L$, $\zeta' \in L'$. Then an isomorphism $K \longrightarrow K'$ can be extended to an isomorphism $K(\zeta) \longrightarrow K'(\zeta')$ that maps ζ onto ζ', provided the elements ζ and ζ' are transcendental over K and K', respectively (see 2.1 and 7; 1.15 and 2.8), or they are roots of irreducible polynomials that correspond to each other via the induced isomorphism $K[X] \longrightarrow K[X']$ (see [1a], Chapter X, §5). In particular, if $\zeta', \zeta'' \in L$ are roots of an irreducible polynomial in $K[X]$, then there exists a K–isomorphism $\varphi : K(\zeta') \longrightarrow K(\zeta'')$ such that $\varphi(\zeta') = \zeta''$.

For every monic polynomial $P \in K[X]$ there exists a finite extension K' of the field K, such that P can be factorized in $K'[X]$ into linear factors: $P = (X - \zeta_1) \ldots (X - \zeta_n)$, where $\zeta_1, \ldots, \zeta_n \in K'$. (See [1a], Chapter X, §4; or [2], Chapter II, §1.1).

§6. Factorial rings

1. We say that two elements of a ring are *associated* if they differ by an invertible factor or, equivalently, if they generate the same ideal. An integral domain is said to be *factorial* or a *unique factorization domain* if every non-zero non-invertible element can be written as a product of irreducible elements, and this representation is unique to within order and association. In particular, every field is a unique factorization domain.

Let A be a factorial ring.

If $a = a_1, \ldots, a_k$, where the a_i's are irreducible, then each non-invertible divisor of the element a is associated with an element of the form $a_{\alpha_1} \ldots a_{\alpha_s}$, where $\alpha_1 < \ldots < \alpha_s$. An irreducible divisor of a product of elements must divide one of them. Thus an element is irreducible if and only if it is prime (see 1.13). If $c_1, \ldots, c_k$ are all the distinct factors, i.e., non-associated irreducible divisors of a non-zero element a, then $a = dc_1^{s_1} \ldots c_k^{s_k}$, where d is an invertible element and $s_i > 0$. The number s_i is called the *multiplicity of the factor c_i* and is uniquely determined by a. If $s_i > 1$, we say that c_i is a *multiple factor* of the element a. The element a has multiple factors if and only if it is divisible by the square of a non-invertible element. Note that rad $Ac_1^{s_1} \ldots c_k^{s_k} = Ac_1, \ldots, c_k$. Thus the element a does not have multiple factors precisely when rad $Aa = Aa$. Any finite system of elements has a greatest common divisor which is unique up to association. If not all of the elements are equal to zero, and $c_1, \ldots, c_k$ are all non-associated common irreducible divisors, then it is of the form $dc_1^{s_1} \ldots c_k^{s_k}$, where d is an invertible element and $s_i > 0$. We say that elements are *relatively prime* if their greatest common divisor is the identity or, equivalently, if they do not have common irreducible divisors. An element d is the greatest common divisor of the elements $a_1, \ldots, a_k$ if and only if $a_i = db_i$, $i = 1, \ldots, k$, where $b_1, \ldots, b_k$ are relatively prime. (See [1a], Chapter IX, §§2-4).

Note also that if B is a subring of A such that all divisors in A of every non-zero element of B belong to B, then B is factorial.

Finally, note that when A is a polynomial ring over an integral domain, then in the decomposition of a homogeneous polynomial into irreducible factors each of the factors has to be homogeneous (see 2.3).

2. Let A be a factorial ring. A non-zero polynomial $P \in A[X]$ is said to be *primitive* if its coefficients are relatively prime [38] .

In particular, every monic polynomial is primitive. Each polynomial $Q \in A[X]$ can be written as cP, where $c \in A$ and $P \in A[X]$ is a primitive polynomial. In such a setting, c must be the greatest common divisor of the coefficients of the polynomial Q.

THE GAUSS LEMMA . *The product of primitive polynomials is primitive.* (See [1a], Chapter IX, §8).

COROLLARY . *Let $P, Q \in A[X]$, and let us assume that P is primitive. Let K denote the field of fractions of the ring A. If P is a divisor of Q in $K[X]$, then P is also a divisor of Q in $A[X]$.*

Indeed, excluding the trivial case when $Q = 0$, we have $bQ = cGP$, where $b, c \in A \setminus 0$ and $G \in A[X]$ is primitive. By applying the Gauss lemma, we see that GP is primitive, and hence c is the greatest common divisor of the coefficients of the polynomial bQ. Therefore b divides c, and hence $Q = aGP$, where $a \in A$.

THE GAUSS THEOREM . *If A is a factorial ring, then so is the ring $A[X_1, \ldots, X_n]$.* (See [1a], Chapter IX, §8).

In particular, *the ring of polynomials over a field is factorial.*

3. Let A be a factorial ring of characteristic zero, and let P be a monic polynomial from $A[X]$.

PROPOSITION . *The following conditions are equivalent.*

---

[38] In the case of a polynomial of degree zero, being primitive means the same as being an invertible constant.

(1) *The polynomial P has multiple factors (or, equivalently, it is divisible by the square of a polynomial of positive degree).*

(2) *The polynomial P and its derivative P' have a common divisor of positive degree.*

(3) *The discriminant of the polynomial P is zero.*

In particular, the discriminant of an irreducible monic polynomial is different from zero.

PROOF. If $G \in A[X]$ and G^2 is a divisor of P, then G is a common divisor of P and P'; hence (1) $\Longrightarrow$ (2). Conversely, it follows from the condition (2) that P and P' have a common irreducible divisor, say G. Thus $P = GH$ and $P' = GF$ for some $F, H \in A[X]$. Hence $G'H = G(F - H')$. Therefore G divides H and G^2 divides P. Let K be the field of fractions for the ring A, and let L be an extension of K such that P can be factorized in $L[X]$ into linear factors: $P = (X - \zeta_1) \ldots (X - \zeta_n)$. Then condition (3) means that $P'(\zeta_r) = 0$ for some r. Hence (2) yields (3). Conversely, note that there is a greatest common divisor of the polynomials P and P' in $K[X]$ which is primitive in $A[X]$; it is their common divisor in $A[X]$ and their greatest common divisor in $L[X]$. Therefore it is of positive degree if condition (3) is fulfilled.

§7. Primitive element theorem

A *primitive element of an extension L of a field K* is an element $\zeta \in L$ which is algebraic over K and such that $L = K(\zeta)$. Notice that in such a case, for any $c \in K \setminus 0$ the element $c\zeta$ is then primitive.

THE PRIMITIVE ELEMENT THEOREM. *Every finite extension L of a field K of characteristic zero has a primitive element. Moreover, if $Z_1, \ldots, Z_r$ are infinite subsets of the field K and $L = K(\eta_1, \ldots, \eta_r)$, then there exists a primitive element*

$$\zeta \in \sum_1^r Z_i \eta_i.$$

PROOF. First of all, it is enough to show the existence of a primitive element of the form $\eta_1 + c_2 \eta_2 + \ldots + c_r \eta_r$ for arbitrarily given infinite sets $Z_2, \ldots, Z_r \subset K$, where $c_i \in Z_i$ [39]. Secondly, it is sufficient to prove this

---

[39] For then we can take $c \in Z_1 \setminus 0$ and the primitive element $\zeta' = \eta_1 + c_2 \eta_2 + \ldots + c_r \eta_r$ (with $c_i \in c^{-1} Z_i$) and define $\zeta = c\zeta'$.

statement for $r = 2$ (the general case would follow by induction). Let $L = K(\alpha, \beta)$, where $\alpha, \beta \in L$ are algebraic elements over K. Let $f, g \in K[X]$ be their minimal polynomials, and let L' be an extension of the field L such that f, g can be factorized in $L'[X]$ into linear factors: $f = (X - \alpha_1)\ldots(X - \alpha_m)$ and $g = (X - \beta_1)\ldots(X - \beta_n)$, where $\alpha_1 = \alpha$ and $\beta_1 = \beta$. Since L' is of characteristic zero and f, g are irreducible in $K[X]$, the α_i's, as well as the β_j's, are mutually distinct. Now, there is $c \in Z_2$ such that $\alpha_i + c\beta_j \neq \alpha + c\beta$ for $i = 1, \ldots m$ and $j = 2, \ldots, n$ (because for each such pair i, j the equality could be true for at most one such c). Put $\zeta = \alpha + c\beta$. Clearly, $K(\zeta) \subset K(\alpha, \beta)$. In order to prove the opposite inclusion, consider the polynomial $h = f(\zeta - cX) \in K(\zeta)[X]$. Obviously, $h(\beta) = 0$ and $h(\beta_j) \neq 0$ for $j \geq 2$ (because $\zeta - c\beta_j \neq \alpha_i$ for $i = 1, \ldots, m$, for such j's). Therefore among the factors $X - \beta_j$, only $X - \beta$ is a common divisor of the polynomials g and h in $L'[X]$. It follows that it is their greatest common divisor in $L'[X]$. Let d denote their greatest common divisor in $K(\zeta)[X]$. Then d is also their greatest common divisor in $L'[X]$, and hence $d = a(X - \beta)$, where $a \in L' \setminus 0$. This implies that $a, a\beta \in K(\zeta)$. Thus $\beta \in K(\zeta)$ and $\alpha = \zeta - c\beta \in K(\zeta)$. Hence $K(\alpha, \beta) \subset K(\zeta)$.

An extension L of a field K is said to be *algebraic (over K)* if each element of L is algebraic over K.

COROLLARY ([40]) . *Let L be an algebraic extension of a field K of characteristic zero. Then L is a finite extension of K if and only if the degree n_x of x over K is a bounded function on L. Then its maximum n is equal to $\dim_K L$, and $\{x \in L : n_x = n\}$ is the set of all primitive elements of the extension.*

In fact, if the extension is finite, then $n_x \leq \dim_K L$. Now, assume that $n = \sup n_x < \infty$ and $n_x = n$. For any $z \in L$ we have $K(x) \subset K(x, z) = K(w)$ for some $w \in L$, and $\dim_K K(w) \leq n = \dim_K K(x)$ (see 5.3), which implies that $K(x) = K(x, z)$, i.e., $z \in K(x)$. Therefore $L = K(x)$.

§8. Extensions of rings

Let A be a ring, and let B be a ring extension of A.

1. We say that an element of the ring B is *integral over A* if it is a root of a monic polynomial from $A[X]$. The integral elements of B over A constitute

---

([40]) Observed by A. Płoski and T. Winiarski. Cf. [44] chap. 6, lemma 8J.

a subring (containing A), which is called the *integral closure of the ring A in the ring B*. If the ring A is equal to its integral closure in B, it is said to be *integrally closed in the ring B*. The ring B is *finite over A*, i.e., finitely generated, as a module over A [41] exactly when, as a ring, it is generated over A by a finite number of integral elements (i.e., when $B = A[\zeta_1, \ldots, \zeta_r]$, where $\zeta_i \in B$ are integral over A). In such a case, the ring B is *integral over A*, i.e., each of its elements is integral over A.

If B is integral over A, then every element of an extension C of the ring B which is integral over B is also integral over A. In particular, if C is integral over B, then it is also integral over A.

(See [1], Chapter III, §1; or [4], 5.1 – 5.4).

It follows from the Mather-Nakayama lemma (see 1.16) that if B is an integral domain and I is a non-zero finitely generated ideal in A, then each element $x \in B$ such that $Ix \subset I$ is integral over A [42].

A ring A is said to be *integrally closed* if it is an integral domain which is integrally closed in its field of fractions. Every factorial ring is integrally closed. (See [1], Chapter III, §1; or [2], Chapter V, §1.1; or [3], Chapter IX, §1.) Obviously, if the A_ι are integrally closed rings which are subrings of a given ring, then their intersection $\bigcap A_\iota$ is also an integrally closed ring.

Note also that the Cartesian product of rings $A_1 \times \ldots \times A_k$ is integrally closed in its ring of fractions if and only if the rings $A_1, \ldots, A_k$ are integrally closed in their rings of fractions [43]. (This follows directly from the definition of an integral element; see 1.15a.)

2. Assume now that A is factorial and B is an integral domain. For any element $\zeta \in B$ which is integral over A, there exists a unique monic polynomial in $A[X]$ of minimal degree such that ζ is its root. It is called the *minimal polynomial of ζ over A*. It is irreducible in $A[X]$ and divides (in $A[X]$) every polynomial in $A[X]$ with root ζ. It is also equal to the minimal polynomial of ζ (regarded as an element of the field of fractions of the ring B) over the field of fractions K of the ring A.

Indeed, let $g \in K[X]$ be the minimal polynomial for ζ over K. It is enough to show that $g \in A[X]$ (because in such a case, since g is primitive, every polynomial in $A[X]$ which is divisible by g in $K[X]$ is also divisible by

---

[41] If A and B are fields, then B is a finite extension of the field A.

[42] For in this case I is a module over $A[x]$, and hence $Iz = 0$ for some $z = x^k + a_1 + \ldots + a_k$, $a_i \in A$. Consequently, $z = 0$. (See also [1], (3.1.6); or [4], 5.1.)

[43] The Cartesian product of two or more non-zero rings is never an integral domain: $(1, 0, \ldots, 0) \cdot (0, 1, 0, \ldots, 0) = (0, \ldots, 0)$.

g in $A[X]$). We have $g = cg_0$ for some $c \in K \setminus 0$, where g_0 is a primitive polynomial from $A[X]$. It is sufficient to show that the leading coefficient a of the polynomial g_0 is invertible in A (since, in view of $ac = 1$, this would yield that $c \in A$). Now, $h(\zeta) = 0$ for some monic polynomial $h \in A[X]$. Thus g_0 is a divisor of h in $K[X]$ and hence also in $A[X]$. This implies that a is invertible in A.

More generally, for the above to hold, it suffices to assume that A is integrally closed (and B remains an integral domain).

In fact, we have $g = (X - \zeta_1)\ldots(X - \zeta_r)$ in $L[X]$ for some extension L of K (see 5.3). Then each ζ_i is integral over A, because there is a K–isomorphism $\varphi : K(\zeta) \longrightarrow K(\zeta_i)$ such that $\varphi(\zeta) = \zeta_i$ (see 5.3). It follows that the coefficients of g are integral over A, and so $g \in A[X]$. Next, suppose that $f \in A[X]$ and $f(\zeta) = 0$. We have $f = qg + r$ with $q, r \in A[X]$ and $\deg r < \deg g$ (see 2.4). Then $r(\zeta) = 0$, which implies that $r = 0$, as g is minimal for ζ in $K[X]$. Therefore g divides f.

3. In what follows, let us assume that B is an integral domain. Let $K \subset L$ be the field of fractions of the rings A and B, respectively.

If $B = A[\zeta_1, \ldots, \zeta_r]$, where $\zeta_i \in B$ are algebraic over A, then $L = K[\zeta_1, \ldots, \zeta_r]$. Therefore

LEMMA . *If B is finite over A, then L is a finite extension of the field K. Each element of L is of the form b/a, where $b \in B$ and $a \in A \setminus 0$, and for any $\xi \in B \setminus 0$ there is an $\eta \in B$ such that $\xi\eta \in A \setminus 0$.*

(Indeed, $\xi^{-1} = \eta/a$, where $\eta \in B$ and $a \in A \setminus 0$.)

REMARK ([44]) . For the second part of the lemma it is enough to suppose that B is integral over A.

(It suffices to observe that if $t \in B$, then $tu \in A$ for some $u \in B \setminus 0$: we have $t^r + a_1 t^{t-1} + \ldots + a_r = 0$ with some $a_i \in A$ and r as small as possible, so $u = t^{r-1} + a_1 t^{r-1} + \ldots + a_{r-1} \in B \setminus 0$ and $tu = -a_r$.)

PROPOSITION . *If A is of characteristic zero and B is finite and free as a module over A, then $g_A(B) = \dim_K L$.*

PROOF . Put $m = \dim_K L$ and $l = g_A(B)$. Let $z_1, \ldots, z_n$ be a basis of B regarded as a module over A. The elements $z_1, \ldots, z_n$ are linearly independent over A and hence over K. Therefore $m \geq n \geq l$. On the other hand, L is a finite extension of K and has a primitive element $\zeta \in B$ (of the form $\zeta = \sum_1^m a_i z_i$, where $a_i \in A$, because $L = K(z_1, \ldots, z_n)$ and A is infinite). Hence $1, \zeta, \ldots, \zeta^{m-1} \in B$ is a basis for L over K. Furthermore, B, as a module

---

([44]) Observed by A. Płoski

over A, has a system of generators $x_1, \ldots, x_l$ which also generates L regarded as a vector space over K. (For the elements ζ^ν are linear combinations of the elements x_i with coefficients in A.) Hence $l \geq m$.

An element $\zeta \in B$ is said to be a *primitive element of the extension B of the ring A* if it is a primitive element of the extension L of the field K or, equivalently, if it is algebraic over A and satisfies the condition:

$$\text{if } x \in B, \text{ then } \alpha x \in A[\zeta] \text{ for some } \alpha \in A \setminus 0 .$$

THE PRIMITIVE ELEMENT THEOREM . *Assume that the ring A is of characteristic zero. If $B = A[\eta_1, \ldots, \eta_r]$, where $\eta_i \in B$ are algebraic over A and $Z_1, \ldots, Z_r$ are infinite subsets of the ring A, then there exists a primitive element of the form $\sum_1^r c_i \eta_i$, where $c_i \in Z_i$. If A is factorial, B is finite over A and ζ is a primitive element of B, then*

$$\delta B \subset A[\zeta],$$

where δ is the discriminant of the minimal polynomial $P \in A[X]$ of the element ζ over A ([45]) .

PROOF . It is enough to prove the second part of the theorem, as the first part is an immediate corollary of the primitive element theorem for fields. Let L' be an extension of the field L such that P can be factorized in $L'[X]$ into linear factors $P = (X - \zeta_1) \ldots (X - \zeta_k)$, where $\zeta_i \in L'$ and $\zeta_1 = \zeta$. Then the ζ_i are mutually distinct and integral over A. Now let $x \in B$. Then $x = R(\zeta)$, where $R \in K[X]$ (since $L = K[\zeta]$). The elements $x_i = R(\zeta_i)$, $i = 1, \ldots, k$, are also integral over A. (Indeed, by taking a monic polynomial $g \in A[X]$ such that $g(x) = 0$ and a K–isomorphism $\varphi : K(\zeta) \longrightarrow K(\zeta_i)$ such that $\varphi(\zeta) = \zeta_i$, we get $\varphi(x) = x_i$ and thus $g(x_i) = 0$.) Let us consider the following polynomial from $L'[X]$:

$$Q = \sum_1^k x_i \prod_{\nu \neq i} (X - \zeta_\nu) = \sum_1^k R(\zeta_i) \prod_{\nu \neq i} (X - \zeta_\nu) .$$

Its coefficients are values taken at $(\zeta_1, \ldots, \zeta_n)$ by symmetric polynomials from $K[X_1, \ldots, X_k]$. Hence (by the theorem on symmetric polynomials) they must belong to K. Moreover, they are integral over A (because x_i and ζ_ν are integral over A). Therefore, since A is a factorial ring, they belong to A, i.e., $Q \in A[X]$. Consequently, $xP'(\zeta) = Q(\zeta) \in A[\zeta]$. Since $GP + HP' = \delta$ for some $G, H \in A[X]$, we conclude that $\delta = H(\zeta)P'(\zeta)$, and hence $x\delta \in A[\zeta]$.

([45]) It also follows from the proof that $P'(\zeta)B \subset A[\zeta]$.

§9. Noetherian rings

1. A ring is said to be *noetherian* if it satisfies one of the following equivalent conditions

(1) each of its ideals is finitely generated;

(2) each increasing sequence of its ideals is stationary (i.e., constant from some term on);

(3) every non-empty family of ideals has a maximal element (with respect to inclusion).

(See [1], Chapter II, §1; or [3], Chapter VI, §1; or [2], Chapter V, §2.1.) In particular, every field is a noetherian ring.

Let A be a noetherian ring.

2. An ideal (of the ring A) that is generated by a set E always has a finite system of generators belonging to E. (This is because a maximal element of the family of all the ideals generated by finite subsets of E must contain E, and therefore is equal to the ideal generated by E.)

We have

$$I \subset \operatorname{rad} J \Longleftrightarrow (I^s \subset J \text{ for some } s)$$

for ideals I, J of the ring A. (Indeed, if $I \subset \operatorname{rad} J$ and $a_1, \ldots, a_k$ generate I then, for some l, we have $a_j^l \in J\, i = 1, \ldots, k$, hence $I^{kl} \subset J$.)

If the noetherian ring A is such that every two ideals of A are comparable (i.e., $I \subset J$ or $J \subset I$), then A is a principal ideal domain. This is so because if an ideal I is generated by the elements $x_1, \ldots, x_k$, one of the ideals $Ax_1, \ldots, Ax_k$, e.g., Ax_s must contain the others. Therefore $I = Ax_s$.

3. By a *primary ideal* of a ring we mean a proper ideal I such that $xy \in I \Longrightarrow (x \in \operatorname{rad} I \text{ or } y \in \operatorname{rad} I)$. If I is primary, then its radical $\operatorname{rad} I$ is prime. The inverse image of a primary ideal under a ring homomorphism is a primary ideal ([46]) .

Every proper ideal I of the noetherian ring A has an *irreducible primary decomposition*, i.e., $I = J_1 \cap \ldots \cap J_k$, where J_i are primary ideals such that none of them contains the intersection of the remaining ones ([47]) and the ideals $I_i = \operatorname{rad} J_i$ are mutually distinct. The ideals I_i are prime and they are uniquely determined by I. They are called the *ideals associated with I*.

([46]) Since $f^{-1}(\operatorname{rad} I) = \operatorname{rad} f^{-1}(I)$ for any ring homomorphism f.

([47]) In such a case we say that the *intersection $J_1 \cap \ldots \cap J_k$ is irreducible*.

The minimal elements of the set $\{I_1, \ldots, I_k\}$ are called the *isolated ideals* for I. They are precisely the minimal elements of the family of prime ideals containing I. Every prime ideal that contains I also contains an isolated ideal for I. (See [1], Chapter II, §3; or [4], Chapters 4,7.)

Note that if the radical of an ideal J is prime, then it is an isolated ideal for J. (For if there were a prime ideal I such that $J \subset I \subset \operatorname{rad} J$, we would have $I = \operatorname{rad} J$.)

The set of zero divisors in the ring A is equal to the union of all ideals associated with the zero ideal. Indeed, let $0 = \bigcap_1^k J_i$ be an irreducible primary decomposition. If x is a zero divisor, i.e., $ax = 0$ for some $a \neq 0$, then $a \notin J_s$ for some s, and so $x \in \operatorname{rad} J_s$. Each of the ideals J_s consists of zero divisors. Indeed, if $x \in J_s$, then (excluding the trivial case $k = 1$), there exists $a \in \bigcap_{i \neq s} J_i \setminus J_s$. Hence $ax \in \bigcap_1^k J_i$ and thus $ax = 0$. Therefore $\operatorname{rad} J_s$ consists also of zero divisors (see 1.6).

The same argument proves that the union of all ideals associated with an ideal I of the ring A is equal to the set of zero divisors in A/I.

3a. By a *ring without nilpotents* we mean a ring in which $t^k = 0 \Longrightarrow t = 0$ (48).

PROPOSITION . *If a noetherian ring A is without nilpotents, then its ring of fractions R is isomorphic to a finite Cartesian product of fields* (49).

PROOF . Take an irreducible primary decomposition of zero. Since $\operatorname{rad} 0 = 0$, we have (see 1.5) $0 = \bigcap_1^r I_i$, where I_i are prime and (by deleting some of the I_i's) we may assume that it is an irreducible primary decomposition. Therefore $N = \bigcup_1^r I_i$ is the set of all zero divisors (see n° 3). The sets $\bar{I}_i = \{y/x : y \in I_i, x \in A \setminus N\}$ are prime ideals in the ring R (50). Moreover, $\bar{I}_i \cap A = I_i$ (51), $\bigcap_1^r \bar{I}_i = 0$ (52), and the intersection is irreducible. The set $\bigcup_1^r \bar{I}_i$ contains all non-invertible elements of the ring R (53). This implies that the ideals $\bar{I}_i$ are maximal. Indeed, if an ideal $J \supset \bar{I}_i$ does not contain invertible elements, then $J \subset \bigcup_1^r \bar{I}_\nu$, and so $J \subset \bar{I}_s$ for some s (see 1.11).

---

(48) i.e., 0 is the only nilpotent.

(49) See 1.1.

(50) If $(y'/x')(y''/x'') = y/x$, where $y \in I_i$ and $x, x', x'' \in A \setminus N$, then $y'y''x \in I_i$ and hence $y' \in I_i$ or $y''x \in I_i$, i.e., $y'/x' \in \bar{I}_i$ or $y''/x'' = y''x/x''x \in \bar{I}_i$.

(51) If $u = y/x \in A$, $y \in I_i$, $x \in A \setminus N$, then $ux \in I_i$, and so $u \in I_i$.

(52) If $y_1/x = \ldots = y_r/x$, where $y_i \in I_i$, $x \in A \setminus N$, then $y_1 = \ldots = y_r \in \bigcap_1^r I_i = 0$.

(53) If y/x is non-invertible, then $y \in N$ and hence $y \in I_s$ for some s. Thus $y/x \in \bar{I}_s$.

Hence $s = i$ and $J = \bar{I}_i$. Therefore

$$\bar{I}_i + \bigcap_{\nu \neq i} \bar{I}_\nu = R \quad \text{for} \quad i = 1, \ldots, r .$$

Let $h_i : R \longrightarrow R/\bar{I}_i$ be the natural homomorphisms. It is enough to show that the homomorphism $h = (h_1, \ldots, h_r) : R \longrightarrow R/\bar{I}_1 \times \ldots \times R/\bar{I}_r$ is an isomorphism. Now, $\ker h = \bigcap_1^r \bar{I}_\nu = 0$. In order to verify surjectivity, take $z_i \in R/\bar{I}_i$, $i = 1, \ldots, r$. Then $z_i = h_i(x_i)$, where $x_i \in \bigcap_{\nu \neq i} \bar{I}_\nu$. Thus $h_\nu(x_i) = 0$ for $\nu \neq i$. Taking $x = x_1 + \ldots + x_r$, we get $h(x) = \big(h_1(x_1), \ldots, h_r(x_r)\big) = (z_1, \ldots, z_r)$.

4. For any ideal I, the ring A/I is also noetherian. Note that for ideals $J_0 \supset J \supset I$ we have the equivalence

$$(J_0 \text{ is isolated for } J) \Leftrightarrow (J_0/I \text{ is isolated for } J/I) .$$

If A is a noetherian ring, then so is its image $h(A)$ under a ring homomorphism h (since $h(A)$ is isomorphic to $A/\ker h$).

HILBERT'S BASIS THEOREM . *If a ring A is noetherian, then so is the ring* $A[X_1, \ldots, X_m]$ (See [1], Chapter II, §2; or [3], Chapter VI, §2.)

In particular, any ring of polynomials over a field is noetherian.

If $x_1, \ldots, x_k$ are elements of an extension of the noetherian ring A, then the ring $A[X_1, \ldots, X_k]$ is noetherian. (For the latter is the image of a ring of polynomials over A under a homomorphism.)

4a. Now we are going to prove the Artin-Rees lemma (in the special case of ideals ([54])).

THE ARTIN-REES LEMMA . *Let* $\mathfrak{m}, I$ *be ideals of a noetherian ring A. Then there exists k such that*

$$\mathfrak{m}^n \cap I = \mathfrak{m}^{n-k}(\mathfrak{m}^k \cap I) \text{ for } n \geq k .$$

PROOF ([55]) . Let $x_1, \ldots, x_r$ be generators of the ideal $\mathfrak{m}$. The subring of the ring $A[T]$ given by

$$B = A[x_1 T, \ldots, x_n T] = \left\{ \sum a_\nu T^\nu : a_\nu \in \mathfrak{m}^\nu \right\}$$

([54]) For the case of modules, see e.g. [1], Chapter II, §5; or [4], 10.9; or [7], Chapter VII, §2; or [3], p.155.

([55]) See [3], p.155.

(see 1.7) is noetherian. Notice that if there exist ideals $\mathfrak{m}_\nu \subset \mathfrak{m}^\nu$ of the ring A which satisfy the condition $\mathfrak{m}\mathfrak{m}_\nu \subset \mathfrak{m}_{\nu+1}$ [56], then $J = \{\sum a_\nu T^\nu : a_\nu \in \mathfrak{m}_\nu\}$ is an ideal of the ring B. Obviously, the ideals $\mathfrak{m}_\nu$ are uniquely determined by the ideal J. In particular, since $\mathfrak{m}(\mathfrak{m}_\nu \cap I) \subset \mathfrak{m}^{\nu+1} \cap I$, we conclude that

$$J_n = \left\{\sum a_\nu T^\nu : a_\nu \in \mathfrak{m}^\nu \cap I \text{ for } \nu \leq n, \ a_\nu \in \mathfrak{m}^{\nu-n}(\mathfrak{m}^n \cap I) \text{ for } \nu \geq n\right\}$$

are ideals of the ring B and constitute an increasing sequence. Hence there exists k such that $J_n = J_k$ for $n \geq k$, and so $\mathfrak{m}^n \cap I = \mathfrak{m}^{n-k}(\mathfrak{m}^k \cap I)$.

COROLLARY . *Let $\mathfrak{m}, I, J$ be ideals of a noetherian ring A. Then there exists k such that $(J + \mathfrak{m}^n) \cap I \subset J + \mathfrak{m}^{n-k}I$ for $n \geq k$.*

Indeed, it is sufficient to apply the Artin-Rees lemma to the ring A/J and the images $\varphi(\mathfrak{m}), \varphi(I)$ under the natural homomorphism $\varphi : A \longrightarrow A/J$. Namely, for $n \geq k$, we have

$$\varphi\big((J + \mathfrak{m}^n) \cap I\big) \subset \varphi(\mathfrak{m})^n \cap \varphi(I) \subset \varphi(\mathfrak{m})^{n-k}\varphi(I) = \varphi(\mathfrak{m}^{n-k}I) \ ,$$

which yields the required inclusion.

5. For a noetherian ring, the property of being factorial can be characterized as follows:

A noetherian integral domain A is factorial if and only if each of its irreducible elements is prime [57] .

Indeed, the condition is necessary (see 6.1). Assume now that it is satisfied. In order to prove existence of a decomposition into irreducible factors, denote by $\mathcal{N}$ the set of all non-zero non-invertible elements that do not have such a decomposition. Note that if $x \in \mathcal{N}$, then $Ax \subsetneq Ay$ for some $y \in \mathcal{N}$. (Indeed, we must have $x = yz$ for some non-zero non-invertible y, z. For if $y \in \mathcal{N}$, then it must follow that $y \notin Ax$. Otherwise $y = ax = ayz$ for some $a \in A$, and thus $az = 1$, in contradiction with the fact that z is not invertible.) If we had $\mathcal{N} \neq \emptyset$, then by taking $x \in \mathcal{N}$, we would obtain an infinite sequence of ideals $Ax \subsetneq Ax_1 \subsetneq Ax_2 \subsetneq \ldots$, contrary to our assumption that the ring is noetherian. Uniqueness of decomposition follows directly from the fact that every irreducible (and hence prime) element which divides the product of elements must divide one of them (see 1.13).

6. Similarly, in the case of modules, the conditions (1), (2), (3) are equivalent – provided that ideals are replaced by submodules – and a module satisfying any of these conditions is said to be *noetherian*. Also (see n° 2), a

---

[56] This implies that $\mathfrak{m}^s\mathfrak{m}_\nu \subset \mathfrak{m}_{\nu+s}$.

[57] Another characterization will be given in 12.2.

submodule of a noetherian module generated by a set E always has a finite sequence of generators from E. A finite module over a noetherian ring is always noetherian. (See [1], Chapter II, §1; or [3], Chapter VI, §1.)

It follows that an extension B of a noetherian ring is flat (see 1.16a) if and only if the kernel of any A–homomorphism $h : B^n \longrightarrow B^r$ is the range of some A–homomorphism $B^s \longrightarrow B^n$. (This is so because $A^n \cap \ker h \subset A^n$ is a finite submodule over A.)

In the definition of flatness of an extension B of a noetherian ring A it is enough to take $r = 1$.

Indeed, assume that all A–homomorphisms $B^n \longrightarrow B$ display the above property. Let $r > 1$, and suppose that all A–homomorphisms $B^n \longrightarrow B^{r-1}$ have this property. If $h : B^n \longrightarrow B^r$ is an A–homomorphism, then $h = (h', h'')$, where $h' : B^n \longrightarrow B^{r-1}$, $h'' : B^n \longrightarrow B$ are A–homomorphisms. Thus $\ker h'' = \operatorname{im} \lambda$, where $\lambda : B^s \longrightarrow B^n$ is an A–homomorphism. But $h' \circ \lambda : B^s \longrightarrow B^{r-1}$ is also an A–homomorphism, so $\ker (h' \circ \lambda) = \operatorname{im} \mu$, where $\mu : B^q \longrightarrow B^s$ is an A–homomorphism. Hence $\ker h = \operatorname{im} (\lambda, \mu)$.

Let M be a finite module over a noetherian ring A.

PROPOSITION . *The set of zero divisors in M is a finite union of prime ideals. If an ideal I contains only zero divisors in M, then $Ix = 0$ for some $x \in M \setminus 0$.*

PROOF . For any $x \in M \setminus 0$ define $J_x = \{t \in A : tx = 0\}$. Then J_x is an ideal of A (being the kernel of the epimorphism $A \ni t \longrightarrow tx \in Ax$). Hence $Ax \approx A/J_x$. Let $\mathcal{A}_M$ be the subfamily consisting of the prime ideals from the family $\mathcal{R}_M = \{J_x\}_{x \in M \setminus 0}$ [58] . Clearly, $\mathcal{A}_M = \mathcal{A}_N$ if $M \approx N$. If I is a prime ideal, then $\mathcal{A}_{A/I} = \{I\}$. (Indeed, if $x \in A \setminus I$, then $J_{\bar{x}} = I : x = I$.)

We will prove that $\mathcal{A}_M$ is finite. The family of submodules L such that $\mathcal{A}_L$ is finite is non-empty (because it contains 0). Hence there is a maximal submodule N in this family. Now, if $J_x \in \mathcal{A}_M$ and $N \cap Ax \neq 0$, then $J_x \in \mathcal{A}_N$. (In such a case, $0 \neq ux \in N$ for some $u \in A$, so $u \notin J_x$ and $J_x = J_x : u = J_{ux}$). Thus it is enough to prove the implication $x \in M \setminus N$, $J_x \in \mathcal{A}_M \implies N \cap Ax \neq 0$, since it implies that $\mathcal{A}_M \subset \mathcal{A}_N$. If the implication were false, there would exist $c \in M \setminus N$ for which $J_c \in \mathcal{A}_M$ and $N \cap Ac = 0$. Hence $L = N + Ac \neq N$ and $L/N \approx Ac \approx A/J_c$. We claim that then, contrary to the fact that N is maximal, we would have $\mathcal{A}_L \subset \mathcal{A}_N \cup \{J_c\}$. Indeed, let $J_x \in \mathcal{A}_x$, where $x \in L \setminus N$. If $N \cap Ax \neq 0$, then $J_x \in \mathcal{A}_N$. If $N \cap Ax = 0$, then $J_x = \{t : tx \in N\} = J_{\bar{x}}$. Hence $J_x \in \mathcal{A}_{L/N} = \mathcal{A}_{A/J_c}$ and $J_x = J_c$.

---

[58] The ideals in the family $\mathcal{A}_M$ are said to be *associated* with M, and $\mathcal{A}_M$ is usually denoted by Ann M.

The set D of zero divisors for M is obviously equal to the union of the family $\mathcal{R}_M$, and hence to the union of its maximal ideals. Now, each maximal ideal J_x from this family must be prime. For if $tu \in J_x$ and $u \notin J_x$, then $ux \neq 0$ and $t \in J_{ux}$. But $J_x \subset J_{ux}$, so $J_x = J_{ux}$, hence $t \in J_x$. Consequently, D is the union of the family $\mathcal{A}_M$. It follows that if $I \subset D$, then $I \subset J_x$ for some $J_x \in \mathcal{A}_M$, and so $Ix = 0$.

7. Let R be the ring of fractions of a noetherian ring A. Then, an element $x \in R$ is integral over A if and only if $ax^n \in A$ for some $a \in A$ which is not a zero divisor in A and for all $n \in \mathbf{N}$.

Indeed, if the condition is fulfilled, then $A[X] \subset Aa^{-1}$. But Aa^{-1} is a noetherian module (see n° 6), and hence $A[X]$ is finite over A. This implies (see 8.1) that x is integral. Conversely, if an element $x = v/u \in R$ ($v \in A$, $u \in A$ is not a zero divisor) is integral over A, then it is a root of a monic polynomial of degree n from $A[T]$. It follows that $x^\nu \in A + \ldots + Ax^{n-1}$ for each ν. By putting $a = v^{n-1}$, we get $ax^\nu \in A$ for each ν.

§10. Local rings

1. A ring A is said to be *local* if it has only one maximal ideal $\mathfrak{m}$. (In such a case, every proper ideal is contained in $\mathfrak{m}$.) This is the same as saying that the set of all non-invertible elements of A is an ideal (and then it must be exactly the maximal ideal $\mathfrak{m}$ ([59])). Note that in this case each element of the set $1 + \mathfrak{m}$ is invertible. All the ideals $\mathfrak{m}^k$, $k \geq 1$, are primary ([60]). (See [2], Chapter I, §2.5.)

Let $h : A \longrightarrow B$ be a homomorphism of local rings A, B with the maximal ideals $\mathfrak{m}, \mathfrak{n}$, respectively. Then $h^{-1}(\mathfrak{n}) \subset \mathfrak{m}$ (because the ideal $h^{-1}(\mathfrak{n})$ does not contain the identity).

The homomorphism h is said to be *local* if $h(\mathfrak{m}) \subset \mathfrak{n}$. Then $h(\mathfrak{m}^\nu) \subset \mathfrak{n}^\nu$ for $\nu = 1, 2, \ldots$ ([61]) and $h^{-1}(\mathfrak{n}) = \mathfrak{m}$. Thus, if h is a local epimorphism, then $h(\mathfrak{m}) = \mathfrak{n}$.

([59]) Hence a ring A is local with the maximal ideal $\mathfrak{m}$ if and only if $\mathfrak{m}$ is a proper ideal containing all non-invertible elements.

([60]) Since if $xy \in \mathfrak{m}^k$ and $x^n \notin \mathfrak{m}^k$ (for each n), then $x \notin \mathfrak{m}$. Hence x is invertible, and so $y \in \mathfrak{m}^k$.

([61]) In a local ring one can introduce the *Krull topology* by taking the sequence of all powers of the maximal ideal $\{\mathfrak{m}^\nu\}$ as a neighbourhood basis of 0. (It is easy to see that the algebraic operations are continuous in this topology.) Thus local homomorphisms are continuous in the Krull topology.

Let A be a local ring with the maximal ideal $\mathfrak{m}$.

2. It follows from the Mather-Nakayama lemma (see 1.16) that if M is a finite module over A and $M \subset \mathfrak{m}M$, then $M = 0$. (Indeed, in this case M is finite over $\mathfrak{m}$, so by taking $\xi = 1$ we get $\eta M = 0$ for some $\eta \in 1 + \mathfrak{m}$.) As a consequence, we get

NAKAYAMA'S LEMMA . *Let N and L be submodules of the module M over A. Then $N \subset L + \mathfrak{m}N \Longrightarrow N \subset L$, provided that N is finitely generated.*

Indeed, by taking the images under the natural homomorphism $M \longrightarrow M/L$, we get $(N + L)/L \subset \mathfrak{m}\big((N + L)/L\big)$, and hence $(N + L)/L = 0$, i.e., $N \subset L$.

3. For every proper ideal I, the ring A/I is also local and its maximal ideal is $\mathfrak{m}/I$.

4. Let M be a finite module over A, and let I be a proper ideal in A. We have the equivalence

$$(x_1, \ldots, x_k \text{ generate } M) \Leftrightarrow (\bar{x}_1, \ldots, \bar{x}_k \text{ generate } M/IM \text{ over } A/I) .$$

(This is true because the right hand side implies that $M \subset \sum Ax_i + IM$, and hence $M = \sum_1^k Ax_i$ by Nakayama's lemma.) Therefore $x_1, \ldots, x_k$ is a system of generators for M which is minimal or has the least possible number of elements (^{62}) precisely in the case when $\bar{x}_1, \ldots, \bar{x}_k$ is a system of generators for M/IM over A/I. Thus we have

$$g_A(M) = g_{A/I}(M/IM) .$$

The field $K = A/\mathfrak{m}$ is called the *residual field* of the ring A. If M is a module over A, then $M/\mathfrak{m}M$ is a vector space over K.

Let M be a finite module over A. The linear space $M/\mathfrak{m}M$ is finite dimensional and $x_1, \ldots, x_n$ is a minimal system of generators for the module M if and only if $\bar{x}_1, \ldots, \bar{x}_n$ is a basis for the space $M/\mathfrak{m}M$. This implies that

$$g_A(M) = \dim_K(M/\mathfrak{m}M) .$$

All minimal systems of generators for M have the same number of elements $g_A(M)$. The elements of a minimal system of generators belong to $M \setminus \mathfrak{m}M$

(^{62}) Both conditions are equivalent, since they can be reduced to the case of a vector space (see below).

and each element of $M \setminus \mathfrak{m}M$ belongs to a minimal system of generators. Note also that the system of generators $x_1, \ldots, x_n$ is minimal exactly in the case when $\sum_1^n t_i x_i = 0 \Longrightarrow t_1, \ldots, t_n \in \mathfrak{m}$. (The condition is sufficient, for if the system were not minimal, it would contain a redundant x_s and we would have $x_s - \sum_{i \neq s} t_i x_i = 0$ for some t_i.)

Let $M = L+N$ be a direct sum of submodules. If $x_1, \ldots, x_k$ and $y_1, \ldots, y_l$ are minimal systems of generators for L and N, respectively, then obviously so is $x_1, \ldots, x_k, y_1, \ldots, y_l$ for M. Therefore $g(M) = g(L) + g(N)$, provided that L and N are finite over A.

Assume, in addition, that A is noetherian.

If the module M is free, then its minimal system of generators is always its basis. In fact, let $a_1, \ldots, a_n$ be such a system. Then $\varphi : A^n \ni (t_1, \ldots, t_n) \longrightarrow \sum t_i a_i \in M$ is an epimorphism, hence (see 1.16) there is a monomorphism $\psi : M \longrightarrow A^n$ such that $\varphi \circ \psi = \mathrm{id}_M$. It follows (see 1.8) that $A^n = \ker \varphi + \operatorname{im} \psi$ is a direct sum. But $g(A^n) = n = g(M) = g(\psi(M))$, so we must have $g(\ker \varphi) = 0$, hence $\ker \varphi = 0$, i.e., $a_1, \ldots, a_n$ is a basis of M.

When the module M is free, then, if it is a direct sum of two submodules, they are both free. Indeed, any of their minimal systems of generators $x_1, \ldots, x_k$ and $y_1, \ldots, y_l$ are respective bases, hence $x_1, \ldots, x_k, y_1, \ldots, y_l$ is a basis of M (since it is then a minimal system of generators of M).

In particular, $\mathfrak{m}/\mathfrak{m}^2$ is a linear space over K which is called the *cotangent space* of the ring A [63] . Its dimension is called the *embedding dimension* of A [64] and is denoted by edim A. Thus we have edim $A = g(\mathfrak{m})$. For any ideal $I \subset \mathfrak{m}$ we have the subspace $(I + \mathfrak{m}^2)/\mathfrak{m}^2$, which is the image of I under the natural epimorphism $\mathfrak{m} \longrightarrow \mathfrak{m}/\mathfrak{m}^2$ (see 1.10). Its dimension is called the *rank of the ideal I*:

$$\operatorname{rank} I = \dim_K (I + \mathfrak{m}^2)/\mathfrak{m}^2 = g_A \big((I + \mathfrak{m}^2)/\mathfrak{m}^2 \big) \quad [65] .$$

It is thus equal to the minimal number k of generators of I modulo $\mathfrak{m}^2$, i.e., of elements $x_1, \ldots, x_k \in I$ such that $I \subset \sum A x_i + \mathfrak{m}^2$. When $\mathfrak{m}$ is finitely generated (in particular, when A is noetherian), we have

$$(*) \qquad\qquad \operatorname{rank} I = \operatorname{edim} A = \operatorname{edim} A/I .$$

---

[63] In the case of the ring $\mathcal{O}_A$ of an analytic germ A, it is the dual of the Zariski tangent space for A (see IV. 4.4 below).

[64] In the case of the ring $\mathcal{O}_A$ of an analytic germ A, it is the minimal dimension of manifolds which contain a germ isomorphic to A.

[65] In the case of the ring $\mathcal{O}_A$, it is equal to $\operatorname{rank}_a(f_1, \ldots, f_s)$, where $f_1, \ldots, f_s$ are representatives of generators of I (see I. 1.8).

In fact, edim $A/I = \dim_K(\mathfrak{m}/I)/(\mathfrak{m}/I)^2$ (see 1.9), and we have the isomorphisms

$$(\mathfrak{m}/I)/(\mathfrak{m}/I)^2 = (\mathfrak{m}/I)/((I+\mathfrak{m}^2)/I) \approx \mathfrak{m}/(I+\mathfrak{m}^2) \approx (\mathfrak{m}/\mathfrak{m}^2)/((I+\mathfrak{m}^2)/\mathfrak{m}^2)$$

of modules over A (see 1.4, 10 and 9), and hence of linear spaces over K (see 1.9). Therefore edim $A/I = $ edim $A - \operatorname{rank} I$ (see 1.17).

4a. Let $C \subset A$ be a field. We say that A is a *local C-ring* if the natural homomorphism $C \ni t \longrightarrow \bar{t} \in A/\mathfrak{m}$ is an isomorphism or, equivalently (see A 1.11), an epimorphism. This means that $A = C + \mathfrak{m}$, and the sum is direct (as vector subspaces over C). In this case, if M is a module over A, $M/\mathfrak{m}M$ is a vector space both over K and over C. Moreover, $tx = \bar{t}x$ for $t \in C$ and $x \in M/\mathfrak{m}M$. Thus, in both cases, we have the same subspaces, bases, dimensions, and codimensions.

If I is a proper ideal of a local C-ring A, then the quotient ring A/I is also a local C-ring, after the natural identification $C \longrightarrow A/I$. A homomorphism of quotient C-rings induced (see 1.3) by a C-homomorphism is a C-homomorphism, as well.

If A is a local C-ring and $x_1, \ldots, x_n$ is a sequence of generators for its maximal ideal $\mathfrak{m}$, then $A = C[x_1, \ldots, x_n] + \mathfrak{m}^\nu$, $\nu = 1, 2, \ldots$ ([66]) .

Indeed, in view of the equality $A = C + \mathfrak{m}$, we have

$$\mathfrak{m}^\nu \subset \sum_{|p|=p} Ax^p \subset C[x_1, \ldots, x_n] + \mathfrak{m}^{\nu+1} .$$

Every C-homomorphism of local C-rings is local.

To see this, let $h : A \longrightarrow B$ be a C-homomorphism of the local C-rings A, B with the maximal ideals $\mathfrak{m}$ and $\mathfrak{n}$, respectively. If $z \in \mathfrak{m}$, then (in view of the inclusion $h^{-1}(\mathfrak{n}) \subset \mathfrak{m}$) there is a $c \in C$ for which $z \in h^{-1}(c+\mathfrak{n}) \subset c+\mathfrak{m}$. This implies that $c = 0$, and so $h(z) \in \mathfrak{n}$.

5. Let A be a noetherian local ring.

Then $A/\mathfrak{m}$, $\mathfrak{m}/\mathfrak{m}^2$, $\mathfrak{m}^2/\mathfrak{m}^3, \ldots$ are finite dimensional vector spaces over K.

Note that if $I_1, \ldots, I_k$ are prime ideals and $\mathfrak{m} \setminus M^2 \subset I_1 \cup \ldots \cup I_k$, then $\mathfrak{m} = I_s$ for some s.

In fact, it is enough to show that $\mathfrak{m} \subset \bigcup I_\nu$. If $\mathfrak{m} = \mathfrak{m}^2$, then $\mathfrak{m} = 0$, according to Nakayama's lemma. Otherwise, we take $a \in \mathfrak{m} \setminus \mathfrak{m}^2$. Now, if

---

([66]) Thus the subring $C[x_1, \ldots, x_n]$ is dense in A with Krull's topology.

$t \in \mathfrak{m}$, then $a + t^\nu \in \mathfrak{m} \setminus \mathfrak{m}^2$ for $\nu \geq 2$. Hence some I_r must contain the elements $a + t^\varrho$ and $a + t^\sigma$ for some $\sigma > \varrho > 0$, and thus it contains the element $t^\sigma - t^\varrho = t^\varrho(t^{\sigma - \varrho} - 1)$ as well. Therefore $t^\varrho \in I_r$ and $t \in I_r$.

An ideal I is said to be *defining* (or an *ideal of definition*), if it satisfies one of the following equivalent conditions

(1) rad $I = \mathfrak{m}$;

(2) $\mathfrak{m}$ is an isolated ideal for I;

(3) $\mathfrak{m}^s \subset I \subset \mathfrak{m}$ for some s [67].

It is clear that (1)$\Longleftrightarrow$(3) and (1)$\Longrightarrow$(2) (see 9.2-3). Now, (2) implies that $\mathfrak{m}$ is the only ideal that is associated with I, and so the irreducible primary representation of the ideal I consists of I only. Thus I is primary and rad $I = \mathfrak{m}$.

Note also that:

If an ideal I is defining, then every ideal $J \subset \mathfrak{m}$ such that $I \subset$ rad J is also defining.

If I is a proper ideal, then an ideal $J \supset I$ is defining if and only if J/I is defining in A/I.

KRULL'S INTERSECTION THEOREM . *We have* $\bigcap_1^\infty \mathfrak{m}^\nu = 0$ [68] .

PROOF . Set $I = \bigcap_1^\infty \mathfrak{m}^\nu$. Let $\mathfrak{m}I = \bigcap J_\nu$ be the irreducible primary decomposition. According to Nakayama's lemma, it is enough to show that $I \subset \mathfrak{m}I$, i.e., that $I \subset J_\nu$ for each ν. Take $x \in I$. Then $\mathfrak{m}x \subset J_\nu$. If we assumed that $x \notin J_\nu$, then it would yield $\mathfrak{m} \subset$ rad J_ν. So J_ν would be defining, and consequently $x \in \mathfrak{m}^s \subset J_\nu$ for some s. This contradiction shows that $x \in J_\nu$.

COROLLARY . *For any ideal I we have* $\bigcap_1^\infty (I + \mathfrak{m}^\nu) = I$.

Indeed, if $I \not\subset A$, then by considering the images of ideals under the natural homomorphism $A \longrightarrow A/I$ (see 1.4 and 10), we get

$$\left(\bigcap_1^\infty (\mathfrak{m}^\nu + I) \right) / I \subset \bigcap_1^\infty ((\mathfrak{m}^\nu + I)/I) = \bigcap_1^\infty (\mathfrak{m}/I)^\nu = 0 \ .$$

REMARK . More generally, *if M is a finite module over A, then $N = \bigcap_{\nu=1}^\infty (N + \mathfrak{m}^\nu M)$ for any submodule $N \subset M$.*

We are going to give a direct proof this result [68a]. As before, it is enough

---

[67] This condition is necessary and sufficient for $\{I^\nu\}$ – regarded as a base of neighbourhoods of 0 – to generate Krull's topology (see footnote [61]).

[68] This means that in a noetherian local ring, Krull's topology is Hausdorff.

[68a] See [17b] pp. 211-212.

to prove the statement $\bigcap_1^\infty \mathfrak{m}^\nu M = 0$. Set $Z = \bigcap_1^\infty \mathfrak{m}^\nu M$ and take a maximal submodule $L \subset M$ satisfying $L \cap Z = \mathfrak{m}Z$ (there are submodules that satisfy this equality, e.g., $\mathfrak{m}Z$). It suffices to show that for any $a \in \mathfrak{m}$ the inclusion $a^p M \subset L$ holds, because then we can take p uniform for each element of a system of, say, k generators of $\mathfrak{m}$. This gives $\mathfrak{m}^{kp} M \subset L$, $Z = \mathfrak{m}Z$, and so $Z = 0$ by Nakayama's lemma (in n° 2). Now, the increasing sequence of submodules $M_\nu = \{x : a^\nu x \in L\}$ must be stationary, hence $M_p = M_{p+1}$ for some p. Take $L' = L + a^p M$. We claim that $L' \cap Z = L \cap Z$, which implies $L = L' \supset a^p M$, owing to the maximality of L. Indeed, if $x \in L' \cap Z$, then $x \in a^p y + L$ for some y which must belong to $M_{p+1} = M_p$ (because $a^{p+1} y \in ax + aL \subset L$); therefore $ya^p \in L$, and so $x \in L$. Thus $L' \cap Z \subset L$, which gives $L' \cap Z = L \cap Z$.

Krull's theorem implies

PROPOSITION . *If B is a noetherian local C-ring and $C \subset A$, then every C-homomorphism on the ring A with values in B is determined by its values taken on a system of generators for the ideal $\mathfrak{m}$.*

To see this, let $h : A \longrightarrow B$, $h' : A \longrightarrow B$ be C-homomorphisms and assume that $h = h'$ on the set of generators $\{x_1, \ldots, x_n\}$ of the ideal $\mathfrak{m}$. Then $h = h'$ in $C[x_1, \ldots, x_n]$. Now, if $z \in A$, then for each $\nu \geq 1$, $z = u + v$, where $u \in C[x_1, \ldots, x_n]$ and $v \in \mathfrak{m}^\nu$ (see n° 4a). Thus $h(z) - h'(z) = h(v) - h'(v) \in \mathfrak{n}^\nu$, where $\mathfrak{n}$ denotes the maximal ideal of B. This is so because the homomorphisms h, h' are local (see n° 4a). Hence, by applying Krull's theorem, we get $h(z) = h'(z)$.

6. A ring A is said to be a *discrete valuation ring* if it is an integral domain and its only proper ideals are 0 and the powers of one ideal $\mathfrak{m}$ [69] . Then the ring A is noetherian and local with the maximal ideal $\mathfrak{m}$. It is also a principal ideal domain (see 9.2). In particular, $\mathfrak{m} = Ax$ and $\mathfrak{m}^n = Ax^n$ for $n \in \mathbf{N}$. The function $\nu : A \setminus 0 \longrightarrow \mathbf{N}$ defined (in view of the Krull intersection theorem) by the condition $\nu = s$ on $\mathfrak{m}^s \setminus \mathfrak{m}^{s+1}$ for $s = 0, 1, 2, \ldots$ is called the *valuation on the ring A* [70] . It displays the following properties: (i) $\nu(uv) = \nu(u) + \nu(v)$ and (ii) $(\nu(u) \leq \nu(v)) \Longleftrightarrow (u$ is a divisor of $v)$ for $u, v \in A \setminus 0$. These properties follow from the fact that for each $u \in A \setminus 0$ we have $u = cx^k$, where $c \in A \setminus \mathfrak{m}$. Thus $k = \nu(u)$.

---

[69] Authors often assume that $\mathfrak{m} \neq 0$, i.e., that A is not a field (excluding the trivial valuation).

[70] It is the restriction of a discrete valuation on the field of fractions of the ring A. (See [1], Chapter III, §6; or [4], Chapter 9.)

Obviously, $\nu(u) \geq k \iff u \in \mathfrak{m}^k$ (for $u \in A \setminus 0$, $k \in \mathbf{N}$).

Notice that, in the definition of a discrete valuation ring, the requirement that the ring is an integral domain can be replaced by the condition that it does not have nilpotents. The latter condition implies that the ring is an integral domain because, by the Krull intersection theorem, every element of $A \setminus 0$ is, as before, of the form cx^ν, where $c \in A \setminus \mathfrak{m}$.

Finally (in view of Nakayama's lemma – see n° 2), a discrete valuation ring is precisely a field or a ring in which $\mathfrak{m} \underset{\neq}{\supset} \mathfrak{m}^2 \underset{\neq}{\supset} \ldots$ are the only proper non-zero ideals.

PROPOSITION . *If A is a noetherian local ring with the maximal ideal $\mathfrak{m}$, then the following conditions are equivalent*

(1) *A is a discrete valuation ring,*

(2) *$\mathfrak{m}$ is principal and A is without nilpotents,*

(3) *A is integrally closed of dimension ≤ 1 ([71]) .*

PROOF . Assume (2). Then $\mathfrak{m} = Ax$ for some $x \in A$. Thus $\mathfrak{m}^\nu = Ax^\nu$, $\nu = 0, 1, 2, \ldots$. If I is a non-zero proper ideal, then, by the Krull intersection theorem, $I \subset \mathfrak{m}^s$, $I \not\subset \mathfrak{m}^{s+1}$. Take an element $y \in I \setminus \mathfrak{m}^{s+1}$. Then $y = ax^s$ for some $a \in A \setminus \mathfrak{m}$, and hence $x^s \in I$, i.e., $\mathfrak{m}^s \subset I$. So $I = \mathfrak{m}^s$. Therefore A is a discrete valuation ring. It does not have prime ideals different from 0 and $\mathfrak{m}$, since if $\mathfrak{m} \neq 0$, then none of the ideals $\mathfrak{m}^k$, $k \geq 2$, is prime. (In view of Nakayama's lemma (see n° 2) there are $x \in \mathfrak{m} \setminus \mathfrak{m}^2$, $y \in \mathfrak{m}^{k-1} \setminus \mathfrak{m}^k$ such that $xy \in \mathfrak{m}^k$.) We are going to show that A is integrally closed. Now, let an element u/v of the field of fractions, where $u, v \in A \setminus 0$, be integral over A. Then $a(u/v)^n$ for all $n \in \mathbf{N}$ and some $a \in A \setminus 0$ (see 9.7); that is, $au^n = bv^n$, where $b \in A \setminus 0$. Taking the valuation ν on A, we have $\nu(a) + n\nu(u) = \nu(b) + n\nu(v)$, which implies that $\nu(u) = \nu(v)$. In particular, u is divisible by v, i.e., $u/v \in A$. Thus we have $(1) \iff (2) \implies (3)$.

Now, suppose that (3) is true. We may assume that $0 \subsetneq Ac \subsetneq \mathfrak{m}$ for some c. Then $\mathfrak{m}^{s+1} \subset Ac$, $\mathfrak{m}^s \not\subset Ac$ for some s, since the ideal Ac must be defining, by (3) (see n° 5). Let us take $d \in \mathfrak{m}^s \setminus Ac$. Then $\eta = d/c \notin A$, and so η is not integral over A. Thus $\mathfrak{m}\eta \not\subset \mathfrak{m}$ (see 8.1). But $\mathfrak{m}\eta \subset A$ (because $\mathfrak{m}d \subset Ac$). Therefore $\mathfrak{m}\eta$ is an ideal of the ring A and thus $\mathfrak{m}\eta = A$. Hence $\mathfrak{m} = A\eta^{-1}$, $\eta^{-1} \in \mathfrak{m}$. Thus $(3) \implies (2)$ and the proof is complete.

We will also prove

LEMMA . *A noetherian integrally closed local ring A in which the maximal ideal is associated with a principal ideal is a discrete valuation ring.*

_____

([71]) This means that there are no prime ideals different from 0 and $\mathfrak{m}$. (See 12.1 and 3 below.)

Without loss of generality we may assume that $\mathfrak{m} \neq 0$. Let Au be a principal ideal associated with the ideal $\mathfrak{m}$. Let K denote the field of fractions of the ring A. Then $N = \{x \in K : x\mathfrak{m} \subset A\}$ is a module over A and $\mathfrak{m}N \subset A$ is an ideal. We have

$$\mathfrak{m}N = A .$$

For if it were not the case, one would have $\mathfrak{m}N \subset \mathfrak{m}$, and so $N \subset A$ (since A is integrally closed – see 8.1). On the other hand $Au : \mathfrak{m} \subset Au$ (for if $u \neq 0$, then $x\mathfrak{m} \subset Au$ implies that $x/u \in N \subset A$). This would yield by induction that $Au : \mathfrak{m}^s \subset Au$ for each s (see 1.5). Now, taking the irreducible primary decomposition $Au = \bigcap J_i$ in which $\mathrm{rad}\, J_1 = \mathfrak{m}$, and s such that $\mathfrak{m}^s \subset J$ (see 9.2), one would get the inclusion $(\bigcap_{i>1} J_j)\mathfrak{m}^s \subset As$ and thus $\bigcap_{i>1} J_i \subset Au : \mathfrak{m}^s \subset Au$ (because the decomposition is irreducible). According to the proposition, it is enough to show that $\dim A \leq 1$. Now let us take a prime ideal I such that $0 \subset I \subsetneqq \mathfrak{m}$. We have $IN \subset I : M$ (because $IN\mathfrak{m} \subset I$). But $I : \mathfrak{m} \subset I$ (see 1.11), and so $IN \subset I$. Hence $I \subset Im N \subset \mathfrak{m}I$, and by Nakayama's lemma (see 10.2), $I = 0$.

§11. Localization

1. Let A be an integral domain, and let $\mathfrak{n}$ be a prime ideal in A. Then

$$A\mathfrak{n} = \{x/y : x \in A, \ y \in A \setminus \mathfrak{n}\}$$

is a subring of the field of fractions of the ring A, which is called the *localization of the ring A to the ideal* $\mathfrak{n}$. Note that each element in $A \setminus \mathfrak{n}$ is invertible in $A\mathfrak{n}$.

Every ideal I of the ring A generates an ideal in $A\mathfrak{n}$, which is called the *extension of the ideal I* and denoted by $\bar{I}$ ([72]) . We have

$$\bar{I} = \{x/y : x \in I, \ y \in A \setminus \mathfrak{n}\} .$$

Of course, if a set E generates the ideal I in A, then it also generates the ideal $\bar{I}$ in $A\mathfrak{n}$.

Now $A\mathfrak{n}$ *is a local ring with the maximal ideal* $\bar{\mathfrak{n}}$. Indeed (see 10.1, footnote ([59])), $1 \notin \bar{\mathfrak{n}}$ and each element of $A\mathfrak{n} \setminus \bar{\mathfrak{n}}$ is invertible.

([72]) In general, given a ring homomorphism $h : A \longrightarrow B$, by the extension of the ideal I one means the ideal generated by the image $h(I)$. It is usually denoted by I^c.

The following properties are easy to check. If I, J are ideals of the ring A, then $\overline{I \cap J} = \bar{I} \cap \bar{J}$, $\overline{IJ} = \bar{I}\bar{J}$, $\overline{\text{rad } I} = \text{rad } \bar{I}$ and $\overline{Ad} = A_{\mathfrak{n}}\bar{d}$, $\overline{Id} = \bar{I}\bar{d}$ for $d \in A$. Also, $I = A_{\mathfrak{n}} \iff I \not\subset \mathfrak{n}$. Finally, the ideal $\bar{I} \cap A$ is equal to the union of the ideals $I : x$, where $x \in A \setminus \mathfrak{n}$.

For every ideal J of the ring $A_{\mathfrak{n}}$, we have $\overline{J \cap A} = J$. (Indeed, if $x/y \in J$, where $x \in A$ and $y \in A \setminus \mathfrak{n}$, then $x \in J \cap A$.)

Therefore, if the ring A is noetherian, then so is its localization $A_{\mathfrak{n}}$.

If the ring A is integrally closed, then so is its localization $A_{\mathfrak{n}}$. This is so because if z is an element of the field of fractions ([73]) that satisfies the equation

$$z^k + (a_1/b)z^{k-1} + \ldots + (a_k/b) = 0 ,$$

where $a_i \in A$, $b \in A \setminus \mathfrak{n}$, then

$$(bz)^k + a_1(bz)^{k-1} + \ldots + a_k b^{k-1} = 0 ;$$

this implies that $bz \in A$, and hence $z \in A_{\mathfrak{n}}$.

If $I \subset \mathfrak{n}$ is a primary ideal in the ring A, then $\bar{I}$ is a primary ideal in the ring $A_{\mathfrak{n}}$ and $\bar{I} \cap A = I$. Furthermore, if I is prime, then so is $\bar{I}$.

Indeed, the ideal $\bar{I}$ is primary. For if $(x'/y')/(x''/y'') = x/y$, where $x', x'' \in A$, $x \in I$, and $y', y'', y \in A \setminus \mathfrak{n}$, then $x'x''y = y'y''x \in I$. But $y^n \in A \setminus \mathfrak{n} \subset A \setminus I$ for each n, and so $x'x'' \in I$, which implies that $x' \in I$ or $(x'')^n \in I$ for some k. Next, we have the inclusion $\bar{I} \cap A \subset I$, i.e., the inclusions $I : x \subset I$ for $x \in A \setminus \mathfrak{n}$; this is so because $x^n \in A \setminus \mathfrak{n} \subset A \setminus I$ for all n, hence if $zx \in I$, then $z \in I$. Finally, if I is prime, then $\bar{I} = \overline{\text{rad } I} = \text{rad } \bar{I}$ is prime, too.

In conclusion, $I \longrightarrow \bar{I}$ and $J \longrightarrow J \cap A$ are mutually inverse bijections between the set of primary ideals of the ring A contained in $\mathfrak{n}$ and the set of primary ideals of the ring $A_{\mathfrak{n}}$ ([74]) . These bijections establish a one–to–one correspondence between the prime ideals in both rings. Clearly, these bijections preserve inclusion. It follows that $\dim A_{\mathfrak{n}} = h(\mathfrak{n})$, provided that A is noetherian (see §12).

When A is a noetherian ring, it follows that if $I = \bigcap J_i$ is the irreducible primary decomposition of an ideal $I \subset \mathfrak{n}$ of the ring A, then

$$\bar{I} = \bigcap \{\bar{J_i} : J_i \subset \mathfrak{n}\}$$

([73]) The rings A and $A_{\mathfrak{n}}$ have the same field of fractions.

([74]) Clearly, if $J \not\subset A_{\mathfrak{n}}$, then $J \cap A \subset \mathfrak{n}$ (since $J \subset \bar{\mathfrak{n}}$).

is the irreducible primary decomposition of the ideal $\bar{I}$ of the ring $A_\mathfrak{n}$. Therefore, for the ideals $I \subset I_0 \subset \mathfrak{n}$ of the ring A such that I_0 is prime, we obtain the equivalences

$$(I_0 \text{ is associated with } I) \Longleftrightarrow (\bar{I}_0 \text{ is associated with } \bar{I})$$
$$(I_0 \text{ is isolated for } I) \Longleftrightarrow (\bar{I}_0 \text{ is isolated for } \bar{I}) \ .$$

The ideals $\mathfrak{n}^{(k)} = \bar{\mathfrak{n}}^k \cap A$ $({}^{75})$ are primary $({}^{76})$ with the radical equal to $\mathfrak{n}$ $({}^{77})$. We have $\mathfrak{n} = \mathfrak{n}^{(1)} \supset \mathfrak{n}^{(2)} \supset \dots$. Also, $\mathfrak{n}^{(k)} \mathfrak{n}^{(l)} \subset \mathfrak{n}^{(k+l)}$. The ideal $\mathfrak{n}^{(k)}$ coincides with the union of the ideals $\mathfrak{n}^k : x$, where $x \in A \setminus \mathfrak{n}$.

If $A_\mathfrak{n}$ is a discrete valuation ring, then every primary ideal $0 \neq I \subset \mathfrak{n}$ in the ring A has the form $I = \mathfrak{n}^{(k)}$, where $k > 0$. (Owing to the fact that $\bar{I} \neq 0$, $\bar{I} = \bar{\mathfrak{n}}^k$, and hence $I = \bar{\mathfrak{n}}^k \cap A = \mathfrak{n}^{(k)}$.)

2. Now, let A be an arbitrary ring, and let $\mathfrak{n}$ be one of its prime ideals. The set

$$\mathcal{J}_\mathfrak{n} = \{x \in A : \ xt = 0 \text{ for some } t \in A \setminus \mathfrak{n}\}$$

is an ideal of the ring A, containing only zero divisors and contained in each primary ideal $I \subset \mathfrak{n}$ $({}^{78})$. Moreover, $\mathcal{J}_\mathfrak{n} : u = \mathcal{J}_\mathfrak{n}$ for $u \in A \setminus \mathfrak{n}$.

We can introduce a relation on the set $A \times (A \setminus \mathfrak{n})$ by the formula

$$(x, u) \equiv (y, v) \Longleftrightarrow xv - yu \in \mathcal{J}_\mathfrak{n} \ .$$

It is an equivalence relation $({}^{79})$. The equivalence class of a pair (x, u) is denoted by x/u. It is easy to verify that the quotient $A_\mathfrak{n}$ (with respect to this equivalence relation), with the operations well-defined by the formulae $x/u + y/v = (xv + yu)/(uv)$ and $(x/u)(y/v) = (xy)/(uv)$, is a ring (with unity $1/1$ and zero $0/1$). The ring $A_\mathfrak{n}$ which we have just defined is called the *localization of the ring A to the ideal* $\mathfrak{n}$. We have the natural homomorphism $A \ni x \longrightarrow \bar{x} = x/1 \in A$, and the following conditions are satisfied:

$$x \in A \setminus \mathfrak{n} \Longrightarrow (\bar{x} \text{ is invertible}) \ ({}^{80}) \ ,$$

$({}^{75})$ Those ideals are called the *symbolic powers* of the ideal $\mathfrak{n}$.

$({}^{76})$ Since $\bar{\mathfrak{n}}^k$ are primary (see 10.1 and 9.3).

$({}^{77})$ This is due to the fact that rad $(\bar{\mathfrak{n}}^k \cap A) = \bar{\mathfrak{n}} \cap A = \mathfrak{n}$ (see 9.3, footnote $({}^{51})$).

$({}^{78})$ For suppose that $x \in \mathcal{J}_\mathfrak{n}$ (and take t as in the definition of $\mathcal{J}_\mathfrak{n}$). Then $xt \in I$ and $t^n \notin I$ for each n. So $x \in I$.

$({}^{79})$ To check transitivity $((x, u) \equiv (y, v) \equiv (z, w)) \Longrightarrow (x, u) \equiv (z, w)$ we use the fact that $\mathcal{J}_\mathfrak{n} : v = \mathcal{J}_\mathfrak{n}$.

$({}^{80})$ We have $\bar{x}^{-1} = 1/x$.

every element from $A_{\mathfrak{n}}$ is of the form $\bar{x}\bar{u}^{-1}$, where $x \in A$ and $u \in A \setminus \mathfrak{n}$ ([81]),

$$\bar{x} = 0 \Longleftrightarrow x \in \mathcal{J}_{\mathfrak{n}}.$$

Conversely, the above conditions determine, up to an isomorphism, the ring $A_{\mathfrak{n}}$ together with the homomorphism $A \ni x \longrightarrow \bar{x} \in A_{\mathfrak{n}}$ ([82]) More generally, the same construction works for any multiplicative set $M \subset A \setminus 0$ ([83]) (containing 1) taken in place of $A \setminus \mathfrak{n}$ ([84]). In the case when M is the complement of the set of zero divisors (then $\mathcal{J}_M = 0$), it is exactly the same construction as that used for the ring of fractions of the ring A (see 1.15a).

If A is an integral domain and K is its field of fractions, then $\mathcal{J}_{\mathfrak{n}} = 0$, and thus – after the identification through the well-defined monomorphism $A_{\mathfrak{n}} \ni x/u \longrightarrow x/u \in K$ – the ring $A_{\mathfrak{n}}$ coincides with the localization introduced in n° 1.

Denote by χ the natural homomorphism $A \ni x \longrightarrow \bar{x} \in A_{\mathfrak{n}}$. For any ideal I of the ring A, the ideal of the ring $A_{\mathfrak{n}}$ generated by $\chi(I)$ is called the *extension of the ideal* I and denoted by $\bar{I}$ ([85]). We have

$$\bar{I} = \{\bar{x}\bar{y}^{-1} : x \in I, \; y \in A \setminus \mathfrak{n}\} \, .$$

Certainly, if a set E generates an ideal I in A, then $\chi(E)$ generates the ideal $\bar{I}$ in $A_{\mathfrak{n}}$. In the case of an integral domain A, $\bar{x} = x$ and the ideal $\bar{I}$ is the same as that defined in n° 1.

Now $A_{\mathfrak{n}}$ *is a local ring with the maximal ideal* $\bar{\mathfrak{n}}$. In fact (see footnote ([59])), $1 \notin \bar{\mathfrak{n}}$ and every element of $A_{\mathfrak{n}} \setminus \bar{\mathfrak{n}}$ is invertible.

We have the following properties. If I, J are ideals of the ring A, then $\overline{IJ} = \bar{I}\bar{J}$ and $\overline{Ad} = A_{\mathfrak{n}}\bar{d}$, $\overline{Id} = \bar{I}\bar{d}$, for $d \in A$. On the other hand $\overline{I \cap J} = \bar{I} \cap \bar{J}$ and $\overline{\mathrm{rad}\, I} = \mathrm{rad}\, \bar{I}$, provided $I \supset \mathcal{J}_{\mathfrak{n}}$ or $I \not\subset \mathfrak{n}$. In particular, the latter is true if I is primary. Also, $\bar{I} = A_{\mathfrak{n}} \Longleftrightarrow I \not\subset \mathfrak{n}$ and, finally, the ideal $\chi^{-1}(\bar{I})$ is equal to the union of the ideals $I : x$, $x \in A \setminus \mathfrak{n}$.

([81]) We have $\bar{x}\bar{u}^{-1} = x/u$.

([82]) If a ring B and a homomorphism $A \ni x \longrightarrow \bar{x} \in B$ satisfy these conditions, then there exists a unique isomorphism $B \longrightarrow A_{\mathfrak{n}}$, such that the following diagram commutes (see [4], 3.2)

$$
\begin{array}{ccc}
 & A & \\
\swarrow & & \searrow \\
B & \longrightarrow & A_{\mathfrak{n}}
\end{array}
$$

([83]) That is, a set such that $x, y \in M \Longrightarrow xy \in M$. In particular, $A \setminus \mathfrak{n}$ is such a set.

([84]) The resulting ring is known as the *ring of fractions with respect to* M. (See [1], Chapter 1, §4; or [4], Chapter 3.)

([85]) Cf. footnote ([72]).

The last property follows from the fact that the condition $\bar{z} \in \bar{I}$ (which means that $(zv - u)t = 0$ for some $t, v \in A \setminus \mathfrak{n}$ and $u \in I$) is equivalent to the condition that $zx \in I$ with some $x \in A \setminus \mathfrak{n}$. As for the fourth equality, it suffices to check the inclusion $\bar{I} \cap \bar{J} \subset \overline{I \cap J}$ [86]. Let $\bar{x}\bar{u}^{-1} = \bar{y}\bar{v}^{-1}$, where $x \in I$, $y \in I$, $u, v \in A \setminus \mathfrak{n}$. Then $xv - yu \in \mathcal{J}\mathfrak{n}$. Assuming that $\mathcal{J}\mathfrak{n} \subset I$, we have $yu \in I \cap J$ and hence $\bar{y}\bar{v}^{-1} = \bar{y}\bar{u}(\bar{u}\bar{v})^{-1} \in \overline{I \cap J}$; however if $I \not\subset \mathfrak{n}$, then by taking $z \in \mathfrak{n} \setminus I$ we have $\bar{y}\bar{v}^{-1} = \bar{y}\bar{z}(\bar{v}\bar{z})^{-1} \in \overline{I \cap J}$, because $yz \in I \cap J$. As far as the fifth equality is concerned, let us verify the inclusion $\mathrm{rad}\,\bar{I} \subset \overline{\mathrm{rad}\,I}$. If $I \not\subset \mathfrak{n}$, both sides are equal to $A\mathfrak{n}$. Now, let $I \supset \mathcal{J}\mathfrak{n}$, and let $(xu^{-1})^n = yv^{-1}$, where $y \in I$, $x \in A$, $u, v \in A \setminus \mathfrak{n}$. Then $(xv)^n - yuv^{n-1} \in \mathcal{J}\mathfrak{n}$, and so $xv \in \mathrm{rad}\,I$ and $\bar{x}\bar{u}^{-1} = \bar{x}\bar{v}(\bar{u}\bar{v})^{-1} \in \overline{\mathrm{rad}\,I}$. The opposite inclusion, as well as the remaining properties, can be shown by using the same argument as that employed in the case of integral domains (see n° 1).

We have $\overline{\chi^{-1}(J)} = J$ for every ideal J of the ring $A\mathfrak{n}$. (From $\chi(\chi^{-1}(J)) \subset J$ we derive the inclusion $\subset$. Conversely, if $\bar{x}\bar{y}^{-1} \in J$, then $\bar{x} \in J$, i.e., $x \in \chi^{-1}(J)$. As a result, $\bar{x}\bar{y}^{-1} \in \overline{\chi^{-1}(J)}$.) Therefore:

If the ring A is noetherian, then so is its localization $A\mathfrak{n}$.

If the ring A does not have nilpotents, then neither does its localization $A\mathfrak{n}$. (For if $(\bar{x}\bar{y}^{-1})^n = 0$, then $\bar{x}^n = 0$, i.e., $x^n t = 0$, where $t \in A \setminus \mathfrak{n}$. Hence $(xt)^n = 0$. Thus $xt = 0$, and so $\bar{x} = 0$.)

If $I \subset \mathfrak{n}$ is a primary ideal of the ring A, then $\bar{I}$ is a primary ideal of the ring $A\mathfrak{n}$, and $\chi^{-1}(\bar{I}) = I$. Furthermore, if I is prime, then so is $\bar{I}$. These two properties can be shown exactly as for an integral domain (see n° 1).

Accordingly, $I \longrightarrow \bar{I}$ and $J \longrightarrow \chi^{-1}(J)$ are mutually inverse bijections between the set of primary ideals of the ring A which are contained in $\mathfrak{n}$, and the set of primary ideals of the ring $A\mathfrak{n}$ [87]. These bijections establish a one-to-one correspondence between the prime ideals of the two rings. Obviously, both mappings preserve inclusion. It follows that $\dim A\mathfrak{n} = h(\mathfrak{n})$, provided that A is noetherian.

When A is noetherian, it follows that if $I = \bigcap J_i$ is the irreducible primary decomposition of the ideal I of the ring A, then

$$\bar{I} = \bigcap\{\bar{J}_i : J_i \subset \mathfrak{n}\}$$

is the irreducible primary decomposition of the ideal $\bar{I}$ of the ring $A\mathfrak{n}$. Hence, for the ideals $I \subset I_0 \subset \mathfrak{n}$ in the ring A, where I_0 is prime, we get the following

---

[86] The opposite inclusion follows from the fact that $I \longrightarrow \bar{I}$ is increasing.

[87] Of course, if $J \not\subset A\mathfrak{n}$, then $\chi^{-1}(J) \subset \mathfrak{n}$ (for $J \subset \bar{\mathfrak{n}}$).

equivalences:

$$(I_0 \text{ is associated with } I) \iff (\bar{I}_0 \text{ is associated with } \bar{I})$$
$$(I_0 \text{ is isolated for } I) \iff (\bar{I}_0 \text{ is isolated for } \bar{I}).$$

The ideals $\mathfrak{n}^{(k)} = \chi^{-1}(\mathfrak{n}^k)$ [88] are primary [89] , with radical $\mathfrak{n}$ [90] .

We have $\mathfrak{n} = \mathfrak{n}^{(1)} \supset \mathfrak{n}^{(2)} \supset \dots$. Also, $\mathfrak{n}^{(k)}\mathfrak{n}^{(l)} \subset \mathfrak{n}^{(k+l)}$. The ideal $\mathfrak{n}^{(k)}$ is equal to the union of the ideals $\mathfrak{n}^k : x, \; x \in A \setminus \mathfrak{n}$.

If $A_\mathfrak{n}$ is a discrete valuation ring, then every primary ideal $\mathcal{J}\mathfrak{n} \subsetneq I \subset \mathfrak{n}$ in the ring A is of the form $I = \mathfrak{n}^{(k)}$, where $k > 0$. Indeed, $\bar{I} \neq 0$ (otherwise we would have $I = \chi^{-1}(0) = \mathcal{J}\mathfrak{n})$, and so $\bar{I} = \bar{\mathfrak{n}}^k$. Thus $I = \chi^{-1}(\bar{\mathfrak{n}}^k) = \mathfrak{n}^{(k)}$.

Clearly, an isomorphism of rings $h : A \longrightarrow B$ induces the isomorphism of their localisations $A_\mathfrak{n} \longrightarrow B_{h(\mathfrak{n})}$ (given by $x/y \longrightarrow h(x)/h(y)$).

Finally, observe that for any ideal $I \subset \mathfrak{n}$,

$$A_\mathfrak{n}/\bar{I} \approx (A/I)_{\mathfrak{n}/I} \ .$$

Indeed, let $A \ni x \longrightarrow \tilde{x} \in A/I$ be the natural epimorphism. Then $A_\mathfrak{n} \ni x/y \longrightarrow \tilde{x}/\tilde{y} \in (A/I)_{\mathfrak{n}/I}$ is an epimorphism and it is easy to check that $\bar{I}$ is its kernel.

3. The lemma in 10.6 yields the following

PROPOSITION . *If A is an integrally closed noetherian ring, then the localization A_I of any ideal I associated with a principal ideal is a discrete valuation ring.*

For suppose that the ideal I is associated with a principal ideal, say Au. The localization A_I is an integrally closed noetherian ring (see n° 1), and the ideal $\bar{I}$ is associated with the principal ideal $A_I u$ (see n° 1). By the lemma, the ring A_I is a discrete valuation ring.

A partial converse to the above proposition is given by the following criterion.

THE ZARISKI-SAMUEL LEMMA . *Let A be a noetherian local ring without nilpotents. Assume that there is a universal denominator $u \in A$ for the integral closure B of the ring A in its ring of fractions [91] such that, if u is*

---

[88] These are the so called *symbolic powers of the ideal* $\mathfrak{n}$. In the case when A is an integral domain, this definition reduces to that given in n° 1.

[89] See n° 1, footnote [76].

[90] Since rad $\chi^{-1}(\bar{\mathfrak{n}}^k) = \chi^{-1}(\bar{\mathfrak{n}}) = \mathfrak{n}$ (see 9.3, footnote [46]).

[91] This means that u is not a zero divisor and $uB \subset A$ (see 1.15a).

non-invertible, the localizations A_{I_i} to the ideals I_i associated with the ideal Au are discrete valuation rings. Then the ring A is integrally closed ([92]) .

First we prove

LEMMA . *If a noetherian local ring A does not have nilpotents and is integrally closed in its ring of fractions R, then it must be an integral domain.*

Indeed, according to the proposition in 9.3a, the ring R is isomorphic to the Cartesian product of fields $K_1 \times \ldots \times K_r$. Thus it is enough to show that $r = 1$. Now, if $r > 1$, then the image $z \in R$ of the element $(1, 0, \ldots, 0) \in K_1 \times \ldots \times K_r$ would satisfy the equation $z^2 - z = 0$, and thus $z \in A$. But $z \neq 0, 1$ and $z(1 - z) = 0$ mean together that the elements $z, 1 - z \in A$ are zero divisors, and hence they would be non-invertible, which is impossible in a local ring.

PROOF of the Zariski-Samuel lemma. We may assume that u is non-invertible by excluding the trivial case (in view of the lemma). Then we have a primary decomposition

$$Au = \bigcap_i I_i^{(k_i)} , \text{ where } k_i > 0 ,$$

owing to the fact that u is not a zero divisor (see n° 2). On account of the lemma, it is sufficient to show that $B = A$. Let $x/y \in B$, where $x, y \in A$ and y is not a zero divisor. We have $c(x/y) \in A$ for all $n \in \mathbf{N}$ and some $c \in A$ that is not a zero divisor (see 9.7); moreover, $u(x/y) \in A$. This means that

$$ux = vy \text{ and } cx^n = d_n y^n \text{ for } n \in \mathbf{N} ,$$

where $v, d_n \in A$. It is enough to prove that $v \in I_i^{(k_i)}$ for each i, as in this case $v \in Au$, i.e., $x/y = v/u \in A$. Fix i and consider the localization A_{I_i}. It now suffices to check that $\bar{v} \in \bar{I}_i^{(k_i)}$. We have $\bar{u}\bar{x} = \bar{v}\bar{y}$ and $\bar{c}\bar{x}^n = \bar{d}_n \bar{y}^n$ for $n \in \mathbf{N}$, where $\bar{u}, \bar{y}, \bar{c} \neq 0$ (as u, y, c are not zero divisors – see n° 2). Obviously we may assume that $\bar{v} \neq 0$ and then $\bar{x}, \bar{d}_n \neq 0$. Consider the valuation ν on A_{I_i}

([92]) In the general case, there is the following criterion due to Serre (see [37], IV. 4, th.11): A local noetherian ring A is integrally closed if and only if the following implications hold for its prime ideals I:

(1) $h(I) \leq 1 \Longrightarrow A_I$ is a discrete valuation ring,

(2) $h(I) \geq \Longrightarrow \text{prof } A_I \geq 2$ (see §14 below).

In the case of an integral domain and when there exists a universal denominator (for the integral closure in the ring of fractions), one can delete the second implication. This follows from the Zariski-Samuel lemma (in view of Krull's principal ideal theorem – see 12.1 below).

(see 10.6). We have $\nu(\bar{u}) + \nu(\bar{x}) = \nu(\bar{v}) + \nu(\bar{y})$ and $\nu(\bar{c}) + n\nu(\bar{x}) \geq n\nu(\bar{y})$ for $n \in \mathbb{N}$. Hence $\nu(\bar{x}) \geq \nu(\bar{y})$, which gives $\nu(\bar{u}) \leq \nu(\bar{v})$. But $\nu(\bar{u}) \geq k_i$, since $\bar{u} \in \bar{I}_i^{k_i}$. Therefore $\nu(\bar{v}) \geq k_i$, and so $\bar{v} \in \bar{I}_i^{k_i}$.

§12. Krull's dimension

1. A module is said to be *artinian* if every decreasing sequence of its submodules is stationary. In particular, a ring A is *artinian* (as a module over A) precisely when every decreasing sequence of its ideals is stationary. Note that finite dimensional spaces are artinian. Thus, if M is a finite module over a local ring A with maximal ideal $\mathfrak{m}$, then the module $M/\mathfrak{m}M$ is artinian (since its submodules over A are precisely its subspaces over $A/\mathfrak{m}$).

LEMMA 1. *Let N be a submodule of the module M. If N and M/N are artinian, then so is M.*

Indeed, take a decreasing sequence L_ν of submodules of the module M. The sequences $L_\nu \cap N$ and $L_\nu + N$ are constant after some index p. Hence, for $\nu \geq p$, we have $L_{\nu+1} \subset L_\nu \subset L_{\nu+1} + N$, and thus $L_\nu \subset L_{\nu+1} + (L_\nu \cap N) \subset L_{\nu+1}$.

LEMMA 2. *If A is a noetherian local ring and I is a defining ideal, then A/I is artinian.*

PROOF. We may assume that $I = 0$ (replacing A by A/I if necessary). Then $\mathfrak{m}^s = 0$ for some s. The ideals $A, \mathfrak{m}, \mathfrak{m}^2, \ldots$ are finitely generated, and consequently, the modules $A/\mathfrak{m}, \mathfrak{m}/\mathfrak{m}^2, \mathfrak{m}^2/\mathfrak{m}^3, \ldots$ are artinian. On the other hand, $A/\mathfrak{m}^k \approx (A/\mathfrak{m}^{k+1})/(\mathfrak{m}^k/\mathfrak{m}^{k+1})$, and hence, by lemma 1, if $A/\mathfrak{m}^k$ is artinian, then so is $A/\mathfrak{m}^{k+1}$. Therefore all $A/\mathfrak{m}^k$ are artinian. In conclusion, $A \approx A/\mathfrak{m}^s$ is artinian.

If two ideals $I \subset J$ are prime, we say that J is *immediate to* I if there is no prime ideal I' such that $I \subsetneq I' \subsetneq J$ [93].

KRULL'S PRINCIPAL IDEAL THEOREM. *In any noetherian integral domain, every isolated ideal associated with a principal ideal is immediate to 0.*

---

[93] By passing to the quotient ring A/I_0 (or to the localization $A_\mathfrak{n}$), we have the following equivalences for the prime ideals $I \subset J$ containing I_0 (or contained in $\mathfrak{n}$, respectively):

$$(I \text{ is immediate to } J) \Leftrightarrow (J/I_0 \text{ is immediate to } I/I_0)$$

$$(J \text{ is immediate to } I) \Leftrightarrow (\bar{J} \text{ is immediate to } \bar{I}).$$

(See 1.10 and 11.2.)

PROOF. Let A be a noetherian integral domain, and let J be an isolated ideal for Ac, where $c \in A$. We may assume that the ring A is local with the maximal ideal $\mathfrak{m} = J$ (since in the ring A_J the ideal $\bar{J}$ is isolated for $\overline{Ac} = A_J c$; and if $\bar{J}$ is immediate to 0 in A_J, then so is J to 0 in A). Thus Ac is defining.

Suppose that there is a prime ideal I such that $0 \subsetneqq I \subsetneqq \mathfrak{m}$. Then $c \in \mathfrak{m} \setminus I$. The ring A_I is local and noetherian. We will prove that the sequence of ideals $I_\nu = \bar{I}^\nu \cap A$ is stationary. In view of lemma 2, the ring A/Ac is artinian. Consequently, the sequence $(I_\nu + Ac)/Ac$, and thus also $I_\nu + Ac$, is constant from some index p on. Let $\nu \geq p$. Since $I_{\nu+1} \subset I_\nu \subset I_{\nu+1} + Ac$, we have $I_\nu \subset I_{\nu+1} + (Ac \cap I_\nu)$. Hence, by Nakayama's lemma, it is enough to show that $Ac \cap I_\nu \subset \mathfrak{m} I_\nu$. Now if $ac \in I_\nu$, where $a \in A$, then $ac \in \bar{I}^\nu$. But $c \notin \operatorname{rad} \bar{I}^\nu = \bar{I}$ (otherwise we would have $c \in \bar{I} \cap A = I$). Since $\bar{I}^\nu$ is prime, $a \in \bar{I}^\nu$, and hence $a \in I_\nu$. This means that $ac \in \mathfrak{m} I_\nu$. We have just shown that the sequence I_ν is constant for $\nu \geq p$. Therefore, because of the Krull intersection theorem, we conclude that $I^p \subset I_p = \bigcap_1^\infty I_\nu = \bigcap_1^\infty \bar{I}_\nu = 0$, i.e., $I^p = 0$, which is impossible because $I \neq 0$ and A is an integral domain.

COROLLARY. *Let A be a noetherian ring. If I is a prime ideal and J is an isolated ideal for $I + Ac$, where $c \in A$, then J is immediate to I.*

Indeed, it is enough to notice that the ring A/I is a noetherian integral domain, whereas J/I is an isolated ideal for the principal ideal $(I + Ac)/I$ (generated by $\bar{c}$).

KRULL'S HEIGHT THEOREM. *Let A be a noetherian ring, and let I be an isolated ideal for an ideal generated by r elements. If the ideals $I_0 \subsetneqq \ldots \subsetneqq I_s = I$ are prime, then $s \leq r$.*

PROOF. For $r = 0$, the theorem is obvious. Assume now that $r > 0$, and suppose that the theorem is true for $r - 1$. Let the ideal J be generated by the elements $x_1, \ldots, x_r$, and let I be an isolated ideal for J. We may assume that A is a local ring whose maximal ideal $\mathfrak{m} = I$. (Owing to the fact that in the local ring A_I the maximal ideal $\bar{I}$ is isolated for the ideal $\bar{J}$ that is generated by $\bar{x}_1, \ldots, \bar{x}_r$, and the ideals $\bar{I}_0 \subsetneqq \ldots \subsetneqq \bar{I}_s = \bar{I}$ are prime.) Then the ideal J is defining and $I_s = \mathfrak{m}$. We may assume that $s > 0$ and that I_s is immediate to I_{s-1} (this can be achieved by adding, if necessary, the maximal ideal of the family of prime ideals I such that $I_{s-1} \subsetneqq I \subsetneqq I_s$). Since $\mathfrak{m}$ is an isolated ideal for J, we may also assume that $x_r \notin I_{s-1}$. Then the ideal $I_{s-1} + Ax_r$ is defining, which implies that for some k we have $x_i^k \in y_i + Ax_r$, with $y_i \in I_{s-1}$ ($i = 1, \ldots, r - 1$). There is an ideal $I' \subset I_{s-1}$ which is isolated for $J' = \sum_1^{s-1} Ay_i \subset I_{s-1}$. As $x_i^k \in I' + Ax_r$ for $i = 1, \ldots, r$, it follows that $J \subset \operatorname{rad}(I' + Ax_r)$, and consequently, $I' + Ax_r$ is defining. Hence, according to the corollary of the Krull principal ideal theorem, $\mathfrak{m}$ is immediate to I'. So we must have $I' = I_{s-1}$. The induction hypothesis implies that $s - 1 \leq r - 1$, and thus $s \leq r$.

2. In a noetherian ring A we define the *height of a prime ideal* I by the formula $h(I) = \max\{k : I_0 \subsetneqq \dots \subsetneqq I_k = I\}$, where $I_1, \dots, I_k$ are prime ideals. (Note that, in view of the Krull height theorem, $k \leq g(I)$ for every such sequence.) Naturally, $I \subsetneqq J \Longrightarrow h(I) < h(J)$ for prime ideals I, J. Note also that if $\mathfrak{n}$ is a prime ideal, then $h(\bar{I}) = h(I)$, where $\bar{I}$ is the extension in $A_\mathfrak{n}$ of a prime ideal $I \subset \mathfrak{n}$.

Krull's height theorem can be restated as follows.

If (in a noetherian ring) an ideal I is isolated for an ideal J, then $h(I) \leq g(J)$.

We have the following

THEOREM 1'. *For every prime ideal I of a noetherian ring A*

$$h(I) = \min\{g(J) : I \text{ is isolated for the ideal } J\}.$$

PROOF . First, we will prove that if $0 \leq k \leq h(I)$, then, for some $x_1, \dots, x_k \in I$, every isolated ideal for $\sum_1^k Ax_i$ is of height $\geq k$. This is obvious for $k = 0$. Suppose that it holds for $k - 1$, where $0 < k \leq h(I)$. Then there exist elements $x_1, \dots, x_{k-1} \in I$ such that all the isolated ideals for $\sum_1^{k-1} Ax_i$ are of height $\geq k - 1$. Let $I_1, \dots, I_s$ denote all the isolated ideals whose height is $k - 1$. Hence $I \not\subset I_\nu$ and so $I \not\subset \bigcup I_\nu$. Thus there is an element $x_k \in I \setminus \bigcup I_\nu$. Therefore each isolated ideal for $\sum_1^k Ax_i$ is of height $\geq k$, because it must contain an isolated ideal I' for $\sum_1^{k-1} Ax_i$, and the inclusion is strict if I' is among the ideals I_ν.

Now, if $k = h(I)$, then I is an isolated ideal for $J = \sum_1^k Ax_i$ (since taking an isolated ideal $I^* \subset I$ for J, we have $h(I^*) \geq k$ and hence $I^* = I$), and obviously $g(J) \leq h(I)$.

In particular, if A is also an integral domain, we have the equivalence

$$h(I) = 1 \Longleftrightarrow (I \text{ is isolated for a non-zero principal ideal}).$$

This gives us the following characterization of noetherian rings that are factorial.

A noetherian integral domain A is factorial if and only if each of its prime ideals of height 1 is principal.

Indeed, the condition is sufficient. To see this, let x be an irreducible element, and let I be an isolated ideal for Ax. Then $h(I) = 1$ and hence $I = Az$ for some element z

which must be non-invertible. So $x = az$ for some element a which must be invertible. Thus $I = Ax$, i.e., the element x is prime. Therefore (see 9.5) the ring A is factorial. Now assume that A is factorial, and let I be one of its prime ideals of height 1. Then, for some $c \neq 0$, the ideal I is isolated for Ac, and thus the element c must be non-invertible. Hence $c = x_1 \ldots x_k$, where x_i are prime (see 6.1). But then $x_s \in I$ for some s (since $c \in I$), and so the ideal I is isolated for Ax_s. Therefore $I = Ax_s$.

3. Let A be a noetherian local ring with the maximal ideal $\mathfrak{m}$. We define the *(Krull) dimension of the ring* A by the formula

$$\dim A = \max\{k : I_0 \subsetneq \ldots \subsetneq I_k\} = h(\mathfrak{m}) ,$$

where I_ν are prime ideals $(^{94})$. Thus we have

THEOREM 1". *The dimension of a noetherian local ring A is equal to the minimum of the numbers of generators of the defining ideals of A, i.e.,*

$$\dim A = \min\{g(I) : \ I \text{ is defining}\} .$$

Every system of generators of a defining ideal which realizes this minimum, i.e., every system of $(\dim A)$ elements which generates a defining ideal, is called a *system of parameters* of the ring A. Thus every noetherian local ring has a system of parameters.

PROPOSITION . *If an element* $x \in \mathfrak{m}$ *is not a zero divisor, then* $\dim A/Ax = \dim A - 1$.

PROOF . Set $s = \dim A/Ax$. There are prime ideals $I_1, \ldots, I_{s+1}$ of the ring A such that $Ax \subset I_1 \subsetneq \ldots \subsetneq I_{s+1} = \mathfrak{m}$. But the ideal I_1 is not isolated for 0, as it contains x (see 9.3). Thus there exists a prime ideal $I_0 \subsetneq I_1$. Consequently, $\dim A \geq s + 1$. On the other hand, there is a defining ideal $J \supset Ax$, such that J/Ax is generated by some elements $\bar{x}_1, \ldots, \bar{x}_s$, where $x_1, \ldots, x_s \in J$. But then J must be generated by $x_1, \ldots, x_s, x$, and thus $\dim A \leq s + 1$.

Observe also that if $h : A \longrightarrow B$ is an epimorphism of noetherian local rings, then $\dim B \leq \dim A$.

We also show the following

LEMMA . *Let* $I \subset \mathfrak{n} \subset A$ *be ideals, and suppose that* $\mathfrak{n}$ *is prime. Then the dimension of the rings*

$$A_\mathfrak{n}/\bar{I} \approx (A/I)_{\mathfrak{n}/I}$$

(see 11.2) *is equal to*

$$\max\{k : \ I \subset I_0 \subsetneq \ldots \subsetneq I_k \subset \mathfrak{n}\} ,$$

where I_i are prime ideals.

$(^{94})$ Therefore $\dim A/I \leq \dim A$ for every proper ideal I (see 1.10–11).

Indeed, the prime ideals of $A_{\mathfrak{n}}/\bar{I}$ are precisely those of the form $\bar{J}/\bar{I}$, where $I \subset \mathfrak{n}$ is a prime ideal satisfying the inclusion $\bar{I} \subset \bar{J}$ (see 1.10-1.11 and 11.2), which is equivalent to $I \subset J$ ([95]).

§13. Modules of syzygies and homological dimension

1. Let A be a local ring with the maximal ideal $\mathfrak{m}$.

Let M be a finite module over A. For any minimal system of generators $a = (a_1, \ldots, a_n)$ for the module M we define a submodule of the module A^n by

$$\mathrm{syz}_a M = \left\{ (t_1, \ldots, t_n) \in A^n : \sum_1^n t_i a_i = 0 \right\}.$$

This submodule is equal to the kernel of the epimorphism

$$A^n \ni (t_1, \ldots, t_n) \longrightarrow \sum_1^n t_i a_i \in M.$$

Therefore $M \approx A^n/\mathrm{syz}_a M$. Each of the modules $\mathrm{syz}_a M$ is called a *module of syzygies* for the module M. Notice that $\mathrm{syz}_a M \subset \mathfrak{m}A^n = \mathfrak{m} \times \ldots \times \mathfrak{m}$ (since $\sum t_i a_i = 0 \Longrightarrow t_1, \ldots, t_n \in \mathfrak{m}$ in view of the minimality of the system a).

PROPOSITION. *If M and N are finite modules over A that are isomorphic, then any two of their modules of syzygies are isomorphic. In particular, all modules of syzygies of the module M are mutually isomorphic.*

PROOF. Let $\varphi : M \longrightarrow N$ be an isomorphism, and let $a = (a_1, \ldots, a_n)$, $b = (b_1, \ldots, b_n)$ be minimal systems of generators for the modules M and N, respectively. We have $\mathrm{syz}_a M = \ker \alpha$ and $\mathrm{syz}_b M = \ker \beta$, where $\alpha : A^n \ni (t_1, \ldots, t_n) \longrightarrow \sum_1^n t_i a_i \in M$ and $\beta : A^n \ni (t_1, \ldots, t_n) \longrightarrow \sum_1^n t_i b_i \in N$. Also, $\varphi(a_j) = \sum_{i-1}^n \lambda_{ij} b_i$ for some $\lambda_{ij} \in A$. The homomorphism $\lambda : A^n \longrightarrow A^n$ given by the matrix $[\lambda_{ij}]$ is an isomorphism, because $\det \lambda_{ij}$ is invertible. Indeed, the latter belongs to $A \setminus \mathfrak{m}$ because $\overline{\det \lambda_{ij}} = \det \bar{\lambda}_{ij} \neq 0$ (since $\overline{\varphi(a_i)} = \sum_{i=1}^n \bar{\lambda}_{ij} \bar{b}_i$ and, owing to the fact that $\{\varphi(a_i)\}$ is also a minimal system of generators for N, the sequences $\{\bar{b}_i\}$ and $\{\overline{\varphi(a_i)}\}$ are two bases of

---

([95]) For if $\bar{I} \subset \bar{J}$ and $x \in I$, then $x = u/v$, with $u \in J$ and $v \in A \setminus J$. Hence $xv \in J$, and so $x \in J$.

the space $N/\mathfrak{m}N$ over $A/\mathfrak{m}$). Therefore $\det \lambda_{ij} \in A \setminus \mathfrak{m}$. As the diagram

$$
\begin{array}{ccc}
A^n & \xrightarrow{\ \alpha\ } & M \\
\downarrow{\scriptstyle \lambda} & & \downarrow{\scriptstyle \varphi} \\
A^n & \xrightarrow{\ \beta\ } & N
\end{array}
$$

commutes $(^{96})$, $\ker \beta = \lambda(\ker \alpha)$ and hence $\mathrm{syz}_a M \approx \mathrm{syz}_b N$.

By syz M or $\mathrm{syz}_A M$ we will denote any of the modules of syzygies for the module M. (Moreover, we assume that syz $0 = 0$.) We have syz $M \subset \mathfrak{m}A^n$ and $M \approx A^n/\mathrm{syz}\ M$. Notice also that syz $M = 0$ precisely when M is free.

LEMMA 1 $(^{97})$. *Let M be a finite module over A. If $t \in \mathfrak{m}$ is not a zero divisor in M, then*

$$
\mathrm{syz}_{A/tA}(M/tM) \approx (\mathrm{syz}\ M)/(t\ \mathrm{syz}\ M)
$$

(both sides are regarded here as modules over A/tA) $(^{98})$.

PROOF . Let $a = (a_1, \ldots, a_n)$ be a minimal system of generators of the module M. Then $\bar{a} = (\bar{a}_1, \ldots, \bar{a}_n)$ is a minimal system of generators for the module M/tM over A/tA. Consider the A–module homomorphism $\chi : A^n \ni (u_1, \ldots, u_n) \longrightarrow (\bar{u}_1, \ldots, \bar{u}_n) \in (A/tA)^n$. We have $\chi(\mathrm{syz}_a M) = \mathrm{syz}_{\bar{a}}(M/tM)$ $(^{99})$. (The inclusion $\subset$ is obvious. On the other hand, if $\sum \bar{u}_i \bar{a}_i = 0$, where $u_i \in A$, then $\sum u_i a_i \in tM$. That is, $\sum u_i a_i = t \sum v_i a_i$, where $v_i \in A$, and hence $\sum (u_i - tv_i)a_i = 0$. But $\bar{u}_i = \overline{u_i + tv_i}$, and so $(\bar{u}_1, \ldots, \bar{u}_n) \in \chi(\mathrm{syz}_a M)$.) Thus the restriction $\chi_0 : \mathrm{syz}_a M \longrightarrow \mathrm{syz}_{\bar{a}}(M/tM)$ of the homomorphism χ is an epimorphism. Now $\ker \chi_0 = t\ \mathrm{syz}_a M$. (The inclusion $\supset$ is clear. However, if $\sum u_i a_i = 0$ and $\bar{u} = 0$, then $u_i = tv_i$ with $v_i \in A$, and so $\sum v_i a_i = 0$, as t is not a zero divisor in M.) Therefore the epimorphism χ_0 induces the A–module isomorphism $(\mathrm{syz}_a M)/t\mathrm{syz}_a M \longrightarrow \mathrm{syz}_{\bar{a}}(M/tM)$. It is, at the same time, an A/tA–module isomorphism.

Fix an element $s \in \mathfrak{m}$. By an s–module we will mean any submodule M of any A^k which satisfies the condition $sA^k \subset M \subset \mathfrak{m}A^k$. Then we put $\bar{M} = M/sA^k$. This is also an (induced) module over $\bar{A} = A/sA$.

$(^{96})$ Because $\varphi(\alpha(t_1, \ldots, t_n)) = \sum t_i \varphi(a_j) = \sum_{i,j} \lambda_{ij} t_j b_i = \beta(\lambda(t_1, \ldots, t_n))$.

$(^{97})$ See [31].

$(^{98})$ See [31]. Note that, both in the numerator and in the denominator of the right hand side, we must take the same module of syzygies syz M.

$(^{99})$ Observe that the module $\mathrm{syz}_{\bar{a}}(M/tM)$ is induced by itself viewed as a submodule of the module $(A/tA)^n$.

LEMMA 2 ([100]) . *Assume that $s \in \mathfrak{m} \setminus \mathfrak{m}^2$ is not a zero divisor. Then, for every s–module M that is finite over A, there exists an s–module $S \approx$ syz M, such that $\mathrm{syz}_{\bar{A}} \bar{M} \approx \bar{S}$ (as modules over $\bar{A}$).*

PROOF . Let $sA^k \subset M \subset \mathfrak{m}A^k$, and let $\bar{a} = (\bar{a}_1, \ldots, \bar{a}_l)$ be a minimal system of generators of the module $\bar{M}$ over $\bar{A}$ and hence over A, too, where $a_i = (a_{i1}, \ldots, a_{ik}) \in M$ for $i = 1, \ldots, l$. Then $a = (a_1, \ldots, a_l, se_1, \ldots, se_k)$, where $e_1, \ldots, e_k$ is the canonical basis in A^k, is a minimal system of generators of M. Indeed, if $\sum t_i a_i + \sum u_j se_j = 0$, where $t_i, u_j \in A$, then $\sum t_i \bar{a}_i = 0$, and so $t_i \in \mathfrak{m}$. Thus $u_j s = -\sum t_i a_{ij} \in \mathfrak{m}^2$ and therefore $u_j \in \mathfrak{m}$ (for otherwise $s \in \mathfrak{m}^2$). The mapping

$$\varphi : \ \mathrm{syz}_a M \ni (t_1, \ldots, t_l, u_1, \ldots, u_k) \longrightarrow (t_1, \ldots, t_l) \in A^l$$

is a monomorphism. (Because, if $\sum u_j se_j = 0$, then $u_j s = 0$ and hence $u_j = 0$.) Accordingly, $S = \varphi(\mathrm{syz}_a M) \approx \mathrm{syz}_a M$ and S is an s–module (as $sA^l \subset S$; this is so because, if $(t_1, \ldots, t_l) \in A^l$, then $\sum_1^l st_i a_i + \sum_1^k u_j se_j = 0$, where $u_j = -\sum_i t_i a_{ij}$, and so $(st_1, \ldots, st_l, u_1, \ldots, u_k) \in \mathrm{syz}_a M$). Now the condition $(t_1, \ldots, t_l) \in S$ implies that $\sum t_i a_i + \sum u_j se_j = 0$, for some $u_j \in A$, and $\sum t_i a_i \in A^k$, $(\bar{t}_1, \ldots, \bar{t}_l) \in \mathrm{syz}_{\bar{a}} \bar{M}$. Thus we have the A–module epimorphism $s \ni (t_1, \ldots, t_l) \longrightarrow (\bar{t}_1, \ldots, \bar{t}_l) \in \mathrm{syz}_{\bar{a}} \bar{M}$ ([101]) . As sA^l is its kernel, it induces an A–module isomorphism $\bar{S} \longrightarrow \mathrm{syz}_{\bar{a}} \bar{M}$ which is, at the same time, an $\bar{A}$–module isomorphism.

2. Now let A be a noetherian local ring.

For any finite A–module M we define recursively its *n-th modules of syzygies* for $n = 0, 1, 2, \ldots$. The 0-th module is defined to be M and the n-th one (for $n > 0$) is defined to be any module of syzygies of an $(n-1)$-st module of syzygies. This can be done, because in each step we obtain submodules of some A^k, and hence finite modules over A. It follows from the proposition that the n-th modules of syzyzgies of isomorphic modules are also isomorphic. In particular, all of the n-th modules of syzygies of a module M are mutually isomorphic. We denote any of them by $\mathrm{syz}^n M$. We have $\mathrm{syz}^0 M = M$, $\mathrm{syz}^1 M \approx \mathrm{syz}\, M$ and $\mathrm{syz}^{n+1} M \approx \mathrm{syz}(\mathrm{syz}^n M)$ for $n \geq 0$. Thus $\mathrm{syz}^{n+m} M \approx \mathrm{syz}^n(\mathrm{syz}^m M)$ for $n, m \geq 0$ ([102]) . Clearly, if $n > 0$, then $\mathrm{syz}^n M \subset \mathfrak{m}A^k$ for some k.

([100]) See [31].

([101]) Note that the module $\mathrm{syz}_{\bar{a}} \bar{M}$ is induced by itself when regarded as a submodule of the module $(A/sA)^l$ over A.

([102]) The equalities remain true for an arbitrary choice of modules of syzygies on the right hand side and for some module of syzygies on the left hand side, and vice versa.

For a module M (that is finite over A) we define its *homological dimension* by the formula

$$\text{hd } M = \text{hd}_A M = \begin{cases} \infty, & \text{if } \text{syz}^n \neq 0 \text{ for each } n; \\ \min\{n \in \mathbf{N} : \text{syz}^{n+1} M = 0\}, & \text{otherwise.} \end{cases}$$

Consequently, hd $M = 0$ precisely when the module M is free. If $M \neq 0$, then hd $M = n$ means that $\text{syz}^n M$ is a free non-zero module. Of course, hd $M = \text{hd}(\text{syz } M) + 1$, provided that hd $M > 0$.

PROPOSITION 1. *Let M be a finite module over A. If $t \in \mathfrak{m}$ is neither a zero divisor in A nor in M, then*

$$\text{hd}_{A/tA} M/tM = \text{hd } M \ .$$

PROOF. If follows from lemma 1 that, since t is not a zero divisor in any A^k and thus in any $\text{syz}^n M$, we have $\text{syz}^n_{A/tA} M/tM \approx (\text{syz}^n M)/(t\text{syz}^n M)$. Therefore it is sufficient to note that for an arbitrary finite module N over A, $N = 0 \iff N/tN = 0$ (which follows from Nakayama's lemma).

PROPOSITION 2. *If $s \in \mathfrak{m} \setminus \mathfrak{m}^2$ is not a zero divisor, then*

$$\text{hd}\mathfrak{m} < \infty \iff \text{hd}_{A/sA} \mathfrak{m}/sA < \infty \ .$$

PROOF. Let $n > 0$. According to lemma 2, for every s–module M, there exists an s–module $S \approx \text{syz}^n M$ such that $\text{syz}^n_A \bar{M} \approx \bar{S}$. In particular, for $M = \mathfrak{m}$, we obtain $S \approx \text{syz}^n \mathfrak{m}$ and $\text{syz}^n_{A/sA} \mathfrak{m}/sA \approx S/sA^k$ (for some k). Consequently, if $\text{syz}^n \mathfrak{m} = 0$, then $\text{syz}^n_{A/sA} \mathfrak{m}/sA = 0$, whereas if $\text{syz}^n_{A/sA} \mathfrak{m}/sA = 0$, then $S = sA^k$ is a free module (with the basis $se_1, \ldots, se_k$), and so $\text{syz}^{n+1} \mathfrak{m} = 0$.

Let M be a finite module over A. A *finite free resolution* of M is an exact sequence of homomorphisms of finite modules over A:

(#) $$0 \longrightarrow L_k \longrightarrow \ldots \longrightarrow L_0 \longrightarrow M \longrightarrow 0$$

in which the L_i's are free. The exactness of the sequence (#) means that every pair of its consecutive homomorphisms f, g satisfy the condition im $f = \ker g$. The number k is called the *length* of the resolution (#).

THEOREM . *We have hd$M < \infty$ if and only if M has a finite free resolution. Then hd$M$ is equal to the minimum of the length of finite free resolutions of M.*

PROOF ([103]). First, we prove two lemmas. Let L be a finite free module over A.

LEMMA 1. syz $(M \times L) \approx$ syz M.

For if $M+L$ is a direct sum (of submodules of one module), $a = (a_1, \ldots, a_k)$ is a minimal system of generators of M, and $b_1, \ldots, b_l$ is a basis of L, then $c = (a_1, \ldots, a_k, b_1, \ldots, b_l)$ is a minimal system of generators of $M + L$ (see 10.4) and $\operatorname{syz}_c(M + L) = (\operatorname{syz}_a M) \times 0$.

LEMMA 2. If $f : L \longrightarrow M$ is an epimorphism, then $\ker f \approx (\operatorname{syz} M) \times A^k$ for some k.

In fact, we have syz $M = \ker \varphi$, where $\varphi : A^n \longrightarrow M$ is an epimorphism. There is a homomorphism $h : L \longrightarrow A^n$ such that $f = \varphi \circ h$. (We take h to be defined by the condition $h(b_i) = f(c_i)$, where $b_1, \ldots, b_r$ is a basis of L and $\varphi(c_i) = b_i$; see 1.16.) Then h is an epimorphism. Indeed, since $\varphi(A^n) = \varphi(\operatorname{im} h)$, we have $A^n \subset \operatorname{im} h + \ker \varphi \subset \operatorname{im} h + \mathfrak{m}A^n$, and so $\operatorname{im} h = A^n$ by Nakayama's lemma (see 10.2). It follows that $L = \operatorname{im} \psi + \ker h$ is a direct sum (see 1.9), and hence $\ker h$ is free (see 10.4). Thus it suffices to check that $\ker f = \psi(\ker \varphi) + \ker h$. Now, the inclusion $\supset$ follows from $f = \varphi \circ h$ and $f \circ \psi = \varphi$. On the other hand, if $f(x) = 0$, $x \in L$, then $x = \psi(u) + z$, $h(z) = 0$, which gives $C = f(\psi(u) + z) = \varphi(u)$, i.e., $u \in \ker \varphi$.

PROOF of the theorem. Suppose that $k = \operatorname{hd}M < \infty$. The module $\operatorname{syz}^{i+1} M$ is the kernel of an epimorphism $A^{m_i} \longrightarrow \operatorname{syz}^i M$, $i = 0, 1, \ldots$, and hence, if $i > 0$, of a homomorphism $\varphi_i : A^{m_i} \longrightarrow A^{m_{i-1}}$ whose range is $\operatorname{syz}^i M$; however, if $i = 0$, it is the kernel of an epimorphism $\varphi_0 : A^{m_0} \longrightarrow M_0$. Moreover, $\operatorname{syz}^{k+1} M = 0$. Therefore $0 \longrightarrow A^{m_k} \xrightarrow{\varphi_k} \ldots \xrightarrow{\varphi_1} A^{m_0} \xrightarrow{\varphi_0} M \longrightarrow 0$ is a finite free resolution of M. Conversely, assume that there exists a finite free resolution $0 \longrightarrow A^{n_k} \xrightarrow{f_k} \ldots \xrightarrow{f_1} A^{n_0} \xrightarrow{f_0} M \longrightarrow 0$. It would be sufficient to show that $\ker f_i \approx (\operatorname{syz}^{i+1} M) \times A^{s_i}$ for some s_i, $i = 0, \ldots, k$; for then, by lemma 1 and in view of $\ker f_k = 0$, we would have $\operatorname{syz}^{k+1} M = 0$, i.e., $\operatorname{hd}M \leq k$. Now, if $i = 0$, the required isomorphism is given in lemma 2. Assume now that it is true for $i - 1$ ($i > 0$), i.e., in view of exactness, that $\operatorname{im} f_i \approx (\operatorname{syz}^i M) \times A^s$ for some s. Then, by lemmas 2 and 1, we have $\ker f_i \approx \operatorname{syz}(\operatorname{im} f_i) \times A^p \approx (\operatorname{syz}^{i+1} M) \times A^p$ for some p.

§14. The depth of a module

Let A be a noetherian local ring with the maximal ideal $\mathfrak{m}$, and let M be a finite module over A.

1. A sequence of elements $t_1, \ldots, t_k \in \mathfrak{m}$ is said to be an M-sequence if, for $i = 1, \ldots, k$, the element t_i is not a zero divisor in $M / \sum_1^{i-1} t_\nu M$ (i.e., $t_i x \in \sum_1^{i-1} t_\nu M \implies x \in \sum_1^{i-1} t_\nu M$). In particular, for $i = 1$, this condition ([104]) means that t_1 is not a zero divisor in M. Therefore M-sequences do not exist only in the case when $\mathfrak{m}$ consists only of zero divisors in

---

([103]) See [17b], III. 2.1-3.

([104]) We assume that $\sum_1^0 t_\nu M = 0$.

M. An M–sequence $t_1, \ldots, t_k$ is said to be *maximal* if it cannot be extended, i.e., if $\mathfrak{m}$ contains only zero divisors in $M/\sum_1^k t_\nu M$ [105].

PROPOSITION 1. *If $M \neq 0$, then every M–sequence can be extended to a maximal one* [106].

PROOF . Otherwise we would be able to choose an infinite sequence of elements $t_\nu \in \mathfrak{m}$, such that $t_1, \ldots, t_n$ is an M–sequence for each n. In order to get a contradiction, it is enough to show that the sequence of modules $M_n = \sum_1^n t_\nu M$ is strictly increasing. Now, if we had $M_{n-1} = M_n$, then $t_n M \subset M_{n-1}$, hence $M \subset M_{n-1}$ and Nakayama's lemma would imply that $M = 0$.

LEMMA 1. *A sequence $t_1, \ldots, t_k, u_1, \ldots, u_l$ is a (maximal) M–sequence if and only if $t_1, \ldots, t_k$ is an M–sequence and $u_1, \ldots, u_l$ is a (maximal) L–sequence, where $L = M/\sum_1^k t_\nu M$.*

Indeed, $\sum_1^i u_\nu L = \left(\sum_1^k t_\nu M + \sum_1^i u_\mu M\right)/\sum_1^k t_\nu M$, and hence

$$L/\sum_1^i u_\mu L \approx M/\left(\sum_1^k t_\nu M + \sum_1^i u_\mu M\right) ,$$

which yields the equivalence of both conditions.

LEMMA 2. *If there is a one-element maximal M–sequence, then each element $v \in \mathfrak{m}$ which is not a zero divisor in M is a maximal M–sequence.*

PROOF . Assume that $u \in \mathfrak{m}$ is a maximal M–sequence, i.e., it is not a zero divisor in M and $\mathfrak{m}$ contains only zero divisors in M/uM. Thus $\mathfrak{m}\bar{x} = 0$ for some $x \in M \setminus uM$ (see 9.6), which means that $\mathfrak{m}x \subset uM$, and hence $vx = uy$ for some $y \in M$. Then $y \notin vM$ (for otherwise $vx \in uvM$, which would imply that $x \in uM$) and $\mathfrak{m}y \subset vM$ (because $u\mathfrak{m}y = v\mathfrak{m}x \subset vuM$). This shows that $\mathfrak{m}$ consists of zero divisors in M/vM, and so v is a maximal sequence.

PROPOSITION 2. *Every permutation of an M–sequence is an M–sequence.*

PROOF . It would be enough to prove the proposition for sequences consisting of two terms only, because – in view of lemma 1 – it would imply that any two consecutive terms of an M–sequence could be interchanged. Let t, u

---

[105] Observe that if $\mathfrak{n} \supset A$ is a prime ideal and $t_1, \ldots, t_k$ is an A–sequence, then $\bar{t}_1, \ldots, \bar{t}_k$ is an $A_\mathfrak{n}$–sequence. Indeed, let $\bar{t}_i \bar{x} \bar{y}^{-1} \in \sum_1^{i-1} A_\mathfrak{n} \bar{t}_\nu = \bar{I}$, where $I = \sum_1^{i-1} At_\nu$, $x \in A$, $y \in A \setminus \mathfrak{n}$. Then we have $\bar{t}_i \bar{x} \in \bar{I}$, which means that $(t_i xv - u)vw = 0$ with some $u \in I$ and $v, w \in A \setminus \mathfrak{n}$. It follows that $t_i xvw \in I$, hence $xvw \in I$, and so $\bar{x}\bar{y}^{-1} \in \bar{I}$ (since $\bar{v}$ is invertible).

[106] If $M = 0$, then each finite sequence of elements of $\mathfrak{m}$ is an M–sequence.

be an M–sequence, i.e., t is not a zero divisor in M and $ux \in tM \implies x \in tM$. The submodule $L = \{x \in M : ux = 0\}$ is contained in tM. Thus $L \subset tL$ (since if $L \ni x = ty$, then $uty = 0$ and hence $y \in L$). By Nakayama's lemma, $L = 0$, which means that u is not a zero divisor in M. Moreover, $tx \in uM \implies x \in uM$. Indeed, if $tx = uy$, then $uy \in tM$, and so $y \in tM$. This implies that $tx \in utM$, hence $x \in uM$. Therefore u, t is an M–sequence.

COROLLARY . *Every permutation of a maximal M–sequence is a maximal M–sequence.*

PROPOSITION 3. *All maximal M–sequences have the same number of terms.*

PROOF . We will prove that if $t_1, \ldots, t_k$ and $s_1, \ldots, s_l$, where $k \leq l$, are maximal M–sequences, then $k = l$. If $k = 1$, then, by lemma 2, the element s_1 is a maximal M–sequence, and so $l = 1$. Suppose $k > 1$ and the implication is true for $k-1$. Now, the sets E, F of the zero divisors in $T = M/\sum_1^{k-1} t_i M$ and $S = M/\sum_1^{l-1} s_i M$, respectively, can be expressed as finite unions of prime ideals (see 9.6), and moreover, $E \not\subseteq \mathfrak{m}$ and $F \not\subseteq \mathfrak{m}$ (as $t_k \notin E$ and $s_l \notin F$). Hence $E \cup F \not\subseteq \mathfrak{m}$. Take $u \in \mathfrak{m} \setminus (E \cup F)$. According to lemmas 1 and 2, the element u is a maximal T–sequence and, at the same time, a maximal S–sequence. Therefore, by lemma 2 and the corollary from proposition 2, the sequences $t_1, \ldots, t_{k-1}$ and $s_1, \ldots, s_{l-1}$ are maximal M/uM–sequences. Thus $k - 1 = l - 1$, and so $k = l$.

2. Let $M \neq 0$.

We define the *depth* prof $M = \text{prof}_A M$ of a module M as follows. If there are no M–sequences, i.e., if $\mathfrak{m}$ contains only zero divisors in M, we put prof $M = 0$. Otherwise, by proposition 1, there are maximal sequences and we define prof M as the number of terms (common to all of them, by proposition 3) in a maximal sequence. Clearly, if $M \approx N$, then prof $M = $ prof N.

PROPOSITION 4. *If $t \in \mathfrak{m}$ is not a zero divisor in M* ([107]) *, then* prof $M/tM = $ prof $M - 1$.

Indeed, in view of proposition 1, there exists a maximal M–sequence $t = t_1, \ldots, t_k$, and hence prof $M = k$. If $k = 1$, then $\mathfrak{m}$ contains only zero divisors in M/tM and, consequently, prof $M/tM = 0$. If $k > 1$, then, according to lemma 1, the sequence $t_2, \ldots, t_k$ is maximal in M/tM and so prof $M/tM = k - 1$.

PROPOSITION 5. *If I is a proper ideal such that $IM = 0$, then* $\text{prof}_{A/I} M = \text{prof}_A M$. *A sequence $t_1, \ldots, t_k \in \mathfrak{m}$ is an M–sequence if and only if $\bar{t}_1, \ldots, \bar{t}_k \in \mathfrak{m}/I$ is an M–sequence.*

([107]) Then $M/tM \neq 0$ by Nakayama's lemma.

THEOREM. *Let B be an extension of the ring A such that B is noetherian, local, and finite over A. If $L \neq 0$ is a finite module over B, then $\operatorname{prof}_A L = \operatorname{prof}_B L$.*

The following lemmas are needed in the proof.

LEMMA α. *Let B be an integral domain that is an extension of the ring A. Then every prime ideal $I \supset \mathfrak{m}$ in B must be maximal.*

PROOF. Suppose $J \underset{\neq}{\supset} I$ is an ideal of the ring B. Take $x \in J \setminus I$. We have $a_n x^n + \ldots + a_0 = 0$ for some $a_i \in A$ such that $a_n = 1$. Let us take the smallest possible k such that $a_k \notin I$. Then a_k is invertible. Therefore $a_n x^n + \ldots + a_k x^k \in I$, and so $a_n x^{n-k} + \ldots + a_k \in I$. Thus $a_k \in J$, which implies that $J = B$.

LEMMA β. *Let $\mathfrak{n}$ be the maximal ideal in the ring B. Under the hypotheses of the theorem, $\mathfrak{m} \subset \mathfrak{n}$. Furthermore, if $\mathfrak{m}$ contains only zero divisors in L, then so does $\mathfrak{n}$.*

PROOF. We have $\mathfrak{m} \subset \mathfrak{m}B \subset \mathfrak{m}$ (for otherwise $\mathfrak{m}B = B$, which is impossible, since $B \neq 0$, in view of Nakayama's lemma). The ideal $\mathfrak{m}B$ is defining, because by lemma α the ideal $\mathfrak{n}$ is isolated for $\mathfrak{m}B$. Now, if $\mathfrak{m}$ has only zero divisors in L, then $\mathfrak{m}z = 0$ for some $z \in L \setminus 0$. Thus $(\mathfrak{m}B)z = 0$, which shows that $\mathfrak{m}B$ contains only zero divisors in L. Hence $\mathfrak{n} = \operatorname{rad}(\mathfrak{m}B)$ contains only zero divisors.

PROOF of the theorem. If $t_1, \ldots, t_k$ is an L–sequence for L regarded as a module over A, then the same is true for L considered as a module over B, since $\mathfrak{m} \subset \mathfrak{n}$, according to lemma β. Therefore it suffices to apply lemma β to the module L or to $L / \sum_1^k t_i L$.

3. Let I be a proper ideal of the ring A. Observe that $\operatorname{prof}_A(A/I) = \operatorname{prof}_{A/I}(A/I)$ (by proposition 5).

PROPOSITION 6. *For every ideal J associated with I we have $\operatorname{prof} A/I \leq \dim A/J$.*

PROOF ([108]). Let $t_1, \ldots, t_k \in \mathfrak{m}$ be a maximal (A/I)–sequence, i.e., $k = \operatorname{prof} A/I$. Consider the ideals $I_s = I + \sum_1^s At_\nu$, $s = 0, \ldots, k$. Then it is easy to check that

$$(\#) \qquad \begin{aligned} &t_1, \ldots, t_s, t \text{ is an } (A/I) - \text{sequence} \\ &\Longleftrightarrow t \text{ is not a zero divisor in } A/I_s. \end{aligned}$$

Now it suffices to find a sequence of prime ideals $J = J_0 \subsetneq \ldots \subsetneq J_k$ of the ring A (see 1.10-11). We will define it by induction, requiring that it satisfies the condition: J_s is associated with I_s. The ideal $J_0 = J$ is associated with $I_0 = I$. Assuming that $J_0, \ldots, J_s$, $(s < k)$ are already defined and satisfy the above condition, we claim that there exists an ideal $J_{s+1} \supset J_s$ associated with I_{s+1}. For otherwise (see 1.11 and 9.3) there would exist an

---

([108]) See [17b], III. 1.1, prop. 3.

element $t \in J_s$ which is not a zero divisor in A/I_{s+1}. Then, by (#) (in view of the proposition 2), $t_1, \ldots, t_s, t$ would be an (A/I)–sequence, and hence, by (#) (see 9.3), we would have $t \notin J_s$. Thus the sequence $J_0, \ldots, J_k$ is defined. Finally, $t_{s+1} \in J_{s+1} \setminus J_s$ (because $t_{s+1} \in I_{s+1}$, and by (#)).

Therefore (see 12.3):

$$\text{prof } A/I \leq \dim A/J \leq \dim A/I \quad \text{for each } J \text{ associated with } I \,.$$

In particular (taking $I = 0$), we have

$$(*) \qquad\qquad\qquad \text{prof } A \leq \dim A \,.$$

The ring A is said to be a *Cohen-Macaulay ring* if prof $A = \dim A$. Thus, if A/I is Cohen-Macaulay, then all the rings A/J, where J is associated with I, have the same dimension (equal to $\dim A/I$). Hence we have

COROLLARY 1. *If A/I is Cohen–Macaulay, then every ideal associated with I is isolated for I.*

(For if $J_1 \subsetneq J_2$ are prime, then $\dim A/J_1 > \dim A/J_2$; see 1.10-11.)

Proposition 4 from n° 2 and the proposition from 12.3 imply

COROLLARY 2. *If A is Cohen-Macaulay and $t \in \mathfrak{m}$ is not a zero divisor, then A/At is also Cohen-Macaulay* ([109]).

PROPOSITION 7. *The ring A is Cohen-Macaulay if and only if one of its systems of parameters is an A–sequence. Then each of its systems of parameters is a maximal A–sequence.*

PROOF . In view of (*), the condition is sufficient. Thus it is enough to prove that if A is Cohen-Macaulay, then each of its systems of parameters is an A–sequence. Set $n = \dim A$. If $n = 0$, then, by theorem 1" from 12.3, the zero ideal is defining, and so $\mathfrak{m}^k = 0$. It follows that $\mathfrak{m}$ contains only zero divisors, i.e., prof $A = 0$. Now, let $n > 0$, and suppose the statement is true for $n - 1$. Let A be Cohen-Macaulay, and let $t_1, \ldots, t_n$ be a system of parameters. Set $\bar{A} = A/At_n$. It is easy to check (see 1.4 and 10.3 and 5) that $\bar{t}_1, \ldots, \bar{t}_{n-1}$ generate a defining ideal in $\bar{A}$. We claim that t_n is not a zero divisor. If it were, then (see 9.3) t_n would belong to an ideal associated with 0 and we would have $\dim A \geq \dim A/J \geq n$ (see 12.3), in view of proposition 6. This is impossible because $\dim \bar{A} \leq n - 1$. Therefore, by corollary 2 and the proposition from 12.3, $\bar{A}$ is Cohen-Macaulay of dimension $n - 1$. Consequently, $\bar{t}_1, \ldots, \bar{t}_{n-1}$ is a system of parameters of $\bar{A}$

([109]) This implies a more general statement: If A is Cohen-Macaulay and $t_1, \ldots, t_k \in \mathfrak{m}$ is an A–sequence, then $A/\sum_1^k At_i$ is also Cohen-Macaulay. For if $B = A/\sum_1^{k-1} At_i$ is Cohen-Macaulay, then (since we have $B\bar{t}_k = \sum_1^k At_i / \sum_1^{k-1} At_i$) so is $A/\sum_1^k At_i \approx B/B\bar{t}_k$, because $\bar{t}_k$ is not a zero divisor in B. It follows that the *localization $A_{\mathfrak{n}}$ of a Cohen-Macaulay ring A to a prime ideal $\mathfrak{n}$ is Cohen-Macaulay*. In fact, in view of proposition 1, there exists an A–sequence which cannot be extended (by an element of $\mathfrak{n}$). Then (see footnote ([103])), $\bar{t}_1, \ldots, \bar{t}_k \in \bar{\mathfrak{n}}$ is an A–sequence and the ring A/I, where $I = \sum_1^k At_i$, is Cohen-Macaulay. Hence, by corollary 1, all ideals associated with I are isolated. One of them must contain $\mathfrak{n}$ (for otherwise – see 1.11 and 9.3 – the ideal $\mathfrak{n}$ would contain a non-zero divisor in A/I, contrary to the inextendability of the sequence t_i). Hence it must coincide with $\mathfrak{n}$, and so, by theorem 1' from 12.2, we get prof $A_{\mathfrak{n}} \geq k \geq h(\mathfrak{n}) = \dim A_{\mathfrak{n}}$ (see 11.2). Therefore, in view of (*), it follows that $A_{\mathfrak{n}}$ is a Cohen-Macaulay ring.

and hence an $\bar{A}$–sequence. By lemma 1 and proposition 5 it follows (in view of proposition 2) that $t_1, \ldots, t_n$ is an A–sequence.

§15. Regular rings

Let A be a noetherian local ring with maximal ideal $\mathfrak{m}$.

We always have edim $A = g(\mathfrak{m}) \geq \dim A$ ([110]) . The ring A is said to be *regular* if edim $A = g(\mathfrak{m}) = \dim A$, i.e., $\mathfrak{m}$ has n generators, where $n = \dim A$. Then, obviously, each system of n generators is minimal and vice versa. Thus all elements of such a system belong to $\mathfrak{m} \setminus \mathfrak{m}^2$ and each element of $\mathfrak{m} \setminus \mathfrak{m}^2$ belongs to some such system. Also, $\dim A > 0 \iff \mathfrak{m} \neq 0 \iff \mathfrak{m} \setminus \mathfrak{m}^2 \neq \emptyset$. A regular ring is zero dimensional if and only if it is a field.

THEOREM 1. *Let A be a regular ring of dimension $n > 0$, and let $x = (x_1, \ldots, x_n)$ be a system of generators of the ideal $\mathfrak{m}$. Then, for each $k > 0$, the collection $\{x^p\}_{|p|=k}$ is a minimal system of generators for the ideal $\mathfrak{m}^k$.*

The theorem yields yet another characterization of regular rings (which can be easily seen to be equivalent to the theorem). Namely, $A_0 = A/\mathfrak{m}$, $A_1 = \mathfrak{m}/\mathfrak{m}^2, \ldots, A_k = \mathfrak{m}^k/\mathfrak{m}^{k+1}, \ldots$ are vector spaces over $K = A/\mathfrak{m}$, and we have the natural bilinear mappings $A_k \times A_l \ni (\bar{t}, \bar{u}) \longrightarrow \overline{tu} \in A_{k+l}$. The direct sum $G(A) = \bigoplus_0^\infty A_\nu$ with the multiplication defined by $(\sum x_i)(\sum y_i) = \sum x_j y_j$ (where $x_\nu, y_\nu \in A_\nu$) is a graded ring, which is said to be *associated* with A. (A ring B is said to be *graded* if it possesses a decomposition $B = \bigoplus_0^\infty B_\nu$ into the direct sum of subgroups of the additive group of B, such that $B_k B_l \subset B_{k+l}$ – see [1], Chapter 1, §5; or [4], Chapter 10.) We have a natural gradation in the ring of polynomials $K[X_1, \ldots, X_n] = \bigoplus_0^\infty H_\nu$, where H_ν is the vector space that consists of forms of degree ν. Let $n = \dim A$.

The ring A is regular precisely when

$$G(A) \approx K[X_1, \ldots, X_n],$$

where $\approx$ denotes an isomorphism of graded rings, i.e., when there is an isomorphism $\varphi : K[X_1, \ldots, X_n] \longrightarrow G(A)$ such that $\varphi(H_\nu) = A_\nu$.

For if such an isomorphism exists, we may assume it is a K–isomorphism (by replacing it by $\varphi \circ \psi^{-1}$, where ψ is the automorphism of the ring $K[X_1, \ldots, X_n]$ induced by $\varphi|_K : K \longrightarrow K$). Then $\varphi|_{H_1} : H_1 \longrightarrow A_1$ is an isomorphism of vector spaces over K, and hence $\dim A_1 = n$, which means that the ring A is regular. Conversely, if $x = (x_1, \ldots, x_n)$ is a system of generators of the ideal $\mathfrak{m}$, then $\bar{x}_1, \ldots, \bar{x}_n \in \mathfrak{m}/\mathfrak{m}^2 = A \subset G(A)$ and

$$\varphi : K[X_1, \ldots, X_n] \ni P \longrightarrow P(\bar{x}_1, \ldots, \bar{x}_n) \in G(A)$$

([110]) See 10.4 and 12.3, theorem 1".

is a K–linear homomorphism, as $K = A_0 \subset G(A)$. According to theorem 1, the system $\{\bar{x}^p\}_{|p|=k}$ is a basis for the space A_k and, owing to the fact that $\varphi(X^p) = \bar{x}^p = \overline{x^p}$, where $\bar{x} = (\bar{x}_1, \ldots, \bar{x}_n)$, the restrictions $\varphi_{H_k} : H_k \longrightarrow A_k$ are bijective. Hence φ is an isomorphism of graded rings.

PROOF (in the case when $A/\mathfrak{m}$ is of characteristic zero) ([111]) . Assume that $\sum_{|p|=k} a_p x^p = 0$. It is enough to show that $a_p \in \mathfrak{m}$. Suppose it is not so. Then the form $P = \sum \bar{a}_p X^p \in (A/\mathfrak{m})[X_1, \ldots, X_n]$ is non-zero, and thus $\sum \bar{a}_p \bar{c}^p \neq 0$ for some $\bar{c} = (\bar{c}_1, \ldots, \bar{c}_n) \neq 0$. Here $c = (c_1, \ldots, c_n) \in A^n$. Clearly we may assume that $\bar{c}_1 \neq 0$. Hence the elements $\sum a_p c^p$ and c_1 are invertible. The matrix $[c_{ij}]$, where $c_{i1} = c_i$ and $c_{ij} = \delta_{ij}$ for $i = 1, \ldots, n$ and $j = 2, \ldots, n$, has the determinant equal to c_1. So it is invertible. Let $[d_{ij}]$ be the inverse matrix. By setting $y_i = \sum_1^n d_{ij} x_j$, $(i = 1, \ldots, n)$, we conclude that $x_i = \sum_1^n c_{ij} y_j = c_i y_1 + \sum_2^n c_{ij} y_j$ $(i = 1, \ldots, n)$. Therefore $y_1, \ldots, y_n$ generate the ideal $\mathfrak{m}$, and for $|p| = k$ we have $x^p \in c^p y_1^k + J$, where J is the ideal generated by $y_2, \ldots, y_n$. It follows that $0 = \sum a_p x^p \in (\sum a_p c^p) y_1^k + J$, and hence rad $J = \mathfrak{m}$. Therefore J is a defining ideal generated by $n-1$ elements, which contradicts our assumption that $\dim A = n$.

PROOF in the general case. Set $B = A[X_1, \ldots, X_n]$. For every ideal I of the ring A we have the ideal $I_* = \{\sum a_p X^p \in B : a_p \in I\}$ of the ring B and the mapping $I \longrightarrow I_*$ which is an injection that preserves inclusion. Obviously, $(I_*)^k \subset (I^k)_*$. Notice that any system of generators of the ideal I also generate the ideal I_*. We will prove the following lemmas.

LEMMA α. If I is prime, then I_* is prime.

Indeed, I_* is the kernel of the epimorphism $A[X_1, \ldots, X_n] \longrightarrow (A/I)[X_1, \ldots, X_n]$ induced by the natural epimorphism $A \longrightarrow A/I$. Hence $B/I_* \approx (A/I)[X_1, \ldots, X_n]$ is an integral domain.

LEMMA β. If I is defining, then $\mathfrak{m}_*$ is an isolated ideal for I_*.

This is because $\mathfrak{m}^k \subset I \subset \mathfrak{m}$ for some k, and thus $(\mathfrak{m}_*)^k \subset (\mathfrak{m}^k)_* \subset I_* \subset \mathfrak{m}_*$, which implies that $\mathfrak{m}_* = \mathrm{rad}\,(I_*)$.

LEMMA γ. We have $h(\mathfrak{m}_*) = h(\mathfrak{m})$.

PROOF . Put $r = h(\mathfrak{m})$ and $s = h(\mathfrak{m}_*)$. Clearly, $r \leq s$ by lemma α. On the other hand, some defining ideal J of the ring A is generated by r elements. They generate the ideal J_* in B. According to lemma β, $\mathfrak{m}_*$ is isolated for J_*. Hence $s \leq r$.

PROOF of theorem 1. We may assume that the ring A is an integral domain. Indeed, if $I_0 \subsetneq \ldots \subsetneq I_n$ is a sequence of prime ideals of the ring A, then A/I_0 is an integral domain of dimension n ([112]) and $\bar{x} = (\bar{x}_1, \ldots, \bar{x}_n)$ is a system of generators for its maximal ideal $\mathfrak{m}/I_0$. Now, if $\sum a_p x^p = 0$, then $\sum \bar{a}_p \bar{x}^p = 0$, and so $\bar{a}_p \in \mathfrak{m}/I_0$. Thus $a_p \in \mathfrak{m}$.

([111]) Only this case will be needed in the following chapters.

([112]) See footnote ([94]).

The ring $B_{\mathfrak{m}_*}$ is noetherian and local with the maximal ideal $\overline{\mathfrak{m}_*}$ generated by the elements $x_1, \ldots, x_n$ (since they generate $\mathfrak{m}_*$). In view of lemma γ, we have $\dim B_{\mathfrak{m}_*} = h(\overline{\mathfrak{m}_*}) = h(\mathfrak{m}_*) = h(\mathfrak{m}) = n$. The matrix $[c_{ij}]$, where $c_{i1} = X_i$ and $c_{ij} = \delta_{ij}$ for $i = 1, \ldots, n$ and $j = 2, \ldots, n$, has determinant equal to $X_1 \notin \mathfrak{m}_*$, and thus the matrix is invertible (as a matrix with entries from $B_{\mathfrak{m}_*}$). Let $[d_{ij}]$ be the inverse matrix. By putting $y_i = \sum_1^n d_{ij} x_j$ $(i = 1, \ldots, n)$, we can see that $x_i = \sum_1^n c_{ij} y_j = X_i y_1 + \sum_2^n c_{ij} y_j$ $(i = 1, \ldots, n)$. Thus $y_1, \ldots, y_n$ generate the ideal $\bar{\mathfrak{m}}_*$ and for $|p| = k$ we have $x^p \in X^p y_1^k + J$, where J is the ideal of the ring $B_{\mathfrak{m}_*}$ generated by $y_2, \ldots, y_n$. Now suppose that $\sum_{|p|=k} a_p x^p = 0$, where $a_p \in A$. Then $0 = \sum a_p x^p \in (\sum a_p X^p) y_1^k + J$. If the element $\sum a_p X^p$ did not belong to $\mathfrak{m}_*$, then it would be invertible in $B_{\mathfrak{m}_*}$. So $y_1^k \in J$, which would mean that $y_1, \ldots, y_n \in \operatorname{rad} J$. Therefore $\operatorname{rad} J = \overline{\mathfrak{m}_*}$, and so J would be a defining ideal generated by $n - 1$ elements, contradicting $\dim B_{\mathfrak{m}_*} = n$. Accordingly, $\sum a_p X^p \in \mathfrak{m}_*$, and hence $a_p \in \mathfrak{m}$.

PROPOSITION . *Every regular ring is an integral domain.*

PROOF . Let A be a regular ring of dimension $n > 0$. The natural homomorphism $A \longrightarrow A/\mathfrak{m}$ induces the homomorphism

$$A[X_1, \ldots, X_n] \ni R \longrightarrow \bar{R} \in (A/\mathfrak{m})[X_1, \ldots, X_n] \,,$$

where $\bar{R} = \sum \bar{a}_p X^p$ corresponds to $R = \sum a_p X^p$. Take a system of generators $x = (x_1, \ldots, x_n)$ of the ideal $\mathfrak{m}$. Theorem 1 implies that if $R \in A[X_1, \ldots, X_n]$ is a form, then $\bar{R} \neq 0 \implies R(x) \neq 0$. Now, let $u, v \in A \setminus 0$. It follows from Krull's intersection theorem that $u \in \mathfrak{m}^k \setminus \mathfrak{m}^{k+1}$ and $v \in \mathfrak{m}^l \setminus \mathfrak{m}^{l+1}$ for some k and l (we put $\mathfrak{m}^0 = A$). Then $u = P(x)$ and $v \in Q(x)$ for some forms $P, Q \in A[X_1, \ldots, X_n]$ (of degrees k and l, respectively) such that $\bar{P} \neq 0$ and $\bar{Q} \neq 0$. Thus $\overline{PQ} \neq 0$, and so $uv = (PQ)(x) \neq 0$.

THEOREM 2. *Let A be a regular ring of dimension n, and let I be an ideal in A. Then the ring A/I is regular of dimension k if and only if $0 \leq k \leq n$ and the ideal I is generated by $n - k$ elements of some system of n generators of the ideal $\mathfrak{m}$* [113] .

LEMMA . *If B and C are noetherian local integral domains and have the same dimension, then every epimorphism from B onto C is an isomorphism.*

---

[113] It follows from theorem 2 and the proposition that if A is regular and of dimension $n > 0$, then every system of n generators of the ideal $\mathfrak{m}$ is a maximal A-sequence. In other words, *every regular ring is Cohen–Macaulay*. Indeed, let $x_1, \ldots, x_n$ generate $\mathfrak{m}$. The ring $A/\sum_1^{k-1} A x_i$ is an integral domain and $\bar{x}_k \neq 0$ (because $x_k \notin \sum_1^{k-1} A x_i$). Thus $\bar{x}_k$ is not a zero divisor in $A/\sum_1^{k-1} A x_i$, and hence the same is true about x_k $(k = 1, \ldots, n)$. Moreover, $\mathfrak{m}$ contains only zero divisors in $A/\sum_1^n A x_i = A/\mathfrak{m}$. The converse is not true. It is enough to extend x_1^2 to a maximal A-sequence. Obviously, if a system $x_1, \ldots, x_n$ of generators of $\mathfrak{m}$ is an A-sequence, then $l = n$; (observe that the sequence is minimal since $x_s \notin \sum_{i \neq s} A x_i$, $s = 1, \ldots, l$).

To show this, consider an epimorphism $\varphi : B \longrightarrow C$ and a sequence of prime ideals $0 = I_0 \subsetneqq \ldots \subsetneqq I_k$ of the ring C, where $k = \dim B = \dim C$. If $\ker \varphi \neq 0$, we would have the sequence of prime ideals $0 \subsetneqq \varphi^{-1}(I_0) \subsetneqq \ldots \ldots \subsetneqq \varphi^{-1}(I_k)$ in the ring B, contradicting the assumption that $\dim B = k$.

PROOF of theorem 2. First observe that, in view of the equality $(*)$ from 10.4, the ring A/I is regular of dimension k precisely when rank $I = n - k$ (and then one must have $0 \leq k \leq n$). Now the condition of the theorem is sufficient. Indeed, let $x_1, \ldots, x_n$ be a system of generators of the ideal $\mathfrak{m}$, and let $I = \sum_{k+1}^{n} A x_i$; then $\bar{x}_1, \ldots, \bar{x}_n$ is a basis of the cotangent space $\mathfrak{m}/\mathfrak{m}^2$. Hence $\bar{x}_{k+1}, \ldots, \bar{x}_n$ is a basis of the subspace $(I + \mathfrak{m}^2)/\mathfrak{m}^2$, and so rank $I = n - k$ (see 10.4). Suppose now that rank $I = n - k$. Then there is a basis $\bar{x}_{k+1}, \ldots, \bar{x}_n$ of $(I + \mathfrak{m}^2)/\mathfrak{m}^2$, where $x_{k+1}, \ldots, x_n \in I$, which can be extended to a basis $\bar{x}_1, \ldots, \bar{x}_n$ of $\mathfrak{m}/\mathfrak{m}^2$, where $x_1, \ldots, x_k \in \mathfrak{m}$. Hence $x_1, \ldots, x_n$ generate the ideal $\mathfrak{m}$ (see 10.4). According to the lemma, the natural epimorphism $A/\sum_{k+1}^{n} A x_i \longrightarrow A/I$ is an isomorphism since, in view of the sufficiency we have just proved, both rings are regular of dimension $n-k$. This implies that $I = \sum_{k+1}^{n} A x_i$. Therefore the condition is necessary.

COROLLARY. If the ring A is regular of dimension $n > 0$ and $x \in \mathfrak{m} \setminus \mathfrak{m}^2$, then the ring A/Ax is regular of dimension $(n-1)$.

LEMMA 1. If $x \in \mathfrak{m}$ is not a zero divisor and the ring A/Ax is regular, then the ring A is regular.

Indeed (see the proposition in 12.3), $(\dim A) - 1 = \dim A/Ax = g(\mathfrak{m}/Ax) \geq g(\mathfrak{m}) - 1$, and so $\dim A = g(\mathfrak{m})$.

LEMMA 2. Let $M_1, \ldots, M_k$ be non-zero finite modules over A. If for each i the ideal $\mathfrak{m}$ contains an element that is not a zero divisor in M_i, then $\mathfrak{m} \setminus \mathfrak{m}^2$ contains an element which is not a zero divisor in M_i.

To see this, note that if $\mathfrak{m} \setminus \mathfrak{m}^2 \subset \bigcup D_i$, where D_i denotes the set of zero divisors in M_i, then, since D_i is a finite union of prime ideals I_{ij} (see the proposition in 9.6), we have $\mathfrak{m} = I_{sj}$ for some s, j (see 10.5). Hence $\mathfrak{m} \subset D_s$.

LEMMA 3. If hd $\mathfrak{m} < \infty$, then

$$(\dim A = 0) \Longleftrightarrow (A \text{ is a field}) \Longleftrightarrow (\mathfrak{m} \text{ contains only zero divisors}).$$

Indeed, if $\dim A = 0$, then $\mathfrak{m}$ is an isolated ideal for 0, and so it contains only zero divisors. Conversely, assume now that $\mathfrak{m}$ contains only zero divisors. Then, for some $t \neq 0$, $t\mathfrak{m} = 0$. Hence $t \, \mathrm{syz}^k \mathfrak{m} = 0$ for each $k \in \mathbb{N}$. Set $r = \mathrm{hd} \, \mathfrak{m}$. If $\mathfrak{m} \neq 0$, then $\mathrm{syz}^r \mathfrak{m} \approx A^l$ for some $l > 0$. Consequently, $tA^l = 0$, which is impossible. In conclusion, $\mathfrak{m} = 0$, i.e., A is a field.

THE HILBERT-SERRE THEOREM . *The ring A is regular if and only if* hd $\mathfrak{m} < \infty$ ([114]) .

PROOF . Set $n = \dim A$. If $n = 0$, the theorem is a consequence of lemma 3. Let $n > 0$ and assume that the theorem is true for $(n-1)$–dimensional rings. Then, in accordance with lemmas 3 and 2, there is an element $s \in \mathfrak{m} \setminus \mathfrak{m}^2$, which is not a zero divisor, provided hd$A < \infty$. Naturally, this is also the case when the ring A is regular. In view of the corollary from theorem 2 and lemma 1, the regularity of the ring A is equivalent to the regularity of the ring A/As. Since $\dim A/As = n - 1$ and because of proposition 2 from 13.2, this is equivalent to the condition hd $\mathfrak{m} < \infty$.

THE SYZYGY THEOREM . *If the ring A is regular, then for every finite module $M \neq 0$ over A we have*

$$(*) \qquad\qquad \text{hd } \mathfrak{m} + \text{prof } M = \dim A \quad (^{115}) .$$

First, we prove that the hypotheses of the theorem imply

LEMMA . *If $\dim A > 0$ and prof $M = 0$, then* syz $M \neq 0$ *and*

$$\text{prof(syz } M) = 1.$$

PROOF . By lemma 3, there is an element $u \in \mathfrak{m}$ which is not a zero divisor in A. Hence it is not a zero divisor in A^k. Therefore prof $A^k > 0$, provided that $k > 0$. As a result, M cannot be free, i.e., $N = $ syz $M \neq 0$. As $N \subset A^r$ for some $r > 0$, u is not a zero divisor in N. Since $\mathfrak{m}$ contains only zero divisors in $A^r/N \approx M$, we have $\mathfrak{m}a \subset N$ for some $a \in A^r \setminus N$. Now u is a maximal N–sequence. Indeed, $\mathfrak{m}(ua) \subset uN$ and $ua \in N \setminus uN$ (because $a \notin N$), which means that $\mathfrak{m}$ contains only zero divisors in N/uM. Therefore prof $N = 1$.

PROOF of the theorem. Put $n = \dim A$. If $n = 0$, then A is a field, and so hd $M = $ prof $M = 0$. Suppose $n > 0$ and the theorem is true for $(n - 1)$–dimensional rings.

We may assume that prof $M > 0$. For if prof $M = 0$, then, in view of the lemma, syz $M \neq 0$ and prof syz $M = 1$. Thus (assuming validity of the

---

([114]) Then hd $M < \infty$ for each finite module M over A, according to the syzygy theorem below. In addition, $\dim A = \max\{\text{hd } M : M$ is a finite module over $A\}$ (see the next footnote). The regularity of the ring A can be also characterized by the condition hd $A/\mathfrak{m} < \infty$, because syz $A/\mathfrak{m} = \mathfrak{m}$.

([115]) Thus, if A is regular, then hd $M \leq \dim A$ for every finite module M over A. In addition, the equality holds for $M = A/\mathfrak{m}$ (as prof $A/\mathfrak{m} = 0$).

theorem in the case when $\dim A = n$ and for modules of depth > 0) we have hd syz $M + \text{prof syz } M = n$, and hence $(*)$ follows (as hd syz $M = \text{hd}\mathfrak{m} - 1$).

Then lemma 2 (in view of lemma 3) yields the existence of an element $s \in \mathfrak{m} \setminus \mathfrak{m}^2$ which is neither a zero divisor in A nor in M. According to the corollary from theorem 2, the ring A/As is regular and of dimension $(n-1)$. Moreover, by Nakayama's lemma, $M/sM \neq 0$. Hence $\text{hd}_{A/As}M/sM + \text{prof}_{A/As}M/sM = n - 1$. This implies $(*)$, since proposition 1 from 13.2 gives $\text{hd}_{A/As}M/sM = \text{hd } M$ and by propositions 5 and 4 from 14.2 we have $\text{prof}_{A/As}M/sM = \text{prof}_A M/sM = \text{prof } M - 1$.

COROLLARY . *If the ring A is regular, then prof $A = \dim A$.*

In other words, *every regular ring is Cohen–Macaulay* ([116]) .

We will also prove

THEOREM 3. *If the ring A is regular, then so is its localization to any prime ideal $\mathfrak{n}$.*

PROOF . For any submodule L of any module A^k define the submodule $\bar{L} = \{(x-1/y, \ldots, x_k/y): (x_1, \ldots, x_k) \in L, y \in A \setminus \mathfrak{n}\}$ of $A_{\mathfrak{n}}^k$. In particular, $\bar{\mathfrak{n}}$ is the maximal ideal of $A_{\mathfrak{n}}$ (in accordance with 1.11). For any homomorphism $h : A^r \longrightarrow A^s$ let $\bar{h} : A_{\mathfrak{n}}^r \longrightarrow A_{\mathfrak{n}}^s$ denote the homomorphism that has the same matrix as h (see 1.16). It is easy to check that

$$(*) \qquad\qquad \ker \bar{h} = \overline{\ker h} \quad \text{and} \quad \text{im } \bar{h} = \overline{\text{im } h} .$$

In view of the syzygy theorem, we have $\text{hd}\mathfrak{n} < \infty$. Consequently, by the theorem from 13.2, there exists a finite free resolution $0 \longrightarrow A^{r_k} \xrightarrow{h_k} \ldots \longrightarrow A_{\mathfrak{n}}^{r_0} \xrightarrow{h_0} \mathfrak{n} \longrightarrow 0$. The relations $(*)$ imply that $0 \longrightarrow A_{\mathfrak{n}}^{r_k} \xrightarrow{h_k} \ldots \longrightarrow A_{\mathfrak{n}}^{r_0} \xrightarrow{h_0} \bar{\mathfrak{n}} \longrightarrow 0$ is an exact sequence and hence it is a finite free resolution of $\bar{\mathfrak{n}}$. Therefore, by the theorem from 13.2, we have $\text{hd}\bar{\mathfrak{n}} < \infty$. This implies that, according to the Hilbert-Serre theorem, $A_{\mathfrak{n}}$ is regular.

([116]) See footnote ([113]).

CHAPTER B

TOPOLOGY

§1. Some topological properties of sets and families of sets

Let X be a topological space.

If a set $G \subset X$ is open, then, for any $E \subset X$, the closure of the set $E \cap G$ in G is $\bar{E} \cap G$.

Let $\{G_\iota\}$ be an open cover of the space X. The set E is – respectively - an open, closed, dense, nowhere dense set if for each ι the same is true about the set $E \cap G$ in G_ι.

Let $\{E_\iota\}$ be a *locally finite family of subsets* of the space X (i.e., a family such that each point $x \in E$ has a neighbourhood U such that $E_\iota \cap U \neq \emptyset$ for only a finite number of indices ι). Then any compact subset of the space X intersects E_ι only for a finite number of indices ι. (In the case when X is locally compact, the last condition characterizes local finiteness of the family $\{E_\iota\}$.) The family $\{\bar{E}_\iota\}$ is also locally finite. Moreover, $\overline{\bigcup E_\iota} = \bigcup \bar{E}_\iota$. If all the sets E_ι are closed, then so is $\bigcup E_\iota$. If all the sets E_ι are nowhere dense, then so is $\bigcup E_\iota$. If for each ι the set F_ι is nowhere dense in E_ι, then the set $\bigcup F_\iota$ is nowhere dense in $\bigcup E_\iota$. For any continuous mapping $f : Z \longrightarrow X$ (where Z is a topological space), the family $\{f^{-1}(E_\iota)\}$ is also locally finite. If $\{F_\kappa\}$ is a locally finite family of subsets of a topological space Y, then the family $\{E_\iota \times F_\kappa\}$ is locally finite in $X \times Y$. If the space X has a countable base, then the family $\{E_\iota\}$ is countable.

PROPOSITION . *Let $\mathcal{R}$ be a locally finite family of closed connected subsets of the space X, and let E denote the union of the family $\mathcal{R}$. Then every* *connected component of the set E is the union of some equivalence class of the relation $\sim$ in $\mathcal{R}$ defined by*

$$F' \sim F'' \iff \left(\begin{array}{c} \textit{there exists a sequence } F' = F_0, \ldots, F_r = F'' \\ \textit{of sets in the family } \mathcal{R} \textit{ such that} \\ F_{i-1} \cap F_i \neq \emptyset, \; i = 1, \ldots, r \end{array} \right).$$

To see this, note that all such unions of equivalence classes are connected and pairwise disjoint. Thus it is enough to show that they are open in E. Let $Z = \bigcup \{F : F \in \Theta\}$, where Θ is an equivalence class, and let $a \in Z$. Then a belongs to some $F' \in \Theta$. As $\mathcal{R}$ is a locally finite family of closed sets, there is a neighbourhood U of the point a such that, if $F \in \mathcal{R}$ has a non-empty intersection with U, then it contains a. Now, if $x \in U \cap E$, then x belongs to some $F \in \mathcal{R}$, and so $F \cap U \neq \emptyset$. Therefore $a \in F$, which implies that $F \sim F'$, $F \in \Theta$, and $x \in Z$. Accordingly, $U \cap E \subset Z$.

A set $E \subset X$ is said to be *locally closed* if it satisfies one of the following equivalent conditions:

(1) Each point $x \in E$ has a neighbourhood U such that $E \cap U$ is closed in U.

(2) E is the intersection of an open set and a closed set [1] .

(3) $\bar{E} \setminus E$ is closed (or, in other words, E is open in $\bar{E}$).

If E is locally closed, then the set $\partial E = \bar{E} \setminus E$ is called the *border of the set E*. The latter is always closed and nowhere dense in $\bar{E}$.

If the space X is locally compact, then a subset of X is locally closed if and only if it is a locally compact subspace.

Naturally, every closed set is locally closed. For a locally closed set E we have the equivalence: (E is nowhere dense)$\iff$ int $E = \emptyset$.

Let $E \subset Y \subset X$. If E is locally closed in X, then it is also locally closed in Y. If E is locally closed in Y and Y is locally closed in X, then E is locally closed in X.

Let X be a locally compact space with a countable base. Then a subset of X is an F_σ-set (i.e., is a countable union of closed sets) precisely when it is a countable union of compact sets. Any locally closed set is an F_σ-set. If $f : X \longrightarrow Y$ is a continuous mapping into a Hausdorff space Y, then the image of an F_σ-set is also an F_σ-set.

---

[1] Some other equivalent characterizations: E is closed in the relative topology of an open subset of X; E is open in the relative topology of a closed subset of X.

§2. Open, closed, and proper mappings

Let M and N be topological spaces.

1. We say that a mapping $f : M \longrightarrow N$ is *open* if the image of every open set in M is an open set in N. This is the same as to say that for any point $a \in M$ the image of any neighbourhood of a is a neighbourhood of $f(a)$.

If f is open, then so is the restriction $f_G : G \longrightarrow N$ to any open set G in M and also the mapping $f_{f^{-1}(E)} : f^{-1}(E) \longrightarrow E$ for any $E \subset N$.

The composition and the Cartesian product of open mappings is an open mapping.

If $g : P \longrightarrow M$ is a continuous surjection on a topological space P and $f : M \longrightarrow N$, then ($f \circ g$ is open) $\Longrightarrow$ (f is open).

A bijection $f : M \longrightarrow N$ is a homeomorphism precisely when it is open and continuous.

The natural projection $M \times N \longrightarrow M$ is always open and continuous.

If a mapping $f : M \longrightarrow N$ is open, then the inverse image of a dense subset of N is dense in M (²).

If a mapping $f : M \longrightarrow N$ is open and M is Hausdorff, then the function $N \ni z \longrightarrow \# f^{-1}(z)$ is lower–semicontinuous. Indeed, if $\# f^{-1}(z) \geq k$, then there are k distinct points $x_1, \ldots, x_k \in M$ such that $z = f(x_\nu)$. Take pairwise disjoint neighbourhoods $V_1, \ldots, V_k$ of these points. Then $W = f(V_1) \cap \ldots \cap f(V_k)$ is a neighbourhood of the point z. Now, if $z' \in W$, we have $z' = f(x'_\nu)$ with $x'_\nu \in V_\nu$, so the points $x'_1, \ldots, x'_k$ must be distinct and, consequently, $\# f^{-1}(z') \geq k$.

2. If the mapping $f : M \longrightarrow N$ is open and continuous, then the inverse image of a nowhere dense set in N is nowhere dense in M (³).

Note that in this case $\overline{f^{-1}(E)} = f^{-1}(\bar{E})$ for every $E \subset N$ (⁴).

Let $f : M \longrightarrow N$ be an open continuous surjection. Then a subset of the space N is – respectively – open, closed, dense, nowhere dense if the same is true of its inverse image. The image of the family of all neighbourhoods

---

(²) Because $f(G) \cap E \neq \emptyset \Longrightarrow G \cap f^{-1}(E) \neq \emptyset$.

(³) For if $Z \subset N$, then the inverse image of a set which is open, dense, and disjoint from Z, is open, dense, and disjoint from $f^{-1}(Z)$.

(⁴) Indeed, for every neighbourhood U of a point $x \in f^{-1}(\bar{E})$ we have $f(U) \cap E \neq \emptyset$, and so $U \cap f^{-1}(E) \neq \emptyset$.

of a point $x \in M$ is the family of all neighbourhoods of the point $f(x)$ in N. The image of a base for the topology of M is a base for the topology of N. A mapping $\varphi : E \longrightarrow L$, where $E \subset N$ and L is a topological space ($L = \mathbf{R}$, respectively), is continuous (upper or lower semicontinuous, respectively) precisely when the same holds for $\varphi \circ f$ ([5]).

3. Assume that both M and N are Hausdorff spaces.

We say that a mapping $f : M \longrightarrow N$ is *closed* if the image of any closed subset of M is closed in N. This holds precisely when $\overline{f(E)} \subset f(\bar{E})$ for all $E \subset M$. Therefore a mapping f is both continuous and closed if and only if $\overline{f(E)} = f(\bar{E})$ for $E \subset M$.

For a mapping $f : M \longrightarrow N$ and the natural projection $\pi_f : f \longrightarrow M$ we have the following equivalent conditions

$$(f \text{ is continuous}) \Longleftrightarrow (\pi_f \text{ is a homeomorphism}) \Longleftrightarrow$$

$$\Longleftrightarrow (\pi_f \text{ is open}) \Longleftrightarrow (\pi_f \text{ is closed}) \quad ([6]) .$$

If the space N is compact, the natural projection $\pi : M \times N \longrightarrow M$ is closed. For if $F \subset M \times N$ is a closed set and $a \in M \setminus \pi(F)$, then $(a \times N) \cap F = \emptyset$. Therefore $(U \times N) \cap F = \emptyset$ for some neighbourhood U of the point a, which implies that $U \subset M \setminus \pi(F)$. (See also n° 4 below).

Consequently, if the mapping $f : M \longrightarrow N$ is *locally bounded* (in the sense that each point of the space M has a neighbourhood whose image is relatively compact), then

$$(f \text{ is continuous}) \Longleftrightarrow (f \text{ is a closed set in } M \times N) .$$

Indeed, assuming that the set f is closed if $\overline{f(U)}$ is compact, then $f_U : U \longrightarrow \overline{f(U)}$ is continuous, as the natural projection $f_U \longrightarrow U$ is closed (because the set f_U is closed in $U \times \overline{f(U)}$).

4. Now, let M and N be locally compact spaces.

We say that a continuous mapping $f : M \longrightarrow N$ is *proper* if the inverse image of any compact subset of N is compact in M. Equivalently, f is proper

---

([5]) It suffices to note that if $(\varphi \circ f)^{-1}(H)$ is open in $f^{-1}(E)$, then $f^{-1}(\varphi^{-1}(H)) = (\varphi \circ f)^{-1}(H) = G \cap f^{-1}(E)$ for some open set G in M, and hence $\varphi^{-1}(H) = f(G) \cap E$ is open in E.

([6]) The last equivalence follows from the fact that π_f is bijective.

if every point of the space N has a neighbourhood whose inverse image is relatively compact. Naturally, if the space M is compact, then every continuous mapping $f : M \longrightarrow N$ is proper.

If f is proper, then so is its restriction $f_E : E \longrightarrow N$ to any closed set $E \subset M$. In addition, the restriction $f_{f^{-1}(H)} : f^q(H) \longrightarrow H$ is proper for any locally closed set H in N [7] .

The composition and the Cartesian product of proper mappings is a proper mapping.

If $f : M \longrightarrow N$ and $g : N \longrightarrow P$ are continuous mappings of locally compact spaces, then

$$(g \circ f \text{ is proper}) \Longrightarrow (f \text{ is proper}) .$$

(For if $E \subset N$ is compact, then $f^{-1}(E)$ is a closed subset of the compact set $f^{-1}\Big(g^{-1}(g(E))\Big)$.)

If the space N is compact, then the natural projection $M \times N \longrightarrow M$ is proper.

A continuous mapping $f : M \longrightarrow N$ is proper if and only if it is closed and has compact fibres. In fact, assume that f is proper. Take any closed $E \subset M$ and any $z \in \overline{f(E)}$. Take a compact neighbourhood U of z. Then $z \in \overline{U \cap f(E)}$. But $U \cap f(E) = f(f^{-1}(U) \cap E)$ is compact, so $z \in f(E)$. Conversely, suppose that f is closed and has compact fibres. Take any compact $F \subset N$ and any centred family $\{E_\iota\}$ of closed subsets of $f^{-1}(F)$ [8] . Then $\{f(E_{\iota_1} \cap \ldots \cap E_{\iota_s})\}$ is a centred family of closed subsets of E, and hence its intersection contains a point a. It follows that the family $\{E_\iota \cap f^{-1}(a)\}$ is centred, and therefore its intersection is non-empty, which gives $\bigcap E_\iota \neq \emptyset$.

If a mapping $f : M \longrightarrow N$ is proper, then the image under f of a locally finite family of subsets of M is also locally finite.

If a mapping $f : M \longrightarrow N$ is proper and $b \in N$, then for any open set $G \supset f^{-1}(b)$ there exists a neighbourhood Δ of the point b such that $f^{-1}(\Delta) \subset G$. In particular, if the fiber $f^{-1}(b)$ consists of a single point a, the inverse image of a base of neighbourhoods of the point b is a base of neighbourhoods of the point a. Otherwise, there would be an open set

---

[7] Note that if the restriction $f_G : G \longrightarrow H$ is proper, where $G \subset M$ is open and dense (and $f(G) \subset H$), then we must have $G = f^{-1}(H)$. (Indeed, in this case G must be closed in $f^{-1}(H)$ – see below.)

[8] A family of sets $\mathcal{R}$ is said to be centred (or to have the finite intersection property) if $E_1 \cap \ldots \cap E_k \neq \emptyset$ for any finite collection $E_1, \ldots, E_k \in \mathcal{R}$.

$G \supset f^{-1}(b)$ such that the family of compact sets $f^{-1}(\Delta) \setminus G$, where Δ varies over all compact neighbourhoods of the point b, would be centred; hence its intersection $f^{-1}(b) \setminus G$ would be non-empty.

Note also that if the restriction of a continuous mapping $f : M \longrightarrow N$ to a locally closed set E is proper, then the set E must be closed. (Otherwise, by taking a point $a \in \bar{E} \setminus E$ and compact neighbourhood V of the point $f(a)$, the set $(f_E)^{-1}(V) = f^{-1}(V) \cap E$ would not be closed.)

§3. Local homeomorphisms and coverings

Let M and N be Hausdorff spaces.

1. Let the mapping $f : M \longrightarrow N$ be a *local homeomorphism* (i.e., for each $a \in M$ the restriction of f to a neighbourhood of the point a is a homeomorphism onto a neighbourhood of $f(a)$).

Then f is continuous and open. Hence it is enough to consider only open neighbourhoods in the definition of a local homeomorphism.

If the mapping f is bijective, it is a homeomorphism.

Suppose now that continuous mappings $h_i : E \longrightarrow M$ satisfy $f(h_i(z)) = z$ on the set $E \subset N$, for $i = 1, 2$. If $h_1(c) = h_2(c)$ for some $c \in E$, then $h_1 = h_2$ in a neighbourhood in E of the point c. Thus, if the set E is connected, then either $h_1 = h_2$ in E or $h_1 \neq h_2$ in E. (Due to the fact that the set $\{h_1(z) = h_2(z)\}$ is open and closed in E.)

LEMMA . *If M is connected and there is a continuous mapping $h : N \longrightarrow M$ such that $f(h(z)) = z$ in N, then f must be a homeomorphism.*

Indeed, it is enough to show that h is surjective. Now, owing to the fact that $h(N) = \{x : h(f(x)) = x\}$, the set $h(N)$ is closed. Let $a \in h(N)$. The restriction f_U of f to a neighbourhood U of the point a is a homeomorphism onto the neighbourhood $f(U)$ of the point $c \in f(a)$; since $h(c) = a = (f_U)^{-1}(c)$, we have $h = (f_U)^{-1}$ in a neighbourhood $V \subset f(U)$ of the point c, and then $h(V) = (f_U)^{-1}(V)$ is a neighbourhood of the point a which is contained in $h(N)$. Thus the set $h(N)$ is open. Hence $h(N) = M$.

2. A mapping $f : M \longrightarrow N$ is said to be a *covering* if every point of the space N has an open neighbourhood V such that $f^{-1}(V)$ is the union of some open sets U_ι which are pairwise disjoint and for which the mappings $f_{U_\iota} : U_\iota \longrightarrow V$ are homeomorphisms (^{9}) .

Every covering is a local homeomorphism (and hence an open and continuous mapping).

Let $f : M \longrightarrow N$ be a covering. For any point $c \in N$ we can find – as in the definition – a neighbourhood V and sets U_ι such that there are exactly $\#f^{-1}(z)$ sets U_ι for every $z \in V$; hence the number $\#f^{-1}(z)$ depends only on c. It is called the *multiplicity of the covering f at the point c*. When regarded as a function of c, it is locally constant. In the case when it is constant in N, its only value p is said to be the *multiplicity of the covering f* and we also say that the covering f is *p–sheeted*. In particular, this is the case (for some p) when the space N is connected. Every 1–sheeted covering is a homeomorphism. We say that a covering f is *finite* when its multiplicity is finite at each point of the space N.

If $f : M \longrightarrow N$ is a covering, then for $E \subset N$ the restriction $f_{f^{-1}(E)} : f^{-1}(E) \longrightarrow E$ is also a covering. The Cartesian product of coverings is a covering. Clearly, the composition of a covering with a homeomorphism (in any order) is a covering.

If M and N are locally connected, then the restriction of a (finite) covering $f : M \longrightarrow N$ to any closed open subset of the space M (in particular, to a connected component of the space M) is a (finite) covering (^{10}) .

PROPOSITION 1. *If M and N are locally compact, then for any mapping $f : M \longrightarrow N$ we have the following equivalence:*

$$(f \text{ is a finite covering}) \Longleftrightarrow$$

$$(f \text{ is a local homeomorphism and a proper mapping}) .$$

PROOF . Suppose f is a finite covering. Then any point of the space N has a neighbourhood whose inverse image is compact. (It is enough to take a compact neighbourhood V such that $f^{-1}(V) = K_1 \cup \ldots \cup K_s$, where the K_i's are homeomorphic to V.) Thus the mapping f is proper.

Assume now that f is a local homeomorphism and a proper mapping. Let $c \in N$. Then the set $f^{-1}(c)$ is compact and discrete (for if $a \in f^{-1}(c)$, f_U is injective on a neighbourhood U of a). Hence $f^{-1}(c)$ is finite, say $f^{-1}(c) = \{a_1, \ldots, a_r\}$. Take disjoint neighbourhoods $U_1, \ldots, U_r$ of the points $a_1, \ldots, a_r$,

---

(^{9}) In such a case, the same property is displayed by any open neighbourhood $V' \subset V$ together with the sets $U_\iota' = U_\iota \cap f^{-1}(V')$.

(^{10}) Compare with the previous footnote.

respectively, such that f_{U_i} is a homeomorphism and $f(U_i)$ is a neighbourhood of the point c for $i = 1, \ldots, r$. Take a compact neighbourhood V^* of c. The set $T = f\big(f^{-1}(V^*) \backslash (U_1 \cup \ldots \cup U_r)\big)$ is compact and does not contain the point c. Therefore there is an open neighbourhood U of the point c which is disjoint from T and such that $V \subset V^* \cap \bigcap f(U_i)$. Then $f^{-1}(V) \subset U_1 \cup \ldots \cup U_r$ (since $x \in f^{-1}(V) \backslash (U_1 \cup \ldots \cup U_r)$ would imply that $f(x) \in T \cap V$). Thus $f^{-1}(V) = U_1' \cup \ldots \cup U_r'$, where $U_i' = U_i \cap f^{-1}(V)$ are open and disjoint. Moreover, the mappings $f_{U_i'} : U_i' \longrightarrow f(U_i') = f(U_i) \cap V = V$ are homeomorphisms. Hence the mapping f is a finite covering.

PROPOSITION 2. *If $M \neq \emptyset$ is connected and N is homeomorphic to $\mathbf{R}^n$, then every covering $f : M \longrightarrow N$ is one–sheeted, i.e., is a homeomorphism.*

PROOF . Without loss of generality we may assume that $N = \mathbf{R}^n$. By the lemma, it is enough to show the existence of a continuous mapping $h : N \longrightarrow M$ satisfying $f\big(h(z)\big) = z$ in N. Since the multiplicity of f is constant, f must be surjective, and so there is an $a \in M$ such that $f(a) = 0$. Let $\mathcal{T}$ be the class of all M–valued continuous mappings g such that their domains are open and star–shaped with respect to 0, $g(0) = a$, and $f\big(g(x)\big) = x$ in the domain of g. The class $\mathcal{T}$ is non-empty, as it contains $\big(f_U\big)^{-1}$ for a suitably chosen neighbourhood U of the point a. Now $h = \bigcup\{g : g \in \mathcal{T}\}$ is a mapping and it belongs to $\mathcal{T}$. (Any two of the mappings in $\mathcal{T}$ coincide on the intersection of their domains, since the latter is connected and contains 0.) It suffices to show that the domain H of h is equal to N. Suppose that it is not true and $H \subsetneq N$. Then there exists a point $c \in N \setminus H$ such that $[0, c) \subset H$. Let us take an open convex neighbourhood V of the point c for which $f^{-1}(V)$ is the union of open disjoint sets U_ι and $f_{U_\iota} : U_\iota \longrightarrow V$ are homeomorphisms ([11]) . Consider $b \in [0, c) \cap V$ and an open convex set W such that $[0, b] \subset W \subset H$. We have $h(b) \in U_\kappa$ for some κ, and hence $h(b) = \bar{h}(b)$, where $\bar{h} = \big(f_{U_\kappa}\big)^{-1}$; hence $f\big(\bar{h}(z)\big) = z$ in V. Thus $h = \bar{h}$ in $W \cap V$, and therefore $h_W \cup \bar{h}$ is a continuous mapping on the set $W \cup V \supset [0, c]$ and its restriction to an open convex set W_0, such that $[0, c] \subset W_0 \subset W_0 \cup V$, belongs to $\mathcal{T}$. In conclusion, $W_0 \subset H$ and hence $c \in H$, which is a contradiction.

([11]) See footnote ([8]).

§4. Germs of sets and functions

Let S, T, U be topological spaces.

1. Let $a \in S$. By the *germs of sets* at the point a (in the space S) we mean the equivalence classes of the equivalence relation

"$E' \cap V = E'' \cap V$ for some neighbourhood V of the point a "

in the set of all subsets of the space S. The equivalence class of a set E is called the *germ of E at the point a* and is denoted by E_a.

The relation of inclusion, the operation of taking the finite union or intersection of sets, the difference of sets or the complement of a set, together with their elementary properties from the algebra of sets, carry over in a natural way to germs at a ([12]) . In this new context, the role of the empty set and the whole space are played, respectively, by the *empty germ* $\emptyset$, i.e., the germ of the empty set and the *full germ*, i.e., the germ of the whole space (the representatives of the latter germ are precisely neighbourhoods of the point a). The above operations are well-defined by the formulae $E_a \cup F_a = (E \cup F)_a$, $E_a \cap F_a = (E \cap F)_a$, etc. The inclusion of germs $A \subset B$ is defined by the condition that $\tilde{A} \subset \tilde{B}$ for some representatives $\tilde{A}$, $\tilde{B}$ of those germs. Thus, for sets $E, F \subset S$, the inclusion $E_a \subset F_a$ means that $E \cap V \subset F \cap V$ for some neighbourhood V of the point a.

Note also that if $a \in E$, then every neighbourhood in E of the point a is a representative of the germ E_a and every representative of the germ E_a contains a neighbourhood in E of the point a.

For a germ A at the point a and a set $E \subset S$ we write $A \subset E$ when $A \subset E_a$, i.e., when A is the germ of a subset of the set E. We define also $E \cap A = A \cap E = A \cap E_a$. The germ of the set $\{a\}$ at the point a will be denoted simply by a. We will also write that $a \in A$ if A is the germ of a set containing the point a.

If $a \in S$ and $b \in T$, then the formula $E_a \times F_b = (E \times F)_{(a,b)}$, where $E \subset S$ and $F \subset T$ defines the *Cartesian product of germs*. (The same works for an arbitrary finite number of germs.)

Let h be a homeomorphism of a neighbourhood of the point a onto a neighbourhood of the point $b = h(a) \in T$. The *image of a germ at a* (under h) is well-defined, as a germ at b, by the formula $h(E_a) = h(E)_b$, where $E \subset S$. Obviously, $h(A \cup B) = h(A) \cup h(B)$, $h(A \cap B) = h(A) \cap h(B)$,

---

([12]) In what follows, all the necessary properties can be checked easily.

and $A \subset B \iff h(A) \subset h(B)$ for germs A, B at a. If $h(A) = C$, then $h^{-1}(C) = A$. For any homeomorphism g of a neighbourhood of the point b onto a neighbourhood of the point $g(b) \in U$, we have $g(h(A)) = (g \circ h)(A)$ for every germ A at a.

If $a \in S \subset T$ is a subspace and $E \subset S$, then the germ of the set E in the space S at the point a is identified with the germ of this set in the space T at a. (This identification is compatible with inclusion, finite union, and finite intersections of germs, but not with taking complements.)

2. Let A be a germ of a set at the point $a \in S$, and let X be an arbitrary set. By the *germs of functions* from A to X we mean the equivalence classes with respect to the following equivalence relation

$$\text{``}F' = F'' \text{ on a representative of the germ } A\text{''}$$

in the set of all X–valued functions defined on representatives of the germ A. The equivalence class of such a function F is called the *germ of F on A* and denoted by F_A. Also, for any function F whose domain contains A, the germ $F_A = (F_{\tilde{A}})_A$ is well-defined, where $\tilde{A}$ is a representative of the germ A. We will use the symbol $f : A \longrightarrow X$ to express the fact that f is the germ on A of an X–valued function.

In the case when X is a ring (or a module over a ring R), the above relation agrees with the multiplication and addition of functions (^{13}) (or multiplication of a function by elements from R, respectively). As a result, all those operations can be defined on the set of the germs on A of X–valued functions (^{14}) and they furnish this set with a structure of a ring (or a module over R, respectively).

The *restriction of the germ $f : A \longrightarrow X$ to a germ $C \subset A$* is well-defined by the formula $f_C = \tilde{f}_C$, where $\tilde{F}$ is a representative of the germ f. Then, for any germ $D \subset C$, we have $f_D = (f_C)_D$. If $a \in A$, then the value of the germ f at a is well-defined by the formula $f(a) = \tilde{f}(a)$.

Let B be the germ of a set at the point $b \in T$. The germ $f : A \longrightarrow T$ is said to be the *germ of a mapping of the germ A into the germ B.* We write $f : A \longrightarrow B$ if every representative of the germ B contains the range of some representative of the germ f. In the case when $a \in A$, this happens only if f is the germ of a mapping from a representative of the germ A to a representative

---

(^{13}) Note that $f + g$ and fg are defined on the intersection of the domains of f and g.

(^{14}) The operations are well-defined by the formulae $F_a + G_a = (G + G)_a$ and $(F_a G_a) = (FG)_a$ (or $\zeta F_a = (\zeta F)_a$, respectively).

of the germ B which is continuous at a and assumes the value b at a (^{15}).
Then, if B' is the germ of a set at the point b, we have $f : A \longrightarrow B'$ precisely
when f has a representative whose range is contained in some representative
of the germ B'. In particular, $f : A \longrightarrow B'$ if $B' \supset B$. Sometimes we will
use the symbol $f^{B'}$ instead of f, to indicate that the germ f is treated as
$f : A \longrightarrow B'$. If h is the germ of an X–valued function defined on the germ
C of a set in U and $g : B \longrightarrow C$, then the *composition* $h \circ g : B \longrightarrow X$ is
well–defined by the formula $h \circ g = (\tilde{h} \circ \tilde{g})_B$, where $\tilde{h}, \tilde{g}$ are representatives of
of the germs h, g. We have $g \circ f : A \longrightarrow C$ and the composition of germs is
associative: $h \circ (g \circ f) = (h \circ g) \circ f$. When $g : B \longrightarrow D \subset C$, then $h_D \circ g = h \circ g$.
We denote by e_A the germ of the identity mapping on a representative of A.
Clearly, $e_A : A \longrightarrow A'$ if $A \subset A'$. We put $e_A^{A'} = (e_A)^{A'}$ and $\varrho_A = (e_A)^{S_a}$.
Obviously, $f \circ e_A = f$ and $f = e_B \circ f$. If $C \subset A$, then $(e_A)_C = e_C$ and, for
any germ h of a function on A, $h_C = h \circ e_C$. Therefore $(g \circ f)_C = g \circ f_C$ for
any germ g of a function on B.

The *diagonal product of germs* $g_1, \dots, g_k$ of functions on B with values in $Y_1, \dots, Y_n$, respectively, is well-defined by the formula $(g_1, \dots, g_k) =$
$(\tilde{g}_1, \dots, \tilde{g}_k)_B$, where $\tilde{g}_1, \dots, \tilde{g}_k$ are representatives of these germs with a common domain. We have $(g_1, \dots, g_k) \circ f = (g_1 \circ f, \dots, g_k \circ f)$ and, in particular,
$(g_1, \dots, g_k)_D = ((g_1)_D, \dots, (g_k)_D)$ if $D \subset B$. Similarly, we can define the
Cartesian product of germs of functions.

Assume now that $a \in A$ or $A = \emptyset$, and $b \in B$ or $B = \emptyset$. The *germ of
a homeomorphism of the germ A onto the germ B* is the germ f on A, of a
homeomorphism $\tilde{f}$ of a representative of the germ A onto a representative of
the germ B taking the value b at a, if A and B are non-empty. (Naturally,
$f : A \longrightarrow B$.) In this case, the inverse germ f^{-1} is well-defined as the germ
on B of the homeomorphism $\tilde{f}^{-1}$ and we have $f^{-1} \circ f = e_A$, $f \circ f^{-1} = e_B$,
and $(f^{-1})^{-1} = f$. If $C \subset A$, the *image of the germ C* is well-defined by the
formula $f(C) = \tilde{f}(C)$. If $D = f(C)$, then $C = f^{-1}(D)$, and moreover, if $C \ni a$
or $C \neq \emptyset$, then the restriction $f_C : C \longrightarrow D$ is the germ of a homeomorphism
of C onto D. The composition of the germs of homeomorphisms (of A onto
B and of B onto C) is the germ of a homeomorphism (of A onto C). If
$f : A \longrightarrow B$ and $g : B \longrightarrow C$ are the germs of continuous mappings such
that $g \circ f = e_A$ and $f \circ g = e_B$, then f and g are mutually inverse germs of
homeomorphisms (of A onto B and of B onto A) (^{16}).

(^{15}) Note that, in general, the ranges of representatives of the germ f (even if the domains
are sufficiently small) need not represent the same germ of a set at the point b. To see this,
take the germ at 0 of the mapping $\mathbf{R}^2 \ni (x, y) \longrightarrow (x, xy) \in \mathbf{R}^2$.

(^{16}) Indeed, assuming that $a \in A$, $b \in B$, and taking representatives $\tilde{f} : \tilde{A} \longrightarrow T$, $\tilde{g} :$

If G is a homeomorphism that maps a neighbourhood of the point $b \in T$ onto a neighbourhood of a point c in U and C is the germ of a set at the point c, then G_B, where $B = G^{-1}(C)$ is the germ of a set at b, is the germ of a homeomorphism of the germ B onto the germ C. A *substitution* into the germ h of a function on C is defined by $h \circ G = h \circ G_B$. Thus, if F is a homeomorphism of a neighbourhood of the point a in S onto a neighbourhood of the point b in T, then we have the associative law $(h \circ G) \circ F = h \circ (G \circ F)$.

If X is a ring (resp., module) and $f : A \longrightarrow B$, then we have a ring (resp., module) homeomorphism $\eta \longrightarrow \eta \circ f$ of the ring (resp., module) of germs of X–valued functions on B into the ring (resp., module) of germs of X-valued functions on A. It becomes an isomorphism if f is the germ of a homeomorphism of the germ A onto the germ B. Similarly, the mappings $\eta \longrightarrow \eta_C$, where $a \in A$, are homomorphisms of the ring (resp., module) of germs of X–valued functions on A into the ring (resp., module) of germs of X–valued functions on C or into the ring (resp., module) X.

3. In the case of the full germ $A = S_a$, the germ F_A is said to be the *germ of the function F at the point a* and is denoted by F_a. Hence it is the equivalence class of the function F defined in a neighbourhood of a, with respect to the equivalence relation (in the set of all the X–valued functions that are defined in a neighbourhood of the point a) given by: "$F' = F''$" in a neighbourhood of the point a".

Note that if $a \in E \subset S$, then the set of germs of functions on $A = E_a$ can be identified with the set of germs of functions at a in the space E via the natural bijection $f_A \longrightarrow f_a$ (where the f are functions on neighbourhoods of the point a in the space E). In the case when X is a ring (resp., module), the bijection becomes a ring (resp., module) isomorphism.

4. Let $f : S \longrightarrow X$ and $g : S \longrightarrow X$. Sometimes we will write $f \equiv g$ instead of $f = g$ in S (and $f \equiv g$ in E instead of $f = g$ in E for any $E \subset S$). The symbol $f \not\equiv g$ will be used to denote the fact that $f_x \neq g_x$ for all $x \in S$ (i.e., that $f \equiv g$ is not true in any open non-empty subset or, equivalently, that the set $\{f(x) \neq g(x)\}$ is dense). Hence, if X is a Hausdorff topological space and f, g are continuous, $f \not\equiv g$ precisely in the case when the set $\{f(x) = g(x)\}$ is nowhere dense. If $E \subset S$, then $f \not\equiv g$ in E will indicate that $f_E \not\equiv g_E$.

$\tilde{B} \longrightarrow S$ for sufficiently small neighbourhoods U, V of the points a, b in $\tilde{A}, \tilde{B}$, respectively, we have $\tilde{g}(\tilde{f}(x)) = x$ in U and $\tilde{f}(\tilde{g}(y)) = y$ in V. But $\tilde{f}(U)$ is always a neighbourhood of the point b in $\tilde{B}$ (because $U \supset \tilde{g}(W)$ for some neighbourhood $W \subset V$, and so $\tilde{f}(U) \supset \tilde{f}(\tilde{g}(W)) = W$). Therefore we may assume that $V = \tilde{f}(U)$; then $\tilde{f}_U : U \longrightarrow V$ and $\tilde{g}_V : V \longrightarrow U$ are mutually inverse homeomorphisms.

§5. The topology of a finite dimensional vector space
(over C or R)

1. Let X be an n–dimensional real or complex vector space (i.e., over **R** or over **C**). Then X has a (unique) *natural topology* such that some (and hence each) linear coordinate system on X is a homeomorphism ([17]) . Every norm on X defines the same natural topology. Thus every two norms are equivalent and X, regarded as a normed space, is complete. Every vector subspace of the space X is closed. If $E \subset X$, then

$$(E \text{ is compact}) \Longleftrightarrow (E \text{ is closed and bounded}),$$

$$(E \text{ is relatively compact}) \Longleftrightarrow (E \text{ is bounded}).$$

A set E is said to be bounded if it is contained in a ball with respect to a norm on X. (This condition is independent of the choice of the norm and the centre of the ball.) Finally, the space X is locally compact.

The set of all linearly independent sequences $(x_1, \ldots, x_n) \in X^k$ is open in X^k and the set of all sequences $(x_1, \ldots, x_r) \in X^r$ that generate X^r is open in X^r. In particular, the set of all bases of the space X is open in X^n.

In the space X, the closure of a cone is a cone. Any compact set which does not contain the origin generates a closed cone. (For the cone generated by a compact non-empty set $E \subset X \setminus 0$ is $\pi(\{(z,w) \in X \times E : z \wedge w = 0\})$, where the natural projection $\pi : X \times E \longrightarrow X$ is closed.)

If $S, T \subset X$ are cones, then

$$T_0 \subset S_0 \Longleftrightarrow T \subset S,$$

and hence

$$T_0 = S_0 \Longleftrightarrow T \subset S.$$

(In other words, a cone is determined by its germ at 0.)

2. Let X, Y, Z, X', Y' be finite dimensional (real or complex) vector spaces.

Every linear mapping from the space X to the space Y is continuous. In the case when it is surjective, it is also open.

([17]) It is, at the same time, the only Hausdorff topology on X under which X is a topological vector space (i.e., the algebraic operations on X are continuous). If X is a vector space over **C**, then, naturally, it is a vector space over **R** and both natural topologies coincide.

Every polynomial mapping $P : X \longrightarrow Y$ is continuous. If $P = 0$ on a non-empty open set, then $P \equiv 0$ [18]. Therefore the condition $P \not\equiv 0$ is the negation of the condition $P \equiv 0$. If P is non-zero, the set $\{P = 0\}$ is closed and nowhere dense, whereas the set $\{P \neq 0\}$ is open and dense.

In particular, the operations $+$ and $\cdot$ are continuous [19].

The set $L(X, Y)$ of all linear mappings on the space X into Y is a finite dimensional vector space and $\dim L(X, Y) = (\dim X)(\dim Y)$. The composition

$$L(X, Y) \times L(Y, Z) \ni (\varphi, \psi) \longrightarrow \psi \circ \varphi \in L(X, Z) ,$$

the diagonal product

$$L(X, Y) \times L(X, Z) \ni (\varphi, \psi) \longrightarrow (\varphi, \psi) \in L(X, Y \times Z),$$

and the Cartesian product

$$L(X, Y) \times L(X', Y') \ni (\varphi, \varphi') \longrightarrow \varphi \times \varphi' \in L(X \times X', Y \times Y')$$

are continuous mappings [20]. Note also that the mapping

$$L(X, Y) \times X \ni (f, x) \longrightarrow f(x) \in L$$

is continuous [21].

The function $L(X, Y) \ni \varphi \longrightarrow \operatorname{rank} \varphi \in \mathbf{N}$ is lower semicontinuous.

Indeed, if $\operatorname{rank} \varphi_0 = r$, then $\varphi_0(a_1), \ldots, \varphi_0(a_r)$ are linearly independent for some $a_1, \ldots, a_r$. Thus, for each mapping φ from some neighbourhood of the mapping φ_0, the elements $\varphi(a_1), \ldots, \varphi(a_r)$ are also linearly independent, and hence $\operatorname{rank} \varphi \geq r$.

Accordingly, the sets $\{\varphi \in L(X, Y) : \operatorname{rank} \varphi \geq k\}$, $k \in \mathbf{N}$, are open. In particular, both the set of all monomorphisms in $L(X, Y)$ and the set of all epimorphisms in $L(X, Y)$ are open [22].

---

[18] It is sufficient to prove the statement in the case of a polynomial $P : \mathbf{K}^n \longrightarrow \mathbf{K}$, where $\mathbf{K} = \mathbf{C}$ or $\mathbf{K} = \mathbf{R}$. Now, the property holds true when $n = 1$. Let $n > 1$ and assume that it is also true in the case of $n - 1$ variables. It follows that $P = 0$ in $G \times H$ for some non-empty open sets $G \subset \mathbf{K}^{n-1}$ and $H \subset \mathbf{K}$. Thus $P = 0$ in $\mathbf{K}^{n-1} \times H$, and so $P = 0$ in $\mathbf{K}^{n-1} \times \mathbf{K}$.

[19] On X^2 and $\mathbf{C} \times X$ or $\mathbf{R} \times X$, respectively.

[20] The same holds for the composition, the diagonal product, and the Cartesian product of a finite number of mappings.

[21] Since the above mappings are linear (the first three) or bilinear (the last one) and hence polynomial.

[22] By taking $k = \dim X$ and $k = \dim Y$, respectively, we can obtain these two sets.

If $\dim X = \dim Y$, then the set of all isomorphisms $L_0(X,Y) \subset L(X,Y)$ is open and dense. Moreover, the mapping $L_0(X,Y) \ni \varphi \longrightarrow \varphi^{-1} \in L_0(X,Y)$ is open and dense in the vector space of the $n \times n$–matrices (as $C \longrightarrow \det C$ is a non-zero polynomial) and the mapping $\{\det C \ne 0\} \ni C \longrightarrow C^{-1} \in \{\det C \ne 0\}$ is continuous.

A linear mapping $f : X \longrightarrow Y$ is proper precisely in the case when $|f(x)| \longrightarrow \infty$ as $|x| \longrightarrow \infty$. (Notice that it would be equivalent to each of these two conditions to require that the inverse image of any bounded set is bounded.)

Let M be a locally compact topological space, and let $E \subset M \times Y$ be a locally closed set. Then the natural projection $E \longrightarrow M$ is proper if and only if E is closed and each point of the space M has a neighbourhood U such that the fibres $E_x = \{y \in Y : (x,y) \in E\}$, for $x \in U$, are *uniformly bounded*, i.e., they are all contained in a common ball (with respect to a norm in Y).

3. Let $z^n + a_1 z^{n-1} + \ldots + a_n$ be a monic polynomial with complex coefficients. Observe that $\zeta_1, \ldots, \zeta_n$ is a complete sequence of its roots ([23]) precisely when $a_j = \sigma_j(\zeta_1, \ldots, \zeta_n)$ for $j = 1, \ldots, n$. Then

$$(|a_i| \le r, i = 1, \ldots, n) \Longrightarrow (|\zeta_j| \le 2r, j = 1, \ldots, n) .$$

(For if $|\zeta| > 2r$, then

$$|\zeta^n + a_1 \zeta^{n-1} + \ldots + a_n| \ge |\zeta|^n \left(1 - \frac{r}{|\zeta|} - \ldots - \left(\frac{r}{|\zeta|}\right)^n \right) > 0 \, .)$$

Consequently, the polynomial mapping

$$\sigma : \mathbf{C}^n \ni (z_1, \ldots, z_n) \longrightarrow (\sigma_1(z_1, \ldots, z_n), \ldots, \sigma_n(z_1, \ldots, z_n)) \in \mathbf{C}^n$$

is a proper surjection. (Since the inverse image of any bounded set is bounded in view of the fundamental theorem of algebra.) Observe that $\sigma(z_1, \ldots, z_n) = 0$ if and only if $z_1 = \ldots = z_n = 0$.

THE THEOREM ON CONTINUITY OF ROOTS. *Let $\zeta_1, \ldots, \zeta_n$ be a complete sequence of roots of the polynomial $z^n + a_1 z^{n-1} + \ldots + a_n$. Then for each $\varepsilon > 0$ there exists $\delta > 0$ such that if $|c_i - a_i| < \delta$, $i = 1, \ldots, n$, then $|z_j - \zeta_j| < \varepsilon$, $j = 1, \ldots, n$, for a (suitably ordered) complete sequence of roots $z_1, \ldots, z_n$ of the polynomial $z^n + c_1 z^{n-1} + \ldots + c_n$.*

---

[23] i.e., $z^n + a_1 z^{n-1} + \ldots + a_n = (z - \zeta_1) \ldots (z - \zeta_n)$.

PROOF . The set $E = \bigcap_\alpha \bigcup_i \{z \in \mathbf{C}^n : |z_{\alpha_i} - \zeta_i| \geq \varepsilon\}$, where $\alpha = (\alpha_1, \ldots, \alpha_n)$ varies over all permutations of $\{1, \ldots, n\}$, is symmetric and closed. Furthermore, $\sigma^{-1}(\sigma(E)) = E$. Indeed, if $\sigma(z) = \sigma(w)$ and $w \in E$, then z is a permutation of w (since both z and w are sequences of the roots of the same polynomials), and so $z \in E$. Hence we have $\sigma(\backslash E) = \sigma\big(\backslash\sigma^q(\sigma(E))\big) = \sigma\big(\sigma^{-1}(\backslash\sigma(E))\big) = \backslash\sigma(E)$ [24] . But $\sigma(E)$ is closed, because σ, being a proper mapping, is closed. Therefore the set

$$\sigma(\backslash E) = \sigma\left(\bigcup_\alpha \bigcap_i \{z \in \mathbf{C}^n : |z_{\alpha_i} - \zeta_i| < \varepsilon\}\right)$$

is open. Since it contains the point $(a_1, \ldots, a_k) = \sigma(\zeta_1, \ldots, \zeta_n)$, it must also contain the set $\{(c_1, \ldots, c_n) : |c_i - a_i| < \delta\}$ for some $\delta > 0$. Thus, if $|c_i - a_i| < \delta$, $i = 1, \ldots, n$, then $c_1, \ldots, c_n$ is the sequence of the coefficients of a monic polynomial whose roots are $z_1, \ldots, z_n$, and which is such that $|z_{\alpha_i} - \zeta_i| < \varepsilon$, $i = 1, \ldots, n$, for some permutation α.

§6. The topology of the Grassmann space

Let X be a complex n–dimensional vector space [25] .

1. For any $k \in \mathbf{N}$, the *Grassmann space* $\mathbf{G}_k(X)$ is the set of all k–dimensional subspaces of the space X. In particular, $\mathbf{P}(X) = \mathbf{G}_1(X)$, i.e., the set of all lines passing through 0 is called the *projective space of dimension* $n-1$ and we let $\dim \mathbf{P}(X) = \dim X - 1$ [26] . Clearly, $\mathbf{G}_k(X) = \emptyset$ for $k > n$.

From now on we will assume that $0 \leq k \leq n$.

Let $B_k(X)$ denote the set of all linearly independent sequences $z \in X^k$. The set is open in X^k. In particular, $B(X) = B_n(X)$ is the set of all bases of the space X.

In the Grassmann space $\mathbf{G}_k(X)$ we can introduce the following topology. The open sets are defined to be the sets whose inverse images under the surjection

$$\alpha = \alpha_k = \alpha^X = \alpha_k^X : B_k(X) \ni x = (x_1, \ldots, x_k) \longrightarrow \sum_1^k \mathbf{C}x_i \in \mathbf{G}_k(X)$$

---

[24] The symbol $\backslash E$ denotes the complement of the set E.

[25] In the present context, the real case does not differ from the complex one.

[26] In Chapter VII we will give $\mathbf{P}(X)$ the structure of an $(n-1)$–dimensional complex manifold (see VII. 2.1).

are open in $B_k(X)$ (^{27}) . Then the surjection α is continuous and open. (Indeed, if $G \subset B_k(X)$, then $\alpha^{-1}(\alpha(G)) = \bigcup A(G)$, where

$$A : \ X^k \in (x_1, \ldots, x_k) \longrightarrow \left(\sum_1^k a_{1j} x_j, \ldots, \sum_1^k a_{kj} x_j \right) \in X^k \ , \ \det a_{ij} \neq 0,$$

are isomorphisms and hence also homeomorphisms.)

The space $\mathbf{G}_k(X)$ is compact. To see this, note first that it is a Hausdorff space. This is so since the diagonal in $\mathbf{G}_k(X)^2$, i.e., the set $\{(U,V) : \ U = V\}$, is closed, as its inverse image under the open surjection $\alpha \times \alpha$ (see 2.1) is the set

$$\{(x,y) \in B_k(X)^2 : \ x_1 \wedge \ldots \wedge x_k \wedge y_i = 0, \ i = 1, \ldots, k\} \ ,$$

which is closed in $B_k(X)^2$. Secondly, the space $\mathbf{G}_k(X)$ is the image under α of the compact set of all orthonormal sequences in X^k (with respect to a fixed Hermitian product in X).

The space $\mathbf{G}_k(X)$ is connected (see n° 8 below).

The topology of the space $\mathbf{G}_k(X)$ has a countable basis (furnished by the image under α of any countable basis in $B_k(X)$). Thus the notions of open sets, closed sets, etc. can be characterized in $\mathbf{G}_k(X)$ in terms of the convergence of sequences (^{28}) .

If $Z \subset X$ is a vector subspace, then the topology of the space $\mathbf{G}_k(Z)$ coincides with that induced by the topology of the space $\mathbf{G}_k(X)$. (Indeed, the mapping $\mathbf{G}_k(Z) \hookrightarrow \mathbf{G}_k(X)$ and its inverse are both continuous; see 2.2.) The set $\mathbf{G}_k(Z)$ is closed in $\mathbf{G}_k(X)$. Moreover, it is nowhere dense, provided that $k > 0$ and $Z \subsetneq X$ (see 2.2).

Note that if $G \subset X$ is an open set, the set $\{l \in \mathbf{G}_k(X) : \ L \cap G \neq \emptyset\}$ is open in $\mathbf{G}_k(X)$ (as it is equal to $\alpha(\{x \in B_k(X) : \ x_1 \in G\})$). Thus, if $F \subset X$ is a closed set, then the set $\{L \in \mathbf{G}_k(X) : \ L \subset F\}$ is closed in $\mathbf{G}_k(X)$.

2. In the space $\mathbf{G}_k(X) \times \mathbf{G}_l(X)$, where $k + l \leq n$, the set $\{(U,V) : \ U \cap V = 0\}$ is open, and the mapping

$$s : \ \{U \cap V = 0\} \ni (U,V) \longrightarrow U + V \in \mathbf{G}_{k+1}(X)$$

(^{27}) Or, equivalently, in X^k.

(^{28}) Note that, in the space $\mathbf{G}_k(X)$, a sequence U_ν converges to U if and only if for some (and hence any) basis x in U there exists a sequence $x_\nu \longrightarrow x$, where x_ν is a basis in U_ν for $\nu = 1, 2, \ldots$. This is a consequence of the fact that the image under α of a base of neighbourhoods of an x in $B_k(X)$ is a base of neighbourhoods of $\alpha(x)$ in $\mathbf{G}_k(X)$ (see 2.2).

is continuous. (Indeed, $\alpha_k \times \alpha_l$ is a continuous and open surjection, the set $(\alpha_k \times \alpha_l)^{-1}(\{U \cap V = 0\})$ is open, and the mapping $s \circ (\alpha_k \times \alpha_l) \subset \alpha_{k+l}$ is continuous – see 2.2.)

3. We have the following

LEMMA . *Let Y be a finite dimensional complex vector space. The mapping $B(X) \times Y^n \ni (x,y) \longrightarrow f_{xy} \in L(X,Y)$ is continuous, where f_{xy} is given by $f_{xy}(x_i) = y_i$ $(i = 1, \ldots, n)$.*

Indeed, the mapping

$$X^n \ni x \longrightarrow F_x \in L\big(L(X,Y), Y^n\big) \ ,$$

where $F_x(f) = \big(f(x_1), \ldots, f(x_n)\big)$ is linear and hence continuous. Now, if $(x,y) \in B(X) \times Y^n$, then $F_x(f_{xy}) = y$, F_x is an isomorphism and we have $f_{xy} = F_x^{-1}(y)$. This yields the continuity of the mapping $(x,y) \longrightarrow f_{xy}$ (see 5.2).

4. With every k–dimensional subspace V of the space X one can associate the $(n-k)$–dimensional subspace $V^\perp = \{\varphi \in X^* : \varphi V = 0\}$ of the dual space X^*. We have the following properties

$$(V_1 \cap \ldots \cap V_r)^\perp = V_1^\perp + \ldots + V_r^\perp, \ (V_1 + \ldots + V_r)^\perp = V_1^\perp \cap \ldots \cap V_r^\perp \ ,$$
$$U \subset V \Longleftrightarrow U^\perp \supset V^\perp, \ 0^\perp = X^*, \ X^\perp = 0 \ .$$

Moreover, if $\varphi : X \longrightarrow Y$ is an isomorphism of vector spaces, then

$$\varphi^*(V^\perp) = \varphi^{-1}(V)^\perp \text{ for any subspace } V \subset Y \ .$$

We will prove that the bijection

$$\tau = \tau_k = \tau^X = \tau_k^X : \ \mathbf{G}_k(X) \ni V \longrightarrow V^\perp \in \mathbf{G}_{n-k}(X^*)$$

is a homeomorphism. In fact, since $\mathbf{G}_k(X)$ is compact, it is enough to show that the mapping τ is continuous. Set $\alpha^* = \alpha_{n-k}^{X^*}$. Let $x \in B_k(X)$. Fix $t = (t_{k+1}, \ldots, t_n) \in X^{n-k}$ in such a way that $(x,t) \in B(X)$. Then $(z,t) \in B(X)$ for z in some neighbourhood W of the point x. Now, for $z \in W$, the forms $f_z^i \in X^*$, $i = k+1, \ldots, n$, given by $f_z^i(z_\nu) = 0$, $\nu = 1, \ldots, k$, and $f_z^i(t_j) = \delta_{ij}$, $j = k+1, \ldots, n$, constitute a basis of the space $\alpha(z)^\perp$. Hence $\alpha(z)^\perp = \alpha^*(f_z)$, where $f_z = (f_z^{k+1}, \ldots, f_z^n)$. This implies the continuity of the mapping $(\tau \circ \alpha)_W$, because the mapping $W \ni z \longrightarrow f_z$ is continuous by the lemma. Therefore the composition $\tau \circ \alpha$ is continuous and so is the mapping τ (see 2.2).

5. For any subspace $U \subset X$ of dimension $\leq k$ we define the *Schubert cycle* :

$$\mathbf{S}^k(U) = \mathbf{S}^k(U, X) = \{V \in \mathbf{G}_k(X) : V \supset U\} \ .$$

Clearly (see n° 4),

$$(\#) \qquad \begin{array}{l} \tau\big(\mathbf{S}^k(U)\big) = \mathbf{G}_{n-k}(U^{\perp}) \text{ for any subspace } U \text{ of dimension } \leq k \ , \\ \tau\big(\mathbf{G}_k(V)\big) = \mathbf{S}^{n-k}(V^{\perp}) \text{ for any subspace } V \text{ of dimension } \geq k \ . \end{array}$$

Accordingly, the Schubert cycle $\mathbf{S}^k(U)$ is a closed subset of the space $\mathbf{G}_k(X)$. In addition, it is nowhere dense provided that $k < n$ and $U \neq 0$ (see n° 1).

6. The function

$$\mathbf{G}_{p_1}(X) \times \ldots \times \mathbf{G}_{p_r}(X) \ni (V_1, \ldots, V_r) \longrightarrow \dim(V_1 + \ldots + V_r) \in \mathbf{N}$$

is lower semicontinuous. Indeed (see 2.2), by composing it with $\alpha_{p_1} \times \ldots \times \alpha_{p_r}$ we obtain a restriction of the lower semicontinuous function $X^p \ni c = (c_1, \ldots, c_p) \longrightarrow \operatorname{rank} g_c \in \mathbf{N}$ [29] , where $p = p_1 + \ldots + p_r$ and $g_c : \mathbf{C}^p \ni z = (z_1, \ldots, z_p) \longrightarrow \sum_1^p z_i c_i \in X$ (see 5.2).

Therefore the function $\mathbf{G}_{k_1}(X) \times \ldots \times \mathbf{G}_{k_r}(X) \ni (V_1, \ldots, V_r) \to \dim(V_1 \cap \ldots \cap V_r) \in \mathbf{N}$ is upper semicontinuous (as $\dim(V_1 \cap \ldots \cap V_r) = n - \dim(V_1^{\perp} + \ldots + V_r^{\perp})$).

Consequently,

The set $\{(V_1, \ldots, V_r) : \dim(V_1 \cap \ldots \cap V_r) \leq s\}$ *is open in the space* $\mathbf{G}_{k_1}(X) \times \ldots \times \mathbf{G}_{k_r}(X)$ *for all* $s \in \mathbf{N}$.

This implies (see A.1.18) that the set of sequences of subspaces $(V_1, \ldots, V_r) \in \mathbf{G}_{k_1}(X) \times \ldots \times \mathbf{G}_{k_r}(X)$ which intersect transversally is open.

Next, the set $\{(U, V) : U \subset V\}$ in the space $\mathbf{G}_p(X) \times \mathbf{G}_k(X)$, where $0 \leq p \leq k$ is closed (since it is equal to $\{(U, V) : \dim(U \cap V) \geq p\}$).

7. Let $\Phi \subset \mathbf{G}_k(X)$ be a closed set. For $0 \leq p \leq k$, the set $\bigcup\{\mathbf{G}_p(V) : V \in \Phi\}$ is closed in $\mathbf{G}_p(X)$. (This is so, because it is the image of the compact set $\big(\mathbf{G}_p(X) \times \Phi\big) \cap \{(U, V) : U \subset V\}$ under the projection $(U, V) \longrightarrow U$.)

---

[29] Because the linear mapping $X^p \ni c \longrightarrow g_c \in L(\mathbf{C}^p, X)$ is continuous.

Let $\Omega \subset \mathbf{G}_k(X)$ be an open set, and let $0 \le p \le k$. The set $\{\mathbf{G}_p(V) : V \in \Omega\}$ is open in $\mathbf{G}_p(X)$. (Since its inverse image under α_k is the open set $\pi(\alpha_k^{-1}(\Omega))$, where $\pi : B_k(X) \ni (x_1, \ldots, x_k) \longrightarrow (x_1, \ldots, x_p) \in B_p(X)$.)

Let $k \le l \le n$. The set $\{V \in \mathbf{G}_l(X) : \mathbf{G}_k(V) \cap \Omega \ne 0\}$ is open in $\mathbf{G}_l(X)$. (Indeed, its image under τ_l is the set $\{\mathbf{G}_{n-l}(W) : W \in \tau_k(\Omega)\}$ which, according to the previous property, is open.)

Let $k \le l \le n$. If $\mathcal{K} \subset \mathbf{G}_k(X)$ and if for each V from a dense subset $\mathcal{L}$ of $\mathbf{G}_l(X)$ the set $\mathcal{K} \cap \mathbf{G}_k(V)$ is dense in $\mathbf{G}_k(V)$, then the set $\mathcal{K}$ itself is dense in $\mathbf{G}_k(X)$. For otherwise there would exist an open non-empty set $\Omega \subset \mathbf{G}_k(X) \backslash \mathcal{K}$, and then the open non-empty set $\{V \in \mathbf{G}_l(X) : \mathbf{G}_k(V) \cap \Omega \ne \emptyset\}$ would contain some $V \in \mathcal{L}$. Therefore the set $\Omega \cap \mathbf{G}_k(V)$ would be non-empty, open in $\mathbf{G}_k(V)$, and disjoint with $\mathcal{K} \cap \mathbf{G}_k(V)$, contrary to our assumption.

8. For any $(n - k)$–dimensional subspace $V \subset X$, the set $\Omega(V)$ of linear complements of V is open and dense in $\mathbf{G}_k(X)$. (Indeed, its inverse image under α is the open and dense set $\{(x_1, \ldots, x_k) \in B_k(X) : \omega \wedge x_1 \wedge \ldots \wedge x_k \ne 0\}$, where ω is a direction vector of the space V.)

Note that a finite number of sets $\Omega(V)$ is sufficient to cover $\mathbf{G}_k(X)$. Namely, if $e_1, \ldots, e_n$ is a basis of the space X, then $\mathbf{G}_k(X) = \bigcup \Omega(V_\nu)$, where $V_\nu = \sum_1^{n-k} \mathbf{C}e_{\nu_i}$ and $\nu = (\nu_1, \ldots, \nu_{n-k})$, $1 \le \nu_1 \le \ldots \le \nu_{n-k} \le n$. In fact, if $U \in \mathbf{G}_k(X)$, then by taking a direction vector ω in the space U we have $\omega \wedge e_{\nu_1} \wedge \ldots \wedge e_{\nu_{n-K}} \ne 0$ for some ν, and hence $U \in \Omega(V_\nu)$.

Let $U \in \mathbf{G}_k(X)$, and let V be a linear complement of the subspace U. Then the bijection

$$\varphi_{UV} : L(U,V) \ni f \longrightarrow \hat{f} \in \Omega(V) ,$$

where $\hat{f} = \{u + f(u) : u \in U\}$, is a homeomorphism $(^{30})$.

To show this, fix a base $e_1, \ldots, e_k$ in the space U. If $f \in L(U,V)$, then $x_f = (e_1 + f(e_1), \ldots, e_k + f(e_k))$ is a basis for the subspace $\hat{f}$, and therefore $\varphi_{UV}(f) = \alpha(x_f)$. So the continuity of the mapping φ_{UV} follows from the continuity of the mapping $L(U,V) \ni f \longrightarrow x_f$. Now, let $t' \in U$, $t'' \in V$ be the components of a vector $t \in X$. If $x = (x_1, \ldots, x_k) \in \alpha^{-1}(\Omega(V))$, then x is a basis of the subspace $\alpha(x) = \hat{f}$, where $f \in L(U,V)$, and we

---

$(^{30})$ Thus the Grassmann space $\mathbf{G}_k(X)$ is locally homeomorphic to $\mathbf{C}^{k(n-k)}$. In Chapter VI we will prove that $\{\varphi_{UV}\}$ is a complex atlas on the space $\mathbf{G}_k(X)$ which defines the structure of a complex manifold of dimension $k(n - k)$ (see VII. 4.1).

have $f(x_i') = x_i''$ $(i = 1, \ldots, k)$, $x' = (x_1', \ldots, x_k') \in B(U)$ [31] , and $x'' = (x_1'', \ldots, x_k'') \in V^k$. Hence $f = f_{x'x''}$, which implies that $\varphi_{UV}^{-1}((\alpha(x))) = f_{x'x''}$. Therefore, by the lemma from n° 3, the composition $\varphi_{UV}^{-1} \circ \alpha$ is continuous. Accordingly, the mapping φ_{UV}^{-1} is continuous (see 2.2).

It follows that the space $\mathbf{G}_k(X)$ is connected (since its dense subset $\Omega(V)$ is connected).

Note also that (for any norm on X) the family of sets

$$\{T \in \mathbf{G}_k(U) : \ T \subset \{u + v : |v| \leq \varepsilon|u|\}\}_{\varepsilon > 0} \ ,$$

where $u \in U$, $v \in V$, constitutes a base of neighbourhoods of the element U in the space $\mathbf{G}_k(X)$. (This is due to the fact that this family is the image under φ_{UV} of the collection of all closed balls in $L(U, V)$ centred at 0.)

9. Let Y be a vector space that is isomorphic to X. If $\varphi \in L_0(X, Y)$, then the bijection

$$\tilde{\varphi} = \varphi_{(k)} : \ \mathbf{G}_k(X) \ni T \longrightarrow \varphi(T) \in \mathbf{G}_k(Y)$$

is called the *isomorphism of the Grassmann spaces* $\mathbf{G}_k(X)$ and $\mathbf{G}_k(Y)$ *induced by* φ. It is a homeomorphism. (This follows from the equality $\tilde{\varphi} \circ \alpha^X = \alpha^Y \circ (\varphi \times \ldots \varphi)_{B_k(X)}$ – see 2.2.) Naturally, the inverse of such an isomorphism is also an isomorphism, and so is the composition of two such isomorphisms. Moreover, $\tilde{\varphi}^{-1} = \widetilde{\varphi^{-1}}$ and $\widetilde{\varphi \circ \psi} = \tilde{\varphi} \circ \tilde{\psi}$. Finally, we have the equality (see n° 4)

$$(\varphi^*)_{(n-k)} = \tau_k^X \circ (\varphi^{-1})_{(k)} \circ (\tau_k^Y)^{-1} \ .$$

10. Now let us consider the projective space $\mathbf{P}(X)$. The mapping

$$\alpha = \alpha_1^X : \ X \setminus 0 \ni z \longrightarrow \mathbf{C}z \in \mathbf{P}(X)$$

is an open continuous surjection (see 1).

For any affine hyperplane $H \subset X$, we denote by H_* the unique vector hyperplane that is parallel to H.

If $H \not\ni 0$, there exists a unique form $\lambda_H \in X^*$ such that $H = \{z : \lambda_H(z) = 1\}$. Its kernel is H_*. Hence we obtain two mutually inverse homeomorphisms

$$\alpha_H : \ H \ni z \longrightarrow \mathbf{C}z \in \mathbf{P}(X) \setminus \mathbf{P}(H_*)$$

$$\alpha_H^{-1} : \ \mathbf{P}(X) \setminus \mathbf{P}(H_*) \ni \mathbf{C}z \longrightarrow z/\lambda_H(z) \in H \ .$$

---

[31] Because $\hat{f} \ni t \longrightarrow t' \in U$ is an isomorphism.

(The continuity of the mapping α_H^{-1} follows from the continuity of the composition $\alpha_H^{-1} \circ \alpha : \; X \setminus H^* \ni z \longrightarrow z/\lambda_H(z) \in H$; see 2.2.)

With every subset E of the projective space $\mathbf{P}(X)$ one can associate a cone in the space X, given by

$$\tilde{E} = \bigcup \{\lambda : \; \lambda \in E\} \cup 0 \, ,$$

which is called the *cone of the set* E. Conversely, to each cone S in the space X, there corresponds a subset of the projective space $\mathbf{P}(X)$, given by

$$S^{\sim} = \{\lambda \in \mathbf{P}(X): \; \lambda \subset S\} = \{\mathbf{C}z : \; z \in S \setminus 0\} \quad (^{32}).$$

Then $E \longrightarrow E^{\sim}$ and $S \longrightarrow S^{\sim}$ are mutually inverse bijections. They preserve unions and intersections (of arbitrary families) as well as inclusion. Furthermore, $(E \setminus F)^{\sim} = (E^{\sim} \setminus f^{\sim}) \cup O$ for $E, F \subset \mathbf{P}(X)$. Also, $\emptyset^{\sim} = 0$ and $0^{\sim} = \emptyset$.

Moreover, $\alpha^{-1}(E) = E^{\sim} \setminus 0$ and $\alpha(E^{\sim} \setminus 0) = E$ for $E \subset \mathbf{P}(X)$.

This implies that a set $E \subset \mathbf{P}(X)$ is closed precisely in the case when its cone $E^{\sim}$ is closed (see n° 1).

We have the equality $\overline{(E^{\sim})} = (\bar{E})^{\sim}$ for $E \subset \mathbf{P}(X)$, and similarly, $\overline{(S^{\sim})} = (\bar{S})^{\sim}$ for $S \subset X$. (Since the closure of a cone S coincides with the intersection of all the closed cones containing S; see 5.1.) Therefore, if $E \subset F \subset \mathbf{P}(X)$, the set E is dense in F if and only if its cone $E^{\sim}$ is dense in the cone $F^{\sim}$.

Finally, if $\varphi : \; X \longrightarrow Y$ is an isomorphism of vector spaces, then

$$\tilde{\varphi}(E)^{\sim} = \varphi(E^{\sim}) \; \text{ for } \; E \subset \mathbf{P}(X) \, .$$

11. Let $0 \leq k \leq n$, and let $\mathbf{G}'_k(X)$ denote the set of all k–dimensional affine subspaces of the space X. In particular, $\mathbf{P}'(X) = \mathbf{G}_1(X)$ denotes the set of all affine lines in the space X.

For any $L \in \mathbf{G}'_k(X)$, let $L_* \in \mathbf{G}_k(X)$ denote the unique vector subspace that is parallel to L $(^{33})$.

The (Hausdorff) topology in $\mathbf{G}'_k(X)$ is defined by transfer via the bijection

$$\chi : \; \mathbf{G}_{k+1}(\mathbf{C} \times X) \setminus \mathbf{G}_{k+1}(0 \times x) \ni N \longrightarrow L \in \mathbf{G}'_k(X) \, ;$$

$(^{32})$ Hence, for a cone $S \subset X$ and $z \in X \setminus 0$, we have $z \in S \Longleftrightarrow \mathbf{C}z \in S^{\sim}$.

$(^{33})$ This agrees with the notation used in n° 10.

this is well-defined by the condition $1 \times L = N \cap (1 \times X)$. Then

$$(\#) \qquad\qquad N = (0 \times L_*) + \mathbf{C}(1, c) \text{ for any } c \in L .$$

The mapping

$$\beta : \; X \times B_k(X) \ni (z_0, \ldots, z_k) \longrightarrow z_0 + \sum_1^k \mathbf{C} z_i \in \mathbf{G}'_k(X)$$

is a continuous and open surjection. Indeed, the composition

$$\chi^{-1} \circ \beta : \; X \times B_k(X) \ni (z_0, \ldots, z_k) \longrightarrow \mathbf{C}(1, z_0) + \sum_1^k \mathbf{C}(0, z_i) \in \mathbf{G}_{k+1}(\mathbf{C} \times X)$$

is continuous (see $(\#)$ and n° 1). It is also open. Indeed, the mapping

$$\gamma : \; ((\mathbf{C} \setminus 0) \times X) \times (\mathbf{C} \times X)^k \ni ((t_0, w_0), \ldots, (t_k, w_k))$$
$$\longrightarrow (w_0/t_0, w_1 - (t_1/t_0) w_0, \ldots, w_k - (t_k/t_0) w_0) \in X^{k+1}$$

is a continuous surjection (see 2.1) and so is the restriction $\gamma_G : \; G \longrightarrow X \times B_k(X)$, where $G = \gamma^{-1}(X \times B_k(X))$. The composition

$$(\chi^{-1} \circ \beta) \circ \gamma_G : \; G \ni ((t_0, w_0), \ldots, (t_k, w_k)) \longrightarrow \mathbf{C}(1, w_0/t_0) +$$
$$+ \sum_1^k \mathbf{C}(0, w_i - (t_i/t_0) w_0) = \sum_0^k \mathbf{C}(t_i, w_i) \in \mathbf{G}_{k+1}(\mathbf{C} \times X) \quad (^{34})$$

is open (see n° 1).

Therefore the open sets in $\mathbf{G}'_k(X)$ are exactly the sets whose inverse images under the mapping β are open $(^{35})$.

The space $\mathbf{G}_k(X)$ is a closed subset of the space $\mathbf{G}'_k(X)$; it is also nowhere dense, provided that $k < n$. Its topology is the same as the induced topology. Indeed, $\mathbf{G}_k(X)$ is a compact space and $\iota : \; \mathbf{G}_k(X) \hookrightarrow \mathbf{G}'_k(X)$ is a continuous injection, as the composition $\iota \circ \alpha = \beta \circ (z \longrightarrow (0, z))$ is continuous (see 2.1).

$(^{34})$ Thus $G \subset B_{k+1}(\mathbf{C} \times X)$.

$(^{35})$ Similarly as in footnote $(^{28})$, it follows that, for $L = x_0 + \sum_1^k \mathbf{C} z_i \in \mathbf{G}'_k(X)$ and $L_\nu \in \mathbf{G}'_k(X)$, we have $L_\nu \longrightarrow L$ precisely when $L_\nu = z_{0\nu} + \sum_{i=1}^k \mathbf{C} z_{i\nu}$ for some $z_{0\nu} \longrightarrow z_0, \ldots, z_{k\nu} \longrightarrow z_k$.

Moreover, when $k < n$, the inverse image $\beta^{-1}(\mathbf{G}_k(X)) = \{z_0 \wedge \ldots \wedge z_k = 0\}$ is nowhere dense (see 5.2 and A. 3.1).

The mapping

$$\nu = \nu_k : \ \mathbf{G}'_k(X) \ni L \longrightarrow L_* \in \mathbf{G}_K(X)$$

is a continuous open surjection. This is the case, since the composition $\nu \circ \beta = \alpha \circ \big((z_0, z) \longrightarrow z\big)$ is continuous and open (see 2.1–2).

Note that the set of all the sequences of subspaces $(L_1, \ldots, L_r) \in \mathbf{G}'_{k_1}(X) \times \ldots \times \mathbf{G}'_{k_r}(X)$ which intersect transversally is open. Namely, it is equal to the inverse image under $\nu_{k_1} \times \ldots \times \nu_{k_r}$ of the open subset of $\mathbf{G}_{k_1}(X) \times \ldots \times \mathbf{G}_{k_r}(X)$ consisting of all sequences of subspaces which intersect transversally (see n° 6 and A. 1.18).

Also, the mapping

$$\nu = \nu_k : \ \mathbf{G}'_k(X) \ni L \longrightarrow L_* \in \mathbf{G}_k(X)$$

is an open continuous surjection. This is so because the composition $\sigma \circ (\mathrm{id}_X \times \alpha) = \beta$ is continuous and open and $\mathrm{id}_X \times \alpha$ is a continuous and open surjection.

Observe that if $G \subset X$ is an open set, then the set $\{L \in \mathbf{G}'_k(X) : \ L \cap G \neq \emptyset\}$ is open in $\mathbf{G}'_k(X)$. (This is because the latter is equal to $\beta(\{x_0 \in G\})$.) Accordingly, if $F \subset X$ is a closed set, then the set $\{L \in \mathbf{G}'_k(X) : \ L \subset F\}$ is closed in $\mathbf{G}'_k(X)$.

12. Set $\mathbf{P} = \mathbf{P}(X)$. By a k–*dimensional projective subspace* of the space $\mathbf{P}$ (where $-1 \leq k \leq n-1$) we mean any subset of $\mathbf{P}$ of the form $\mathbf{P}(L)$, where $L \in \mathbf{G}_{k+1}(X)$ [36] . Obviously, $\mathbf{P}(L) = L\tilde{}$. Denote by $\mathbf{G}_k(\mathbf{P})$ the set of all k–dimensional projective subspaces of the space $\mathbf{P}$. We have the bijection

$$\omega = \omega^{\mathbf{P}} : \ \mathbf{G}_{k+1}(X) \ni L \longrightarrow L\tilde{} \in \mathbf{G}_k(\mathbf{P}) \ .$$

By *(projective) lines* we mean one–dimensional projective subspaces and the name *(projective) hyperplanes* is used for $(\dim \mathbf{P} - 1)$–dimensional subspaces of the projective space $\mathbf{P}$. The latter are exactly the sets of the form $\mathbf{P}(H) = H\tilde{}$, where H is a hyperplane in the space X. Note that each k–dimensional subspace, where $k < \dim \mathbf{P}$, is the intersection of a $(k + 1)$–dimensional subspace and a hyperplane (see n° 10).

---

[36] Compare with footnote [28]. It is a k–dimensional submanifold of the manifold $\mathbf{P}$ (see VII. 2.2).

It is easy to see that for any two distinct points in $\mathbf{P}$ there is a unique line passing through them, and that a set $E \subset \mathbf{P}$ is a projective subspace if and only if for any two distinct points of the set E the line passing through these points is contained in E.

Since the intersection of any family of projective subspaces is a projective subspace, it follows that for each set $E \subset \mathbf{P}$ there exists a projective subspace generated by E, i.e., the smallest projective subspace containing E. A set E generates a projective subspace T precisely when its cone $E^\sim$ generates the subspace $T^\sim$ (see n° 10).

Clearly, every isomorphism between projective spaces maps projective subspaces onto projective subspaces (preserving their dimensions). Moreover, the subspace generated by a set is mapped onto the subspace generated by the image of this set.

Let Y be an n–dimensional vector space. If $H \subset \mathbf{P}(X)$ and $L \subset \mathbf{P}(Y)$ are hyperplanes, and if $\lambda \in \mathbf{P}(X) \setminus H$, and $\mu \in \mathbf{P}(Y) \setminus L$, then there is an isomorphism $\psi : \mathbf{P}(X) \longrightarrow \mathbf{P}(Y)$ such that $\psi(H) = L$ and $\psi(\lambda) = \mu$ [37]. (For we have the direct sums $X = H^\sim + \lambda$, $Y = L^\sim + \mu$, and so there exists an isomorphism $\varphi : X \longrightarrow Y$ such that $\varphi(H^\sim) = L^\sim$ and $\varphi(\lambda) = \mu$. It suffices to set $\psi = \tilde{\varphi}$; see n° 10.)

Note also that if $U \subset X$ is a subspace and $V \subset X$ is a complementary subspace, then (given a norm on X) the family of sets

$$\{u + v : u \in U, \ v \in V, |v| \leq \varepsilon|u|\}^\sim, \ \varepsilon > 0,$$

is a neighbourhood basis of the projective subspace $\mathbf{P}(U)$ in $\mathbf{P}$ [38]. Indeed, each of the sets in the family contains the set $\{\mathbf{C}(u+v) : |v| < \varepsilon|u|\} \supset \mathbf{P}(U)$ (with the same ε), which is open since the mapping α is open. Furthermore, any open set that contains $\mathbf{P}(U)$ must also contain one of the sets from the family; since these sets are compact, they constitute a filter–base and their intersection is equal to $\mathbf{P}(U)$.

The (Hausdorff) topology on $\mathbf{G}_k(\mathbf{P})$ is defined by transfer via the bijection ω. Observe that then the mapping

$$\mu : \ \mathbf{G}'_k(X) \setminus \mathbf{G}_k(X) \ni T \longrightarrow (\mathbf{C}T)^\sim \in \mathbf{G}_k(\mathbf{P}) \quad [39]$$

is a continuous and open surjection. Indeed, since both the composition $(\omega^{-1} \circ \mu) \circ \beta_{B_{k+1}(X)} = \alpha_{k+1}$ and the mapping $\beta_{B_{k+1}(X)} : \ B_{k+1}(X) \longrightarrow$

---

[37] More generally, if $U, U' \in \mathbf{P}(X)$ are disjoint projective subspaces of dimensions k and k', respectively, and the same is true about $V, V' \subset \mathbf{P}(Y)$, then there is an isomorphism $\psi : \mathbf{P}(X) \longrightarrow \mathbf{P}(Y)$ such that $\psi(U) = V$ and $\psi(U') = V'$.

[38] A family of subsets $\{B_\iota\}$ of a topological space is said to be a *neighbourhood basis of a subset* E of the space if $E \subset \text{int} B_\iota$ and every open set containing E contains at least one of the B_ι's.

[39] The set $\mathbf{C}T$ is then equal to the vector subspace generated by T.

$\mathbf{G}'_k(X) \setminus \mathbf{G}_k(X)$ are continuous and open surjections, we conclude that so is $(\omega^{-1} \circ \mu)$, and this implies our assertion (see n° 11 and 2.1–2).

Note also that if $F \subset \mathbf{P}$ is a closed set, then the set $\{L \in \mathbf{G}_k(\mathbf{P}) : L \cap F = \emptyset\}$ is open. In fact, it is enough to check (see n° 10) that the set $\Omega = \{U : U \cap F^\sim = 0\} \subset \mathbf{G}_{k+1}(X)$ is open. Let $U \in \Omega$, i.e., $U \cap F^\sim = 0$. Let V be a linear complement of the space U. Fix a norm on X. Since $F^\sim$ is a closed cone (see n° 10), there exists $\varepsilon > 0$ such that $|v| \geq \varepsilon |u|$ for $u + v \in F^\sim$, $u \in U$, $v \in V$. Therefore the set $\{T : T \subset \{|v| \leq \varepsilon |u|\}\} \subset \mathbf{G}_{k+1}(X)$ is a neighbourhood of the element U (see n° 8) contained in Ω.

CHAPTER C

COMPLEX ANALYSIS

§1. Holomorphic mappings

1. Every vector space over $\mathbf{C}$ is also a vector space over $\mathbf{R}$ [1].

Let X and Y be vector spaces over $\mathbf{C}$. A mapping $\varphi : X \longrightarrow Y$ is said to be $\mathbf{C}$–*linear* (or $\mathbf{R}$– *linear*) if it is linear as a morphism of the complex (resp., real) vector spaces X,Y. An $\mathbf{R}$–linear mapping $\varphi : X \longrightarrow Y$ is $\mathbf{C}$–linear if and only if $\varphi(ix) = i\varphi(x)$ for all $x \in X$.

The space $\mathbf{C}^n$, regarded as a vector space over $\mathbf{R}$, is identified with $(\mathbf{R}^2)^n = \mathbf{R}^{2n}$. When an n-dimensional complex vector space is regarded as a vector space over $\mathbf{R}$, it is 2n-dimensional.

In what follows, we will be assuming that all vector spaces (over $\mathbf{C}$ or $\mathbf{R}$) are finite-dimensional.

2. Let f be a function defined in a neighbourhood of a point $a \in \mathbf{C}$, with values in a complex vector space Y. Then the *complex derivative*

$$f'(a) = \lim_{z \to 0} \frac{f(a+z) - f(a)}{z}$$

---

[1] It is enough to restrict multiplication to $\mathbf{R} \times X$. On the other hand, if X is a vector space over $\mathbf{R}$, it can be endowed with the structure of a complex vector space (whose underlying real structure coincides with the original one), provided that X is even dimensional. This can be achieved by introducing an endomorphism $x \longrightarrow ix$, such that $i(ix) = -x$, which would play the role of multiplication by i. Consequently, the multiplication by complex numbers would be given by the formula $(\alpha + \beta i)x = \alpha x + \beta(ix)$.

exists if and only if f is differentiable at a ([2]) and the *differential* $d_a f$ is
C–linear ([3]) . In this case, $d_a f(z) = f'(a)z$. (This is because the condition
that $f((a + z) - f(a))/\, z \longrightarrow c$ as $z \longrightarrow 0$ is equivalent to the condition
that $f(a + z) - f(z) = cz + o(z)$ as $z \longrightarrow 0$.) Moreover, $\frac{\partial f}{\partial x}(a) = f'(a)$ and
$\frac{\partial f}{\partial y}(a) = if'(a)$ (because $d_a f(x, y) = f'(a)x + if'(a)y$).

Now let f be a function defined on a neighbourhood of a point $a \in \mathbf{C}^n$
and with values in a complex vector space Y. If f is differentiable at the
point a ([4]) , then the differential $d_a f$ is C–linear if and only if the partial
derivatives $\frac{\partial f}{\partial z_\nu}(a), \nu = 1, \dots, n$, exist. Then

$$d_a f(z) = \sum_{\nu=1}^{n} \frac{\partial f}{\partial z_\nu}(a)z_\nu, \ z = (z_1, \dots, z_n).$$

Indeed, we have $d_a f(z) = \sum_{\nu=1}^{n} {}^\nu d_a f(z)$, where ${}^\nu d_a f$ denotes the differential
of the mapping $\zeta \longrightarrow f(\dots, a_{\nu-1}, \zeta, a_{\nu+1}, \dots)$ at a_ν and $a = (a_1, \dots, a_n)$.
Therefore each of the above conditions is equivalent to C–linearity of the
differentials ${}^\nu d_a f, \ \nu = 1, \dots, n$.

3. We say that the series $\sum c_p$ (or more precisely, $\sum_{\mathbf{N}^n} c_p$), where $c_p \in \mathbf{C}$
and $p \in \mathbf{N}^n$, is *convergent* and its sum is equal to $c \in \mathbf{C}$ (this is also written
as $c = \sum c_p$) if, for each $\varepsilon > 0$, there exists a finite subset $Z_0 \subset \mathbf{N}^n$ such
that for each finite subset $Z \supset Z_0$ we have $|\sum_Z c_p - c| < \varepsilon$ ([5]) . Then the
series $\sum_{|p| \geq k} c_p$ ([6]) is also convergent and $\sum c_p = \sum_{|p| < k} c_p + \sum_{|p| \geq k} c_p$.
Moreover, for any partition of $\mathbf{N}^n$ into finite sets $Z_\nu, \ \nu = 0, 1, \dots$, the series
$\sum_{\nu=0}^{\infty} \sum_{Z_\nu} c_p$ is convergent and its sum is equal to c.

A series $\sum c_p$ is said to be *absolutely convergent* if the series $\sum |c_p|$ is
convergent. (This definition is equivalent to that given in [6], §2.) Then the

---

([2]) Where $\mathbf{C}$ and Y are regarded as vector spaces over $\mathbf{R}$.

([3]) This condition is equivalent to the *Cauchy-Riemann equations*. (Since an $\mathbf{R}$–linear
endomorphism of $\mathbf{C} = \mathbf{R}^2$ given by the matrix $\begin{bmatrix} a & b \\ c & d \end{bmatrix}$ is C–linear if and only if $a = d$
and $b = -c$.)

([4]) $\mathbf{C}^n$ and Y are considered as real vector spaces.

([5]) The convergence of the series $\sum c_p$ can be characterized by Cauchy's condition: for
each $\varepsilon > 0$ there exists a finite subset $Z_0 \subset \mathbf{N}^n$ such that $|\sum_Z c_p| < \varepsilon$ for each finite
subset Z disjoint from Z_0.

([6]) That is, the series $\sum c_p'$, where $c_p' = 0$ for $|p| < k$, and $c_p' = c_p$ for $|p| \geq k$.

series $\sum c_p$ is convergent ([7]) .

4. The power series $\sum a_p z^p$ with coefficients $a_p \in \mathbf{C}$ (where $z = (z_1, \ldots$ $\ldots, z_n) \in \mathbf{C}^n$, $p = (p_1, \ldots, p_n) \in \mathbf{N}^n$, and $z^p = z_1^{p_1} \ldots z_n^{p_n}$) is said to be *convergent* if it is absolutely convergent in a neighbourhood of the origin in $\mathbf{C}^n$ (i.e., convergent for every z from the neighbourhood). If the convergence holds in a neighbourhood of the form $\{|z_\nu| \leq \delta_\nu, \nu = 1, \ldots, n\}$, then the series $\sum_{\nu=0}^{\infty} f_\nu(z)$ of forms $f_\nu(z) = \sum_{|p|=\nu} a_p z^p$ is absolutely and uniformly convergent in this neighbourhood and $\sum a_p z^p = \sum_{\nu=0}^{\infty} f_\nu(z)$. (Indeed, in this case the series $\sum |a_p| \delta^{|p|}$ is convergent and $|f_\nu(z)| \leq \sum_{|p|=\nu} |a_p| \delta^{|p|}$ in $\{|z_s| \leq \delta, \ s = 1, \ldots, n\}$.)

5. We say that a complex function f defined on a neighbourhood of a point $c \in \mathbf{C}^n$ is *holomorphic* at c if in a neighbourhood of zero $f(c + z)$ can be expressed as the sum of a convergent power series: $f(c + z) = \sum a_p z^p$. In other words, $f(z) = \sum a_p (z - c)^p$ in a neighbourhood of c. The last expression is called the power series expansion of f at c. The function f displays the above property if and only if f is continuous and has all first order partial derivatives in a neighbourhood of the point c. If this is the case, f has the partial derivatives of all orders in a neighbourhood of c and

$$a_p = \frac{1}{p!} \frac{\partial^{|p|} f}{\partial z^p}(c) \ \text{ for } \ p \in \mathbf{N}^n.$$

(Recall that $p! = p_1! \ldots p_n!$ and $\frac{\partial^{|p|} f}{\partial z^p} = \frac{\partial^{|p|} f}{\partial z_1^{p_1} \ldots \partial z_n^{p_n}}$.) The coefficients of the expansion of a holomorphic function at c are uniquely determined by the function. In particular, if a function f is holomorphic at c and $\frac{\partial^{|p|} f}{\partial z^p}(c) = 0$ for $p \in \mathbf{N}^n$, then $f = 0$ in a neighbourhood of c. (See [6], §§3 and 4.)

We say that a complex function defined on an open subset $G \subset \mathbf{C}^n$ is *holomorphic* if it is holomorphic at each point of G ([8]) . This is equivalent to the condition that f is continuous in G and holomorphic with respect to each variable separately. Then all complex partial derivatives $\frac{\partial^{|p|} f}{\partial z^p}$, $p \in \mathbf{N}^n$, exist and are holomorphic in G. Furthermore, if a polydisc $P = \{z : |z_\nu - c_\nu| \leq r, \nu = 1, \ldots, n\}$ with centre at c is contained in G, then $f(z) = \sum a_p (z - c)^p$

([7]) And vice versa. (It is enough to notice that the series $\sum \mathrm{Re}\, c_p$ and $\sum \mathrm{Im}\, c_p$ are convergent, and to use Cauchy's condition.) Therefore the fact that $\sum_{\nu=0}^{\infty} c_\nu$ is convergent (in the usual sense) does not imply that $\sum_N c_\nu$ is convergent.

([8]) In particular, polynomials on $\mathbf{C}^n$ are holomorphic.

in P, where $a_p = \frac{1}{p!} \frac{\partial^{|p|} f}{\partial z^p}(c)$, and the series is absolutely convergent in P([9]).
Moreover, the following Cauchy estimates hold

$$|a_p| \le \frac{M}{r^{|p|}} \quad \text{for} \quad p \in \mathbf{N}^n \quad \text{if} \quad |f(z)| \le M \text{ in } P .$$

Notice also that a function f is holomorphic at a point c if and only if it is holomorphic in a neighbourhood of c. Hence, if the power series $\sum a_p z^p$ is convergent, then the function $f(z) = \sum a_p(z - c)^p$ is holomorphic in a neighbourhood of the point c. (See [6], §§3 and 4.) ([10]) .

Assume that G is connected. If for some $c \in G$ the germ f_c is zero, i.e., $f = 0$ in a neighbourhood of c, then $f \equiv 0$. (Since the set $\{z \in G : f_z = 0\} = \{z \in G : \frac{\partial^{|p|} f}{\partial z^p}(z) = 0, \ p \in \mathbf{N}^n\}$ is open and closed in G.) Thus the condition that $f \not\equiv 0$ is the negation of the condition $f \equiv 0$ (see B.4.4).

A holomorphic function is a $\mathcal{C}^\infty$–function when regarded as a function of real variables. For if f is holomorphic, then

$$\frac{\partial^{|p+q|} f}{\partial x^p \partial y^q} = i^{|q|} \frac{\partial^{|p+q|} f}{\partial z^{p+q}} \quad \text{for} \quad p, q \in \mathbf{N} ,$$

where $z_\nu = x_\nu + i y_\nu$, $x = (x_1, \ldots, x_n)$, $y = (y_1, \ldots, y_n)$.

Therefore, for a function f which is holomorphic at 0, we have

$$f(0) = o(|z|^k) \iff \frac{\partial^{|p|} f}{\partial z^p}(0) = 0 \quad \text{for} \quad |p| \le k.$$

This yields the *Taylor formula*

$$f(z) = \sum_{|p| \le k} \frac{1}{p!} \frac{\partial^{|p|} f}{\partial z^p}(0) z^p + o(|z|^k).$$

It follows that if f is a polynomial of degree $\le k$ and $f(z) = o(|z|^k)$, then $f = 0$.

([9]) It is also uniformly convergent in P. (One can define uniform convergence in a similar manner to that in which ordinary convergence was defined in n° 3.)

([10]) Recall that a function $f : G \longrightarrow \mathbf{C}$ – where G is an open subset in $\mathbf{R}^n$ – is said to be *analytic* if for each point $c \in G$ it can be expanded into a power series $f(c + x) = \sum a_p x^p$ which is convergent in a neighbourhood of zero in $\mathbf{R}^n$. (Then also $\mathrm{Re}f$ and $\mathrm{Im}f$ are analytic. Note that absolute convergence of the series $\sum a_p z^p$ in a neighbourhood of 0 in $\mathbf{C}^n$ is equivalent to absolute convergence in a neighbourhood of 0 in $\mathbf{R}^n$.) This holds exactly when f can be extended to a function that is holomorphic in a neighbourhood of the set G in $\mathbf{C}^n$.

We have a *maximum principle* for holomorphic functions, i.e., if f is holomorphic at $c \in \mathbf{C}^n$ and the function $z \longrightarrow |f(z)|$ attains its maximum at c, then f is constant in a neighbourhood of c (see[6], §5).

6. A mapping $f = (f_1, \ldots f_m) : G \longrightarrow \mathbf{C}^m$, where $G \subset \mathbf{C}^n$ is an open set, is said to be *holomorphic* if and only if the functions $f_1, \ldots, f_m$ are holomorphic. Equivalently, f is holomorphic if and only if it is differentiable in G and for each $a \in G$ the differential $d_a f$ is **C**–linear. For any $a \in G$, the matrix $\left[\frac{\partial f_\mu}{\partial z_\nu}(a)\right]$ is then the complex Jacobi matrix of f at a, i.e., it is the matrix of the differential $d_a f$ regarded as a **C**–linear mapping:

$$d_a f(z) = \left(\sum_{\nu=1}^n \frac{\partial f_1}{\partial z_\nu}(a) z_\nu, \ldots, \sum_{\nu=1}^n \frac{\partial f_m}{\partial z_\nu}(a) z_\nu \right) \text{ for } z = (z_1, \ldots, z_n) \in \mathbf{C}^n.$$

Indeed, if f is differentiable at a, **C**–linearity of $d_a f$ is equivalent to the existence of $\frac{\partial f}{\partial z_\nu}(a)$, $\nu = 1, \ldots, n$, and then $d_a f(z) = \sum_{\nu=1}^n \frac{\partial f}{\partial z_\nu}(a) z_\nu$ (see n° 2).

Let X and Y be vector (or affine) spaces over **C**, and let $G \subset X$, $H \subset Y$ be open sets. A mapping $f : G \longrightarrow H$ is said to be *holomorphic* if it is differentiable in G and for each $a \in G$ the differential $d_a f$ is **C**–linear [11]. The definition yields the following properties. The composition of holomorphic mappings is holomorphic; the Cartesian product and the diagonal product of holomorphic mappings are holomorphic. A linear combination, a product or a quotient (with the denominator different from zero everywhere) of holomorphic functions is a holomorphic function. Obviously, a mapping $f : G \longrightarrow Y$ is holomorphic if and only if it is holomorphic in some (and hence in all) complex linear (or affine) coordinate systems in X and Y [12]. If a mapping $f : G \longrightarrow H$ is holomorphic and $f(U) \subset V$ for some open subsets $U \subset G$ and $V \subset Y$, then the mapping $f_U : U \longrightarrow V$ is also holomorphic. Holomorphic mappings are of class $\mathcal{C}^\infty$ [13].

---

[11] When $X = \mathbf{C}^n$, we say that f is a holomorphic mapping of n variables. If $Y = \mathbf{C} = H$, we say that f is a holomorphic function. Naturally, $f : G \longrightarrow H$ is holomorphic if and only if $f : G \longrightarrow Y$ is holomorphic.

[12] This is because **C**–linear mappings are holomorphic.

[13] A mapping $f = (f_1, \ldots, f_m) : G \longrightarrow \mathbf{R}^m$, where $G \subset \mathbf{R}^n$ is open, is said to be *analytic* if $f_1, \ldots, f_m$ are analytic (see n° 5, footnote [10]). A mapping defined on an open subset of a real vector space and with values in a real vector space is called analytic if it is analytic in some (and hence in all) linear coordinate systems in these vector spaces. Analytic mappings are of class $\mathcal{C}^\infty$. Every holomorphic mapping $f : G \longrightarrow Y$ is analytic if X and Y are regarded as vector spaces over **R**. (This follows from the fact that if $X = \mathbf{C}^n$ and $Y = \mathbf{C}$, then the function f is analytic because $f(z_1 + iw_1, \ldots, z_n + iw_n)$ is holomorphic as a function of the complex variables $z_1, w_1, \ldots, z_n, w_n$.)

Note also that if $f : G \longrightarrow Y$ is a holomorphic mapping ($G \subset X$ and X, Y are vector spaces), then also the mapping $G \ni z \longrightarrow d_z f \in L(X,Y)$ is holomorphic. (Indeed, if $X = \mathbf{C}^n$, $Y = \mathbf{C}^m$, $\mathbf{C}_m^n$ denotes the vector space of $(m \times n)$–matrices, and $f = (f_1, \ldots, f_m)$, then $G \ni z \longrightarrow \left[\frac{\partial f_i}{\partial z_j}(z) \right] \in \mathbf{C}_m^n$ is holomorphic.)

If f is a holomorphic mapping of n complex variables with values in a vector space, then $d_a f(z) = \sum_{\nu=1}^n \frac{\partial f}{\partial z_\nu}(a) z_\nu$.

If $f_1, \ldots, f_m$ are holomorphic functions in a neighbourhood of a point $a \in \mathbf{C}^n$ and h is a holomorphic mapping in a neighbourhood of $(f_1(a), \ldots, f_m(a))$ in $\mathbf{C}^m$ with values in a vector space then

$$\frac{\partial}{\partial z_\nu} \left(h \circ (f_1, \ldots, f_m) \right)(a) = \sum_{j=1}^n \frac{\partial h}{\partial w_j} \left(f_1(a), \ldots, f_m(a) \right) \frac{\partial f_j}{\partial z_\nu}(a), \quad \nu = 1, \ldots, n.$$

(This is true because, by putting $f = (f_1, \ldots, f_m)$, we get

$$d_a(h \circ f)(z) = \left((d_{f(a)} h) \circ (d_a f_1, \ldots, d_a f_m) \right)(z_1, \ldots, z_n) =$$

$$= \sum_j \frac{\partial h}{\partial w_j}(f(a)) \sum_\nu \frac{\partial f_j}{\partial z_\nu}(a) z_\nu.)$$

7. Let X be a complex vector space. Note that if f is a holomorphic function in a neighbourhood of zero in X, then, in some neighbourhood of zero, $f(x) = \sum_0^\infty f_\nu(x)$, where f_ν is a homogeneous polynomial of degree $\nu = 0, 1, 2, \ldots$, and the series is absolutely and uniformly convergent (see n° 4) (^{14}) . Moreover, the polynomials f_ν are uniquely determined by the function f (^{15}) .

Let $f_1, \ldots, f_p$ be functions that are holomorphic in a neighbourhood of 0 in X and whose germs do not vanish. Then, for every z belonging to an open dense subset of X, the germs $\left(t \longrightarrow f_i(tz) \right)_0$, $i = 1, \ldots, p$, are different from zero.

(^{14}) If f is holomorphic in X, then the equality and absolute convergence of the series hold in X.

(^{15}) For if $\sum_0^\infty f_\nu = 0$ in a neighbourhood of the origin in X, then, for each $z \in X$, we have $\sum_0^\infty f_\nu(z) t^\nu = \sum_0^\infty f_\nu(tz) = 0$ in a neighbourhood of zero in $\mathbf{C}$. Hence $f_\nu(z) = 0$ for $\nu = 0, 1, \ldots$.

In fact, it is enough to show this for a single function f. In a neighbourhood of zero, $f(z) = \sum_r^\infty f_\nu(z)$ and the series is absolutely convergent. Here f_ν is a homogeneous polynomial of degree ν for $\nu = r, r+1, \ldots$, and f_r is non-zero. Hence the set $\{f_r \neq 0\}$ is open and dense in X (see B. 5.2). Now, if $z \in \{f_r \neq 0\}$, then the function $t \longrightarrow f(tz) = \sum_r^\infty f_\nu(z)t^\nu$ is holomorphic in a neighbourhood of zero and it is never identically zero on any neighbourhood of zero in $\mathbf{C}$.

8. Let X and Y be complex vector spaces. All polynomial mappings on X into Y are holomorphic $(^{16})$. We have the following

LIOUVILLE THEOREM . *A holomorphic function f on X is a polynomial of degree $\leq k$ if and only if, for some $M > 0$ and for some norm in X, $|f(x)| \leq M(1 + |x|^k)$ in X. A holomorphic mapping $f : X \longrightarrow Y$ is polynomial if and only if (for some norms in X and Y) it holds that $|f(x)| \leq M(1 + |x|^k)$, for some M, $k > 0$.*

To prove the theorem, we may assume that $X = \mathbf{C}^n$ and $Y = \mathbf{C}$. We have $f(z) = \sum a_p z^p$ in $\mathbf{C}^n$. If the above inequality holds, then it follows from Cauchy's estimates (see n° 5) that $|a_p| \leq \frac{M(1+(nr)^k)}{r^{|p|}}$ for arbitrary $r > 0$. Therefore $a_p = 0$ when $|p| > k$.

COROLLARY . *A mapping $f : X \longrightarrow Y$ that is holomorphic and homogeneous must be polynomial $(^{17})$.*

9. Let G be an open subset of a complex vector space X.

If $X = \mathbf{C}^n$ and a sequence of holomorphic functions $\{f_\nu\}$ in G is *almost uniformly convergent in G* $(^{17a})$, then its limit f is holomorphic in G, and for each $p \in \mathbf{N}^n$ the sequence $\left\{\dfrac{\partial^{|p|} f_\nu}{\partial z^p}\right\}$ is almost uniformly convergent to $\dfrac{\partial^{|p|} f}{\partial z^p}$ in G.

If a sequence of holomorphic functions $\{f_\nu\}$ in G is locally bounded and convergent in G, then its limit is a holomorphic function.

$(^{16})$ It suffices to note that this is the case when $X = \mathbf{C}^n$ and $Y = \mathbf{C}^m$.

$(^{17})$ This can also be shown directly. When $Y = \mathbf{C}$, we have $f = \sum_0^\infty f_\nu$, where f_ν is a homogeneous polynomial of degree ν (see n° 7, footnote $(^{15})$). Hence, if f is homogeneous of degree r, then $f(x)t^r = \sum_0^\infty f_\nu(x)t^\nu$ for $x \in X$ and $t \in \mathbf{C}$. Hence $f(x) = f_r(x)$ for $x \in X$.

$(^{17a})$ i.e., it is uniformly convergent in a neighbourhood of every point of G, or, equivalently, it is uniformly convergent on every compact subset of G.

In fact, we may assume that $X = \mathbf{C}^n$. The sequence $\{f_\nu\}$ is equicontinuous in G (see [5], Chapter V, §17; it follows from the proof of Montel's theorem that if $|f_\nu(z)| \leq M$ in the polydisc $\{|z_\nu - a_\nu| < 3r : \nu = 1, \ldots, n\}$, then $|f_\nu(z'') - f_\nu(z')| \leq \frac{M}{r} \sum_1^n |z_\nu'' - z_\nu'|$ for z'', z' belonging to the polydisc $\{|z_\nu - a_\nu| < r : \nu = 1, \ldots, n\}$). Thus the sequence is almost uniformly convergent, and hence its limit is holomorphic.

Observe that the above argument contains a proof of Montel's theorem for several complex variables:

A locally bounded family of holomorphic functions in G is equicontinuous. Therefore each sequence of functions from this family contains a subsequence that is almost uniformly convergent to a holomorphic function in G.

Let L be a rectifiable arc in $\mathbf{C}$, and let $f(z, w)$ be a locally bounded complex function on $G \times L$ which is holomorphic with respect to z and continuous with respect to w (e.g., continuous in $G \times L$ and holomorphic with respect to z). Then the function

$$g(z) = \int_L f(z, w)dw \text{ for } z \in G$$

is holomorphic.

Indeed, $g(z)$ is the limit of the sequence of Riemann sums, which – when regarded as functions of z – form a locally bounded sequence of holomorphic functions in G.

10. Let V be an open convex neighbourhood of the origin in $\mathbf{C}^n$, and let G be an open subset of a complex vector space X. We have the following

HADAMARD LEMMA . *Let $f(z, w)$ be a holomorphic function on $V \times G$. If $f(0, w) = 0$ in G, then $f(z, w) = \sum_{\nu=1}^n z_\nu f_\nu(z, w)$ in $V \times G$, where f_ν are holomorphic functions on $V \times G$* [18] .

Indeed, if $z \in V$ and $w \in G$, the function $\varphi_{zw}(t) = f(tz, w)$ is holomorphic in a neighbourhood of the line segment $L = [0, 1] \subset \mathbf{C}$. Thus we get

$$f(z, w) = \varphi_{zw}(1) = \int_L \varphi_{zw}'(t)dt = \sum_{\nu=1}^n z_\nu f_\nu(z, w) ,$$

---

[18] Then $f_\nu(0, w) = \frac{\partial f}{\partial z_\nu}(0, w)$ in G, $\nu = 1, \ldots, n$. (It suffices to differentiate the equality with respect to z_ν.)

where the functions $f_\nu(z,w) = \int_L \frac{\partial f}{\partial z_\nu}(tz,w)dt$ are holomorphic in $V \times G$.

In particular, if f is a holomorphic function in a neighbourhood of $0 \in \mathbf{C}^n$ and $f(0) = 0$, then $f(z) = \sum_{\nu=1}^n f_\nu(z)z_\nu$ in a neighbourhood of zero, with some coefficients f_ν that are holomorphic in a neighbourhood of zero. If, in addition, $f(z) = o(|z|^{r+1})$, where $r \geq 0$, then we can require that $f_\nu(z) = o(|z|^r)$. (This follows from the above formulae for f_ν, since $\frac{\partial f}{\partial z_\nu}(z) = o(|z|^r)$.) By induction we can derive, for $k \geq 1$, the following corollary:

A holomorphic function f in a neighbourhood of $0 \in \mathbf{C}^n$ satisfies $f(z) = o(|z|^{k-1})$ if and only if $f(z) = \sum_{|p|=k} f_p(z)z^p$ in a neighbourhood of zero, for some f_p that are holomorphic in a neighbourhood of zero $(^{19})$.

11. Let Z be a closed subset of a topological space T. We say that a function defined on $T \setminus Z$ with values in a vector space (over $\mathbf{C}$ or $\mathbf{R}$) is *locally bounded near* Z if each point $z \in Z$ has a neighbourhood U such that the restriction $f_{U\setminus Z}$ is bounded. In the case when f is continuous and T is locally compact, it means precisely that f is *locally bounded on* T, i.e., each point of T has a neighbourhood U such that f_U (i.e., $f_{U\setminus Z}$) is bounded.

Let X be a complex vector space.

RIEMANN'S LEMMA . *Let Z be a closed subset of the open set $G \subset X \times \mathbf{C}$ such that for every $z \in X$ the set $Z_z = \{w : (z,w) \in Z\}$ is a discrete subset of $G_z = \{w : (z,w) \in G\}$. Then every holomorphic function f in $G \setminus Z$ that is locally bounded near Z (or – more generally – such that for each $z \in X$ the function $w \longrightarrow f(z,w)$ is locally bounded near Z_z) has a (unique) holomorphic extension in G.*

Indeed, in view of Riemann's theorem, for any $z \in X$ the function $G_z \setminus Z_z \ni w \longrightarrow f(z,w) \in \mathbf{C}$ has an extension $\tilde{f}_z$ which is holomorphic in G_z. Then the function $\tilde{f}(z,w) = \tilde{f}_z(w)$ is defined in G, it is holomorphic with respect to w, and extends the function f. It is enough to show that it is holomorphic at an arbitrary point $(z_0,w_0) \in Z$. Now, if $| \cdot |$ is a norm in X, there exists $\varepsilon > 0$ such that $\{|z - z_0| < \varepsilon, |w - w_0| \leq \varepsilon\} \subset G$ and $\{z = z_0, |w - w_0| = \varepsilon\} \subset G \setminus Z$. Since Z is closed, we have $\{|z - z_0| <$

$(^{19})$ This implies Taylor's formula with Hadamard's remainder for any holomorphic function f in a neighbourhood of 0 in $\mathbf{C}^n$:

$$f(z) = \sum_{|p| \leq k} \frac{1}{p!} \frac{\partial^p f}{\partial z^p}(0)z^p + \sum_{|p|=k+1} f_p(z)z^p$$

in a neighbourhood of zero for some f_p which are holomorphic in the neighbourhood .

$\delta, |w - w_0| = \varepsilon\} \subset G \setminus Z$ for some δ such that $0 < \delta < \varepsilon$. Therefore

$$\tilde{f}(z, w) = \frac{1}{2\pi i} \int\limits_{|s-w_0|=\varepsilon} \frac{f(z, s)}{s - w} ds$$

is holomorphic in the set $\{|z - z_0| < \delta, \ |w - w_0| < \varepsilon\}$ (see n° 9).

HARTOGS' LEMMA . *Let B be an open connected subset of X, and let V be a non-empty open subset of B. Let $0 \leq r < R$, and let*

$$G = (B \times \{w \in \mathbf{C} : \ r < |w| < R\}) \cup (V \times \{w \in \mathbf{C} : \ |w| < R\})$$

and

$$H = B \times \{w \in \mathbf{C} : \ |w| < R\}.$$

Then every holomorphic function on G has a (unique) holomorphic extension to H.

Indeed, let f be a holomorphic function on G. Let $r < \varrho < R$. The function $\tilde{f}(z, w) = \frac{1}{2\pi i} \int_{|s|=\varrho} \frac{f(z,w)}{s-w} ds$ is holomorphic in $H_0 = B \times \{|w| < \varrho\}$ (see n° 9). It coincides with the function f in the set $V \times \{|w| < \varrho\}$ and hence also in $H_0 \cap G = (B \times \{r < |w| < \varrho\}) \cup (V \times \{|w| < \varrho\})$, because the latter is connected (see n° 5). Thus $f \cup \tilde{f}$ is a holomorphic function in $H_0 \cup G = H$.

COROLLARY . *Let $G \subset X$ be an open set, $k \geq 0$, $U = \{z \in \mathbf{C}^k : \ |z_\nu| < R\}$, $E = \{z \in \mathbf{C}^k : \ |z_\nu| \leq r_\nu\}$, where $0 \leq r_\nu < R_\nu \ (\nu = 1, \ldots, k)$. Then every function that is holomorphic in $G \times (U \setminus E)$ has a (unique) holomorphic extension to $G \times U$.*

It is enough to consider the case when G is connected, and then to note that

$$U \setminus E = \left(U_* \times \{r_k < |t| < R_k\}\right) \cup \left((U_* \setminus E_*) \times \{|t| < R_k\}\right),$$

where

$$U_* = \{u \in \mathbf{C}^{k-1} : \ |u_\nu| < R_\nu\} \text{ and } E_* = \{u \in \mathbf{C}^{k-1} : \ |u_\nu| \leq r_\nu\}.$$

12. Let X and Y be complex vector (or affine) spaces. A *biholomorphic* mapping between open subsets of X and Y, respectively, is a bijection f of X and Y such that f and f^{-1} are holomorphic. Then we must have

$\dim X = \dim Y$ (provided that $f \neq \emptyset$). Obviously, the inverse of a biholomorphic mapping and the composition of biholomorphic mappings are biholomorphic. A holomorphic diffeomorphism is biholomorphic $(^{20})$. If f is a holomorphic mapping from a neighbourhood of a point $a \in X$ to Y, such that the differential $d_a f$ is an isomorphism, then, for some open neighbourhood U of a, the set $f(U)$ is open and the restriction $f_U : U \longrightarrow f(U)$ is biholomorphic.

Note that the mapping $L_o(X,Y) \ni \varphi \longrightarrow \varphi^{-1} \in L_o(X,Y)$ (see B. 5. 2) is biholomorphic. Indeed, it is sufficient to use the fact that linear isomorphisms and the mapping $A \longrightarrow A^{-1}$ of the set of non-singular matrices onto itself are biholomorphic.

13. Let X, Y, Z be complex vector spaces with $\dim Y = \dim Z$. We have the following

IMPLICIT FUNCTION THEOREM . *Let F be a holomorphic mapping on a neighbourhood of a point $(a, b) \in X \times Y$ with values in Z, such that $F(a, b) = 0$ and the differential $''d_{ab}F$ of the mapping $\eta \longrightarrow F(a, \eta)$ at the point b is an isomorphism. Then we have equality of the germs of sets*

$$\{F(x, y) = 0\}_{(a,b)} = \{y = \varphi(x)\}_{(a,b)}$$

for some holomorphic mapping φ defined on a neighbourhood of the point a with values in Y. Furthermore,

$$F\big(x, \varphi(x)\big) = 0 \text{ in a neighbourhood of } a, \ \varphi(a) = b,$$

and the germ of a continuous function satisfying these conditions is uniquely determined.

Moreover, if F is a holomorphic Z-valued mapping in a neighbourhood of the graph of a continuous Y-valued mapping φ defined on an open set $G \subset X$, then the conditions

$$F\big(x, \varphi(x)\big) = 0 \text{ and } ''d_{x\varphi(x)}F \text{ is an isomorphism for } x \in G$$

imply that the mapping φ is holomorphic.

To see this, denote by $'d_{cd}F$ the differential of the mapping $\zeta \longrightarrow F(\zeta, d)$ at the point c. Now it is enough to apply the real case of the implicit function theorem and to note that the differential $d_x \varphi = -(''d_{x\varphi(x)}F)^{-1} \circ {}'d_{x\varphi(x)}F$ is **C**–linear (for each x in a neighbourhood of a or for each x in G, respectively).

$(^{20})$ This is so, because the differentials of the inverse mapping are **C**–linear as inverses of the differentials of the mapping.

§2. The Weierstrass preparation theorem

1. Consider a monic polynomial P in $w \in \mathbf{C}$ whose coefficients $a_1, \ldots a_n$ are holomorphic functions in an open subset B of a complex vector space, i.e., the function

$$P(z, w) = w^n + a_1(z)w^{n-1} + \ldots + a_n(z).$$

The function P is holomorphic in $B \times \mathbf{C}$. The function $D(z) = D_n\big(a_1(z), \ldots$
$\ldots, a_n(z)\big)$ is holomorphic in B and is called the *discriminant* of the polynomial P. It is defined by

$$D(z) = \prod_{i<j}(w_i - w_j)^2 = (-1)^{\binom{n}{2}} \prod_{j=1}^{n} \frac{\partial P}{\partial w}(z, w_j)$$

for $z \in B$, where $w_1, \ldots, w_n$ is the complete sequence of roots of the polynomial $w \longrightarrow P(z, w)$ (see A. 4.3).

Set $Z = \{(z, w) \in B \times \mathbf{C} : P(z, w) = 0\}$. The natural projection $Z \longrightarrow B$ is open. (This follows from the theorem on continuity of roots; see B. 5.3). Therefore, if $D \not\equiv 0$, the set $Z \cap \{D \neq 0\}$ is dense in Z (see B. 2.1).

2. Now, let the set B be a neighbourhood of zero in $\mathbf{C}^n$. Then the function $P(z, w)$ is said to be a *distinguished polynomial* with respect to w of degree n if $a_\nu(0) = 0$, $\nu = 1, \ldots, n$. This means precisely that the polynomial $w \longrightarrow P(0, w)$ does not have zeros different from 0. Then (see B. 5.3) for each $\varepsilon > 0$ there exists $\delta > 0$ such that

$$\big(P(z, w) = 0, \ |z| < \delta\big) \Longrightarrow |w| < \varepsilon.$$

If $P(z, w)$ and $Q(z, w)$ are monic polynomials with holomorphic coefficients defined in a neighbourhood of the origin in $\mathbf{C}^m$, then PQ is a distinguished polynomial if and only if P and Q are distinguished polynomials.

3. We say that a holomorphic function $f(z, w)$ in a neighbourhood of zero in $\mathbf{C}^m \times \mathbf{C}$ is *w–regular* if $f(0, w) \not\equiv 0$ in a neighbourhood of zero in $\mathbf{C}$, i.e., if $\big(w \longrightarrow f(0, w)\big)_0$ is a non-zero germ. Then, for $k \in \mathbf{N}$, we have $f(0, w) = c(w)w^k$ in a neighbourhood of zero, where c is a holomorphic function that never vanishes. We say that the function f is *w–regular of order k*.

If the functions $f_1, \ldots, f_p$ are holomorphic in a neighbourhood of zero in a complex m–dimensional vector space X and such that their germs at 0 are non-zero, then for every linear coordinate system from an open dense subset of $L_0(X, \mathbf{C})$ the functions are z_m–regular $(^{21})$.

Indeed, there exists an open dense subset G of X such that for any $z \in G$ the germs $\big(t \longrightarrow f_i(tz)\big)$, $i = 1, \ldots p$, are non-zero (see 1.7). The mapping $\Phi : L(\mathbf{C}^m, X) \ni \psi \longrightarrow \psi(0, \ldots, 0, 1) \in X$ is an epimorphism, and so it is open and continuous (see B. 5.2). Therefore (see B. 2.2) the set $\Phi^{-1}(G)$ is open and dense in $L(\mathbf{C}^m, X)$, and hence the set $H = \Phi^{-1}(G) \cap L_0(\mathbf{C}^m, X)$ is open and dense in $L_0(\mathbf{C}^m, X)$. Now, if $\psi \in H$, then $\psi(0, \ldots, 0, 1) \in G$. Thus the germs of the functions $t \longrightarrow (f_i \circ \psi)(0, \ldots, 0, t) + f_i\big(t\psi(0, \ldots, 0, 1)\big)$, $i = 1, \ldots, p$, are non-zero $(^{22})$.

4. We have the following

WEIERSTRASS PREPARATION THEOREM (CLASSICAL VERSION). *If a function $f(z, w)$ is holomorphic in a neighbourhood of zero in $\mathbf{C}^m \times \mathbf{C}$ and w–regular of order k, then*

$$f = hP$$

in a neighbourhood of zero, where P is a distinguished polynomial in w and h is a non-vanishing holomorphic function in a neighbourhood of zero $(^{23})$. The germs P_0 and h_0 are uniquely determined by f. The degree of P is k.

PROOF . The uniqueness of the germs can be shown as follows. If P', h' is another such pair, we take a neighbourhood of zero $U = \{|z| < \delta, \ |w| < \varepsilon\}$ such that $hP = h'P'$, $h \neq 0$, and $h' \neq 0$ in U. Moreover, we require that $P(z, w) = 0$, $|z| < \delta \Longrightarrow |w| < \varepsilon$, as well as $P'(z, w) = 0$, $|z| < \delta \Longrightarrow |w| < \varepsilon$. Then, for $|z| < \delta$, the monic polynomials $w \longrightarrow P(z, w)$ and $w \longrightarrow P'(z, w)$ have the same zeros (with the same multiplicities), that is, the same complete sequence of roots, and hence they are equal. Therefore $P = P'$ in U, and so $h = h'$ in U (because $\{P \neq 0\}$ is dense in U).

To prove the existence part of the theorem, take $\varepsilon > 0$ such that the function f is holomorphic in a neighbourhood of the set $\{|z| \leq \varepsilon, \ |w| \leq \varepsilon\}$ and the function $w \longrightarrow f(0, w)$ has no zeros in the set $\{0 < |w| \leq \varepsilon\}$. Since k is the multiplicity of the zero of the latter function at 0, Rouché's theorem (see [5], Chapter VI, §8) implies that there is $0 < \delta < \varepsilon$ such that $f(z, w) \neq 0$ on the circle $\{|w| = r\}$ for $|z| < \delta$, and the function $w \longrightarrow f(z, w)$ has

---

$(^{21})$ See A. 3.1, footnote $(^{34})$.

$(^{22})$ This completes the proof, as $L_0(X, \mathbf{C}^m) \ni \varphi \longrightarrow \varphi^{-1} \in L_0(\mathbf{C}^m, X)$ is a homeomorphism (see B. 5.2).

$(^{23})$ Therefore, in a neighbourhood of the origin, the sets $\{f = 0\}$ and $\{P = 0\}$ coincide.

exactly k zeros $w_1(z), \ldots, w_k(z)$ (counted with their multiplicities) in the disc $\{|w| < \varepsilon\}$.

For $|z| < \delta$ and $w \in \mathbf{C}$, set

$$P(z, w) = \big(w - w_1(z)\big) \ldots \big(w - w_k(z)\big) = w^k + a_1(z) w^{k-1} + \ldots + a_k(z) ,$$

where $a_\nu(z) = \sigma_\nu\big(w_1(z), \ldots, w_k(z)\big)$, $\nu = 1, \ldots, k$ (see A. 4.1). It is enough to show that the function P is a distinguished polynomial, i.e., that the functions $a_1, \ldots, a_k$ are holomorphic in $|z| < \delta$ (clearly, the functions $a_1, \ldots, a_k$ vanish at 0). Indeed, this is sufficient because for $|z| < \delta$ the holomorphic functions $w \longrightarrow f(z, w)$ and $w \longrightarrow P(z, w)$ have the same zeros (with the same multiplicities) in $\{|w| < \varepsilon\}$. Consequently, in view of Riemann's lemma (see 1.11), the function $\frac{f(z, w)}{P(z, w)}$ on $U \setminus \{P = 0\}$ (where $U = \{|z| < \delta, \ |w| < \delta\}$) has a holomorphic extension h in U which is different from zero in U, and $f = hP$ in U (since $\{P \neq 0\}$ is dense in U).

Now, the functions $b_1, \ldots, b_k$ defined for $|z| < \delta$ by the formulae

$$b_\nu(z) = \frac{1}{2\pi i} \int\limits_{|w| = \varepsilon} w^\nu \frac{\frac{\partial f}{\partial w}(z, w)}{f(z, w)} dw = w_1(z)^\nu + \ldots + w_k(z)^\nu, \ \nu = 1, \ldots, k,$$

are holomorphic (see 1.9); the second equality follows from the residue theorem for the logarithmic derivative (see [5], Chapter VI, §6.3). Since $\sigma_\nu(v) = Q_\nu\big(s_1(v), \ldots, s_k(v)\big)$ in $\mathbf{C}^k$, $\nu = 1, \ldots, k$, where Q_ν are polynomials and $s_\nu(v) = v_1^\nu + \ldots + v_k^\nu$ (see A. 4.2), we have $a_\nu(z) = Q_\nu\big(b_1(z), \ldots, b_k(z)\big)$ for $|z| < \delta$, $\nu = 1, \ldots, k$. Therefore the functions a_ν are holomorphic in $\{|z| < \delta\}$.

5. Now we are going to prove

THE WEIERSTRASS PREPARATION THEOREM (DIVISION VERSION). *Let $f(z, w)$ be a holomorphic function in a neighbourhood of the origin in $\mathbf{C}^m \times \mathbf{C}$ which is w-regular of order k. Then, for every function $g(z, w)$ that is holomorphic in a neighbourhood of zero,*

$$g = Qf + R$$

in a neighbourhood of zero, where Q is a holomorphic function in a neighbourhood of zero and R is a polynomial in w of degree $< k$ with coefficients which are holomorphic in a neighbourhood of zero in $\mathbf{C}^m$. The germs Q_0 and R_0 are uniquely determined by the functions f and g [24].

---

[24] By taking $g = w^k$ we can see that the classical version of the Weierstrass theorem follows.

PROOF . In view of the classical version of the preparation theorem, we may assume that f is a distinguished polynomial of degree k. There exist $\delta, \varepsilon > 0$ such that the functions f and g are holomorphic in a neighbourhood of the set $\{|z| \leq \delta, \; |w| \leq \varepsilon\}$ and $f(z,w) = 0, \; |z| < \delta \Longrightarrow |w| < \varepsilon$. Then the function

$$Q(z,w) = \frac{1}{2\pi i} \int\limits_{|s|=\varepsilon} \frac{g(z,s)}{f(z,s)} \frac{1}{s-w} ds$$

is holomorphic in the set $U = \{|z| < \delta, \; |w| < \varepsilon\}$ (see 1.9). Since

$$g(z,w) = \frac{1}{2\pi i} \int\limits_{|s|=\varepsilon} \frac{g(z,s)}{s-w} ds$$

in U, we have

$$(g - Qf)(z,w) = \frac{1}{2\pi i} \int\limits_{|s|=\varepsilon} \frac{g(z,s)}{f(z,s)} \frac{f(z,s) - f(z,w)}{s-w} ds$$

in U. On the other hand, for $s \neq w$ and $|z| < \delta$ the function $\frac{f(z,s)-f(z,w)}{s-w}$ is a polynomial in w of degree $< k$ and has coefficients that are holomorphic in the set $\{|z| < \delta, \; s \in \mathbf{C}\}$. Thus it follows that, in U, the function $g - Qf$ is a polynomial in w of degree $< k$ and has holomorphic coefficients in $\{|z| < \delta\}$.

To show the uniqueness, suppose that $Qf + R = 0$ in a neighbourhood of the origin. Then the same is true in some $U = \{|z| < \delta, \; |w| < \varepsilon\}$ (where $\delta, \varepsilon > 0$), the functions Q, f are holomorphic there, and the coefficients of the polynomial R are holomorphic in the set $\{|z| < \delta\}$. In view of the classical version of the preparation theorem, ε and δ can be chosen in such a way that, for $|z| < \delta$, the function $w \longrightarrow f(z,w)$ has exactly k zeros (counted with their multiplicities) in the disc $\{|w| < \varepsilon\}$. Then the polynomial $w \longrightarrow R(z,w)$ (of degree $< k$) has at least k zeros. Therefore $R = 0$ in U, and hence $Q = 0$ in U.

§3. Complex manifolds

In this section we will define complex manifolds, holomorphic mappings between manifolds, and state their basic properties. The proofs of these properties are trivial and almost all of them are the same as in the case of *smooth* (i.e., $\mathcal{C}^\infty$–differentiable) or *(real) analytic* (^{25}) *manifolds*. Hence we will only give some necessary hints concerning proofs.

Let $n \in \mathbf{N}$.

1. A *complex manifold of dimension n* is a topological Hausdorff space M furnished with a *complex atlas* (*modelled on* $\mathbf{C}^n$), i.e., with a family of homeomorphisms $\varphi_\iota : G_\iota \longrightarrow V_\iota$ between open sets $G_\iota \subset M$, $V_\iota \subset \mathbf{C}^n$ such that

(∗) the mappings $\varphi_\kappa \circ \varphi_\iota^{-1} : \varphi_\iota(G_\iota \cap G_\kappa) \longrightarrow \varphi_\kappa(G_\iota \cap G_\kappa)$ are holomorphic (and hence biholomorphic), and $M = \bigcup G_\iota$.

Two complex atlases (modelled on $\mathbf{C}^n$) are said to be *equivalent* if their union is also an atlas. It is assumed that equivalent atlases define the same structure of a complex manifold on M, and hence the subsequent definitions should be independent of the choice of equivalent atlases (^{26}) . Each atlas which is equivalent to the atlas $\{\varphi_\iota\}$ is said to be an *atlas of the manifold* M. The dimension of the manifold M will be denoted by $\dim M$.

Obviously, a complex atlas defines the structure of a smooth manifold (^{27}) on M of dimension $2n$ (see 1.1). A *chart* or a *(local) coordinate system* of the manifold M is a homeomorphism φ of an open set in M onto an open subset of $\mathbf{C}^n$, such that $\{\varphi_\iota\} \cup \{\varphi\}$ is an atlas. In particular, the φ_ι are charts. (Notice that the restriction of a chart to an open set, as well as the composition of a chart with a biholomorphic mapping between open sets in $\mathbf{C}^n$, is also a chart.) The atlases of the manifold M are precisely the families of charts on M whose domains cover M. Let $a \in M$. A chart φ whose domain contains a point a such that $\varphi(a) = 0$ is called a *chart* or a *(local) coordinate system for the manifold M at the point a*.

2. The definition of an atlas can be made slightly more general by assuming that each V_ι is an open subset of a complex vector (or affine) space

---

(^{25}) See 1.6, footnote (^{13}).

(^{26}) This means, in fact, that the structure of a complex manifold on the space M should be regarded as a maximal class of equivalent atlases. Indeed, any given atlas is equivalent to the unique *maximal atlas* which is the union of all atlases equivalent to the given one, and consists of all charts on M (see below). A maximal atlas A can be characterized by the condition: "Each homeomorphism ψ between open sets in M and $\mathbf{C}^n$, respectively, such that $\psi \circ \varphi^{-1}$ is a biholomorphic for each $\varphi \in \mathcal{A}$, belongs to A." Thus we can define a complex manifold as a Hausdorff space with a maximal complex atlas and therefore avoid the notion of equivalent atlases in the definition.

(^{27}) Also a real analytic one. (See 1.6, footnote (^{13}).)

X_ι of dimension n. We will, in some cases, call such an atlas a *generalized atlas* (modelled on X when all X_ι are the same space X). It defines on M the structure of a complex manifold given by the atlas $\{\chi_\iota \circ \varphi_\iota\}$ modelled on $\mathbf{C}^n$, where $\chi_\iota : X_\iota \longrightarrow \mathbf{C}^n$ are isomorphisms (the structure is independent of the choice of the isomorphisms) ([28]). As before, two atlases are said to be equivalent if their union is an atlas. Two equivalent atlases define the same complex manifold structure on M. Every generalized atlas that is equivalent to the atlas $\{\varphi_\iota\}$ is called a generalized atlas of the manifold M.

If X is a complex vector (or affine) space of dimension n, then by a (generalized) chart on M modelled on X we mean a homeomorphism φ from an open set in M onto an open set in X, such that $\{\varphi_\iota\} \cup \{\varphi\}$ is a generalized atlas ([29]). In particular, φ_ι is a chart modelled on X_ι. Atlases (generalized atlases) on M are precisely the collections of charts (generalized charts) on M whose domains cover M.

Let φ be a chart with domain G, modelled on a normed affine space X ([30]). Then we have a metric on G (transferred from X by φ): $\varrho_\varphi(x,y) = |\varphi(y) - \varphi(x)|$. If φ and ψ are charts whose domains contain a point a and which are modelled on a normed affine space, then there exist a neighbourhood U of the point a and a constant $M > 0$ such that $\varrho_\varphi(x,y) \leq M\varrho_\psi(x,y)$ and $\varrho_\psi(x,y) \leq M\varrho_\varphi(x,y)$ for all $x, y \in U$.

3. An *inverse atlas* (generalized, modelled on X) is a family of mappings $\{\psi_\iota\}$ such that $\{\psi_\iota^{-1}\}$ is an atlas (generalized, modelled on X). In other words, it is a family of homeomorphisms $\psi_\iota : V_\iota \longrightarrow G_\iota$ of open sets $V_\iota \subset \mathbf{C}^n$ (or $V_\iota \subset X_\iota$, or $V_\iota \subset X$) and $G_\iota \subset M$, such that $M = \bigcup G_\iota$ and the mappings

$$\psi_\kappa^{-1} \circ \psi_\iota : \psi_\iota^{-1}(G_\iota \cap G_\kappa) \longrightarrow \psi_\kappa^{-1}(G_\iota \cap G_\kappa)$$

are holomorphic. Naturally, it defines – via the atlas $\{\psi_\iota^{-1}\}$ – the structure of a complex manifold on M. A mapping ψ such that ψ^{-1} is a chart (modelled on X) is called an *inverse chart* (modelled on X). Thus a family of inverse charts on a manifold M whose ranges cover M is an inverse atlas of the manifold M.

([28]) More generally, it can be assumed that V_ι are complex n–dimensional manifolds (see n° 8 and 17 for definitions of holomorphic and biholomorphic mappings between manifolds). Then there exists a unique structure of a complex manifold on M such that the φ_ι's are biholomorphic. (The collection $\{\chi_{\iota\kappa} \circ \varphi_\iota\}$, where $\{\chi_{\iota\kappa}\}$ is an atlas on V_ι for each ι, is an atlas of the manifold M.)

([29]) Thus the charts defined in n° 1 are exactly charts modelled on $\mathbf{C}^n$.

([30]) By a *normed affine space* we mean an affine space X whose underlying vector space is normed. Then $X^2 \ni (x, y) \longrightarrow |y - x| \in \mathbf{R}$ is a metric on X.

4. In what follows, we will be assuming that all complex manifolds have countable atlases. This means precisely that M has a countable basis for its topology.

Any complex manifold is locally compact, and hence it is a Baire space. It is also locally connected, and so its connected components are open.

Notice that an atlas of a manifold determines the topology of the manifold. Namely, if M is a set which is not endowed with topology, we can define an atlas as a family of bijections $\varphi_\iota : G_\iota \longrightarrow V_\iota$ from subsets G_ι of M onto open sets $V_\iota \subset X_\iota$ (where the X_ι are n–dimensional complex vector spaces), such that the conditions $(*)$ are fulfilled. For such an atlas, there exists a unique topology in M for which the sets G_ι are open and the mappings φ_ι are homeomorphisms. The open sets in this topology are precisely the sets $H \subset M$ such that $\varphi_\iota(H)$ are open for all ι. It remains to impose the condition that M is a Hausdorff space. This is equivalent to the condition that the graph of the mapping $\varphi_\kappa \circ \varphi_\iota^{-1}$ is closed in $V_\iota \times V_\kappa$ for each ι and κ ([31]) . Then $\{\varphi_\iota\}$ is an atlas in the sense of the former definition and it defines on M the structure of a complex manifold. This gives an alternative method of defining a complex manifold ([32]) .

5. Any n–dimensional complex vector (or affine) space X is, naturally, a complex manifold of dimension n. Its atlas is given by any linear (or affine) coordinate system. Every such system is a chart for the manifold. The identity mapping on $\mathbf{C}^n$ is called the *canonical coordinate system* in $\mathbf{C}^n$.

6. For the Cartesian product $M \times N$ of two manifolds M and N of dimensions m and n, respectively, (with the product topology) the structure of a complex manifold (of dimension $m + n$) is given by the atlas $\{\varphi_\iota \times \psi_\kappa\}$, where $\{\varphi_\iota\}$ and $\{\psi_\kappa\}$ are atlases for M and N, respectively. (The manifold structure on $M \times N$ is independent of the choice of atlases on M and N.)

7. Let M be an n–dimensional complex manifold. A set $Z \subset M$ is said to be a *(complex) submanifold* of M if there exists $k \geq 0$ such that for each $z \in Z$ there is a chart $\varphi = \varphi_z$ whose domain contains z and for which $\varphi(Z)$ is an open subset of a k–dimensional subspace $L = L_z$ of $\mathbf{C}^n$. (Note that $\varphi : G \longrightarrow V$ can always be taken as a coordinate system at z ([33]) . Moreover, the chart can be chosen so that $\varphi(Z \cap G) = L \cap V$ ([34])

([31]) This follows from the fact that M is a Hausdorff space if and only if the diagonal $\{(z, w) \in M^2 : z = w\}$ is closed in M^2.

([32]) See n° 2, footnote ([26]).

([33]) By composing φ with a suitable translation.

([34]) The sets V and G can be made smaller, if necessary.

and $V = \mathbf{C}^k \times 0$ $(^{35})$.) Then $\{\varphi_z|_Z\}$ is a complex (generalized) atlas for Z, and it defines on Z the structure of a complex manifold of dimension k (which is independent of the choice of the charts φ_z). It is called the *induced structure* (of dimension k) on Z. The set Z with the induced structure is regarded as a manifold of dimension k and is also called a (complex k–dimensional) *submanifold*.

Of course, we could have an equivalent definition of a submanifold by taking generalized charts $\varphi = \varphi_z$ and assuming that the set $\varphi(Z)$ is an open subset of a k–dimensional affine subspace. Then $\{\varphi_z|_Z\}$ would be a (generalized) atlas for Z.

If $Z \neq \emptyset$, the number k is uniquely determined (see 1.12 and 1.6). It is also called the *dimension of the submanifold* Z regarded as a subset, and denoted by $\dim Z$. The number $n - k$ is called the *codimension* of Z and is denoted by $\operatorname{codim} Z$. In addition, we assume that $\dim \emptyset = -\infty$, i.e., the submanifold $\emptyset$ (regarded as a subset) is of dimension $-\infty$. (When regarded as a manifold, $\emptyset$ is a submanifold of every dimension $k \leq n$, and hence the above definition of dimension differs only for the empty submanifold.) Clearly, $\dim Z \leq \dim M$. Submanifolds of dimension n of M, when regarded as subsets, coincide with non-empty open sets in M. When regarded as manifolds, they coincide with open sets in M (endowed with the induced n–dimensional structure). In particular, connected components of the manifold M are n–dimensional submanifolds of M (see n° 4).

Notice that every submanifold (of a manifold M) is a locally closed set, and hence it is open in its closure (see B. 1). Submanifolds are F_σ–sets.

We say that a set $Z \subset M$ is a *submanifold* (of dimension k and codimension $n - k$) *at a point* $a \in Z$ if the germ Z_a is the germ of a k–dimensional submanifold containing a. In other words, it means that some neighbourhood of the point a in the set Z is a k–dimensional submanifold. A manifold $Z \subset M$ is a k–dimensional submanifold if and only if it is a k–dimensional submanifold at each of its points. Therefore the non-empty open subsets of a k–dimensional submanifold (in particular, its components) are submanifolds of dimension k. If a set $Z \subset M$ is the union of a family of open subsets of Z, each of which, when regarded as a manifold, is a k–dimensional submanifold, then Z regarded as a manifold is also a k–dimensional submanifold.

If Z and W are submanifolds of the manifolds M and N, respectively, then $Z \times W$ is a submanifold of the manifold $M \times N$. Moreover, the manifold structure induced on $Z \times W$ coincides with the structure of the Cartesian product of the manifolds Z and W.

$(^{35})$ By composing φ with a suitable linear automorphism of $\mathbf{C}^n$.

Note also that for an n–dimensional manifold the diagonal $\{(z, w) \in M^2 : z = w\}$ is an n–dimensional submanifold of the manifold M^2.

Any k–dimensional affine subspace of a complex affine space is a k–dimensional submanifold.

An affine space does not contain any compact submanifolds of positive dimensions (see the remark in n° 9 below).

8. Let M and N be complex manifolds with atlases $\{\varphi_\iota\}$ and $\{\psi_\kappa\}$, respectively. We say that a mapping $f : M \longrightarrow N$ is *holomorphic* if all mappings $\psi_\kappa \circ f \circ \varphi_\iota^{-1}$ are holomorphic (^{36}) . In the case when $N = \mathbf{C}$, we say that f is a *holomorphic function*. If $\{G_\iota\}$ is an open cover of the manifold M, the mapping $f : M \longrightarrow N$ is holomorphic if and only if all the restrictions f_{G_ι} are holomorphic. The composition of holomorphic mappings is holomorphic. The diagonal product (as well as the Cartesian product) of holomorphic mappings is holomorphic. The restriction of a holomorphic mapping to a submanifold is holomorphic. If $W \subset N$ is a submanifold, then a mapping $f : M \longrightarrow W$ is holomorphic precisely when the mapping $f : M \longrightarrow N$ is holomorphic. If $Z \subset M$ is a submanifold, then a mapping $f : Z \longrightarrow N$ is holomorphic if and only if each point of the submanifold Z has an open neighbourhood U in M such that the restriction $f_{Z\cap U}$ is the restriction of a holomorphic mapping on U with values in N. The natural projections $M \times N \longrightarrow M$ and $M \times N \longrightarrow N$ are holomorphic.

If a set $Z \subset M$ is a submanifold at a point $a \in Z$, then we say that a mapping $f : Z \longrightarrow N$ is *holomorphic at the point* a if its restriction to a neighbourhood (which is a submanifold) in Z of the point a is holomorphic. Therefore, if $Z \subset M$ is a submanifold, then the mapping $f : Z \longrightarrow N$ is holomorphic if and only if it is holomorphic at each point of the submanifold Z.

9. In particular, a complex function $f : M \longrightarrow \mathbf{C}$ is holomorphic precisely in the case when all the functions $f \circ \varphi_\iota^{-1}$ are holomorphic. Any linear combination, product, and quotient (with non-vanishing denominator) of holomorphic functions is a holomorphic function. We denote by $\mathcal{O}_M$ the ring of all holomorphic functions on M.

Assume now that the manifold M is connected and f is a holomorphic function on M. In this case, if a germ of the function f is zero, then $f = 0$ in M (for the set $\{z \in M : F_z = 0\}$ is open and closed in M; see 1.5 and

---

(^{36}) For each ι, κ, the domain $\varphi_\iota\big(f^{-1}(\text{domain of } \psi_\kappa)\big)$ of this mapping *should be open* and this is obviously equivalent to the *continuity* of f. This definition coincides with the previous one when M and N are open subsets of complex vector spaces (see 1.6).

B. 1.); it follows that $\mathcal{O}_M$ is an integral domain. Hence the condition that $f \not\equiv 0$ is the negation of the condition $f \equiv 0$ (see B. 4.4). Two holomorphic functions on M are equal if and only if their germs at some point are equal. In particular, a holomorphic function on M which is constant on a non-empty open set must be constant in M. This implies the following

MAXIMUM PRINCIPLE. *If M is connected and the function $M \ni z \longrightarrow |f(z)|$ attains its maximum, then the function f must be constant.*

(This is true, since in view of the maximum principle in $\mathbf{C}^n$ (see 1.5) f must be constant in a neighbourhood of the point at which $|f(z)|$ attains its maximum. It follows that if a set $E \subset M$ is compact and $K > 0$, then

$$|f(z)| \leq K \text{ in } M \setminus E \Longrightarrow |f(z)| \leq K \text{ in } M.$$

REMARK. It follows from the maximum principle that *there are no compact submanifolds of $\mathbf{C}^n$ of positive dimension* [37] . Indeed, if Z were a connected component of such a submanifold, it would be a compact connected submanifold (see n° 7), and then the functions $Z \ni z \longrightarrow z_i \in \mathbf{C}$ would be constant. Thus the set Z would consist of a single point.

This shows an essential difference between the complex case and the case of manifolds of class $\mathcal{C}^k$, $1 \leq k \leq \infty$, or that of real analytic manifolds. According to Whitney's and Grauert's embedding theorems, every manifold (whose topology has a countable basis) is isomorphic to a submanifold of $\mathbf{R}^n$ for some n. In such cases, the theory of manifolds can be reduced to a study of submanifolds of $\mathbf{R}^n$. In contrast, in the case of complex manifolds, the notion of an "abstract" manifold (i.e., one defined via an atlas) is considerably more significant. Even the Riemann sphere $\bar{\mathbf{C}}$ (see n° 19 below), the simplest among the compact complex manifolds of positive dimension, is not isomorphic (i.e., biholomorphic; see n° 10) to any submanifold of the space $\mathbf{C}^n$. Other important complex compact manifolds, namely the projective spaces and Grassmann manifolds, will be discussed in Chapter VII.

10. A mapping $f : M \longrightarrow N$ is said to be *biholomorphic* if it is bijective and the mappings f and f^{-1} are holomorphic. If there is a biholomorphic mapping $f : M \longrightarrow N$, we say that the manifolds M and N are *biholomorphic* or *isomorphic*. Then $\dim M = \dim N$ (see 1.12). Any chart on the manifold M modelled on X is a biholomorphic mapping on an open subset of M onto an open set in X. Clearly, the inverse of a biholomorphic mapping is a biholomorphic mapping and the composition of biholomorphic

---

[37] This is a special case of the theorem stating that *every compact analytic subvariety of $\mathbf{C}^n$ is finite*. The theorem follows from the maximum principle for analytic varieties (see IV. 5 below).

mappings is biholomorphic. Any holomorphic diffeomorphism is biholomorphic (see n° 8 and 1.12). If $f : M \longrightarrow N$ is biholomorphic and Z is a submanifold of the manifold M, then $f(Z)$ is a submanifold of the manifold N and $f_Z : Z \longrightarrow f(Z)$ is biholomorphic.

Two structures of a complex manifold on a set E coincide if and only if the identity mapping on the set E is biholomorphic, when regarded as a mapping from E with the first structure to E with the second one. Observe that it is enough to check whether the identity mapping is holomorphic (see below, V.1 corollary 2 from theorem 2).

Thus a manifold structure on a submanifold $N \subset M$ is the same as the induced one precisely when the mapping $N \hookrightarrow M$ is holomorphic (N is furnished with the former structure).

If $f : M \longrightarrow E$ is bijective, then there exists a unique complex manifold structure on the set E such that f is a biholomorphic mapping ([38]) . This structure is given by the atlas $\{\varphi_\iota \circ f^{-1}\}$, where $\{\varphi_\iota\}$ is an atlas of the manifold M. It is called the *complex manifold structure transferred by the bijection f*.

If a set $Z \subset M$ is a submanifold at a point $a \in Z$, we say that a mapping $f : Z \longrightarrow N$ is *biholomorphic at a point a* if there are open neighbourhoods Δ in Z and Ω in N of the points a and $f(a)$, respectively, such that Δ is a submanifold and the mapping $f_\Delta : \Delta \longrightarrow \Omega$ is biholomorphic. A mapping $f : M \longrightarrow N$ is said to be *locally biholomorphic* if it is biholomorphic at each point of the manifold M. (Then, obviously, the mapping is a holomorphic local homeomorphism.) Both the composition and the Cartesian product of locally biholomorphic mappings are locally biholomorphic. A locally biholomorphic mapping is biholomorphic if and only if it is a bijection.

Let $f : M \longrightarrow N$ be a locally biholomorphic surjection. Then there is no other manifold structure on N for which f is a locally biholomorphic mapping. A subset $Z \subset N$ is a submanifold if and only if its inverse image $f^{-1}(Z) \subset M$ is a submanifold. In this case, $f_{f^{-1}(Z)} : f^{-1}(Z) \longrightarrow Z$ is locally biholomorphic. A mapping h on the manifold N into an arbitrary manifold is holomorphic if and only if the composition $h \circ f$ is holomorphic ([39]).

Indeed, it is enough to note that if $\{\varphi_\iota\}$ is an inverse atlas on M such that the restrictions of f to the ranges of the φ_ι are injective, then $\{f \circ \varphi_\iota\}$ is an inverse atlas.

([38]) It is the unique structure for which the bijection f is holomorphic (see the above remark).

([39]) See 4.2. Analogous properties are true when M and N are (real) analytic manifolds. In particular, if a surjection $f : M \longrightarrow N$ is a local analytic isomorphism, then a mapping h on the manifold N, with values in another manifold, is analytic if and only if the composition $h \circ f$ is analytic.

10a. Let $\{M_\iota\}$ be a family of manifolds. A manifold M together with a family of biholomorphic mappings $f_\iota : M_\iota \longrightarrow M$ onto open sets in M which cover M is called a *gluing* of the family $\{M_\iota\}$ [40] . (Then all the M_ι's must have the same dimension.)

Each of the compositions $f_{\iota\kappa} = f_\kappa^{-1} \circ f_\iota$ is a biholomorphic mapping of an open subset of the manifold M_ι onto an open subset of the manifold M_κ. We say that the manifold M – or more precisely the pair M, $\{f_\iota : M_\iota \longrightarrow M\}$ – is a *gluing of the manifolds M_ι by the biholomorphic mappings $f_{\iota\kappa}$*. The family $\{f_{\iota\kappa}\}$ satisfies the conditions

$$(\#)\begin{cases} f_\iota \text{ is the identity on } M_\iota; \quad f_{\kappa\iota} = f_{\iota\kappa}^q; \quad f_{\kappa\lambda} \circ f_{\iota\kappa} \subset f_{\iota\lambda}; \\[2mm] \text{the graph of } f_{\iota\kappa} \text{ is closed in } M_\iota \times M_\kappa; \quad \text{(for all } \iota, \kappa, \lambda). \end{cases}$$

(Note that the graph of $\{f_{\iota\kappa}\}$ is the set $\{f_\iota(z) = f_\kappa(w)\}$.)

Conversely, given a family $\{f_{\iota\kappa}\}$ that satisfies the conditions $(\#)$, where $f_{\iota\kappa}$ is a biholomorphic mapping of an open subset of the manifold M_ι onto an open subset of the manifold M_κ, there exists a gluing of the manifolds M_ι by the biholomorphic mappings $f_{\iota\kappa}$ [41] . Indeed, assuming that the M_ι's are disjoint [42] , we define the following relation in $M' = \bigcup M_\iota :$ "$f_{\iota\kappa}(z) = w$ for some ι, κ". It is an equivalence relation because of the first three conditions in $(\#)$. Let M be the quotient space with respect to this relation, and let $\psi : M' \longrightarrow M$ denote the natural surjection. Then the injections $f_\iota = \psi_{M_\iota} : M_\iota \longrightarrow M$ satisfy $f_\kappa^{-1} \circ f_\iota = f_{\iota\kappa}$, and so (see n° 4) there exists a unique complex manifold structure on M such that the mappings f_ι are biholomorphic.

If $\{f_{\iota\kappa}\}_{\iota\neq\kappa}$ is a family (of biholomorphic mappings) which satisfies the last three conditions in $(\#)$, then, after adding to the family the identity mappings $f_{\iota\iota}$ on M_ι, we obtain a family fulfilling all the conditions in $(\#)$.

---

[40] The manifolds M_ι are identified, via f_ι, with their images in M.

[41] The gluing is unique up to an isomorphism. This means that if M, $\{f_\iota : M_\iota \longrightarrow M\}$ and N, $\{g_\iota : M_\iota \longrightarrow N\}$ are gluings by the biholomorphic mappings $f_{\iota\kappa}$ and $g_{\iota\kappa}$, respectively, then there exist a (unique) biholomorphic mapping $\varphi : M \longrightarrow N$ such that the diagrams

$$\begin{array}{ccc} & M & \\ f_\iota\nearrow & & \searrow g_\iota \\ M & \xrightarrow{\ \varphi\ } & N \end{array}$$

commute. Indeed, owing to the identities $f_\kappa^{-1} \circ f_\iota = g_\kappa^{-1} \circ g_\iota$, the families $\{g_\iota \circ f_\iota^{-1}\}$ and $\{f \circ g_\iota^{-1}\}$ can be glued to mutually inverse holomorphic mappings between M and N.

[42] This can be achieved by replacing M by $L \times M_\iota$ (with the structure of complex manifolds transferred via the natural bijections $z \longrightarrow (\iota, z)$).

Therefore there exists a gluing M. In such a case, we also say that M is a *gluing of the manifolds M_ν by the biholomorphic mappings $f_{\iota\kappa}$, $\iota \neq \kappa$.*

11. Let M be an n–dimensional complex manifold, and let $a \in M$. The *tangent space $T_a M$ at the point a to M* (regarded as a $2n$–dimensional differential manifold) has the natural structure of a complex vector space transferred from $\mathbf{C}^n$ by the differential $d_a\varphi : T_a M \longrightarrow \mathbf{C}^n$ of any chart φ whose domain contains a. (The structure is independent of the choice of the chart φ.) If $f : M \longrightarrow N$ is a holomorphic mapping between complex manifolds, then the differential $d_a f : T_a M \longrightarrow T_{f(a)} M$ is $\mathbf{C}$–linear $(^{43})$. If $g : M \longrightarrow P$ is also a holomorphic mapping between complex manifolds, then we have the chain rule $d_a(g \circ f) = d_{f(a)} g \circ d_a f$. If f is a biholomorphic mapping, then the differential $d_a f$ is an isomorphism (of complex vector spaces). If $Z \subset M$ is a submanifold and $a \in Z$, then $T_a Z$ can be identified with a subspace of the tangent space $T_a M$ (and the differential of the inclusion $Z \hookrightarrow M$ at the point a is the inclusion $T_a Z \hookrightarrow T_a M$). Moreover, if $f : M \longrightarrow N$ is a holomorphic mapping of complex manifolds, $W \subset N$ is a submanifold, and $f(Z) \subset W$, then $d_a(f_Z) = (d_a f)_{T_a Z}$ $(^{44})$ and $(d_a f)(T_a Z) \subset T_{f(a)} W$. In particular, if f is constant on Z, then $T_a Z \subset \ker d_a f$, and if f is biholomorphic at a, we have $T_{f(a)} f(Z) = (d_a f)(T_a Z)$. In the case when M is an open subset of a vector (or affine) space X, the tangent space $T_a M$ can be identified with X (or with the underlying vector space X). If M is an open subset of an affine subspace L of a vector space X, then the tangent space can be identified with L_* (see B. 6.11). The tangent space $T_{(a,b)}(M \times N)$ to the Cartesian product of complex manifolds M and N is identified with the Cartesian product $T_a M \times T_b N$ of the tangent spaces to these manifolds. If $f : M \longrightarrow N$, $g : M' \longrightarrow N$ are holomorphic mappings of complex manifolds, then $d_{(a,b)}(f \times g) = (d_a f) \times (d_b g)$, and if $M = M'$, then $d_a(f,g) = (d_a f, d_a g)$ $(^{45})$.

If a set $Z \subset M$ is a submanifold at a point $a \in Z$, i.e., if some neighbourhood in Z of the point a is a submanifold, then by $T_a Z$ we mean the tangent space to this submanifold at a. If, in addition, $f : Z \longrightarrow N$ is a mapping with values in a manifold N and f is holomorphic at the point a, then by $d_a f$ we mean the differential of the restriction of f to a neighbourhood of a in Z that is a submanifold.

We say that subsets $Z_1, \ldots, Z_r \subset M$ *intersect transversally at a point $a \in$*

$(^{43})$ Conversely, if at each point of the manifold M the differential of a $\mathcal{C}^1$–mapping $f : M \longrightarrow N$ is $\mathbf{C}$–linear, then the mapping must be holomorphic.

$(^{44})$ This holds both for $f_Z : Z \longrightarrow N$ and $f_Z : Z \longrightarrow W$.

$(^{45})$ The same holds for any finite number of components.

$\bigcap_1^r Z_i$ if they are submanifolds at this point and $T_a Z_1, \ldots, T_a Z_r$ intersect each other transversally (see A. 1.18). We will prove (see n° 16 below) that in such a case the set $\bigcap_1^r Z_i$ is a submanifold at the point a. Clearly, if $h : M \longrightarrow N$ is a biholomorphic mapping, then also $h(Z_1), \ldots, h(Z_r)$ intersect transversally at $h(a)$. In the case when $r = 2$ and the sets Z_1 and Z_2 have complementary dimensions (i.e., the sum of the dimensions is equal to n), then the condition for transversality of their intersection is $(T_a Z_1) \cap (T_a Z_2) = 0$. Similarly, in the case of $r = n$ and Z_i's of codimension 1, the condition is $\bigcap_1^n T_a Z_i = 0$. We say that subsets $Z_1, \ldots, Z_r$ intersect each other transversally if they do so at each point $a \in \bigcap_1^n T_a Z_i$.

12. For any holomorphic mapping $f : M \longrightarrow N$ between complex manifolds we can define the rank of f at a point z and the rank of f, respectively, by the formulae: $\operatorname{rank}_z f = \operatorname{rank} d_z f$ [46] and $\operatorname{rank} f = \max\{\operatorname{rank}_z f : z \in M\}$ [47]. The function $M \ni z \longrightarrow \operatorname{rank}_z f \in \mathbf{N}$ is lower semicontinuous. (Since this holds when M and N are open subsets of complex vector spaces; see B. 5.2 and 1.6.) Thus, for any $k \in \mathbf{N}$, the set $\{z \in M : \operatorname{rank}_z f > k\}$ is open. In particular, the set of all points $z \in M$ at which $\operatorname{rank}_z f$ attains its maximum, is open. Therefore $\operatorname{rank} f = \max\{\operatorname{rank}_z f : z \in E\}$, provided that $E \subset M$ is a dense set.

13. Let M and N be complex manifolds. If a set $Z \subset M$ is a submanifold at a point $a \in Z$, then a mapping $f : Z \longrightarrow N$ is biholomorphic at the point a if and only if it is holomorphic at the point a and the differential $d_a f$ is an isomorphism. Consequently, a mapping $f : M \longrightarrow N$ is locally biholomorphic if and only if it is holomorphic and, for each $z \in M$, the differential $d_z f$ is an isomorphism.

The implicit function theorem (see 1.13) carries over literally to the case when X, Y, Z are complex manifolds.

14. Let M and N be complex manifolds. By an *embedding* of the manifold N into the manifold M we mean any biholomorphic mapping of the manifold N onto a submanifold of M. We say that a holomorphic mapping $\varphi : N \longrightarrow M$ is an immersion at a point $c \in N$ if a restriction of φ to an open neighbourhood of the point c is an embedding. Equivalently, it means that the differential $d_c \varphi$ is injective. If this is the case at each point $c \in \mathbf{N}$, then the mapping φ is said to be an *immersion*. Hence a holomorphic mapping $\varphi : N \longrightarrow M$ is an embedding if and only if it is an immersion and

---

[46] Obviously, $d_z f$ is regarded here as a linear mapping of complex vector spaces.

[47] If $M = \emptyset$, we put $\operatorname{rank} f = -\infty$.

the mapping $\varphi : N \longrightarrow \varphi(N)$ is a homeomorphism (see n° 7 and 8) ([48]).
Naturally, every locally biholomorphic mapping is an immersion. The composition of embeddings is an embedding and the composition of immersions
is an immersion. A set $Z \subset M$ is a submanifold precisely when it admits
a manifold structure for which the inclusion $Z \hookrightarrow M$ is an embedding. It
follows that if $Z \subset M$ is a submanifold and $W \subset Z$, then W is a submanifold
of the manifold Z if and only if it is a submanifold of M.

Note also that an embedding is proper if and only if its range is closed
(see B. 2.4).

15. Let M be a complex manifold of dimension m. The property that
$Z \subset M$ is a k–dimensional submanifold at a point $a \in Z$ (where $k \leq m$) can
be characterized by each of the following conditions:

(a) $Z_a = \{\Phi = c\}_a$, where Φ is a holomorphic mapping from a neighbourhood
of the point a to an $(n - k)$-dimensional manifold N and is such that
$\Phi(a) = c$ and $d_a\Phi$ is surjective. (We can always take $\mathbf{C}^{m-k}$ and 0 as N
and c, respectively.) Then $T_a Z = \ker\ d_a\Phi$.

(b) $Z_a = \varphi(N)_a$ and $a = \varphi(c)$, where $\varphi : N \longrightarrow M$ is an embedding of a
k-dimensional manifold N and $c \in N$. (Here, we can take $\mathbf{C}^k$ and 0 as
N and c, respectively.) Then $T_a Z = \operatorname{im} d_c\varphi$.

If M is a complex vector space, we have yet another characterization:

(c) $Z_a = \{z + f(z) : z \in G\}_a$ for some direct sum decomposition $M = X + Y$,
where $\dim X = k$, and for a holomorphic mapping $f : G \longrightarrow Y$ on an
open neighbourhood G of a point c in X such that $a = c + f(c)$. In this
case, $T_a Z = \{z + d_a f(z) : z \in X\}$.

The characterization given in (a) yields the following two corollaries:

If $\Phi : M \longrightarrow N$ *is a holomorphic mapping into a k–dimensional manifold* N, $c \in N$, *and the differential* $d_z\Phi$ *is surjective for each z such that*
$\Phi(z) = c$, *then* $k \leq m$ *and the set* $\{\Phi(z) = c\}$ *is an $(m - k)$-dimensional
submanifold of the manifold* M.

If $\varphi_1, \ldots, \varphi_k$, $k \leq m$, *are holomorphic functions in a neighbourhood of
a point* $a \in M$ *and the differentials* $d_a\varphi_1, \ldots, d_a\varphi_k$ *are linearly independent,
then for some neighbourhood U of the point a the set* $\{z \in U : \varphi_1(z) = \ldots =
\varphi_k(z) = 0\}$ *is an $(m - k)$-dimensional submanifold of the manifold* M ([49]).

16. Let $Z_1, \ldots, Z_r \subset M$ be submanifolds of codimensions $k_1, \ldots, k_r$,

---

([48])   Therefore every embedding is an immersion (but not vice versa).

([49])   This is so because if $\lambda_1, \ldots, \lambda_k$ are independent linear forms on a complex vector
space X, then the mapping $(\lambda_1, \ldots, \lambda_k) : X \longrightarrow \mathbf{C}^k$ is surjective.

respectively, and let $a \in \bigcap_1^r Z_i$.

The submanifolds $Z_1, \ldots, Z_r$ intersect transversally at the point a if and only if in some coordinate system at a they are subspaces which intersect transversally, i.e., if there exists a coordinate system $\Phi : U \longrightarrow \Omega$ at a such that $\Phi(Z_i) = T_i \cap \Omega$, where $T_1, \ldots, T_r \subset \mathbf{C}^m$ are subspaces which intersect transversally. Then the set $\bigcap_1^r Z_i$ is a submanifold at a, of codimension $k = k_1 + \ldots + k_r$, and $T_a\left(\bigcap_1^r Z_i\right) = \bigcap_1^r T_a Z_i$. Moreover, $Z_1, \ldots, Z_r$ intersect transversally at every point of some neighbourhood of a in $\bigcap_1^r Z_i$. More generally, at every point z in a neighbourhood of a, those of the submanifolds Z_i which contain z intersect transversally at z.

In fact, suppose that $Z_1, \ldots, Z_r$ intersect transversally at a. Take a submanifold $Z_{r+1} \ni a$ such that $T_a Z_{r+1}$ is a linear complement of $\bigcap_1^r T_a Z_i$. Hence its dimension is equal to $k_{r+1} = m - k$. By 15(a), we have $Z_i \cap U = \{\Phi_i = 0\}$, where $\Phi_i : U \longrightarrow \mathbf{C}^{k_i}$ are holomorphic mappings of an open neighbourhood U of a, whose differentials $d_a \Phi_i$ are surjective. Then the differential $d_a \Phi$ of the mapping $\Phi = (\Phi_1, \ldots, \Phi_{r+1}) : U \longrightarrow \mathbf{C}^m$ is an isomorphism (as $\ker d_a \Phi = \bigcap_1^{r+1} T_a Z_i = 0$). Hence, after making U small enough, the mapping $\Phi : U \longrightarrow \Omega = \Phi(U)$ is a coordinate system at a. Then we have $\Phi^{-1}(T_i) = Z_i \cap U$ with the subspaces $T_i = \{z : u_i = 0\} \subset \mathbf{C}^m$ (where $z = (u_1, \ldots, u_{r+1})$, $u_i \in \mathbf{C}^{k_i}$) intersecting transversally (see A. 1.18).

Note also that in this case there exists a biholomorphic mapping of a neighbourhood of the point $a \in M$ onto a neighbourhood of zero in $T_a M$ which takes a to 0 and maps the traces of the Z_I's on the domain of the mapping onto the traces of $T_a Z_i$'s in the range. Indeed, the composition $(d_a \Phi)^{-1} \circ \phi$ satisfies the above requirements.

Hence, if the submanifolds $Z_1, \ldots, Z_r$ intersect transversally, then their intersection $\bigcap_1^r Z_i$ is a submanifold of codimension $k_1 + \ldots + k_r$.

If submanifolds $Z, W \subset M$ are such that $\dim Z + \dim W = m$ and they intersect transversally at a point $a \in Z \cap W$, then a is an isolated point of their intersection $Z \cap W$. Similarly, if submanifolds $Z_1, \ldots, Z_m \subset M$ of codimension 1 intersect transversally at a point $a \in \bigcap_1^m Z_i$, then a is an isolated point of their intersection $\bigcap_1^r Z_i$.

Let X be a complex vector space of dimension n.

PROPOSITION . *If a subspace $L_0 \in \mathbf{G}'_k(X)$ intersects transversally with a submanifold $T \subset X$ of dimension $(n-k)$* [50] *at a point of T, then the same holds for each subspace L belonging to some neighbourhood of the subspace L_0 in $\mathbf{G}'_k(X)$.*

---

[50] It is easy to prove that the proposition is true without the assumption $\dim \Gamma = n - k$.

As a matter of fact, it is sufficient to prove the proposition for $\mathbf{G}_k(X)$ instead of $\mathbf{G}'_k(X)$. Indeed, take the homeomorphism χ defined in B. 6.11. Setting $L = \chi(N)$, we have the equivalence: (L intersects Γ transversally at a) $\Longleftrightarrow$ (N intersects $1 \times \Gamma$ transversally at $(1, a)$). This is true because, for $T \in \mathbf{G}_{n-k}(X)$, we have : $L_* \cap T = 0 \Longleftrightarrow N \cap (0 \times T) = 0$. The last equivalence, in turn, follows from the fact that $N \cap (0 \times X) = 0 \times L_*$ (due to the equality ($\#$) in B. 6.11).

Now, since we have the homeomorphisms φ_{UV} from B. 6.8, we may assume that X is the Cartesian product of vector spaces M and N of dimensions k and $n - k$, respectively. We may also assume that $L_0 \in L(M, N)$, and it would be enough to prove the proposition with $L(M, N)$ in place of $\mathbf{G}_k(X)$.

Suppose that the graph L_0 intersects Γ transversally at the point $a \in \Gamma \cap L_0$. We may assume (see n° 15 condition (a) and n° 12) that $\Gamma = \{\Phi(u, v) = 0\}$, where $\Phi : U \longrightarrow \mathbf{C}^k$ is a holomorphic mapping on an open neighbourhood $U \subset M \times N$ of the point a such that the differential $d_{u,v}\Phi$ is surjective for each $(u, v) \in \Gamma$. Define $\Psi_L(u, v) = v - L(u)$ for $L \in L(M, N)$ and apply the implicit function theorem to the system of equations

$$(*) \qquad \Phi(u, v) = 0, \ \Psi_L(u, v) = 0 .$$

Note that, for any triple u, v, L satisfying the above system of equations, we have the equivalences

$$(\#) \qquad \begin{aligned} &\bigl(L \text{ intersects } \Gamma \text{ transversally at } (u, v)\bigr) \\ &\Longleftrightarrow (L \cap \ker d_{u,v}\Phi = 0) \\ &\Longleftrightarrow (d_{u,v}(\Phi, \Psi_L) \text{ is an isomorphism}) . \end{aligned}$$

Thus the differential $d_a(\Phi, \Psi_{L_0})$ is an isomorphism. Hence there exists a continuous mapping $\Omega \ni L \longrightarrow (u_L, v_L) \in X$ on a neighbourhood Ω of the element L_0, such that $(u_{L_0}, v_{L_0}) = a$, and (u_L, v_L) satisfies ($*$), i.e., $(u_L, v_L) \in \Gamma \cap L$ for $L \in \Omega$. Furthermore, we may assume (taking a smaller Ω) that the differential $d_{u_L, v_L}(\phi, \Psi_L)$ is an isomorphism for $L \in \Omega$. Consequently, ($\#$) yields that each $L \in \Omega$ intersects Γ transversally at (u_L, v_L).

17. Let M and N be complex manifolds, and let $m = \dim M$. The graph of a holomorphic mapping $f : M \longrightarrow N$ is an m–dimensional submanifold of the manifold $M \times N$. Moreover, the natural projection $f \longrightarrow M$ is a biholomorphic mapping. Conversely, every complex submanifold $\Gamma \subset M \times N$, such that the natural projection $\Gamma \longrightarrow M$ is a biholomorphic mapping, is the graph of a holomorphic mapping from M to N.

A *topographic submanifold in the Cartesian product* $M \times N$ is the graph of a holomorphic mapping on a non-empty open subset of the manifold M with values in the manifold N. Equivalently, a non-empty submanifold $Z \subset M \times N$ is topographic if its natural projection into M is a biholomorphic mapping onto an open subset of M. Clearly $\dim Z = m$. Note that the image of a topographic submanifold in $M \times (N \times L)$ under the natural projection $M \times (N \times L) \longrightarrow M \times N$ is a topographic manifold in $M \times N$.

A subset $E \subset M \times N$ is said to be a *topographic submanifold of a point* $a \in E$ if some neighbourhood of a in E is a topographic submanifold of $M \times N$, or – equivalently – if E is a submanifold at the point a and the natural projection $\pi : E \longrightarrow M$ is biholomorphic at a. Yet another equivalent characterization is that E is an m–dimensional submanifold at a and the submanifold $\pi(a) \times N$ intersects the set E transversally at a (^{51}) . Clearly, the set of such points a is open in E. If this is the case for each $a \in E$, then the set E is an m–dimensional submanifold and is called a *locally topographic submanifold* of $M \times N$. Therefore a submanifold $Z \subset M \times N$ is locally topographic if and only if the natural projection $Z \longrightarrow M$ is locally biholomorphic.

Let $\chi : L \longrightarrow M$ be a holomorphic mapping of complex manifolds. If E is a (locally) topographic submanifold in $M \times N$, then $(\chi \times e)^{-1}(E)$ – where $e = \mathrm{id}_N$ – is a (locally) topographic submanifold of $L \times N$ (^{52}) . If $\varphi : M \longrightarrow M'$ and $\psi : N \longrightarrow N'$ are biholomorphic mappings, then the set $E \subset M \times N$ is a topographic submanifold (at a point $a \in E$) precisely when the set $(\varphi \times \psi)(E) \subset M' \times N'$ is a topographic submanifold (at the point $(\varphi \times \psi)(a)$).

A k–dimensional submanifold of the space $\mathbf{C}^n$ is said to be *(locally) topographic* if it is (locally) topographic in the Cartesian product $\mathbf{C}^k \times \mathbf{C}^{n-k}$. We say that a set $E \subset \mathbf{C}^n$ is a *topographic* (or k–*topographic*) *submanifold at a point* $a \in E$ if a neighbourhood of a in E is a topographic submanifold (of dimension k). This is the case precisely when E is a k–dimensional submanifold at a and the subspace $a + N$, where $N = \{z_1 = \ldots = z_k = 0\}$, intersects E transversally at a. We say that a subset E of the manifold M is a *topographic* (k–*topographic*) *submanifold at the point* a *in the coordinate system* φ whose domain contains a if the image $\varphi(E)$ is a topographic (k–topographic) submanifold at the point $\varphi(a)$.

18. Let M be a complex manifold, and let N be a complex vector space.

(^{51}) This follows from condition (a) in n° 15 and the implicit function theorem.

(^{52}) For if $E \cap (\Delta \times \Omega)$ is the graph of the mapping f, then $(\chi \times e)^{-1}(E \cap (\Delta \times \Omega)) = (\chi \times e)^{-1}(E) \cap (\chi^{-1}(\Delta) \times \Omega)$ is the graph of the mapping $(\chi \times e)^{-1}(f) = f \circ \chi$.

Note that if $g(z, w)$ is a holomorphic function on an open subset $G \subset M \times \mathbf{C}^k$, then its partial derivatives $\partial^{|q|} g / \partial w^q$, where $q \in \mathbf{N}^k$, are also holomorphic (in G) $(^{53})$.

A holomorphic function $f(z, w)$ on $M \times N$ is said to be N–*homogeneous of degree* r if for each $z \in M$ the function $N \ni w \longrightarrow f(z, w)$ is homogeneous of order r or, equivalently, a form of degree r (see 1.8, corollary of Liouville's theorem). For such a function, in any coordinate system $\varphi : N \longrightarrow \mathbf{C}^k$, the function $g(z, w) = \big(f(z, \varphi^{-1}(w)\big)$ is a form of degree r with respect to w, and its coefficients are holomorphic on M $(^{54})$.

If f is a holomorphic function in a neighbourhood of the point $(a, 0) \in M \times N$, then $f = \sum_0^\infty f_\nu$ and the series is absolutely and uniformly convergent in an open neighbourhood $U \times \Delta$ of that point, where each f_ν is a holomorphic N–homogeneous function of degree ν in $U \times N$, $\nu = 0, 1, \ldots$. Then the functions f_ν are uniquely determined by the function f $(^{55})$.

Indeed, we may assume that $M = \mathbf{C}^n$, $a = 0$, and $N = \mathbf{C}^k$. Then the function f is holomorphic, $|f(z, w)| \le L$ for some $L > 0$, $f(z, w) = \sum_q \frac{1}{q!} \frac{\partial^{|q|} f}{\partial w^q}(z, 0) w^q$, and the series is absolutely convergent in a neighbourhood of zero of the form $U \times \{|w_j| < 2r\}$, where U is a neighbourhood of zero in M. Accordingly, $f(z, w) = \sum_0^\infty f_\nu(z, w)$ in $U \times \{|w_j| \le 2r\}$, where $f_\nu(z, w) = \sum_{|q|=\nu} \frac{1}{q!} \frac{\partial^{|q|} f}{\partial w^q}(z, 0) w^q$ in $U \times \mathbf{C}^k$. From the Cauchy estimates (see 1.5) we deduce that $|f_\nu(z, w)| \le K \sum_{|q|=\nu} (\frac{1}{2})^{|q|}$ in the neighbourhood $U \times \{|w_j| < r\}$, where $K > 0$ is a constant. This fact, combined with the convergence of the series $\sum_q (\frac{1}{2})^{|q|}$ $(^{56})$, implies that the series $\sum_0^\infty f_\nu$ is absolutely and uniformly convergent in the neighbourhood.

A function $f(z, w)$ which is holomorphic in $M \times N$ is called an N–*polynomial* if it is a polynomial with respect to $w \in N$ for arbitrarily fixed

---

$(^{53})$ Indeed, if ψ is a chart on M and the chart domain is Ω, then

$$\frac{\partial^{|q|} g}{\partial w^q}(z, w) = \frac{\partial^{|q|} h}{\partial w^q}(\psi(z), w)$$

in $G \cap (\Omega \times \mathbf{C}^k)$, where $h = g \circ (\psi \times \mathrm{id}_{\mathbf{C}^k})^{-1}$.

$(^{54})$ As they are equal to $\frac{1}{q!} \frac{\partial^{|q|} g}{\partial w^q}(z, 0)$, $|q| = r$.

$(^{55})$ If $\sum f_\nu = 0$ in $U \times \Delta$, then for each $z \in U$ the forms $w \longrightarrow f_\nu(z, w)$ are identically zero (see 1.7), and so $f_\nu = 0$, $\nu = 0, 1, \ldots$.

$(^{56})$ This follows from the fact that the series $\sum w^q$ – being the power series expansion of the holomorphic function $\frac{1}{(1-w_1)\ldots(1-w_k)}$ – is convergent in the polydisc $P = \{|w_j| < 1\}$.

z in M. Then, if the manifold M is connected, the function f, in any linear coordinate system $\varphi : N \longrightarrow \mathbf{C}^k$ (i.e., the function $g(z,w) = f(z,\varphi^{-1}(w))$, is a polynomial with holomorphic coefficients on M. (Indeed, in such a case we have $M = \bigcup_1^\infty Z_\nu$, where $Z_\nu = \left\{ \frac{\partial^{|q|} g}{\partial w^q}(z,0) = 0 \text{ if } |q| > \nu \right\}$ are closed sets in M; thus $\operatorname{int} Z_r \neq \emptyset$ for some r, and therefore g must be a polynomial of degree $\leq r$ with coefficients $\frac{1}{q!}\frac{\partial^{|q|} g}{\partial w^q}(z,0)$ that are holomorphic in M.) Note also that in this case there is a holomorphic $(\mathbf{C} \times N)$-homogeneous function F on $M \times (\mathbf{C} \times N)$ such that $f(z,w) = F(z,1,w)$ in $M \times N$ ([57]).

19. By the *extended plane* we mean the set $\bar{\mathbf{C}} = \mathbf{C} \cup \{\infty\}$ (i.e., $\mathbf{C}$ with the "infinity point" ∞ added) endowed with the following topology. The family of open sets consists of the open sets in $\mathbf{C}$ and the complements in $\bar{\mathbf{C}}$ of the compact subsets of $\mathbf{C}$. Thus the sets $\{z \in \mathbf{C} : |z| > \nu\} \cup \{\infty\}$, $\nu = 1,2,\ldots$, constitute a base of neighbourhoods of the point ∞. The space $\bar{\mathbf{C}}$ is compact and $\mathbf{C}$ is an open and dense subset.

A complex manifold structure on $\bar{\mathbf{C}}$ can be defined by an atlas consisting of two charts: $\varphi : \bar{\mathbf{C}} \setminus \infty \ni z \longrightarrow z \in \mathbf{C}$ and $\psi : \bar{\mathbf{C}} \setminus 0 \ni z \longrightarrow \frac{1}{z} \in \mathbf{C}$ (where $\frac{1}{\infty} = 0$). (Therefore $\mathbf{C}$ – whether regarded as an open subset of the manifold $\bar{\mathbf{C}}$ or a complex vector space – carries the same complex manifold structure.) The compact complex manifold $\bar{\mathbf{C}}$ with the above structure is called the *Riemann sphere* (or the *extended plane*) ([58]).

Notice that the mapping $\varrho : \bar{\mathbf{C}} \ni z \longrightarrow \frac{1}{z} \in \bar{\mathbf{C}}$ (where $\frac{1}{0} = \infty$ and $\frac{1}{\infty} =$

([57]) If $g(z,w) = \sum_{|q|\leq r} a_q(z)w^q$, it suffices to set $F(z,t,w) = G(z,t,\varphi(w))$, where

$$G(z,t,w) = \sum_{\nu+|q|=r} a_\nu(z)t^\nu w^q.$$

([58]) The space $\bar{\mathbf{C}}$, regarded as a (real) analytic manifold, is isomorphic to the two-dimensional sphere $S = \{(w,t) \in \mathbf{C} \times \mathbf{R} = \mathbf{R}^3 : |w|^2 + t^2 = 1\}$, via the stereographic mapping $\sigma : S \longrightarrow \bar{\mathbf{C}}$ given by $\sigma(w,t) = \frac{2w}{1-t}$ in $S \setminus (0,1)$ and $\sigma(0,1) = \infty$. The mapping σ is an analytic isomorphism (i.e., a bijection such that σ and σ^{-1} are analytic). Indeed, $\sigma^{-1}(z) = (4z,|z|^2 - 4)(|z|^2 + 4)^{-1}$ for $z \in \mathbf{C}$, and $\sigma^{-1}(\infty) = (0,1)$, and the restrictions $\sigma_{S \setminus (0,1)}$, $\sigma_{S \setminus (0,-1)} : (w,t) \longrightarrow \varrho(\frac{\bar{w}}{2(1+t)})$, $(\sigma^{-1})_{\mathbf{C}}$ and $(\sigma^{-1})_{\bar{\mathbf{C}} \setminus 0} : z \longrightarrow (4\bar\varrho(z), 1 - 4|\varrho(z)|^2)(1 + 4|\varrho(z)|^2)^{-1}$ are analytic (see the definition of ϱ below). Note also that the stereographic projections of the sphere S onto the tangent planes T_+, T_- at the "poles" $(0,1)$, $(0,-1)$, respectively, composed with the mappings π_+ and $(z \longrightarrow \bar{z}) \circ \pi_-$ (where $\pi_\pm : T_\pm \longrightarrow \mathbf{C}$ are the natural projections), constitute a complex atlas that defines on S a manifold structure identical to that induced from $\bar{\mathbf{C}}$ through σ. (See [5], Chapter II, §10).

0) is biholomorphic. To see this, observe that $\varrho^{-1} = \varrho$ and the restrictions $\varrho_{\bar{C}\setminus 0}$, $\varrho_{\bar{C}\setminus\infty}$ coincide with the holomorphic mappings $\psi : \bar{C}\setminus 0 \longrightarrow \bar{C}$, $\psi^{-1} : \bar{C}\setminus\infty \longrightarrow \bar{C}$, respectively.

20. Let M be a complex manifold, and let G be a group of biholomorphic mappings of M onto itself that satisfies the condition

(d) Each point in $M \times M$ has a neighbourhood that has a non-empty intersection with at most one of the graphs of the biholomorphic mappings from G ([59]).

Then there exists a unique structure of a complex manifold on M/G ([60]), such that the natural surjection $\pi : M \longrightarrow M/G$ is a local biholomorphism. The mapping becomes then a covering. Obviously, $\dim M/G = \dim M$.

Indeed, there could be at most one such structure (see n° 10). The union $Z \subset M^2$ of all the graphs of the biholomorphic mappings from G is closed. Each point $c \in M$ has an open neighbourhood U_c which is the range of an inverse chart $\varphi_c : G_c \longrightarrow U_c$, such that $U_c \times U_c$ does not intersect any of the graphs $\tau \in G$ different from the identity e. That is, $\tau(U_c) \cap U_c = \emptyset$ for $\tau \in G \setminus e$. Hence the open sets $\tau(U_c)$, $\tau \in G$, are disjoint. For each $c \in M$, the restriction π_{U_c} is injective (if $z, w \in U_c$ and $\pi(z) = \pi(w)$, then $w = \sigma(z)$, $\sigma \in G$, and thus $\sigma = e$, which means that $z = w$). Then the restrictions $\pi_{\tau(U_c)}$, $\tau \in G$, are also injective (as $\pi \circ \tau = \pi$). Furthermore, they all have the same range $\Delta_c = \pi(U_c)$. Now, the family of bijections $\psi_c = \pi_{U_c} \circ \varphi_c : G_c \longrightarrow \Delta_c$ is an inverse atlas on M/G in the sense of n° 4 ([61]). Indeed, the ranges of the mappings ψ_c cover M/G. For any $a, b \in M$, the composition $\psi_b^{-1} \circ \psi_a$ is holomorphic and its graph is closed in $G_a \times G_b$. This is so, because the composition $\pi_{U_b}^{-1} \circ \pi_{U_a}$ is holomorphic and its graph is closed in $U_a \times U_b$, since the graph is equal to $Z \cap (U_a \times U_b)$ and the restrictions $\tau \cap (U_a \times U_b)$ have disjoint domains. Finally, π is a locally biholomorphic covering. Indeed, the mappings $\pi_{U_c} = \psi_c \circ \varphi_c^{-1} : U_c \longrightarrow \Delta_c$ are biholomorphic, Δ_c is an open neighbourhood of the point $\pi(c)$, the inverse image $\pi^{-1}(\Delta_c)$ is the union of the sets $\tau(U_c)$, $\tau \in G$, and the restrictions $\pi_{\tau(U_c)} : \tau(U_c) \longrightarrow \Delta_c$ are bijective.

Now let M be a *commutative (complex) Lie group*, i.e., a complex manifold with the structure of a commutative group whose group operations ([62]) are holomorphic or – equivalently – such that the mapping $r : M^2 \ni (z, w) \longrightarrow z - w \in M$ is holomorphic. In such a case, the mapping $M \ni z \longrightarrow -z \in M$ and the translations $\tau_c : M \ni z \longrightarrow z + c \in M$, $c \in M$, are biholomorphic.

Let $Z \subset M$ be a discrete subgroup (i.e., Z consists of isolated points). Clearly, it

---

([59]) An equivalent condition: the union of all the graphs of the biholomorphic mappings in G is closed and each point of the manifold M has a neighbourhood U such that $\tau(U) \cap U = \emptyset$ for each $\tau \in G$ different from the identity. It is said then that G "acts discretely" on M.

([60]) Let M be a set. Any group G of bijections of M induces the equivalence relation defined by the condition: "z is equivalent to w if $w = \varphi(z)$ for some $\varphi \in G$". By M/G we denote the corresponding quotient space.

([61]) That is, the family $\{\psi^{-1}\}$ is an atlas in the sense of n° 4.

([62]) That is, the mappings $(z, w) \longrightarrow z + w$ and $z \longrightarrow -z$.

is closed (^{63}). The group G of the translations τ_c, $c \in Z$, satisfies condition (d) (^{64}). Obviously, $M/G = M/Z$ and the mapping π is the natural homomorphism. Therefore:

The quotient group M/Z has a unique complex manifold structure such that the natural homomorphism $\pi :\ M \longrightarrow M/Z$ is a locally biholomorphic mapping and hence a covering. The quotient group M/Z is a commutative (complex) Lie group of dimension dim M.

Indeed, the mapping $\tilde{r} :\ (M/Z)^2 \ni (v, w) \longrightarrow u - v \in M/Z$ is holomorphic, since $\tilde{r} \circ (\pi \times \pi) = \pi \circ r$ and $\pi \times \pi$ is a locally biholomorphic surjection (see n° 10).

21. Now we are going to define *complex tori* as quotients of complex linear spaces and lattices.

First, consider a real m–dimensional vector space M. A *lattice* in M is a discrete (additive) subgroup $\Lambda \subset M$ which generates the vector space M. Such a subgroup is always closed (^{65}). Note that (if M is normed) the function $x \longrightarrow \varrho(x, \Lambda)$ is bounded in M. (Indeed, take a basis $a_1, \dots, a_m \in \Lambda$ for M. Then, for any $x \in M$, we have $x = \sum_1^m (k_\nu + \varepsilon_\nu) a_\nu$, $k_\nu \in \mathbf{Z}$, $|\varepsilon_\nu| \leq 1$, and hence $|x - \sum_1^m k_\nu a_\nu| \leq \sum_1^m |a_\nu|$.)

A subspace $N \subset M$ is said to be Λ–*distinguished* if $\Lambda \cap N$ is a lattice in N (i.e., if $\Lambda \cap N$ generates N). Let $p :\ M \longrightarrow M/N$ be the natural mapping. The following equivalence holds

$$(N \text{ is } \Lambda\text{–distinguished}) \Longleftrightarrow (p(\Lambda) \text{ is a lattice in } M/N) .$$

Indeed, introduce norms on M and M/N. Suppose that N is Λ–distinguished. Take a bounded $F \subset M/N$. As $F = p(E)$ for some bounded set $E \subset M$, $p(\Lambda) \cap F$ is finite. Consequently, $p(\Lambda) \cap F \subset p(\Lambda \cap E_0)$, where $E_0 = E + \{x \in N :\ |x| \leq r\}$ is bounded. Hence $\Lambda \cap E_0$ is finite and so is $p(\Lambda) \cap F$. Therefore the subgroup $p(\Lambda)$ is discrete, and so it is a lattice in M/N (as it generates M/N). Conversely, assume that $p(\Lambda)$ is a lattice in M/N. If the subgroup $\Lambda \cap N$ were not a lattice in N, it would generate a subspace $N_0 \subsetneq N$. As $\varrho(x, \Lambda) < d$ in M for some d, there would exist $c \in N$ such that the ball $\{|x - c| \leq 2d\}$ would be disjoint from N_0, and then the same would hold for the balls $\{|x - \nu c| \leq 2d\}$, $\nu = 1, 2, \dots$. Take $a_\nu \in \Lambda$ such that $|a_\nu - \nu c| \leq d$, $\nu = 1, 2, \dots$. If $\nu < \mu$, then $|a_\mu - a_\nu - (\mu - \nu)c| \leq 2d$, and therefore $a_\mu - a_\nu \notin N_0$. Accordingly, $p(a_\mu) \neq p(a_\nu)$ (for otherwise one would have $a_\mu - a_\nu \in N \cap \Lambda \subset N_0$). Thus the points $p(a_\nu) \in p(\Lambda)$ would be distinct. But $|p(a_\nu)| = |p(a_\nu - \nu c)| \leq |p|d$ and, as a consequence, $p(\Lambda)$ would not be discrete.

Every lattice in M is of the form $\sum_1^m \mathbf{Z} a_\nu$, where $a_1, \dots, a_m$ is a basis of M (and, of course, conversely). Then $a_1, \dots, a_m$ is called a *basis* of this lattice.

In fact, the above statement is trivial for $m = 1$. Let $m > 1$, and let us assume that the statement is true for $m - 1$. Take a vector line L that intersects $\Lambda \setminus 0$. The line is, obviously, Λ–distinguished, and so $L = \mathbf{R} a_m$ and $\Lambda \cap L = \mathbf{Z} a_m$ for some $a_m \in \Lambda$. Moreover, $p(\Lambda)$ is a lattice in M/L, where $p :\ M \longrightarrow M/L$ is the natural mapping.

(^{63}) For if there were an $a \in \bar{Z} \setminus Z$, then there would exist distinct $b, c \in Z$, arbitrarily close to a, and hence $b - c \in Z \setminus 0$ could be made arbitrarily small.

(^{64}) For let $a, b \in M$, and let U be a neighbourhood of zero such that $Z \cap ((U - U) - (U - U)) = 0$. If $(z, z + c) \in (a + U) \times (b + U)$, then $c \in b - a + (U - U)$, which can be true for at most one $c \in Z$ only.

(^{65}) See footnote (^{63}).

Therefore $p(\Lambda) = \sum_1^{m-1} \mathbf{Z} p(a_i)$ and $p(a_1), \ldots, p(a_{m-1})$ is a basis of the space M/Λ, where $a_1, \ldots, a_{m-1} \in \Lambda$. Thus it follows that $a_1, \ldots, a_m$ is a basis for the space M. Now, $\Lambda = \sum_1^m \mathbf{Z} a_i$. For if $c \in \Lambda$, then $p(c) = \sum_1^{m-1} k_i p(a_i)$, where $k_i \in \mathbf{Z}$. This means that $c - \sum_1^{m-1} k_i a_i \in \Lambda \cap L$ and implies that $c = \sum_1^m k_i a_i$ for some $k_m \in \mathbf{Z}$.

Note also that every closed subgroup $H \subset M$ must be of the form $H = L + \Omega$, where $L \subset M$ is a subspace, whereas Ω is a discrete subgroup of any of the linear complements of the subspace L.

Indeed, the union L of all the lines (through 0) contained in H is a subspace. Let N be a linear complement of this subspace. Then $H = L + \Omega$, where $\Omega = H \cap N$ is a closed subgroup that does not contain any of the lines (passing through 0). Now, Ω must be a discrete subgroup. Otherwise, after fixing a norm on M, one could choose a sequence $\Omega \setminus 0 \ni u_\nu \longrightarrow 0$ such that $u_\nu / |u_\nu| \longrightarrow v \neq 0$, and then $\mathbf{R} v \subset \Omega$. (Since, if $\lambda \in \mathbf{R}$, then, by taking $k_\nu \in \mathbf{Z}$ such that $|k_\nu - \lambda/|u_\nu|| \leq 1$, we have $k_\nu u_\nu \longrightarrow \lambda v$).

Now let X be an n–dimensional complex vector space, and let Λ be a lattice in X. The space X, regarded as a complex manifold together with its additive group structure, is, naturally, a Lie group and Λ is a discrete subgroup. Accordingly, X/Λ is a Lie group of dimension n and the natural homomorphism is a locally biholomorphic covering.

Every such manifold X/Λ is called a *complex torus*. It is connected ([66]) and compact ([67]). When regarded as a real analytic manifold (or a real Lie group), it is isomorphic to the real torus S^{2n}, where $S = \{z \in \mathbf{C} : |z| = 1\}$ ([68]).

Let X and Y be (complex) vector spaces, and let Λ and Ω be lattices in these spaces, respectively. We have the complex tori X/Λ and Y/Ω and the natural mappings $\pi : X \longrightarrow X/\Lambda$ and $\pi' : Y \longrightarrow Y/\Omega$.

Every linear mapping $F : X \longrightarrow Y$ such that $F(\Lambda) \subset \Omega$ induces a torus homomorphism $\tilde{F} : X/\Lambda \longrightarrow Y/\Omega$; this mapping is well-defined by the formula

$$\tilde{F} \circ \pi = \pi' \circ F \quad (^{69}).$$

It is also a holomorphic mapping (see n° 10).

PROPOSITION . *Every holomorphic mapping of tori* $f : X/\Lambda \longrightarrow Y/\Omega$ *is of the form* $f = c + \tilde{F}$, *where* $c \in Y/\Omega$ *and* $F \in L(X, Y)$, $F(\Lambda) \subset \Omega$.

PROOF . Of course, all mappings of this form are holomorphic. Conversely, assume that the mapping $f : X/\Lambda \longrightarrow Y/\Omega$ is holomorphic. Consider the set $(\pi \times \pi')^{-1}(f) \subset$

([66]) Since it is the image of the space X under π.

([67]) For if $a_1, \ldots, a_{2n}$ is a basis for the lattice Λ, then $X = \Lambda + \sum_1^{2n} [0, 1] a_n u$, and hence $X/\Lambda = \pi(\sum_1^{2n} [0, 1] a_\nu)$.

([68]) Indeed, the surjection $\lambda : X \ni \sum_1^{2n} t_i a_i \longrightarrow (e^{2\pi i t_1}, \ldots, e^{2\pi i t_{2n}}) \in S^{2n}$ is a local analytic isomorphism and also a group epimorphism whose kernel is Λ. Therefore it induces a group isomorphism $\kappa : X/\Lambda \longrightarrow S^{2n}$. Then $\kappa \circ \pi = \chi$ and $\kappa^{-1} \circ \chi = \pi$, from which it follows that κ is an analytic isomorphism (see n° 10, footnote ([39])).

([69]) If $F \in L(X, Y)$ and $\varphi : X/\Lambda \longrightarrow Y/\Omega$ satisfies $\varphi \circ \pi = \pi' \circ F$, then $F(\Lambda) \subset \Omega$ and $\varphi = \tilde{F}$. (Since $\pi' \circ F$ is constant on Λ, it is equal to zero on Λ.) We have $\tilde{F}_1 = \tilde{F}_2 \Longrightarrow F_1 = F_2$ (since if $\pi' \circ F = 0$, then $F(X) \subset \Omega$, and so $F = 0$). Furthermore, $\widetilde{F_1 + F_2} = \tilde{F}_1 + \tilde{F}_2$, $\widetilde{G \circ F} = \tilde{G} \circ \tilde{F}$; finally, if $e = \mathrm{id}_X$, then $\tilde{e} = \mathrm{id}_{X/\Lambda}$.

$X \times Y$. It is a submanifold (see n° 10) and is equal to $(e \times \pi')^{-1}(f \circ \pi)$, where e is the identity mapping on the space X ([70]). Now, the natural projection $(\pi \times \pi')^{-1}(f) \longrightarrow X$ is a locally biholomorphic covering. This is so, because it is the composition of the restriction $(e \times \pi')^{-1}(f \circ \pi) \longrightarrow f \circ \pi$ of the mapping $e \times \pi' : X \times Y \longrightarrow X \times Y/\Omega$, which is a locally biholomorphic covering, with the natural projection $f \circ \pi \longrightarrow X$, which is a biholomorphic mapping (see n° 10 and B. 3.2). Let H be a connected component of the set $(\pi \times \pi')^{-1}(f)$. Then $H \subset X \times Y$ is a submanifold and the natural projection $H \longrightarrow X$ is a covering (see B. 3.2) and hence a homeomorphism (see B. 3.2, proposition 2). But it is also locally biholomorphic, and so it must be biholomorphic. Consequently, it is the graph of a holomorphic mapping $H : X \longrightarrow Y$ (see n° 17) and we have

$$f(\pi(\zeta)) = \pi'(H(\zeta)) \text{ for } \zeta \in X .$$

Accordingly, if $a \in \Lambda$, then $H(\zeta + a) - H(\zeta) \in \Omega$ for $\zeta \in X$, which implies that the mapping $X \ni \zeta \longrightarrow H(\zeta + a) - H(\zeta) \in Y$ must be constant, and hence $d_{\zeta+a} H = d_\zeta H$ for $\zeta \in X$. It follows that the holomorphic mapping $X \ni \zeta \longrightarrow d_\zeta H \in L(X, Y)$ is bounded ([71]) and therefore constant: $d_\zeta H = F \in L(X, Y)$ for each $\zeta \in X$ (see C. 1.8, Liouville's theorem). Thus we have $H = b + F$ for some $b \in Y$, and so $(f - c) \circ \pi = \pi' \circ F$, where $c = \pi'(b)$. This implies (see footnote ([69])) that $F(\Lambda) \subset \Omega$ and $f - c = \tilde{F}$.

COROLLARY 1. *The holomorphic homomorphisms of the tori* $X/\Lambda \longrightarrow Y/\Omega$ *are precisely the induced mappings* $\tilde{F}$, *where* $F \in L(X, Y)$ *and* $F(\Lambda) \subset \Omega$. *All holomorphic mappings of the tori are of the form of the sum of a constant and a holomorphic homomorphism.*

COROLLARY 2. *The biholomorphic mappings of the tori* $f : X/\Lambda \longrightarrow Y/\Omega$ *are exactly the mappings of the form* $f = c + \tilde{F}$, *where* $c \in Y/\Omega$ *and* $F \in L_0(X, Y)$, $F(\Lambda) \subset \Omega$.

Indeed, let f be such a biholomorphic mapping. We may assume that $f(0) = 0$ ([72]). Then $f = \tilde{F}$ and $f^{-1} = \tilde{G}$, where $F \in L(X, Y)$, $G \in L(X, Y)$, $F(\Lambda) \subset \Omega$, and $G(\Omega) \subset \Lambda$. Hence $\tilde{G} \circ \tilde{F} = \tilde{e}$, and so $G \circ F = e$ (see footnote ([69])). Similarly, $F \circ G$ is the identity mapping (on Y). Thus $F = G^{-1}$ is an isomorphism and $F(\Lambda) = \Omega$.

This yields the following

COROLLARY 3. *The tori* X/Λ *and* Y/Ω *are biholomorphic if and only if the lattices* Λ *and* Ω *are* C*-equivalent, i.e., if* $\Omega = F(\Lambda)$ *for some* $F \in L_0(x, Y)$.

As we have noticed, any two complex n–dimensional tori are isomorphic as real analytic manifolds. On the other hand, corollary 3 implies that for any $n > 0$ there exist an infinite (even uncountable) number of different types of n–dimensional complex tori ([73]). This is so, as there is an infinite (uncountable) number of lattices in $\mathbf{C}^n$ which are not C–equivalent ([74]).

([70]) Because $(\pi(z), \pi'(w)) \in f \Longleftrightarrow (z, \pi'(w)) \in f \circ \pi$.

([71]) In view of the equality $X = \sum_1^{2n} [0, 1] a_i + \Lambda$, where $a_1, \ldots, a_{2n}$ is a basis of the lattice Λ.

([72]) By replacing f by its composition with a suitable translation.

([73]) In the sense that two tori are of the same type if they are biholomorphic.

([74]) Indeed, with each lattice Λ in $\mathbf{C}^n$ one associates the set $E_\Lambda = \{(\zeta_1, \ldots, \zeta_n) \in \mathbf{C}^n : \sum_1^n \zeta_i c_i \in \Lambda$ for some $c_1, \ldots, c_n \in \Lambda$ which are linearly independent over $\mathbf{C}\}$. Then the same set corresponds to all C–equivalent lattices. Therefore it is enough to note that the

Let $Z \subset X$ be a subspace. If it is Λ–distinguished, i.e., if $\Lambda \cap Z$ is a lattice in Z, then its image $\pi(Z) \subset X/\Lambda$ is a submanifold which is biholomorphic to the torus $Z/\Lambda \cap Z$. The natural isomorphism

$$Z/\Lambda \cap Z \longrightarrow \pi(Z)$$

is a biholomorphic mapping.

Indeed, let $\iota : Z/\Lambda \cap Z \longrightarrow X/\Lambda$ be the natural monomorphism. We have $\iota \circ \pi_0 = \pi_Z$, where the natural homomorphism $\pi_0 : Z \longrightarrow Z/\Lambda \cap Z$ is a locally biholomorphic mapping. Thus the mapping ι is an immersion (because so is $\pi_Z : Z \longrightarrow X/\Lambda$). It is also a homeomorphism onto its range and hence an embedding.

On the other hand, if the subspace Z is not Λ–distinguished, then its image $\pi(Z)$ is not even locally closed (and hence is not a submanifold).

For otherwise it would be closed $(^{75})$, and so would the set $\pi^{-1}(\pi(z)) = Z + \Lambda$. Now, let $p : X \longrightarrow X/Z$ be the natural mapping, and let Y denote a linear complement of the subspace Z. Then $p_Y : Y \longrightarrow X/Z$ is an isomorphism. Therefore the set $p(\Lambda) = p((Z + \Lambda) \cap Y) \subset X/Z$ would be a closed countable subgroup, and so it would be discrete. Consequently, it would be a lattice and the subspace Z would be Λ–distinguished.

For a subgroup Γ of the torus X/Λ, the following conditions are equivalent

(1) Γ is a submanifold that is biholomorphic to a torus,

(2) Γ is a connected submanifold,

(3) $\Gamma = \pi(Z)$, where Z is a Λ–distinguished subspace.

In such a case, Γ is said to be a *subtorus* of the torus X/Λ.

In fact, the implication (3)$\Longrightarrow$(1) has already been proved and (1)$\Longrightarrow$(2) is trivial. Only (2)$\Longrightarrow$(3) remains to be shown. Assume that the subgroup $\Gamma \subset X/\Lambda$ is a connected submanifold. Then it must be also closed $(^{76})$. Therefore the subgroup $\pi^{-1}(\Gamma) \subset X$ is a closed complex submanifold, and hence its component $Z \subset X$ which contains zero is a complex vector subspace $(^{77})$. Since $\pi_{\pi^{-1}(\Gamma)} : \pi^{-1}(\Gamma) \longrightarrow \Gamma$ is a covering, the same is true for the restriction $\pi_Z : Z \longrightarrow \Gamma$ (see B. 3.2). This implies that $\pi(Z) = \Gamma$. In addition,

---

sets E_Λ are uncountable and they cover $\mathbf{C}^n \setminus \mathbf{R}^n$. In the case when $n = 1$, it is easy to verify (see [10a], Chapter 2, §2) that each lattice is $\mathbf{C}$–equivalent to one of the form $\Lambda_\zeta = \mathbf{Z} + \zeta \mathbf{Z}$, $\zeta \in \mathcal{F}$, where $\mathcal{F} \subset \mathbf{C}$ is the so-called "fundamental domain" :

$$\mathcal{F} = \{|z| \geq 1, \ \frac{-1}{2} \leq Re \ z \leq 0\} \cup \{|z| > 1, \ 0 < Re \ z < \frac{1}{2}\} \ .$$

For different $\zeta, \zeta' \in \mathcal{F}$, the lattices Λ_ζ, $\Lambda_{\zeta'}$ are not equivalent. In this way one can establish a one-to-one correspondence between the types of one–dimensional complex tori and the points of the "fundamental domain" $\mathcal{F}$.

$(^{75})$ A locally closed subgroup $\Gamma \subset X/\Lambda$ must be closed. Indeed, if U is an open neighbourhood of zero in X/Λ such that $\Gamma \cap U$ is closed in U, then $\Gamma \cap (c+U)$ is closed in $c+U$ for $c \in \Gamma$ (as then $c + \Gamma = \Gamma$). It is enough to see that such $c + U$, where $c \in \Gamma$, constitute an open cover of the set $\bar{\Gamma}$: if $a \in \bar{\Gamma}$, then there exists $c \in \Gamma \cap (a - U)$, and so $a \in c + U$. (See B. 1.)

$(^{76})$ See the previous footnote.

$(^{77})$ A real subspace, which is a complex submanifold, is a complex subspace.

the subspace Z must be Λ–distinguished, for otherwise its image $\pi(Z)$ would not be locally closed.

REMARK: The mapping $Z \longrightarrow \pi(Z)$ is a one–to–one correspondence between the set of Λ–distinguished subspaces of X and the set of subtori of X/Λ $(^{78})$.

The image of a subtorus under a holomorphic homomorphism of tori is a subtorus. Indeed (see corollary 2), if Z is a Λ–distinguished subspace, then $F\big(\pi(Z)\big) = \pi'\big(F(Z)\big)$ and the subspace $F(Z)$ is Ω–distinguished, as it is generated by $F(Z \cap \Lambda) \subset F(Z) \cap \Omega$. This yields

COROLLARY 4. *The submanifolds of the torus T which are biholomorphic to tori are precisely the translated subtori $c + S$, where $c \in T$ and S is a subtorus. The image of a translated subtorus under a holomorphic mapping between tori is a translated subtorus.*

Now, we are going to give an example of a 2–dimensional (complex) torus that contains only one–dimensional subtori $(^{79})$.

Let X be a (complex) two–dimensional vector space. It is sufficient to find a lattice Λ in X for which there is only one Λ–distinguished line. Now, it is enough to take the lattice whose basis is $a \neq 0$, $b = ia$, $c \notin L = \mathbf{C}a$, $d = ic + \lambda a$ $(^{80})$, where $\lambda \in \mathbf{C} \setminus \mathbf{Q}(i)$. Then L is the unique Λ–distinguished line. For otherwise there would exist $z_1, z_2 \in \Lambda \setminus L$ such that $z_2 = \mu z_1$, where $\mu \in \mathbf{C} \setminus \mathbf{R}$. Consequently, $z_\nu = \alpha_\nu a + \gamma_\nu c + \delta_\nu \lambda a$, where $\alpha_\nu, \gamma_\nu, \delta_\nu \in \mathbf{Q}(i)$, $\gamma_\nu \neq 0$, and $\delta_\nu, \gamma_\nu - \delta_\nu i \in \mathbf{R}$, for $\nu = 1, 2$. Thus we would have $\gamma_2 = \mu \gamma_1$, and so $\mu \in \mathbf{Q}(i)$. But $a, c, \lambda a$ are linearly independent over $\mathbf{Q}(i)$, hence $\delta_1 = \mu \delta_2$. It follows that $\mu \in \mathbf{R}$, since, if $\delta_1 = \delta_2 = 0$, then $\gamma_1, \gamma_2 \in \mathbf{R}$. Therefore we would have a contradiction.

§4. The rank theorem. Submersions

Let M and N be complex manifolds.

1. Put $m = \dim\ M$. We have the following

RANK THEOREM . *Let $f : \ M \longrightarrow N$ be a holomorphic mapping whose rank is constant and equal to r (i.e., $\mathrm{rank}_z f = r$ for all $z \in M$). Let $a \in M$. Then, for sufficiently small open neighbourhoods U, V of the points $a, f(a)$, respectively, such that $f(U) \subset V$, there exist charts $\varphi : \ U \longrightarrow U'$, $\psi : \ V \longrightarrow V'$, where U', V' are open sets of complex vector spaces X', Y', respectively, such that*

$$\psi \circ f_U \circ \varphi^{-1} = \lambda_{U'}\ ,$$

$(^{78})$ For this mapping is injective: we have $Z_1 \subset Z_2 + \Lambda \Longrightarrow Z_1 \subset Z_2$ (in view of Baire's theorem), which yields $Z_1 + \Lambda = Z_2 + \Lambda \Longrightarrow Z_1 = Z_2$.

$(^{79})$ See [41].

$(^{80})$ It is a basis for X over $\mathbf{R}$, because so is a, ia, c, ic.

where $\lambda : X' \longrightarrow Y'$ is a linear mapping of rank r ([81]). Moreover, $f(U)$ is an r-dimensional submanifold of the manifold N and the non-empty fibres ([82]) of the mapping f are $(m-r)$-dimensional submanifolds of the manifold M.

PROOF. The second part of the theorem follows from the first one. It is enough to observe that $\lambda(U')$ is an open subset of the r-dimensional subspace $\lambda(X')$ (see B. 5.2) and the fibres of the mapping $\lambda_{U'}$ are open sets in subspaces of dimension $(m-r)$. Moreover, $f(U) = \psi^{-1}(\lambda(U'))$ and $f_U^{-1}(c) = \varphi^{-1}\left(\lambda_{U'}^{-1}(\psi(c))\right)$ for $c \in V$.

As for the first part of the theorem, it suffices to find just one pair of neighbourhoods U, V with the required properties. Furthermore, it is enough to consider the case when M is an open subset of a vector space X and N is a vector space. Then $d_a f : X \longrightarrow N$. Let $K = \ker\, d_a f$, $L = \mathrm{im}\; d_a f$ ([83]) and let $p : K \longrightarrow N$, $q : N \longrightarrow L$ be projections. Then the differential $d_a(p, q \circ f) = (p, d_a f) : X \longrightarrow K \times L$ is an isomorphism (since it is injective and $\dim X = \dim(K \times L)$). Therefore there exist an open neighbourhood U of the point a in M, an open connected neighbourhood T of the point $p(a)$ in K, and an open neighbourhood W of the point $q(f(a))$ in L, such that the mapping $\varphi(p, q \circ f)_U : U \longrightarrow T \times W$ is biholomorphic (see 1.12). Then $(q \circ f_U \circ \varphi^{-1})(t, w) = w$ for $(t, w) \in T \times W$. Hence, for each $(t, w) \in T \times W$, we have $\ker\, d_{(t,w)}(f_U \circ \varphi^{-1}) \subset K \times 0$. Since both spaces have the same dimension (since $\dim \ker\, d_z f = \dim K$ for $z \in M$), they must be equal. Consequently, for each $w \in W$, the differential of the mapping $T \ni t \longrightarrow (f_U \circ \varphi^{-1})(t, w) \in N$, at any point $t \in T$, is zero. Therefore the mapping $f_U \circ \varphi^{-1} : T \times W \longrightarrow N$ is independent of t; this means that $f_U(\varphi^{-1}(t, w)) = g(w)$ for $(t, w) \in T \times W$, where $g : W \longrightarrow N$ is a holomorphic mapping. Thus $q(g(w)) = w$ for $w \in W$. Putting $h(w) = g(w) - w$ in W, we have $q(h(w)) = 0$ for $w \in W$. The set $H = q^{-1}(W)$ is an open neighbourhood of the point $f(a)$ and $f(U) \subset H$. The mapping $\psi : H \ni y \longrightarrow y = h(y) \in H$ is biholomorphic (since $H \ni y \longrightarrow y + h(q(y)) \in H$ is the inverse) and $\psi(g(w)) = w$ in W. Hence $\psi\left(f_U(\varphi^{-1}(t, w))\right) = w$ in $T \times W$, which means that the mapping $\psi \circ f_U \circ \varphi^{-1}$ is the restriction to $T \times W$ of the linear mapping $K \times L \ni (t, w) \longrightarrow w \in N$

([81]) It means that f can be locally linearized by a choice of coordinate systems: f_U in the coordinate systems φ and ψ is a restriction (to an open set) of a linear mapping (of rank r).

([82]) By the fibres of a mapping $g : X \longrightarrow Y$ we mean the sets $g^{-1}(c)$, where $c \in Y$.

([83]) It follows that $f^{-1}(b)$, where $b \in f(M)$, are $(m-n)$-dimensional submanifolds (see 3.7).

of rank r.

The rank theorem implies the following

PROPOSITION . *If $\emptyset \neq V \subset M$ and $\emptyset \neq W \subset N$, then:*

$(V \times W$ *is a submanifold of the manifold* $M \times N$ $) \Longleftrightarrow$
$\Longleftrightarrow (V$ *and W are submanifolds of the manifold* M *and* N, *respectively*).

PROOF . Suppose that $\Lambda = V \times W$ is a p–dimensional submanifold. Let $\pi_1 : \Lambda \longrightarrow M$, $\pi_2 : \Lambda \longrightarrow N$ be the natural projections. The functions $\varrho_i : \Lambda \ni u \longrightarrow \mathrm{rank}_u \pi_i$ are lower semicontinuous (see 3.12). Note that if $\varrho_1 = r$ in a neighbourhood of a point $(a, b) \in \Lambda$, then V is an r–dimensional submanifold at a. (Indeed, by the rank theorem, there exist neighbourhoods U, U' of the points a, b in V, W, respectively, such that $U = \pi_1(U \times U')$ is an r–dimensional submanifold.) Similarly, if $\varrho_2 = s$ in a neighbourhood of a point $(a, b) \in \Lambda$, then W is an s–dimensional submanifold at b.

Therefore it is enough to prove that $\varrho_1 + \varrho_2 = p$ on Λ (⁸⁴) . Now we have the inequality $\varrho_1 + \varrho_2 \geq p$ (⁸⁵) . The function $\varrho_1 + \varrho_2$ attains its maximum $p' \geq p$ at a point $(a, b) \in \Lambda$. By the semicontinuity of ϱ_i, we must have $\varrho_i = \varrho_i(a, b)$ in a neighbourhood of (a, b), $i = 1, 2$. It follows that $p' = \varrho_1(a, b) + \varrho_2(a, b) = \dim_a V + \dim_b W = p$, which completes the proof.

2. Let $m = \dim M$ and $n = \dim N$. We say that a holomorphic mapping $f : M \longrightarrow N$ is a *submersion at a point* $a \in M$ if $m \geq n$ and there exists an open neighbourhood U of this point such that the image $f(U)$ is open and there are biholomorphic mappings $\varphi : U \longrightarrow \Delta \times \Omega$, $\psi : f(U) \longrightarrow \Delta$, where $\Delta \subset \mathbf{C}^n, \Omega \subset \mathbf{C}^{m-n}$ are open sets such that the mapping

$$\psi \circ f \circ \varphi^{-1} : \Delta \times \Omega \longrightarrow \Delta$$

(⁸⁴) For (in view of the semicontinuity) the functions ϱ_i must then be continuous, which shows that the sets V, W are submanifolds at each of their points. Moreover, $\varrho_1(z, w) = \dim_z V$, $\varrho_2(z, w) = \dim_w W$, and hence the functions $z \longrightarrow \dim_z V$, $w \longrightarrow \dim_w W$ must be constant.

(⁸⁵) Due to the fact that the differentials $d_u \pi_1 : T_u \Lambda \longrightarrow T_z M$ and $d_u \pi_2 : T_u \Lambda \longrightarrow T_w N$ are the natural projections (see 3.11), it is enough to observe that if T is a vector subspace of the Cartesian product $X \times Y$ of vector spaces and $p : T \longrightarrow X$ and $q : T \longrightarrow Y$ are the natural projections, then

$$\dim T \leq \dim p(T) + \dim q(T)$$

(as $T \subset p(T) \times q(T)$).

is the natural projection. In view of the rank theorem, this is the same as saying that the differential $d_a f$ is surjective [86]. (Indeed, if $d_a f$ is surjective, then $\operatorname{rank}_z f = n$ in a neighbourhood of the point a (see 3.12), and now one can apply the rank theorem. In this case, λ is an epimorphism, and hence (see A. 1.19) one can assume that $\lambda : \mathbf{C}^m \longrightarrow \mathbf{C}^n$ is the natural projection. Making the neighbourhoods U, V smaller, if necessary, one gets the required result.)

A holomorphic mapping $f : M \longrightarrow N$ is said to be a *submersion* if it is a submersion at each point of the manifold M – or equivalently – if for each point $z \in M$ the differential $d_z f$ is an epimorphism. Of course, every submersion $f : M \longrightarrow N$ is an open mapping and all fibres of f are $(m - n)$–dimensional submanifolds of the manifold M, by the rank theorem [87].

Any locally biholomorphic mapping is a submersion. Epimorphisms of vector spaces are submersions.

The Cartesian product of submersions is a submersion.

The composition of submersions is a submersion. If $f : M \longrightarrow N$ is a holomorphic mapping and the composition $f \circ g$ with a holomorphic surjection $g : L \longrightarrow M$ (where L is a manifold) is a submersion, then f is a submersion. (This follows from the formula $d_u(f \circ g) = d_{g(u)} f \circ d_u g$.)

Let $f : M \longrightarrow N$ be a submersion, and let $Z \subset N$. If Z is a submanifold, then so is $f^{-1}(Z)$, and the restriction $f_{f^{-1}(Z)} : f^{-1}(Z) \longrightarrow Z$ is a submersion. Moreover,

$$\dim f^{-1}(Z) = \dim Z + (m - n),$$

provided that $f^{-1}(Z) \neq \emptyset$, and for $c \in f^{-1}(Z)$ we have

$$T_c f^{-1}(Z) = (d_c f)^{-1}(T_{f(c)} Z) .$$

If, in addition, f is surjective, then

$$(Z \text{ is a submanifold}) \iff (f^{-1}(Z) \text{ is a submanifold})$$

---

[86] This can also be shown directly. Suppose that M, N are open subsets of vector spaces Z, X, respectively. Take the mapping $(f, \pi) : M \longrightarrow X \times Y$, where $\pi : M \longrightarrow Y = \ker d_a f$ is the restriction to M of a linear projection. The differential $d_a(f, \pi) : Z \longrightarrow X \times Y$ is an isomorphism. Therefore there is an open neighbourhood U of the point a and open sets $\Delta \subset X, \Omega \subset Y$, such that the mapping $\chi = (f, \pi)_U : U \longrightarrow \Delta \times \Omega$ is biholomorphic. Then the mapping $f \circ \chi^{-1} : \Delta \times \Omega \longrightarrow \Delta$ is the natural projection.

[87] Or directly from the definition.

and for mappings $h : Z \longrightarrow L$ between the manifolds we have the equivalence

$$(h : Z \longrightarrow L \text{ is holomorphic}) \Longleftrightarrow (h \circ f : f^{-1}(Z) \longrightarrow L \text{ is holomorphic}) .$$

These properties follow directly from the definition of a submersion (because of the equality $f^{-1}(Z) \cap U = f_U^{-1}(Z \cap V)$, where $V = f(U)$). Indeed, they hold for the natural projection $\Delta \times \Omega \longrightarrow \Delta$ and hence also for its compositions with biholomorphic mappings.

Note that the *rank theorem* can be reformulated as follows.

If $f : M \longrightarrow N$ is a holomorphic mapping whose rank is constant and equal to r, then each point of the manifold M has an open neighbourhood U such that the image $f(U)$ is an r–dimensional submanifold and the mapping $f_U : U \longrightarrow f(U)$ is a submersion.

Finally, we have

LEMMA . *Let $f : M \longrightarrow N$ be a submersion, and let $L \subset M$ be an n–dimensional submanifold. If a fibre $f^{-1}(c)$ (where $c \in N$) has a transversal intersection with L at $a \in M$, then, for some neighbourhoods Ω, U of the points c, a, respectively, each fibre $f^{-1}(w)$, $w \in \Omega$, intersects $L \cap U$ in one point only and the intersection is transversal.*

Indeed, we may assume (see 3.15) that $L = \varphi^{-1}(0)$, where $\varphi : M \longrightarrow \mathbf{C}^{m-n}$ is a submersion. Then (ker $d_a f$)∩(ker $d_a \varphi$) = 0 (see 3.15–16), and so $d_a(f, \varphi)$ is an isomorphism. Consequently, there is an open neighbourhood U of the point a such that $(f, \varphi)_U$ maps U biholomorphically onto a neighbourhood of the point $(c, 0)$. Therefore there is a neighbourhood Ω of the point c such that if $w \in \Omega$, then $f^{-1}(w) \cap L \cap U = (f, \varphi)_U^{-1}(c, 0) = \{b\}$ and $d_b(f, \varphi)$ is an isomorphism. This implies that (ker $d_b f$)∩(ker $d_b \varphi$) = 0. The latter means that $f^{-1}(w)$ and L intersect transversally at b.

COMPLEX ANALYTIC GEOMETRY

CHAPTER I

RINGS OF GERMS OF HOLOMORPHIC FUNCTIONS

§1. **Elementary properties. Noether and local properties. Regularity**

1. Let M be a complex manifold (¹) of dimension n, and let $a \in M$. The set of germs at a of functions that are holomorphic at a is a ring (see B. 4.2 and 3). It is denoted by $\mathcal{O}_a$ or $\mathcal{O}_a(M)$ and called the *ring of germs of holomorphic functions at a* (²).

In particular, $\mathcal{O}_n = \mathcal{O}_0(\mathbf{C}^n)$ denotes the *ring of germs of holomorphic functions at $0 \in \mathbf{C}^n$* (³). (The ring $\mathcal{O}_0$ is identified with $\mathbf{C}$.) By z_i we will also denote the germ of the function z_i at 0, i.e., the germ $(z \longrightarrow z_i)_0 \in \mathcal{O}_n$, $i = 1, \ldots, n$.

Of course, the ring $\mathcal{O}_a$ contains $\mathbf{C}$ as a subring (⁴). Hence it is also a complex vector space and each of its ideals is a vector subspace (⁵).

(¹) In what follows we will often omit the adjective "complex" in the expressions like "complex manifold ", "complex vector space", etc.

(²) Clearly, the mapping $\mathcal{O}_M \ni f \longrightarrow f_a \in \mathcal{O}_a$ is a homeomorphism.

(³) Note that the ring $\mathcal{O}_n$ can be identified with the ring $\mathbf{C}\{Z_1, \ldots, Z_n\}$ of convergent power series via the isomorphism which sends the germ of a holomorphic function at $0 \in \mathbf{C}^n$ to its power series expansion at 0.

(⁴) After the identification of $\mathbf{C}$ with the subring of constant germs in the ring $\mathcal{O}_a$.

(⁵) The spaces $\mathcal{O}_a$ and their non-zero ideals are infinite dimensional (if $n > 0$). Indeed, (see below) the space $\mathcal{O}_a$ is isomorphic to the space $\mathcal{O}_n$, in which the subspace of germs of polynomials (which is isomorphic to $\mathbf{C}[Z_1, \ldots, Z_n]$) is infinite dimensional. Now if I is a non-zero ideal, then, by taking $h \in I \setminus 0$, we have the monomorphism of vector spaces $\mathcal{O}_n \ni f \longrightarrow hf \in I$ (see 3).

Note that if M is a vector space, then $\mathcal{O}_0$ contains the subring (vector subspace) of germs of polynomials on M (^{6}). It is identified with the ring (vector space) of polynomials on M via the isomorphism that associates with each polynomial its germ at 0 (see B. 5.2).

If φ is a holomorphic mapping in a neighbourhood of the point a with values in a complex manifold N and $b = \varphi(a)$, then the mapping $\mathcal{O}_b \ni f \longrightarrow f \circ \varphi_a \in \mathcal{O}_a$ is a homomorphism (see B. 4.3). (It is a $\mathbf{C}$–homomorphism and hence also a linear mapping.) If, in addition, φ maps a neighbourhood of the point a biholomorphically onto a neighbourhood of the point b, then the mapping $\mathcal{O}_b \ni f \longrightarrow f \circ \varphi \in \mathcal{O}_a$ is an isomorphism (see B. 4.2 and 3); the image $I \circ \varphi$ of any ideal I of the ring $\mathcal{O}_b$ is an ideal in the ring $\mathcal{O}_a$.

In particular, $\mathcal{O}_a \approx \mathcal{O}_n$. If φ is a coordinate system at a, then $\mathcal{O}_a \ni f \longrightarrow f \circ \varphi^{-1} \in \mathcal{O}_n$ is an isomorphism. Let $f \in \mathcal{O}_a$. By the germ f in the *coordinate system* φ we mean the germ $f \circ \varphi^{-1} \in \mathcal{O}_n$. Similarly, if I is an ideal of the ring $\mathcal{O}_a$, then the ideal $I \circ \varphi^{-1}$ is called the *ideal I in the coordinate system* φ.

The ring $\mathcal{O}_k$, for $k \leq n$, will often be identified with the subring of germs of functions independent of the variables $z_{k+1}, \ldots, z_n$ in the ring $\mathcal{O}_n$ by means of the monomorphism $\mathcal{O}_k \ni f \longrightarrow f \circ \pi_0 \in \mathcal{O}_n$; here $\pi : (z_1, \ldots, z_n) \longrightarrow (z_1, \ldots, z_k)$. Therefore (after these identifications) we have the inclusions

$$\mathbf{C} = \mathcal{O}_0 \subset \mathcal{O}_1 \subset \ldots \subset \mathcal{O}_n .$$

2. For any germ $f \in \mathcal{O}_a$, the following are well-defined: $f(a) = \tilde{f}(a)$, $df = d_a \tilde{f}$, and also $\frac{\partial^{|p|} f}{\partial^p z_i} = \left(\frac{\partial^{|p|} \tilde{f}}{\partial^p z_i} \right)_0$ for $f \in \mathcal{O}_n$, where $\tilde{f}$ is a representative of the germ f.

A germ $f \in \mathcal{O}_a$ is invertible if and only if $f(a) \neq 0$. Consequently, the germs $z_i \in \mathcal{O}_n$ are irreducible (^{7}).

Observe that for a germ $f \in \mathcal{O}_n$ we have $f(a) = 0$ and $df \neq 0$ exactly when f is equal to the germ z_n in some coordinate system (^{8}). In this case, the germ f is irreducible.

(^{6}) If $M = \mathbf{C}^n$, then the subring is $\mathbf{C}[z_1, \ldots, z_n]$.

(^{7}) Otherwise we would have $z_i = f(z)g(z)$ in a neighbourhood of zero, for some holomorphic functions f, g which vanish at 0. This would imply that $d_0 z_i = d_0(fg) = 0$, which is impossible.

(^{8}) This is due to the fact that for a holomorphic function f at $0 \in \mathbf{C}^n$ such that $f(0) = 0$ and $\frac{\partial f}{\partial z_n}(0) \neq 0$, we have $f = z_n \circ \varphi$, where $\varphi = (z_1, \ldots, z_{n-1}, f)$ and the restriction of φ to a neighbourhood of zero is a coordinate system at 0.

For every germ $f \in \mathcal{O}_{ab} = \mathcal{O}_{ab}(M \times N)$, where N is a complex manifold and $b \in N$, the germs $f(z, b) = \left(z \longrightarrow \tilde{f}(z, b)\right)_a$ and $f(a, w) = \left(w \longrightarrow \tilde{f}(a, w)\right)_b$ are well-defined. The symbol $\tilde{f}(z, w)$ here stands for a representative of the germ f. (We have $f(z, b) = f \circ \iota_a$ and $f(a, w) = f \circ \kappa_b$, where $\iota : M \ni z \longrightarrow (z, b) \in M \times N$ and $\kappa : N \ni w \longrightarrow (a, w) \in M \times N$.) The mappings

$$\mathcal{O}_{ab} \ni f \longrightarrow f(z, b) \in \mathcal{O}_a \quad \text{and} \quad \mathcal{O}_{ab} \ni f \longrightarrow f(a, w) \in \mathcal{O}_b$$

are epimorphisms.

Notice also that if $f_1, \ldots, f_k \in \mathcal{O}_a$ and $P \in \mathcal{O}_a[T_1, \ldots, T_k]$, then the function $\tilde{P}\left(z, \tilde{f}_1(z), \ldots, \tilde{f}_k(z)\right)$ represents the germ $P(f_1, \ldots, f_k) \in \mathcal{O}_a$, where $\tilde{f}_i$ are representatives of the germs f_i and $\tilde{P}(z, w_1, \ldots, w_k)$ is a polynomial in w whose coefficients are representatives of the coefficients of the polynomial P.

3. The ring $\mathcal{O}_a$ and every quotient of $\mathcal{O}_a$ by a proper ideal are of characteristic zero. (Because $\mathcal{O}_a$ contains the field $\mathbf{C}$; see A. 1.14.)

The ring $\mathcal{O}_a$ is an integral domain. Indeed, if $f, g \in \mathcal{O}_a \setminus 0$, then they have holomorphic representatives $\tilde{f} \not\equiv 0$, $\tilde{g} \not\equiv 0$ in a connected neighbourhood U of the point a. Consequently, the set $\{\tilde{f}\tilde{g} = 0\} = \{\tilde{f} = 0\} \cup \{\tilde{g} = 0\}$ is nowhere dense in U, which means that $\tilde{f}\tilde{g} \not\equiv 0$, and so $fg = (\tilde{f}\tilde{g})_a \neq 0$.

4. A germ from $\mathcal{O}_n$ is said to be *regular* (or z_n-*regular*) *of order* k if it is the germ of a holomorphic function which is z_n-regular of order k (see C. 2.3) ([9]). Note that such a germ is not invertible precisely when $k > 0$. The product of germs from $\mathcal{O}_n$ is regular if and only if each of the germs involved is regular (see n° 3).

If M is a vector space and $f_1, \ldots, f_p \in \mathcal{O}_0(M)$ are non-zero germs, then in any linear coordinate system from an open and dense subset of $L_0(M, \mathbf{C}^n)$ these germs are regular (see C. 2.3). In particular, this implies that:

Any non-zero germ in $\mathcal{O}_a$ is regular in some coordinate system at a.

Furthermore, for any germs $f_1, \ldots, f_p \in \mathcal{O}_n \setminus 0$ there exists an automorphism $\chi : \mathcal{O}_n \longrightarrow \mathcal{O}_n$ such that the germs $\chi(f_1), \ldots, \chi(f_p)$ are regular ([10]).

([9]) Thus a germ $f \in \mathcal{O}_n$ is regular if and only if $f(0, z_n) \neq 0$.

([10]) It is enough to take $\chi : \mathcal{O}_n \ni g \longrightarrow g \circ L \in \mathcal{O}_n$ with an $L \in L_0(\mathbf{C}^n, \mathbf{C}^n)$ such that the germs $f_1 \circ L, \ldots, f_p \circ L$ are regular.

The *Weierstrass preparation theorem – division version* (see C. 2.5) can be stated as follows:

If a germ $f \in \mathcal{O}_n$ is regular of order k, then every germ $g \in \mathcal{O}_n$ can be represented in the form

$$g = qf + \sum_{\nu=0}^{k-1} a_\nu z_n$$

with a unique q and unique germs $a_0, \ldots, a_{k-1} \in \mathcal{O}_{n-1}$.

5. Consider the subring $Q_n = \mathcal{O}_{n-1}[z_n]$ of the ring $\mathcal{O}_n$. It is the set of the germs of polynomials in z_n whose coefficients are holomorphic at $0 \in \mathbf{C}^{n-1}$.

We have $Q \approx \mathcal{O}_{n-1}[X]$ – the mapping $\mathcal{O}_{n-1}[X] \ni p \longrightarrow p(z_n) \in Q_n$ establishes an isomorphism. Each non-zero element of Q_n can be uniquely written in the form $a_0 z_n^k + \ldots + a_k$, where $a_0, \ldots, a_k \in \mathcal{O}_{n-1}$ and $a_0 \neq 0$. Here, a_0 is called the *leading coefficient* and k its *degree*. The leading coefficient (respectively, the degree) of the product of non-zero elements of $\mathcal{O}_n$ is the product of the leading coefficients (respectively, the sum of the degrees) of those elements.

An element from Q_n of the form $z_n^k + a_0 z_n^{k-1} + \ldots + a_k$ (where $a_i \in \mathcal{O}_{n-1}$) is said to be *monic* and $D_k(a_1, \ldots, a_k) \in \mathcal{O}_{n-1}$ is called its *discriminant*. (Thus it equals $p(z_n) \in Q_n$, where $p \in \mathcal{O}_{n-1}[X]$ is monic, and then $p(z_n)$ and p have the same discriminant.) The monic elements of Q_n are given by the germs at 0 of monic polynomials with respect to z_n (whose coefficients are holomorphic at $0 \in \mathbf{C}^{n-1}$). If F is such a polynomial and D is its discriminant (see B. 2.1), then the germ D_0 is the discriminant of the monic germ F_0. Monic elements are obviously regular.

Let φ be a coordinate system at $0 \in \mathbf{C}^{n-1}$. Then $\psi : (u, z_n) \longrightarrow (\varphi(u), z_n)$ is a coordinate system at $0 \in \mathbf{C}^n$. If f is a monic germ from Q_n, then so is $f \circ \psi$. Moreover, if δ is the discriminant of the germ f, then $\delta \circ \varphi$ is the discriminant of the germ $f \circ \psi$ ([11]).

6. PROPOSITION 1. *The ring $\mathcal{O}_a$ is noetherian.*

PROOF . It is sufficient to consider the case of $\mathcal{O}_N$. Now, the ring $\mathcal{O}_0 = \mathbf{C}$ is noetherian (see A. 9.1). Let $n > 1$, and let us assume that the ring $\mathcal{O}_{n-1}$ is noetherian. Then the ring Q_n is noetherian because it is isomorphic to $\mathcal{O}_{n-1}[X]$, which is noetherian by the Hilbert basis theorem (see A. 9.4). Now, let I be an ideal of the ring $\mathcal{O}_n$. We may assume that $I \neq 0$ and that

---

([11]) Indeed, $\alpha : \mathcal{O}_{n-1} \ni g \longrightarrow g \circ \varphi \in \mathcal{O}_{n-1}$ is an isomorphism, and hence $\alpha(D_k(a_1, \ldots \ldots, a_k)) = D_k(\alpha(a_1), \ldots, \alpha(a_k))$.

I contains a regular element g (see n° 4 (^{12})). Since $I' = I \cap Q_n$ is an ideal in the ring Q_n, it follows that I' is finitely generated, and so $I' \subset \sum_1^s \mathcal{O}_n g_i$, where $g_1, \ldots, g_s \in I$. Now, $g, g_1, \ldots, g_s$ generate the ideal I. Indeed, if $f \in I$, then in view of the division version of the Weierstrass preparation theorem (see n° 4), we have $f = qg + r$, where $q \in \mathcal{O}_n$ and $r \in Q_n$. On the other hand, $r = f - qg \in I$, so $r \in I'$, and therefore $f \in \mathcal{O}_n g + \sum_1^s \mathcal{O}_n g_i$.

7. The ring $\mathcal{O}_a$ is local with maximal ideal $\mathfrak{m} = \mathfrak{m}_a = \{f \in \mathcal{O}_a : f(a) = 0\}$. Indeed, the set of all non-invertible elements coincides with $\mathfrak{m}$ (see n° 2), and so it is an ideal; see A. 1.10. Obviously, if φ maps a neighbourhood of the point a biholomorphically onto a neighbourhood of a point b of a manifold N and $\varphi(a) = b$, then $\mathfrak{m}_a = \mathfrak{m}_b \circ \varphi$.

We will denote by $\mathfrak{m}_n$ the maximal ideal of the ring $\mathcal{O}_n$.

It follows from Hadamard's lemma (see C. 1.10) that the germs $z_1, \ldots, z_n$ generate the maximal ideal $\mathfrak{m} = \mathfrak{m}_n$ in $\mathcal{O}_n$. Consequently, $\{z^p\}_{|p|=k}$ is a system of generators for the ideal $\mathfrak{m}^k$ (see A. 1.17). Therefore the ideal $\mathfrak{m}^k$ is the set of germs of holomorphic functions f at 0 such that $f(z) = o(|z|^{k-1})$ (see C. 1.10), or, equivalently, whose derivatives of order $< k$ vanish at 0 (see C. 1.5), or whose power series expansion at 0 is of the form $\sum_{|p| \geq k} a_p z^p$. As a result, we have the direct sum $\mathcal{O}_n = P_{k-1} + \mathfrak{m}^k$, where P_{k-1} denotes the vector subspace of germs of polynomials of degree $\leq k - 1$. The subspace P_{k-1} is finite dimensional (see n° 1).

Thus each of the ideals $\mathfrak{m}^k$, regarded as a vector subspace of $\mathcal{O}_a$, is of finite codimension.

The following equivalence holds for the ideal I of the ring $\mathcal{O}_a$:

$$\text{codim } I < \infty \iff I \supset \mathfrak{m}^k \text{ for some } k .$$

Therefore, the defining ideals of the ring $\mathcal{O}_a$ are precisely its proper ideals of finite codimension (see A. 10.5).

To see this, suppose codim $I < \infty$. Then the sequence of natural numbers $\{\text{codim } (I + \mathfrak{m}^k)\}_{k=1,2,\ldots}$ is increasing and bounded, and hence it is stationary. Accordingly, the sequence $\{I + \mathfrak{m}^k\}$ is also stationary (see A. 1.17). Therefore for some k we have $\mathfrak{m}^k \subset I + \mathfrak{m}^{k+1}$ and so, by the Nakayama lemma, $\mathfrak{m}^k \subset I$.

The ring $\mathcal{O}_a$ is a local **C**–ring (see A. 10.4a), because $\mathcal{O}_a = \mathbf{C} + \mathfrak{m}_a$ is a direct sum. Thus the residual field $\mathcal{O}_a / \mathfrak{m}_a$ of the ring $\mathcal{O}_a$ can be identified with **C** via the isomorphism $\mathbf{C} \ni \zeta \longrightarrow \tilde{\zeta} \in \mathcal{O}_a / \mathfrak{m}_a$, where $\tilde{\zeta}$ denotes the

---

(^{12}) If $\chi(I)$ is finitely generated for an automorphism χ, then so is the ideal I.

equivalence class of the germ of the constant ζ (see A. 10.4a). For every module M over $\mathcal{O}_a$, the module $M/(\mathfrak{m}_a M)$ is a vector space over both $\mathcal{O}_a/\mathfrak{m}_a$ and $\mathbf{C}$. (Moreover, $\zeta x = \tilde{\zeta} x$ for $\zeta \in \mathbf{C}$ and $x \in M/(\mathfrak{m}_a M)$.) In both cases we have the same subspaces with the same bases, dimensions, and codimensions (see A. 10.4a).

8. Proposition 2. *The ring $\mathcal{O}_a$ is regular of dimension $n = \dim M$.*

Proof . It is enough to consider the ring $\mathcal{O}_n$. The ring $\mathcal{O}_0 = \mathbf{C}$ is a field, and hence is regular and 0–dimensional (see A. 15). Let $n > 0$. Assume that $\mathcal{O}_{n-1}$ is regular and $(n-1)$–dimensional. By Hadamard's lemma (see C. 1.10), the kernel of the epimorphism $\mathcal{O}_n \ni f \longrightarrow f(z_1, \ldots, z_{n-1}, 0) \in \mathcal{O}_{n-1}$ (see n° 2) is equal to the ideal $\mathcal{O}_n z_n$. Therefore $\mathcal{O}_n/(\mathcal{O}_n z_n) \approx \mathcal{O}_{n-1}$, which implies that the ring $\mathcal{O}_n$ is regular of dimension n (see A. 15 lemma 1, and the proposition in A. 12.3).

Note that the germs $f_1, \ldots, f_n$ in the ideal $\mathfrak{m}_a$ generate $\mathfrak{m}_a$ if and only if their differentials $d_a f_1, \ldots, d_a f_n$ are linearly independent.

Indeed, it suffices to consider the case of $\mathcal{O}_n$. The mapping $\mathfrak{m} \ni g \longrightarrow d_0 f \in L = (\mathbf{C}^n)^*$ is a linear epimorphism whose kernel is $\mathfrak{m}^2$ (see n° 7). Thus $\mathfrak{m}/\mathfrak{m}^2 \ni \tilde{f} \longrightarrow d_0 f \in L$ is an isomorphism, where $\tilde{f}$ denotes the equivalence class of the germ $f \in \mathfrak{m}$. But the germs f_i generate $\mathfrak{m}$ if and only if $\tilde{f}_i$ generate the linear space $\mathfrak{m}/\mathfrak{m}^2$ (see A. 10.4) or, equivalently, if the differentials $d_0 f_i$ generate L. The last condition means that the differentials are linearly independent $(^{13})$.

Observe that the rank of an ideal $I \subset \mathfrak{m}_a$ of $\mathcal{O}_a$ (see A. 10.4) can be expressed by means of the generators $g_1, \ldots, g_k$ of I as follows:

$$\operatorname{rank} I = \operatorname{rank}_a(g_1, \ldots, g_k) \quad (^{14}).$$

$(^{13})$ Here is an alternative proof. If f_j generate $\mathfrak{m} = {}_n\mathfrak{m}$, then $z_i = \sum_j a_{ij} f_j$ for some $a_{ij} \in \mathcal{O}_n$. Thus $\delta_{ij} = \sum_j a_{ij}(0)\frac{\partial f_j}{\partial z_i}(0)$, and so $\det \frac{\partial f_i}{\partial z_j}(0) \neq 0$. Conversely, if $\det \frac{\partial f_i}{\partial z_j} \neq 0$, then $\sum_j c_{ij} \frac{\partial f_j}{\partial z_k}(0) = \delta_{ik}$ for some $c_{ij} \in \mathbf{C}$. Therefore

$$z_i = \sum_j c_{ij}\left(\sum_k \frac{\partial f_j}{\partial z_k}(0)z_k\right) = \sum_j c_{ij} f_j(z) + o(|z|) ,$$

which implies that $\mathfrak{m} \subset \sum_j \mathcal{O}_n f_j + \mathfrak{m}^2$. Finally, in view of the Nakayama lemma,

$$\mathfrak{m} \subset \sum_j \mathcal{O}_n f_j .$$

See also II. 4.2.

For, in the case of $\mathcal{O}_n$, the image of $(I+\mathfrak{m}^2)/\mathfrak{m}^2$ by the isomorphism $\mathfrak{m}/\mathfrak{m}^2 \longrightarrow L$ induced by the epimorphism $\mathfrak{m} \ni f \longrightarrow d_0 f \in L$ is equal to the image of I by this epimorphism, i.e., to the subspace $\sum \mathbb{C} d_0 g_i$ whose dimension is $\mathrm{rank}_0(g_1, \ldots, g_k)$.

§2. Unique factorization property

1. A germ from Q_n is said to be *distinguished of degree k* if it is the germ of a distinguished polynomial in z_n of degree k; in other words, if it is monic, of degree k and with all coefficients (but the leading one) belonging to $\mathfrak{m}_{n-1}$ (see C. 2.2). It is, clearly, regular of order k. Furthermore, it is non-invertible in $\mathcal{O}_n$ (or, equivalently, in Q_n) precisely when $k > 0$ [15] .

The *classical version of the Weierstrass preparation theorem* (see C. 2.4) can now be stated as follows:

Any germ from $\mathcal{O}_n$ which is regular of order k is associated in $\mathcal{O}_n$ with a unique distinguished germ. Moreover, the degree of the distinguished germ is k.

Consequently, distinguished germs that are associated in $\mathcal{O}_n$ must coincide.

Note also that if a germ $g \in Q_n$ is distinguished and $h \in \mathcal{O}_n$, then $gh \in Q_n \Longrightarrow h \in Q_n$.

Indeed, we have $gh = qg + r$, where $q, r \in Q_n$ and the degree of r is less than that of g (see A. 2.4). But the degree of the germ g is equal to the order of g. Thus, in view of the uniqueness part of the preparation theorem (see 1.4), h must be equal to q.

Therefore, if a distinguished germ divides in $\mathcal{O}_n$ a germ from Q_n, then the same is true in Q_n.

2a. A distinguished germ of positive degree is reducible in $\mathcal{O}_n$ if and only if it is a product of distinguished germs of positive degree.

Indeed, suppose that a distinguished germ c is the product of germs g_1 and g_2 which are non-invertible in $\mathcal{O}_n$. Then g_1, g_2 are regular and of positive order (see n° 1 and 1.4). By the preparation theorem, we have $g_i = h_i c_i$, where h_i are invertible in $\mathcal{O}_n$. However the c_i are distinguished and of positive

[14] The right hand side is defined as $\mathrm{rank}_a(\tilde{g}_1, \ldots, \tilde{g}_k)$, where the $\tilde{g}_i$'s are representatives of the g_i's.

[15] The only distinguished germ of degree 0 is 1.

degree. Thus $c = (h_1 h_2)(c_1 c_2)$. Because of the uniqueness in the preparation theorem, $c = c_1 c_2$.

2b. A distinguished germ is reducible in $\mathcal{O}_n$ precisely when it is reducible in Q_n.

To see this, take a distinguished germ $c \in \mathcal{O}_n$. We may assume that c is of positive degree (for otherwise $c = 1$ is irreducible in both $\mathcal{O}_n$ and Q_n). Now, in view of 2a, if c is reducible in $\mathcal{O}_n$, then it is also reducible in Q_n. Conversely, if c is the product of two non-invertible germs in Q_n, we may assume that they are monic, since the product of their leading coefficients is 1. Then they are distinguished of positive degree and hence are non-invertible in $\mathcal{O}_n$ (see n° 1).

3. PROPOSITION . *The ring $\mathcal{O}_a$ is a unique factorization domain.*

PROOF . It is enough to consider the ring $\mathcal{O}_n$. Now, the ring $\mathcal{O}_0 = \mathbf{C}$ is a unique factorization domain (see A. 6.1). Let $n > 0$, and assume that the ring $\mathcal{O}_{n-1}$ is a unique factorization domain. Then, by the Gauss theorem (see A. 6.2), the ring Q_n – being isomorphic to $\mathcal{O}_{n-1}[X]$ – is a unique factorization domain.

In view of A. 9.5, it suffices to show that if an irreducible germ $f \in \mathcal{O}_n$ divides the product of germs $g, h \in \mathcal{O}_n$, then it must divide one of them. By the preparation theorem (see n° 1), we may assume that the germ f is distinguished and $g, h \in Q_n$ (see 1.4 ([16])). But then the germ f is also irreducible in Q_n (see n° 2a) and is a divisor in Q_n of the product gh (see n° 1). Therefore f is a divisor in Q_n of g or of h, and so the same is true in $\mathcal{O}_n$.

REMARK. We have also proved that the ring Q is a unique factorization domain.

As a corollary we have the following property:

In the ring $\mathcal{O}_n$, every non-zero non-invertible germ f has a decomposition $f = f_1^{k_1} \dots f_r^{k_r}$, where $k_i > 0$ and the f_i's are irreducible and mutually non-associated ([17]) .

According to the proposition from A. 6.3, we get (see 1.5) the corollary:

The discriminant of a monic germ $p \in Q_n$ is zero if and only if p is divisible by the square of a monic germ of positive degree.

This implies that:

([16]) For if $\chi : \mathcal{O}_n \longrightarrow \mathcal{O}_n$ is an automorphism, then, if f is irreducible, so is $\chi(f)$, and if $\chi(f)$ divides the product $\chi(g)\chi(h)$, then f divides the product gh.

([17]) It is sufficient to notice that if a germ $g \in \mathcal{O}_n$ is invertible and $k > 0$, then $g = h^k$ for some invertible germ $h \in \mathcal{O}_n$.

A regular germ from $\mathcal{O}_n$ has no multiple factors precisely when the discriminant of its associated distinguished germ (see n° 1) *is non-zero.* (See A. 6.1, 1.4, and n° 1.)

In particular,

The discriminant of a distinguished irreducible germ (see n° 2b) *is non-zero.*

4. Let $p \in \mathcal{O}_n$ be a distinguished germ. If the degree of p is positive, then p can be represented in a unique fashion as $p = p_1^{k_1} \ldots p_r^{k_r}$, where the p_i's are distinguished, irreducible (and hence of positive degree; see n° 1), mutually distinct, and the k_i's are all positive. Then the discriminant of the germ p is non-zero precisely when $k_1 = \ldots = k_r = 1$.

Indeed, $p = q_1 \ldots q_s$, where the germs q_j are irreducible in Q_n. We may assume that they are monic (as the product of their leading coefficients is 1). Then they have to be distinguished (see n° 1). This yields the above decomposition and proves its uniqueness (see n° 1).

We define red $p = p_1 \ldots p_r$, and, in addition, let red $p = 1$ if $p = 1$. For any regular element $f \in \mathcal{O}_n$, we define red $f = $ red p, where p is the distinguished element associated with f (via the preparation theorem). Accordingly, the discriminant of red f is always non-zero ([18]).

§3. The Preparation Theorem in Thom-Martinet version

LEMMA . *Let M be a finite module over $\mathcal{O}_{n+k}$, and let N be finitely generated submodule of M regarded as a module over $\mathcal{O}_n$. Then $M = N + \mathfrak{m}_n M$ implies that $M = N$.*

PROOF . By the Nakayama lemma (see A. 10.2), it is enough to show the implication:

(∗)   *If $M = N + \mathfrak{m}_n M$, then M is finite over $\mathcal{O}_n$.*

Suppose $k = 1$. Let $m = N + \mathfrak{m}_n M$. Then M is finite over the ring $S = \mathcal{O}_n + \mathfrak{m}_n \mathcal{O}_{n+1}$. In view of the Mather-Nakayama lemma (see A. 1.16),

---

([18]) Obviously, the zero sets of representatives of the germs f and red f coincide in a neighbourhood of zero. Then *the zero set of a holomorphic function $\not\equiv 0$ is – in a suitably chosen local coordinate system – the zero set of a distinguished polynomial with non-zero discriminant.*

there exists a germ $\eta = z_{n+}^r + \gamma_1 z_{n+1}^{r-1} + \ldots + \gamma_r$, where $\gamma_i \in \mathcal{S}$, such that $\eta M = 0$. Now, the germ η is regular (since the $\gamma_i(0, \ldots, 0, z_{n+1})$ are constant germs). Denote its order by p. Let $m_1, \ldots, m_s$ be generators of the module M over $\mathcal{O}_{n+1}$. If $x \in M$, then $x = \sum_1^s f_i m_i$, where $f_i \in \mathcal{O}_{n+1}$, but the preparation theorem in division version gives $f_i = g_i \eta + \sum_{j=0}^{p-1} \alpha_{ij}(z_{n+1})^j$, where $g_i \in \mathcal{O}_{n+1}$ and $\alpha_{ij} \in \mathcal{O}_n$. Hence $x = \sum_{ij} \alpha_{ij}(z_{n+1})^j m_i$. Thus the elements $(z_{n+1})^j m_i$ $(i = 1, \ldots, s; j = 0, \ldots, p-1)$ generate M over $\mathcal{O}_n$.

Now let $k > 0$ and let us assume that our implication holds for $k - 1$. Suppose $M = N + \mathfrak{m}_n M$. Then $M = \mathcal{O}_{n+k-1} N + \mathfrak{m}_{n+k-1} M$. Since $\mathcal{O}_{n+k-1} N$ is a finitely generated submodule of M regarded as a module over $\mathcal{O}_{n+k-1}$, the already verified implication $(*)$ for $k = 1$ (with $n + k - 1$ replacing n) shows that the module M is finite over $\mathcal{O}_{n+k-1}$. Hence, by the induction hypothesis, M is finite over $\mathcal{O}_n$.

COROLLARY . *Assume the hypothesis of the lemma, and let L be a submodule of M. Then $M = L + N + \mathfrak{m}_n M$ implies $M = L + N$.*

Indeed, it is sufficient to apply the lemma to the images under the natural homomorphism $M \longrightarrow M/L$ of $\mathcal{O}_n$–modules.

Denote by $\mathcal{O}_v$ the ring of germs at 0 of functions that are holomorphic with respect to the variables $v = (z_{n+1}, \ldots, z_{n+k})$. If I is an ideal of the ring $\mathcal{O}_{n+k}$, then $I(0, v) = \{f(0, v) : f \in I\}$ is an ideal of the ring $\mathcal{O}_v$. It is equal to the image of the ideal I under the epimorphism $\chi : \mathcal{O}_{n+k} \ni f \longrightarrow f(0, v) \in \mathcal{O}_v$ (see 1.2).

THE PREPARATION THEOREM IN THOM-MARTINET VERSION. *Let I be an ideal of the ring $\mathcal{O}_{n+k}$, and let $a_1, \ldots, a_r \in \mathcal{O}_{n+k}$. Then*

$$\mathcal{O}_v = I(0, v) + \sum_1^r \mathbf{C} a_i(0, v) \iff \mathcal{O}_{n+k} = I + \sum_i^r \mathcal{O}_n a_i \quad (^{19}) .$$

PROOF . By considering the images under χ, one can see that the left hand side follows from the right hand side. Now note that, according to Hadamard's lemma (see C. 1.10), we have $\ker \chi = \mathfrak{m}_n \mathcal{O}_{n+k}$. Assuming the equality on the left hand side, we have $\chi(I + \sum_1^r \mathcal{O}_n a_i) = \mathcal{O}_v$, and so

---

(^{19}) This version implies the division version (except for the uniqueness property). Indeed, if a germ $f \in \mathcal{O}_n$ is regular of order k, then (taking $v = z_{n+1}$) we have $\mathcal{O}_v = \sum_{i=0}^{k-1} \mathbf{C} v^i + \mathcal{O}_v f(0, v)$ (in view of Taylor's formula – since $f(0, v)$ generates $\mathfrak{m}_1^k$). Therefore $\mathcal{O}_n = \sum_0^{k-1} \mathcal{O}_{n-1} z_n^i + \mathcal{O}_n f$.

$$\mathcal{O}_{n+k} = I + \sum_{1}^{r} \mathcal{O}_n a_i + \mathfrak{m}_n \mathcal{O}_{n+k} \quad (^{20}) \ .$$

Thus the corollary implies the equality on the right hand side above.

---

(20) If $\chi : M \longrightarrow N$ is an epimorphism of commutative groups, $E \subset M$, and $\chi(E) = N$, then $M = E + \ker \chi$.

CHAPTER II

ANALYTIC SETS, ANALYTIC GERMS, AND THEIR IDEALS

§1. Dimension

1. Let M be a complex n–dimensional manifold. We define the (complex) *dimension of a subset $E \subset M$* by the formula

$$\dim E = \sup\{\dim \Gamma : \ \Gamma \text{ is a submanifold contained in } E\} \quad (^1) \ .$$

In the case E is a submanifold, this definition is consistent with the one used previously (see C. 3.7). If $E \subset N \subset M$, where N is a submanifold, then $\dim E$ does not change when E is regarded as a subset of the manifold N. Biholomorphic mappings preserve the dimension: if $h : \ M \longrightarrow N$ is a biholomorphic mapping between complex manifolds M, N, then $\dim h(E) = \dim E$ for any $E \subset M$.

We define also the *codimension of a subset $E \subset M$* by codim $E = n - \dim E$.

2. Clearly, $\dim \emptyset = -\infty$. The dimension of any countable non-empty set is equal to 0. Note that $\dim E = n \iff \text{int } E \neq \emptyset$. Thus, if E is nowhere dense, $\dim E < n$.

3. We have $E \subset F \implies \dim E \leq \dim F$.

$(^1)$ We assume here that the maximum on the empty set is equal to $-\infty$. This definition of the dimension (which we use for any subset of the manifold M as a technical convenience) is useful only for a rather narrow class of sets that contains all analytic sets. For the latter, it is equivalent to standard definitions of dimension used in analytic geometry.

If subsets E_ι are open in their union $\bigcup E_\iota$, then $\dim(\bigcup E_\iota) = \max \dim E_\iota$ [2]. Also, for any countable family of F_σ–subsets E_i, we have

$$\dim\left(\bigcup E_i\right) = \max \dim E_i \ .$$

Indeed, if a submanifold Γ is contained in $\bigcup E_i$, then – due to the fact that Γ is a Baire space (see C. 3.4) – at least one of the sets E_i contains a non-empty open set in Γ.

4. Let $E \subset M \times N$ (where N is a complex manifold), let $\pi : E \longrightarrow M$ be the natural projection, and let $k \in \mathbf{N}$.

If $\dim \pi^{-1}(z) \leq k$ for $z \in \pi(E)$, then $\dim E \leq k + \dim \pi(E)$.

Indeed, let $\Gamma \subset M \times N$ be a non-empty submanifold contained in E. Take a point $c \in \Gamma$ at which the rank of the mapping $\pi : \Gamma \longrightarrow M$ is maximal. Then π has constant rank in a neighbourhood of the point c in Γ (see C. 3.12), and hence, by the rank theorem (see C. 4.1), there is a neighbourhood Γ_0 of the point c in Γ such that $\pi(\Gamma_0)$ is a submanifold and the non-empty fibres of the mapping π_{Γ_0} are submanifolds of dimension $\dim \Gamma_0 - \dim \pi(\Gamma_0) \leq k$. Thus $\dim \Gamma \leq k + \dim \pi(E)$.

It follows that

$$\dim(E \times F) = \dim E + \dim F$$

for non-empty subsets E and F of complex manifolds.

5. The *dimension of a subset* $E \subset M$ *at a point* $a \in M$ is defined as $\dim_z E = \dim(E \cap U)$, where U is a sufficiently small neighbourhood of the point a (so that the right hand side is independent of U). Therefore $\dim E = \max_{z \in E} \dim_z E = \max_{z \in M} \dim_z E$ [3]. We have: $E \subset F \Longrightarrow \dim_a E \leq \dim_a F$. If $E_1, \ldots, E_k$ are F_σ–subsets, then $\dim_a(E_1 \cup \ldots \cup E_k) = \max_i \dim_a E_i$. If $E \subset M, F \subset N$ (where N is a complex submanifold), then $\dim_{(a,b)}(E \times F) = \dim_a E + \dim_b F$. If $h : M \longrightarrow N$ is biholomorphic mapping of complex manifolds, then $\dim_{h(a)} h(E) = \dim_a E$ for $E \subset M$ and $a \in M$. Note also that the function $M \ni z \longrightarrow \dim_z E$ is upper semicontinuous.

We say that a subset $E \subset M$ is of *constant dimension* k if $\dim_z E = k$ for $z \in E$ [4]. In such a case, also $\dim_z E = k$ for $z \in \bar{E}$. If F_σ–subsets

---

[2] See footnote [1].

[3] See footnote [1].

[4] Then, obviously, $\dim E = k$, provided that $E \neq \emptyset$. (The set $\emptyset$ is of constant dimension k for any $k \leq n$.)

E_ι are of constant dimension k and constitute a locally finite family, then their union $\bigcup E_\iota$ is of constant dimension k. Observe that if a subset F is of constant dimension and $E \subset F$, then $\dim E < \dim F \implies \operatorname{int}_F E = \emptyset$.

6. Let $a \in M$. The *dimension of a germ* A of a set at a point a is defined by $\dim A = \dim_a \tilde{A}$, where $\tilde{A}$ is any representative of the germ A (the right hand side is independent of the choice of representative). Thus $\dim_a E = \dim E_a$ for $E \subset M$. If A and B are germs of sets at a, then $A \subset B \implies \dim A \leq \dim B$. If $A_1, \ldots, A_k$ are germs of F_σ-sets at a, then $\dim(A_1 \cup \ldots \cup A_k) = \max \dim A_i$. If A and B are germs of sets at the points $a \in M$ and $b \in N$, respectively (where N is a complex manifold), then $\dim(A \times B) = \dim A + \dim B$. If $h : M \longrightarrow N$ is a biholomorphic mapping between complex manifolds, then $\dim h(A) = \dim A$ for any germ A of a set at any point $a \in M$. If $A \subset N \subset M$, where $N \ni a$ is a submanifold, then the dimension $\dim A$ remains the same when A is regarded as a germ in N.

If M is a vector space and $S \subset M$ is a cone, then $\dim S = \dim S_0$. Indeed, $S = \bigcup_{k=1}^{\infty} k(S \cap U)$, where U is an open neighbourhood of zero (see n° 3 and 1).

We define also the *codimension of a germ* A by $\operatorname{codim} A = n - \dim A$.

§2. Thin sets

1. Let M be a complex manifold. A subset $Z \subset M$ is said to be *thin* (in M) if it is closed, nowhere dense, and for any open set $\Omega \subset M$ every holomorphic function on $\Omega \setminus Z$ which is locally bounded near $Z \cap \Omega$ (i.e., on Ω; see C. 1.11) extends to a holomorphic function on Ω. In such a case, every holomorphic mapping of the set $\Omega \setminus Z$ into a complex vector space which is locally bounded near $Z \cap \Omega$ (i.e., on Ω) has a holomorphic extension to Ω. If $h : M \longrightarrow N$ is a biholomorphic mapping between complex manifolds, then the set $h(Z)$ is also thin (in N).

2. Any set satisfying the assumptions of Riemann's lemma (see C. 1.11) is thin in G.

In particular, if B is an open subset of a vector space and P is the restriction of a monic polynomial, with coefficients continuous in B, to an open subset $G \subset B \times \mathbf{C}$, then the set $\{P = 0\}$ is thin in G.

3. Any closed subset of a thin set is thin.

If $Z \subset M$ is thin and $\Omega \subset M$ is open, then $Z \cap \Omega$ is thin in Ω. If $Z \subset M$ is closed, $\Omega \subset M$ is open, and $Z \subset \Omega$, then the set Z is thin in M if and only if it is thin in Ω. If $\{\Omega_\iota\}$ is an open cover of the manifold M, then the set Z is thin in M precisely when for each ι the set $Z \cap \Omega_\iota$ is thin in Ω_ι.

If Z is thin in M and W is thin in $M \setminus Z$, then $Z \cup W$ is thin in M.

The union of a locally finite family of thin subsets is a thin subset. (Indeed, in the case of two such sets Z_1, Z_2, the set $Z_2 \setminus Z_1$ is thin in $M \setminus Z_1$.)

4. If the manifold M is connected and Z is a thin subset of M, then the set $M \setminus Z$ is connected. (For otherwise there would exist a locally constant function on $M \setminus Z$ that takes only two values. It would be holomorphic, bounded, and would not extend to a holomorphic function on M; see C. 3.9.)

5. Let N be a complex manifold. A subset of N is said to be *l–bounded* if it is relatively compact in the domain of a chart. Clearly, every subset of such a set is also *l*–bounded. If Z is a closed subset of the manifold M, a mapping $f : M \setminus Z \longrightarrow N$ is said to be *locally l–bounded near Z* if each point of Z has a neighbourhood U in M such that the set $f(U \setminus Z)$ is *l*–bounded [5] . If Z is a thin set, it follows that:

Every holomorphic mapping $f : M \setminus Z \longrightarrow N$ which is locally *l*–bounded near Z has a holomorphic extension to M [6] .

In particular:

If a holomorphic mapping $f : M \setminus Z \longrightarrow N$ extends to a continuous mapping on M, then this extension is holomorphic.

§3. Analytic sets and germs

1. Let M be a complex manifold. By a *globally analytic subset* of the manifold M we mean any set of the form

$$V(f_1, \ldots, f_k) = \{ z \in M : f_1(z) = \ldots = f_k(z) = 0 \} ,$$

where $f_1, \ldots, f_k$ are holomorphic functions on M. (Obviously, it is a closed set.) We say also that this subset is *defined* (in M) *by the functions* $f_1, \ldots, f_k$.

---

[5] Clearly, if N is a vector space, then local boundedness near Z implies local *l*–boundedness near Z.

[6] One cannot replace the *l*–boundedness by relative compactness, e.g., the holomorphic mapping $\mathbf{C}^2 \setminus 0 \ni (z, w) \longrightarrow w/z \in \bar{\mathbf{C}}$ does not have even a continuous extension to $\mathbf{C}^2$.

A globally analytic subset that can be defined by a single function is said to be *principal*. Any finite union and finite intersection of globally analytic subsets in M is a globally analytic subset in M. Furthermore, if the set Z_i is defined by the functions $f_1^{(i)}, \ldots, f_{s_i}^{(i)}$ $(i = 1, \ldots, k)$, then the set $Z_1 \cup \ldots \cup Z_k$ is defined by all the functions $f_{\nu_1}^{(1)} \ldots f_{\nu_k}^{(k)}$ $(\nu_i = 1, \ldots, s_i)$. The Cartesian product of a globally analytic subset of the manifold M and a globally analytic subset of the manifold N is a globally analytic subset of $M \times N$. If $f : M \longrightarrow N$ is a holomorphic mapping between manifolds, then the inverse image of a globally analytic subset of N is a globally analytic subset of M [7]. If $N \subset M$ is a submanifold and Z is a globally analytic subset of M, then $Z \cap N$ is a globally analytic subset of N.

Note that if the manifold M is connected, then every globally analytic proper subset of M is nowhere dense. Indeed, its interior must be empty (see C. 3.9).

2. If M is a complex vector (or affine) space, then a globally analytic subset defined by polynomials on M, i.e., of the form $V(f_1, \ldots, f_k)$, where $f_1, \ldots, f_k \in \mathcal{P}(M)$, is called an *algebraic subset* of the space M. The properties of globally analytic subsets listed above (see n° 1) have obvious counterparts for algebraic subsets. In particular, the inverse image of an algebraic subset under a polynomial mapping is also an algebraic subset. If $N \subset M$ is an affine subspace, then a subset is algebraic in N precisely when it is algebraic in M. Any proper algebraic subset is nowhere dense.

If M and N are vector spaces, the sets $\{f \in L(M, N) : \operatorname{rank} f \leq k\}$, $k = 0, 1, \ldots$, are algebraic in $L(M, N)$. (For this holds when $M = \mathbf{C}^m$ and $N = \mathbf{C}^n$.) In particular, if $\dim M = \dim N$, the set $L_0(M, N) \subset L(M, N)$ is the complement of a nowhere dense algebraic subset, and hence it is open and dense in $L(M, N)$.

3. Let M be a complex manifold. By an *analytic germ at a point $a \in M$* we mean the germ at a of a globally analytic subset of an open neighbourhood of a. If this is the germ at a of a principal subset of a neighbourhood of a, it is said to be *principal*. The finite union and intersection of analytic germs at a are analytic germs at a. The Cartesian product of an analytic germ at a point a of the manifold M with an analytic germ at a point b of a manifold N is an analytic germ at the point (a, b) of the manifold $M \times N$. If $N \subset M$ is a submanifold, then the germ of a set at $a \in N$, which is contained in N, is analytic in N if and only if it is analytic in M. (It is enough to check this for a vector space M and a subspace N.)

---

[7] Naturally, biholomorphic mappings of manifolds preserve global analyticity of subsets.

If $h : M \longrightarrow N$ is a biholomorphic mapping of manifolds, then the image of an analytic germ at a is an analytic germ at $h(a)$. If A is a germ at a and φ is a coordinate system at a, then by the *germ A in the coordinate system* φ we mean the germ $\varphi(A)$ (at $0 \in \mathbf{C}^n$, where $n = \dim M$). (Its analyticity is clearly equivalent to that of the germ A.)

Note that the only analytic germ of dimension $n = \dim M$ is the germ of M (see 1.2 and n° 1).

A *smooth* germ at a is the germ at a of a submanifold containing a. Obviously, its dimension is equal to that of the submanifold. It is an analytic germ (see C. 3.15(a)). If C is a smooth germ at a and A is an analytic one at a, then $A \subsetneq C \Longrightarrow \dim A < \dim C$ [8] . For if it were not the case, A would be the full germ in a manifold representing C.

If M is a vector space, we have the following

CARTAN-REMMERT-STEIN LEMMA . *Every cone $S \subset M$ whose germ S_0 is analytic is an algebraic subset defined by homogeneous polynomials.*

PROOF . There is an open neighbourhood U of zero such that $S \cap U = \{f_1 = \ldots = f_k = 0\}$ for some functions f_i that are holomorphic in U, and $f_i = \sum_{\nu=0}^{\infty} f_{i\nu}$ in U, where $f_{i\nu}$ is a form of degree ν. Since the ring $\mathcal{P}(M)$ is noetherian (see A. 9.4), there exists an s such that for each i the ideal generated by the set $\{f_{i0}, f_{i1}, \ldots\}$ is generated by the forms $f_{i0}, \ldots, f_{is}$, hence each form $f_{i\nu}$, $\nu > s$, is a combination of those s forms. Thus $\tilde{S} \cap U \subset S \cap U$, where $\tilde{S} = \{f_{i\nu} = 0 : i = 1, \ldots, k, \nu = 0, \ldots, s\}$ is an algebraic cone, and so $\tilde{S} \subset S$ (see B. 5.1). Now, let $z \in S$. There is an $\varepsilon > 0$ such that if $|t| < \varepsilon$, then $tz \in U$, and hence $f_i(tz) = \sum_{\nu=0}^{\infty} f_{i\nu}(z)t^{\nu} = 0$. Consequently, $f_{i\nu}(z) = 0$, which means that $z \in \tilde{S}$. Therefore $S = \tilde{S}$ is an algebraic subset defined by homogeneous polynomials.

4. A subset Z of a manifold M is called an *analytic subset* (of M or in M) if its germ at any point of the manifold M is analytic or, equivalently, if every point of the manifold M has an open neighbourhood U such that the set $Z \cap U$ is a globally analytic subset of U [9] . In particular, any closed submanifold of the manifold M is an analytic subset. Any subset of an analytic subset which is closed and open in the induced topology is analytic. The union and the intersection of a locally finite family of analytic subsets of

---

[8] More generally, the implication is true if C is a simple germ; see IV.3, prop. 2.

[9] Obviously, a globally analytic subset of M is analytic in M. Generally speaking, the converse is not true. For instance, the only globally analytic subsets of a compact connected manifold M are $\emptyset$ and M. This is so because, owing to the maximum principle (see C. 3.9), every holomorphic function on M is constant.

M is an analytic subset of M. The Cartesian product of analytic subsets of the manifolds M and N, respectively, is an analytic subset of the manifold $M \times N$ $(^{10})$. If $f : M \longrightarrow N$ is a holomorphic mapping of manifolds, then the inverse image of an analytic subset of N is an analytic subset of M. If $N \subset M$ is a submanifold and Z is an analytic subset of M, then $Z \cap N$ is an analytic subset of N. If $N \subset M$ is a closed submanifold, then a subset of N is analytic in N if and only if it is analytic in M. If $\{G_\iota\}$ is an open cover of the manifold M, then a set $Z \subset M$ is analytic in M precisely when, for each ι, the set $Z \cap G_\iota$ is analytic in G_ι. Every analytic subset is closed (see B. 1).

Analytic subsets of open subsets of the manifold M are called *locally analytic subsets* of M (or in M). Therefore, a set $Z \subset M$ is a locally analytic subset (of M) if and only if its germ at any of its points is analytic or, in other words, if each of its points has an open neighbourhood U such that $Z \cap U$ is globally analytic in U. In particular, every submanifold of the manifold M is a locally analytic subset of M. For subsets Z, M, we have the equivalence

$$(Z \text{ is analytic}) \Longleftrightarrow (Z \text{ is locally analytic and closed}) .$$

Any open subset in a locally analytic subset is locally analytic. If $N \subset M$ is a submanifold, then: if Z is locally analytic in M, the set $Z \cap N$ is locally analytic in N. If $Z \subset N$, then

$$(Z \text{ is locally analytic in } N) \Longleftrightarrow (Z \text{ is locally analytic in } M) .$$

The Cartesian product of locally analytic subsets of the manifolds M and N, respectively, is a locally analytic subset of the manifold $M \times N$ $(^{11})$. If $f : M \longrightarrow N$ is a holomorphic mapping of manifolds, then the inverse image of a locally analytic subset of N is a locally analytic subset of M. The intersection of a locally finite family of locally analytic subsets is a locally analytic subset (whereas the union of two or more locally analytic subsets is not necessarily a locally analytic subset). Clearly, every locally analytic subset is locally closed.

If V and W are non-empty subsets of the manifolds M and N, respectively, then:

$$(V \times W \text{ is (locally) analytic in } M \times N) \Longleftrightarrow$$

$$\Longleftrightarrow (V \text{ and } W \text{ are (locally) analytic in } M \text{ and } N, \text{ respectively}) .$$

(For, e.g., V is the inverse image of the set $V \times W$ under the mapping $M \ni z \longrightarrow (z, b) \in M \times N$, where $b \in W$.)

$(^{10})$ The same holds for any finite number of factors.

$(^{11})$ The same holds for any finite number of factors.

If $f : M \longrightarrow N$ is a surjective submersion, the manifolds M and N are of dimension m and n, respectively, and $Z \subset N$, then

$$\big(Z \text{ is a (locally) analytic subset (of constant dimension)}\big) \Longleftrightarrow$$

$$\Longleftrightarrow \big(f^{-1}(Z) \text{ is a (locally) analytic subset (of constant dimension)}\big).$$

In this case

$$\dim f^{-1}(Z) = \dim Z + (m - n),$$

provided that $Z \neq \emptyset$. This follows directly from the definition of a submersion (similarly as in C. 4.2).

Naturally, all biholomorphic mappings between manifolds preserve (local) analyticity of subvarieties.

5. PROPOSITION . *Every nowhere dense analytic subset $Z \subset M$ is thin in* M.

This is a consequence of the following lemma:

LEMMA . *If Z is a representative of a non-full analytic germ at a point $a \in M$, then there is a coordinate system $\varphi : U \longrightarrow W \subset \Omega \times \mathbf{C}$ at the point a, and a distinguished polynomial P (with respect to z_n, where $n = \dim M$) whose coefficients are holomorphic in Ω and such that $Z \cap U \subset \varphi^{-1}(V(P))$.*

Indeed, if $Z \subset M$ is analytic and nowhere dense, then for any point $a \in M$ and any sufficiently small neighbourhood U chosen as in the lemma, the set $Z \cap U$ is thin in U since (see 2.2) the set $V(P) \cap W$ is thin in W (see 2.3 and 2.1).

In order to prove the lemma, let us take an open neighbourhood U of the point a such that the set $Z \cap U$ is defined by holomorphic functions. One of them, say f, has a non-zero germ f_a, for otherwise $a \in \text{int } Z$. Therefore, taking a smaller U, we can see that there is a coordinate system $\varphi : U \longrightarrow W$, where W is a neighbourhood of zero in $\mathbf{C}^n$, such that the function $g = f \circ \varphi^{-1}$ is z-regular (see I.1.4). Taking smaller U and W, we may assume, in view of the preparation theorem (see C. 2.4), that $V(g) = V(P) \cap W$, where P is a distinguished polynomial with coefficients holomorphic in Ω. We may also suppose that $W \subset \Omega \times \mathbf{C}$. Consequently, we have $Z \cap U \subset V(f) = \varphi^{-1}(V(g)) = \varphi^{-1}(V(P))$.

COROLLARY . *If M is connected, then the ring $\mathcal{O}_M$ is integrally closed.* (See C. 3.9.)

Indeed, let $(f/g)^r + a_1(f/g)^{r-1} + \ldots + a_r = 0$, $f, g, a_i \in \mathcal{O}_M$, $g \neq 0$. Then the equality holds for f/g regarded as a holomorphic function on $\{g \neq 0\}$.

But f/g must then be locally bounded on M (se B. 5.3), hence it extends to a holomorphic function $h \in \mathcal{O}_M$. This means that $f/g \in \mathcal{O}_M$ (since $f = gh$).

Using the notation of C. 2.1 and assuming that $D \not\equiv 0$, as another corollary we have the following

HENSEL'S LEMMA ([11a]). *Let $a \in B$ and $P(a,t) = (t - c_1)^{n_1} \ldots (t - c_r)^{n_r}$ with mutually distinct $c_1, \ldots, c_r$. Then $P = Q_1 \ldots Q_r$ in $U \times \mathbf{C}$, where Q_i are monic polynomials (in t) with holomorphic coefficients in an open neighbourhood $U \subset B$ of a, such that $Q_i(a,t) = (t - c_i)^{n_i}$.*

Indeed, let $\Omega_1, \ldots, \Omega_r$ be disjoint open neighbourhoods of $c_1, \ldots, c_r$, respectively. By the theorem of continuity of roots (see B. 5.3), there is an open neighbourhood $U \subset B$ of a such that for each $z \in U' = U \cap \{D \neq 0\}$ the polynomial $t \longrightarrow P(z,y)$ has precisely n_i distinct roots $w_1, \ldots, w_{n_i}$ in Ω_i. Then

$$Q_i'(z,t) = (t - w_1) \ldots (t - w_{n_i})$$

defines a monic polynomial (in t) whose coefficients (by the implicit function theorem) are holomorphic in U'. The latter are locally bounded on U (see B. 5.3), hence, by the proposition, they extend holomorphically on U. So we obtain polynomials $Q_i \subset Q_i'$ which satisfy the required conditions.

6. *If the manifold M is connected, then every proper analytic subset Z of M is nowhere dense and its complement $M \setminus Z$ is connected (and open).*

Indeed, owing to the fact that proper globally analytic subsets of a connected manifold are nowhere dense (see n° 1), int Z must be closed and hence empty. Therefore the set Z is nowhere dense and $M \setminus Z$ is connected, by the proposition in n° 5 (see 2.4).

7. Let $f : M \longrightarrow N$ be a holomorphic mapping of complex manifolds. Then the sets $\{z \in M : \mathrm{rank}_z f \leq k\}$, $k = 0, 1, \ldots$, are analytic because they are analytic in the case when M is an open subset of a vector space and N is a vector space (see n° 2, n° 1, and C.1.6).

Hence, if the manifold M is connected, then the set $\{z \in M : \mathrm{rank}_z f < \mathrm{rank}\, f\}$ is analytic and nowhere dense, whereas the set $\{z \in M : \mathrm{rank}_z f = \mathrm{rank}\, f\}$ is open, dense, and connected.

8. We are going to prove the following

PROPOSITION . *If Z is a representative of an analytic germ at a point $a \in M$, then there is an arbitrarily small open connected neighbourhood U*

([11a])   The assumption $D \not\equiv 0$ is not essential. The general case follows by taking the factorization of $P_a = T^n + (a_1)_a T^{n-1} + \ldots + (a_n)_a$ into irreducible monic factors in $\mathcal{O}_a[T]$ (see I. 2.3 and A. 6.2). See also [17a] pp. 44-45.

of the point a such that every function f that is holomorphic in $U \setminus Z$ and extends holomorphically across a (^{12}) *extends to a holomorphic function in U.*

To prove the proposition, one may assume that $a \notin \text{int } Z$. Note that the neighbourhood U from the lemma in n° 5 can be chosen arbitrarily small such that the set $Z \cap U$ is analytic and nowhere dense in U. At the same time, one can require the neighbourhood W to be of the form $\Omega \times \{|z_n| < 2r\}$ with a connected Ω, and such that $V(P) \subset \Omega \times \{|z_n| < r\}$ (see C. 2.2). For the triple consisting of a manifold N, a closed subset F of N, and a point $c \in N$, consider the condition:

Every holomorphic function on $N \setminus F$ that has a holomorphic extension on a neighbourhood of the point c, extends to a function which is holomorphic on N.

Now, according to the Hartogs lemma, the triple $(W, V(P), 0)$ satisfies this condition (see C. 1.11 and C. 2.2). Consequently, the condition is satisfied by the triples $\left(U, \varphi^{-1}(V(P)), a\right)$ and $(U, Z \cap U, a)$.

COROLLARY . *Let $f \in \mathcal{O}_a$. There exists an arbitrarily small open connected neighbourhood U of the point a, such that if $F \in \mathcal{O}_U$ is a representative of the germ f, then every function $G \in \mathcal{O}_U$ whose germ G_a is divisible by f is divisible by F in $\mathcal{O}_U$.*

In fact, we may assume that $f \neq 0$. Let $\tilde{f} \in \mathcal{O}_{\tilde{U}}$ be a representative of the germ f, and let $U \subset \tilde{U}$ be a neighbourhood of the point a chosen for the set $Z = V(\tilde{f})$ as in the proposition. Then $F = \tilde{f}_U$. Now, if the germ G_a of a $G \in \mathcal{O}_U$ is divisible by f, then the holomorphic function $U \setminus Z \ni z \longrightarrow G(z)/F(z)$ extends holomorphically across a. Hence it has a holomorphic extension H on U, which implies $G = HF$.

9. *If $Z \subset M$ is a nowhere dense analytic subset, then every sequence of holomorphic functions on M which is almost uniformly convergent in $M \setminus Z$ is also almost uniformly convergent in M.*

In fact, fix $a \in Z$. First, note that the neighbourhood W in the lemma from n° 5 can be taken to be of the form $\Omega \times \{|z_n| < 4r\}$ such that $V(P) \subset$

(^{12}) That is, there exists an open neighbourhood V of the point a such that $f_{V \setminus Z}$ has a holomorphic extension on V. Note that *if $\Sigma \subset M$ is a closed nowhere dense set, then it is easy to check that every holomorphic function on $M \setminus \Sigma$ which extends holomorphically across each point of the set Σ has a holomorphic extension to M.*

$\Omega \times \{|z_n| < r\}$ (see C. 2.2). For the pair consisting of a manifold N and its closed subset F, consider the condition

($\#$) Every sequence of holomorphic functions in N which is almost uniformly convergent in $N \setminus F$ is almost uniformly convergent in N.

In view of the maximum principle, the pair $\left(W, V(P)\right)$ fulfils the condition ($\#$). Indeed, if a sequence of holomorphic functions on W is almost uniformly convergent in $W \setminus V(P)$ and $E \subset \Omega$ is a compact subset, then the sequence – being uniformly convergent in the set $E \times \{2r < |z_n| < 3r\}$ – satisfies Cauchy's condition there. Therefore it also satisfies Cauchy's condition in the set $E \times \{|z_n| < 3r\}$ (see C. 3.9), and hence it is uniformly convergent there. Consequently, the pair $\left(U, \varphi^{-1}(V(P))\right)$ satisfies the condition ($\#$) and so does the pair $(U, Z \cap U)$. This implies – since the point $a \in Z$ was chosen arbitrarily – that the condition is also satisfied by the pair (M, Z).

§4. Ideals of germs and the loci of ideals. Decomposition into simple germs

Let M be a complex manifold, and let $a \in M$.

1. For any germ $f \in \mathcal{O}_a$ we define its *locus* $V(f) = V(\tilde{f})_a$, where $\tilde{f}$ is a holomorphic representative of the germ f. (The right hand side is independent of the choice of representative $\tilde{f}$.) Therefore the principal analytic germs (at a) are precisely the germs of the form $V(f)$, where $f \in \mathcal{O}_a$. Of course, $V(f) = V(g)$ if the germs $f, g \in \mathcal{O}_a$ are associated (see I. 1.2). We have

$$V(f_1 \ldots f_k) = V(f_1) \cup \ldots \cup V(f_k) \quad for \ f_1, \ldots, f_k \in \mathcal{O}_a .$$

For any germs $f_1, \ldots, f_k \in \mathcal{O}_a$, we define their *locus*

$$V(f_1, \ldots, f_k) = V(f_1) \cap \ldots \cap V(f_k) = V(\tilde{f}_1, \ldots, \tilde{f}_k)_a ,$$

where $\tilde{f}_1, \ldots, \tilde{f}_k$ are holomorphic representatives (with a common domain) of these germs. Every analytic germ (at a) is of this form, so it is a finite intersection of principal germs.

2. Let A be an analytic germ (at a). We say that a germ $f \in \mathcal{O}_a$ vanishes on A or $f = 0$ on A if $f_A = 0$, i.e., if a representative of the germ f vanishes on a representative of the germ A, or, equivalently, if $A \subset V(f)$. The set

$$\mathcal{I}(A) = \mathcal{I}(A, M) = \{f \in \mathcal{O}_a : f = 0 \text{ on } A\}$$
$$= \{f \in \mathcal{O}_a : f_A = 0\} = \{f \in \mathcal{O}_a : A \subset V(f)\}$$

is an ideal of the ring $\mathcal{O}_a$ and is called the *ideal of the analytic germ A*. Obviously,

$$A \subset B \Longrightarrow \mathcal{I}(A) \supset \mathcal{I}(B) \quad \text{and} \quad \mathcal{I}(A_1 \cup \ldots \cup A_k) = \mathcal{I}(A_1) \cap \ldots \cap \mathcal{I}(A_k),$$

for all analytic germs A, B, and A_i. Observe that

$$\operatorname{rad} \mathcal{I}(A) = \mathcal{I}(A) .$$

If A is a smooth germ of dimension k and $n = \dim M$, then for $f_1, \ldots$ $\ldots, f_s \in \mathcal{I}(A)$ we have $\operatorname{rank}(df_1, \ldots, df_s) \leq n - k$, *and equality holds if and only if $f_1, \ldots, f_s$ generate $\mathcal{I}(A)$. Thus we have the equivalence*

$$\big(f_1, \ldots, f_s \text{ generate } \mathcal{I}(A)\big) \Longleftrightarrow \big(df_1, \ldots, df_s \text{ generate } (TA)^{\perp}\big)$$

$$\Longleftrightarrow \bigcap_1^s \ker df_i = \emptyset ,$$

where TA denotes the tangent space at a to a representative of the germ A.

Without loss of generality we may assume that $M = \mathbf{C}^n, a = 0$ and $A = (0 \times \mathbf{C}^k)_0$. Put $r = n - k$. If $f \in \mathcal{I}(A)$, then, in view of the Hadamard lemma, $f = \sum_1^r a_i z_i$, where $a_i \in \mathcal{O}_n$. Thus the germs $z_1, \ldots, z_r$ generate the ideal $\mathcal{I}(A)$. Moreover (by differentiating), we get $df = \sum_1^r a_i(0) z_i$. Therefore

$$(*) \qquad\qquad f - (df)_0 \in \mathfrak{m}\mathcal{I}(A) \quad \text{for} \quad f \in \mathcal{I}(A) ,$$

and $\sum \mathbf{C} df_i \subset \sum_1^r \mathbf{C} z_i$. So $\dim\big(\sum \mathbf{C} df_j\big) \leq r$ and equality occurs precisely when

$$(**) \qquad\qquad z_i = \sum_j c_{ij} df_i \quad (i = 1, \ldots, r) \quad \text{for some } c_{ij} \in \mathbf{C} .$$

Now, if f_j generate $\mathcal{I}(A)$, then $z_i = \sum_j a_{ij} f_j$ $(i = 1, \ldots, r)$, where $a_{ij} \in \mathcal{O}_n$. By differentiating both sides we get $(**)$. Conversely, the condition $(**)$, combined with $(*)$, implies that $z_i \in \sum \mathcal{O}_n f_j + \mathfrak{m}\mathcal{I}(A)$ $(i = 1, \ldots, r)$, which gives $\mathcal{I}(A) \subset \sum \mathcal{O}_n f_j + \mathfrak{m}\mathcal{I}(A)$. Hence, by Nakayama's lemma, $\mathcal{I}(A) = \sum \mathcal{O}_n f_j$.

3. Let I be an ideal of the ring $\mathcal{O}_a$. The analytic germ (at a)

$$V(I) = V(f_1, \ldots, f_k),$$

where the f_i generate the ideal I $(^{13})$, is independent of the choice of the generators f_i $(^{14})$. It is called the *locus of the ideal I*. Naturally, $V(f) = V(\mathcal{O}_a f)$, i.e., the principal germs (at a) are the same as the loci of the principal ideals (of the ring $\mathcal{O}_a$). Obviously,

$$V(I_1 + \ldots + I_k) = V(I_1) \cap \ldots \cap V(I_k)$$

for any collection of ideals $I_1, \ldots, I_k$ of the ring $\mathcal{O}_a$. Moreover,

$$I \subset J \Longrightarrow V(I) \supset V(J)$$

for ideals I, J of the ring $\mathcal{O}_a$. In particular, $V(I) \subset V(f)$ for $f \in I$. Consequently,

$$V(I_1 \cap \ldots \cap I_k) = V(I_1) \cup \ldots \cup V(I_k),$$

for any ideals I_j of $\mathcal{O}_a$. This is so, because $V(I \cap J) \subset V(I) \cup V(J)$; indeed, by taking generators f_i of I and generators g_j of J, we get $f_i g_j \in I \cap J$, hence $V(I \cap J) \subset \bigcap_{i,j} V(f_i g_j) = \left(\bigcap_i V(f_i) \right) \cup \left(\bigcap_j V(g_j) \right) = V(I) \cup V(J)$.

We also have

$$V(\mathrm{rad}\ I) = V(I) .$$

Indeed, by taking generators g_i of rad I, we get $g_i^{p_i} \in I$ for some $p_i > 0$. Therefore $V(g_i) = V(g_i^{p_i}) \supset V(I)$, and hence $V(\mathrm{rad}\ I) \supset V(I)$. Note also that I is proper precisely when $V(I) \neq \emptyset$, i.e., when $V(I) \supset a$.

4. If φ maps a neighbourhood of a point a of the manifold M biholomorphically onto a neighbourhood of a point b of the manifold N and $\varphi(a) = b$, then it is easy to verify that

$$V(f \circ \varphi^{-1}) = \varphi(V(f)), \quad \mathcal{I}(\varphi(A)) = \mathcal{I}(A) \circ \varphi^{-1} \quad \text{and} \quad V(I \circ \varphi^{-1}) = \varphi(V(I))$$

for $f \in \mathcal{O}_a$, an analytic germ A at a, and an ideal I of the ring $\mathcal{O}_a$ $(^{15})$. If $g \in \mathcal{O}_a$, then

$$\left(g = 0 \text{ on } V(f) \text{ (on } V(I)) \right) \Longleftrightarrow \left(g \circ \varphi^{-1} = 0 \text{ on } \varphi(V(f)) \left(\text{on } \varphi(V(I)) \right) \right) .$$

$(^{13})$ The ideal I is finitely generated, as the ring $\mathcal{O}_a$ is noetherian (see I. 1.6).

$(^{14})$ It is enough to observe that $\bigcap_1^k V(f_i) \subset V\left(\sum_1^k a_i f_i \right)$ for $a_i, f_i \in \mathcal{O}_a$.

$(^{15})$ In particular, if φ is a coordinate system at a, then the locus of the "germ f (the ideal I) in the coordinate system φ" means the same as "the locus of the germ f (the ideal I) – in the coordinate system φ", and the ideal of the "germ A in the coordinate system φ" means the same as the "ideal of the germ A – in the coordinate system φ".

5. For any analytic germ A, we have

$$V\big(\mathcal{I}(A)\big) = A .$$

Indeed, by taking generators $g_1, \ldots, g_l$ of the ideal $\mathcal{I}(A)$, we get $V\big(\mathcal{I}(A)\big) = \bigcap_i V(g_j) \supset A$. On the other hand, $A = V(f_1) \cap \ldots \cap V(f_k)$ for some $f_i \in \mathcal{O}_a$; thus $f_i \in \mathcal{I}(A)$, hence $V(f_i) \supset V\big(\mathcal{I}(A)\big)$, and so $A \supset V\big(\mathcal{I}(A)\big)$.

The above equality yields the following equivalences

$$A \subset B \Longleftrightarrow \mathcal{I}(A) \supset \mathcal{I}(B) \quad \text{and} \quad A = B \Longleftrightarrow \mathcal{I}(A) = \mathcal{I}(B)$$

for analytic germs A, B (see n° 2). Since the ring $\mathcal{O}_a$ is noetherian (see I. 1.6), we conclude that

Every decreasing sequence of analytic germs is stationary.

For any ideal I of the ring $\mathcal{O}_a$, we have the inclusion

$$I \subset \mathcal{I}\big(V(I)\big) \quad (^{16}) .$$

Indeed, if $f \in I$, then $V(f) \supset V(I)$, and so $f \in \mathcal{I}\big(V(I)\big)$.

6. A non-empty analytic germ A is said to be *irreducible* or *simple* if it cannot be written as the union of two analytic germs that are strictly contained in A $(^{17})$. It follows (by induction) that such a germ is not a finite union of analytic germs strictly contained in A.

If an analytic germ A is irreducible, then

$$(*) \qquad A \subset B_1 \cup \ldots \cup B_k \Longrightarrow (A \subset B_j \text{ for some } j)$$

for analytic germs $B_1, \ldots, B_k$. (Indeed, $A = \bigcup(A \cap B_i)$, and so $A = A \cap B_j$ for some j.)

PROPOSITION 1. *Every analytic germ A has a unique decomposition into a finite union of simple germs A_i such that $A_i \not\subset A_j$ for $i \neq j$.*

PROOF . We will prove that the class N of analytic germs which are not finite unions of simple germs is empty. First, note that every germ $B \in N$

$(^{16})$ By the Hilbert Nullstellensatz (see III. 4.1 below), we have $\mathcal{I}(V(I)) = \mathrm{rad}\, I$. Therefore equality holds when the ideal I is prime (see A. 1.11).

$(^{17})$ Hence an analytic germ is not simple if and only if it is empty or it is *reducible* (e.g., it is the union of two analytic germs that are strictly smaller than this germ).

strictly contains the same germ $C \in N$. Indeed, if $B \in N$, then B is reducible, hence $B = C_1 \cup C_2$, where $C_1 \subsetneqq B$ and $C_2 \subsetneqq B$. Thus, either $C_1 \in N$ or $C_2 \in N$. Now, supposing that $N \neq \emptyset$, one could find strictly decreasing (infinite) sequence of analytic germs, which is impossible. In conclusion, $A = A_1 \cup \ldots \cup A_k$, where each A_i is simple. By removing from this union each germ which is contained in any of the remaining ones, we obtain the condition $A_i \not\subset A_j$ for $i \neq j$. Now let $A = B_1 \cup \ldots \cup B_l$ be another such decomposition of the germ A. It is enough to show that each A_i is equal to some B_j and vice versa. Now, any A_i is contained in some B_j, which is contained in some A_s. Hence $i = s$ and $A_i = B_j$.

Such a representation of the germ A is called its *decomposition into simple germs*, and the germs A_i are called *simple components* of the germ A. If all the simple components of A are of the same dimension k, we will say that A is of *constant dimension* k. (In such a case $\dim A = k$, provided that $A \neq \emptyset$.)

Note that if a germ A is the union of simple germs $A_1, \ldots, A_k$, then its simple components are precisely the maximal elements of the set of these germs.

Every smooth germ is irreducible. For if it were not true, there would exist a connected non-empty submanifold C (representing this germ) such that $\tilde{C} = \tilde{A} \cup \tilde{B}$, where $\tilde{A}, \tilde{B}$ would be proper analytic subsets of $\tilde{C}$, and hence they would be nowhere dense in $\tilde{C}$ (see 3.6), which would be impossible.

PROPOSITION . *An analytic germ A is simple if and only if its ideal $\mathcal{I}(A)$ is prime.*

Indeed, assume that A is irreducible. Then if $fg \in \mathcal{I}(A)$, we have $A \subset V(fg) = V(f) \cup V(g)$, hence $A \subset V(f)$, and so $f \in \mathcal{I}(A)$. Moreover, $\mathcal{I}(A)$ is proper (as $1 \notin \mathcal{I}(A)$). Conversely, if $\mathcal{I}(A)$ is prime, then if $A = B \cup C$, we have $\mathcal{I}(A) = \mathcal{I}(B) \cap \mathcal{I}(C)$ (see n° 2), and hence $\mathcal{I}(A) = \mathcal{I}(B)$ (see A. 1.11). Consequently, $A = B$ (see n° 5). Obviously, $A \neq \emptyset$.

§5. Principal germs

Let M be a complex manifold of dimension n, and let $a \in M$.

1. Note first that if $f \in \mathcal{O}_n$ is a regular germ, then (see I.2.4, footnote (^{17}))

$$V(f) = V(\text{red f}) .$$

Let $f \in \mathcal{O}_a$. The germ $V(f)$ is full or, equivalently, of dimension n (see 3.3) if and only if $f = 0$. It is empty precisely when f is invertible.

If a germ $f \in \mathcal{O}_a$ is non-zero and non-invertible, then $\dim V(f) = n - 1$.

Indeed, we may assume that $\mathcal{O}_a = \mathcal{O}_n$, and that the germ f is regular (see I. 1.4 and 4.4). Now, the distinguished germ red f has a non-zero discriminant and red $f \neq 1$ (as $V(f) \neq \emptyset$). Hence one of its representatives P is a distinguished polynomial of positive degree and with discriminant $D \neq 0$. Therefore, in any neighbourhood of zero (in $\mathbf{C}^n$), there is a point $z = (u, z_n)$ such that $P(z) = 0$ and $D(u) \neq 0$ (see C. 2.2), hence $\frac{\partial P}{\partial z_n}(z) \neq 0$ (see C. 2.1), and so (by the implicit function theorem) the set $\{P = 0\}$ is an $(n-1)$–dimensional submanifold at the point z. In conclusion, the dimension of the germ $V(f) = V(\text{red } f) = \{P = 0\}_0$ is $n - 1$.

2. If a germ $f \in \mathcal{O}_a$ satisfies the conditions $f(a) = 0$, $d_a f \neq 0$, then every germ from $\mathcal{O}_a$ that vanishes on $V(f)$ is divisible by f. In other words, $\mathcal{I}(V(f)) = \mathcal{O}_a f$ [18].

It is enough to verify the statement for $f = z_n \in \mathcal{O}_n$ (see I. 1.2 and 4.4), and this is a direct consequence of Hadamard's lemma (see C. 1.10).

The above is a special case (see I. 1.2) of the following theorem, which itself is a special case of the *Hilbert Nullstellensatz* (see III. 4.1 below, and A. 6.1).

THEOREM . *If a germ $p \in \mathcal{O}_a$ is irreducible or, more generally, does not have multiple factors, then every germ $f \in \mathcal{O}_a$ which vanishes on $V(p)$ is divisible by p, hence $\mathcal{I}(V(p)) = \mathcal{O}_a p$* [19] .

PROOF . We may assume that $\mathcal{O}_a = \mathcal{O}_n$, the germ p is regular (see I. 1.4 and 4.4), and (see the preparation theorem in I. 2.1) that it is DISTINGUISHED of degree $k > 0$ and with the discriminant $\delta \neq 0$ (see I. 2.3). By the division version of the preparation theorem (see I. 1.4), we have

$$f = gp + r ,$$

where $g \in \mathcal{O}_n$ and $r \in Q_n$ is of degree $< k$. There exist representatives P, R, Δ, F, G of the germs p, r, δ, f, g with the following properties. P and Q are polynomials in z_n with holomorphic coefficients in an open connected neighbourhood B of zero in $\mathbf{C}^{n-1}$, whereas F and G are holomorphic in $U = B \times \{|z_n| < \varepsilon\}$. Moreover, R is of degree $< k$, P is distinguished of

---

[18] See 4.5 and 4.3.

[19] See 4.5 and 4.3.

degree k, and one may assume that $\{P = 0\} \subset U$ (see C. 2.2). Finally,

$$F = GP + R \quad \text{in} \quad U \quad \text{and} \quad F = 0 \quad \text{on} \quad \{P = 0\}.$$

It follows that $R = 0$. Indeed, for any u from the subset $\{\Delta \neq 0\}$ which is dense in U (see 3.6), the polynomial $t \longrightarrow P(u,t)$ has k distinct roots (see C. 2.1) at which the polynomial $t \longrightarrow R(u,t)$ of degree $< k$ must vanish. This implies $R = 0$. Thus $r = 0$, and the proof is complete.

The theorem yields (see I. 2.3 and A. 6.1) the following corollaries.

COROLLARY 1. *If the germs* $p, q \in \mathcal{O}_a$ *do not have multiple factors, then*

$$V(p) = V(q) \Longleftrightarrow (p \text{ and } q \text{ are associated}).$$

In view of proposition 2 from 4.6, we have

COROLLARY 2. *If* $p \in \mathcal{O}_a$ *is irreducible, then* $V(p)$ *is simple. If* $f \in \mathcal{O}_a$ *is a non-zero non-invertible germ and* $f = p_1^{k_1} \dots p_r^{k_r}$ *is its decomposition into irreducible factors (where* $k_i > 0$ *and the* p_i*'s are not mutually associated), then* $V(f) = V(p_1) \cup \dots \cup V(p_r)$ *is the decomposition of the germ* $V(f)$ *into simple germs.*

This implies

COROLLARY 3. *If* $f, g \in \mathcal{O}_a$, *and* g *is irreducible, then*

$$V(f) = V(g) \Longleftrightarrow (f = hg^k, \text{ where } h \in \mathcal{O}_a \text{ is invertible and } k > 0).$$

Finally, corollary 2 implies that $\mathcal{I}(V(f)) = \text{rad } \mathcal{O}_a f$ (see A. 6.1 [20]) and this yields (see A. 6.1)

COROLLARY 4. *A non-zero germ from* $\mathcal{O}_a$ *has no multiple factors if and only if it generates the ideal of its locus.*

3. Assume now that $n > 0$.

LEMMA . *If* $p \in \mathcal{O}_n$ *is an irreducible distinguished germ and* P *is its representative which is a distinguished polynomial whose coefficients are holomorphic in an open connected neighbourhood* B *of zero, then the discriminant* Δ *of* P *is not zero. Moreover the set* $\Lambda = \{P = 0, \Delta \neq 0\}$, *which is open and dense in* $\{P = 0\}$, *is an* $(n-1)$*-dimensional connected submanifold.*

PROOF . We have $\Delta \not\equiv 0$, because p has a non-zero discriminant (see I. 2.3, 1.5). The set Λ is open and dense in $\{P = 0\}$ by the theorem on

---

[20] Since $\mathcal{O}_a p_1 \dots p_r = \mathcal{O}_a p_1 \cap \dots \cap \mathcal{O}_a p_r$. See also Hilbert's theorem in III. 4.1 below.

continuity of roots (see B. 5.3). It is a locally topographic submanifold in $H \times \mathbf{C}$, where $H = \{\Delta \neq 0\}$ (and hence is of dimension $n - 1$). This is due to the implicit function theorem, as $\frac{\partial P}{\partial z_n} \neq 0$ on Λ (see C. 2.1). Therefore the natural projection $\pi : \Lambda \longrightarrow H$ is a local homeomorphism (see C. 3.17), it is proper (see B. 5.2 and B. 5.3), and hence is a finite covering (see B. 3.2, proposition 1). Suppose now that the submanifold Λ is not connected. We have a partition $\Lambda = \Lambda_1 \cup \Lambda_2$, where the Λ_i's are non-empty, closed, and open in Λ. Then $\pi_i = \pi_{\Lambda_i} : \Lambda_i \longrightarrow H$ is an r_i–sheeted covering with $r_i > 0$, since the set H is connected (see B. 3.2 and 3.6). The function $\tilde{P}_i$ is well-defined by the formula

$$\tilde{P}_i(z) = (z_n - w_1) \ldots (z_n - w_{r_i}) \,,$$

where

$$\pi_i^{-1}(u) = u \times \{w_1, \ldots, w_{r_i}\} \quad \text{and} \quad u = (z_1, \ldots, z_{n-1}) \in H \,.$$

It is a monic polynomial whose coefficients are locally bounded near $\{\Delta = 0\}$ (see B. 5.3). They are holomorphic in H. Indeed, if $u_0 \in H$, then $\pi_i^{-1}(u_0) = u_0 \times \{w_1^0, \ldots, w_{r_i}^0\}$ and (by the implicit function theorem) there exist holomorphic functions $\omega_1, \ldots, \omega_{r_i}$ on an open connected neighbourhood U of the point u_0, such that $\omega_\nu(u_0) = w_\nu^0$, $P(u, \omega_\nu(u)) = 0$ for $u \in U$, and the $\omega_\nu(u)$'s are mutually distinct. Their graphs, being connected, must be contained in Λ_i, and hence for $u \in U$ we have $\pi_i^{-1}(u) = u \times \{\omega_1(u), \ldots, \omega_{r_i}(u)\}$. In other words $\tilde{P}_i(z) = (z_n - \omega_1(u)) \ldots (z_n - \omega_{r_i}(u))$. Since the set $\{\Delta = 0\}$ is thin in B (see 3.5), the coefficients extend to holomorphic functions on B, which means that $\tilde{P}_i$ has an extension P_i that is a monic polynomial of degree r_i with coefficients which are holomorphic in B. Obviously, $P = \tilde{P}_1 \tilde{P}_2$ in $H \times \mathbf{C}$, hence $P = P_1 P_2$ in $B \times \mathbf{C}$, which implies that the germ p is the product of two monic germs of positive degrees (see I. 1.5). Thus p is reducible (see I. 2.1).

We are going to prove the following property

If non-zero germs $f, g \in \mathcal{O}_a$ are relatively prime, then $\dim(V(f) \cap V(g)) < n - 1$.

In fact, we may assume that the germs f, g are non-invertible and also (see 4.1 and A. 6.1) that they are irreducible. Furthermore, we may assume that $\mathcal{O}_a = \mathcal{O}_n$ and (in view of the preparation theorem) that the germ g is distinguished (see I. 1.4 and 4.4). Let P be a representative of the germ g chosen as in the lemma. We may take the neighbourhood $W = B \times \{|z_n| < \varepsilon\}$ arbitrarily small and such that it contains the set $\{P = 0\}$ (see C. 2.2). In particular, we may assume that the germ f has a representative that is holomorphic in W. Suppose now that $\dim(V(f) \cap V(g)) = n - 1$. Then there exists a non-empty submanifold $\Gamma \subset \{F = 0\} \cap \{P = 0\}$ of dimension $n - 1$. Now, $\dim\{\Delta = 0\} \leq n - 2$ (see 1.2), and hence the set $\{P = 0, \ \Delta = 0\}$ is of dimension $\leq n - 2$, as the fibres of its natural projection onto $\mathbf{C}^{n-1}$ are finite (see 1.4). Therefore the above set cannot contain the submanifold Γ. Consequently, the set $\Gamma_0 = \Gamma \cap \Lambda$, open in Γ, is non-empty. It is an $(n - 1)$–dimensional submanifold, and so it is open in Λ. As F vanishes on Γ_0, it also vanishes on Λ and on $\{P = 0\}$. Thus the germ f vanishes on $V(g)$, and hence, by the theorem in n° 2, it is divisible by g, which is impossible.

If $f_1, \ldots, f_k \in \mathcal{O}_a$, then

$$(*) \qquad\qquad V(f_1) \cap \ldots \cap V(f_k) = V(g) \cup B ,$$

where g is the greatest common divisor of the germs f_i and B is a germ of dimension $\leq n - 2$.

Indeed, omitting the trivial cases when one of the germs f_i is invertible or $f_1 = \ldots = f_k = 0$ and removing the zero germs, we may assume that the germs f_i are non-zero and non-invertible. Let $g_1, \ldots, g_l$ be all distinct (up to association) irreducible divisors of the germs $f_1, \ldots, f_k$. Then $V(f_i) = \bigcup \{V(g_j) : \ g_j$ is a divisor of $f_i\}$. It follows that $V(f_1) \cap \ldots \cap V(f_k)$ is the union of all the germs of the form $V(g_{\alpha_1}) \cap \ldots \cap V(g_{\alpha_k})$, where g_{α_i} is a divisor of f_i $(i = 1, \ldots, k)$. Those among them for which $\alpha_1 = \ldots = \alpha_k$ coincide with the germs $V(g_j)$, where g_j is a common divisor of the germs $f_1, \ldots, f_k$, and hence their union is the germ $V(g)$ (see A. 6.1). The remaining ones are of dimension $\leq n - 2$, according to the previously described property.

For an analytic germ A (at a), we have the equivalences:

$$(A \text{ is simple of dimension } n - 1) \iff (A = V(f), \text{ where } f \in \mathcal{O}_a$$

$$\text{is irreducible}). \ (^{21})$$

$(^{21})$ This, combined with the lemma, implies that every simple germ of dimension $(n - 1)$ has an arbitrarily small representative in which there is an open dense subset that is a connected $(n - 1)$–dimensional submanifold. This follows also from proposition 1 in IV. 3 that characterizes irreducible germs (in view of corollary 3 from proposition 2 in IV. 2.8).

$(A$ is of constant dimension $n-1) \Longleftrightarrow (A = V(f)$, where $f \in \mathcal{O}_a$ is non-zero

(moreover, f can be chosen to be

without multiple factors) $(^{22}))$.

Clearly, the right hand sides imply the left hand sides (see n° 2, corollary 2; and n° 1). Conversely, suppose A is of constant dimension $n-1$. According to $(*)$ (see 4.1), we have $A = V(g) \cup B$, where $\dim B \leq n-2$. Each simple component of B must be contained in $V(g)$, for otherwise (see 4.6) it would be a simple component of A. Thus $A = V(g)$. Obviously, the germ g is non-zero (and one can replace g by a germ that does not have multiple factors). Assume, in addition, that A is irreducible of dimension $n-1$. By taking the decomposition of g into irreducible factors we conclude (see n° 2, corollary 2) that $A = V(f)$ for some irreducible germ $f \in \mathcal{O}_a$.

The second of the above equivalences can be restated as follows.

$(A$ is principal$) \Longleftrightarrow (A$ is of constant dimension $n-1$ or $n)$.

From the representation $(*)$ we get the following equivalence for germs $f_1, \ldots$ $\ldots, f_k$:

$(f_1, \ldots, f_k$ are relatively prime$) \Longleftrightarrow \left(\dim(V(f_1) \cap \ldots \cap V(f_k)) \leq n-2\right)$.

Indeed, both sides are equivalent to the condition $V(g) = \emptyset$ (see n° 1).

Note also that, for non-invertible germs $f, g \in \mathcal{O}_a$ we have:

$(f, g$ are relatively prime$) \Longleftrightarrow \dim(V(f) \cap V(g)) = n-2$.

This is a corollary of the inequality

$$\mathrm{codim}(V(f_1) \cap \ldots \cap V(f_k)) \leq k ,$$

i.e.,

$$\dim(V(f_1) \cap \ldots \cap V(f_k)) \geq n-k ,$$

for any non-invertible germs $f_i \in \mathcal{O}_a$ (see III. 4.6, inequality $(*)$ below, and n° 1) $(^{23})$.

$(^{22})$ Then, in view of the theorem from n° 2, the germ f is unique up to association.

$(^{23})$ This inequality implies also that the germ $V(f) \cap V(g)$ must be of constant dimension $n-2$ (see IV. 3.1, the corollary from proposition 4 below).

§6. One-dimensional germs. The Puiseux theorem

1. Let $H(z, w)$ be a polynomial in $w \in \mathbf{C}$ which is monic of degree p and has holomorphic coefficients in a neighbourhood of zero in $\mathbf{C}$ (see C. 2.1). Then the germ $H_0 \in Q_2$ can be identified with a polynomial from $\mathcal{O}_1[T]$ (via the natural isomorphism $\mathcal{O}_1[T] \longrightarrow Q_2$; see I. 1.5).

THE PUISEUX THEOREM (FIRST VERSION). *If H_0 is irreducible in $\mathcal{O}_1[T]$, then there is a holomorphic function on the disc $\Omega = \{|z| < \delta\}$, such that*

$$H(z^p, w) = \prod_{\nu=0}^{p-1} \left(w - h(e^{2\pi i \nu / p} z) \right) \quad in \quad \Omega \times \mathbf{C} .$$

Moreover, if H is a distinguished polynomial, then $h(0) = 0$.

The following lemma will be used in the proof of the theorem.

LEMMA . *If a bounded holomorphic function η defined on the half-plane $P = \{\operatorname{im} z > \beta\}$, where $\beta \in \mathbf{R}$, satisfies the condition $\eta(z + 1) = \eta(z)$ in P, then*

$$\eta(z) = \gamma(e^{2\pi i z}) \quad in \quad P$$

for some holomorphic function γ on $\{|w| < e^{-2\pi\beta}\}$.

Indeed, the above equation uniquely determines a bounded function γ in the range $S = \{0 < |w| < e^{-2\pi\beta}\}$ of the mapping $P \ni z \longrightarrow e^{2\pi i z}$ and the function is holomorphic, in view of the implicit function theorem [24] .

PROOF of the theorem. The discriminant of the polynomial H_0 is not zero (see the propositions in A. 6.3 and I. 2.3). Since it coincides with the germ of the discriminant D of the polynomial H, there exists $\varrho > 0$ such that the coefficients of the polynomial H are bounded and holomorphic in the disc $\{|z| < \varrho\}$ and $D(z) \neq 0$ in the annulus $\{0 < |z| < \varrho\}$. Thus $\frac{\partial H}{\partial w}(z, w) \neq 0$, provided that $0 < |z| < \varrho$ and $H(z, w) = 0$ (see C. 2.1). Hence the function $G(t, w) = H(e^{2\pi i t}, w)$ is holomorphic in $P \times \mathbf{C}$, where $P = \{\operatorname{im} t > \beta\}$ and $e^{-2\pi\beta} = \varrho$. It is a monic polynomial of degree p the coefficients of which are holomorphic and bounded in P; we have $\frac{\partial G}{\partial w}(t, w) \neq 0$ in the set $Z = \{G = 0\}$ and all the roots of the polynomial $w \longrightarrow G(t, w)$ are distinct for each $t \in P$. So, in view of the implicit function theorem, the set Z is a locally topographic submanifold of $P \times \mathbf{C}$. Since the natural projection $\pi : Z \longrightarrow P$ is proper (see B. 5.2 and 3), it is a finite covering (see B. 3.2 proposition 1) of multiplicity p.

---

[24] For if $w_0 = e^{2\pi i z_0}$, $z_0 \in P$, then $e^{2\pi i \zeta(w)} = w$ in U and $\zeta(U) \subset P$, where ζ is a holomorphic function in a neighbourhood U of the point z_0. Hence $\gamma(w) = \eta(\zeta(w))$ in U.

Any topological component ζ of the submanifold Z is also locally topographic and the restriction $\pi_\zeta : \zeta \longrightarrow P$ is also a covering (see B. 3.2). It must be one-sheeted (see B. 3.2, proposition 2), so ζ is a holomorphic function in P. Therefore $Z = \eta_1 \cup \ldots \cup \eta_p$, where the η_i are mutually distinct holomorphic functions on P. Consequently, $G(t,w) = \prod_1^p (w - \eta_i(t))$ in $P \times \mathbf{C}$ and the functions η_i are bounded (see B. 5.3). Let $\Lambda = \{\eta_1, \ldots, \eta_p\}$. Each of the functions $\eta_i(t+l)$, $l \in \mathbf{Z}$, belongs to Λ, because $Z - (l,0) \subset Z$, hence each equivalence class of the relation "$\eta_i(t+l) = \eta_j(t)$ in P for some $l \in \mathbf{Z}$" must be of the form $\{\eta_\nu(t), \eta_\nu(t+1), \ldots, \eta_\nu(t+p_\nu - 1)\}$. Moreover, $\eta_\nu(t+p_\nu) = \eta_\nu(t)$ in P. Changing the order of indices, we get

$$\Lambda = \{\eta_1(t), \ldots, \eta_1(t + p_1 - 1), \ldots, \eta_k(t), \ldots, \eta_k(t + p_k - 1)\} .$$

This means that $G(t,w) = \prod_{s=1}^k G_s(t,w)$, where $G_s(t,w) = \prod_{\nu=0}^{p_s - 1}(w - \eta_s(t + \nu))$ in $P \times \mathbf{C}$. Now, the coefficients of the polynomial G_s satisfy the hypotheses of the lemma (since $G_s(t + 1, w) = G_s(t, w)$). Hence $G_s(t,w) = H_s(e^{2\pi it}, w)$ in $P \times \mathbf{C}$, where H_s is a polynomial of degree $p_s > 0$ with holomorphic coefficients in the disc $\{|z| < \varrho\}$, and we have $H = H_1 \ldots H_k$ in $\{|z| < \varrho\}$. It follows that $k = 1$ and $p_1 = p$, for otherwise H_0 would be reducible in $\mathcal{O}_1[T]$ (see A. 1.13 and A. 2.3). Putting $\eta = \eta_1$, we obtain $G(t,w) = \prod_{\nu=0}^{p-1}(w - \eta(t + \nu))$ in $P \times \mathbf{C}$. On the other hand, the function $\frac{1}{p}P = \{\text{im } t > \beta/p\} \ni t \longrightarrow \eta(pt)$ satisfies the hypotheses of the lemma, and therefore $\eta(pt) = h(e^{2\pi it})$ in $\frac{1}{p}P$, where h is a holomorphic function on the disc $\{|z| < \varrho^{1/p}\}$. Thus

$$H(e^{2\pi ipt}, w) = G(pt, w) = \prod_{\nu=0}^{p-1}\left(w - h(e^{2\pi i(t + \nu/p)})\right) \quad \text{in} \quad \left(\frac{1}{p}P\right) \times \mathbf{C} ,$$

which implies that

$$H(z^p, w) = \prod_{\nu=0}^{p-1}\left(w - h(e^{2\pi i\nu/p}z)\right) \quad \text{in} \quad \{|z| < \varrho^{1/p}\} \times \mathbf{C} .$$

Since $\mathcal{O}_1[T]$ is factorial (by Gauss' theorem in A. 6.2 and the proposition in I. 2.3), we obtain the following

COROLLARY . *If $H(z, w)$ is a polynomial with respect to $w \in \mathbf{C}$ which is monic of degree p and whose coefficients are holomorphic in a neighbourhood* *of zero in* **C**, *then there exist an integer exponent* $k > 0$ *and holomorphic functions* $h_1, \ldots, h_p$ *in the disc* $\Omega = \{|z| < \delta\}$ *such that*

$$H(z^k, w) = \prod_1^p (w - h_i(z)) \quad \text{in} \quad \Omega \times \mathbf{C} \;.$$

2. In this section we make an exception and use some material from Chapters III–V.

The Puiseux theorem implies the following parametric description of a simple one-dimensional germ in $\mathbf{C}^2$ (a special case of the second version of Puiseux's theorem; see below).

PROPOSITION . *If an analytic germ A at $0 \in \mathbf{C}^2$ is simple and 1-regular* $(^{25})$, *then it has a representative of the form*

$$(\#) \qquad\qquad V = \{(t^p, h(t)) : \ |t| < \delta\} \;.$$

Here $p \in \mathbf{N} \setminus 0$ and h is a holomorphic function in the disc $\{|t| < \delta\}$ such that $h(0) = 0$. Also the mapping $\{|t| < \delta\} \ni t \longrightarrow (t^p, h(t)) \in V$ is a homeomorphism.

Indeed, $A = V(f)$, where $f \in \mathcal{O}_2$ is irreducible (see 5.3). Since $A \cap \{z = 0\} = 0$ (see the proposition from III. 4.4), the germ f must be w–regular. The Weierstrass preparation theorem implies that $A = V(H_0)$ (see I. 2.1 and 4.1), where H is a distinguished polynomial of degree p whose germ H_0 is irreducible in $\mathcal{O}_2$ and hence also in Q_2 (see I. 2.2b). According to the Puiseux theorem $H(t^p, w) = \prod_{\nu=0}^{p-1} \left(w - h\left(e^{2\pi i \nu / p} t \right) \right)$ in $\{|t| \leq \delta\} \times \mathbf{C}$, where $\delta > 0$ and h is a holomorphic function in a neighbourhood of the set $\{|t| \leq \delta\}$ such that $h(0) = 0$. Now, the representative $V = \{H(z, w) = 0, \ |z| < \delta^p\}$ of the germ A is equal to the right hand side of $(\#)$. Furthermore, δ can be chosen so small that the discriminant of the polynomial H does not vanish in the set $\{0 < |z| \leq \delta^p\}$. Then for $0 < |t| \leq \delta$ the values $h(e^{2\pi i \nu / p} t)$, $\nu = 0, \ldots, p-1$, are mutually distinct, from which it follows easily that the mapping $\{|t| \leq \delta\} \ni t \longrightarrow (t^p, h(t))$ is injective. Hence the latter is a homeomorphism onto its range, and so the mapping $\{|t| < \delta\} \ni t \longrightarrow (t^p, h(t)) \in V$ is also a homeomorphism.

Now let $n \geq 2$.

$(^{25})$ See III. 2.5. Therefore, up to a linear change of coordinates, it is an arbitrary one-dimensional simple germ (see the proposition from III. 4.2, and III. 2.7).

THE PUISEUX THEOREM (SECOND VERSION). *The germ A at $0 \in \mathbf{C}^n$ is analytic, simple, and 1-regular precisely when it has a representative of the form*

$$(*) \qquad\qquad V = \{(t^p, h(t)) : |t| < \delta\} ,$$

where $p \in \mathbf{N} \setminus 0$ and $h : \{|t| < \delta\} \longrightarrow \mathbf{C}^{n-1}$ is a holomorphic mapping that fixes 0. Moreover, p and h can be chosen in such a way that the mapping $\{|t| < \delta\} \ni t \longrightarrow (t^p, h(t)) \in V$ is a homeomorphism. The representative V defined by $(*)$ is a locally analytic representative. By making $\delta > 0$ smaller, we obtain locally analytic representatives of the germ A that form a base of open neighbourhoods of zero in V.

A germ A at a point a of an n-dimensional complex manifold is analytic, simple, and 1-dimensional if and only if in some coordinate system at a it has a representative of the form $(*)$. Then, in any representative of A, there is a base of open neighbourhoods of a consisting, in some coordinate system at a, of representatives $(*)$ with all sufficiently small $\delta > 0$ and some fixed p, h.

PROOF . The second part of the theorem is a consequence of the first part (see the proposition in III. 4.2, and III. 2.7).

Let V be a set defined by $(*)$. Since the mapping $\{|t| < \delta\} \ni t \longrightarrow (t^p, h(t)) \in \{|s| < \delta^p\} \times \mathbf{C}^{n-1}$ is proper $(^{26})$, Remmert's theorem (see V. 5.10) implies that the set V is analytic in $\{|s| < \delta^p\} \times \mathbf{C}^{n-1}$. It is one-dimensional (see II. 1.4 and III. 4.3) and irreducible (as the image of the disc $\{|t| < \delta\}$; see V. 4.6 and IV. 2.8). When δ is replaced by δ_0 such that $0 < \delta_0 \leq \delta$, then $(*)$ also defines an analytic one-dimensional and irreducible set, which can be made arbitrarily small. Being equal to $V \cap \{|z_1| < \delta_0^p\}$, it is an open neighbourhood of 0 in V. Thus the germ V_0 is irreducible (see IV. 3.1 proposition 1), one-dimensional, and hence 1-regular, because $V \cap \{z_1 = 0\} = 0$ (see the corollary in III. 4.4).

Suppose now that A is an analytic germ at $0 \in \mathbf{C}^n$ that is simple and 1-regular (and hence one-dimensional). It remains to prove that it has a representative of the form $(*)$ such that the mapping $\{|t| < \delta\} \ni t \longrightarrow (t^p, h(t)) \in V$ is a homeomorphism.

In view of the classic descriptive lemma (see III. 3.1–2), there is an open neighbourhood of 0 in $\mathbf{C}$ and a representative $\tilde{V}$ of the germ A which is

---

$(^{26})$ Because its composition with the projection onto $\{|s| < \delta^p\}$ is the proper mapping $\{|t| < \delta\} \ni t \longrightarrow t^p \in \{|t| < \delta^p\}$ (see B. 2.4).

analytic in $\tilde{\Omega} \times \mathbf{C}^{n-1}$, bounded, such that $\tilde{V} \cap \{z_1 = 0\} = 0$, and

$$\tilde{V} \setminus 0 = \{z_1 \in \tilde{\Omega} \setminus 0, \ \tilde{p}(z_1, z_2) = 0, \ z_j = g_j(z_1, z_2), \ j = 3, \ldots, n\} \ .$$

Here, $\tilde{p}$ is a distinguished polynomial of degree $p > 0$ with holomorphic coefficients in $\tilde{\Omega}$ and with the discriminant $\tilde{\delta}(z_1) \neq 0$ in $\tilde{\Omega} \setminus 0$. Moreover, the germ $\tilde{p}_0$ is irreducible in $\mathcal{O}_1[T]$ (see III. 3.1, III. 2.5, and A. 8) and the functions $g_3, \ldots, g_n$ are holomorphic in $(\tilde{\Omega} \setminus 0) \times \mathbf{C}$ $(^{27})$. By the first version of the Puiseux theorem, we have

$$\tilde{p}(t^p, z_2) = \prod_{\nu=0}^{p-1} \left(z_2 - h_2\left(e^{2\pi i \nu/p} t\right) \right) \quad \text{for } |t| \leq \delta \ ,$$

where $\delta > 0$, $\{|z_1| \leq \delta^p\} \subset \tilde{\Omega}$, and h_2 is a holomorphic function in a neighbourhood of the set $\{|t| \leq \delta\}$ such that $h_2(0) = 0$. It follows easily that

$$\tilde{V} \cap \{0 < |z_1| \leq \delta^p\} =$$
$$\{z: \ z_1 = t^p, \ z_2 = h_2(t), \ z_j = g_j(t^p, h_2(t)), \ j = 3, \ldots n, \ \text{where } 0 < |t| \leq \delta\},$$

and (as in the proof of the proposition) that the mapping $\{|t| \leq \delta\} \ni t \longrightarrow (t^p, h_2(t))$ is injective. The functions $g_j(t^p, h_2(t))$, $j = 3, \ldots, n$, must be bounded in $\{0 < |t| \leq \delta\}$, and hence they have holomorphic extensions h_j in a neighbourhood of the set $\{|t| \leq \delta\}$. Furthermore, since the set $\tilde{V}$ is closed in $\tilde{\Omega} \times \mathbf{C}^{n-1}$, we have $(0, 0, h_3(0), \ldots, h_n(0)) \in \tilde{V}$, and so $h_3(0) = \ldots = h_n(0) = 0$. Consequently, we have

$$\tilde{V} \cap \{|z_1| \leq \delta^p\} = \{z_1 = t^p, \ z_j = h_j(t), \ j = 2, \ldots, n, \ \text{where } |t| \leq \delta\}.$$

The mapping $\{|t| \leq \delta\} \ni t \longrightarrow (t^p, h_2(t), \ldots, h_n(t))$ is injective, hence it is a homeomorphism onto its range. Therefore the mapping $\{|t| < \delta\} \ni t \longrightarrow (t^p, h_2(t), \ldots, h_n(t))$ is a homeomorphism onto the representative $V = \tilde{V} \cap \{|z_1| < \delta^p\}$ of the germ A.

We will also give a direct variant of the second part of the proof (not based on the first version of the Puiseux theorem).

By proposition 1 from IV. 1.4, the germ A has a normal triple of the form $(\Omega, 0, V)$, where $\Omega = \{|z_1| < \varrho\}$ and $V \subset \Omega \times \{|v| < R\}$ with $R > 0$ (see IV. 1.2–2). Then $V \setminus 0 = V_{\Omega \setminus 0}$ is a locally topographic closed submanifold of

---

$(^{27})$ It is enough to take $\tilde{\Omega} \subset \subset \Omega$ sufficiently small and $\tilde{V} = V_{\tilde{\Omega}}$.

$(\Omega \setminus 0) \times \mathbf{C}^{n-1}$ and the natural projection $\pi : V \setminus 0 \longrightarrow \Omega \setminus 0$ is a finite covering. Consider the holomorphic mapping $g : P \times \mathbf{C}^{n-1} \ni (\zeta, v) \longrightarrow (e^{2\pi i \zeta}, v) \in (\Omega \setminus 0) \times \mathbf{C}^{n-1}$, where $P = \{\operatorname{im}\zeta > \beta\}$ and $e^{-2\pi\beta} = \varrho$. It is surjective. The set $W = g^{-1}(V \setminus 0)$ is a locally topographic closed submanifold of $P \times \mathbf{C}^{n-1}$ (see C. 3.17). Since $W \subset P \times \{|v| < R\}$, the natural projection $\tilde{\pi} : W \longrightarrow P$ is proper (see B. 5.2) and hence is a finite covering (see B. 3.2, proposition 1). Moreover,

$$(\#) \qquad\qquad W_\zeta = V_{e^{2\pi i \zeta}} \quad \text{for} \quad \zeta \in P .$$

Thus (as in the proof of the first version of the Puiseux theorem),

$$(\#\#) \qquad\qquad W = \eta_1 \cup \ldots \cup \eta_q ,$$

where $\eta_\nu : P \longrightarrow \mathbf{C}^{n-1}$ are bounded holomorphic mappings mutually distinct at each point of P. Each of the mappings $\eta_\nu(\zeta + l)$ (i.e., $P \ni \zeta \longrightarrow \eta_\nu(\zeta+l) \in \mathbf{C}^{n-1}$), where $l \in \mathbf{Z}$, coincides with some η_μ, since $W - (l, 0) \subset W$. Thus, for any $\nu = 1, \ldots, q$, there is a p_ν such that the mappings $\eta_\nu(\zeta)$, $\eta_\nu(\zeta + 1), \ldots, \eta_\nu(\zeta + p_\nu - 1)$ differ at each point and $\eta_\nu(\zeta + p_\nu) = \eta_\nu(\zeta)$ in P. Now, take a collector $P_\nu : (\mathbf{C}^{n-1})^{p_\nu + 1} \longrightarrow \mathbf{C}^{r_\nu}$ (see III. 1.1) and consider the holomorphic mapping $G_\nu(\zeta, v) = P_\nu(\eta_\nu(\zeta), \ldots, \eta_\nu(\zeta + p_\nu - 1), v)$ of the set $P \times \mathbf{C}^{n-1}$ to the space $\mathbf{C}^{r_\nu}$. Its components are polynomials in v whose coefficients satisfy the hypotheses of the lemma from n° 1 (since $G_\nu(\zeta + 1, v) = G_\nu(\zeta, v)$). Hence $G_\nu(\zeta, v) = F_\nu(e^{2\pi i \zeta}, v)$, where $F_\nu : \Omega \times \mathbf{C}^{n-1} \longrightarrow \mathbf{C}^{r_\nu}$ is a holomorphic mapping. Since

$$\{v : \ G_\nu(\zeta, v) = 0\} = \{\eta_\nu(\zeta), \ldots, \eta_\nu(\zeta + p_\nu - 1)\}$$

for $\zeta \in P$, the equality $(\#\#)$ implies that

$$W_\zeta = \bigcup_1^q \{v : \ G_\nu(\zeta, v) = 0\} .$$

Thus, in view of $(\#)$, we have $V_{z_1} = \bigcup_1^q \{v : \ F_\nu(z_1, v) = 0\}$ for $z_1 \in \Omega \setminus 0$. This means that $V \setminus 0 = \bigcup_1^q \{F_\nu = 0\}_{\Omega \setminus 0}$. Therefore $V \subset \bigcup_1^q \{F_\nu = 0\}$ (see IV. 1.1, property (4)). But V is irreducible (see IV. 2.8, corollary 5 from proposition 2), and so $V \subset \{F_s = 0\}$ for some s (see IV. 2.8). Therefore $V \setminus 0 = \{F_s = 0\}_{\Omega \setminus 0}$. Set $p = p_s$ [28] and $\eta = \eta_s$. Since $\eta(\zeta + p) = \eta(\zeta)$, the components of the mapping

$$\frac{1}{p} P = \{\operatorname{im}\zeta > \beta/p\} \ni \zeta \longrightarrow \eta(p\zeta) \in \mathbf{C}^{n-1}$$

---

[28] It is easy to check that $p_s = q$.

satisfy the hypotheses of the lemma from n° 1, and hence $\eta(p\zeta) = h(e^{2\pi i\zeta})$ in $\frac{1}{p}P$, where $h : \{|t| < \varrho^{1/p}\} \longrightarrow \mathbf{C}^{n-1}$ is a holomorphic mapping. Since

$$\{v : \ G_s(p\zeta) = 0\} = \{\eta(p\zeta + \nu) : \ \nu = 0, \ldots, p-1\}$$

for $\zeta \in \frac{1}{p}P$, it follows that if $0 < |t| < \varrho^{1/p}$ (i.e., if $t = e^{2\pi i\zeta}$ for some $\zeta \in \frac{1}{p}P$), then

$$\{v : \ F_s(t^p, v) = 0\} = \{h(te^{2\pi i\nu/p}) : \ \nu = 0, \ldots, p-1\}$$

and $h(te^{2\pi i\nu/p})$ are mutually distinct. This easily implies the equality

$$V \setminus 0 = \{F_s(z_1, v) = 0, \ 0 < |z_1| < \varrho\} = \{(t^p, h(t)) : \ 0 < |t| < \varrho^{1/p}\}$$

and also, since $h(0) = 0$ $(^{29})$, the equality $V = \{(t^p, h(t)) : \ |t| < \varrho^{1/p}\}$. We may also infer the injectivity of the mapping $\{|t| < \varrho^{1/p}\} \ni t \longrightarrow (t^p, h(t))$. Finally, take $0 < \delta < \varrho^{1/p}$. The mapping $\{|t| \le \delta\} \ni t \longrightarrow (t^p, h(t))$ is a homeomorphism onto its range. Therefore the mapping $\{|t| < \delta\} \ni t \longrightarrow (t^p, h(t))$ is a homeomorphism onto the representative $V \cap \{|z_1| < \delta^p\}$ of the germ A.

COROLLARY 1. *If V is an analytic subset of a complex manifold, then, for each point $a \in V$ at which $\dim_a V = 1$, there is an open neighbourhood in V which is homeomorphic to the topological space obtained from a finite union of disjoint open discs by identification of their centres. Namely, $V \cap U = W_1 \cup \ldots \cup W_r$, where U is an open neighbourhood of the point a and W_i are analytic sets in U which are homeomorphic to an open disc and such that $W_i \cap W_j = a$ for $i \neq j$.*

Indeed, $V \cap U' = V_1 \cup \ldots \cup V_r$, where U' is an open neighbourhood of the point a and V_i are analytic sets in U' with germs $(V_i)_a$ irreducible, one-dimensional, and mutually distinct. We have $(V_i \cap V_j)_a = a$ for $i \neq j$ (see IV. 3.1, proposition 2; and III. 4.3). Therefore there exist open neighbourhoods W_i of the point a in V_i, respectively, such that $W_i \cap W_j = a$ for $i \neq j$, and there are homeomorphisms $h_i : \{|t| < \delta_i\} \longrightarrow W_i$. It is easy to find an open neighbourhood U of the point a such that the sets W_j are analytic in U $(^{30})$.

COROLLARY 2. *At each point a of a locally analytic subset V of a vector space M at which the germ V_a is irreducible and one-dimensional, there is a "tangent line"*

$$\lambda = \lim_{V \setminus a \ni z \to a} \mathbf{C}(z - a) .$$

$(^{29})$ Because $(0, h(0)) \in V$; see IV. 1.1, property (2).

$(^{30})$ It is sufficient to take $U = U_0 \cup \bigcup_{i=1}^{r}(U_i \setminus \bigcup_{\nu \neq i} \overline{W}_\nu)$, where $U_0, \ldots, U_r$ are open neighbourhoods of the point a such that W_i is analytic in U_i and $U_0 \subset \bigcap_1^r U_i$.

Given a norm on M, the set of limits $\lim \frac{z_\nu - a}{|z_\nu - a|}$, where $V \setminus a \ni z_\nu \longrightarrow a$, coincides with $\{z \in \lambda : |z| = 1\}$.

Indeed, one may assume that $M = \mathbf{C}^n$, $a = 0$, and the germ V_0 is 1–regular. Now, it is enough to observe that, for the representative (∗), the following limit exists and is different from zero:

$$\lim_{t \to 0} \left(t^p, h(t)\right)/t^r, \quad \text{where } r = \min(p, p_2, \dots, p_n)$$

and p_j are the multiplicities of the zeros at 0 of the components of the mapping h.

COROLLARY 3. *If V is an analytic subset of a complex manifold and $\dim_a V = 1$ ([31]), then V contains a simple arc of class C^1 with an endpoint a.*

Indeed, one may assume that V is locally analytic in $\mathbf{C}^n$, and that the germ V_0 is simple and 1–regular. Now, for the representative (∗) and a sufficiently small $\tau > 0$, the arc $\{(t, h(t)) : 0 \le t \le \tau\}$ is simple and of class C^1, because the limit $\lim_{t \to 0} \frac{d}{dt}\left(pt^{p-1}, h'(t)\right)/t^{r-1}$ exists and is different from zero. Here $r = \min(p, p_2, \dots, p_n)$ and the p_j are the multiplicities of the zeros at 0 of the components of the mapping h.

([31]) It is enough to assume that $\dim_a V \ge 1$, since then V_a contains a one–dimensional analytic germ. (See e.g. the proof of the lemma in IV. 4.3.)

CHAPTER III

FUNDAMENTAL LEMMAS

§1. Lemmas on quasi-covers

1. Let X be a complex vector space, and let G be a finite subgroup of the group of linear automorphisms of X. A subset Z of the space X (respectively, a mapping f defined on X) is said to be *invariant with respect to G* or *G–invariant* if $\varphi(Z) = Z$ (respectively, $f \circ \varphi = f$) for each $\varphi \in G$.

LEMMA 0. *Every G–invariant algebraic subset of X can be defined by G–invariant polynomials.*

Indeed, let $G = \{\varphi_1, \ldots, \varphi_l\}$, and let $Z \subset X$ be a G–invariant algebraic subset defined by the polynomials $f_1, \ldots, f_k$. Then

$$Z = \{g_{ij} = 0; \ i = 1, \ldots, k; \ j = 1, \ldots, l\} = \{F_{i\nu} = 0; \ i = 1, \ldots, k;$$
$$\nu = 1, \ldots, l\} ,$$

where $g_{ij} = f_i \circ \varphi_j$, $F_{i\nu} = \sigma_\nu \circ (g_{i1}, \ldots, g_{il})$, and $\sigma_1, \ldots, \sigma_l$ are the basic symmetric polynomials of l variables (see B. 5.3). Hence it is enough to check that the polynomials $F_{i\nu}$ are G–invariant. Let us take an arbitrary φ_s. As the mapping $G \ni \varphi \longrightarrow \varphi \circ \varphi_s \in G$ is bijective, we have $\varphi_j \circ \varphi_s = \varphi_{\beta_j}$, where $j = 1, \ldots, k$ and $(\beta_1, \ldots, \beta_l)$ is a permutation of the set $\{1, \ldots, l\}$. Thus $g_{ij} \circ \varphi_s = g_{i\beta_j}$ and we get $F_{i\nu} \circ \varphi_s = \sigma_\nu \circ (g_{i\beta_1}, \ldots, g_{i\beta_l}) = F_{i\nu}$.

In particular, if X and Y are vector spaces and $p \in \mathbf{N} \setminus 0$, then the mappings

$$\Pi_\alpha : \ X^p \times Y \ni (x_1, \ldots, x_p, y) \longrightarrow (x_{\alpha_1}, \ldots, x_{\alpha_p}, y) \in X^p \times Y ,$$

where $\alpha = (\alpha_1, \ldots, \alpha_p)$ are permutations of the set $\{1, \ldots, p\}$, form a finite subgroup of the group of linear automorphisms of the vector space $X^p \times Y$. Subsets of this space and mappings defined on this space which are invariant with respect to the above subgroup are said to be *symmetric* with respect to $x = (x_1, \ldots, x_p)$. In other words, a set $Z \subset X^p \times Y$ (respectively, a mapping f defined on $X^p \times Y$) is symmetric with respect to x if for every permutation α, $\Pi_\alpha(Z) = Z$ (respectively, $f \circ \Pi_\alpha = f$).

Lemma 0 yields the following

LEMMA 1. *If a set $Z \subset X^p \times Y$ is algebraic and symmetric with respect to x, then it can be defined by polynomials that are symmetric with respect to x, i.e., there exists a polynomial mapping $P : X^p \times Y \longrightarrow \mathbf{C}^s$ that is symmetric with respect to x and such that $Z = P^{-1}(0)$.*

Let X be a vector space. A polynomial mapping $P(\eta_1, \ldots, \eta_p, v)$ on the space X^{p+1} with values in a vector space is called a *collector* if it is symmetric with respect to $\eta = (\eta_1, \ldots, \eta_p)$ and

$$P^{-1}(0) = \{v = \eta_1\} \cup \ldots \cup \{v = \eta_p\} .$$

For example, the polynomial $\mathbf{C}^{p+1} \ni (\eta, v) \longrightarrow (v - \eta_1) \ldots (v - \eta_p) \in \mathbf{C}$ is a collector. It follows from lemma 1 that for every vector space X and $p \in \mathbf{N}$ there exists a collector $P : X^{p+1} \longrightarrow \mathbf{C}^s$.

LEMMA 2. *There is a collector $P : X^{p+1} \longrightarrow \mathbf{C}^r$ such that, for each $\eta = (\eta_1, \ldots, \eta_p) \in X^p$ satisfying the condition $\eta_i \neq \eta_1$ for $i > 1$, the differential $d_{\eta_1}(v \longrightarrow P(\eta, v))$ is injective.*

PROOF . Take an arbitrary $r \geq n = \dim X$. There exist linear forms $\varphi_1, \ldots, \varphi_r \in X^*$, such that every set of n of these forms is linearly independent. (They can be chosen by induction: if a sequence $\varphi_1, \ldots, \varphi_k \in X^*$, where $k \geq n$, has this property, we take φ_{k+1} from the complement of the union of all hyperplanes generated by $n - 1$ elements of this sequence $(^1)$.) Observe that the following property holds when $a_{ij} \in \mathbf{C}$ for $i \in I$ and $j \in J$, where I and J are finite sets:

$$(\#I > (\#J)(n-1), \prod_{j \in J} a_{ij} \text{ for } i \in I) \implies (\#\{i : a_{ij} = 0\} \geq n$$

(*)

$$\text{for some } j \in J).$$

(Otherwise, since $I \subset \bigcup_j \{i : a_{ij} = 0\}$, we would get $\#I \leq (\#J)(n-1)$.) Assume that $r > p(n-1)$. Then $P = (P_1, \ldots, P_r)$, where $P_i(\eta, v) = \varphi_i(v - $

$(^1)$ This is possible, as hyperplanes are nowhere dense.

$\eta_1)\ldots\varphi_i(v-\eta_p)$, is a collector. Indeed, by $(*)$, the equation $P(\eta,v)=0$ implies that $\varphi_i(v-\eta_j)=0$ for some j and for n distinct values of the index i, and hence $v=\eta_j$. Now fix η such that $\eta_i\neq\eta_1$ for $i>1$. We have to show that the differential $d_{\eta_1}P_\eta$ is injective, where $P_\eta=(P_{\eta_1},\ldots,P_{\eta_r})$ and $P_{\eta_i}:\ v\longrightarrow\varphi_i(v-\eta_1)\ldots\varphi_i(v-\eta_r)$. We have $d_{\eta_1}P_{\eta_i}=c_i\varphi_i$, where $c_i=\prod_{j>1}\varphi_i(\eta_1-\eta_j)$. Observe that $\#\{i:\ c_i=0\}\leq(p-1)(n-1)$. For if this were not the case, then, in view of $(*)$, there would exist an index $j>1$ such that $\varphi_i(\eta_1-\eta_j)=0$ for at least n values of the index i, and hence η_i would be equal to η_1, contrary to our assumption about η. Thus, if we assume that $r\geq(p-1)(n-1)+n$, then $c_i\neq0$ for at least n distinct values of the index i, which implies that the differential $d_{\eta_1}P_\eta=(c_1\varphi_1,\ldots,c_r\varphi_r)$ is injective.

2. Let M be a connected manifold, and let X be a vector space. By a *quasi-cover* in the Cartesian product $M\times X$ we mean a pair (Z,Λ) in which Z is a thin subset of M, and Λ is a closed locally topographic submanifold of $(M\setminus Z)\times X$ (see C. 3.17) such that the natural projection $\bar\pi:\ \bar\Lambda\longrightarrow M$ is proper (or, equivalently, each point of the manifold M has a neighbourhood U such that the fibres $\Lambda_u=\{v\in X:\ (u,v)\in\Lambda\}$ $(^2)$, for $u\in U$, are uniformly bounded (see B. 5.2)). If (Z,Λ) is a quasi-cover, then the natural projection $\pi:\ \Lambda\longrightarrow M\setminus Z$ is a finite covering (see B. 3.2, proposition 1; and B. 2.4). Since $M\setminus Z$ is connected (see II. 2.4), then the covering has a multiplicity (see B. 3.2), say k, which is called the *multiplicity* of the quasi-cover (Z,Λ) and we say that the quasi-cover is k–sheeted. The set $\bar\Lambda$ is called the *adherence* of the quasi-cover.

Notice that if (Z,Λ) is a quasi-cover in $M\times X$, then so is (Z,Λ') for any closed and open subset Λ' of Λ; in particular, it is true when Λ' is a component of Λ. Moreover, $(Z\cap G,\Lambda_G)$ is a quasi-cover in $G\times X$, where G is an open and connected subset of M.

3. Let (Z,Λ) be an s–sheeted quasi-cover in the product $M\times X$, where $s>0$.

LEMMA 3. *If* $R(\eta_1,\ldots,\eta_p,v)$ *is a polynomial mapping on* X^{p+1}, *with values in* $\mathbf{C}^q$, *which is symmetric with respect to* $\eta=(\eta_1,\ldots,\eta_p)$, *then there is a unique holomorphic mapping* $H:\ M\times X\longrightarrow\mathbf{C}^q$ *such that for any* $(u,v)\in(M\setminus Z)\times X$,

$$H(u,v)=R(\eta_1,\ldots,\eta_p,v),\quad\text{where}\quad\{\eta_1,\ldots,\eta_p\}=\Lambda_u\ .$$

$(^2)$ Throughout the book we will be using the following notation. If $Z\subset M\times N$, then $Z_x=\{y\in N:\ (x,y)\in Z\}$ for $x\in M$, and $Z_E=Z\cap(E\times N)=\pi^{-1}(E)$ for $E\subset M$, where $\pi:\ Z\longrightarrow N$ is the natural projection.

Indeed, furnishing X with a linear coordinate system, we can write $R(\eta, v) = \sum_s a_s(\eta) v^s$, where $a_s : X^p \longrightarrow \mathbf{C}^q$ are symmetric polynomial mappings. Then the mappings $c_s^0 : M \backslash Z \longrightarrow \mathbf{C}^q$, well-defined by the formula $c_s^0(u) = a_s(\eta_1, \ldots, \eta_p)$ for $u \in M \backslash Z$, where $\{\eta_1, \ldots, \eta_p\} = \Lambda_u$, are holomorphic and locally bounded near Z. Therefore they have holomorphic extensions $c_s : M \longrightarrow \mathbf{C}^q$ (see II. 2.1). The mapping defined by $H(u, v) = \sum_s c_s(u) v^s$ has the required property. The uniqueness of H follows from the fact that $(M \backslash Z) \times X$ is dense in $M \times X$.

PROPOSITION 1. *The projection* $\bar{\pi} : \bar{\Lambda} \longrightarrow M$ *is open. Therefore the function* $M \ni u \longrightarrow \# \bar{\Lambda}_u$ *is lower semi-continuous* (see B. 2.1) *and finite:* $\# \bar{\Lambda}_u \le p$ [3].

THE FIRST LEMMA ON QUASI-COVERS. *The adherence* $\bar{\Lambda}$ *is analytic* [4] *in* $M \times X$. *If* $P : X^{p+1} \longrightarrow \mathbf{C}^r$ *is a collector, then there exists a unique holomorphic mapping* $F = F_P : M \times X \longrightarrow \mathbf{C}^q$ *such that*

$$F(u, v) = P(\eta_1, \ldots, \eta_p, v), \quad \text{where } \{\eta_1, \ldots, \eta_p\} = \Lambda_u,$$
$$\text{for } (u, v) \in (M \backslash Z) \times X.$$

Then $\bar{\Lambda} = F^{-1}(0)$.

PROOF of proposition 1 and the first lemma on quasi-covers. The existence and uniqueness of the mapping F follows from lemma 3. Let $\tilde{\pi} : F^{-1}(0) \longrightarrow M$ be the natural projection. Then $\tilde{\pi}^{-1}(M \backslash Z) = \Lambda$. It is sufficient to show that $\tilde{\pi}$ is open because, from the fact that $M \backslash Z$ is dense, we have $\bar{\Lambda} = F^{-1}(0)$ (see B. 2.1), and consequently, $\bar{\pi} = \tilde{\pi}$. If it were not true, there would exist a compact neighbourhood $U \times V$ of some point $(u_0, v_0) \in F^{-1}(0)$ and a sequence $u_\nu \longrightarrow u_0$ such that $u_\nu \notin \tilde{\pi}(F^{-1}(0) \cap (U \times V))$; moreover, as Z is nowhere dense, one can require that $u_\nu \notin U \backslash Z$. Then we would have $P(\eta_1^\nu, \ldots, \eta_p^\nu, v) = F(u_\nu, v) \neq 0$ for each ν and $v \in V$, where $\eta_i^\nu \in \Lambda_{u_\nu}$. This would imply $\eta_i^\nu \notin V$. In view of the uniform boundedness of the fibres Λ_{u_ν}, one could choose convergent subsequences $\eta_i^{\alpha_\nu} \longrightarrow \eta_i \neq v_0$. Taking the limit, we would get $F(u_0, v_0) = P(\eta_1, \ldots, \eta_p, v_0) \neq 0$, in contradiction with the choice of (u_0, v_0).

According to lemma 1, there is a polynomial mapping $Q(\eta_1, \ldots, \eta_p, v)$ of X^{p+1} to $\mathbf{C}^s$ which is symmetric with respect to $\eta_1, \ldots, \eta_p$ and such that

$$(*) \qquad\qquad Q^{-1}(0) = \bigcup_{i \neq j} \{v = \eta_i = \eta_j\}.$$

---

[3] This is because $M \backslash Z$ is dense and $\# \Lambda_u \le p$.

[4] It is even globally analytic.

By lemma 3, there exist a holomorphic mapping $G : M \times X \longrightarrow \mathbf{C}^s$ such that

$(**)$ $G(u,v) = Q(\eta_1, \ldots, \eta_p, v),$ where $\{\eta_1, \ldots, \eta_p\} = \Lambda_u,$

for $(u,v) \in (M \setminus Z) \times X.$

We have the following lemmas

LEMMA 4. *If at a point (u_0, v_0) the adherence $\bar{\Lambda}$ is a topographic submanifold, then $G(u_0, v_0) \neq 0$.*

LEMMA 5. *If $P : X^{p+1} \longrightarrow \mathbf{C}^r$ is a collector, $F = F_P$, and the adherence $\bar{\Lambda}$ is a topographic submanifold at a point (u_0, v_0), then there are $\eta_1, \ldots, \eta_p \in X$ such that $\eta_i \neq \eta_1 = v_0$ for $i > 1$, and*

$$F(u_0, v) = P(\eta_1, \ldots, \eta_p, v) \quad \text{for some} \quad v \in X \ .$$

PROOF of lemmas 4 and 5. We can choose neighbourhoods U and V of the points u_0 and v_0, respectively, in such a way that the set $\bar{\Lambda} \cap (U \times V)$ is the graph of a continuous mapping $\varphi : U \longrightarrow V$ (with $\varphi(u_0) = v_0$). Take a sequence $u_\nu \longrightarrow u_0$ such that $u_\nu \in U \setminus Z$. Then for each ν we have $\Lambda_{u_\nu} = \{\eta_1^\nu, \ldots, \eta_p^\nu\}$, where $\eta_1^\nu = \varphi(u_\nu)$ and $\eta_i^\nu \notin V$ for $i > 1$. Furthermore, $G(u_\nu, v) = Q(\eta_1^\nu, \ldots, \eta_p^\nu, v)$ and $F(u_\nu, v) = P(\eta_1^\nu, \ldots, \eta_p^\nu, v)$ for $v \in X$. In view of the uniform boundedness of the fibres Λ_{u_ν}, it is possible to choose convergent subsequences $\eta_i^{\alpha_\nu} \longrightarrow \eta_i$. Taking the limit, we obtain $\eta_1 = \varphi(u_0) = v_0 \neq \eta_i$ for $i > 1$, $F(u_0, v) = P(\eta_1, \ldots, \eta_p, v)$ for $v \in X$, and $G(u_0, v_0) = Q(\eta_1, \ldots, \eta_p, v_0) \neq 0$, in view of $(*)$.

THE SECOND LEMMA ON QUASI-COVERS. *The set Σ of points of the adherence $\bar{\Lambda}$ at which it is not a topographic manifold is an analytic subset* [5] *in $M \times X$.*

PROOF . By lemma 4, it is enough to prove that $\Sigma \subset G^{-1}(0)$, since then $\Sigma = \bar{\Lambda} \cap G^{-1}(0)$, which shows — in view of the first lemma on quasi-covers — that Σ is analytic in $M \times X$.

Let $(u_0, v_0) \in \Sigma$ be fixed. There is no neighbourhood of (u_0, v_0) in $\bar{\Lambda}$ which is the graph of a function. In fact, assume the contrary. Since, according to proposition 1, the projection $\bar{\pi}$ is open, there would exist an open neighbourhood U of u_0 and a compact neighbourhood V of v_0 such that $\bar{\Lambda} \cap (U \times V)$ would be the graph of a mapping $\varphi : U \longrightarrow V$, which must be continuous (see B. 2.3). Since $\varphi_{U \setminus Z} \subset \Lambda$ and Λ is locally topographic, this

---

[5] It is even globally analytic.

would imply that $\varphi_{U\setminus Z}$ is holomorphic and hence so is φ (see II. 2.3 and 5), which means that $\bar{\Lambda}$ would be a topographic submanifold at (u_0, v_0).

Thus there exist sequences $(u_\nu, \eta_1^\nu), (u_\nu, \eta_2^\nu) \in \bar{\Lambda}$ which are convergent to (u_0, v_0) and such that $\eta_1^\nu \neq \eta_2^\nu$. Furthermore, due to the fact that Z is nowhere dense and the projection $\bar{\pi}$ is open, we may assume that $u_\nu \in M\setminus Z$. Therefore $\Lambda_{u_\nu} = \{\eta_1^\nu, \ldots, \eta_p^\nu\}$ and by $(**)$ we get $G(u_\nu, v_0) = Q(\eta_1^\nu, \ldots, \eta_p^\nu, v_0)$. In view of the uniform boundedness of the fibres Λ_{u_ν}, we can choose convergent subsequences $\eta_i^{\alpha_\nu} \longrightarrow \eta_i$. Taking limits, we get $\eta_1 = \eta_2 = v_0$ and $G(u_0, v_0) = Q(\eta_1, \ldots, \eta_p, v_0)$. By $(*)$, this implies that $G(u_0, v_0) = 0$.

PROPOSITION 2. *There is a collector* $P : X^{p+1} \longrightarrow \mathbf{C}^r$ *such that*

$$T_z \bar{\Lambda} = \ker d_z F$$

(where $F = F_P$*) holds for each point at which the adherence* $\bar{\Lambda}$ *is a topographic submanifold.*

Indeed, let P be a collector from lemma 2, and let $z = (u_0, v_0)$. According to lemmas 2 and 5, the mapping $v \longrightarrow d_z F(0, v)$ is injective, as it is the differential of the mapping $v \longrightarrow F(u_0, v) = P(\eta_1, \ldots, \eta_p, v)$ at the point $v_0 = \eta_1$. Hence

$$\dim(\ker d_z F) \leq \dim T_{u_0} M = \dim T_z \bar{\Lambda} .$$

But $T_z \bar{\Lambda} \subset \ker d_z F$ (because $F = 0$ on $\bar{\Lambda}$; see C. 3.11), and thus $T_z \bar{\Lambda} = \ker d_z F$.

§2. Regular and k–normal ideals and germs

1. Let I be an ideal in the ring $\mathcal{O}_n$, and let $\chi : \mathcal{O}_n \longrightarrow \mathcal{O}_n/I$ be the natural epimorphism. We put $\hat{f} = \chi(f)$ for $f \in \mathcal{O}_n$ and $\hat{\mathcal{O}}_l = \chi(\mathcal{O}_l)$ for $l = 1, \ldots, n$. In particular, $\hat{\mathcal{O}}_n = \mathcal{O}/I$. We have (see A. 2.1) the induced epimorphisms

$$\mathcal{O}_n[T_1, \ldots, T_s] \ni P \longrightarrow \hat{P} \in \hat{\mathcal{O}}_n[T_1, \ldots, T_s] ,$$

where $\hat{P} = \sum \hat{a}_p T^p$ for $P = \sum a_p T^p$. If $P \in \mathcal{O}_n[T_1, \ldots, T_s]$, then (see A. 2.2):

$$P(g_1, \ldots, g_s)\hat{} = \hat{P}(\hat{g}_1, \ldots, \hat{g}_s) \quad \text{for} \quad g_1, \ldots, g_s \in \mathcal{O}_n ,$$
$$P(Q_1, \ldots, Q_s)\hat{} = \hat{P}(\hat{Q}_1, \ldots, \hat{Q}_s) \quad \text{for} \quad Q_1, \ldots, Q_s \in \mathcal{O}_n[S_1, \ldots, S_r] .$$

2. For $0 \leq k \leq n$, the ideal I is said to be *k–normal*, if it satisfies one of the following equivalent conditions:

(1) I contains a regular germ from $\mathcal{O}_l$ for $l = k + 1, \ldots, n$.

(2) I contains a distinguished germ from $\mathcal{O}_k[z_l]$ for $l = k + 1, \ldots, n$.

(3) $\hat{\mathcal{O}}_n = \hat{\mathcal{O}}_k[\hat{z}_{k+1}, \ldots, \hat{z}_n]$ and $\hat{z}_{k+1}, \ldots, \hat{z}_n$ are integral over $\hat{\mathcal{O}}_k$.

(4) $\hat{\mathcal{O}}_n$ is finite over $\hat{\mathcal{O}}_k$ (and hence it is also integral over $\hat{\mathcal{O}}_k$; see A. 8.1)

We define (-1)–normality by condition (1); the only (-1)–normal ideal is $\mathcal{O}_n$. Of course, if an ideal I is k–normal, then it is also l–normal for $k \leq l \leq n$ and every ideal J containing I is also k–normal.

One checks the equivalences of (1) – (4) as follows. The implication (2)$\Longrightarrow$(1) is trivial and we have already proved (3)$\Longrightarrow$(4) (see A. 8.1). Assume that condition (1) is satisfied, and let $k + 1 \leq l \leq n$. According to the preparation theorem (see I. 2.1), the ideal I contains a monic germ from $\mathcal{O}_{l-1}[z_l]$, and hence the element $\hat{z}_l$ is integral over $\hat{\mathcal{O}}_{l-1}$. In view of the division version of the preparation theorem (see I. 1.4), we have $\mathcal{O}_l \subset I + \mathcal{O}_{l-1}[z_l]$ and thus $\hat{\mathcal{O}}_l = \hat{\mathcal{O}}_{l-1}[\hat{z}_l]$. Now (3) follows by induction (see A. 8.1 and A. 2.7). Finally, assume (4). Then $\hat{z}_l$ is integral over $\hat{\mathcal{O}}_k$, $l = k+1, \ldots, n$. This implies that the ideal I contains a monic – and hence z_l–regular – germ from $\mathcal{O}_k[z_l]$. Therefore, by the preparation theorem (see I. 2.1), the ideal I contains a distinguished germ from $\mathcal{O}_k[z_l]$, which is (2).

If the ideal I is k–normal, where $k \geq 0$, then it is generated by a finite subset of $\mathcal{O}_k[z_{k+1}, \ldots, z_n]$.

To see this, consider $B = \mathcal{O}_k[z_{k+1}, \ldots, z_n]$. It is sufficient to prove that I is generated by $B \cap I$ (see A. 9.2). According to (2), the ideal I' generated by $B \cap I$ is k–normal, and in view of (3) applied to I', we have $\mathcal{O}_n = B + I'$. Therefore $I' \subset I \subset I' + B$. Hence $I \subset I' + (B \cap I) \subset I'$ (see A. 1.4), which gives $I = I'$.

Notice that an ideal I is k–normal if and only if rad I is k–normal as well. (This follows from (1).)

3. If $P_j(u, z_j)$ are distinguished polynomials with respect to z_j whose coefficients are holomorphic in a connected open neighbourhood Ω of zero in $\mathbf{C}^k$ ($j = k + 1, \ldots, n$ and $u = (z_1, \ldots, z_k)$), then we define the *Weierstrass set*

$$W = W(P_{k+1}, \ldots, P_n)$$
$$= \{z \in \mathbf{C}^n : \ u \in \Omega, \ P_{k+1}(u, z_{k+1}) = \ldots = P_n(u, z_n) = 0\} \ .$$

The Weierstrass set W is an analytic nowhere dense subset of $\Omega \times \mathbf{C}^{n-k}$. The natural projection $\pi : W \longrightarrow \Omega$ is proper (see B. 5.2–3) and $\pi^{-1}(0) \subset 0$. Moreover, $\dim W \leq k$. (For the fibres $\pi^{-1}(u)$ are finite; see II. 1.4.)

Observe that if $\Delta \subset \Omega$ is an open and connected neighbourhood of zero, then $W(P_{k+1}, \ldots, P_n)_\Delta = W((P_{k+1})_{\Delta \times \mathbf{C}}, \ldots, (P_n)_{\Delta \times \mathbf{C}})$, and if $W \neq \emptyset$ (i.e., if the degrees of the P_j's are positive), then $0 \in W$ and the sets W_Δ form a base of neighbourhoods of zero in W (see C. 2.2).

Assume now that the discriminants of the polynomials P_j are $\not\equiv 0$. Then we denote the union of the zero sets of these discriminants by $Z = Z(P_{k+1}, \ldots, P_n)$. Then Z is an analytic nowhere dense subset of Ω. In view of the implicit function theorem, the set $W_{\Omega \setminus Z}$ is a locally topographic submanifold of $(\Omega \setminus Z) \times \mathbf{C}^{n-k}$ (see C. 2.1). Hence $(Z, W_{\Omega \setminus Z})$ is a quasi-cover in $\Omega \times \mathbf{C}^{n-k}$.

4. An analytic germ A at $0 \in \mathbf{C}^n$ is said to be k–normal if its ideal $\mathcal{I}(A)$ is k–normal. Then every analytic germ $B \subset A$ is also k–normal (see II. 4.2 and n° 2).

Every analytic and k–normal germ A at $0 \in \mathbf{C}^n$, where $k \geq 0$, is contained in some Weierstrass set $W(P_{k+1}, \ldots, P_m)$. Moreover, one may choose the P_j's with discriminants $\not\equiv 0$.

Indeed, for such A, the ideal $\mathcal{I}(A)$ contains distinguished germs $p_j \in \mathcal{O}_k[z_j]$, $j = k+1, \ldots, n$ (see n° 2, condition (2)). It is enough to take as P_j's representatives of those germs (or of the germs red p_j) that are distinguished polynomials with holomorphic coefficients in an open connected neighbourhood Ω of zero in $\mathbf{C}^k$. Indeed, in this case,

$$A = V(\mathcal{I}(A)) \subset \bigcap_{k+1}^{n} V(p_j) \subset W(P_{k+1}, \ldots, P_n)$$

(or $A = V(\mathcal{I}(A)) \subset \bigcap_{k+1}^{n} V(p_j) = \bigcap_{k+1}^{n} V(\text{red } p_j) = W(P_{k+1}, \ldots, P_n)$; see II. 5.1. Then the discriminants of the polynomials P_j are $\not\equiv 0$; see I. 2.4 and I. 1.5).

This yields the following corollaries.

The dimension of any k–normal analytic germ is $\leq k$.

Any k–normal analytic germ (where $k \geq 0$) has a representative $Z \subset \mathbf{C}^k \times \mathbf{C}^{n-k}$ for which the fibres Z_u, $u \in \mathbf{C}^k$, are finite.

If Z is a representative of an $(n-2)$–normal germ, then there is an arbitrarily small open neighbourhood of zero U (in which $Z \cap U$ is closed) *such that every holomorphic function on $U \setminus Z$ extends to a holomorphic function on U.*

In fact, one can find an arbitrarily small neighbourhood U of the form $\Omega \times \{|z_{n-1}| < 2r, \ |z_n| < 2r\}$, where Ω in an open connected set, and a Weierstrass set $W(P_{n-1}, P_n)$ with polynomials P_j whose coefficients are holomorphic in Ω, such that $Z \cap U$ is analytic in U and $Z \cap U \subset W(P_{n-1}, P_n) \subset \Omega \times \{|z_{n-1}| < r, \ |z_n| < r\}$ (see C. 2.2 and n° 3). Then, in view of the Hartogs lemma, every holomorphic function on $U \setminus Z$ has a holomorphic extension in U (see the corollary in C. 1.11 and II. 3.6).

5. An ideal I of the ring $\mathcal{O}_n$ is said to be *k-regular*, where $-1 \le k \le n$, if it is k-normal and $\mathcal{O}_k \cap I = 0$ in the case when $k \ge 0$. (We define $-\infty$-regularity in the same way as $-\infty$-normality.) If such a k exists, then the ideal I is also said to be *regular*. Then (by (1) in n° 2)

$$(I \text{ is } l - \text{normal}) \Longleftrightarrow l \ge k .$$

Thus such a k is unique.

If an ideal I is k-regular, where $k \ge 0$, then the mapping $\chi_{\mathcal{O}_k} : \mathcal{O}_k \longrightarrow \hat{\mathcal{O}}_k$ is an isomorphism $(^6)$ and we have the induced isomorphisms $\mathcal{O}_k[T_1, \ldots, T_s] \ni P \longrightarrow \hat{P} \in \hat{\mathcal{O}}_k[T_1, \ldots, T_s]$.

Notice that *an ideal I is k-regular (regular) only if so is* rad I $(^7)$.

An analytic germ A at $0 \in \mathbf{C}^n$ is said to be *k-regular (regular)*, if its ideal $\mathcal{I}(A)$ is k-regular (regular).

6. Let M be a complex n-dimensional manifold, and let φ be a coordinate system at a point $a \in M$. Then an ideal I of the ring $\mathcal{O}_a$ or an analytic germ A at a is called, respectively, *k-normal, k-regular, regular in the coordinate system φ* if the ideal or the germ, respectively, in this coordinate system (i.e., the ideal $I \circ \varphi^{-1}$ or the germ $\varphi(A)$) is k-normal, k-regular, or regular.

A linear change of the coordinates $z_1, \ldots, z_k$, as well as a linear change of the coordinates $z_{k+1}, \ldots, z_n$, preserves k-normality and k-regularity of an ideal of the ring $\mathcal{O}_n$ or of an analytic germ at $0 \in \mathbf{C}^n$. (This indicates that if φ is an automorphism and e' is the identity mapping of $\mathbf{C}^k$, ψ is an automorphism and e'' is the identity mapping of $\mathbf{C}^{n-k}$, then if an ideal of $_n\mathcal{O}$

$(^6)$ The condition $\mathcal{O}_k \cap I = 0$ means precisely that it is an isomorphism.

$(^7)$ See n° 2. If $\mathcal{O}_k \cap I = 0$ and $f \in \mathcal{O}_k \cap$ rad I, then $f^s \in \mathcal{O}_k \cap I$ for some $s > 0$, and hence $f = 0$.

or an analytic germ at $0 \in \mathbf{C}^n$ is k–normal or k–regular, it is also k–normal or k–regular, respectively, both in the coordinate systems $\varphi \times e''$ and $e' \times \psi$.)

Indeed, it is enough to verify the claim for ideals (see II. 4.4). Now, let $\chi = \varphi \times e''$ or $\chi = e' \times \psi$. Then $\mathcal{O}_k \circ \chi^{-1} = \mathcal{O}_k$. The condition (4) of k–normality of an ideal I (see n° 2) is equivalent to the condition $\mathcal{O}_n \sum_i \mathcal{O}_k g_i + I$ for some $g_i \in \mathcal{O}_n$. Thus, if I is k–normal, then, by passing to the images under the automorphism $f \longrightarrow f \circ \chi^{-1}$ of the ring $\mathcal{O}_n$, we conclude that the ideal $I \circ \chi^{-1}$ is k–normal. Moreover, if $\mathcal{O}_n \cap I = 0$, then $\mathcal{O}_k \cap (I \circ \chi^{-1}) = 0$.

In particular, this implies that:

If a k–normal ideal of the ring $\mathcal{O}_n$ is r–regular after a linear change of the coordinates $z_1, \ldots, z_k$ (i.e., in the coordinate system $\varphi \times e''$, as above), then $r \leq k$. (See n° 5.)

7. Let $0 \leq k \leq n$, and let e'' denote the identity mapping of the space $\mathbf{C}^{n-k}$.

PROPOSITION . *If ideals $I_1, \ldots, I_m$ of the ring $\mathcal{O}_n$, or analytic germs $A_1, \ldots, A_m$ at $0 \in \mathbf{C}^n$, are k–normal, then all of them are regular after any change of the coordinates $z_1, \ldots, z_k$ that belong to a dense subset of $L_0(\mathbf{C}^k, \mathbf{C}^k)$, i.e., in any coordinate system $\varphi \times e''$ for any φ from a dense subset of $L_0(\mathbf{C}^k, \mathbf{C}^k)$.*

PROOF [8] . It suffices to consider the case of ideals. For $0 \leq r \leq n$, consider the condition:

(c_r) The ideal I is r–normal or s–regular for some $s \geq r$.

Denote by e_r the identity mapping of $\mathbf{C}^{n-r}$. Let $r > 0$. In view of the previous property (see n° 6) and condition (1) for $(r-1)$–and r–normality (see n° 2), if the ideals $I_1, \ldots, I_m$ satisfy the condition (c_r), then they satisfy the condition (c_{r-1}) in the coordinate system $\varphi \times e_r$ for any φ from a dense subset of $L_0(\mathbf{C}^r, \mathbf{C}^r)$. (Indeed, each of those ideals which is r–normal but not r–regular contains a non-zero germ from $\mathcal{O}_r$; see n° 5. These germs are regular in any coordinate system φ from a dense subset of $L_0(\mathbf{C}^r, \mathbf{C}^r)$; see I. 1.4.) Therefore, for any φ_k from the dense subset of $L_0(\mathbf{C}^k, \mathbf{C}^k)$, one can choose successively $\varphi_{k-1}, \ldots, \varphi_1 \in L_0(\mathbf{C}^k, \mathbf{C}^k)$ arbitrarily close to the identity mapping such that the ideals $I_1, \ldots, I_m$ in the coordinate system $(\varphi_1 \circ \ldots \circ \varphi_k) \times e''$ satisfy condition (c_0). Owing to the fact that the composition is continuous (see B. 5.2), it follows that the automorphisms $\varphi \in L_0(\mathbf{C}^k, \mathbf{C}^k)$ for which the

---

[8] The set of such φ is open. Namely, it is the complement of a nowhere-dense algebraic subset of $L(\mathbf{C}^k, \mathbf{C}^k)$. (See corollary 2 from the Cartan-Remmert theorem in V. 3.3.)

ideals $I_1, \ldots, I_m$ in the coordinate system $\varphi \times e''$ satisfy condition (c_0) form a dense subset. Finally, it is enough to observe that any 0–normal ideal is 0–or (-1)–regular.

The proposition implies the following corollaries.

Every k–normal analytic ideal (of $\mathcal{O}_n$) or germ (at $0 \in \mathbf{C}^n$) becomes r–regular, with some $r \leq k$, after a suitable change of the coordinates $z_1, \ldots, z_k$. (See n° 6 and II. 4.4.)

If M is an n–dimensional manifold and $a \in M$, then arbitrary ideals $I_1, \ldots, I_m$ of the ring $\mathcal{O}_a$ or arbitrary analytic germs $A_1, \ldots, A_m$ at a are simultaneously regular in a suitable coordinate system at a (see II. 4.4). In the case when M is a vector space and $a = 0$, such coordinate systems that, in addition, are linear, constitute a dense set in $L_0(M, \mathbf{C}^n)$ (see footnote ([8])).

If A is a k–dimensional analytic germ at $0 \in \mathbf{C}^n$ $(0 \leq k \leq n$ or $k = -\infty)$, then: (A is k–regular) $\Longleftrightarrow$ (A is k–normal).

Indeed, assuming that $k \geq 0$, if A is k–normal, then, after a suitable change of the coordinates $z_1, \ldots, z_k$, it becomes r–regular, where $r \leq k$. Thus $\dim A \leq r$ (see n° 4), and so $k = r$. Hence A is k–regular (see n° 6).

§3. Rückert's descriptive lemma

Let $0 \leq k \leq n$.

1. Let I be a k–regular prime ideal of the ring $\mathcal{O}_n$. Then the ring $\hat{\mathcal{O}}_n$ is an integral domain (see A. 1.11 and 2.1) and $\mathcal{O}_k$ is factorial (see I. 2.3 and 2.5). Thus, for each $\hat{z}_j$, $j = k+1, \ldots, n$, we have a minimal polynomial $\hat{p}_j \in \hat{\mathcal{O}}_k[T]$, where $p_j \in \mathcal{O}_k[T]$ (see A. 8.2, 2.2, and 2.5).

LEMMA 1. *The germs $p_j(z_j)$, $j = k+1, \ldots, n$, belong to the ideal I, are distinguished (in $\mathcal{O}_k[z_j]$, respectively), and have non-zero discriminants.*

Indeed, since $\widehat{p_j(z_j)} = \hat{p}_j(\hat{z}_j) = 0$ (see 2.1), the germ $p_j(z_j)$ belongs to I. It has a non-zero discriminant, because p_j is irreducible (see A. 8.2, the proposition in A. 6.3, and I. 1.5). It is distinguished. Indeed, the preparation theorem (see I. 2.1) implies that it is associated with a distinguished germ $q_j(z_j)$, where $q_j \in \mathcal{O}_k[T]$. Thus $q_j(z_j) \in I$, which means that $\hat{q}_j(\hat{z}_j) = 0$, and hence $\hat{p}_j$ (as the minimal polynomial for $\hat{z}_j$) is a divisor of $\hat{q}_j$. Accordingly, the germ $p_j(z_j)$ is a divisor in $\mathcal{O}_k[z_j]$ of the distinguished germ $q_j(z_j)$, and so it is itself distinguished (see I. 2.1).

LEMMA 2. *The Weierstrass set* $W(\tilde{p}_{k+1},\ldots,\tilde{p}_n)$, *where* $\tilde{p}_j$ *are representatives of the germs* $p_j(z_j)$ *and have non-zero discriminants, can be made arbitrarily small.* (*See* I. 1.5 *and* 2.3.)

RÜCKERT'S DESCRIPTIVE LEMMA . *Let* $A = V(I)$ *be the locus of a* k-*regular prime ideal* I *of the ring* $\mathcal{O}_n$. *In particular,* A *can be an irreducible* k-*regular analytic germ at* $0 \in \mathbf{C}^n$ [9] . *Then there exist:*

an open connected neighbourhood Ω *of zero in* $\mathbf{C}^k$,

an analytic nowhere dense subset Z *in* Ω,

a representative V *of the germ* A, *which is analytic in* $\Omega \times \mathbf{C}^{n-k}$,

such that

(1) *the natural projection* $\pi : V \longrightarrow \Omega$ *is proper,*

(2) $\pi^{-1}(0) = 0$,

(3) *the set* $V' = V_{\Omega \setminus Z}$ *is a non-empty locally topographic submanifold of* $\Omega \times \mathbf{C}^{n-k}$.

Thus (*see* B. 3.2, *proposition* 1; B. 2.4; *and* II. 3.6) *the projection* $\pi_{V'} : V' \longrightarrow \Omega \setminus Z$ *is a finite covering of multiplicity* > 0.

Note that, in such a case, (Z, V') is a quasi-cover in $\Omega \times \mathbf{C}^{n-k}$ (see B. 2.4).

Under the assumptions of the descriptive lemma, we have the following corollaries (notation as in the lemma).

COROLLARY 1. *The sets* $V_{\Omega'}$, *where* $\Omega' \subset \Omega$ *are open neighbourhoods of the origin in* $\mathbf{C}^k$, *form a base of neighbourhoods of zero in* V, *and* $\pi(V_{\Omega'}) = \Omega'$. *Therefore the image under the projection onto* $\mathbf{C}^k$ *of any representative of the germ* A *is a neighbourhood of zero.*

Indeed, in view of (1) and (2), the sets $V_{\Omega'}$ can be made arbitrarily small (see B. 2.4). The projection $\pi_{V_{\Omega'}} \longrightarrow \Omega'$ is proper and hence closed. Therefore, since $\pi(V_{\Omega' \setminus Z}) = \Omega' \setminus Z$, we have $\pi(V_{\Omega'}) \supset \Omega'$ (see B. 2.4 and B. 2.3).

COROLLARY 1a. *For any open neighbourhood* $\Omega' \subset \Omega$ *of zero, the triple* $\Omega', Z \cap \Omega', V_{\Omega'}$ *satisfies the conditions* (1) – (3). *For any sufficiently small* Ω', *the set* $V_{\Omega' \setminus Z}$ *is dense in* $V_{\Omega'}$.

---

[9] See II. 4.6, proposition 2; and II. 4.5. From Hilbert's Nullstellensatz (see 4.1) it follows that these two conditions are equivalent.

To see this, note that the first lemma on quasi-covers (see 1.3) implies that the set $\tilde{V} = \overline{V_{\Omega \setminus Z}} \cap (\Omega \times \mathbf{C}^{n-k})$ is analytic in $\Omega \times \mathbf{C}^{n-k}$. Also, the set V_Z is analytic in $\Omega \times \mathbf{C}^{n-k}$ and we have $V = \tilde{V} \cup V_Z$; so, $A = \tilde{V}_0 \cup (V_Z)_0$. But $A \neq (V_Z)_0$ by corollary 1. Hence $A = \tilde{V}_0$ ([10]). Thus, in view of corollary 1, we have $V_{\Omega'} \subset \overline{V_{\Omega' \setminus Z}}$, provided that Ω' is sufficiently small.

COROLLARY 2. *The multiplicity of the covering* $\pi_{V'}$ *depends exclusively on the germ* A *and is independent of any linear change of the coordinates* $z_{k+1}, \ldots, z_n$ ([11]) *(cf. 2.6).*

In order to see this, take $\chi = e' \times \psi$, where e' is the identity mapping of $\mathbf{C}^k$ and $\psi \in L_0(\mathbf{C}^{n-k}, \mathbf{C}^{n-k})$. Let $\tilde{A} = \chi(A)$. Let $V, \tilde{V}$ be representatives of the germs $A, \tilde{A}$, respectively, and $\Omega, \tilde{\Omega}$ be neighbourhoods of zero, chosen as in the descriptive lemma. Note that $\chi(V)$ is also a representative of the germ $\tilde{A}$. By corollary 1, the sets $\tilde{V}_{\Omega'}$ form a base of neighbourhoods of zero in $\tilde{V}$, whereas the sets $\chi(V)_{\Omega'} = \chi(V_{\Omega'})$ form a base of neighbourhoods of zero in $\chi(V)$. Therefore we must have $\tilde{V}_{\Omega'} = \chi(V)_{\Omega'}$ for some $\Omega' \subset \Omega \cap \tilde{\Omega}$. Hence $\tilde{V}_u = \chi(V)_u = \psi(V_u)$ for $u \in \Omega'$. Thus the multiplicities of the coverings $\pi_{V'}$ and $\pi_{\tilde{V}'}$ – being equal to $\#V_u$ and $\#\tilde{V}_u$, respectively, for $u \in \Omega'$ from the complement of a nowhere dense subset of Ω' – must coincide.

PROOF of the descriptive lemma. There is a system of generators

$$f_1(z_{k+1}, \ldots, z_n), \ldots, f_\nu(z_{k+1}, \ldots, z_n)$$

for the ideal I, where $f_i \in \mathcal{O}_k[X_{k+1}, \ldots, X_n]$ (see 2.2). In view of lemma 1, we may assume that this system contains the germs $p_j(z_j)$, $j = k+1, \ldots, n$. By lemma 2, there exist arbitrarily small open connected neighbourhoods of zero in $\Omega \subset \mathbf{C}^k$ and $\Delta \subset \mathbf{C}^{n-k}$, and a Weierstrass set $W = W(\tilde{p}_{k+1}, \ldots, \tilde{p}_n) \subset \Omega \times \Delta$, where the $\tilde{p}_j$, which are representatives of the $p_j(z_j)$, have coefficients that are holomorphic in Ω. We may also assume that the germs $f_i(z_{k+1}, \ldots, z_n)$ have representatives $\tilde{f}_i$ which are polynomials in $v = (z_{k+1}, \ldots, z_n)$ with coefficients holomorphic in Ω. Then each of the germs $\tilde{p}_j$ is equal to some $\tilde{f}_i$. Hence the set $V = \{\tilde{f}_1 = \ldots = \tilde{f}_r = 0\}$ is an analytic subset of $\Omega \times \mathbf{C}^{n-k}$ that represents A and is contained in W. As a result,

$$(\#) \qquad\qquad V \subset \Omega \times \Delta,$$

([10]) Due to the fact the germ A is irreducible – see footnote ([9]). In this chapter, corollary 1a will be used only in the proof of the classic descriptive lemma, which is not used in the proof of Hilbert's Nullstellensatz.

([11]) It is also independent of any linear change of the coordinates $z_1, \ldots, z_k$.

the natural projection $\pi : V \longrightarrow \Omega$ is proper, and $\pi^{-1}(0) = 0$ (see 2.3 and B. 2.4; we have $\tilde{f}_i(0) = 0$, for otherwise $I = \mathcal{O}_n$). Now the set $W_{\Omega \setminus Z^*}$, where $Z^* = Z(\tilde{p}_{k+1}, \ldots, \tilde{p}_n)$, is a locally topographic submanifold of $\Omega \times \mathbf{C}^{n-k}$ (see lemma 2, and 2.3). Consequently, it is enough to show that for suitably chosen Ω and Δ there is an analytic set $Z \supset Z^*$, that is nowhere dense in Ω and such that

$$\begin{cases} V' = V_{\Omega \setminus Z} \neq \emptyset, \\ \text{for each } z \in V', \text{ the germ } V'_z \text{ contains a smooth germ of dimension } k. \end{cases}$$

For if $z \in V'$, then we must have $V'_z = W_z$ (otherwise, since $V' \subset W_{\Omega \setminus Z^*}$, one would have $\dim V'_z < k$; see II. 3.3), and so V' is a topographic submanifold at z.

According to the primitive element theorem for integral domains (see A. 8.3), there is a primitive element $\hat{w}$ of the extension $\hat{\mathcal{O}}_n$ of the ring $\hat{\mathcal{O}}_k$, where $w \in \hat{\mathcal{O}}_n$ (see 2.2 condition (3)). Therefore $\hat{\delta}\hat{z}_j = \hat{Q}_j(\hat{w})$, where $\delta \in \mathcal{O}_k \setminus 0$ and $Q_j \in \mathcal{O}_k[T]$ $(j = k+1, \ldots, n)$. Thus $\delta z_j - Q_j(w) \in I$, and so

$$(\alpha) \qquad \delta z_j - Q_j(w) = \sum_i a_{ij} f_i(z_{k+1}, \ldots, z_n), \quad j = k+1, \ldots, n ,$$

where $a_{ij} \in \mathcal{O}_n$. Since the element $\hat{w}$ is integral over $\hat{\mathcal{O}}_k$ (see (4) in 2.2), it has a minimal polynomial $\hat{G} \in \hat{\mathcal{O}}_k[T]$, where $G \in \mathcal{O}_k[T]$. Then the polynomial G is irreducible (see A. 8.2 and 2.5), and hence its discriminant $\delta_0 \in \mathcal{O}_k$ is non-zero (see the proposition in A. 6.3). Thus we have $\hat{G}(\hat{w}) = 0$, and so $G(w) \in I$. Therefore

$$(\beta) \qquad\qquad G(w) = \sum_i b_i f_i(z_{k+1}, \ldots, z_n) ,$$

where $b_i \in \mathcal{O}_n$. For some m, we have

$$(\gamma) \qquad \delta^m f_i(z_{k+1}, \ldots, z_n) = F_i(\delta z_{k+1}, \ldots, \delta z_n), \quad i = 1, \ldots, r ,$$

for some $F_i \in \mathcal{O}_n[X_{k+1}, \ldots, X_n]$. Then

$$\hat{F}_i(\hat{Q}_{k+1}, \ldots, \hat{Q}_n)(\hat{w}) = \hat{F}_i(\hat{\delta}\hat{z}_{k+1}, \ldots, \hat{\delta}\hat{z}_n) = \big(\delta^m f_i(z_{k+1}, \ldots, z_n)\big)^{\hat{}} = 0$$

(see A. 2.2 and 2.1), and hence $\hat{G}$ must be a divisor of $\hat{F}_i(\hat{Q}_{k+1}, \ldots, \hat{Q}_n)$ in $\hat{\mathcal{O}}_k[T]$ (see A. 8.2); i.e., $\hat{F}_i(\hat{Q}_{k+1}, \ldots, \hat{Q}_n) = \hat{G}\hat{H}_i$, for some $H_i \in \mathcal{O}_k[T]$.

Therefore $F_i(Q_{k+1}, \ldots, Q_n) = GH_i$ (see 2.1 and 2.5) and, substituting the germ $t \in \mathcal{O}_{ut} \supset \mathcal{O}_u = \mathcal{O}_k$ of the function $(u,t) \longrightarrow t$, we obtain (see A. 2.2):

$$(\delta) \qquad F_i\big(Q_{k+1}(t), \ldots, Q_n(t)\big) = G(t)H_i(t), \quad i = 1, \ldots, r \ .$$

Now the neighbourhood $\Omega \times \Delta$ can be chosen so small that the germs w, a_{ij}, b_i have representatives $\tilde{w}, \tilde{a}_{ij}, \tilde{b}_i$ which are holomorphic in $\Omega \times \Delta$, the germs δ, δ_0 have representatives $\tilde{\delta} \not\equiv 0$, $\tilde{\delta}_0 \not\equiv 0$ which are holomorphic in Ω, and the coefficients of the polynomials Q_j, G, H_i, F_i have holomorphic representatives in Ω. Let $\tilde{Q}_j(u,t), \tilde{G}(u,t), \tilde{H}_i(u,t), \tilde{F}_i(u,v)$ be polynomials with respect to t and v, respectively, whose coefficients are holomorphic representatives in Ω of the coefficients of the polynomials Q_j, G, H_i, F_i, respectively. Then the equalities (α)–(δ) imply

$$(a) \quad \tilde{\delta}(u)z_j - \tilde{Q}_j\big(u, \tilde{w}(z)\big) = \sum_{i=1}^{r} \tilde{a}_{ij}(z)\tilde{f}_i(z) \quad \text{in} \quad \Omega \times \Delta, \quad j = k+1, \ldots, n,$$

$$(b) \quad \tilde{G}\big(u, \tilde{w}(z)\big) = \sum_{1}^{r} \tilde{b}_i(z)\tilde{f}_i(z) \quad \text{in} \quad \Omega \times \Delta,$$

$$(c) \quad \tilde{\delta}(u)^m \tilde{f}_i(z) = \tilde{F}_i\big(u, \tilde{\delta}(u)z_{k+1}, \ldots, \tilde{\delta}(u)z_n\big) \quad \text{in} \quad \Omega \times \mathbf{C}^{n-k}, \ i = 1, \ldots, r,$$

$$(d) \quad \tilde{F}_i\big(u, \tilde{Q}_{k+1}(u,t), \ldots, \tilde{Q}_n(u,t)\big) = \tilde{G}(u,t)\tilde{H}_i(u,t) \quad \text{in} \quad \Omega \times \mathbf{C},$$

$$i = 1, \ldots, r.$$

Indeed, it is easy to check that the germs at 0 of the left and right hand sides of the equalities (a)–(d) are equal, respectively, to the left and right hand sides of the equalities (α)–(δ) [12] . Note that $\tilde{\delta}_0$ is the discriminant of

---

[12] For instance, one can take the polynomials $Q_j^*, G^*, H_i^* \in \mathcal{O}_\Omega[T]$, and $F_i^* \in \mathcal{O}_\Omega[X_{k+1},$ $\ldots, X_n]$ whose coefficients coincide with those of $\tilde{Q}_j, \tilde{G}_j, \tilde{H}_i$, and $\tilde{F}_i$, respectively. Then – after the identifications $\mathcal{O}_\Omega \subset \mathcal{O}_{\Omega \times \Delta}$ and $\mathcal{O}_\Omega \subset \mathcal{O}_{\Omega \times \mathbf{C}}$ – the equalities (a)–(d), with (c) restricted to $\Omega \times \Delta$, will become

$$(a) \quad \tilde{\delta}z_j - Q_j^*(\tilde{w}) = \sum_{i=1}^{r} \tilde{a}_{ij}\tilde{f}_i, \quad j = k+1, \ldots, n,$$

$$(b) \quad G^*(\tilde{w}) - \sum_{1}^{r} \tilde{b}_i\tilde{f}_i,$$

$$(c) \quad \tilde{\delta}^m f_i = F_i^*(\tilde{\delta}z_{k+1}, \ldots, \tilde{\delta}z_n), \quad i = 1, \ldots, r,$$

$$(d) \quad F_i^*\big(Q_{k+1}^*(t), \ldots, Q_n^*(t)\big) = G^*(t)H_i^*(t), \quad i = 1, \ldots, r,$$

where z_j denote the functions $\Omega \times \Delta \ni z \longrightarrow z_j \in \mathbf{C}$ and t denotes the function $\Omega \times \mathbf{C} \ni (z,t) \longrightarrow t \in \mathbf{C}$. Now, the images of the left and right hand sides of the above equalities

$\tilde{G}$ (¹³) .

Now we set $Z = Z^* \cup \{\tilde{\delta}\tilde{\delta}_0 = 0\}$. Then

$$V' = \{(u,v) \in \mathbf{C}^n : u \in \Omega \setminus Z, \ \tilde{f}_i(u,v) = 0, \ i = 1,\ldots,r\} .$$

The set

$$\Lambda = \{(u,v,t) \in \mathbf{C}^{n+1} : u \in \Omega \setminus Z, \ \tilde{G}(u,t) = 0, \ \tilde{\delta}(u)z_j = \tilde{Q}_j(u,t),$$
$$j = k+1,\ldots,n\}$$

is non-empty (because G and also $\tilde{G}$ are of degree > 0). In view of the implicit function theorem (see C. 2.1), it is a locally topographic submanifold of $\Omega \times \mathbf{C}^{n-k+1}$. Let $\pi^* : \ \mathbf{C}^{n+1} \ni (z,t) \longrightarrow z \in \mathbf{C}^n$. The relations (a)–(d) imply that $V' = \pi^*(\Lambda)$. (The inclusion $V' \supset \pi^*(\Lambda)$ can be derived from the equalities (c) and (d), while the inclusion $V' \subset \pi^*(\Lambda)$ follows from the relations (#), (a), and (b) with $t = \tilde{w}(z)$.) Consequently, the set V' is non-empty and each of its points belongs to a k–dimensional submanifold which, in turn, is contained in this set (see C. 3.17). This completes the proof of the descriptive lemma.

REMARK. It follows from the above argument that by taking $Z = \{\tilde{\delta} = 0\}$ we get also $V' = \pi(\Lambda)$.

2. Using the second part of the primitive element theorem for integral domains, we can obtain a more precise description of the submanifold V'. First note that:

Under the hypothesis of the descriptive lemma (assuming $k < n$) and after a suitable change of the coordinates $z_{k+1},\ldots,z_n$ (see 2.6), $\hat{z}_{k+1}$ becomes a primitive element of the extension $\hat{\mathcal{O}}_n$ of the ring $\hat{\mathcal{O}}_k$.

Indeed, the set $\chi(\mathbf{C} \setminus 0) \subset \hat{\mathcal{O}}_k$ is infinite and hence there is a primitive element of the form $\sum_{k+1}^n \hat{c}_j \hat{z}_j$, where $c_j \in \mathbf{C} \setminus 0$ (see A. 8.3 and (3) in 2.2). Let us take the coordinate system $\varphi = \left(z_1,\ldots,z_k,\sum_{k+1}^n c_j z_j, z_{k+2},\ldots,z_n\right)$. Consider the ideal $I' = I \circ \varphi^{-1}$. We have the corresponding ring $\hat{\mathcal{O}}'_n = \mathcal{O}_n/I'$. Its subring $\hat{\mathcal{O}}'_k$ and its elements $\hat{z}'_j$ are the images of the ring $\mathcal{O}_k$ and the germs

---

under the homomorphisms $\mathcal{O}_{\Omega \times \Delta} \in f \longrightarrow f_0 \in \mathcal{O}_n$ and $\mathcal{O}_{\Omega \times \mathbf{C}} \ni h \longrightarrow h_0 \in \mathcal{O}_{ut}$ are, respectively, the left and right hand sides of the equalities (α)–(δ). This is so because the images of the polynomials Q_j^*, G^*, H_i^*, F_i^* via the induced homomorphisms are the polynomials Q_j, G, H_i, F_i (see A. 2.2).

(¹³) Because $\tilde{\delta}_0$ is the discriminant of G^* (see A. 4.3 and I. 1.5).

z_j, via the natural epimorphism $\mathcal{O}_n \longrightarrow \hat{\mathcal{O}}'_n$. Now, the ideal I' is the image of the ideal I under the isomorphism $\mathcal{O}_n \ni f \longrightarrow f \circ \varphi^{-1} \in \mathcal{O}_n$, whereas the germ z'_{k+1} is the image of the germ $\sum_{k-1}^{n} c_j z_j$. Therefore we have the induced isomorphism $\hat{\mathcal{O}}_n \longrightarrow \hat{\mathcal{O}}'_n$ under which the image of the subring $\hat{\mathcal{O}}_k$ is the subring $\hat{\mathcal{O}}'_k$, and the image of the element $\sum_{k+1}^{n} \hat{c}_j \hat{z}_j$ is the element $\hat{z}'_{k+1}$. This implies that $\hat{z}'_{k+1}$ is a primitive element of the extension $\hat{\mathcal{O}}'_n$ of the ring $\hat{\mathcal{O}}'_k$.

Suppose now that $\hat{z}_{k+1}$ is a primitive element of the extension $\hat{\mathcal{O}}_n$ of the ring $\hat{\mathcal{O}}_k$ (and that the assumptions of the descriptive lemma are satisfied). Then $G = p_{k+1}$ in the proof of the descriptive lemma. According to the second part of the primitive element theorem for integral domains (see A. 8.3; $\hat{\mathcal{O}}_n$ is finite over $\hat{\mathcal{O}}_k$, by (4) in 2.2), we can set $\delta = \delta_0$. Moreover, we may assume that $Q_{k+1} = \delta T$. Now, taking $Z = \{\tilde{\delta} = 0\}$, we have $V' = \pi^*(\Lambda)$, by the remark that follows the proof of the descriptive lemma. But then

$$\Lambda = \{(u,v,t) \in \Omega \times \mathbf{C}^{n-k+1} : \tilde{\delta}(u) \neq 0, \tilde{p}_{k+1}(u,t) = 0, z_{k+1} = t,$$
$$\tilde{\delta}(u)z_j = \tilde{Q}_j(u,t), j = k+2,\ldots,n\},$$

and so

$$V' = \{(u,v) \in \Omega \times \mathbf{C}^{n-k} : \tilde{\delta}(u) \neq 0, \tilde{p}_{k+1}(u,z_{k+1}) = 0, \tilde{\delta}(u)z_j =$$
$$= \tilde{Q}_j(u,z_{k+1}), j = k+2,\ldots,n\}.$$

The implicit function theorem implies that the set V' is a non-empty locally topographic submanifold in $\Omega \times \mathbf{C}^{n-k}$. Therefore (taking into account corollary 1a) we have

RÜCKERT'S CLASSIC DESCRIPTIVE LEMMA . *Suppose that the assumptions of the descriptive lemma, with $k < n$, are satisfied. Assume that $\hat{z}_{k+1}$ is a primitive element of the extension $\hat{\mathcal{O}}_n$ of the ring $\hat{\mathcal{O}}_k$; this can be achieved by a suitable change of the coordinates $z_{k+1},\ldots,z_n$. Then there is a triple Ω, Z, V that satisfies the conclusions of the descriptive lemma, with V' dense in V and such that the following Rückert formula holds:*

$$(R) \quad V' = V_{\Omega \setminus Z} = \{(u,v) \in \Omega \times \mathbf{C}^{n-k} : \tilde{\delta}(u) \neq 0, \tilde{p}_{k+1}(u,z_{k+1}) = 0,$$
$$\tilde{\delta}(u)z_j = \tilde{Q}_j(u,z_{k+1}), \; j = k+2,\ldots,n\}.$$

Here, $\tilde{p}_{k+1}, \tilde{Q}_{k+2},\ldots,\tilde{Q}_n$ are polynomials in z_{k+1} whose coefficients are holomorphic in Ω. Moreover, $\tilde{p}_{k+1}$ is a representative of the germ $p_{k+1}(z_{k+1})$ [13a], and is a distinguished polynomial with discriminant $\tilde{\delta} \not\equiv 0$.

[13a]   The polynomial $\hat{p}_{k+1}$ is the minimal polynomial for $\hat{z}_{k+1}$ over $\hat{\mathcal{O}}_k$; see n° 1.

Finally, $Z = \{u \in \Omega : \ \tilde{\delta}(u) = 0\}$. ([14])

REMARK 1. The above argument concerning the choice of a coordinate system

$$\varphi = \Big(z_1, \ldots, z_k, \sum_{k+1}^{n} c_j z_j, z_{k+2}, \ldots, z_n\Big)$$

for a k–regular prime ideal I shows (see the primitive element theorem in A. 8.3) that if $Z_{k+1}, \ldots, Z_n \subset \mathbf{C} \setminus 0$ are infinite subsets, then there exist $c_j \in Z_j$, $j = k+1, \ldots, n$, such that the ideal $I \circ \varphi^{-1}$ satisfies the assumptions of the classic descriptive lemma.

REMARK 2. Replacing the polynomials $\tilde{Q}_i$ by the remainders from their division by $\tilde{p}_{k+1}$ (see A. 2.4), one may assume that the degrees of the $\tilde{Q}_j$ are smaller than the degree of p_{k+1}, i.e., the multiplicity of the covering $\pi_{V'}$ (see the corollary below).

COROLLARY . *The multiplicity of the covering $\pi_{V'}$ in the descriptive lemma is equal to the dimension of the field of fractions of the ring $\hat{\mathcal{O}}_k$. In the classic descriptive lemma, it coincides with the degree of the polynomial p_{k+1}.*

In fact, this is the case in the classic descriptive lemma. For if $\tilde{\delta}(u) \neq 0$, then all roots of the polynomial $t \longrightarrow p_{k+1}(u, t)$ are distinct, and hence their number $\# V'_u$ coincides with the degree of the polynomial p_{k+1}; the latter is equal to the dimension of the field of fractions of the ring $\hat{\mathcal{O}}_n$ over the field of fractions of the ring $\hat{\mathcal{O}}_k$. This is so because $\hat{p}_{k+1}$ is the minimal polynomial

---

([14]) As a result, we have also (after reducing the size of Ω)

(#) $V' = \{(u, v) \in \Omega \times \mathbf{C}^{n-k} : \ \tilde{\delta}(u) \neq 0, \tilde{p}_{k+1}(u, z_{k+1}) = 0,$

$$\frac{\partial \tilde{p}_{k+1}}{\partial z_{k+1}}(u, z_{k+1}) z_j = \tilde{R}_j(u, z_{k+1}), \ j = k+2, \ldots, n\},$$

where $\tilde{R}_j$ are polynomials in z_{k+1} with coefficients that are holomorphic in Ω. We may also assume (as in remark 2 below) that the degrees of $\tilde{R}_j$ are smaller than the degree of p_{k+1}, i.e., smaller than the multiplicity of the covering $\pi_{V'}$.

Indeed, we have $\hat{p}_{k+1}(\hat{z}_{k+1})\hat{z}_j = R_j(\hat{z}_{k+1})$, where $R_j \in \mathcal{O}_k[T]$, $j = 2, \ldots, n$. (See A. 8.3 footnote ([38]).) Thus $p_{k+1}(z_{k+1}) z_j - (z_{k+1}) \in I$, and hence (after a suitable reduction of size of Ω) $\frac{\partial \tilde{p}_{k+1}}{\partial z_{k+1}}(u, z_{k+1}) z_k - \tilde{R}_j(u, z_{k+1}) = 0$ on V, where $\tilde{R}_j$ are polynomials with respect to z_{k+1} with coefficients holomorphic in Ω. Here the $\tilde{R}_j$ represent the coefficients of the polynomials R_j, respectively. Let V'' denote the right hand side of the equality (#). Then $V' \subset V''$. But (except for the trivial case $k = n - 1$) the sets V', V'' are graphs of mappings of the set $\{(u, z_{k+1} \in \Omega \times \mathbf{C} : \ \tilde{\delta}(u) \neq 0, \tilde{p}_{k+1}(u, z_{k+1}) = 0\}$ to the space $\mathbf{C}^{n-k-1}$. Hence $V' = V''$.

for $\hat{z}_{k+1}$ (see A. 5.3 and A. 8.2). Now the dimension is unaffected by any linear change of the coordinates $z_{k+1}, \ldots, z_n$.

Indeed, consider the ideal $I' = I \circ \chi$, where $\chi = e' \times \varphi$, e' is the identity mapping of $\mathbf{C}^k$, and $\varphi \in L_0(\mathbf{C}^{n-k}, \mathbf{C}^{n-k})$. Consider the corresponding ring $\hat{\mathcal{O}}'_n = \mathcal{O}_n / I'$ and its subring $\hat{\mathcal{O}}'_k$ – the image of $\mathcal{O}_k$ under the natural epimorphism $\mathcal{O}_n \longrightarrow \hat{\mathcal{O}}'_n$. The isomorphism $\mathcal{O}_n \ni f \longrightarrow f \circ \chi^{-1} \in \mathcal{O}_n$, which maps I onto I', induces an isomorphism $\hat{\mathcal{O}}_n \longrightarrow \hat{\mathcal{O}}'_n$, mapping $\hat{\mathcal{O}}_k$ onto $\hat{\mathcal{O}}'_k$. The mapping is such that its extension to the fields of fractions maps the field of fractions of the ring $\hat{\mathcal{O}}_k$ onto the field of fractions of the ring $\hat{\mathcal{O}}'_k$ (see A. 1.15).

Thus (except for the trivial situation when $k = n$) the general case follows in view of corollary 2 of the descriptive lemma.

§4. Hilbert's Nullstellensatz and other consequences (concerning dimension, regularity, and k–normality)

1. Let M be a complex manifold, and let $a \in M$.

HILBERT'S NULLSTELLENSATZ. *For any ideal I of the ring $\mathcal{O}_a$ we have* $\mathcal{I}(V(I)) = \operatorname{rad} I$. *If a germ f vanishes on $V(I)$, then $f^m \in I$ for same* $m > 0$ [15]. *In particular $\mathcal{I}(V(I)) = I$ if the ideal I is prime.* (See A. 1.11).

REMARK. This is obviously equivalent to the following classical version of Hilbert's Nullstellensatz:

If $f, g_1, \ldots, g_r$ are holomorphic functions in a neighbourhood U of the point a and $f = 0$ on the set $\{g_1 = \ldots = g_r = 0\}$, then there is an exponent $m > 0$ such that $f^m = \sum_1^r c_i g_i$ in a neighbourhood $W \subset U$ of the point a for some holomorphic functions c_i in W.

PROOF of Hilbert's theorem. First suppose that the ideal I is prime. One may assume that $M = \mathbf{C}^n$, $a = 0$, and I is k–regular for some $k \geq 0$ (see 2.7, I. 1.1 and II. 4.4). In view of corollary 1 from the descriptive lemma (see 3.1), we have

$$(*) \qquad \mathcal{O}_k \cap \mathcal{I}(V(I)) = 0 \quad [16].$$

---

[15] The inclusion $\mathcal{I}(V(I)) \supset \operatorname{rad} I$ is trivial (see II. 4.3 and II. 4.5).

[16] For every germ from $\mathcal{O}_k \cap \mathcal{I}(V(I))$ has a representative which is independent of $z_{k+1}, \ldots, z_n$, vanishes on a representative of the germ $V(I)$, and hence vanishes in a neighbourhood of zero.

Since $I \subset \mathcal{I}(V(I))$ (see II. 4.5), it is enough to show the opposite inclusion. Now, let $f \in \mathcal{O}_n \setminus I$. Then $\hat{f} \in \hat{\mathcal{O}}_n \setminus 0$, and so $\hat{f}\hat{g} \in \mathcal{O}_k \setminus 0$ for some $g \in \mathcal{O}_n$ (see A. 8.1, lemma 3.1; and (4) in 2.2). Thus $fg \in h + I \subset h + \mathcal{I}(V(I))$ for some $h \in \mathcal{O}_k \setminus I$. One must have $h \notin \mathcal{I}(V(I))$ in view of $(*)$. Therefore $fg \notin \mathcal{I}(V(I))$, and so $f \notin \mathcal{I}(V(I))$.

Consider now an arbitrary ideal I. One may assume that $I \neq \mathcal{O}_a$. Then $I = \mathcal{I}_1 \cap \ldots \cap \mathcal{I}_k$, where $\mathcal{I}_i$ are primary ideals (see A. 9.3). Since the ideals rad $\mathcal{I}_i$ are prime (see A. 9.3), we have $\mathcal{I}(V(I)) = \bigcap_i \mathcal{I}(V(\mathcal{I}_i)) = \bigcap_i \mathcal{I}(V(\text{rad } \mathcal{I}_i)) = \bigcap_i \text{rad } \mathcal{I}_i = \text{rad } I$ (see II. 4.2–3 and A. 1.5), which completes the proof.

Consequently, we have *the mutually inverse bijections* $A \longrightarrow \mathcal{I}(A)$ *and* $I \longrightarrow V(I)$ *between the set of simple analytic germs at a and the set of prime ideals of the ring $\mathcal{O}_a$* (see II. 4.6, proposition 2; and II. 4.5).

If the ideal I is primary, then the germ $V(I)$ is simple (see II. 4.3 and A. 9.3).

If I is a proper ideal and $I_1, \ldots, I_r$ are all its isolated ideals, then $V(I) = V(I_1) \cup \ldots \cup V(I_r)$ is the decomposition of the germ $V(I)$ into simple germs. (The equality follows from the irreducible primary decomposition of I; see A. 9.3 and II. 4.3. Next, we have $V(I_i) \not\subset V(I_j)$ if $i \neq j$, for otherwise the Nullstellensatz would imply $I_i \supset I_j$; see II. 4.2).

If $I = J_1 \cap \ldots \cap J_r$ is the irredundant primary decomposition, then $V(I) = V(J_1 \cup \ldots \cup V(J)_r)$ is not necessarily the decomposition into simple germs. A counter-example is furnished by the primary ideals $J_1 = \mathcal{O}_2 w$, $J_2 = \mathfrak{m}^2$ in the ring $\mathcal{O}_2$ for which $V(J_1) = \{w = 0\}_0$ and $V(J_2) = 0$.

An ideal I of the ring $\mathcal{O}_n$ is k–normal, k–regular, or regular if and only if the germ $V(I)$ is, respectively, k–normal, k–regular, regular. (This follows from the Nullstellensatz; see 2.4 and 2.5.)

2. Let $0 \leq k \leq n$ or $k = -\infty$.

PROPOSITION . *Any k–regular analytic germ at $0 \in \mathbf{C}^n$ is k–dimensional.*

PROOF . One may assume that $k \geq 0$. For an irreducible germ, the proposition follows, since $\dim A \leq k$ (see 2.4), from the descriptive lemma and corollary 1; for in this case each representative of the germ A must contain a k–dimensional submanifold. In the general case, the germ is of the form $V(I)$, where I is a k–regular ideal of the ring $\mathcal{O}_n$ (see II. 4.5). Because $k \geq 0$, the ideal I is proper, and hence $I = J_1 \cap \ldots \cap J_r$, where J_i are primary ideals (see A. 9.3). Each of these ideals is k–normal (see 2.2) and so, after a suitable change of the coordinates $z_1, \ldots, z_k$, each of them becomes k_i–regular, where $k_i \leq k$ (see the proposition in 2.7, and 2.6). Since the $V(J_i)$ (after this change of coordinates) are simple and k_i–regular, respectively (see n° 1), we have $\dim V(J_i) = k_i$. Therefore $\dim V(I) = \max k_i$, because $V(I) = V(J_1) \cup \ldots \cup V(J_r)$ (see II. 1.6 and II. 4.3). Now, if we had $k_i < k$, $i = 1, \ldots, r$, then non-zero germs $g_i \in \mathcal{O}_k \cap J_i$ would exist (see (1) in 2.2); then $0 \neq g_1, \ldots, g_k \in I \cap \mathcal{O}_k$, contrary to the fact that I is k–regular. Hence $k = \max k_i = \dim V(I)$.

Let M be an n–dimensional manifold, and let $a \in M$. The proposition implies that:

An analytic germ at a is k–dimensional precisely when it is k–regular in some coordinate system at a. If M is a vector space and $a = 0$, such coordinate systems (which, in addition, are linear) constitute a dense subset of $L_0(M, \mathbf{C}^n)$. (See 2.7.)

An analytic germ at a is of dimension $\leq k$ if and only if it is k–normal in some coordinate system at a. This coordinate system can be linear if M is a vector space and $a = 0$. (See 2.4 and 2.5.)

This implies the following

HARTOGS' THEOREM . *If $Z \subset M$ is an analytic subset of dimension $\leq n - 2$, then every holomorphic function on $M \setminus Z$ extends to a holomorphic function on M.*

Indeed, for any $a \in Z$, the germ Z_a is $(n-2)$–normal in some coordinate system at a, and thus (see 2.4) the function f extends holomorphically across the point a (see II. 3.8 footnote ([12])).

REMARK. Clearly, Hartogs' theorem holds for holomorphic mappings on the set $M \setminus Z$ with values in a vector space.

The vector space cannot be replaced by a manifold. A counter-example is provided by $(\pi^{\mathbf{C}^2 \setminus 0})^{-1} : \mathbf{C}^2 \setminus 0 \longrightarrow \Pi_2$, where $\pi : \Pi_2 \longrightarrow \mathbf{C}^2$ is the blow-up of $\mathbf{C}^2$ at the origin (see VII. 5.2 below).

Note also that:

If A is a k–regular analytic germ at $0 \in \mathbf{C}^n$, then the image under the projection onto $\mathbf{C}^k$ of any representative of the germ A is a neighbourhood of zero.

Indeed, by the proposition, the germ A is k–dimensional and therefore so is one of its simple components, say A_0 (see II. 1.6). But A_0 is also k–normal (see 2.4) and hence k–regular (see 2.7). Consequently, by corollary 1 from the descriptive lemma (see 3.1), our claim is true for A_0, and hence also for A.

3. If M is a manifold and $a \in M$, then, for any ideal I of the ring $\mathcal{O}_a$, we have the equivalences:

$$\operatorname{codim} I < \infty \Longleftrightarrow (I \supset \mathfrak{m}^s \text{ for some } s) \Longleftrightarrow$$
$$\Longleftrightarrow \dim V(I) \le 0 \Longleftrightarrow V(I) \subset a .$$

In fact, numbering these conditions, we already have the equivalence $1 \Longleftrightarrow 2$ (see I. 1.7), whereas the implications $2 \Longrightarrow 4 \Longrightarrow 3$ are trivial. It remains to show that $3 \Longrightarrow 2$. Now, if $\dim V(I) \le 0$, then, in some coordinate system at a, the ideal I is 0–normal (see n° 2 and n° 1). Therefore the ideal I – in this coordinate system – contains (see (2) in 2.2) distinguished elements from $\mathbf{C}[z_i]$, $(i = 1, \ldots, n = \dim M)$. These elements will be of the form $z_j^{s_i}$. Therefore the ideal will also contain the elements z^p for $|p| = s = s_1 + \ldots + s_n$. Thus it contains the ideal $\mathfrak{m}^s$ (see I. 1.7) (^{17}) .

In particular, for an analytic germ A at a we have: $\dim A \le 0 \Longleftrightarrow A \subset a$ (see II. 4.5). Hence

An analytic subset of a manifold M is of dimension ≤ 0 precisely when it is discrete. (See II. 1.5.)

The next characterization of systems of parameters for $\mathcal{O}_a$ also follows:

$$(f_1, \ldots, f_n \text{ is a system of parameters of } \mathcal{O}_a) \Longleftrightarrow (V(f_1, \ldots, f_n) = a)$$

for $f_1, \ldots, f_n \in \mathcal{O}_a$, where $n = \dim M$.

Indeed, since $\dim \mathcal{O}_a = n$ (by proposition 2 in I. 1.8), each side is equivalent to the condition: $\mathfrak{m}_a^s \subset \sum \mathcal{O}_a f_i \subset \mathfrak{m}_a$ for some s (see A. 10.5).

4. Let $N = \{z \in \mathbf{C}^n : z_1 = \ldots = z_k = 0\}$, where $0 \le k \le n$, and let A be an analytic germ at 0 in $\mathbf{C}^n$. We have the following geometric characterization of k–normality:

PROPOSITION . *(The germ A is k–normal)* $\Longleftrightarrow A \cap N \subset 0$.

PROOF . Set $I = \mathcal{I}(A)$. Let $\mathcal{O}_v$ denote the ring of germs (at 0) of holomorphic functions of the variables $v = (z_{k+1}, \ldots, z_n)$; then $I(0, v)$ is an ideal of the ring $\mathcal{O}_v$ (see I. §3). Note that $A \cap N = 0 \times V(I(0, v))$ (^{18}) . Now

---

(^{17}) Since the implication $2 \Longrightarrow 1$ is trivial, instead of using the equivalence $1 \Longleftrightarrow 2$, it suffices to show $1 \Longrightarrow 3$. If $\dim V(I) > 0$, then $I' = I \circ \varphi^{-1}$ is k–regular, where $k > 0$, for some coordinate system φ at a. Hence $\mathcal{O}_k \cap I' = 0$, and so $\operatorname{codim} I = \operatorname{codim} I' = \infty$. Note also that the implication $4 \Longrightarrow 2$ follows from the Nullstellensatz.

(^{18}) For take representatives $F_1, \ldots, F_r$ of generators of the ideal I. Then $F_i(0, v)$ are representatives of generators of the ideal $I(0, v)$ and we have $\{F_1 = \ldots = F_r = 0\} \cap N = 0 \times \{F_1(0, v) = \ldots = F_r(0, v) = 0\}$. Since $A = V(I)$ (see II. 4.5), it is enough to take the germs at 0 in the former equality.

k–normality of the germ A, i.e., k–normality of the ideal I, means (see (4) in 2.2) that $\mathcal{O}_n = \sum \mathcal{O}_k a_i + I$ for some $a_i \in \mathcal{O}_n$. In view of the Thom-Martinet version of the preparation theorem (see I. §3), this is equivalent to the condition $\mathcal{O}_v = \sum \mathbf{C} a_i(0, v) + I(0, v)$ for some $a_i \in \mathcal{O}_n$, and hence to the condition codim $I(0, v) < \infty$. This, in turn, is equivalent (see n° 3) to the inclusion $V\big(I(0, v)\big) \subset 0$, hence to the inclusion $A \cap N \subset 0$.

COROLLARY . *If* $\dim A = k$, *then: (A is k–regular)* $\Longleftrightarrow A \cap N = 0$. (See 2.7.)

The proposition implies also that:

If $A \cap N = 0$, then a representative of the germ A is defined by polynomials with respect to $z_{k+1}, \ldots, z_n$ whose coefficients are holomorphic in a neighbourhood of zero in $\mathbf{C}^k$. (For $A = V\big(\mathcal{I}(A)\big)$, and $\mathcal{I}(A)$ is k–normal; see 2.2.)

5. Let M be an n–dimensional vector space, and let A be a non-empty germ at 0.

The proposition in n° 4 implies the following formula

$$(\#) \quad \text{codim } A = \max\{\dim N : N \text{ is a subspace such that } N \cap A = 0\} \quad (^{19}).$$

Indeed, if $0 \leq l \leq n$, then the condition codim $A \geq l$, i.e., $\dim A \leq k = n - l$, is equivalent (see n° 2 and the proposition in n° 4) to the existence of a linear coordinate system in which $A \cap N = 0$, where $N = \{z_1 = \ldots = z_k = 0\}$. In other words, it is equivalent to the existence of an l–dimensional subspace N such that $N \cap A = 0$.

Notice that codim $A = \dim N$ for any maximal subspace N such that $N \cap A = 0$.

Namely, in such a case, $N = \{z_1 = \ldots = z_k = 0\}$ in some (linear) coordinate system with $k = \text{codim } N$. According to the proposition in n° 4, the germ A is k–normal, and therefore, after a suitable change of the coordinates $z_1, \ldots, z_k$, it is r–regular, where $0 \leq r \leq k$ (see 2.7; we have

---

(¹⁹) It follows from formula (∗) in n° 6 that for an analytic germ $A \neq \emptyset$ at a point a of the manifold M one has: codim $A \geq l \Longleftrightarrow A \cap C = a$ for some l–dimensional germ C which is analytic (or even smooth) at a (where $0 \leq l \leq \dim M$). This yields the formulae

$$\text{codim } A = \max\{\dim C : C \text{ is an analytic germ at } a \text{ such that } C \cap A = a\} =$$
$$= \max\{\dim C : C \text{ is a smooth germ at } a \text{ such that } C \cap A = a\}.$$

$A \neq \emptyset$). Thus $\dim A = r$ (see the proposition in n° 2) and, in view of the corollary from n° 4, we have $N_0 \cap A = 0$, where $N_0 = \{z_1 = \ldots = z_r = 0\}$. Hence $r = k$, since N is maximal, which gives $\dim A = k$.

If $M = H + N$ is a direct sum of subspaces and $A \cap N = 0$, then:

$$\dim A = \dim H \iff \text{(the projection onto } H \text{ of any representative}$$
$$\text{of the germ } A \text{ is a neighbourhood of zero).}$$

Indeed, one may assume that $M = \mathbf{C}^n$, $H = \{z_{k+1} = \ldots = z_n = 0\}$, and $N = \{z_1 = \ldots = z_k = 0\}$, where $k = \dim H$. Now, the right hand side of the above equivalence implies that N is a maximal subspace such that $N \cap A = 0$, and so codim $A = \dim N$, i.e., $\dim A = k$. Conversely, if $\dim A = k$, then the germ A is k–regular (see the corollary in n° 4), which implies the right hand side of the equivalence (see n° 2).

6. Let M be a manifold. Then, for any analytic germs $A_1, \ldots, A_k$ at a point $a \in M$, we have the inequality

$$(*) \qquad \operatorname{codim}(A_1 \cap \ldots \cap A_k) \leq \operatorname{codim} A_1 + \ldots + \operatorname{codim} A_k \quad (^{20}) .$$

Indeed, one may assume that M is an n-dimensional vector space and $a = 0$. Obviously, it is enough to prove that $\operatorname{codim}(A \cap B) \leq \operatorname{codim} A + \operatorname{codim} B$ for analytic germs A, B at 0. We may assume that $A \neq \emptyset$ and $B \neq \emptyset$. First suppose that B is the germ of a subspace L. Clearly $A \cap L \neq \emptyset$. Now the formula ($\#$) from n° 5 shows that there is a subspace N such that $\operatorname{codim}(A \cap L) = \dim N$ and $A \cap L \cap N = 0$. Hence, by the same formula, codim $A \geq \dim(L \cap N)$ and, since $\dim N \leq \operatorname{codim} L + \dim(L \cap N)$ $(^{21})$, one derives $\operatorname{codim}(A \cap L) \leq \operatorname{codim} L + \operatorname{codim} A$. In the general case we have $h(A \cap B) = (A \times B) \cap D$, where D is the diagonal of M^2 and h is

---

$(^{20})$ This inequality implies that $\dim A + \dim B \leq \dim C + \dim(A \cap B)$ for analytic germs A, B contained in a smooth germ C. The last inequality is not true without the assumption of smoothness of the germ C, even if the latter is irreducible. To see this, it is enough to take $M = \mathbf{C}^4$ and (denoting the variables by s, t, x, y) the germs $A = \{s = t = 0\}_0$, $B = \{x = y = 0\}_0$, $C = \{sy + tx = 0\}_0$. The last germ is irreducible and 3–dimensional, since in the analytic subset $\{sy + tx = 0\}$ there is an arbitrarily small open neighbourhood V of 0 such that its set V^0 of regular points is a connected 3–dimensional submanifold (see IV. 3.1, proposition 1; and IV. 2.8 proposition 2).

$(^{21})$ Because $\dim N + \dim L = \dim(N + L) + \dim(N \cap L) \leq n + \dim(N \cap L)$.

the biholomorphic mapping $M \ni z \longrightarrow (z, z) \in D$. Hence $\dim(A \cap B) = \dim((A \times B) \cap D)$ (see II. 1.6). Therefore, by the case already proved,

$$2n - \dim(A \cap B) = \mathrm{codim}((A \times B) \cap D) \leq$$
$$\leq \mathrm{codim}(A \times B) + \mathrm{codim}\, D = \mathrm{codim}\, A + \mathrm{codim}\, B + n$$

(see II. 1.6), which implies the required inequality.

It follows from the inequality $(*)$ that if $Z_1, \ldots, Z_k \subset M$ are analytic subsets of constant dimension (see II. 1.5) with non-empty intersection, then

$$\mathrm{codim}(Z_1 \cap \ldots \cap Z_k) \leq \mathrm{codim} Z_1 + \ldots + \mathrm{codim} Z_k \ .$$

7. Let M be a manifold of dimension n, and let $a \in M$.

Let $f_1, \ldots, f_p \in \mathfrak{m}_a$. We always have the inequality codim $V(f_1, \ldots$ $\ldots, f_p) \leq p$ (by $(*)$ in n° 6; see II. 5.1). We say that the germs $f_1, \ldots, f_p$ *comprise a set-theoretic complete intersection* if codim $V(f_1, \ldots, f_p) = p$. An analytic germ A at a which can be comprised of such germs $f_1, \ldots, f_p$ (i.e., $A = V(f_1, \ldots, f_p)$ and $p = \mathrm{codim}\, A$) is said to be a *set-theoretic complete intersection*.

PROPOSITION . *If an ideal I of $\mathcal{O}_a$ is generated by germs which comprise a set-theoretic complete intersection, then the quotient ring $\mathcal{O}_a/I$ is Cohen-Macaulay.*

Indeed, suppose that codim $V(f_1, \ldots, f_p) = p$ for a system of generators $f_1, \ldots, f_p$ of I. In view of $(\#)$ in n° 5, one may assume that $M = \mathbf{C}^n$, $a = 0$, and $V(z_1, \ldots, z_{n-p}, f_1, \ldots, f_p) = 0$. Then the sequence $z_1, \ldots, z_{n-p}, f_1, \ldots, f_p$ is a system of parameters of $\mathcal{O}_n$ (see n° 3). By proposition 7 in A. 14.3, it is also an $\mathcal{O}_n$-sequence, as $\mathcal{O}_n$ is Cohen-Macaulay (see A. 15, corollary of the syzygy theorem; and I. 1.8, proposition 2 $(^{22})$). Therefore, as is easy to check (see A. 1.11, A. 10.3, and A. 10.5), the elements $\bar{z}_1, \ldots, \bar{z}_{n-p}$ of $\mathcal{O}_n/I$ generate a defining ideal of $\mathcal{O}_n/I$. Moreover, by lemma 1 and proposition 5 in A. 14, they form an $(\mathcal{O}_n/I)$-sequence. It follows that $\dim \mathcal{O}_n/I \leq n - p \leq \mathrm{prof}\, \mathcal{O}_n/I$ (see A. 12.3, theorem $1''$), which proves (see $(*)$ in A. 14.3) that $\mathcal{O}_n/I$ is Cohen-Macaulay.

$(^{22})$ In view of proposition 7 in A. 14.3, this can be shown directly by checking that $z_1, \ldots, z_n$ is an $\mathcal{O}_n$-sequence.

CHAPTER IV

GEOMETRY OF ANALYTIC SETS

§1. Normal triples

Let $0 \leq k \leq n$.

1. A *normal triple* of dimension k (in $\mathbf{C}^n$) is a triple (Ω, Z, V), where Ω is an open connected neighbourhood of zero in $\mathbf{C}^k$, Z is a nowhere dense analytic subset of Ω, and V is an analytic subset of $\Omega \times \mathbf{C}^{n-k}$ such that the following conditions are satisfied:

(1) the natural projection $\pi : V \longrightarrow \Omega$ is proper,

(2) $\pi^{-1}(0) = 0$,

(3) the set $V_{\Omega \setminus Z}$ is a locally topographic submanifold in $\Omega \times \mathbf{C}^{n-k}$,

(4) the set $V_{\Omega \setminus Z}$ is dense in V.

In such a case, the projection $V_{\Omega \setminus Z} \longrightarrow \Omega \setminus Z$ is a finite covering. Its multiplicity is > 0 (see B. 3.2, B. 2.4, and II. 3.6 $(^1)$) and is called the *multiplicity of the normal triple* (Ω, Z, V). The set V is said to be the *crown of the normal triple* (Ω, Z, V).

A normal triple (Ω, Z, V) whose crown V is a representative of an analytic germ A at 0 in $\mathbf{C}^n$ is called a *normal triple of the germ A*. Thus such a triple satisfies the assumptions of Rückert's descriptive lemma (see III. 3.1), and moreover, the set V' is dense in V $(^{1a})$.

$(^1)$ In view of (2) and (4), we have $V_{\Omega \setminus Z} \neq \emptyset$.

$(^{1a})$ See III. 3.1, corollary 1a.

2. Let (Ω, Z, V) be a k–dimensional normal triple of the germ A.

If $\tilde{Z} \supset Z$ is a nowhere dense analytic subset of Ω, then $(\Omega, \tilde{Z}, V)$ is also a k–dimensional normal triple of the germ A.

If $\Omega' \subset \Omega$ is an open connected neighbourhood of zero in $\mathbf{C}^k$, then $(\Omega', Z \cap \Omega', V_{\Omega'})$ is also a k–dimensional normal triple of the germ A. Furthermore, the sets $V_{\Omega'}$ form a base of open neighbourhoods of zero in V (see n° 1, conditions (1),(2); and B. 2.4).

As a consequence, the multiplicity of a normal triple of a given germ is the same for all normal triples of the germ. (Cf. n° 7 below.)

3. Let (Ω, Z, V) be a k–dimensional normal triple, and let $\pi : V \longrightarrow \Omega$ be the natural projection.

We have $\pi(V) = \Omega$. (For $\pi(V_{\Omega \backslash Z}) = \Omega \backslash Z$ and π is closed; see B. 2.3–4.)

The pair $(Z, V_{\Omega \backslash Z})$ is a quasi-cover in $\Omega \times \mathbf{C}^{n-k}$ with multiplicity equal to the multiplicity of the normal triple (Ω, Z, V) whose adherence is the crown V (see III. 1.2; and n° 1, conditions (3), (4), (1) (^{2})).

Therefore we have the following corollaries:

The projection $\pi : V \longrightarrow \Omega$ is open, the function $\Omega \ni u \longrightarrow \#V_u$ is lower semicontinuous, and the fibres V_u, $u \in \Omega$ are finite. (See III. 1.3, proposition 1.)

Moreover, $\dim V_Z < k$. (See II. 1.2–4.) Consequently,

The crown V is of constant dimension k (^{3}) .

In view of the second lemma on quasi-covers (see III. 1.3) we have:

The set of points of the crown V at which it is not a k–topographic submanifold is an analytic subset in $\Omega \times \mathbf{C}^{n-k}$.

Finally, it follows from proposition 2 in III. 1.3 that:

The crown V can be defined by holomorphic functions $f_1, \ldots, f_r$ on $\Omega \times \mathbf{C}^{n-k}$ such that, for each point $c \in V$ at which V is a k–topographic submanifold, we have $T_c V = \bigcap_1^r \ker d_c f_i$.

4. Now we are going to prove the following:

(^{2}) Conversely, in view of the first lemma on quasi-covers (see III. 1.3), if Ω is an open connected neighbourhood of zero in $\mathbf{C}^k$ and (Z, Λ) is a quasi-cover in $\Omega \times \mathbf{C}^{n-k}$ that satisfies the condition $\bar{\Lambda}_0 = 0$, then $(\Omega, Z, \bar{\Lambda})$ is a normal triple of dimension k.

(^{3}) Indeed, according to the conditions (3), (4), we have $\dim_z(V_{\Omega \backslash Z}) = k$ for each $z \in V$.

PROPOSITION 1. *An analytic germ at* $0 \in \mathbf{C}^n$ *has a normal triple of dimension* k *if and only if it is* k–*regular and of constant dimension* k $(^4)$.

PROOF . Let A be a k–regular germ. If it is irreducible, then, in accordance with corollary 1a from the descriptive lemma (see III. 3.1), it has a normal triple. Suppose now that A is of constant dimension k. Then $A = A_1 \cup \ldots \cup A_r$, where the A_i are k–dimensional irreducible germs (see II. 4.6). Hence they are k–regular (see III. 2.4 and 7). One can choose a Weierstrass set $W = W(P_{k+1}, \ldots, P_n) \supset A$, where the P_j have holomorphic coefficients in Ω and discriminants $\not\equiv 0$, and also k–dimensional normal triples (Ω, Z, V_i) of the germs A_i $(i = 1, \ldots, r)$, in such a way that $V = V_1 \cup \ldots \cup V_r \subset W$ and $Z \supset Z(P_{k+1}, \ldots, P_n)$ (see III. 2.3–4 and n° 2). Then (Ω, Z, V) is a normal triple (of dimension k) of the germ A. Indeed, $V_{\Omega \setminus Z}$ is dense in V (since $(V_i)_{\Omega \setminus Z}$ are dense in V_i, respectively) and $V_{\Omega \setminus Z}$ is a locally topographic submanifold at z in $\Omega \times \mathbf{C}^{n-k}$: if $z \in V_{\Omega \setminus Z}$, then W is a topographic submanifold (see III. 2.3) and $V_z = W_z$ (for otherwise – see II. 3.3 – one would have $\dim(V_i)_z \leq \dim V_z < k$; this would contradict the condition (3) from n° 1, as $z \in (V_i)_{\Omega \setminus Z}$ for some i).

Conversely, assume that the germ A has a normal triple (Ω, Z, V) of dimension k. Then $\dim A = k$ (see n° 3), $A \cap \{z_1 = \ldots = z_k = 0\} = 0$ (see (2) in n° 1), and so the germ A is k–regular (see the corollary in III. 4.4). Let $A = A_1 \cup \ldots \cup A_s$ be the decomposition into simple germs such that $A_1, \ldots, A_r$ are all the k–dimensional germs in the decomposition. One may assume (see n° 2) that $V = V^1 \cup \ldots \cup V^s$, where V^i are representatives of the germs A_i, respectively, are analytic in $\Omega \times \Delta$, where Δ is an open neighbourhood of zero in $\mathbf{C}^{n-k}$, and $\dim A_i = \dim V^i$. Now, $V_{\Omega \setminus Z} \subset V^1 \cup \ldots \cup V^r$. For if $z \in V_{\Omega \setminus Z}$, then $V_z = V_z^1 \cup \ldots \cup V_z^s$, and hence $V_z = V_z^i$ for some i (since V_z is smooth; see II. 4.6). Therefore $z \in V^i$ and $\dim A_i \geq \dim V_z^i = \dim V_z = k$ (see n° 1, condition (3)), and so $i \leq r$. Consequently, $V \subset V^1 \cup \ldots \cup V^r$ (see n° 1, condition (4)), and hence $A = A_1 \cup \ldots \cup A_r$. This means that A is of constant dimension k.

Now let M be a vector space, and let N be a subspace of M.

COROLLARY . *Every analytic germ at a point* $a \in M$ *satisfying the condition* $A \cap (a + N) = a$ *has a representative* W *such that, for each affine subspace* N' *parallel to* N, *the set* $W \cap N'$ *is finite. Hence, if* V *is a locally analytic subset of the space* M *and* $a \in V$, *then* $V_a \cap (a + N) = 0$ *implies that* $V_z \cap (z + N) = z$ *for each* z *from some neighbourhood of* a *in* V $(^5)$.

_____

$(^4)$ In view of the proposition from III. 4.2, this condition can be restated as follows: A is k–regular and of constant dimension.

$(^5)$ The second conclusion of the corollary follows directly from the proposition in

Indeed, one may assume that $M = \mathbf{C}^n$, $N = \{z_1 = \ldots = z_k = 0\}$, $a = 0$, and – in view of the propositions in III. 4.4 and III. 2.7 – that, for each i, the irreducible components A_i of the germ A are r_i–regular, where $r_i \leq k$ (see III. 2.5). Now it is enough to define W as the union of all the crowns of the normal triples of the germs A_i (see n° 3).

5. Let A be an analytic germ at a point a of an n–dimensional manifold M.

If φ is a coordinate system at the point a and G is the domain of φ, then a set $V \subset G$ is said to be a *normal representative of the germ A with respect to the coordinate system* φ if $\varphi(V)$ is the crown of a normal triple $(\Omega, Z, \varphi(V))$ of the germ $\varphi(A)$. Then the set V is a locally analytic representative of the germ A. Moreover, there is an open neighbourhood $U \subset G$ of the point a such that the sets V and $\varphi(V)$ are analytic subsets of U or $\varphi(U)$, respectively, and $\varphi(U) \subset \Omega \times \mathbf{C}^{n-k}$ (where k is the dimension of the triple) [6] .

A *normal representative of the germ A* is defined as a normal representative of the germ A with respect to some coordinate system at a. It is always of constant dimension (see n° 3) and is non-empty (it contains the point a; see (2) in n° 1).

Proposition 1 implies (see III. 4.2 and n° 2):

PROPOSITION 2. *The germ A has a normal representative if and only if it is non-empty and of constant dimension. In any representative of such a germ, there is a base of open neighbourhoods of the point a that consists of normal representatives of the germ (with respect to a given coordinate system at a in which the germ A is regular; if M is a vector space and $a = 0$, it can be any coordinate system from a dense subset of $L_0(M, \mathbf{C}^n)$).*

6. We have the following proposition.

PROPOSITION 3. *If (Ω, Z, V) is a normal triple of an irreducible germ, then the manifold $V_{\Omega \setminus Z}$ is connected.*

Indeed, if this were not the case, then the manifold $V_{\Omega \setminus Z}$ would be equal to the union of open, disjoint, and non-empty sets Λ^1, Λ^2. Then (see n° 3 and

---

III. 4.4. To see this, take M, N and a as in the proof of the corollary. Then the ideal $\mathcal{I}(V_0)$ is k–normal, hence there exist distinguished polynomials $f_l(z_1, \ldots, z_{l-1}, z_l)$, $l = k+1, \ldots, n$, that vanish on some open neighbourhood of zero in V. Then, for each $c \in W$, the germ $(f_l(c+z))_0$ is a regular element in $\mathcal{O}_l \cap \mathcal{I}(V_c - c)$ ($l = k+1, \ldots, n$), and therefore the germ $V_c - c$ is k–normal. This means that $V_c \cap (c + N) = c$.

[6] It suffices to set $U = \varphi^{-1}(\Omega \times \mathbf{C}^{n-k})$, because for such a set $\varphi(V) \subset \varphi(U) \subset \Omega \times \mathbf{C}^{n-k}$. (See II. 3.4.)

III. 1.2) the pairs (Z, Λ^i) would be quasi-covers in $\Omega \times \mathbf{C}^{n-k}$ [7] with positive multiplicities, and according to the first lemma on quasi-covers (see III. 1.3), their adherences V^i would be analytic in $\Omega \times \mathbf{C}^{n-k}$. Since $V = V^1 \cup V^2$ (see (4) in n° 1), we would have $V_0 = V_0^1 \cup V_0^2$. But $0 \in \bar{\Lambda}^i$ (because $\emptyset \neq \Lambda_\omega^i \subset V_{\Omega'}$ for each neighbourhood Ω' of zero in $\mathbf{C}^k$; see n° 2), hence $V_0^i \not\subset V_0^j$ for $i \neq j$, contradicting the fact that the germ V_0 is irreducible.

Under the assumptions of proposition 3, the crown V is connected. Thus any normal representative of an irreducible germ (at a point of a manifold) is a connected set [8] . Hence we have the following:

THEOREM . *Any locally analytic subset of a complex manifold is locally connected.*

In fact, assume that V is an analytic subset of a complex manifold, and let $a \in V$. There exists an arbitrarily small neighbourhood U of the point a and analytic subsets $V_1, \ldots, V_r$ of U, such that $V \cap U = V_1 \cup \ldots \cup V_r$ and the germs $(V_i)_a$ are irreducible. In view of proposition 2, there is a normal representative W_i of the germ $(V_i)_a$ which is a neighbourhood of a in V_i. Hence $W_1 \cup \ldots \cup W_r \subset U$ is a connected neighbourhood of the point a in V.

COROLLARY . *Connected components of a (locally) analytic subset of a manifold M are (locally) analytic subsets of M.*

(Indeed, they are both open and closed in M; see II. 3.4.)

7. The corollary that follows the classic descriptive lemma (see III. 3.2) implies that:

If (Ω, Z, V) is a normal triple of a k-regular irreducible germ A, then the multiplicity of the covering $V_{\Omega \setminus Z} \longrightarrow \Omega \setminus Z$ is equal to the dimension of the field of fractions of the ring $\hat{O}_n$ over the field of fractions of the ring $\hat{O}_k$; see III. 2.1, *where $I = \mathcal{I}(A)$.*

For $A = V(I)$ (see II. 4.5) and if $\tilde{V}$ is a representative of A as in the descriptive lemma (see III. 3.1; and II. 4.6, proposition 2), we have $\tilde{V}_{\Omega'} = V_{\Omega'}$ for some open neighbourhood Ω' of zero in $\mathbf{C}^k$ (see n° 2; and III. 3.1, corollary 1).

8. Note also that *if a normal triple (Ω, Z, V) has multiplicity 1, then its crown V is the graph of a holomorphic mapping of the set Ω into the space $\mathbf{C}^{n-k}$.*

---

[7] Where k is the dimension of the normal triple.

[8] It is easy to check that a normal representative (of a non-empty germ of constant dimension) is always connected.

Indeed, in such a case $V_{\Omega \setminus Z}$ is the graph of a holomorphic mapping, which is locally bounded near Z (see n° 1, condition (1); and B. 5.2). This mapping has (see II. 3.5) a holomorphic extension on Ω whose graph (in view of condition (4) in n° 1) coincides with V.

§2. Regular and singular points
Decomposition into simple components

Let M be an n–dimensional manifold.

1. Let V be a locally analytic subset of M. The points of V at which V is a complex submanifold (of dimension k) (⁹) are called the *regular points* (of dimension k) of set V. The remaining points of the subset V are said to be its *singular points* (¹⁰) . By $V^*, V^0, V^{(k)}$ we denote the set of the singular points, the regular points, or the regular points of dimension k, respectively, of the set V. Thus we have the decompositions $V = V^* \cup V^0$ and $V^0 = V^{(0)} \cup \ldots \cup V^{(n)}$. Obviously, the set $V^{(k)}$ is a k–dimensional submanifold (as a manifold; see C. 3.7) and is open in V. Therefore V^0 is an open subset of the set V, while V^* is a closed subset of V. Moreover, the submanifolds $V^{(k)}$ are closed and open subsets of V^0. Consequently, the connected components of V^0 are submanifolds, and they are open in V^0 and in V (see C. 3.7 and C. 3.4). Clearly, if $G \subset M$ is an open set, we have

$$(\#) \qquad (V \cap G)^0 = V^0 \cap G \ \text{ and } \ (V \cap G)^* = V^* \cap G$$

(⁹) See C. 3.7.

(¹⁰) An analytic subset can be a topological manifold in a neighbourhood of a singular point. Take, for instance, the set $\{z^3 = w^2\}$, which is an analytic (even algebraic) subset in $\mathbf{C}^2$. Now, the mapping $B \ni t \longrightarrow (t^2, t^3) \in B^2 \cap \{z^3 = w^2\}$, where $B = \{|z| < 1\}$, is a homeomorphism. On the other hand, the set $\{z^3 = w^2\}$ is not a submanifold at the point 0, since (see C. 3.15, condition (a)) for every function f that is holomorphic at 0 and vanishing on this set, its differential $d_0 f$ is equal to zero. Indeed, $f(z, w) = az + bw + \ldots$ and $f(t^2, t^3) = at^2 + bt^3 + o(t^3) = 0$, and so $a = b = 0$ (see C. 1.5).

More generally, the same is true for the set $V = \{z^p = w^q\}$, where $p, q > 1$ are relatively prime. In such a case, the mapping $\{|t| < 1\} \ni t \longrightarrow (t^q, t^p) \in V$ is a surjection onto a neighbourhood of zero in V. Therefore the germ V_0 is irreducible (see II. 6.2) and the point 0 has – in the set V – a neighbourhood homeomorphic to the unit disc. On the other hand, if, e.g., $q < p$, then $\lambda = \{w = 0\}$ is the "tangent line" to V at 0 (see II. 6.2, corollary 2). Hence, if the set V were a submanifold at 0, then the line λ would be its tangent space at 0. But for each neighbourhood W of the point 0 in V, there are fibres of the natural projection $W \longrightarrow \lambda$ containing at least two points.

(and similarly, $(V \cap G)^{(k)} = V^{(k)} \cap G$). If $h : M \longrightarrow N$ is a biholomorphic mapping between two manifolds, then $h(V)^0 = h(V^0)$, $h(V)^* = h(V^*)$, and $h(V)^{(k)} = h(V^{(k)})$.

Observe that if M is a linear space, then

$$(V \text{ is a cone}) \Longrightarrow (V^* \text{ is a cone or } V^* = \emptyset) .$$

Indeed, in such a situation, $\lambda V^* = (\lambda V)^* = V^*$ for $\lambda \in \mathbf{C} \setminus 0$ (as $M \ni z \longrightarrow \lambda z \in M$ is biholomorphic), and if $V^* \neq \emptyset$, then $0 \in V^*$ (since V^* is closed in V).

If V and W are locally analytic subsets of the manifolds M and N, respectively, then

$$(V \times W)^0 = V^0 \times W^0 .$$

(The inclusion $\supset$ is obvious, while the inclusion $\subset$ follows from the proposition in C. 4.1.)

It follows easily (in view of (#); see C. 4.2) that if $f : M \longrightarrow N$ is a submersion, then

$$f^{-1}(W)^0 = f^{-1}(W^0) \text{ and } f^{-1}(W)^* = f^{-1}(W^*)$$

for each locally analytic subset $W \subset N$ $(^{11})$.

Note also that if $\Gamma \subset M$ is a submanifold whose closure $\bar{\Gamma}$ is an analytic subset of M, then $\Gamma \subset (\bar{\Gamma}^0)$ $(^{12})$, and so $(\bar{\Gamma}^*) \subset \bar{\Gamma} \setminus \Gamma$.

PROPOSITION 1. *The set V^0 is open and dense in V, whereas the set V^* is closed and nowhere dense in V.*

In fact, it suffices to show that V^0 is dense in V. First of all, observe that this is true for normal representatives of analytic germs (see 1.1, conditions (3),(4); and 1.5). Now, let $a \in V$. It is enough to prove that $V^0 \cap U \neq \emptyset$ for an arbitrarily small open neighbourhood U of a in M. We have $V_a = A \cup B$, where A, B are analytic germs, A is irreducible, and $A \not\subset B$ (see II. 4.6). Hence, in view of proposition 2 from 1.5, one can find an arbitrarily small open neighbourhood U of a and analytic representatives W, Z of the germs A, B in U such that $V \cap U = W \cup Z$ and the representative W is normal. Then the set W^0 is dense in W. By (#), we get $V^0 \cap U \supset V^0 \cap (U \setminus Z) = (W \setminus Z)^0 = W^0 \setminus Z \neq \emptyset$, since the set $W^0 \setminus Z$ is dense in the non-empty set $W \setminus Z$.

$(^{11})$ We use here the previous formula with $W = N$. This special case of the formula can be verified directly, without use of the proposition from C. 4.1.

$(^{12})$ Since Γ is an open set in $\bar{\Gamma}$ (see C. 3.7 and B. 1).

If W is a normal representative of a k–dimensional analytic germ, then W^0 is a k–dimensional submanifold and $\dim W^* < k$.

Indeed (see 1.5), the crown V of the normal triple (Ω, Z, V) of a germ of dimension k is of constant dimension k and $V^* \subset V_Z$ (see 1.3; and 1.1, condition (3)).

Proposition 3 from 1.6 yields the following:

If W is a normal representative of a k–dimensional simple germ, then W^0 is a connected submanifold of dimension k.

Indeed (see 1.5), for a normal triple (Ω, Z, V) of a simple germ, the set V^0 is connected, as it contains the subset $V_{\Omega \backslash Z}$ which is connected and dense (see 1.1, condition (4)).

2. We have the following lemmas.

LEMMA 1. *If $V, W \subset M$ are analytic subsets and the set V^0 is connected, then*
$$W \subsetneqq V \implies (W \text{ is nowhere dense in } V) \, .$$

This follows from proposition 1, since V^0 is a submanifold (see n° 1), and so $W \cap V^0$ is nowhere dense in V^0, for otherwise (see II. 3.6 and 4) we would have $V^0 \subset W$ and hence $V \subset W$.

LEMMA 2. *Let $V = \bigcup V_i$ be a locally finite union of analytic subsets of M such that $V_i \cap V_j$ is nowhere dense in V_i if $i \neq j$. Then*
$$(*) \qquad V^* = \bigcup_i V_i^* \cup \bigcup_{i \neq j} (V_i \cap V_j) \quad \text{and} \quad V^0 = \bigcup_i \left(V_i^0 \backslash \bigcup_{\nu \neq i} V_\nu \right) .$$

In fact, it is enough to show the second equality $(*)$, as the right hand sides of both equalities complement each other in V. First, observe that $V_i^0 \backslash \bigcup_{\nu \neq i} V_\nu = V^0 \backslash \bigcup_{\nu \neq i} V_\nu$ (see n° 1, the first equality in $(\#)$). Thus the inclusion $\supset$ is obvious. Now, let $a \in V^0$. One can choose an open neighbourhood U of the point a in such a way that $V \cap U = \bigcup_I (V_i \cap U)$, where I is a finite set of indices and $V \cap U$ is a connected submanifold. By lemma 1, we have $V \cap U = V_i \cap U$ for some i, and hence $V_\nu \cap U \subset V_\nu \cap V_i$ for each ν. Consequently, if one had $a \in V_\nu$ for some $\nu \neq i$, then a would be an interior point of the set $V_\nu \cap V_i$ in V_ν. That would contradict the assumptions above. Thus $a \in V_i^0 \backslash \bigcup_{\nu \neq i} V_\nu$.

LEMMA 3. *If $V \subset M$ is an analytic subset and a is one of its points, then there exists an open neighbourhood U of the point a and analytic representatives $V_1, \ldots, V_r$ in U of the simple components of the germ V_a such that $V \cap U = V_1 \cup \ldots \cup V_r$ and $V_i \cap V_j$ is nowhere dense in V_i if $i \neq j$.*

It is clear that there is an open neighbourhood U' of the point a and analytic subsets $V_1', \ldots, V_r'$ in U' representing the simple components $A_1, \ldots, A_r$ of the germ V_a such that $V \cap U' = V_1' \cup \ldots \cup V_r'$. Take normal representatives $V_i'' = V_i' \cap U_i$ of the germs A_i, where $U_i \subset U'$ are open neighbourhoods of the point a (see 1.5, proposition 2). Then if $i \neq j$, by lemma 1, the set $V_i'' \cap V_j'$ is nowhere dense in V_i'', because $A_i \not\subset A_j$ and $(V_i'')^0$ is connected (see n° 1). Therefore it suffices to set $U = \bigcap U_i$ and $V_i = V_i'' \cap U$.

3. Let X be an n–dimensional vector space, and let $0 \leq k \leq n$. For each linear coordinate system $\varphi \in L_0(X, \mathbf{C}^n)$, define

$$\lambda(\varphi) = \varphi^{-1}(e_{k+1}) \wedge \ldots \wedge \varphi^{-1}(e_n) \, ,$$

where $e_1, \ldots, e_n$ is the canonical basis for $\mathbf{C}^n$. A sequence $\varphi_1, \ldots, \varphi_r \in L_0(X, \mathbf{C}^n)$ is said to be k–complete if $\lambda(\varphi_1), \ldots \lambda(\varphi_r)$ generate the space $\bigwedge^{n-k} X$. In such a case, if $\varphi_1', \ldots, \varphi_r' \in L_0(X, \mathbf{C}^n)$ are sufficiently close to $\varphi_1, \ldots, \varphi_r$, then also $\varphi_1', \ldots, \varphi_r'$ is a k–complete sequence (see B 5.1–2). Obviously a k–complete sequence exists in every space (see A. 3, footnote (^{26})).

LEMMA 4. *If a set $E \subset X$ is a k–dimensional submanifold at a point $a \in E$, then, for any k–complete sequence of linear coordinate systems in X, the set E is a k–topographic submanifold at a in one of those coordinate systems.*

To see this, consider a k–complete sequence $\varphi_1, \ldots, \varphi_r \in L_0(X, \mathbf{C}^n)$. Taking a direction vector μ of the subspace $T_a E$, we have $\mu \wedge \lambda(\varphi_s) \neq 0$ for some s. Therefore $(T_a E) \cap \left(\sum_{i=k+1}^n \mathbf{C} \varphi_s^{-1}(e_i) \right) = 0$, and hence $T_{\varphi_s(a)} \varphi_s(E) \cap \{z_1 = \ldots = z_k = 0\} = 0$ (see C. 3.11). In conclusion, the set $\varphi_s(E) \subset \mathbf{C}^n$ is a k–topographic submanifold at $\varphi_s(a)$ (see C. 3.1).

4. Let $V \subset M$ be a locally analytic subset.

THEOREM 1. *The set V^* is a locally analytic subset. Moreover, $\dim V^* < \dim V$ if $V \neq \emptyset$. If V is an analytic subset, then so is V^*.*

PROOF. Let $a \in V$, and set $k = \dim V_a$. It is enough to show that for some open neighbourhood U of the point a, the set $V^* \cap U$ is analytic in U, of dimension $< k$ (see II. 3.4, II. 1.3 and 5–6).

In the case when the germ V_a is irreducible, the above is true for a sufficiently small U, and furthermore, $V^0 \cap U$ is a k–dimensional submanifold. Indeed, one may assume that M is a vector space and $a = 0$ (see n° 1). There exists a k–complete sequence $\varphi_1, \ldots, \varphi_r \in L_0(M, \mathbf{C}^n)$ such that the germ V_a has a representative W_i which is normal with respect to the coordinate system

φ_i, $i = 1, \ldots, r$ (see 1.5, proposition 2; and n° 3). The set W_i and the set Σ_i of the points of W_i at which it is not a k–topographic submanifold in the coordinate system φ_i are analytic in some open neighbourhood U_i of the point a (see 1.5 and 1.3). Now, for a sufficiently small $U \subset \bigcap_i^r U_i$, we have $V \cap U = W_i \cap U$, $i = 1, \ldots, r$. Then $V^0 \cap U = W_i^0$ is a k–dimensional submanifold (see n° 1), and, in view of lemma 4, we have $V^0 \cap U = \bigcup_1^r (W_i \setminus \Sigma_i) \cap U$. Hence $V^* \cap U = \bigcap_1^r (\Sigma_i \cap U)$ is an analytic subset of U. Finally, $\dim(V^* \cap U) < k$ since $V^* \cap U = W_i^* \cap U$ (see n° 1).

In the general case, take a neighbourhood U of the point a and analytic subsets V_i of U, as in lemma 3. According to the first part of the proof, by reducing the size of U, we may assume that the sets V_i^* are analytic subsets of dimension $< k$, while the sets V_i^* are submanifolds of dimension $\leq k$. Hence, by lemma 2 (the first formula in $(*)$), the set $V^* \cap U = (V \cap U)^*$ is analytic in U. It is of dimension $< k$, since the sets $V_i \cap V_j$, $i \neq j$, are also of dimension $< k$. This is so because, setting $W = V_i \cap V_j$, we have $W \subset (W \cap V_i^0) \cup V_i^*$ and the set $W \cap V_i^0$, which is nowhere dense in V_i^0, is of dimension $< k$.

5. Let $V \subset M$ be a locally analytic subset. Theorem 1 implies the following corollaries.

Every locally analytic set is a finite union of submanifolds.

It is always true that $\dim V^* \leq n - 2$, *provided that* $n \geq 1$. Indeed, this is so if M is non-empty and connected, because, in such a case, the only n–dimensional analytic subset is M (see II. 3.6 and II. 1.2). This implies the general case (see n° 1, the second formula in $(\#)$).

Let $W \subset N$ *be a locally analytic subset of* M. *If* $V \neq \emptyset$, *then*

$$(W \text{ is nowhere dense in } V) \Longrightarrow (\dim W < \dim V).$$

In fact, $W \subset \bigcup_{\nu=0}^k (W \cap V^{(\nu)}) \cup V^*$, where $k = \dim V$. Also $W \cap V^{(\nu)}$ is of dimension $< \nu$, since it is nowhere dense in $V^{(\nu)}$ (see n° 1 and II. 1.2). It follows that

$$(W \text{ is nowhere dense in } V) \Longleftrightarrow (\dim_z W < \dim_z V \text{ for } z \in W)$$

(see B. 1). In particular, *if the analytic set* V *is non-empty and of constant dimension, we have*

$$(W \text{ is nowhere dense in } V) \Longleftrightarrow \dim W < \dim V .$$

It follows from theorem 1 that

$$\dim V = \dim V^0 .$$

Therefore (see n° 1, proposition 1)

$$\dim V = \max_{x \in E} \dim_x V \quad \text{if } E \text{ is a dense subset of } V \text{ ,}$$

and, in particular,

$$\dim V = \dim Z \quad \text{if } Z \text{ is an open dense subset of } V \text{ .}$$

Thus

$$(V \text{ is of constant dimension } k) \Longleftrightarrow (\dim_z V = k \text{ for } z \in E)$$

if E is a dense subset of V. In particular,

$$(V \text{ is of constant dimension } k) \Longleftrightarrow (Z \text{ is of constant dimension } k)$$

if Z is an open dense subset of V, and moreover,

$$(V \text{ is of constant dimension } k) \Longleftrightarrow$$

$$(V^0 \text{ is a submanifold of dimension } k \text{ as a manifold}).$$

6. Let $V \subset M$ be a locally analytic subset, and let $a \in V$.

THEOREM 2. *If $f_1, \ldots, f_r$ are representatives of generators of the ideal $\mathcal{I}(V_a)$ and U is a sufficiently small open neighbourhood of the point a, then f_i are holomorphic in U, we have $V \cap U = \{z \in U : f_1(z) = \ldots = f_r(z) = 0\}$, and*

$$(\#) \qquad T_c V = \bigcap_1^r \ker d_c f_i \quad \text{for } c \in V^0 \cap U \text{ .}$$

PROOF. One may assume that M is a vector space and $a = 0$. It is enough to show that there is an open neighbourhood U of a and $f_1, \ldots, f_r$ satisfying the conditions in the conclusion of the theorem. Indeed, let $g_1, \ldots, g_s$ be representatives of generators of the ideal $\mathcal{I}(V_a)$. If $U_0 \subset U$ is a sufficiently small open neighbourhood of a, then the functions f_i, g_i are holomorphic in U_0. Thus we have $V \cap U_0 = \{g_1 = \ldots = g_s = 0\} \cap U_0$ (since the set $\{g_1 = \ldots = g_s = 0\}$ is a representative of the germ $V(\mathcal{I}(V_a)) = V_a$; see II. 4.5) and $f_i = \sum_j a_{ij} g_j$ in U_0 for some a_{ij} that are holomorphic in U_0 (since $(f_i)_a \in \mathcal{I}(V_a)$). Therefore, for $c \in V^0 \cap U_0$, we get the equality $T_c V = \bigcap_1^s \ker d_c g_j$; the inclusion $\subset$ follows because $g_i = 0$ on $V \cap U_0$ (see C. 3.11), whereas the inclusion $\supset$ is true because $d_c f_i = \sum_j a_{ij}(c) d_c g_j$.

Now, suppose that the germ V_a is irreducible. Set $k = \dim V_a$. There exists a k–complete sequence $\varphi_1, \ldots, \varphi_r \in L_0(M, \mathbf{C}^n)$ such that the germ V_a has a normal representative W_i in the coordinate system φ_i, $i = 1, \ldots, r$ (see 1.5, proposition 2; and n° 3). We have $W_i = \{f_{i1} = \ldots = f_{is_i}\}$ and $T_c W_i = \bigcap_\nu \ker d_c f_{i\nu}$ for $c \in W_i'$ and for some functions $f_{i\nu}$ that are holomorphic in a neighbourhood U_i of zero. Here W_i' denotes the set of points of W_i at which W_i is a k–topographic submanifold in the coordinate system φ_i (see 1.3 and 1.5). Now, take an open neighbourhood of the origin $U \subset \bigcap U_i$ such that $V \cap U = W_i \cap U$, $i = 1, \ldots, r$. Then $V^0 \cap U = W_i^0 \cap U$ and $T_c V = T_c W_i$ for $c \in V^0 \cap U$, $i = 1, \ldots, r$. Therefore $V^0 \cap U$ is a k–dimensional submanifold (see n° 1) and, by lemma 4, we have $V^0 \cap U \subset \bigcup W_i'$. Hence, if $c \in V^0 \cap U$, then (in view of the fact that $d_c f_{i\nu} = 0$ on $T_c V$) we have $T_c V = \bigcap_{i,\nu} \ker d_c f_{i\nu}$. Clearly, $V \cap U = \{z \in U : f_{i\nu} = 0; \ \nu = 1, \ldots, s_i; \ i = 1, \ldots, r\}$.

In the general case, consider a neighbourhood U of the point a and analytic subsets of U as in lemma 3. In light of the first part of the proof, and reducing the size of U, we may assume that $V_i = \{g_{i1} = \ldots = g_{is_i} = 0\}$, $g_{i\nu}$ are holomorphic functions on U, and $T_c V_i = \bigcap_\nu \ker d_c g_{i\nu}$ for $c \in V_i^0$. Then $V \cap U = \{g_{1\nu_1} \ldots g_{r\nu_r} = 0 : \ \nu_i = 1, \ldots, s_i; \ i = 1, \ldots, r\}$ and it suffices to show the equality $T_c V = \bigcap \ker d_c(g_{1\nu_1} \ldots g_{r\nu_r})$ for $c \in V^0 \cap U$. Obviously, the inclusion $\subset$ is true (see C. 3.11), and so the opposite one remains to be verified. Now, by lemma 2, we have $c \in V_i^0 \setminus \bigcup_{j \neq i} V_j$ for some i. Thus on the one hand, $T_c V = T_c V_i$ (since $V_c = (V_i)_c$); on the other hand, if $j \neq i$, then $g_{j\alpha_j}(c) \neq 0$ for some α_j, and so $\ker d_c(g_{1\alpha_1} \ldots g_{i\nu} \ldots g_{r\alpha_r}) = \ker d_c g_{i\nu}$, $\nu = 1, \ldots, s_i$, which implies the inclusion $\supset$.

REMARK. The condition (#) means exactly that if $c \in V^0 \cap U$, then the germs $(f_i)_c$ generate the ideal $\mathcal{I}(V_c)$ (see II. 4.2). In fact, we have a much stronger result, *Cartan's theorem*, according to which there are holomorphic functions $f_1, \ldots, f_s$ on an open neighbourhood U of the point a such that the germs $(f_1)_c, \ldots, (f_s)_c$ generate the ideal $\mathcal{I}(V_c)$ for each $c \in U$. (See VI. 1.3–4 below.)

The case of a principal germ is particularly simple. Namely, if $V_a = V(f_a)$, where f is a function which is holomorphic in a neighbourhood of the point a, then one may assume (see II. 5.3) that f_a has no multiple factors. Then for each c from a neighbourhood of the point a, the germ f_a generates the ideal $\mathcal{I}(V_c)$. (See 3.1 , corollary 1 from proposition 5 below.)

7. Let $V \subset M$ be an analytic set.

THEOREM 3. *The connected components of the set V^0 form a locally finite family and their closures are analytic subsets of M.*

PROOF . Let $\{\Lambda_j\}$ be the family of connected components of the set V^0. It is sufficient to prove that each point $a \in M$ has an open neighbourhood U such that the number of non-empty sets $\Lambda_j \cap U$ is finite and the sets $\bar{\Lambda}_j \cap U$ are analytic in U. One may assume that $a \in V$. Take an open neighbourhood U' of a and analytic subsets $V_1, \ldots, V_r$ of U' as in lemma 3. The set $V^* \cap V_i$ is nowhere dense in V_i since, by lemma 2, it is contained in the set $V_i^* \cup \bigcup_{\nu \neq i}(V_\nu \cap V_i)$ (see n° 1, proposition 1). In view of proposition 2 from 1.5, there is a normal representative W_i of the germ $(V_i)_a$ which is an open neighbourhood of a in V_i. Then the set $V^* \cap W_i^0$ is nowhere dense in W_i^0 and $W_i^* \subset V_i^* \subset V^*$ (see lemma 2), which implies that the set $\Gamma_i = W_i \setminus V^* = W_i^0 \setminus (V^* \cap W_i^0)$ is connected (see II. 3.6 and n° 1) and dense in W_i, $i = 1, \ldots, r$. Now, if $U \subset U'$ is a sufficiently small open neighbourhood of the point a, then the sets $W_i \cap U$ are analytic in U, and $V \cap U = \bigcup_1^r (W_i \cap U)$. Hence $V^0 \cap U = \bigcup(\Gamma_i \cap U) = \bigcup(\Lambda_j \cap U)$. But since each Γ_i is a connected subset of the set V^0, it must be contained in some Λ_j, i.e., $\Gamma_i \cap U \subset \Lambda_j \cap U$. It follows that the number of non-empty sets of the form $\Lambda_j \cap U$ is finite and each of them is the union of some of the sets $\Gamma_i \cap U$. Thus each of the sets $\bar{\Lambda}_j \cap U$ is the union of some of the sets $\bar{\Gamma}_i \cap U = W_i \cap U$, and hence is analytic in U.

8. A non-empty analytic set $V \subset M$ is said to be *irreducible* or *simple* if it cannot be expressed as the union of two analytic subsets of M properly contained in V ([13]) . It follows (by induction) that such a set V cannot be expressed as a finite union of analytic subsets properly contained in V. Then, if V_i are analytic subsets of M, it follows that

$$(V \subset V_1 \cup \ldots \cup V_k) \Longrightarrow (V \subset V_j \text{ for some } j) .$$

(Since $V = \bigcup(V \cap U_i)$, and so $V = V \cap V_j$ for some j.) Obviously, any closed connected non-empty submanifold is an irreducible analytic set. (See II. 3.6 and 4.)

Note that the image of an irreducible analytic set under a holomorphic mapping is also irreducible, provided that it is analytic.

Irreducibility of an analytic subset of an open set G does not depend on the set G (see II. 3.4). This allows one to define *irreducibility of a locally analytic set* (see II. 3.4).

Lemma 1, combined with theorem 3, implies (see B. 1 and II. 3.4):

PROPOSITION 2. *A non-empty analytic set $V \subset M$ is irreducible precisely when the set V^0 is connected.*

([13]) Hence an analytic set is not irreducible precisely when it is empty or *reducible*, i.e., is the union of two analytic subsets properly contained in it.

Hence, in view of proposition 1, we have the following corollaries.

COROLLARY 1. *An irreducible analytic set is of constant dimension and the set of its regular points V^0 is a connected submanifold of dimension* dim V. (See n° 1 and n° 5.)

COROLLARY 2. *An analytic set is irreducible if and only if it contains a non-empty dense connected subset of regular points.*

COROLLARY 3. *An analytic set is irreducible (of dimension k) if and only if it is the closure of a non-empty connected submanifold (of dimension k).* (See C. 3.7.)

COROLLARY 4. *Any irreducible analytic set is connected* ([14]) .

COROLLARY 5. *A normal representative of a simple germ is irreducible.* (See n° 1.)

COROLLARY 6. *If V, W are non-empty analytic subsets of the manifolds M, N, respectively, then*

$$(V \times W \text{ is irreducible}) \Longleftrightarrow (V \text{ and } W \text{ are irreducible}) .$$

(See n° 1.) ([15])

Let $V, W \subset M$ be analytic sets.

PROPOSITION 3. *If $W \subset V$ and V is irreducible, then*

$$W \subsetneqq V \Longleftrightarrow (W \text{ is nowhere dense in } V) \Longleftrightarrow \dim W < \dim V .$$

Indeed, numbering the above conditions, we see that the implication 3 $\Longrightarrow$ 1 is trivial; we have already shown 2 $\Longrightarrow$ 3 (see n° 5), and 1 $\Longrightarrow$ 2 follows from lemma 1 and proposition 2.

COROLLARY 1. *If V is irreducible and $V \not\subset W$, then $V \setminus W$ and $V^0 \setminus W$ are dense in V and connected.*

In fact, $W \cap V$ is nowhere dense in V; hence also the set $W \cap V^0$ is nowhere dense in V^0. Therefore, in view of proposition 2 (see II. 3.6), the set $V^0 \setminus W$ is connected and dense in V^0. By proposition 1, it is dense in V, and also the set $V \setminus W$ is connected and dense in V.

By proposition 2, we have:

COROLLARY 2. *If the set V is irreducible and $V \not\subset W$, then also the analytic subset $V \setminus W$ of $M \setminus W$ is irreducible.*

([14]) Obviously, the converse is false: a connected analytic set can be reducible.

([15]) The Cartesian product of two non-empty sets is connected if and only if both of the compact sets are connected.

PROPOSITION 4. *Let* $a \in V \cap W$. *If* V *is irreducible, then* $V_a \subset W_a \implies$
$V \subset W$. *Hence, if the* V, W *are irreducible, then* $V_a = W_a \implies V = W$. *This*
means that an irreducible analytic set is uniquely determined by its germ at
any of its points.

Indeed, let $V_a \subset W_a$. Then $V_a = (V \cap W)_a$. If we had $V \not\subset W$, then
$W \cap W \subsetneq V$ and, by proposition 3, we would have $\dim V_a = \dim(V \cap W)_a \leq$
$\dim(V \cap W) < \dim V$, contradicting the fact that V is of constant dimension
(see corollary 1 from proposition 2).

REMARK (RITT'S LEMMA). Instead of the inclusion $V_a \subset W_a$, it suffices
to assume that a simple component of the germ V_a is contained in W_a. (See
3.1, proposition 4 below.)

COROLLARY . *If* V *is irreducible and* $\{W_i\}$ *is a countable family of ana-*
lytic subsets of M, *then*

$$V \subset \bigcup W_i \implies (V \subset W_j \text{ for some } j) \,.$$

Indeed, V is a Baire space and $V = \bigcup(W_i \cap V)$. Therefore some set
$W_j \cap V$ has an interior point a (with respect to V), and hence $V_a \subset (W_j)_a$.
This implies that $V \subset W_j$.

9. The *decomposition of an analytic set* $V \subset M$ *into simple components*
is the representation of V as the union of a locally finite family of analytic
subsets V_i of M which are irreducible and such that $V_i \not\subset V_j$ for $i \neq j$. (Such
a family is always countable; see B. 1.) According to theorem 4 below, such
a decomposition always exists and is unique. The sets V_i are called *simple*
components of the set V.

The decomposition into simple components of an analytic subset of an
open set G does not depend on the set G (see n° 8). Therefore the *decompo-*
sition into simple components of a locally analytic set is well defined (see II.
3.4).

THEOREM 4 (On the decomposition into simple components). *Every an-*
alytic set $V \subset M$ *has a unique decomposition into simple components:* $V =$
$\bigcup V_i$. *Then*

$$V_i \setminus V^* = V_i^0 \setminus V^* = V_i^0 \setminus \bigcup_{\nu \neq i} V_\nu$$

and all these sets are submanifolds, open and dense in V_i^0 *(and hence also in*
V_i), *and form the family of all connected components of the set* V^0. *Therefore*
the simple components of the analytic set V *are precisely the closures of the*

connected components of the set V^0. We have the formulae

$$V^* = \bigcup_i V_i^* \cup \bigcup_{i \neq j}(V_i \cap V_j) \quad \text{and} \quad V^0 = \bigcup_i (V_i^0 \setminus \bigcup_{\nu \neq i} V_\nu) .$$

PROOF. Let $\{\Lambda_i\}$ be the family of connected components of the set V^0. Then – in view of theorem 3 – we have $V = \bigcup \bar{\Lambda}_i$ (see proposition 1 and B. 1), and the sets $\bar{\Lambda}_i$ are analytic. This is a decomposition into simple components, as the $\bar{\Lambda}_i$ are irreducible (see corollary 3 from proposition 2) and $\bar{\Lambda}_i \not\subset \bar{\Lambda}_j$ for $i \neq j$ (since $\Lambda_i \cap \bar{\Lambda}_j = \emptyset$). Now let $V = \bigcup V_i$ be a decomposition into simple components. Owing to proposition 3, for $i \neq j$, the set $V_i \cap V_j$ is nowhere dense in V_i, and hence lemma 2 yields the formulae $(*)$. Obviously $V^0 = \bigcup(V_i \setminus V^*)$ and (because of $(*)$) the sets $V_i \setminus V^*$ are disjoint and $V_i \setminus V^* = V_i^0 \setminus V^*$. Since the set V_i^0 is a connected submanifold (see corollary 1 from proposition 2), while the set $V_i^0 \cap V^* = \bigcup_{\nu \neq i} V_\nu \cap V_i^0$ (see $(*)$) is nowhere dense in V_i^0, the set $V_i^0 \setminus V^*$ is dense in V_i and connected (see II. 3.6 and proposition 1). Moreover, $V_i^0 \setminus V^*$ is open in V^0, as $V_i^0 \setminus V^* = V^0 \setminus \bigcup_{\nu \neq i} V_\nu$ [16]. Thus the sets $V_i^0 \setminus V^*$ form the family of all connected components of the set V^0. Therefore the family $\{V_i\}$ coincides with the family $\{\bar{\Lambda}_i\}$.

With notation as in the theorem, we have the following corollaries.

COROLLARY 1. *A set $Z \subset V$ is nowhere dense in V precisely when for each i the set $Z \cap V_i$ is nowhere dense in V_i.* (See B. 1.)

COROLLARY 2. *We have $V^{(k)} = \bigcup\{V_i \setminus V^* : \dim V_i = k\}$, $k = 0, \ldots, n$.* (See n° 1 and n° 5.)

COROLLARY 3. *The set V is of constant dimension k if and only if all of its simple components are of constant dimension k.* (See n° 5.)

By the second formula in $(*)$, we have:

COROLLARY 4. *A point $z \in V$ is a regular point of V if and only if it belongs to only one of the sets V_i and is a regular point of V_i. Then also $V_z = (V_i)_z$.*

COROLLARY 5. *If H is an open and closed set in V^0, e.g., if H is a connected component of V^0, then the set $V \setminus H$ is analytic. In particular, the sets $V \setminus V^{(k)}$ are analytic. If $V \neq \emptyset$ and $k = \dim V$, then $\dim(V \setminus V^{(k)}) < \dim V$.*

Indeed, $H = \bigcup(V_i' \setminus V^*)$ for some simple components V_i' of V. Denoting the remaining ones by V_j'', we have $V \setminus H = \bigcup V_j'' \cup V^*$.

---

[16] Because of $(*)$, we have $V_i^0 \cap V^0 = V_i^0 \setminus \bigcup_{\nu \neq i} V_\nu = V^0 \setminus \bigcup_{\nu \neq i} V_\nu$.

COROLLARY 6. *If* $V = \bigcup V_i$ *and* $W = \bigcup W_j$ *are the decompositions into simple components of non-empty analytic subsets* V, W *of manifolds* M, N, *respectively, then* $V \times W = \bigcup_{i,j}(V_i \times W_j)$ *is the decomposition of the analytic set* $V \times W$ *into simple components.* (See B. 1; and n° 8, corollary 6 from proposition 2.)

COROLLARY 7. *Assume that* M *is a vector space. If the set* V *is a cone, then so is each of its simple components.* (*In this case the set* V *is algebraic* ([17]) *and so are all of its simple components; their number is finite* ([18]) .)

Indeed, if $z \in V_i \setminus V^*$, then $\mathbf{C}z \subset V_i$. This follows from the corollary of proposition 4, as then $\mathbf{C}z$ is an irreducible analytic set which is contained in V, but not in the union $\bigcup_{\nu \neq i} V_\nu$ (it does not contain the point z). Consequently, since the set $V_i \setminus V^*$ is dense in V_i, we obtain $tV_i \subset V_i$ for $t \in \mathbf{C}$.

COROLLARY 8. *If* G *is an open subset of* V, *then the trace on* G *of any simple component of* V *is a union of some simple components of* G.

(Since $G \cap V^0 = G^0$, the trace on G of a connected component of V^0 is a union of connected components of G^0.)

Theorem 4, combined with its corollaries 2 and 3, yields the following *decomposition of an analytic set* V *into components of constant dimension*

$$V = W_0 \cup \ldots \cup W_n \ ,$$

where (see B. 1)

$$W_k = \bigcup \{V_i : \ \dim V_i = k\} = \overline{V^{(k)}}$$

is an analytic set of constant dimension $(k = 0, \ldots, n)$. (Naturally, the decomposition is uniquely defined in this way ([19]) .)

In view of the proposition from B. 1 (combined with corollary 4 from proposition 2), we have:

PROPOSITION 5. *Any connected component of an analytic set* V *is the union of an equivalence class with respect to the relation* $\approx$, *defined on the set of all simple components of the* V *by the formula*

$$V' \approx V'' \Longleftrightarrow \ (\textit{There exists a sequence } V' = V_0, \ldots, V_r = V'' \textit{ of simple}$$
$$\textit{components of the analytic set } V \textit{ such that } V_{i-1} \cap V_i \neq \emptyset,$$
$$i = 1, \ldots, r).$$

([17]) In view of the Cartan-Remmert-Stein lemma (see II. 3.3).

([18]) Since each of them contains 0. See also VII. 11, proposition 2.

([19]) A decomposition $V = Z_0 \cup \ldots \cup Z_n$, where Z_k is an analytic set of constant dimension $(k = 0, \ldots, n)$, is generally not unique.

A locally analytic set is said to be *locally irreducible* if at each of its points its germ is irreducible. By proposition 5 (in view of proposition 4 from n° 8), we get

COROLLARY . *For a locally irreducible analytic set, its decomposition into simple components coincides with its decomposition into connected components. Hence, if it is connected, it must be irreducible.*

10. Let $V, W \subset M$ be analytic sets.

THEOREM 5. *The closures of any connected component of the set $V \setminus W$, any open and closed set in $V \setminus W$, and, in particular, the set $\overline{V \setminus W}$, are unions of some simple components of V not contained in W, and so they are analytic. The family of the connected components of the set $V \setminus W$ is locally finite.*

Indeed, let V_i be all the simple components of V that are not contained in W. Then $V \setminus W = \bigcup(V_i \setminus W)$ and, by corollary 1 from proposition 3, for each i the set $V_i \setminus W$ is connected and dense in V_i. Therefore each connected component of the set $V \setminus W$ and also each open and closed set in $V \setminus W$ is the union of some of the sets $V_i \setminus W$. Thus (see B. 1) its closure is the union of the corresponding V_i's, and hence it is analytic (see II. 3.4); moreover, it follows that the family of connected components of the set $V \setminus W$ is locally finite.

REMARK. The above argument shows that the family of simple components of the set $\overline{V \setminus W}$ is equal to the family of simple components of the set V that are not contained in W.

We will also prove the following proposition.

PROPOSITION 6. *Let $V \subset M$ be an analytic set, and let f be a holomorphic function in $M \setminus V$. If each $(n-1)$-dimensional simple component of V contains a point across which f extends holomorphically, then f has a holomorphic extension on M.*

One may assume that $\dim V \le n - 1$ [20] . Let G be the set of points of the set V^0 across which f extends holomorphically. This set is obviously open in V^0. It is also closed in V^0. In order to prove this claim, let a be a point of the closure of G in V^0. If $a \in \bigcup_0^{n-2} V^{(i)}$, then $a \in G$ by the Hartogs theorem (see III. 4.2). In the case when $a \in V^{(n-1)}$, it is enough to observe that, according to the Hartogs lemma (see C. 1.11), each holomorphic function on $U \times \{0 < |z_n| < r\}$ – where $U \subset \mathbf{C}^{n-1}$ is an open connected set –

---

[20] By disregarding the connected components of the manifold M which are contained in V.

which extends holomorphically across a point from $U \times 0$, has a holomorphic extension in $U \times \{|z_n| < r\}$. Consequently, $G = V^0$ because, in view of theorem 4 from n° 9, each connected component of V^0 contains a point of G. Thus ([21]) the function f extends to a holomorphic function on $M \setminus V^*$ and hence, by the Hartogs theorem, also on M, since $\dim V^* \leq n - 2$.

REMARK. Proposition 6 is clearly true for holomorphic mappings on $M \setminus V$ with values in a vector space.

§3. Some properties of analytic germs and sets

Let M be a complex manifold.

1. We are going to prove the following:

PROPOSITION 1. *Let A be an analytic germ at a point $a \in M$, and let V be a representative of A. The following conditions are equivalent:*

(1) *The germ A is simple.*

(2) *The germ A has an arbitrarily small representative which is locally analytic and irreducible.*

(3) *There exists an arbitrarily small open neighbourhood U of the point a such that $V \cap U$ is analytic in U and irreducible.*

In fact, one may assume that V is an analytic set containing a. Now the implication (3) $\Longrightarrow$ (2) is trivial, while (1) $\Longrightarrow$ (3) follows from proposition 2 in 1.5 and corollary 5 stated after proposition 2 from 2.8. It remains to show that (2) $\Longrightarrow$ (1). Suppose the germ A is reducible. Then there are analytic subsets V', V'' in an open neighbourhood U of the point a such that $A = (V' \cup V'')_a$ and $V'_a, V''_a \subsetneq A$. If the germ A had an irreducible locally analytic representative $W \subset U$, then W would be an analytic subset in an open set $U_0 \subset U$, and hence $W \subset (V' \cup V'') \cap U_0$ (see 2.8, proposition 4). Therefore, e.g., $W \subset V' \cap U_0$ (see 2.8), and so $A \subset V'_a$. This contradicts the statement $V'_a \subsetneq A$. In conclusion, the neighbourhood U does not contain any irreducible analytic representative of the germ A.

Let A and C be analytic germs at a point $a \in M$.

PROPOSITION 2. *If the germ C is simple, then*

$$A \subsetneq C \Longrightarrow \dim A < \dim C .$$

([21]) See II. 3.8, footnote ([13]).

Indeed, there is an open neighbourhood U of the point a and analytic representatives W, V in U of the germs A, C, respectively, such that $W \subsetneqq V$. By taking a smaller U, we may assume that $\dim V = \dim C$ and – in view of proposition 1 – that V is irreducible. Therefore proposition 3 from 2.8 implies that $\dim A \leq \dim W < \dim V = \dim C$.

PROPOSITION 3a. *If $A = A_1 \cup \ldots \cup A_r$ is the decomposition of the germ A into simple germs, then, for a sufficiently small open neighbourhood U of the point a, there are analytic representatives $V, V_1, \ldots, V_r$ in U of the germs $A, A_1, \ldots, A_r$, respectively, such that $V = V_1 \cup \ldots \cup V_r$ is the decomposition of the subset V into irreducible components.*

PROPOSITION 3b. *If $Z, Z_1, \ldots, Z_r$ are locally analytic sets containing a point a and such that $Z_a = (Z_1)_a \cup \ldots \cup (Z_r)_a$ is the decomposition into simple germs, then there is an arbitrarily small open neighbourhood U' of the point a such that $Z \cap U' = (Z_1 \cap U') \cup \ldots \cup (Z_r \cap U')$ is the decomposition into simple components.*

PROOF of propositions 3a and 3b. Let Z, Z_i be representatives of the germs A, A_i $(i = 1, \ldots, r)$. Then, for a sufficiently small open neighbourhood U of the point a, the sets $Z \cap U$, $Z_i \cap U$ are analytic subsets of U and $Z \cap U = (Z_1 \cap U) \cup \ldots \cup (Z_r \cap U)$. There exists a simple component V_i of the set $Z_i \cap U$ such that $(V_i)_a = A_i$ (see II. 4.6). Then no other simple component contains the point a (for otherwise, such a component would be contained in V_i; see 2.8, proposition 4); for each i, denote by V_i' the union of those components. Hence $V' = V_1' \cup \ldots \cup V_r' \not\ni a$. By taking $V = V_1 \cup \ldots \cup V_r$, we obtain the conclusion of proposition 3a. As far as proposition 3b is concerned, it is enough to take $U' = U \setminus V'$, because then $Z_i \cap U' = V_i \setminus V'$ (see 2.8, corollary 2 from proposition 3).

COROLLARY . *If V' is a simple component of an analytic set $V \subset M$ and $a \in M$, then the germ V_a' is the union of some simple components of the germ V_a. Hence every simple component of V_a' is a simple component of V_a. Conversely, each simple component of V_a is a simple component of some V_a', where V' is a simple component of V.*

Indeed, there is an open neighbourhood U of the point a such that $V \cap U = \bigcup W_i$ is the decomposition into simple components, hence $V_a = \bigcup (W_i)_a$. Then $V' \cap U$ is the union of some of the sets W_i (see 2.9, corollary 3 from theorem 4). The last part follows from $(*)$ in II. 4.6.

PROPOSITION 4. *A locally analytic set $W \subset M$ is of constant dimension k precisely when for each $a \in W$, the germ W_a is of constant dimension k.*

For suppose that W is of constant dimension k. Let $a \in W$, and let $A = W_a = A_1 \cup \ldots \cup A_r$ be the decomposition into simple germs. Choose a neighbourhood U and analytic sets $V, V_1, \ldots, V_r$ in U as specified in proposition 3a. Then $V \cap U' = W \cap U'$ for some open neighbourhood $U' \subset U$ of a. We have $a \in V_i$; hence, because of theorem 4 from 2.9, the set $(V_i^0 \setminus V^*) \cap U'$ is non-empty and open in $V^0 \cap U' = W^0 \cap U'$ (see 2.1). So it is open in W^0 and, clearly, also in V_i^0. Therefore $\dim A_i = \dim V_i = \dim V_i^0 = \dim W^0 = k$ (see 2.8, corollary 1 from proposition 2; and 2.5).

COROLLARY. *An analytic germ is of constant dimension k if and only if it has a locally analytic representative of constant dimension k.* (See 1.5.)

PROPOSITION 5. *If f is a holomorphic function in a neighbourhood of a point $a \in M$, then the following conditions are equivalent:*

(1) *the germ f_a is non-zero and has no multiple factors,*

(2) *$d_z f \neq 0$ on $V(f)^0$ in a neighbourhood of the point a,*

(3) *$d_z f \neq 0$ on a dense subset of $V(f)$ in a neighbourhood of the point a.*

The implication $(1) \Longrightarrow (2)$ follows from corollary 4 in II. 5.2 and theorem 2 in 2.6. It is enough to show $(3) \Longrightarrow (1)$. Assume that (3) is satisfied. Then the germ f_a is non-zero. We may assume that it is non-invertible; then $f = f_1^{s_1} \ldots f_r^{s_r}$ in an open neighbourhood U of the point a, where f_i are holomorphic in U, the germs $(f_i)_a$ are irreducible and mutually non-associated, and $s_i > 0$. One may also require (in view of corollary 2 from II. 5.2 and proposition 3b) that $V(f) \cap U = V(f_1) \cup \ldots \cup V(f_r)$ is the decomposition into simple components, and also $d_z f \neq 0$ on a dense subset of $V(f)^0 \cap U$ (see 2.1, proposition 1). Now, fixing any i, we have $d_z f \neq 0$ for some $z \in V(f)^0 \cap V(f_i)$ (see 2.9, theorem 4) and then $z \notin V(f_\nu)$ for $\nu \neq i$. Thus one must have $s_i = 1$.

It follows (in view of corollary 4 from II. 5.2) that:

COROLLARY 1. *Let f be a holomorphic function in a neighbourhood of a point a, and let $V = V(f)$. If the germ f_a does not have multiple factors, then neither does f_c for all c from a neighbourhood of the point a, and hence it generates the ideal $\mathcal{I}(V_c)$.*

This implies (see II. 5.3) the *Cartan coherence theorem* ([22]) *in the case of analytic subsets of constant dimension 1 of a manifold.*

If $V \subset M$ is an analytic set of constant codimension 1, then for each point of the manifold M there is a holomorphic function f in a neighbourhood U of this point such that $\mathcal{I}(V_z) = \mathcal{O}_z f_z$ for all $z \in U$.

COROLLARY 2. *Let $p : M \times \mathbf{C} \longrightarrow \mathbf{C}$ be a monic polynomial whose coefficients are holomorphic in M and with discriminant $\delta : m \longrightarrow \mathbf{C}$. Then $\delta \not\equiv 0$ if and only if no germ p_c, $c \in M \times \mathbf{C}$, has multiple factors.*

Indeed, if $\delta \not\equiv 0$, then $d_w p \neq 0$ in the set $V(p)_{\{\delta \neq 0\}}$, which is dense in

---

([22]) See VI. 1.3–4 below.

$V(p)$ [23] . Hence the germs p_c, $c \in M \times \mathbf{C}$, have no multiple factors. On the other hand, if $\delta_a = 0$ for some $a \in M$, then (in view of the proposition from A. 6.3 applied to $\mathcal{O}_a$), there is an open neighbourhood U of the point a such that the polynomial $p_{U \times \mathbf{C}}$ is divisible (in $\mathcal{O}_U[t]$) by the square of a monic polynomial q of positive degree. Then $q(a, \tau) = 0$ for some τ, and so the germ $q_{(a,\tau)}$ is non-invertible. But the germ $p_{(a,\tau)}$ is divisible by $q^2_{(a,\tau)}$, which means that it has multiple factors.

2. We have the following theorems:

THEOREM 1. *The intersection of any family of analytic subsets of M is an analytic subset of M.*

THEOREM 2. *If V_ν is a decreasing sequence of analytic subsets of M, then, for every relatively compact set $E \subset M$, the sequence $V_\nu \cap E$ is stationary.*

They are consequences of the following lemma:

LEMMA . *If $\{V^\iota\}$ is a family of analytic subsets of M and $V = \bigcap V^\iota$, then every point $a \in V$ has an open neighbourhood U such that $V \cap U = V^{l_1} \cap \ldots \cap V^{l_r} \cap U$ for some $\iota_1, \ldots, \iota_r$* [24] .

PROOF . Let I be the ideal of $\mathcal{O}_a$ generated by the set $\bigcup \mathcal{I}(V_a^\iota)$. The latter does not contain invertible elements (since $a \in V^\iota$). Hence $I \neq \mathcal{O}_a$, which means that $V(I) \neq \emptyset$. Hence, in view of proposition 3a, the germ $V(I)$ has an analytic representative $\tilde{V}$ in an open neighbourhood $\tilde{U}$ of the point a such that all its simple components W^i contain the point a. Since $I \supset \mathcal{I}(V_a^\iota)$, we have $W_a^i \subset V(I) \subset V_a^\iota$ (see II. 4.3 and 5), so proposition 4 from 2.8 implies that $W^i \subset V^\iota$. This gives

$$(*) \qquad\qquad\qquad \tilde{V} \subset V .$$

On the other hand, the ideal I is generated by a finite number of elements of the set $\bigcup \mathcal{I}(V_a^\iota)$ (see A. 9.2), hence $I \subset \mathcal{I}(V_a^{\iota_1}) + \ldots + \mathcal{I}(V_a^{\iota_r})$ for some $\iota_1, \ldots, \iota_r$. Consequently (see II. 4.3 and 5), we get $V(I) \supset V_a^{\iota_1} \cap \ldots \cap V_a^{\iota_r}$. Therefore there is an open neighbourhood $U \subset \tilde{U}$ of the point a such that $\tilde{V} \cap U \supset V^{\iota_1} \cap \ldots \cap V^{\iota_r} \cap U \supset V \cap U$. Hence, in view of $(*)$, the required equality follows.

---

[23] Because of the theorem on continuity of roots (see B. 5.3).

[24] In connection with theorem 2 it is enough to observe that each point of the closure $\bar{E}$ has an open neighbourhood U such that $V_\nu \cap U = V \cap U$ for sufficiently large ν, where $V = \bigcap V_\nu$.

§4. The ring of an analytic germ. Zariski's dimension

Let M be an n–dimensional manifold.

1. According to the definition of holomorphic mapping, which will be given in Chapter V (see V. 3.1), *holomorphic functions on a locally analytic subset* $V \subset M$ are local restrictions of holomorphic functions on open sets in M. In other words, a function $f : V \longrightarrow \mathbf{C}$ is said to be *holomorphic* if each point of the set V has an open neighbourhood U such that $f_{V \cap U}$ is the restriction of a holomorphic function on U. Obviously, any holomorphic function on V is continuous. Note that the restriction of a holomorphic function on V to an open subset of V is also a holomorphic function.

In the case when V is a submanifold, the above notion is consistent with the previously given definition of a holomorphic function on a manifold. (See C. 3.8.)

2. Let A be a non-empty analytic germ at a point $a \in M$.

The set $\mathcal{O}_A$ of germs on A of holomorphic functions on analytic representatives of A is a ring (see B. 4.2). After the identification of $\mathbf{C}$ with the subring of germs of constants, $\mathcal{O}_A$ is a local $\mathbf{C}$–ring (see A. 10.4a) whose maximal ideal is $\mathfrak{m}_A = \{f \in \mathcal{O}_A : f(a) = 0\}$. Observe that $\mathcal{O}_A \ni f \longrightarrow f(a) \in \mathbf{C}$ is a $\mathbf{C}$–epimorphism with the kernel $\mathfrak{m}_A$. We have the natural $\mathbf{C}$–epimorphism

$$(*) \qquad\qquad \mathcal{O}_a \ni f \longrightarrow f_A \in \mathcal{O}_A$$

(see B. 4.2) which maps $\mathfrak{m}_a$ onto $\mathfrak{m}_A$ (cf. A. 10.1 and 4a); as the kernel of the epimorphism is $\mathcal{I}(A)$ (see II. 4.2), it induces a $\mathbf{C}$–isomorphism (see A. 10.4a)

$$\mathcal{O}_A \overset{C}{\approx} \mathcal{O}_a / \mathcal{I}(A) \ .$$

The ring $\mathcal{O}_A$ is called the *ring of the analytic germ A*. Naturally, it is also a complex vector space and each of its ideals is a subspace of this space. If A is the full germ, then $\mathcal{O}_A = \mathcal{O}_a$.

Let $C \subset A$ be an analytic germ. We define the *ideal of the germ C in the ring $\mathcal{O}_A$*:

$$\mathcal{I}_A(C) = \{f \in \mathcal{O}_A : f_C = 0\} \ .$$

(See B. 4.2.) If $C \neq \emptyset$, then the mapping $\mathcal{O}_A \ni f \longrightarrow f_C \in \mathcal{O}_C$ is a $\mathbf{C}$–epimorphism which maps the ideal $\mathfrak{m}_A$ onto $\mathfrak{m}_C$. Its kernel is $\mathcal{I}_A(C)$, hence (see A. 10.4a)

$$\mathcal{O}_C \overset{C}{\approx} \mathcal{O}_a / \mathcal{I}_A(C) \ .$$

Consider $\mathcal{O}_A$ as a module over $\mathcal{O}_a$ [25]. Its submodules coincide with its ideals and have the same generators. The mapping $(*)$ is a module epimorphism that maps $\mathcal{I}(C)$ onto $\mathcal{I}_A(C)$. Hence we have the natural isomorphism of modules over $\mathcal{O}_a$:

$$\mathcal{I}_A(C) \approx \mathcal{I}(C)/\mathcal{I}(A) \ .$$

In particular, $g\big(\mathcal{I}_A(C)\big) = g\big(\mathcal{I}(C)/\mathcal{I}(A)\big)$.

Obviously, if a locally analytic set $V \subset M$ is a representative of the germ A, then the local C–ring $\mathcal{O}_A$ can be identified with the local C–ring $\mathcal{O}_a(V)$. Here $\mathcal{O}_a(V)$ is the local C–ring of the germs at a of holomorphic functions on open neighbourhoods in V of the point a whose maximal ideal is $\mathfrak{m}_a(V) = \{f \in \mathcal{O}_a(V) : \ f(a) = 0\}$ [26]. The restriction $h_V \in \mathcal{O}_a(V)$ of a germ $h \in \mathcal{O}_a$ is well-defined by the formula $(f_a)_V = (f_V)_a$ [26a], and the epimorphism $(*)$ is

$$\mathcal{O}_a \ni h \longrightarrow h_V \in \mathcal{O}_a(V) \ .$$

Its kernel is $\mathcal{I}(V_a)$ and it maps $\mathfrak{m}_a$ onto $\mathfrak{m}_a(V)$.

In particular, when Z is a submanifold containing a we have $\mathcal{O}_{Z_a} \overset{\mathrm{C}}{\approx} \mathcal{O}_a(Z)$.

3. Let A be a non-empty analytic germ at a point $a \in M$.

Since the ring $\mathcal{O}_a$ is local and noetherian (see I. 1.6–7), the ring $\mathcal{O}_A$ is also local and noetherian (see A. 10. 3 and A. 9.4).

Clearly, the ring $\mathcal{O}_A$ is an integral domain precisely in the case when the germ A is simple (see II. 4.6, proposition 2). Note also that if $A = A_1 \cup \ldots \cup A_r$ is the decomposition into simple germs, then a germ $f \in \mathcal{O}_A$ is not a zero divisor if and only if $f_{A_i} \neq 0$, $i = 1, \ldots, r$ or, equivalently, if it has a representative $\tilde{f} \not\equiv 0$. Indeed, the second condition implies the third one in view of proposition 2 in 2.8 and corollary 1 of proposition 2 in 2.8. Now, if $f_{A_s} = 0$ for some s, then f is a zero divisor because, taking $\varphi \in \mathcal{I}\big(\bigcup_{i \neq s} A_1\big) \setminus \mathcal{I}(A_s)$ (see II. 4.5 and 6), we obtain $f\varphi = 0$.

PROPOSITION 1. *The dimension of the germ A coincides with the Krull dimension* (see A. 12.3) *of the ring $\mathcal{O}_A$:*

$$\dim A = \dim \mathcal{O}_A \ .$$

---

[25] With the multiplication $\mathcal{O}_a \times \mathcal{O}_A \ni (f, \varphi) \longrightarrow f_A \varphi \in \mathcal{O}_A$.

[26] The natural isomorphism is given by $\mathcal{O}_a(V) \ni f_a \longrightarrow f_A \in \mathcal{O}_A$, (where f is a holomorphic function on an open neighbourhood in V of a).

[26a] Where f is holomorphic function on an open neighbourhood of a in M.

PROOF. We have the bijection $C \longrightarrow \mathcal{I}(C)$, which reverses inclusion, of the set of irreducible analytic germs at a onto the set of prime ideals of the ring $\mathcal{O}_a$ (see III. 4.1 and II. 4.5). Consequently, the mapping $C \longrightarrow \mathcal{I}(C)/\mathcal{I}(A)$ of the set of irreducible analytic germs $C \subset A$ is a bijection onto the set of the prime ideals of the ring $\mathcal{O}_a/\mathcal{I}(A) \approx \mathcal{O}_A$. This bijection reverses inclusion (see A. 1.11 and II 4.5). Therefore it is enough to prove the following:

LEMMA. *We have* $\dim A = \max\{k : C_0 \subsetneq \ldots \subsetneq C_k\}$, *where* $C_i \subset A$ *are simple analytic germs.*

PROOF of the lemma. For every such sequence of germs C_i, we have $k \leq \dim A$. This is so because $0 \leq \dim C_0 < \ldots < \dim C_k \leq \dim A$ by proposition 2 from 3.1. Therefore it is enough to prove that any irreducible germ C of dimension $k > 0$ contains a $(k-1)$–dimensional irreducible germ B. Knowing this, we take an irreducible component C_s of the germ A of dimension $s = \dim A$, and we obtain a sequence of irreducible germs $C_s \supsetneq \ldots \supsetneq C_0$. To prove the former statement, one may assume that M is a vector space and $a = 0$. There is a hyperplane $H \not\supset C$ (for otherwise one would have $C \subset 0$). Then $\dim(C \cap H) = k - 1$. Indeed, since $C \cap H \subsetneq C$, we have $\dim(C \cap H) \leq k - 1$ by proposition 2 from 3.1. On the other hand, the inequality (*) from III. 4.6 implies $\dim(C \cap H) \geq k - 1$. So it is sufficient to take as B a simple component of dimension $k - 1$ of the germ $C \cap H$.

PROPOSITION 2. *The germ* A *is smooth of dimension* k *if and only if its ring* $\mathcal{O}_A$ *is regular of dimension* k.

PROOF. The necessity of the condition follows from proposition 2 in I. 1.8 because, if A is the germ of a submanifold Z containing a, then $\mathcal{O}_A \approx \mathcal{O}_a(Z)$ (see n° 2). Assume now that the ring $\mathcal{O}_A \approx \mathcal{O}_a/\mathcal{I}(A)$ is regular of dimension k. Consequently, since the ring $\mathcal{O}_a$ is regular of dimension n (see I. 1.8, proposition 2), theorem 2 from A. 15 implies the existence of generators $f_1, \ldots, f_n$ of the ideal $\mathfrak{m}_a$ such that $f_1, \ldots, f_{n-k}$ generate the ideal $\mathcal{I}(A)$. Let $\tilde{f}_1, \ldots, \tilde{f}_n$ be their holomorphic representatives on an open neighbourhood of the point a. The differentials $d_a \tilde{f}_1, \ldots, d_a \tilde{f}_n$ are linearly independent (see I. 1.8), and so $A = V(\mathcal{I}(A)) = \{\tilde{f}_1 = \ldots = \tilde{f}_{n-k} = 0\}_a$ is the germ of a k–dimensional submanifold containing a (see II. 4.5 and C. 3.15).

3a. Let $A \subset B$ be analytic germs at a point $a \in M$, and assume that A is simple. Let $d_A B$ denote the maximum dimension of simple components of B that contain A. Then the following equality holds

(*) $$d_A B - \dim A = \max\{k : A \subset V_0 \subsetneq \ldots \subsetneq V_k \subset B\},$$

where the V_i are simple germs.

Indeed, this follows as in the proof of the lemma in n° 3. First, in view of $(*)$ in II. 4.2, we get $k \leq d_a B - \dim A$ for every such sequence V_i. Next, it suffices to show that for each simple germ $V \supset A$ of dimension $> \dim A$ there exists a germ $H \supset A$ of codimension 1 such that $H \not\supset V$. Now, for such a germ, we have $\mathcal{I}(V) \subsetneqq \mathcal{I}(A)$ (see II. 4.2 and 5). Thus it is enough to take $H = V(f)$, where $f \in \mathcal{I}(A) \setminus \mathcal{I}(V)$, since codim $H = 1$ (see II. 5.1).

The equality $(*)$, combined with the lemma in A. 12.3, implies

LEMMA $(^{26b})$. *The rings* $(\mathcal{O}_a)_{\mathcal{I}(A)}/\overline{\mathcal{I}(B)} \approx (\mathcal{O}_B)_{\mathcal{I}_B A}$ *are of dimension* $d_A B - \dim A$.

In fact, both rings are isomorphic to $(\mathcal{O}_a/\mathcal{I}(B))_{\mathcal{I}(A)/\mathcal{I}(B)}$. For the image of the ideal $\mathcal{I}(A)/\mathcal{I}(B)$ under the natural isomorphism $\mathcal{O}_a/\mathcal{I}(B) \longrightarrow \mathcal{O}_B$ (see n° 2) is the ideal $\mathcal{I}_B A$. Now, we have the bijection $V \longrightarrow \mathcal{I}(V)$ of the set of simple germs $\{A \subset V \subset B\}$ onto the set of prime ideals $\{\mathcal{I}(B) \subset I \subset \mathcal{I}(A)\}$ (see III. 4.1 and II. 4.5) — notice that inclusion is reversed. Therefore (see A. 1.11 and A. 11.2) the dimension of those rings is $\max\{k : A \subset V_c \subsetneqq \ldots \subsetneqq V_k \subset B\}$, where V_i are simple germs.

4. Let A be a non-empty analytic germ at a point $a \in M$.

We define the *Zariski dimension* of the germ A by the formula

$$\dim\operatorname{zar} A = \operatorname{edim} \mathcal{O}_A = \dim_{\mathbb{C}}(\mathfrak{m}_A/\mathfrak{m}_A^2) \ .$$

Any biholomorphic mapping between analytic sets (see n° 5 below) preserves Zariski's dimension of their germs (see n° 7 below).

Now

$$(**) \hspace{3cm} \dim\operatorname{zar} A = \operatorname{codim} S_A \ ,$$

where $S_A = \{d_a\varphi : \varphi \in \mathcal{I}(A)\}$ is a subspace of the space $(T_a M)^*$. Hence, if $\varphi_1, \ldots, \varphi_r$ generate the ideal $\mathcal{I}(A)$, then

$$(***) \hspace{3cm} \dim\operatorname{zar} A = \dim \ker(d_a\varphi_1, \ldots, d_a\varphi_r) \ .$$

Indeed, according to equality $(*)$ from A. 10.4 (see A. 10.4a), we have (see I. 1.8) $\dim S_A = \operatorname{rank} \mathcal{I}(A) = n - \dim\operatorname{zar} A$, which gives the formula $(**)$.

It follows that

$$(\#) \hspace{3cm} A \subset B \Longrightarrow \dim\operatorname{zar} A \leq \dim\operatorname{zar} B$$

for non-empty analytic germs A, B (see II. 4.2). Moreover,

$$(\#\#) \hspace{3cm} \dim\operatorname{zar} A = \dim A \text{ *if* } A \text{ *is smooth* } .$$

$(^{26b})$ See [7aa].

(Indeed, it is true when A is full, since then $S_A = 0$.)

Furthermore, for non-empty analytic germs $A \subset B$, we have

$$(\#\#\#) \qquad \dim \mathrm{zar}\, B \leq \dim \mathrm{zar}\, A + g(\mathcal{I}_B(A)) \,.$$

To see this, put $r = g(\mathcal{I}_B(A))$. There are germs $\varphi_1, \ldots, \varphi_r \in \mathcal{I}(A)$ such that their equivalence classes generate $\mathcal{I}(A)/\mathcal{I}(B)$ (see n° 2). Hence, if $\psi_1, \ldots, \psi_s$ generate $\mathcal{I}(B)$, then $\varphi_1, \ldots, \varphi_r, \psi_1, \ldots, \psi_s$ generate $\mathcal{I}(A)$ (see A. 1.7). Therefore, in view of $(***)$,

$$\dim \mathrm{zar}\, B = \dim \ker(d_a \psi_1, \ldots, d_a \psi_s) \leq$$
$$\leq \dim \ker(d_a \psi_1, \ldots, d_a \psi_s, d_a \varphi_1, \ldots, d_a \varphi_r) + r$$
$$= \dim \mathrm{zar}\, A + r \quad (^{27}).$$

PROPOSITION 3. *We have*

$$\dim \mathrm{zar}\, A = \min\{\dim \Gamma : \Gamma \supset A\} \quad (^{28}),$$

where $\Gamma \subset M$ are submanifolds.

In fact, set $k = \dim \mathrm{zar}\, A$. Because of $(\#)$ and $(\#\#)$, it is enough to find a submanifold $\Gamma \supset A$ of dimension k. Now, according to $(**)$, we have $\dim S_A = n - k$, hence there exist functions $\varphi_1, \ldots, \varphi_{n-k}$, holomorphic at a, such that $A \subset \{\varphi_i = 0\}$ $(i = 1, \ldots, n - k)$. Notice that the differentials $d_a \varphi_1, \ldots, d_a \varphi_{n-k}$ are linearly independent (see II. 4.2). It follows that for some open neighbourhood U of the point a, the set $\Gamma = \{z \in U : \varphi_1(z) = \ldots = \varphi_{n-k}(z) = 0\}$ is a k–dimensional submanifold containing the germ A (see C. 3.15).

COROLLARY 1. *We have $\dim A \leq \dim \mathrm{zar}\, A \leq \dim M$.*

COROLLARY 2. *A germ A is smooth if and only if $\dim \mathrm{zar}\, A = \dim A$.* (See II. 3.3.)

PROPOSITION 4. *We have the equivalence*

$$\dim \mathrm{zar}\, A = \dim M \iff \mathcal{I}(A) \subset \mathfrak{m}_a^2 \,.$$

Indeed, in view of $(*)$ in A. 10.4 (see A. 10.4a), we have $\dim \mathrm{zar}\, A = \mathrm{edim}\, \mathcal{O}_a/\mathcal{I} = \mathrm{codim}_{\mathfrak{m}_a}(\mathcal{I}(A) + \mathfrak{m}_a^2)$. But $\dim M = \dim \mathrm{zar}\, M_a = \mathrm{codim}_{\mathfrak{m}_a} \mathfrak{m}_a^2$,

---

$(^{27})$ Since, for subspaces L, N of a vector space, we have $\dim L \leq \dim(L \cap N) + \mathrm{codim}\, N$.

$(^{28})$ It is interesting to note that the right hand side does not depend on the way in which the germ A (i.e., its representative) is immersed in a manifold.

hence (see A. 1.17) both conditions are equivalent to the equality $\mathcal{I}(A) + \mathfrak{m}_a^2 = \mathfrak{m}_a^2$.

The subspace $T_A^{zar} = S_A^\perp = \bigcap\{\ker d_a\varphi : \varphi \in \mathcal{I}(A)\} \subset T_a M$ – or the space $(\mathfrak{m}_A/\mathfrak{m}_A^2)^*$, which is naturally isomorphic to the former one ([28a]) – is called the *Zariski tangent space* to the germ A. In accordance with (**), we have

$$\dim \operatorname{zar} A = \dim T_A^{zar}.$$

Thus, combining the implication $A \subset B \Longrightarrow T_A^{zar} \subset T_B^{zar}$ and the equality $T_{\Gamma_a}^{zar} = T_a \Gamma$ for any submanifold $\Gamma \ni a$ (see (##), and C. 3.11 and 15) with proposition 3, one concludes that:

The tangent space at the point a to any submanifold of minimal dimension containing the germ A coincides with the Zariski tangent space T_A^{zar}.

5. Let N be a k–dimensional manifold. According to the definition of holomorphic mappings between locally analytic subsets – which will be given in Chapter V (see V. 3.1) – a mapping $f : V \longrightarrow W$ of locally analytic subsets $V \subset M$, $W \subset V$ is called *holomorphic* if each point of the set V has an open neighbourhood U in M such that $f_{V \cap U}$ is the restriction of a holomorphic mapping of U into N. Such a mapping f is, obviously, continuous. If $Z \subset V$ is a locally analytic set, then the restriction $f : Z \longrightarrow W$ is also holomorphic. The composition of holomorphic mappings is holomorphic. The diagonal product of holomorphic mappings is holomorphic.

A mapping $f : V \longrightarrow W$ is said to be *biholomorphic* (see V. 3.4) if it is bijective and the mappings f, f^{-1} are holomorphic. Naturally, such a mapping is a homeomorphism. The inverse mapping f^{-1} is also biholomorphic. If $Z \subset V$ is a locally analytic set, then $f(Z)$ is a locally analytic set and the restriction $f_Z : Z \longrightarrow f(Z)$ is a biholomorphic mapping.

In the case when V and W are submanifolds, the above definitions coincide with the previously given definitions of holomorphic and biholomorphic mappings of manifolds. (See C. 3.8 and 10.)

6. Let A be an analytic germ at a point a of the manifold M. A germ $f : A \longrightarrow N$ is said to be *holomorphic* if it has a holomorphic representative (on a locally analytic representative of the germ A). If $C \subset A$ is an analytic germ, then f_C is a holomorphic germ. In the case of the full germ $A = M_a$, we define $d_a f = d_a \tilde{f}$, where $\tilde{f}$ is a representative of the germ f. (This definition agrees with that concerning the case $N = \mathbf{C}$; see I. 1.2.) The composition of

---

([28a]) This is because of the natural isomorphisms $S_A^\perp \approx ((T_a M)^*/S_A)^*$ and $\mathfrak{m}_A/\mathfrak{m}_A^2 \approx (T_a M)^*/S_A$ which are induced by the linear mapping $\mathfrak{m}_A \longrightarrow (T_a M)^*/S_A$ sending f_A onto the equivalence class of $d_a f$.

holomorphic germs $f : A \longrightarrow B$ and $g : B \longrightarrow L$, i.e., the germ $g \circ f : A \longrightarrow L$, where B is an analytic germ at the point $b = f(a) \in N$ and L is a manifold, is holomorphic. Moreover, if A and B are full, then $d_a(g \circ f) = d_b g \circ d_a f$. The diagonal product of holomorphic germs $f_i : A \longrightarrow N_i$, where $N_1, \ldots, N_r$ are manifolds, i.e., the germ $(f_1, \ldots, f_r) : A \longrightarrow N_1 \times \ldots \times N_r$, is holomorphic, and moreover, $d_a(f_1, \ldots, f_r) = (d_a f_1, \ldots, d_a f_r)$ if A is full. Obviously, the germ e_A is holomorphic. Note that $e_{(\mathbf{C}^n)_0} = (z_1, \ldots, z_n)$. If $\varphi_1, \ldots, \varphi_r \in \mathfrak{m}_A$, then $(\varphi_1, \ldots, \varphi_r) : A \longrightarrow (\mathbf{C}^r)_0$ and $z_i \circ (\varphi_1, \ldots, \varphi_r) = \varphi_i$, $i = 1, \ldots, r$.

Let A and B be analytic germs at points $a \in M$ and $b \in N$, respectively. A *biholomorphic germ* of the germ A onto the germ B is defined as the germ f on A of a biholomorphic mapping of a locally analytic representative of the germ A onto a locally analytic representative of the germ B, such that $f(a) = b$ when the germs A, B are non-empty; it is, of course, holomorphic. Obviously, it is the germ of a homeomorphism and f^{-1} is a biholomorphic germ. If such a biholomorphic germ exists, we say that the *germs A and B are biholomorphic* or *isomorphic* and we write $A \approx B$. (Naturally, this is an equivalence relation.) If f is a biholomorphic germ of A onto the germ B and $C \subset A$ is an analytic germ, then $D = f(C)$ is an analytic germ, and f_C is a biholomorphic germ of the germ C onto the germ D. If for some analytic germs $D \subset A$, $C \subset B$ such a germ f exists, we write $(A, C) \approx (B, D)$. (This is also an equivalence relation.) The composition of biholomorphic germs (of A onto A' and of A' onto A'') is a biholomorphic germ (of A onto A'').

If $f : A \longrightarrow B$ and $g : B \longrightarrow A$ are holomorphic germs such that $g \circ f = e_A$ and $f \circ g = e_B$, then f and g are mutually inverse germs of biholomorphic mappings of A onto B and of B onto A, respectively.

Indeed, there is a homeomorphism F of a representative $\tilde{A}$ of the germ A onto a representative $\tilde{B}$ of the germ B such that F, F^{-1} are representatives of the germs f, g, respectively (see B. 4.2). Moreover, there are locally analytic representatives $\tilde{\tilde{A}} \subset \tilde{A}$, $\tilde{\tilde{B}} \subset \tilde{B}$ such that the mappings $F_{\tilde{\tilde{A}}}$, $(F^{-1})_{\tilde{\tilde{B}}}$ are holomorphic. Omitting the trivial case of empty germs, take a neighbourhood $U \subset \tilde{\tilde{A}}$ of the point a such that U is open in $\tilde{A}$ and $V = F(U) \subset \tilde{\tilde{B}}$. Then V is a neighbourhood of the point b and V is open in $\tilde{B}$. Furthermore, U and V are open in $\tilde{\tilde{A}}$ and $\tilde{\tilde{B}}$, respectively. Hence they are locally analytic representatives of the germs A and B, respectively. Moreover, the mappings $F_U : U \longrightarrow V$, $(F_U)^{-1} = (F^{-1})_V : V \longrightarrow U$ are holomorphic.

For every holomorphic germ $f : M_a \longrightarrow N_b$, we have the equivalence (see C. 3.13):

$$\left(f \text{ is a biholomorphic germ (of } M_a \text{ onto } N_b) \right) \Longleftrightarrow$$

$$\Longleftrightarrow \left(d_a f : T_a M \longrightarrow T_b N \text{ is an isomorphism} \right).$$

Generally (see II. 4.2 and C.3.14), the germs $\varphi_1, \dots, \varphi_p$ generate $\mathfrak{m}_a$ if and only if $(\varphi_1, \dots, \varphi_p) : M_a \longrightarrow (\mathbf{C}^p)_0$ is an immersion germ, i.e., a biholomorphic germ onto a smooth germ.

Finally, we have the following, easy to verify lemma.

LEMMA 1. *If* $\dim M = \dim N$ *and the germs* A, B *are smooth, then every biholomorphic germ of* A *onto* B *is the restriction* $F_A : A \longrightarrow B$ *of a biholomorphic germ* $F : M_a \longrightarrow N$.

7. Let A, B be non-empty analytic germs at points $a \in M$, $b \in N$, respectively.

For every holomorphic germ $f : A \longrightarrow B$, we have the $\mathbf{C}$–homomorphism of local $\mathbf{C}$–rings $f^* : \mathcal{O}_B \ni \psi \longrightarrow \psi \circ f \in \mathcal{O}_A$. Obviously we have $(g \circ f)^* = f^* \circ g^*$ (if $g : B \longrightarrow D$ is a holomorphic germ). Naturally, e_A^* is the identity mapping of $\mathcal{O}_A$. Hence, if f is a biholomorphic germ of A onto B, then f^* is a $\mathbf{C}$–isomorphism of $\mathcal{O}_B$ onto $\mathcal{O}_A$ and $(f^{-1})^* = (f^*)^{-1}$. Therefore, if $A \approx B$, then $\mathcal{O}_A \overset{\mathbf{C}}{\approx} \mathcal{O}_B$ and $\dim \operatorname{zar} A = \dim \operatorname{zar} B$. If $C \subset A$ is a non-empty analytic germ, then we have the $\mathbf{C}$–epimorphism $(e_C^A)^* : \mathcal{O}_A \ni \varphi \longrightarrow \varphi_C \in \mathcal{O}_C$ (see n° 1).

LEMMA 2. *For holomorphic germs* $f : A \longrightarrow B$ *and* $g : A \longrightarrow B$, *we have the implications* $f^* = g^* \Longrightarrow f = g$.

PROOF . Taking a germ $\Psi = (\psi_1, \dots, \psi_k) : N_b \longrightarrow (\mathbf{C}^k)_0$, where $\psi_i \in \mathcal{O}_b$, we have $\psi_i \circ f = (\psi_i)_B \circ f = (\psi_i)_B \circ g = \psi_i \circ g$. Therefore $\Psi \circ f = \Psi \circ g$, which implies $f = g$.

LEMMA 3. *Let* $F : M_a \longrightarrow N_b$ *be a holomorphic germ. If* $\dim \operatorname{zar} A = \dim M$ *and* $F_A^* : \mathcal{O}_b \longrightarrow \mathcal{O}_A$ *is an epimorphism, then* F *is an immersion germ (i.e., a biholomorphic germ onto a smooth germ at* b).

PROOF . Consider generators $\psi_1, \dots, \psi_k$ of the ideal $\mathfrak{m}_b$. Then $\Psi = (\psi_1, \dots, \psi_k) : N_b \longrightarrow (\mathbf{C}^k)_0$ is a biholomorphic germ (see n° 6). Since F_A^* is an epimorphism, we have $F_A^*(\mathfrak{m}_b) = \mathfrak{m}_A$ (see A. 10.1 and 4a), and therefore the germs $\psi_i \circ F_A = (\varphi_i)_A$, where $\varphi_i = \psi_i \circ F \in \mathfrak{m}_a$, generate the ideal $\mathfrak{m}_A$. These germs correspond under the isomorphism $\mathcal{O}_a/\mathcal{I}(A) \longrightarrow \mathcal{O}_A$ (see n° 2) to the equivalence classes of the germs φ_i; hence the latter generate the ideal $\mathfrak{m}_a/\mathcal{I}(A)$. This means that $\mathfrak{m}_a \subset \sum_1^k \mathcal{O}_a \varphi_i + \mathcal{I}(A)$. Now, by proposition 4, we have $\mathcal{I}(A) \subset \mathfrak{m}_a^2$; it follows that, by Nakayama's lemma, $\mathfrak{m}_a \subset \sum_1^k \mathcal{O}_a \varphi_i$, i.e., $\varphi_1, \dots, \varphi_k$ generate $\mathfrak{m}_a$. Consequently, $\Phi = (\varphi_1, \dots, \varphi_k) : M_a \longrightarrow (\mathbf{C}^k)$

is an immersion germ (see n° 6). But $\Phi = \Psi \circ F$, and hence $F = \Psi^{-1} \circ \Phi$ is an immersion germ.

PROPOSITION 5. *If* $\dim \operatorname{zar} A = \dim M$, *then any holomorphic germ* $F : M_a \longrightarrow N_b$ *whose restriction* F_A *is a biholomorphic germ must be an immersion germ. If* $\dim M = \dim N$, *then every biholomorphic germ of the germ* A *onto the analytic germ* B *is the restriction* $F_A : A \longrightarrow B$ *of a biholomorphic germ* $F : M_a \longrightarrow N_b$.

PROOF . In view of lemma 1 and proposition 3, it is enough to prove the first part of the proposition. Now F_A is a biholomorphic germ onto some germ B, hence $(F_A^B)^* : \mathcal{O}_B \longrightarrow \mathcal{O}_A$ is an isomorphism. But $F_A = \varrho_B \circ F_A^B$, and so $F_A^* = (F_A^B)^* \circ \varrho_B^* : \mathcal{O}_b \longrightarrow \mathcal{O}_A$ is an epimorphism. In view of lemma 3, the proof is complete.

LEMMA 4. *Let* $f : A \longrightarrow B$ *be a holomorphic germ, and let* $C \subset A$, $D \subset B$ *be analytic germs. Then* $f : C \longrightarrow D \iff f^*(\mathcal{I}_B(D)) \subset \mathcal{I}_A(C)$. *If, in addition,* f *is a biholomorphic germ of* A *onto* B, *then* $f(C) = D \iff f^*(\mathcal{I}_B(D)) = \mathcal{I}_A(C)$.

PROOF . The second part follows from the first one, as then f^* is an isomorphism and $(f^*)^{-1} = (f^{-1})^*$. Suppose that $f_C : C \longrightarrow D$. Then, if $\psi \in \mathcal{I}_B(D)$, we have $\psi_D = 0$. Hence $(\psi \circ f)_C = \psi_D \circ f_C = 0$, i.e., $f^*(\psi) \in \mathcal{I}_A(C)$. Conversely, let $\mathcal{I}_B(D) \circ f \subset \mathcal{I}_A(C)$. For some locally analytic representatives $\tilde{A}, \tilde{B}, \tilde{D}$ of the germs A, B, D, respectively, we have $\tilde{B} \supset \tilde{D} = \{\tilde{g}_i = \ldots = \tilde{g}_s = 0\}$, where $\tilde{g}_i$ are holomorphic on $\tilde{B}$. Therefore there is a holomorphic representative $\tilde{f} : \tilde{A} \longrightarrow \tilde{B}$ of the germ f. Then $g_i = (\tilde{g}_i)_B \in \mathcal{I}_B(D)$ (since $(g_i)_D = ((\tilde{g}_i)_{\tilde{D}})_D = 0$). Thus $g_i \circ f \in \mathcal{I}_A(C)$, which means that $(g_i \circ f)_C = 0$. Therefore $\tilde{g}_i \circ \tilde{f} = 0$ on some representative $\tilde{C} \subset \tilde{A}$ of the germ C; hence $\tilde{f}(\tilde{C}) \subset \tilde{D}$, and so $f : C \longrightarrow D$ (see B. 4.2).

PROPOSITION 6. *Every* **C**-*homomorphism* $h : \mathcal{O}_B \longrightarrow \mathcal{O}_A$ *is of the form* $h = f^*$, *where* $f : A \longrightarrow B$ *is a holomorphic germ.*

PROOF . Let us take generators $\psi_1, \ldots, \psi_k$ of the ideal $\mathfrak{m}_b$. Then $\Psi = (\psi_1, \ldots, \psi_k) : N_b \longrightarrow (\mathbf{C}^k)_0$ is the germ of a biholomorphic mapping (see n° 6). We have

$$h((\psi_i)_B) = (\varphi_i)_A, \quad \text{for some } \varphi_i \in \mathfrak{m}_a ,$$

because ϱ_B^* and h are local homomorphisms and $\varrho_A^*(\mathfrak{m}_a) = \mathfrak{m}_A$ (see A. 10.1 and 4a). Consider the holomorphic germs $\Phi = (\varphi_1, \ldots, \varphi_k) : M_a \longrightarrow (\mathbf{C}^k)_0$ and $F = \Psi^{-1} \circ \Phi : M_a \longrightarrow N_b$. Now, we have the equality

$$\varrho_A^* \circ F^* = h \circ \varrho_B^* .$$

This follows from the proposition in A. 10.5, because both sides are **C**–homomorphisms $\mathcal{O}_b \longrightarrow \mathcal{O}_A$ that coincide on ψ_i (since we have $\psi_i \circ F = z_i \circ \Psi \circ F = z_i \circ \Phi = \varphi_i$, and so $(\psi_i \circ F)_A = (\varphi_i)_A = h\big((\psi_i)_B\big)$.) In particular, $\varrho_A^*\Big(F^*\big(\mathcal{I}(B)\big)\Big) = 0$, that is $F^*\big(\mathcal{I}(B)\big) \subset \ker \varrho_A^* = \mathcal{I}(A)$. Therefore $f = F_A : A \longrightarrow B$, by lemma 4. Now, since $\varrho_B \circ f = F \circ \varrho_A$, we conclude that $h \circ \varrho_B^* = \varrho_A^* \circ F^* = f^* \circ \varrho_B^*$. But ϱ_B^* is an epimorphism, and so $h = f^*$.

Proposition 6, combined with lemma 2, implies:

COROLLARY 1. *The mapping* $f \longrightarrow f^*$ *is a bijection of the set of all holomorphic germs* $A \longrightarrow B$ *onto the set of all* **C**–*homomorphisms* $\mathcal{O}_B \longrightarrow \mathcal{O}_A$; *moreover, biholomorphic germs correspond precisely to the* **C**–*isomorphisms* [29] . *We have*

$$A \approx B \Longleftrightarrow \mathcal{O}_A \overset{C}{\approx} \mathcal{O}_B .$$

If $C \subset A$ and $D \subset B$, the above corollary and lemma 4 yield:

COROLLARY 2. *We have*

$$(A,C) \approx (B,D) \Longleftrightarrow \big(\mathcal{O}_A, \mathcal{I}_A(C)\big) \overset{C}{\approx} \big(\mathcal{O}_B, \mathcal{I}(D)\big) \quad [30] .$$

Finally, in view of proposition 5, we have

COROLLARY 3. *If* $\dim M = \dim N$, *then*

$$A \approx B \Longleftrightarrow (M_a, A) \approx (N_b, B) \Longleftrightarrow \mathcal{O}_A \overset{C}{\approx} \mathcal{O}_B \Longleftrightarrow$$

$$\Longleftrightarrow \big(\mathcal{O}_a, \mathcal{I}(A)\big) \overset{C}{\approx} \big(\mathcal{O}_b, \mathcal{I}(B)\big) .$$

§5. The maximum principle

Let V be an analytic subset of the manifold M.

THEOREM 1. *If a holomorphic function* f *on* V *is non-constant on every neighbourhood of a point* a *in* V, *then the set* $f(V)$ *is a neighbourhood*

---

[29] In fact, if f^* is an isomorphism, then f is a biholomorphic germ of A onto B. Indeed, we then have $(f^*)^{-1} = g^*$ for some holomorphic germ $g : B \longrightarrow A$, and so $(g \circ f)^* = f^* \circ g^* = e_A^*$, i.e., $g \circ f = e_A$ (by lemma 2). Similarly, $g \circ f = e_B$. This means that f is a biholomorphic germ of A onto B (see n° 6).

[30] That is, there exists a **C**–isomorphism $h : \mathcal{O}_A \longrightarrow \mathcal{O}_B$ such that $h\big(\mathcal{I}_A(C)\big) = \mathcal{I}_B(D)$.

of $f(a)$ in $\mathbf{C}$. *(Then $f(W)$ is, obviously, a neighbourhood of $f(a)$ for any neighbourhood W of a in V.)*

PROOF . First note that there is an open neighbourhood U of the point a and analytic subsets $V_1, \ldots, V_r$ of U such that the germs $(V_i)_a$ are simple and $V \cap U = V_1 \cup \ldots \cup V_r$. Then, for some s, the function f_{V_s} is non-constant on any neighbourhood of a in V_s. Thus we may assume that the germ V_a is irreducible. Furthermore, one may assume that M is an open neighbourhood of $a = 0$ in a vector space N, f is a holomorphic function on M, and $f(a) = 0$. Set $n = \dim N$ and $k = \dim V_0$. According to the assumption, $C = V_0 \cap \{f = 0\} \subsetneq V_0$. Hence $\dim C \leq -1$, by proposition 2 from 3.1. Therefore, by formula $(\#)$ from III. 4.5, there is an $(n - (k-1))$–dimensional subspace $L \subset N$ such that $C \cap L = 0$. So $\{f_W = 0\} \cap L = 0$ for some neighbourhood W of 0 in V; i.e., in the space $N \times \mathbf{C}$, we have $f_W \cap (L \times 0) = 0$, which gives

$$f_0 \cap (L \times 0) = 0 .$$

We have the direct sum $N \times \mathbf{C} = (L' \times \mathbf{C}) + (L \times 0)$, where L' is a linear complement of L. Moreover, $\dim(L' \times \mathbf{C}) = k$ and $\dim f_W = k$ (as the natural projection $f \longrightarrow M$ is biholomorphic and it maps the set f onto V, and hence the image of the germ f_0 is V_0; see C. 3.17 and II. 1.6). Therefore (see III. 4.5) the image of the set f by projection onto $L' \times \mathbf{C}$ is a neighbourhood of zero. The image of this neighbourhood by the projection onto $\mathbf{C}$ is a neighbourhood of zero in $\mathbf{C}$ (see B. 2.1) and coincides with the image of the set f by the projection onto $\mathbf{C}$, that is, with the set $f(V)$.

COROLLARY . *If the set V is irreducible, then any holomorphic function f on V is either constant or is an open mapping* ([31]) .

Indeed, if the function f is constant in a neighbourhood of a point of V, then – in view of proposition 1 from 2.1 and since V^0 is a connected submanifold (see 2.8, corollary 1 from proposition 2) – the function f is constant on V^0 (see C. 3.9) and hence on V. Otherwise, according to theorem 1, the function f is an open mapping.

Therefore we have:

PROPOSITION . *Every compact analytic subset of an affine space is finite.*

In fact, it is enough to consider the case of the space $\mathbf{C}^n$. Assume first that V is a compact irreducible analytic subset of $\mathbf{C}^n$. Now the function $V \ni z \longrightarrow z_i \in \mathbf{C}$ $(i = 1, \ldots, n)$ is not an open mapping, as the image of the set V is a non-empty compact set in $\mathbf{C}$. Therefore the function must be

---

([31]) For connected analytic sets, this alternative is – in general – not true. For instance, the function $(z, w) \longrightarrow w$ on the set $\{z = 0\} \cup \{w = 0\} \subset \mathbf{C}^2$ is neither constant nor open.

constant (for $i = 1, \ldots, n$), and so the set V consists of a single point. This completes the proof, as the number of simple components of any compact analytic set is finite.

As an application, we will prove the following fixed point theorem for holomorphic mappings (see Hervé [24], p.83):

Let G be an open connected subset of an affine space X. If $f : G \longrightarrow G$ is a holomorphic mapping whose range $f(G)$ is relatively compact in G ([32]), then the mapping has a unique fixed point.

PROOF . We may assume that $X = \mathbf{C}^n$. The set $E = \overline{f(G)} \subset G$ is compact. Consider the sequence of iterations $f_n = \underbrace{f \circ \ldots \circ f}_{n} : G \longrightarrow G$. We have $f_n(G) \subset E$. Now, we apply Montel's theorem twice (see C. 1.9): there exists a subsequence f_{n_ν} which converges (almost uniformly) to a holomorphic mapping $f' : G \longrightarrow G$ and such that $n_{\nu+1} - n_\nu \longrightarrow \infty$. Next, there is a subsequence $f_{n_{\alpha_\nu+1} - n_{\alpha_\nu}}$ that converges (almost uniformly) to a holomorphic mapping $f'' : G \longrightarrow G$, and $f''(G) \subset E$. Since $f_{n_{\alpha_\nu+1}} = f_{n_{\alpha_\nu+1} - n_{\alpha_\nu}} \circ f_{n_{\alpha_\nu}}$, we conclude (by taking limits) that $f' = f'' \circ f'$. Thus $f'(G) \subset \{f''(z) = z\} \subset E$, and so – in view of the proposition – the compact subset $\{f''(z) = z\} \subset \mathbf{C}^n$ must be finite. Hence the set $f'(G)$, which is connected and finite, consists of a single point: $f'(z) = z_0$ in G. Now, z_0 is the unique fixed point of the mapping f. Indeed, by taking limits, we have $f \circ f' = f' \circ f$, and therefore $f(z_0) = f(f'(z_0)) = f'(f(z_0)) = z_0$. Finally, if $f(z) = z$, then $f_n(z) = z$, and by passing to the limit, we obtain $z = f'(z) = z_0$.

THEOREM 2. *If V is connected, then every holomorphic function f on V such that $|f(z)|$ attains its maximum must be constant ([33]) .*

PROOF . Suppose that $|f(z)|$ attains its maximum at $a \in V$. If the set V is irreducible, the function f is not constant, by the corollary from theorem 1. Indeed, f is not an open mapping: since $f(V) \subset \{w : |w| \leq |f(a)|\}$, the set $f(V)$ is not a neighbourhood of the point $f(a)$. Passing to the general case, set $c = f(a)$ and take any point $z_0 \in V$. The points a, z_0 belong to some simple components V', V'' of the set V. According to proposition 5 from 2.9, there is a sequence $V' = V_0, \ldots, V_r = V''$ of simple components of the set V such that $V_{i-1} \cap V_i \neq \emptyset$, $i = 1, \ldots, r$. As $|f_{V_0}(z)|$ attains its maximum at the point a, we have $f(z) = c$ on V_0. But, if $f(z) = c$ on V_{i-1} (where $0 < i \leq r$), then the same is true on V_i. Indeed, by taking $b \in V_{i-1} \cap V_i$, we have $f(b) = c$, and so $|f_{V_i}(z)|$ attains its maximum at the point b. Therefore $f(z) = c$ on V_i. Hence $f(z) = c$ on V_r and, in particular, $f(z_0) = c$.

([32]) That is, the set $\overline{f(G)}$ is contained in G and is compact.

([33]) If the set V is irreducible, then theorem 2 is a direct consequence of the corollary from theorem 1 (see the first part of the proof of theorem 2). The general case requires a different approach (see footnote ([31])).

§6. The Remmert-Stein removable singularity theorem

1. Let M be a p–dimensional manifold, and let X be a vector space.

LEMMA . *If $W \subset M \times X$ is an analytic subset of constant dimension p such that the natural projection $\pi : W \longrightarrow M$ is proper and if Σ is the set of points of W at which W is not a topographic submanifold, then the set $\pi(\Sigma)$ is thin in M and the set $\pi^{-1}(\pi(\Sigma))$ is nowhere dense in W.*

PROOF . First of all we are going to prove that

(∗) every point $c \in W$ has an open neighbourhood U such that the set $\Sigma \cap U$ is nowhere dense in $W \cap U$ and the set $\pi(\Sigma \cap U)$ is thin in $\tilde\pi(U)$.

Here $\tilde\pi : M \times X \longrightarrow M$ is the natural projection. By passing to the images under the chart $\varphi \times \psi$, where φ is a coordinate system of M at the point a, $c = (a, b)$, and ψ is an affine coordinate system in X at b, one may assume that M is an open neighbourhood of zero in $\mathbf{C}^p$, $X = \mathbf{C}^{n-p}$ (for some n), and $c = 0$ (see C. 3.17). Now, according to the proposition from §5, the set $W \cap \{z_1 = \ldots = z_p = 0\} = \pi^{-1}(0)$ is finite, and so $W_0 \cap \{z_1 = \ldots = z_p\} = 0\} = 0$. Therefore the germ W_0 is p–regular (see the corollary from III. 4.4). Since it is of constant dimension p (see 3.1, proposition 4), it has a normal triple (Ω, Z, V) such that $V = W \cap U$, where U is an open neighbourhood of zero (see 1.4, proposition 1; and 1.2); then $\tilde\pi(U) = \Omega$ (see 1.3). Now, $\Sigma \cap U$ is the set of points of V at which the subset V is not a topographic submanifold. Hence the set $\Sigma \cap U$ is closed in V (see C. 3.17) and contained in V_Z (see 1.1, property (3)). Therefore it is nowhere dense in $V = W \cap U$ (see 1.1, property (4)). Since the mapping $\pi_V : V \longrightarrow \Omega$ is proper (see 1.1, property (1)), the set $\pi(\Sigma \cap U)$ is closed in Ω (see B. 2.4), and so – since it is a subset of Z – it must be thin in $\Omega = \tilde\pi(U)$ (see II. 2.3).

Now, let us take any point $a \in M$. By the proposition from §5, the set $\pi^{-1}(a) = W \cap (a \times X)$ is finite: $\pi^{-1}(a) = \{c_1, \ldots, c_r\}$. Let U_i be a neighbourhood of c_i chosen accordingly to (∗) ($i = 1, \ldots, r$). We have $\pi^{-1}(a) \subset U_1 \cup \ldots \cup U_r$, hence there is an open neighbourhood $\Delta \subset \bigcap_1^r \tilde\pi(U_i)$ of the point a such that $W_\Delta = \pi^{-1}(\Delta) \subset U_1 \cup \ldots \cup U_r$ (see B. 2.4). Then $\pi(\Sigma) \cap \Delta = (\pi(\Sigma \cap U_1) \cup \ldots \cup \pi(\Sigma \cap U_r)) \cap \Delta$. Therefore the set $\pi(\Sigma) \cap \Delta$ is thin in Δ (see II 2.3). Consequently, the set $\pi(\Sigma)$ is thin in M (see II. 2.3). It follows from (∗) that Σ is nowhere dense in W (see B. 1). Since the mapping $\pi_{W \setminus \Sigma} : W \setminus \Sigma \longrightarrow M$ is a local homeomorphism (see C. 3.17), the set $\pi^{-1}(\pi(\Sigma)) \setminus \Sigma = (\pi_{W \setminus \Sigma})^{-1}(\pi(\Sigma))$ is nowhere dense in $W \setminus \Sigma$ (see B. 2.2 and B. 3.1). Hence the set $\pi^{-1}(\pi(\Sigma))$ is nowhere dense in W.

Let Z be a thin subset of the manifold M.

PROPOSITION . *If W is an analytic subset of $(M \setminus Z) \times X$ of constant dimension p (recall that $p = \dim M$) such that the natural projection $\overline{W} \longrightarrow M$ is proper, then the set $\overline{W}$ is an analytic subset of $M \times X$.*

PROOF . We may assume that the manifold M is connected. Let Σ be the set of points of W at which W is not a topographic submanifold. As the projection $\pi : W \longrightarrow M \setminus Z$ is also proper (see B. 2.4), the lemma implies that $Z \cup \pi(\Sigma)$ is thin in M (see II. 2.3) and $\pi^{-1}(\pi(\Sigma))$ is nowhere dense in W. Now the set $W_0 = W \setminus \pi^{-1}(\pi(\Sigma)) = W \cap \left((M \setminus Z \setminus \pi(\Sigma)) \times X \right)$ is a locally topographic closed submanifold of $\left(M \setminus (Z \cup \pi(\Sigma)) \right) \times X$ and $\overline{W}_0 = \overline{W}$. Accordingly, the pair $(Z \cup \pi(\Sigma), W_0)$ is a quasi-cover in the Cartesian product $M \times X$. Thus, by the first lemma on quasi-covers (see III. 1.3), the set $\overline{W} = \overline{W}_0$ is an analytic subset of $M \times X$.

2. We have the following

LEMMA 1. *If W is a locally analytic set of dimension $\leq p$ in an n-dimensional vector space M, where $0 \leq p \leq n$, then the set $W \cap L$ is discrete for each L from a dense subset of $\mathbf{G}_{n-p}(M)$.*

Indeed, this is the case if $p = 0$ (see III. 4.3). One may assume (omitting trivial cases) that $n > 1$. Let $p > 0$, and suppose the lemma holds for $p - 1$. Therefore, if for a hyperplane H

$$(*) \qquad\qquad \dim(W \cap H) \leq p - 1 ,$$

the set $W \cap L$ is discrete for each L from a dense subset of $\mathbf{G}_{n-p}(H)$. Thus it suffices to show that the condition $(*)$ is satisfied for every H from a dense subset of $\mathbf{G}_{n-1}(M)$ (see B. 6.7). Now let $\{W_i\}$ be the family of simple components of dimension > 0 of W. Each of the sets $\{H \in \mathbf{G}_{n-1}(M) : H \supset W_i\}$ is nowhere dense in $\mathbf{G}_{n-1}(M)$ (because it is contained in $\mathbf{S}^{n-1}(\lambda_i)$, where λ_i is a straight line that intersects $W_i \setminus 0$; see B. 6.5). Thus the complement Θ of the union of these sets is dense in the space $\mathbf{G}_{n-1}(M)$, as the latter is a Baire space, because it is compact (see B. 6.1). Now, if $H \in \Theta$, then for each i we have $W_i \cap H \subsetneq W_i$, and so $\dim(W_i \cap H) < \dim W_i \leq p$ (see 2.8, proposition 3). Thus $\dim(W_i \cap H) \leq p - 1$, and since the same estimate holds for the remaining simple components of W, we have $\dim(W \cap H) \leq p - 1$ (see II. 1.3).

We say that a *subset E of the manifold M is analytic on an open subset G of M* if $E \cap G$ is analytic in G. If $\{G_\iota\}$ is a family of open subsets of the manifold M, then a subset $E \subset M$ that is analytic on each G_ι is also analytic on $\bigcup G_\iota$ (see II. 3.4).

LEMMA 2. *Let Ω be an open neighbourhood of zero in an n–dimensional vector space M, and let $S \subsetneqq M$ be a subspace. If W is an analytic subset of $\Omega \setminus S$ of constant dimension p, where $0 \leq p \leq n$, such that there is an $(n-p)$–dimensional subspace $L \subset S$ for which the set $\overline{W} \cap \Omega \cap L \setminus 0$ is discrete, then the set $\overline{W}$ is analytic on an open neighbourhood of zero.*

PROOF . Take a linear complement K in M of the subspace L. It is of dimension p. Set $N = S \cap K$. Then $S = N + L$ and $N \subsetneqq K$. Replacing M, S, Ω, and W by their images under the natural isomorphism $M \longrightarrow K \times L$, we may assume that $M = K \times L$, $S = N \times L$, and the set $\overline{W} \cap \Omega \cap (0 \times (L \setminus 0))$ is discrete. Thus (after choosing a norm in L) one can find $\varepsilon > 0$ and an open neighbourhood U of zero in K such that $U \times \{|v| < 2\varepsilon\} \subset \Omega$ and $(U \times \{|v| = \varepsilon\}) \cap \overline{W} = \emptyset$. Now, the set

$$W_0 = W \cap (U \times \{|v| < \varepsilon\}) = W \cap ((U \setminus N) \times \{|v| \leq \varepsilon\})$$

is analytic in $(U \setminus N) \times L$, for it is locally analytic and it is closed in $(U \setminus N) \times L$ (see II. 3.4). It is of constant dimension p and the natural projection $\overline{W}_0 \cap (U \times L) \longrightarrow U$ is proper (see B. 5.2). Therefore, according to the proposition in n° 1, the set $\overline{W}_0 \cap (U \times L)$ is analytic in $U \times L$, and hence the set $\overline{W} \cap (U \times \{|v| < \varepsilon\}) = \overline{W}_0 \cap (U \times \{|v| < \varepsilon\})$ is analytic in $U \times \{|v| < \varepsilon\}$.

LEMMA 3. *Let Ω be an open neighbourhood of zero in an n–dimensional vector space M, and let $N \subset M$ be a subspace of dimension $< p$, where $0 < p \leq n$. If W is an analytic subset of $\Omega \setminus N$ of constant dimension p, then its closure $\overline{W}$ is analytic on some neighbourhood of zero.*

PROOF . Since $\dim N < p$, there are $(n - p)$–dimensional subspaces $L_1, \ldots, L_r \subset M$ such that

$$(**) \quad N \cap L_i = 0 \ (i = 1, \ldots, r) \quad \text{and} \quad \bigcap_1^r (N + L_i) = N \quad (^{34}) .$$

In view of lemma 1, one may require that the sets $W \cap L_i$ are discrete since the set of r–tuples $(L_1, \ldots, L_r)$ satisfying the condition $(**)$ is open in $\mathbf{G}_{n-p}(M)^r$ (see B. 6.2 and B. 6.6 $(^{35})$). Now the set $\tilde{W} = W \cup (N \cap \Omega)$ is closed in Ω and

---

$(^{34})$ It suffices to take $(n - p)$–dimensional subspaces L_i of a linear complement of the subspace N such that $\bigcap_1^r L_i = 0$. (The dimension of the complement is $> n - p$.)

$(^{35})$ Because the condition $\bigcap_1^r (N + L_i) = N$ is equivalent to the condition $\dim \bigcap_1^r (N + L_i) \leq \dim N$.

the sets $\tilde{W} \cap L_i \setminus 0 = W \cap L_i$ are discrete. Consequently, $\overline{W \setminus (N + L_i)} \cap \Omega \subset \tilde{W}$ and the sets $\overline{W \setminus (N + L_i)} \cap \Omega \cap L_i \setminus 0$ are also discrete. Therefore, according to lemma 2, the sets $\overline{W \setminus (N + L_i)}$ are analytic on a neighbourhood of zero. But $(**)$ implies that $W = \bigcup_1^r (W \setminus (N + L_i))$, hence the set $\overline{W}$ is also analytic on a neighbourhood of zero.

3. Let M be an n–dimensional manifold.

For any set $E \subset M$ and any point $a \in E$, we put

$$\underline{\dim}_a E = \min \{ \dim_z E : z \in E \cap U \} ,$$

where U is a sufficiently small neighbourhood of the point a (^{36}) .

THE REMMERT-STEIN THEOREM (on removable singularities). *Let W be an analytic set in $M \setminus V$, where V is an analytic subset of M. If*

(1) $$\underline{\dim}_z W > \dim_z V \quad for \quad z \in \overline{W} \cap V ,$$

and thus in particular if

(2) $$\dim_z W > \dim V \quad for \quad z \in W ,$$

then $\overline{W}$ is an analytic subset of M. Therefore $W \cup V$ is also an analytic subset of M.

PROOF . It is enough to prove the theorem under the assumption (2). Indeed, assuming (1) and applying this version of the theorem to a sufficiently small open neighbourhood U of any point $z \in \overline{W} \cap V$ (and to the sets $W \cap U$, $V \cap U$), we obtain the required conclusion (see II. 3.4). One may assume that W is of constant dimension $p > \dim V$. For, taking the decomposition $W = W_0 \cup \ldots \cup W_r$ into components of constant dimension (see 2.9), we have $W_p = \emptyset$ for $p \leq \dim V$ (since then $W^{(p)} = \emptyset$, in view of (2)). Moreover, one may assume that $0 < p \leq n$ (because if $p = 0$, then $V = \emptyset$). Put $k = \dim V$. The theorem is trivial when $k = -\infty$. Now, let $k \geq 0$, and suppose that the theorem is valid when V is of dimension $< k$. Now, by lemma 3, the set $\overline{W}$ is analytic on a neighbourhood of any point $z \in V^0$ (^{37}) , which implies that

---

(^{36}) Obviously, the right hand side of the equality does not dependent on U, provided that U is sufficiently small. Note that if W is a locally analytic subset of the manifold M, then $\underline{\dim}_a W = \min \{ k : a \in \overline{W^{(k)}} \}$.

(^{37}) We take a coordinate system $\varphi : U \longrightarrow \Omega$ such that $\varphi(V \cap U) = N \cap \Omega$, where $N \subset \mathbf{C}^n$ is a subspace (of dimension $\leq k < p$).

it is analytic on $M \setminus V^*$. Thus the set $\overline{W} \setminus V^*$ is analytic on $M \setminus V^*$. It is of constant dimension p (see 2.5 [38]) and, according to theorem 1 from 2.4, we have $\dim V^* < k$. Therefore the set $\overline{W} = \overline{\overline{W} \setminus V^*}$ [39] is analytic in M.

REMARK 1. The condition (1) could not be weakened by replacing the strict inequality by the weak one. For instance, for $M = \mathbf{C}^2$, $V = \{z = 0\}$, and $W = \{w = e^{1/z}, \ z \neq 0\}$, the set $\overline{W}$ is not analytic in $\mathbf{C}^2$. (In fact, the set $\overline{W} \cap \{w = 1\}$ is not analytic: it consists of the points $(0,1)$, $(1/(2\pi\nu i), 1)$, where $\nu = \pm 1, \pm 2, \ldots$, and hence it is 0–dimensional but not discrete; see III. 4.3.)

REMARK 2. The Hartogs theorem (see III. 4.2) can be derived from the Remmert-Stein theorem as follows. The graph of f is analytic in $(M \times \bar{\mathbf{C}}) \setminus (Z \times \bar{\mathbf{C}})$ and of constant dimension n; also $\dim(Z \times \bar{\mathbf{C}}) < n$, and therefore the closure $\bar{f}$ is analytic in $M \times \bar{\mathbf{C}}$ (of constant dimension n; see 2.5). As $\bar{f} \cap (M \times \infty) \subset Z \times \infty$, we have $\operatorname{codim}(\bar{f} \cap (M \times \infty)) \geq 3 > \operatorname{codim} \bar{f} + \operatorname{codim}(M \times \infty)$, which implies that $\bar{f} \cap (M \times \infty) = \emptyset$ (see III. 4.6). Hence the function f is locally bounded near Z, and so it can be extended holomorphically to M (see the proposition in II. 3.5).

The Remmert-Stein theorem yields the following

PROPOSITION . *A set* $V \subset M$ *is analytic if and only if it has a decomposition into (disjoint) submanifolds*

$$V = \Gamma^n \cup \ldots \cup \Gamma^0 \ ,$$

where Γ^i *is of dimension* i, *as a manifold, and the sets* $\Gamma^k \cup \ldots \cup \Gamma^0$, $k = 0, \ldots, n$, *are closed* [40] .

Indeed, if the set V is analytic, then by corollary 5 from theorem 4 in 2.9, we define the sequence of analytic sets $V = V_n \supset \ldots \supset V_0$, where $V_{k-1} = V_k \setminus V_k^{(k)}$ ($k = n, n-1, \ldots, 1$); then $\dim V_k \leq k$ ($k = 0, \ldots, n$). So, it suffices to set $\Gamma^i = V_i^{(i)}$ ($i = 0, \ldots, n$). Conversely, assume that V has a decomposition with the above properties. Put $Z_k = \Gamma^0 \cup \ldots \cup \Gamma^k$ ($k = 0, \ldots, n$). The set Z_0 is clearly analytic. Let $0 < k \leq n$, and suppose that Z_{k-1} is analytic. Since $\dim Z_{k-1} < k$ and Γ^k, as a closed submanifold of $M \setminus Z_{k-1}$, is analytic in $M \setminus Z_{k-1}$ (see II. 3.4), the Remmert-Stein theorem shows that the set $Z_k = Z_{k-1} \cup \Gamma^k$ is analytic in M. Consequently, the set $V = Z_n$ is analytic in M.

---

[38] The set W is open and dense in $\overline{W} \setminus V^*$.

[39] See footnote [38].

[40] See [30], page 60, where such sets are called *–analytic. A decomposition like this is, in general, not unique.

REMARK. The manifolds Γ^i in the above decomposition are always analytically constructible (see 8.3 below). (This is so because $\Gamma^i = Z_i \setminus Z_{i-1}$ for $i > 0$, and $\Gamma^0 = Z_0$.)

§7. Regular separation

1. We say that a pair of closed sets E, F in an open subset of a vector space M satisfies the condition (S) at a point $a \in G$ if either $a \notin E \cap F$ or $a \in E \cap F$, and (after endowing M with a norm)

$$\varrho(z, E) + \varrho(z, F) \geq c\varrho(z, E \cap F)^p$$

in a neighbourhood of the point a for some $c, p > 0$. (Obviously, the last condition does not depend on the norm on M.) If $U \subset G$ is an open neighbourhood of the point a, then the above condition (S) is satisfied if and only if the pair $E \cap U$, $F \cap U$ satisfies the condition (S) at the point a [41]. (Therefore (S) is, in fact, a condition on the germs E_a, F_a.)

If $h : G \longrightarrow H$ is a biholomorphic mapping between open subsets of vector spaces, then a pair $E, F \subset G$ satisfies the condition (S) at a point $a \in G$ precisely when the pair $h(E), h(F) \subset H$ satisfies the condition (S) at the point $h(a)$ [42]. Thus the condition (S) can be carried over – in a natural way – to the case of manifolds. Namely, we say that a pair of closed subsets E, F of a manifold M satisfies the condition (S) at a point $a \in M$ if, for some (and hence for every) chart φ whose domain Ω contains a, the pair $\varphi(E), \varphi(F)$ satisfies the condition (S) at the point $\varphi(a)$. The latter means that (see C. 3.2) either $a \notin E \cap F$ or $a \in E \cap F$, and $\varrho_\varphi(z, E \cap \Omega) + \varrho_\Omega(z, F \cap \Omega) \geq c\varrho_\varphi(z, E \cap F \cap \Omega)^p$ in a neighbourhood of the point a for some $c, p > 0$.

We say that the pair of closed subsets E, F of a manifold M satisfies the *condition of regular separation* if it satisfies the condition (S) at each point $a \in M$ or, equivalently, at each point $a \in E \cap F$.

It is easy to see that if M is an open subset of a vector space, then a pair E, F of closed subsets of M satisfies the condition of regular separation if and only if either $E \cap F = \emptyset$ or $E \cap F \neq \emptyset$ and (after endowing the space with a norm) we have the inequality

$$\varrho(z, E) + \varrho(z, F) \geq c\varrho(z, E \cap F)^p$$

---

[41] Indeed, if $a \in Z$ and $\{|z| < 2\varepsilon\} \subset U$, then $\varrho(z, Z \cap U) = \varrho(z, Z)$ in $\{|z| < \varepsilon\}$.

[42] This is due to the fact that, for some K, the inequality $|h(z') - h(z)| \leq K|z' - z|$ holds in a neighbourhood of the point a (and the same is true for h^{-1} in a neighbourhood of the point $h(a)$). Obviously, it is sufficient for the mapping h to be a diffeomorphism.

in every compact subset of M for some $c, p > 0$ (that depend on the compact set and the norm).

We have the following:

THEOREM . *Every pair of analytic subsets of a complex manifold satisfies the condition of regular separation* [43] *.*

As we will see, this result follows from the proposition which we will prove in n° 2.

2. We say that a continuous mapping f of an open subset G of a vector space M into a vector space N satisfies the condition (s) at a point $a \in G$ if either $f(a) \neq 0$ or $a \in Z = f^{-1}(0)$, and, after endowing M and N with norms,

$$|f(z)| \geq c\varrho(z, Z)^p \text{ in a neighbourhood of } a$$

for some $p, c > 0$. (Clearly, the condition does not depend on the choice of norms.) If $U \subset G$ is an open neighbourhood of the point a, the above condition (s) holds precisely when the restriction f_U satisfies the condition (s) at the point a [44]. (Therefore (s) is, in fact, a condition for the germ f_a.) If $h : H \longrightarrow G$ is a biholomorphic mapping of open subsets of vector spaces and $c \in H$, then a continuous mapping $f : G \longrightarrow N$ satisfies the condition (s) at a point $h(c)$ if and only if the mapping $f \circ h$ satisfies the condition (s) at the point c [45] .

PROPOSITION . *Every holomorphic mapping f on an open subset G of a vector space M with values in a vector space N satisfies the condition (s) at each point $a \in G$. Hence, if $Z = f^{-1}(0) \neq \emptyset$, then in any compact subset of G (and after endowing M and N with norms) we have the inequality*

$$|f(z)| \geq c\varrho(z, Z)^p$$

for some $c, p > 0$ (that depend on the choice of the compact set and the norms).

REMARK. For holomorphic functions (i.e., if $N = \mathbf{C}$), the proposition is a direct corollary of the Weierstrass preparation theorem. Namely, one may assume that $M = \mathbf{C}^n$, $a = 0$, and that f is z_n-regular (see C. 2.3). Next, one may assume, by the preparation theorem (see C. 2.4 [46]), that f is

---

[43] In a more general setting, the regular separation holds for closed semi-analytic subsets of real–analytic manifolds. (See [27], p. 85.)

[44] See footnote [41].

[45] See footnote [42].

[46] Because $|h(z)| \geq \varepsilon$ in a sufficiently small neighbourhood of zero for some $\varepsilon > 0$.

the restriction to $G = \{|z'| < \delta, \; |z_n| < \varepsilon\}$, where $z' = (z_1, \ldots, z_{n-1})$, of a distinguished polynomial P whose coefficients are holomorphic in $\{|z'| < \delta\}$. Finally, we may assume that $(P(z) = 0 \text{ and } |z'| < \delta) \Longleftrightarrow |z_n| < \varepsilon$ (see C. 2.2). Then, if $z = (z', z_n) \in G$, we have $P(z', t) = (t - \zeta_1) \ldots (t - \zeta_p)$ for $t \in \mathbf{C}$ and $z^{(\nu)} = (z', \zeta_\nu) \in Z$, hence $|f(z)| = |z - z^{(1)}| \ldots |z - z^{(p)}| \geq \varrho(z, Z)^p$.

Now the theorem in n° 1 can be deduced from the proposition. Indeed, it is enough to consider the case of globally analytic subsets V, W of an open subset G of the space $\mathbf{C}^n$, that is, $V = f^{-1}(0)$, $W = g^{-1}(0)$, where $f : G \longrightarrow \mathbf{C}^r$, $g : G \longrightarrow \mathbf{C}^s$ are holomorphic mappings. Then $V \cap W = (f, g)^{-1}(0)$. Now, if $a \in V \cap W$, then – according to the proposition – we have $|(f, g)(z)| \geq c\varrho(z, V \cap W)^p$ in the ball $\{|z - a| < 2\varepsilon\} \subset G$ for some $c, p, \varepsilon > 0$. Hence $c\varrho(z, V \cap W)^p \leq |f(z)| + |g(z)| \leq K\big(\varrho(z, V) + \varrho(z, W)\big)$ in a ball $\{|z - a| < \varepsilon\}$ for some $K > 0$ (the last estimate follows from the mean value theorem).

The proposition is now implied by the following lemma.

LEMMA . *If Z is an analytic subset of an open subset G of $\mathbf{C}^n$, then, for any point $a \in Z$, there is an open neighbourhood $U \subset G$ and a holomorphic mapping $g : U \longrightarrow \mathbf{C}^s$ such that $Z \cap U = g^{-1}(0)$ and $|g(z)| \geq c\varrho(z, Z)^p$ in U, for some $c, p > 0$.*

In fact, to demonstrate the proposition, one may assume that $M = \mathbf{C}^n$ and $N = \mathbf{C}^r$. Let $a \in Z$, and suppose that U and g are chosen according to the lemma. We have $f = (f_1, \ldots, f_r)$, $g = (g_1, \ldots, g_s)$. Hence, by Hilbert's Nullstellensatz (see the remark in III. 4.1), there exists an exponent m such that $g_i^m = \sum_i a_{ij} f_j$ in an open neighbourhood $U_0 \subset U$ of the point a, where a_{ij} are holomorphic functions in U_0. Furthermore, by taking a smaller U_0, one may assume that the functions a_{ij} are bounded. As a consequence, for some $K > 0$ and all $z \in U_0$, we have $c^m \varrho(z, Z)^{pm} \leq s^m \max |g_i(z)|^m \leq K|f(z)|$.

PROOF of the lemma. First consider the special case when the germ Z_a is irreducible. One may assume that $a = 0$ and that the germ Z_0 is k–regular, where $0 \leq k \leq n$. (This can be achieved by a linear change of coordinates; see III. 2.7.) Now, owing to proposition 1 from 1.4, the germ Z_0 has a normal triple (Ω, Σ, V) of dimension k. Moreover, one may require that $V = Z \cap U$, where $U \subset \Omega \times \mathbf{C}^{n-k}$ is an open neighbourhood of zero (see 1.2). As the pair $(\Sigma, V_{\Omega \setminus \Sigma})$ is a quasi-cover in $\Omega \times \mathbf{C}^{n-k}$ whose multiplicity is $p > 0$ and whose adherence is V (see 1.3), lemma I on quasi-covers (see III. 1.3) yields that $V = F^{-1}(0)$, where $F = F_P : \Omega \times \mathbf{C}^{n-k} \longrightarrow \mathbf{C}^s$ and $P : (\mathbf{C}^{n-k})^{p+1} \longrightarrow \mathbf{C}^s$ is any collector. By taking $g = F_U$, we get $Z \cap U = g^{-1}(0)$, and it is enough to choose the collector P in such a way that the second condition in the conclusion of the lemma is satisfied. Now, for each $\alpha = (\alpha_1,\ldots,\alpha_p) \in J = \{1,\ldots,n-k\}^p$, we take the polynomial $P_\alpha(\eta,v) = \varphi_{\alpha_1}(\eta_1-v)\ldots\varphi_{\alpha_p}(\eta_p-v)$, where $\eta = (\eta_1,\ldots,\eta_p) \in (\mathbf{C}^{n-k})^p$, $v \in \mathbf{C}^{n-k}$, and $\varphi_j : \mathbf{C}^{n-k} \ni (v_1,\ldots,v_{n-k}) \longrightarrow v_j \in \mathbf{C}$ $(j = 1,\ldots,n-k)$. Take $P = (P_{\gamma_1},\ldots,P_{\gamma_s})$ as the collector, where $J = \{\gamma_1,\ldots,\gamma_s\}$. Then, for $(u,v) \in (\Omega\backslash\Sigma)\times\mathbf{C}^{n-k}$, we have $V_u = \{\eta_1,\ldots,\eta_p\}$, $F(u,v) = P(\eta_1,\ldots,\eta_s,v)$, and

$$s|F(u,v)| \geq \sum_J |P_\alpha(\eta_1,\ldots,\eta_p,v)| =$$

$$\left(\sum_{\nu=1}^{n-k} |\varphi_\nu(\eta_1-v)|\right) \ldots \left(\sum_{\nu=1}^{n-k} |\varphi_n u(\eta_p-v)|\right) \geq |\eta_1-v|\ldots|\eta_p-v| \geq \varrho((u,v),V)^p.$$

Here we use the fact that $(u,\eta_i) \in V$ $(i = 1,\ldots,p)$. Therefore $s|F(z)| \geq \varrho(z,V)^p$ also holds in $\Omega \times \mathbf{C}^{n-k}$. This means that for $z \in U$ we have $|g(z)| \geq \frac{1}{s}\varrho(z,V)^p \geq \frac{1}{s}\varrho(z,Z)^p$.

Passing to the general case, we have $Z \cap U' = Z_1 \cup \ldots \cup Z_m$ for some open neighbourhood U' of the point a and some analytic subsets Z_i in U' whose germs $(Z_i)_a$ are irreducible. By the first part of the proof, there exists an open neighbourhood $U \subset U'$ of the point a and holomorphic mappings $g_i : U \longrightarrow \mathbf{C}^{s_i}$ such that $Z_i \cap U = g_i^{-1}(0)$ and $|g_i(z)| \geq c\varrho(z,Z_i)^p$ in U, where $c,p > 0$. Consequently, the set Z_i is defined by the functions $g_{i\nu}$, where $g_i = (g_{i1},\ldots,g_{is_i})$, and so the set $Z \cap U$ is defined by the functions $h_\nu = g_{1\nu_1}\ldots g_{n\nu_m}$, where $\nu = (\nu_1,\ldots,\nu_m) \in \Theta = \{\nu : \nu_i = 1,\ldots,s_i$ for $i = 1,\ldots,m\}$ (see II. 3.1). In other words, $Z \cap U = h^{-1}(0)$, where $h = (h_{\alpha_1},\ldots,h_{\alpha_r})$ and $\Theta = \{\alpha_1,\ldots,\alpha_r\}$. Finally, for $z \in U$, we have

$$r|h(z)| \geq \sum_\Theta |h_\nu(z)| = \left(\sum_{\nu=1}^{s_1} |g_{1\nu}(Z)|\right)\ldots\left(\sum_{\nu=1}^{s_m} |g_{m\nu}(z)|\right) \geq$$

$$\geq |g_1(z)|\ldots|g_m(z)| \geq c^m \varrho(z,Z_1)^p\ldots\varrho(z,Z_m)^p \geq c^m \varrho(z,Z)^{pm} .$$

§8. Analytically constructible sets

Let M be an n–dimensional manifold.

1. By an *analytically constructible leaf* (in M) we will mean a non-empty connected submanifold $\Gamma \subset M$ such that the sets $\bar{\Gamma}$ and $\bar{\Gamma} \setminus \Gamma$ are analytic [47].

---

[47] Corollary 1 from proposition 5 in n° 3 justifies this terminology.

Analytically constructible leaves are precisely the sets of the form $V \setminus W$, where V, W are analytic sets, V is irreducible, and $V^* \subset W \subsetneq V$. Obviously, the condition $W \subsetneq V$ can be replaced by $V \not\subset W$.

Indeed, if Γ is an analytically constructible leaf, then $\bar{\Gamma}$ is irreducible (see 2.8, corollary 3 from proposition 2) and $\bar{\Gamma}^* \subset \bar{\Gamma} \setminus \Gamma \subsetneq \bar{\Gamma}$ (as $\Gamma \subset \bar{\Gamma}^0$; see C. 3.7). Conversely, if V, W are analytic sets, V is irreducible, and $V^* \subset W \subsetneq V$, then the set $V \setminus W = V^0 \setminus (W \cap V^0)$ is connected, open, and dense in V^0 (see II. 3.6; and 2.8, proposition 2). Hence it is a non-empty connected submanifold, $\overline{V \setminus W} = V$ (see 2.1, proposition 1) and $\overline{V \setminus W} \setminus (V \setminus W) = W$.

If V, W are analytic sets and $V^* \subset W$, then the connected components of the set $V \setminus W$ are analytically constructible leaves. For every connected component H of the set $V \setminus W$ is a submanifold (since it is an open subset of a connected component of the set V^0) and the set $\bar{H}$ is analytic (see 2.10, theorem 5), and so is the set $\bar{H} \setminus H = \bar{H} \cap W$.

In particular, if V is an analytic set, then the connected components of the set V^0 are analytically constructible leaves.

The Cartesian product of analytically constructible leaves is an analytically constructible leaf ([48]) .

2. Let $\mathcal{A}, \mathcal{B}$ be families of subsets of the same space. We say that the family $\mathcal{A}$ is *compatible* with the family $\mathcal{B}$ if for every $A \in \mathcal{A}$ and $B \in \mathcal{B}$ either $A \subset B$ or $A \subset \setminus B$. In the case when $\mathcal{B} = \{B\}$, we say that the family $\mathcal{A}$ is *compatible* with the set B. If, in addition, $\mathcal{A} = \{A\}$, then we say that the set A is *compatible* with the set B. (Clearly, the family $\mathcal{A}$ is compatible with the family $\mathcal{B}$ if and only if each set $A \in \mathcal{A}$ is compatible with each set $B \in \mathcal{B}$. Any reduction of sizes of the families preserves their compatibility.) If the family $\mathcal{A}$ covers the set B and is compatible with this set, then the set B is the union of some sets from the family $\mathcal{A}$.

A locally finite partition of the manifold M into (disjoint) non-empty connected submanifolds Γ_ν^i such that $\dim \Gamma_\nu^i = i$ and each of the sets $\partial \Gamma_\mu^k$ is the union of some of the sets Γ_ν^i, $i < k$, is called a *complex stratification* of the manifold M. Note that the last condition above (with the other ones satisfied) is fulfilled if and only if

(1) the sets $\bigcup_{i \le k} \Gamma_\nu^i$ $(k = 0, \dots, n)$ are closed,

(2) the set Γ_ν^i is compatible with the set $\overline{\Gamma_\mu^k}$ if $i < k$,

---

([48]) Because $\overline{\Gamma_1 \times \Gamma_2} \setminus (\Gamma_1 \times \Gamma_2) = ((\bar{\Gamma}_1 \setminus \Gamma_1) \times \bar{\Gamma}_2) \cup (\bar{\Gamma}_1 \times (\bar{\Gamma}_2 \setminus \Gamma_2))$.

(3) $\bar{\Gamma}^k_\nu \cap \Gamma^k_\nu = \emptyset$ for $\nu \neq \mu$ $(k = 0, \ldots, n)$. [49]

The Remmert-Stein theorem (see 6.3) implies the following

PROPOSITION 1. *The elements of any complex stratification of M are analytically constructible leaves.*

Indeed, let $\{\Gamma^i_\nu\}$ be a complex stratification of the manifold M. Obviously the sets Γ^0_ν are analytically constructible leaves. Let $0 < k \leq n$ and assume that the Γ^i_ν are analytically constructible leaves if $i < k$. Then the set $V = \bigcup_{i<k} \Gamma^i_\nu = \bigcup_{i<k} \bar{\Gamma}^i_\nu$ (see property (1)) is analytic of dimension $< k$ (see II. 3.4, B. 1, and II. 1.3). Now each set Γ^k_μ is an analytic subset of $M \setminus V$ (as a closed submanifold; see II. 3.4). Hence the set $\bar{\Gamma}^k_\mu$ is analytic in M. Also, $\partial \Gamma^k_\kappa$ is analytic in M because it is the union of some of the $\bar{\Gamma}^i_\nu$. Therefore Γ^k_μ is an analytically constructible branch.

PROPOSITION 2. *For any locally finite family $\{W_j\}$ of analytic subsets (in M), there is a complex stratification of the manifold M which is compatible with this family.*

In the proof of proposition 2 we will need the following lemma. Let W, V be analytic sets. Set $\tau_k(W, V) = \bigcup(W \cap V_i)$, where $\{V_i\}$ is the family of all k–dimensional simple components of V which are not contained in W $(k = 0, \ldots, n)$. Therefore $\tau_k(W, V)$ is an analytic set of dimension $< k$ (see 2.8, proposition 3) contained in $W \cap V$.

LEMMA . *Any open connected subset H of the manifold $V^{(k)}$ which is disjoint with $\tau_k(W, V)$ is compatible with W.*

Indeed, it is enough to show that the set $H \cap W$ is open in H (since it is closed in H), i.e., that if $a \in H \cap W$, then $V_a \subset W_a$. Now such a point a belongs to a k–dimensional simple component $\tilde{V}$ of the set V, and $\tilde{V}_a = V_a$ (see 2.9, corollaries 2 and 4). Since $a \notin \tau_k(W, V)$, one must have $\tilde{V} \subset W$, hence $V_a \subset W_a$.

PROOF of proposition 2. We are going to define a sequence of analytic sets $M = Z_n \supset \ldots \supset Z_{-1}$ satisfying the conditions $\dim Z_i \leq i$ and $Z_i \setminus Z_{i-1} \subset Z_i^{(i)}$. If the sets $Z_n \supset \ldots \supset Z_k$ (where $n \geq k \geq 0$) satisfying these conditions are already defined, we proceed as follows. For each $i = n, \ldots, k+1$, we take the family $\{\Gamma^i_\nu\}$ of the connected components of the set $Z_i \setminus Z_{i-1}$. They are i–dimensional submanifolds (since they are open in $Z_i^{(i)}$) and their closures

---

[49] The conditions (1)–(3) are sufficient, since the family $\{\Gamma^i_\nu\}_{i<k}$ covers the set $\partial \Gamma^k_\mu = \bar{\Gamma}^k_\mu \setminus \Gamma^k_\mu$ and is compatible with this set.

are analytic (see 2.10, theorem 5). Set

$$Z_{k-1} = \bigcup_{\mu} Z_k^{\mu} \cup Z_k^* \cup \bigcup_{i>k} \tau_k(\bar{\Gamma}_{\nu}^i, Z_k) \cup \bigcup_j \tau_k(W_j, Z_k) \,,$$

where $\{Z_{\mu}^k\}$ is the family of simple components of dimension $\leq k$ of the set Z_k. Then $\dim Z_{k-1} \leq k-1$ (see 2.4, theorem 1) and $Z_k \setminus Z_{k-1} \subset Z_k^{(k)}$ (see 2.9, corollary 2). Thus all the Z_i, as well as Γ_{ν}^k, are defined. Now, the family $\{\Gamma_{\nu}^i\}$ is a complex stratification of M, since the conditions (1)–(3) are satisfied. Indeed, the sets $\bigcup_{i \leq k} \Gamma_{\nu}^i = Z_k$ are closed. Since the Γ_{ν}^k are disjoint open subsets of the manifold $Z_k^{(k)}$, we have $\bar{\Gamma}_{\mu}^k \cap \Gamma_{\varrho}^k = \emptyset$ for $\mu \neq \varrho$. Finally, if $k < 1$, then (due to the lemma and the fact that $\Gamma_{\mu}^k \cap \tau_k(\overline{\Gamma_{\nu}^i}, Z_k) = \emptyset$) the set Γ_{μ}^k is compatible with $\overline{\Gamma_{\nu}^i}$. The stratification is compatible with the family $\{W_j\}$, because, in view of the lemma, any of the sets Γ_{ν}^k is compatible with every W_j (because $\Gamma_{\nu}^k \cap \tau_k(W_j, Z_k) = \emptyset$).

COROLLARY . *Every analytic set has a complex stratification, i.e., a locally finite partition into (disjoint) analytically constructible leaves Γ_{ν}^i such that $\dim \Gamma_{\nu}^i = i$ and each set $\partial \Gamma_{\mu}^k$ is the union of some of the sets Γ_{ν}^i, $i < k$.*

3. By *analytically constructible sets* (in M) we mean the elements of the smallest family of subsets of the manifold M that contains all analytic subsets (of M) and is closed with respect to operations of taking the locally finite union of sets and the complement of a set ([50]) . Thus the difference and the finite intersection of analytically constructible sets is also an analytically constructible set.

LEMMA 1. *Let $\mathcal{K}$ denote the family of locally finite unions of analytically constructible leaves. For any locally finite family of sets $E_i \in \mathcal{K}$, there exists a complex stratification of the manifold M which is compatible with this family.*

Indeed, each of the sets E_i is equal to a locally finite union $\bigcup_{\nu}(V_{i\nu} \setminus W_{i\nu})$, where $V_{i\nu}, W_{i\nu}$ are analytic sets and $W_{i\nu}$ is nowhere dense in $V_{i\nu}$. Then $V_{i\nu} \subset \bar{E}_i$ and the family $\{\bar{E}_i\}$, as well as the families $\{V_{1\nu}\}, \{V_{2\nu}\}, \ldots$, are locally finite (see B. 1). This implies that the family of all the sets $V_{i\nu}, W_{j\mu}$ is also locally finite. Now, it is enough – according to proposition 2 – to take a complex stratification compatible with this family.

PROPOSITION 3. *For a subset E of M, the following conditions are equivalent:*

([50]) It is contained in the class of semi-analytic subsets of the manifold M regarded as a real-analytic manifold. (See [27], p. 65–68.)

(1) E is analytically constructible,

(2) E is a locally finite union of analytically constructible leaves,

(3) E is the union of some elements of a complex stratification of the manifold M,

(4) $E = \bigcup(V_i \setminus W_i)$, where $\{V_i\}$ is a locally finite family of irreducible analytic sets and W_i are analytic sets; in addition, one may require that $W_i \subsetneqq V_i$.

Indeed, by lemma 1 and proposition 1, we have (2) $\Longleftrightarrow$ (3). The family $\mathcal{K}$ is the class of sets satisfying condition (2) or, equivalently, condition (3). We have the implication (2) $\Longrightarrow$ (4) (see n° 1 and B. 1), whereas the implication (4) $\Longrightarrow$ (1) is trivial. Finally, the class $\mathcal{K}$ contains all analytic sets (see n° 2, corollary from proposition 2) is closed with respect to the operations of taking the locally finite union of sets and the complement of a set (see condition (3)). Therefore it contains the class of all analytically constructible sets, i.e., (1) $\Longrightarrow$ (2).

Proposition 3 yields the following corollaries.

The connected components of an analytically constructible set are analytically constructible and constitute a locally finite family. (See condition (2).)

The Cartesian product of analytically constructible subsets of manifolds M and N, respectively, is an analytically constructible subset of the manifold $M \times N$. (See condition (2), and then B. 1 and n° 1.)

If $f : M \longrightarrow N$ is a holomorphic mapping of manifolds, then the inverse image (under f) of an analytically constructible set in N is an analytically constructible set in M. (See condition (4) and B. 1.)

Therefore, if $Z \subset M$ is a submanifold, then the trace on Z of an analytically constructible set in M is an analytically constructible set in Z. If Z is a closed submanifold, then a set $E \subset Z$ is analytically constructible in Z precisely when it is analytically constructible in M. (See condition (4) and II. 3.4.)

Each analytically constructible set is an F_σ-set (see condition (2), as well as C. 3.7 and B. 1).

LEMMA 2. *If E is an analytically constructible set (in M) and $E \subset F \subset M$, then*

$$(E \text{ is nowhere dense in } F) \Longleftrightarrow \text{int}_F E = \emptyset .$$

For E is a locally finite union of locally closed sets (see condition (2) and B. 1).

PROPOSITION 4. *Every analytically constructible set is a locally connected space.*

Indeed, let E be an analytically constructible set, and let U be an open neighbourhood of a point $a \in E$. The set $E \cap U$ is analytically constructible in U, and thus the family of its connected components is locally finite. Therefore only a finite number of those components, say $F_1, \ldots, F_r$, have the property that $a \in \bar{F}_i$. Then $\bigcup_1^r F_i \subset U$ is a connected neighbourhood of the point $a \in E$.

Let us prove here a special case of the *curve selection lemma* (Bruhat-Cartan-Wallace):

If E is a constructible set and $a \in \bar{E}$ is not an isolated point of E, then there is a (simple) arc λ of class C^1 with an endpoint a and such that $\lambda \setminus a \subset E$ ([51]).

Indeed, in view of proposition 3, one may assume that $E = V \setminus W$, where V, W are analytic subsets of an open neighbourhood of zero in $\mathbf{C}^n$, $a = 0 \in V$, the germ V_0 is of constant dimension $k > 0$, and $\dim W_0 < k$. (See 3.1 proposition 4; 2.8, corollary 1 from proposition 2; and proposition 3.) Secondly, one may assume that the germs V_0 and W_0 are regular (see III. 2.7). Let $L = \{z_1 = \ldots = z_{k-1} = 0\}$. Then $W_0 \cap L \subset 0$ and, considering a normal triple of V_0 (see proposition 1 in 1.4), we see that $\dim(V_0 \cap L) = 1$. Now it is enough to take – in accordance with corollary 3 from Puiseux's theorem (see II. 6.2) – a sufficiently small arc contained in $V \cap L$ with an endpoint 0.

Condition (2) implies (see B. 1 and II. 3.4):

PROPOSITION 5. *The closure of an analytically constructible set is analytic. The closed analytically constructible sets are precisely the analytic sets.*

REMARK. If E is an analytically constructible set, then the set $\bar{E} \setminus E$ is nowhere dense in $\bar{E}$. (This follows from lemma 2.)

COROLLARY 1. *The analytically constructible leaves coincide exactly with the non-empty connected submanifolds that are analytically constructible.* (See C. 3.7.)

COROLLARY 2. *The interior of an analytically constructible set is an analytically constructible set. (It is even the complement of an analytic subset).*

COROLLARY 3. *An analytically constructible set is locally closed if and only if it is locally analytic.*

Condition (2), combined with lemma 1, yields:

PROPOSITION 6. *For any locally finite family of analytically constructible sets there is a complex stratification of the manifold M compatible with this family.*

4. For any subset E of a topological space X, we define a decreasing sequence of closed sets $V_i = V_i(E)$, $i = 0, 1, \ldots$, as follows. We define by

---

([51]) The lemma is true and useful in the more general situation when E is a semi-analytic subset of a real-analytic manifold. (See [27], p.103.)

recursion a sequence E_i, setting $E_0 = E$ and $E_{i+1} = \bar{E}_i \setminus E_i$, and we put $V_i = \bar{E}_i$. In particular, $V_0 = \bar{E}$. Note that if $G \subset X$ is an open set, then for the subset $E \cap G$ of the space G we have

$$(*) \qquad\qquad V_i(E \cap G) = V_i(E) \cap G .$$

We have the following:

LEMMA 3. *If* $V_{2r} = \emptyset$, *then* $E = (V_0 \setminus V_1) \cup \ldots \cup (V_{2r-2} \setminus V_{2r-1})$.

(Because $E_i = \bar{E}_i \setminus E_{i+1} = (V_i \setminus V_{i+1}) \cup (V_{i+1} \setminus E_{i+1}) = (V_i \setminus V_{i+1}) \cup E_{i+2}$.)

PROPOSITION 7. *A set* $E \subset M$ *is analytically constructible if and only if the sets* $V_i = V_i(E)$ *are analytic and* $V_s = \emptyset$ *for some s. Then* V_{i+1} *is nowhere dense in* V_i, $i = 0, 1, \ldots$, *and we have* $V_i = \emptyset$ *for* $i > n = \dim M$. *If* $2r > n$, *then*

$$(**) \qquad\qquad E = (V_0 \setminus V_1) \cup \ldots \cup (V_{2r-2} \setminus V_{2r-1}) .$$

Indeed, by lemma 3, the condition is sufficient. Now suppose that the set E is analytically constructible. Then – in view of proposition 5 and the remark – the sets V_i are analytic and V_{i+1} is nowhere dense in V_i, $i = 0, 1, \ldots$. Thus $V_i = \emptyset$ for $i > n$ (see 2.5), and by applying lemma 3, we get the formula $(**)$.

COROLLARY 1. *If a set* $E \subset M$ *is analytically constructible, then there exists an analytic set* Z *which is nowhere dense in* $\bar{E}$ *and such that* $\bar{E} \setminus Z \subset E$.

(It is enough to take $Z = V_1$.)

COROLLARY 2. *The analytically constructible sets are precisely the sets of the form*

$$(V_0 \setminus V_1) \cup \ldots \cup (V_{2k} \setminus V_{2k+1}) ,$$

where $V_0 \supset \ldots \supset V_{2k+1}$ *are analytic sets such that* V_{i+1} *is nowhere dense in* V_i $(i = 0, \ldots, 2k)$.

COROLLARY 3. *The class of all analytically constructible sets is the algebra of sets* [52] *generated by the class of all analytic sets.*

Proposition 7, together with formulae $(*)$ and $(**)$, implies:

PROPOSITION 8. *If* $\{G_\iota\}$ *is an open cover of the manifold* M, *then a set* $E \subset M$ *is analytically constructible in* M *if and only if for each* ι *the set* $E \cap G_\iota$ *is analytically constructible in* G_ι.

REMARK. It follows from proposition 8 that a set $E \subset M$ is analytically constructible if and only if $E_z \in \mathcal{K}_z$ for each $z \in M$, where $\mathcal{K}_z$ denotes the

---

[52] That is, it is a class of sets which is closed with respect to the operations of taking the union of two sets or taking the complement of a set.

algebra of germs [53] (of sets at z) generated by the class of analytic germs at z. In other words, E is analytically constructible if it can be described locally by holomorphic functions [54] , i.e., if each point of M has an open neighbourhood U such that $E \cap U = \bigcup_i \bigcap_j E_{ij}$ for some finite family of sets E_{ij} of the form $\{f_{ij} = 0\}$ or $\{f_{ij} \neq 0\}$. Here f_{ij} are holomorphic functions in U.

5. For each analytically constructible set $E \subset M$ we define the sets E_0, E^*, and $E^{(k)}$ ($k = 0, \ldots, n$) in the same fashion as that used for analytic sets (see 2.1). Thus we have the decomposition $E^0 = E^{(0)} \cup \ldots \cup E^{(n)}$, and the set $E^{(i)}$ is a submanifold (of dimension i as a manifold) and is open in E^0 ($i = 0, \ldots, n$). The space E^0 is locally connected and the sets $E^{(i)}$ are unions of its connected components. Each connected component of the set E^0 is a submanifold. We have also the formulae $(E \cap G)^0 = E^0 \cap G$, $(E \cap G)^* = E^* \cap G$, and $(E \cap G)^{(k)} = E^{(k)} \cap G$ if G is an open set. The set E is said to be *smooth* if $E = E^0$; then it is a locally analytic set.

LEMMA 4. *If E is an analytically constructible set, then $E^0 = (V_0)^0 \setminus V_1$, where $V_i = V_i(E)$.*

PROOF . In view of proposition 7, the sets V_i are analytic and we have the formula (**). Hence we have $E \setminus V_1 = V_0 \setminus V_1$, and so $E^0 \setminus V_1 = (V_0)^0 \setminus V_1$. Therefore it is enough to show that $E^0 \cap V_1 = \emptyset$. Now, if there was a point $z \in E^0 \cap V_1$, then, for some neighbourhood U of z, the set $E \cap U$ would be closed in U, and moreover, $(V_1 \setminus V_2) \cap U \neq \emptyset$. Here we use the fact that $z \in V_1$ and the set V_2 is nowhere dense in V_1. Since $E \subset (V_0 \setminus V_1) \cup V_2 = \bar{E} \setminus (V_1 \setminus V_2)$, we would have $E \cap U \subsetneqq \bar{E} \cap U$, contrary to the fact that $E \cap U$ is closed in U.

COROLLARY . *In any analytically constructible set E, the set E^0 is open and dense, while the set E^* is closed and nowhere dense.*

From lemma 4 and proposition 7, we have the following

PROPOSITION 9. *If a set $E \subset M$ is analytically constructible, then so are the sets E^0, E^*, and $E^{(k)}$ ($k = 0, \ldots, n$). Furthermore, the connected components of the set E^0 are analytically constructible leaves.*

Indeed, the set $(V_0)^0$ is analytically constructible (see 2.4, theorem 1), and hence so is the set E^0, as well as its connected components (see n° 3). This implies the analytical constructibility of the sets $E^{(k)}$, as well as the

---

[53] That is, a class of germs (of sets at z) which is closed with respect to the operations of taking the union of two germs and the complement of a germ.

[54] See [27], p. 66.

second part of the proposition (see corollary 1 from proposition 5).

Let $E \subset M$ be an analytically constructible set.

We have $\dim \bar{E} = \dim E$. (By 2.5, this is true for analytically constructible leaves. Then the general case follows by condition (2) in proposition 3; see II. 1.3 and n° 3.)

It follows that $\dim_z \bar{E} = \dim_z E$ for any $z \in M$.

If $k = \dim E \geq 0$, then $\dim(\bar{E} \setminus E^{(k)}) < \dim E$.

Indeed, consider a complex stratification of the manifold M which is compatible with E (see proposition 6). Then the sets E and $\bar{E}$ are unions of some leaves of dimension $\leq k$ of this stratification; they contain exactly the same k–dimensional leaves – all of them are contained in $E^{(k)}$. Therefore the set $\bar{E} \setminus E^{(k)}$ is contained in a union of leaves of dimension $\leq k - 1$.

In particular, if Γ is an analytically constructible leaf, then $\dim \partial \Gamma < \dim \Gamma$ [55].

Now let $E \subset F$ be analytically constructible sets (in M).

If E is nowhere dense in F and $F \neq \emptyset$, then $\dim E < \dim F$. (For $\bar{E}$ is nowhere dense in $\bar{F}$; see 2.5.) In view of lemma 2, one derives the following equivalence:

$$(E \text{ is nowhere dense in } F) \Longleftrightarrow (\dim_z E < \dim_z F \text{ for } z \in E) .$$

---

[55] This follows also from the fact that $\partial \Gamma$ is nowhere dense in $\bar{\Gamma}$ (see 2.5 and C. 3.7).

CHAPTER V

HOLOMORPHIC MAPPINGS

§1. Some properties of holomorphic mappings of manifolds

Let M and N be complex manifolds.

THEOREM 1. *Let* $f : M \longrightarrow N$ *be a holomorphic mapping, and let* $V \subset M$ *be a locally analytic set. Let* $k \in \mathbf{N}$. *If*

$$\text{rank}_z f \leq k \quad for \quad z \in V ,$$

then $f(V)$ *is a countable union of submanifolds of dimension* $\leq k$. *Hence*

$$\dim f(V) \leq k .$$

Indeed, set $m = k + \dim V$. The case when $m = -\infty$ is trivial. Suppose now that $m \geq 0$ and the theorem is true if $k + \dim V < m$. We have

$$V = V^* \cup \bigcup_i (V_0^{(i)} \cup V_1^{(i)}) ,$$

where $V_0^{(i)} = \{z \in V^{(i)} : \text{rank}_z f_{V^{(i)}} = k\}$ and $V_1^{(i)} = \{z \in V^{(i)} : \text{rank}_z f_{V^{(i)}} < k\}$ (see IV. 2.1). Now $V = \emptyset$ or $\dim V^* < \dim V$ (see IV. 2.4, theorem 1), and so the set $f(V^*)$ is a countable union of submanifolds of dimension $\leq k$. Next, if $k = 0$, then $V_1^{(i)} = \emptyset$; if $k > 0$, then, since $V_1^{(i)}$ is an analytic subset (of dimension $\leq \dim V$) of the manifold $V^{(i)}$ (see II. 3.7), the set $f(V_1^{(i)})$ is a countable union of submanifolds of dimension $\leq k - 1$.

Finally, by the rank theorem (see C. 4.1), the set $f(V_0^{(i)})$ is a countable union of submanifolds of dimension k.

COROLLARY 1. *We have* $\dim f(E) \leq \dim E$ *for each analytically constructible set* $E \subset M$. *In particular,* $\dim f(M) \leq \dim M$.

The corollary holds because the statement is true when E is a submanifold. (See IV. 8.3, proposition 3.)

COROLLARY 2 (SARD'S THEOREM). *The set of critical values of the mapping* f (1) *is a countable union of submanifolds of dimension* $< n = \dim N$, *and hence it is a set of dimension* $< n$.

REMARK. It follows that the set is of measure zero and of first category (2).

THEOREM 2. *Let* $U \subset M$. *If the graph of a mapping* $f : U \longrightarrow N$ *is locally analytic in* $M \times N$ *and* $\dim_x f \geq \dim M$ *for* $x \in f$, *then the set* U *is open and the mapping* f *is holomorphic.*

PROOF . It is enough to show that if $(a, b) \in f$, then the mapping f is defined and holomorphic in a neighbourhood of the point a. One may assume that M is a neighbourhood of zero in $\mathbf{C}^n$, $N = \mathbf{C}^k$, and $(a, b) = 0$. Now, as $\dim f \leq n$ (see II. 1.4), the set f is of constant dimension n. Hence its germ f_0 is of constant dimension n (see IV. 3.1, proposition 4); since $f_0 \cap (0 \times N) = 0$, it is n–regular (see III. 4.4). Consequently, by proposition 1 from IV. 1.4, it has a normal triple (Ω, Z, V) and one may assume that the crown V of the triple is a neighbourhood of zero in f (see IV. 1.2). But then the covering $V_{\Omega \setminus Z} \longrightarrow \Omega \setminus Z$ is one–sheeted, which proves that V is the graph of a holomorphic mapping in Ω (see IV. 1.8). Thus the mapping f is defined and holomorphic in Ω.

COROLLARY 1. *If* $f : M \longrightarrow N$ *is a holomorphic injection and* $\dim N \leq \dim M$, *then the set* $f(M)$ *is open in* N *and the mapping* $f : M \longrightarrow f(M)$ *is biholomorphic.*

REMARK. If $\dim N > \dim M$, then f may not be an immersion. For example, the holomorphic mapping $f : \mathbf{C} \ni z \longrightarrow (z^2, z^3) \in \mathbf{C}^2$ is injective, but $d_0 f = 0$.

COROLLARY 2. *Any holomorphic bijection* $f : M \longrightarrow N$ *is a biholomor-*

---

(1) It is the image under f of the set of critical points of f, i.e., of points $z \in M$ for which the differential $d_z f$ is not surjective. Thus it is the image of the analytic set $\{z \in M : \operatorname{rank}_z f < n\}$ (see II. 3.7).

(2) More generally, Sard's theorem says (see, e.g., [13], Chapter XVI, §23; or [43], Chapter VII, §1) that the set of critical values of any $\mathcal{C}^\infty$–mapping between smooth manifolds (with countable bases for topology) is of measure zero (i.e., its image under any chart is of measure zero). Hence, as an F_σ–set, it is also of first category.

phic mapping.

For then $N = f(M)$, and so $\dim N \leq \dim M$ (see corollary 1 of theorem 1).

COROLLARY 3. *Every local holomorphic homeomorphism of manifolds and, in particular, any holomorphic covering of manifolds is locally biholomorphic.*

COROLLARY 4. (THE ANALYTIC GRAPH THEOREM). *Any continuous mapping* $f : M \longrightarrow N$ *whose graph is analytic in* $M \times N$ *is holomorphic.*

In fact, take $(a, b) \in F$. The sets f_Ω, where Ω is any open neighbourhood of the point a, form a base of neighbourhoods of the point (a, b) in f. Therefore, since $\dim f_\Omega \geq \dim M$ (by corollary 1 of theorem 1 applied to the natural projection $\Omega \times N \longrightarrow \Omega$), we have $\dim_{(a,b)} f \geq \dim M$.

REMARK . Instead of continuity, it is enough to assume local boundedness of the mapping f. For the latter property implies continuity, as f is a closed set in $M \times N$ (see B. 2.3). In particular, if the manifold N is compact, the assumption that the mapping f is continuous is redundant. In the general case however, it is necessary. For instance, the graph of the mapping $f : \mathbf{C} \longrightarrow \mathbf{C}$ given by $f(0) = 0$ and $f(z) = z^{-1}$ for $z \neq 0$, is analytic in $\mathbf{C}^2$, but the mapping f is not even continuous.

§2. The multiplicity theorem. Rouché's theorem

Let M and N be manifolds of the same dimension $n > 0$, and let $f : M \longrightarrow N$ be a holomorphic mapping.

1. We say that the mapping f is *light at the point* $a \in M$ if a is an isolated point of its fibre $(^3)$ $f^{-1}(f(a))$. In such a case, we define the *multiplicity of the mapping* f *at the point* a by

$$(*) \qquad\qquad m_a f = \sup\{\#f_\Omega^{-1}(w) : w \in \Delta\} ,$$

where Ω and Δ are sufficiently small neighbourhoods of the points a and $f(a)$, respectively (so that the right hand side is independent of Ω and Δ).

Furthermore, if the neighbourhood Ω is sufficiently small, the equality $(*)$ is true for each neighbourhood Δ. Since, denoting the right hand side by

---

$(^3)$ By the fibre of the point a we mean the fibre of f containing a, i.e., the set $f^{-1}(f(a))$. (See C. 4.1, footnote $(^{79})$).

$m(\Omega, \Delta)$, we have $m_a f = m(\Omega, \Delta)$ if $\Omega \subset \Omega_0$, it follows that $\Delta \subset \Delta_0$ for some neighbourhoods Ω_0 and Δ_0. One may assume that $f(\Omega_0) \subset \Delta_0$. Then, for $\Omega \subset \Omega_0$ and for any Δ, we have $m(\Omega, \Delta) = m(\Omega, \Delta \cap \Delta_0) = m_a f$.

If G is an open neighbourhood of the point $a \in M$ and the image $f(G)$ is contained in an open set $H \subset N$, then the mapping f is light at a if and only if the mapping $f_G : G \longrightarrow H$ is light at a, and we have $m_a f_G = m_a f$. If $\varphi : M \longrightarrow M'$, $\psi : N \longrightarrow N'$ are biholomorphic mappings of manifolds and $\varphi(a) = b$, then the mapping $g = \psi \circ f \circ \varphi^{-1}$ is light at b if and only if the mapping f is light at a. In this case $m_b g = m_a f$.

Note that a holomorphic function f on a neighbourhood of a point $a \in \mathbf{C}$ is light at a if and only if f_a is not a constant germ (see [5], Chapter V, §11). Then the multiplicity $m_a f$ is equal to the multiplicity of the zero of the function $f(z) - f(a)$ at a (see [5], Chapter VI, §9).

If the differential $d_a f$ is an isomorphism, the mapping f is light at a and $m_a f = 1$ (⁴).

PROPOSITION . *If the mapping f is light at the point $a \in M$, then $m_a f < \infty$. Moreover, there are arbitrarily small neighbourhoods Ω, Δ of the points $a, f(a)$, respectively, and a nowhere dense analytic set W in Δ such that*

$$m_a f = f_\Omega^{-1}(w) \quad for \quad w \in \Delta \setminus W ,$$

and each point $w \in \Delta \setminus W$ is a regular value of the mapping f_Ω (⁵) .

First, we will prove the following

LEMMA . *Suppose that M and N are open neighbourhoods of zero in $\mathbf{C}^n$, the mapping f is light at 0, and $f(0) = 0$. Then the germ A at 0 of the submanifold*

$$\{(w, z) \in N \times M : w = f(z)\}$$

is simple and n–regular. Moreover, there exists a normal triple (Δ, W, V) of the germ A such that $V = \{(w, z) \in N \times M : w = f(z), z \in \Omega\}$, where Ω is an open neighbourhood of zero in M, the mapping $f_\Omega : \Omega \longrightarrow N$ is open, and the multiplicity p of the covering $V_{\Delta \setminus W} \longrightarrow \Delta \setminus W$ is equal to $m_a f$. Furthermore, the neighbourhoods Ω and Δ can be made arbitrarily small.

Indeed, the germ A is n–dimensional and smooth (see II. 3.3), hence it is simple (see II. 4.6). It is n–regular, as $A \cap (0 \times M) = 0$ (⁶) (see III. 4.4).

(⁴) This is a consequence of the implicit function theorem. See also the multiplicity theorem (n° 2 below).

(⁵) i.e., it is not a critical value. In this case, it means that for each $z \in f_\Omega^{-1}(w)$ the differential $d_z f$ is an isomorphism.

(⁶) Because 0 is an isolated point of the fibre $f^{-1}(0) = \{z : f(z) = 0\}$.

By proposition 1 from IV. 1.4, it has a normal triple (Δ, W, V) in which the neighbourhood Δ of zero in N can be chosen arbitrarily small. At the same time the crown V is an open (arbitrarily small) neighbourhood of zero in the set $\{(w, z) : w = f(z)\}$ (see IV. 1.2). Hence $V = \{(w, z) : w = f(z), z \in \Omega\}$, where Ω is the image of V by the natural projection onto M, and thus it is an open neighbourhood of zero in M. The mapping $f_\Omega : \Omega \longrightarrow \Delta$ is open because it is the composition of two open mappings: namely, the homeomorphism $\Omega \ni z \longrightarrow (f(z), z) \in V$ and the projection $V \longrightarrow \Delta$ (see IV. 1.3). Therefore the mapping $f_\Omega : \Omega \longrightarrow N$ is also open. Finally, for $w \in \Delta \setminus W$, we have $V_w = f_\Omega^{-1}(w)$, and so $\#f_\Omega^{-1}(w)$ is equal to the multiplicity p of the covering $V_{\Delta \setminus W} \longrightarrow \Delta \setminus W$. Owing to the fact that the function $\Delta \ni w \longrightarrow \#f^{-1}(w)$ is lower semicontinuous (see B. 2.1), the right hand side of the equality $(*)$ is p. Thus, if Δ, V are sufficiently small, we get $m_a f = p$.

Now the proposition follows from the lemma. Indeed, one may assume that M, N, and f are as in the lemma. Then, if $w \in \Delta \setminus W$, it follows that $\#f_\Omega^{-1}(w) = \#V_w = p = m_a f$. If also $z \in f_\Omega^{-1}(w)$, i.e., if $(w, z) \in V_{\Delta \setminus W}$, then V is a topographic submanifold at the point (w, z) (see IV. 1.1, property (3)). Consequently, the restriction of the mapping f to a suitable neighbourhood of the point z is a biholomorphic mapping (onto a neighbourhood of the point w). Thus the differential $d_z f$ must be an isomorphism.

2. For any point $a \in M$, we will denote by $I_a f$ the ideal of the ring $\mathcal{O}_a$ generated by the set $\mathfrak{m}_{f(a)} \circ f_a$. In the case when $N = \mathbf{C}^n$ and $f(a) = 0$, the ideal $I_a f$ is generated by the germs $(f_1)_a, \ldots, (f_n)_a$, where $f = (f_1, \ldots, f_n)$.

The mapping f is light at the point a if and only if $\operatorname{codim} I_a f < \infty$.

Indeed, one may assume that $N = \mathbf{C}^n$ and $f(a) = 0$ (see I. 1.7 and B. 4.3–2), and then the displayed assertion follows from the equivalence in III. 4.3.

We have

THE MULTIPLICITY THEOREM . *If the mapping f is light at a point a, then*

$$m_a f = \operatorname{codim} I_a f .$$

PROOF . One may assume that M, N are open neighbourhoods of the point $a = 0$ in $\mathbf{C}^n$, and that $f(0) = 0$. (See I. 1.7 and 1, B. 4.2–3, and also n° 1.) Then we can apply the lemma from n° 1. Let us use the same notation as in the lemma. Let $\mathcal{O}_z, \mathcal{O}_w, \mathcal{O}_{wz}$ denote the rings of germs at 0 of holomorphic functions in variables w or z, or both w and z, respectively. The ideal $I = \mathcal{I}(A)$ is prime (see II. 4.6, proposition 2) and n–regular. Thus the ring $\hat{\mathcal{O}}_{wz} = \mathcal{O}_{wz}/I$ is an integral domain and is finite over the subring $\hat{\mathcal{O}}_w$ ([7]) (see III. 2.2, property (4)). Moreover, the multiplicity p is equal to the dimension of the field of fractions of the ring $\hat{\mathcal{O}}_{wz}$ over the field of fractions of the subring $\hat{\mathcal{O}}_w$ (see IV. 1.7).

Now, we have the well-defined epimorphism

$$\eta : \mathcal{O}_{wz} \ni \varphi_0 \longrightarrow \Big(\varphi(f(z),z)\Big)_0 \in \mathcal{O}_z$$

(see B. 4.2–3). Its kernel is the ideal I (see II. 4.2). Therefore it induces the isomorphism

$$\kappa : \hat{\mathcal{O}}_{wz} \longrightarrow \mathcal{O}_z ,$$

and we have

$$\kappa(\hat{\mathcal{O}}_w) = \mathcal{R} ,$$

where $\mathcal{R} = \eta(\mathcal{O}_w) = \mathcal{O}_w \circ f_0$. Therefore the ring $\mathcal{O}_z$ is finite over the subring $\mathcal{R}$ and the multiplicity p is equal to the dimension of the field of fractions of the ring $\mathcal{O}_z$ over the field of fractions of the subring $\mathcal{R}$.

Now we are going to prove that $\mathcal{O}_z$ regarded as a module over $\mathcal{R}$ is free. Indeed, the ring $\mathcal{R}$ is regular and n–dimensional, since it is isomorphic to the ring $\mathcal{O}_w$ (see I. 1.8, proposition 2); namely, the mapping $\mathcal{O}_w \ni \varphi \longrightarrow \varphi \circ f_0 \in \mathcal{R}$ is an isomorphism, due to the fact that the mapping f_Ω is open. In accordance with the syzygy theorem (see A. 15), we have $\mathrm{hd}_{\mathcal{R}}\mathcal{O}_z + \mathrm{prof}_{\mathcal{R}}\mathcal{O}_z = n$. But, in view of the theorem in A. 14.2 and the corollary from the syzygy theorem, we also have $\mathrm{prof}_{\mathcal{R}}\mathcal{O}_z = \mathrm{prof}_{\mathcal{O}_z}\mathcal{O}_z = n$, and hence $\mathrm{hd}_{\mathcal{R}}\mathcal{O}_z = 0$. The latter means that $\mathcal{O}_z$ is free as a module over $\mathcal{R}$ (see A. 13.2).

According to proposition in A. 8.3, it follows that

$$p = g_{\mathcal{R}}(\mathcal{O}_z) = \dim_{\mathcal{R}/\mathfrak{n}}(\mathcal{O}_z/\mathfrak{n}\mathcal{O}_z) ,$$

where $\mathfrak{n}$ is the maximal ideal in the ring $\mathcal{R}$ (see A. 10.4). But $\mathfrak{n} = \mathfrak{m}_w \circ f_0$, where $\mathfrak{m}_w$ is the maximal ideal in the ring $\mathcal{O}_w$; hence $I_0 f = \mathfrak{n}\mathcal{O}_z$. As $\mathcal{R} = \mathfrak{n} + \mathbf{C}$ is a direct sum, we conclude (see A. 10.4a) that

$$p = \dim_{\mathbf{C}} \mathcal{O}_z/I_0 f = \mathrm{codim}\, I_0 f ,$$

and so – in view of the lemma in n° 1 – we obtain the conclusion of the theorem.

COROLLARY . *Let* $b = f(a)$, *and let* $\psi : M \times N \longrightarrow L$ *be a holomorphic mapping into an* n–*dimensional manifold* L *such that* $z \longrightarrow \psi(z,b)$ *is constant*

([7]) $\hat{\mathcal{O}}_w$ denotes the image of the subring $\mathcal{O}_w \subset \mathcal{O}_{wz}$.

and the differential $"d_{ab}\psi$ (^{7a}) is an isomorphism. If the mapping f is light at the point a, then so is the mapping $g(z) = \psi(z, f(z))$, and $m_a g = m_a f$.

In fact, one may assume that M, N, L are open convex neighbourhoods of zero in $\mathbf{C}^n$, and that $a = b = \psi(a, b) = 0$. The mapping g is light at a, because, in view of the implicit function theorem, $\{\psi = 0\}_0 = \{w = 0\}_0$ (see C. 1.13). Let $f = (f_1, \ldots, f_n)$, $g = (g_1, \ldots, g_n)$, and $\psi = (\psi_1, \ldots, \psi_n)$. According to Hadamard's lemma (and also footnote (^{18}); see C. 1.10), we have $\psi_i(z, w) = \sum_{j=1}^n h_{ij}(z, w) w_j$, where h_{ij} are holomorphic in $M \times N$ and $\det h_{ij}(0) \neq 0$. Therefore $g_i(z) = \sum_{j=1}^n h_{ij}(z, f(z)) f_j(z)$, $i = 1, \ldots, n$, and so $I_a g = I_a f$.

A different proof of this property will be given after Rouché's lemma (see the remark in n° 3 below).

3. Assume now that N is a vector space.

If the set of zeros of the mapping f is finite, say $f^{-1}(0) = \{a_1, \ldots, a_k\}$, then we define the *number of zeros of the mapping f* (counted with their multiplicities) by the formula

$$\nu(f) = \sum_1^k m_{a_i} f .$$

Note that if $\varphi : M' \longrightarrow M$ is a biholomorphic mapping between manifolds, then $\nu(f \circ \varphi) = \nu(f)$.

The proposition in n° 1 implies the following

LEMMA 1. *Assume that the set $f^{-1}(0)$ is finite and $\inf_{M \setminus E} |f(z)| > 0$ for some compact set $E \subset M$ (after endowing N with a norm). Then there exist an open neighbourhood Δ of zero in N and a nowhere dense analytic subset W of Δ such that*

$$\nu(f) = \#f^{-1}(w) \quad for \quad w \in \Delta \setminus W ,$$

and each point $w \in \Delta \setminus W$ is a regular value of the mapping f (^{8}) .

Indeed, it follows from the proposition that there are mutually disjoint open neighbourhoods $\Omega_1, \ldots, \Omega_k$ of the zeros $a_1, \ldots, a_k$ of the mapping f, an open neighbourhood Δ of zero in N, and a nowhere dense analytic subset

---

(^{7a})   i.e., the differential of $w \longrightarrow \psi(a, w)$ at b.

(^{8}) i.e., for each $z \in f^{-1}(w)$ the differential $d_z f$ is an isomorphism.

W of Δ such that $m_{a_i} f = \# f_{\Omega_i}^{-1}(w)$ for $w \in \Delta \setminus W$. Moreover each point $w \in \Delta \setminus W$ is a regular value of f_{Ω_i} $(i = 1, \ldots, k)$. We obviously have $\inf_{M \setminus \bigcup \Omega_i} |f(z)| > 0$, and so $|f(z)| > \varepsilon$ in $M \setminus \bigcup \Omega_i$ for some $\varepsilon > 0$. Now, by taking a smaller Δ so that $\Delta \subset \{|w| < \varepsilon\}$, we get the conclusion of the lemma because $f^{-1}(w) = \bigcup f_{\Omega_i}^{-1}(w)$ for $w \in \Delta$ $(^9)$.

ROUCHÉ'S LEMMA . *Let*

$$f_t(z) = F(t, z) \quad in \quad M$$

for $t \in H$, where H is a manifold and $F : H \times M \longrightarrow N$ is a holomorphic mapping $(^{10})$. Suppose that for each $t \in H$ the set $f_t^{-1}(0)$ is finite and $f_t(z) \neq 0$ outside a compact set $E \subset M$ (independent of t). Then the function

$$H \ni t \longrightarrow \nu(f_t)$$

is locally constant. If the manifold M is biholomorphic to an open set in a vector space, then the finiteness of the sets $f_t^{-1}(0)$ follows from the other assumptions $(^{11})$.

First, we will prove – under the assumptions of Rouché's lemma – the following

LEMMA 2. *Let w be a regular value of the mapping f_{t_0}. Assume that the set $f_{t_0}^{-1}(w)$ is finite, and that $f_t(z) \neq w$ in $M \setminus E$ for $t \in H$. Then $\# f_t^{-1}(w) = \# f_{t_0}^{-1}(w)$ for each t from a neighbourhood of the point t_0.*

Indeed, we have $f_{t_0}^{-1}(w) = \{a_1, \ldots, a_k\}$. Therefore $F(t_0, a_i) = w$, $i = 1, \ldots, k$, and, by the implicit function theorem, there are mutually disjoint open neighbourhoods $\Omega_1, \ldots, \Omega_k$ of the points $a_1, \ldots, a_k$ and a neighbourhood U of the point t_0 such that if $t \in U$, then $F(t, b_i(t)) = w$ for a unique $b_i(t) \in \Omega_i$ $(i = 1, \ldots, k)$. Now, $f_{t_0}(z) \neq w$ in the compact set $E \setminus \bigcup \Omega_i$; hence $f_t(z) \neq w$ in $E \setminus \bigcup \Omega_i$ for t from a neighbourhood $U_0 \subset U$ of the point t_0. Thus, for $t \in U_0$, we have $f_t^{-1}(w) = \{b_1(t), \ldots, b_k(t)\}$, and so $\# f_t^{-1}(w) = \# f_{t_0}^{-1}(w)$.

PROOF of Rouché's lemma. It is enough to show that in the case when H is an open neighbourhood of zero in a normed vector space L, the function $t \longrightarrow \nu(f_t)$ is constant in some neighbourhood of zero. We may assume (by taking a larger E, and a smaller M and H) that for some $\varepsilon > 0$ (and after

---

$(^9)$ Since $f(z) \neq w$ in $M \setminus \bigcup \Omega_i$.

$(^{10})$ Holomorphic dependence on the parameter t is not essential. See the remarks following Rouché's theorem below.

$(^{11})$ See the proposition in IV. 5.

furnishing N with a norm), we have $|F(t, z)| \geq \varepsilon$ in $H \times (M \setminus E)$, and $B = \{|t| \leq \varepsilon\} \subset H$. Now, consider the holomorphic mapping $\Phi : H \times M \longrightarrow L \times N$ defined by $\Phi(t, z) = (t, F(t, z))$. Then

$$\Phi^{-1}(t, w) = t \times f_t^{-1}(w) \quad \text{for} \quad t \in H, \ w \in N \ ,$$

the set $\Phi^{-1}(0)$ is finite, and we have $|\Phi(t, z)| \geq \varepsilon$ in $(H \times M) \setminus (B \times E)$. Therefore, according to lemma 1, there are connected open neighbourhoods $U \subset H$ and $\Delta \subset \{|w| < \varepsilon\}$ of the zeros in L and N, and a nowhere dense analytic subset Λ in $U \times \Delta$, such that $\#\Phi_s^{-1}(s, w) = \nu(\Phi)$, i.e.,

$$(\#) \qquad \#f_s^{-1}(w) = \nu(\Phi) \quad \text{if} \quad (s, w) \in (U \times \Delta) \setminus \Lambda \ .$$

Let $t \in U$. In view of lemma 1, there is an open neighbourhood $\Delta_0 \subset \Delta$ of zero in N and a nowhere dense analytic subset W_0 of Δ_0 such that

$$(\#\#) \qquad \left. \begin{array}{l} \nu(f_t) = \#f_t^{-1}(w) \text{ and} \\ w \text{ is a regular value of } f_t \end{array} \right\} \quad \text{provided that } w \in \Delta_0 \setminus W_0.$$

Now, the set $\{w \in \Delta : \Lambda^w \text{ is nowhere dense in } U\}$, where $\Lambda^w = \{t \in U : (t, w) \in \Lambda\}$, is dense in Δ. For if it were not, then, due to connectedness of U, we would have $\Lambda \supset U \times \Delta'$ for some non-empty open Δ' (see II. 3.6). Hence there exists a point $w_0 \in \Delta_0 \setminus W_0$ such that Λ^{w_0} is nowhere dense in U. Note that $f_s(z) \neq w_0$ in $M \setminus E$ for $s \in H$ (as $F(s, z) \notin \Delta_0$ when $(s, z) \in H \times (M \setminus E)$). Because of this and $(\#\#)$, we can now apply lemma 2. Thus $\#f_s^{-1}(w_0) = \#f_t^{-1}(w_0)$ for some $s \in U \setminus \Lambda^{w_0}$. Then $(s, w_0) \in (U \times \Delta) \setminus \Lambda$. In view of $(\#)$ and $(\#\#)$, it follows that

$$\nu(f_t) = \#f_t^{-1}(w_0) = \#f_s^{-1}(w_0) = \nu(\Phi) \ .$$

So we have proved that the function $t \longrightarrow \nu(f_t)$ is constant in U.

REMARK . Using Rouché's lemma, one can deduce the corollary from n° 2 as follows. One may assume that M, N, L are open neighbourhoods of zero in a normed vector space and that $a = b = \psi(a, b) = 0$. By the implicit function theorem (see C. 1.13), there is an $\varepsilon > 0$ such that $\{|z| < \varepsilon\} \subset M$, $\{|w| < \varepsilon\} \subset N$, and

$$\big(|z| < \varepsilon, \ |w| < \varepsilon, \ \psi(z, w) = 0\big) \Longrightarrow w = 0 \ .$$

Furthermore, there exists $0 < \delta < \frac{1}{2}\varepsilon$ such that $\big(|z| < \delta, \ f(z) = 0\big) \Longrightarrow z = 0$ and $f(\{|z| < \delta\}) \subset \{|w| < \varepsilon\}$. It follows that if $t \in \mathbf{C}$ and $|t| < 2$, then the only zero of the function $g_t(z) = \psi(tz, f(z))$ in $\{|z| < \delta\}$ is 0. Thus, in view of Rouché's lemma (and because $g_0 = \tilde{\psi} \circ f$, where $\tilde{\psi} : z \longrightarrow \psi(0, z)$ is biholomorphic at 0; see C. 3.13), we conclude that $m_0 g = m_0 g_1 = \nu(g_1) = \nu(g_0) = m_0 g_0 = m_0 f$.

ROUCHÉ'S THEOREM . *Assume that the manifold M is biholomorphic to an open subset of a vector space. If $g : M \longrightarrow N$ is a holomorphic mapping such that (after endowing N with a norm) the following inequality holds*

$$|g(z)| < |f(z)| \quad \text{in} \quad M \setminus E$$

for some compact set $E \subset M$ $(^{12})$, *then*

$$\nu(f + g) = \nu(f) .$$

PROOF . One may assume (by taking a larger E and a smaller M) that $|f(z)| - |g(z)| \geq \varepsilon$ and $|g(z)| \leq C$ in $M \setminus E$ for some positive ε, C. Then, setting $r = 1 + \varepsilon/(2C)$, we have for $|t| < r$ and $z \in M \setminus E$ the estimate $|f(z) + tg(z)| \geq |f(z)| - r|g(z)| > 0$. Therefore one can apply Rouché's lemma to the mapping $f_t(z) = f(z) + tg(z)$ defined in $\{|t| < r\} \times M$. So the function $t \longrightarrow \nu(f_t)$ is constant in the disc $\{|t| < r\}$ (because the latter is connected). Thus $\nu(f + g) = \nu(f_1) = \nu(f_0) = \nu(f)$.

REMARKS. In the case of an arbitrary manifold M, the following version of *Rouché's theorem* is true:

If the set $f^{-1}(0)$ is finite and contained in the interior of a compact set $E \subset M$, then there is $\delta > 0$ such that a holomorphic mapping $g : M \longrightarrow N$ satisfying the conditions: $|g(z) - f(z)| < \delta$ in E and $g(z) \neq 0$ in $M \setminus E$ implies that the set $g^{-1}(0)$ is finite and $\nu(g) = \nu(f)$.

In fact, let M_i be neighbourhoods of the zeros of f such that $M_i \subset E$, the closures of the M_i's are mutually disjoint, and each M_i is biholomorphic to an open ball in $\mathbf{C}^n$. It is enough to apply Rouché's theorem to the restrictions f_{M_i} and to note that $\inf\{|f(z)| : z \in E \setminus \bigcup M_i\} > 0$.

It is now easy to see that in *Rouché's lemma* the holomorphic dependence on the parameter t is not essential. It suffices to assume that $F(t, z)$ is a continuous mapping in $H \times M$ which is holomorphic with respect to t, where H is a topological space (the rest of the assumptions remain unchanged).

$(^{12})$ Obviously, in such a case, the sets $f^{-1}(0)$ and $(f + g)^{-1}(0)$ are finite. (See the proposition in IV. 5.)

§3. Holomorphic mappings of analytic sets

1. Let V and W be locally analytic subsets of manifolds M and N, respectively.

We say that a mapping $f : V \longrightarrow W$ is *holomorphic* if each point of V has an open neighbourhood U in M such that $f_{V \cap U}$ is the restriction of a holomorphic mapping of U into N. In the case when V and W are submanifolds, the above notion coincides with that of a holomorphic mapping of manifolds (see C. 3.8).

Clearly, if $M' \subset M$ and $N' \subset N$ are submanifolds that contain V and W, respectively, then for the mapping $f : V \longrightarrow W$ to be holomorphic it is irrelevant whether V, W are regarded as locally analytic subsets of the manifolds M, N or of the manifolds M', N' (see II. 3.4).

Combining the above definition with basic properties of (locally) analytic sets, analytically constructible sets, and holomorphic mappings of manifolds, one deduces the following properties:

Obviously, every holomorphic mapping between locally analytic sets is continuous. If V is the union of a family of open sets V_ι in V, then a mapping $f : V \longrightarrow W$ is holomorphic if and only if all the restrictions $f_{V_\iota} : V_\iota \longrightarrow W$ are holomorphic. The restriction of a holomorphic mapping $f : V \longrightarrow W$ to a locally analytic subset $Z \subset V$ of M is holomorphic. If $W' \subset W$ is a locally analytic subset of N, then a mapping $f : V \longrightarrow W'$ is holomorphic if and only if $f : V \longrightarrow W$ is holomorphic. The graph of a holomorphic mapping $f : V \longrightarrow W$ is locally analytic in $M \times N$; the inverse image under a holomorphic mapping $f : V \longrightarrow W$ of a locally analytic subset $T \subset W$ of N is locally analytic in M. The natural projections $V \times W \longrightarrow V$ and $V \times W \longrightarrow W$ are holomorphic.

The composition of holomorphic mappings is holomorphic. The Cartesian product and the diagonal product of holomorphic mappings are holomorphic.

Note also that if $g : L \longrightarrow M$ is a surjective submersion, then

$$(f : V \longrightarrow W \text{ is holomorphic}) \Longleftrightarrow (f \circ g : g^{-1}(V) \longrightarrow W \text{ is holomorphic}) .$$

The above follows directly from the definition of submersion (just as in C. 4.2).

Suppose now that V, W are analytic subsets of M and N, respectively, and that $f : V \longrightarrow W$ is a holomorphic mapping. Then the graph of the mapping f is analytic in $M \times N$. The inverse image of an analytic set or an analytically constructible set $T \subset W$ in N is – respectively – analytic or analytically constructible in M. If the image $f(Z)$ of an analytic set $Z \subset V$ in M is analytic in N; then, if the set Z is irreducible, so is its image $f(Z)$. We have the inequality

$$\dim f(E) \leq \dim E$$

for each analytically constructible set $E \subset V$ in M.

2. Let $f : V \longrightarrow W$ be a holomorphic mapping of locally analytic sets V, W in the manifolds M, N, respectively.

If $z \in V^0$, then some open neighbourhood Ω in V of the point z is a submanifold (of the manifold M), and then $\mathrm{rank}_z f$ is well-defined as the rank at the point z of the mapping $f_\Omega : \Omega \longrightarrow N$. Naturally, in the case when V and W are submanifolds, the above definition coincides with that given in C. 3. 12.

THEOREM 1. *If the set V is analytic in M, then for any connected component W of the set V^0 and for any $k \in \mathbf{N}$, the set*

$$\{z \in W : \mathrm{rank}_z f \leq k\}$$

is analytically constructible in M.

PROOF . According to theorem 4 from IV. 2.9, we have $W = V_i^0 \setminus V^*$, where V_i is a simple component of V. Hence

$$\{z \in W : \mathrm{rank}_z f \leq k\} = \{z \in V_i^0 : \mathrm{rank}_z f_{V_i} \leq k\} \setminus V^* .$$

Therefore one may assume that V is of constant dimension, and it is enough to prove that the set

$$E = \{z \in V^0 : \mathrm{rank}_z f \leq k\}$$

is analytically constructible in M (see IV. 2.8, corollary 1 from proposition 2). Since the property of being analytically constructible is local (see IV. 8.4, proposition 8), one may assume that $N = \mathbf{C}^n$, $f = F_V$, where $F = (F_1, \ldots, F_n) : M \longrightarrow \mathbf{C}^n$ is a holomorphic mapping. Moreover, it is enough to show that each point $a \in V$ has an open neighbourhood U in M such that the set $E \cap U$ is analytically constructible in U. Now, theorem 2 from IV. 2.6 says that there is an open neighbourhood U of the point a and holomorphic functions $G_1, \ldots, G_r$ on U such that $T_z V^0 = \bigcap_1^r \ker d_z G_i$ for $z \in V^0 \cap U$. Then, by putting $G = (G_1, \ldots, G_r)$, we have $\ker d_z f_{V^0} = \ker d_z(F, G)$ for $z \in V^0 \cap U$ (see C. 3.11). It follows that

$$E \cap U = \{z \in U : \mathrm{rank}_z(F, G) \leq k + \mathrm{codim}\, V\} \cap V^0 ,$$

which shows that the set $E \cap U$ is analytically constructible in U (see II. 3.7; and IV. 2.4, theorem 1).

THEOREM 2. *Let $k \in \mathbf{N}$. We have the inequalities*

$$\dim V \geq k + \dim f(V) \quad if \quad \dim f^{-1}(w) \geq k \quad for \quad w \in f(V);$$
$$\dim V \leq k + \dim f(V) \quad if \quad \dim f^{-1}(w) \leq k \quad for \quad w \in f(V).$$

PROOF of the first inequality $(^{13})$. One may assume that $N \neq \emptyset$. Next, we may assume that W is an affine space and that the set $f(V)$ is open in W. Indeed, there is a submanifold $\Gamma \subset f(V)$ of dimension equal to $\dim f(V)$ which is biholomorphic to an open subset of an affine space; then it suffices to have the inequality for the mapping $f_{f^{-1}(\Gamma)} : f^{-1}(\Gamma) \longrightarrow \Gamma$ (because $\left(f_{f^{-1}(\Gamma)} \right)^{-1}(w) = f^{-1}(w)$ for $w \in \Gamma$). Clearly, one may assume that V is analytic. Let $l = \dim W$. The case when $l = 0$, is trivial. Suppose that $l > 0$ and the inequality is true for any $(l-1)$–dimensional space W. Take a non-constant affine mapping $\chi : W \longrightarrow \mathbf{C}$. Since the set $(\chi \circ f)(V)$ is open and non-empty, it contains a point c which is not the value of any constant restriction of $\chi \circ f$ to a simple component of V. Then $W_0 = \chi^{-1}(c)$ is an affine space of dimension $l - 1$, while $V_0 = f^{-1}(W_0) = (\chi \circ f)^{-1}(c)$ is a non-empty analytic set of dimension $< \dim V$ (see IV. 2.8, proposition 3). Consider now the mapping $f_{V_0} : V_0 \longrightarrow W_0$. Since the set $f(V_0) = f(V) \cap W_0$ is non-empty and open in W_0, and $f_{V_0}^{-1}(w) = f^{-1}(w)$ for $w \in f(V_0)$, we have

$$\dim V > \dim V_0 \geq k + \dim f(V_0) = k + \dim f(V) - 1 ,$$

and so $\dim V \geq k + \dim f(V)$.

PROOF of the second inequality. Consider the natural projection $\pi : f \longrightarrow N$. We have $\pi(f) = f(V)$ and $\pi^{-1}(w) = f^{-1}(w) \times w$. Hence $\dim \pi^{-1}(w) \leq k$ for $w \in \pi(f)$. Thus $\dim V \leq \dim f \leq k + \dim \pi(f) = k + \dim f(V)$ (see n° 1 and II. 1.4).

COROLLARY 1. *If f is a mapping whose fibres are discrete, and in particular, if f is an injection, then $\dim f(V) = \dim V$. If, in addition, V is of constant dimension, then so is $f(V)$.*

COROLLARY 2. *Let $Z \subset W$ be a locally analytic subset of N. If $\dim f^{-1}(w) = k$ for each $w \in Z$, then $\dim f^{-1}(Z) = k + \dim Z$. Moreover, if the mapping f is open and both the subset Z and all the fibres $f^{-1}(w)$,*

$(^{13})$ Based on an idea due to Narasimhan [33].

$w \in Z$, are of constant dimension, then the subset $f^{-1}(Z)$ is also of constant dimension ([14]).

Indeed, to show the second part of the corollary, let $T = f^{-1}(Z)$, and let Ω be an arbitrary open neighbourhood in V of any point of T. For each $w \in f(T \cap \Omega)$, the dimension of the fibre $(f_{T \cap \Omega})^{-1}(w) = f^{-1}(w) \cap \Omega$ is equal to k and the dimension of the set $f(T \cap \Omega) = Z \cap f(\Omega)$ is equal to $\dim Z$. Therefore $\dim(T \cap \Omega) = k + \dim Z$.

For each $z \in V$, denote by $l_z f$ the germ at z of the fibre of z:

$$l_z f = \left(f^{-1}(f(z)) \right)_z .$$

THEOREM 3 (semicontinuity). *The function*

$$V \ni z \longrightarrow \dim l_z f$$

is upper semicontinuous: for each $a \in V$ we have the inequality $\dim l_z f \leq \dim l_a f$ *in a neighbourhood (in V) of a.*

PROOF . One may assume that M and N are vector spaces. Set $n = \dim M$. Fix an arbitrary $k \in \mathbf{N}$. Now, for any point $z \in V$, the condition $\dim l_z f \leq k$ is equivalent to the condition: for some $(n - k)$–dimensional subspace $L \subset M$, the point z is an isolated point of the set $f^{-1}(f(z)) \cap (z + L)$ (see III. 4.5, formula (#)). The latter occurs if and only if $(z, f(z))$ is an isolated point of the set

$$\left(f^{-1}(f(z)) \cap (z + L) \right) \times f(z) = f \cap \left((z, f(z)) + (L \times 0) \right) .$$

Now, by the corollary from proposition 1 in IV. 1.4, if our condition is satisfied for the point $z = a \in V$, then it is also satisfied for each point z from some neighbourhood (in V) of the point a.

In view of theorem 2, we get the following

COROLLARY . *We have* $\dim l_z f \geq \dim_z V - \dim f(V)$ *for $z \in V$.*

Indeed, if it were not true, then, denoting the right hand side of the inequality by p, we would have an open neighbourhood U (in V) of the point

---

([14]) The assumption that the mapping f is open cannot be omitted even in the case when V and W are submanifolds. For instance, consider the mapping $f : \mathbf{C}^3 \ni (t, z, w) \longrightarrow (tw, zw, t) \in \mathbf{C}^3$ and the set $Z = 0 \times 0 \times \mathbf{C}$ of constant dimension 1. All the fibres $f^{-1}(0) = (0 \times \mathbf{C} \times 0) \cup (0 \times 0 \times \mathbf{C})$ and $f^{-1}(0, 0, c) = c \times \mathbf{C} \times 0$, where $c \neq 0$, are of constant dimension 1, whereas the variety $f^{-1}(Z) = (\mathbf{C}^2 \times 0) \cup (0 \times 0 \times \mathbf{C})$ is not of constant dimension.

z such that $\dim f_U^{-1}(w) \le p - 1$ for $w \in f(U)$, and $\dim_z V = \dim U \le p - 1 + \dim f(U) < p + \dim f(V)$. This contradicts the definition of p.

We define the *rank of the mapping f* by the formula

$$\operatorname{rank} f = \max\{\operatorname{rank}_z f : z \in V^0\} \quad (^{15}) .$$

Clearly, this is an extension of the previously given definition in the case when V and W are submanifolds (see C. 3.12).

If the set V is irreducible, then $\operatorname{rank} f_\Omega = \operatorname{rank} f$ for any non-empty open subset Ω of V; next,

$$(*) \qquad\qquad \dim l_z f \ge \dim V - \operatorname{rank} f \quad \text{for } z \in V ,$$

and hence

$$(**) \qquad\qquad \dim f^{-1}(w) \ge \dim V - \operatorname{rank} f \quad \text{for } w \in f(V)$$

(cf. n° 3 below) $(^{16})$.

Indeed, the set $G = \{z \in V^0 : \operatorname{rank}_z f = \operatorname{rank} f\}$ is open and dense in the submanifold V^0 (see II. 3.7; and IV 2.8, corollary 1 from proposition 2). In view of the rank theorem (see C. 4.1), we have inequality $(*)$ $(^{17})$ for $z \in G$, and hence also for $z \in V$, because of the semicontinuity theorem (theorem 3).

WHITNEY'S LEMMA . *We have* $\operatorname{rank} f_{V^*} \le \operatorname{rank} f$.

PROOF . Let $r = \operatorname{rank} f_{V^*}$. We may assume that $V^* \ne \emptyset$, and then $\operatorname{rank}_z f_{V^*} = r$ in some set $\Delta \ne \emptyset$ which is open in $(V^*)^0$ and is a submanifold (see C. 3.12). Set $k = \dim \Delta$. According to the rank theorem (see C. 4.1), we may assume that $\Gamma = f(\Delta)$ is an r–dimensional submanifold, and that the fibres of the mapping $f_\Delta : \Delta \longrightarrow \Gamma$ are $(k - r)$–dimensional submanifolds. Hence $\dim l_z f_{V^*} = k - r$ for $z \in \Delta$. Now, if we had $\operatorname{rank} f < r$, i.e.,

$$\operatorname{rank}_z f < r \quad \text{for } z \in V^0 ,$$

then we would have $\dim f(V^0) < r$ (see §1, theorem 1), and so $\Gamma \not\subset f(V^0)$. Thus, there would exist a point $a \in \Delta$ such that $f(a) \notin f(V^0)$. This would mean that $f^{-1}(f(a)) \subset V^*$, and hence we would have

$$\dim l_a f = k - r .$$

_____

$(^{15})$ If $V = \emptyset$, we put: $\operatorname{rank} f = -\infty$.

$(^{16})$ Clearly, without the assumption of irreducibility of V, the inequality is no longer true.

$(^{17})$ In fact, we have equality.

It follows from the semicontinuity theorem (theorem 3) that, for some open neighbourhood U of the point a, we have

$$\dim l_z f \leq \dim l_a f \text{ if } z \in U .$$

Set $m = \dim V_a$. As a consequence of the decomposition into components of constant dimension (see IV. 2.9), there is a point $c \in V^{(m)} \cap U$. In addition, it can be chosen so that in a neighbourhood of the point, $\operatorname{rank}_z f_{V^{(m)}}$ is constant (see C. 3.12) and hence equal to some $s < r$. Therefore, by the rank theorem (and because $l_c f_{V^{(m)}} = l_c f$), we would have

$$m = s + \dim l_c f < r + \dim l_a f = k .$$

This is impossible, because $\dim V_a \geq \dim \Delta_a = \dim \Delta$.

THEOREM 4. $\operatorname{rank} f = \dim f(V)$.

Indeed, the inequality $\dim f(V) \geq \operatorname{rank} f$ follows from the fact that $f(V)$ contains (except for the trivial case $V = \emptyset$) non-empty submanifolds of dimension $\operatorname{rank} f$. (See C. 3.12 and the rank theorem in C. 4.1.) Set $k = \dim V$. The opposite inequality is trivial when $k = -\infty$. Assume now that $k \geq 0$ and the inequality is true if $\dim V < k$. Then – in view of theorem 1 from IV. 2.4 and Whitney's lemma – we have $\dim f(V^*) \leq \operatorname{rank} f_{V^*} \leq \operatorname{rank} f$. But also $\dim f(V^0) \leq \operatorname{rank} f$ (see §1, theorem 1). Therefore $\dim f(V) \leq \operatorname{rank} f$.

COROLLARY 1. *If $Z \subset V$ is a locally analytic set (in M), then*

$$\operatorname{rank} f_Z \leq \operatorname{rank} f .$$

COROLLARY 2. *If the set V is irreducible, then the set $f(V)$ is of constant dimension* [18] .

In fact, for any open neighbourhood Δ in W of any point of the set $f(V)$, since $f(V) \cap \Delta = f_{f^{-1}(\Delta)}(f^{-1}(\Delta))$, we have $\dim(f(V) \cap \Delta) = \operatorname{rank} f_{F^{-1}(\Delta)} = \operatorname{rank} f$.

3. Let $f : V \longrightarrow W$ be a holomorphic mapping between analytic subsets V, W of complex manifolds M, N, respectively, and *assume that the set V is irreducible*. (Thus V^0 is a connected submanifold and $\operatorname{rank} f_\Omega = \operatorname{rank} f$ for every open set $\Omega \neq \emptyset$ in V.)

---

[18] Obviously, the assumption of irreducibility of V cannot be replaced by the assumption that V is of constant dimension.

We define the *generic dimension of the fibres of the mapping* f by the formula

$$(*) \qquad\qquad \lambda(f) = \min\{\dim l_z f : z \in V\} .$$

(If $V = \emptyset$, we put $\lambda(f) = -\infty$.) Owing to the semicontinuity theorem (theorem 3 in n° 2), the set V in the equality $(*)$ can be replaced by any dense subset of V. Clearly,

$$\dim f^{-1}(w) \geq \lambda(f) \quad \text{for} \ \ w \in f(V) .$$

Consider the set

$$C(f) = \{z \in V^0 : \ \operatorname{rank}_z f < \operatorname{rank} f\} \cup V^* .$$

It is a nowhere dense subset of V (see II. 3.7; and IV. 2.1, proposition 1), analytic in M (see n° 2, theorem 1; and IV. 8.3, proposition 5).

We have the equalities

$$(1) \qquad\qquad \dim V = \lambda(f) + \operatorname{rank} f$$

and

$$(2) \qquad\qquad \dim l_z f = \lambda(f) \quad \text{for} \ \ z \in V \setminus C(f) .$$

For, according to the rank theorem (see C. 4.1), we have $\dim V = \dim l_z f + \operatorname{rank} f$ for $z \in V \setminus C(f)$.

We also have

$$(3) \qquad \dim f^{-1}(w) = \lambda(f) \text{ in a dense subset of the set } f(V) .$$

For otherwise one would have $\dim f^{-1}(w) > \lambda(f)$ in an open set $\Delta \neq \emptyset$ in $f(\Delta)$. Hence, in view of theorems 2 and 4 from n° 2, we would have

$$\dim V \geq \dim f^{-1}(\Delta) > \lambda(f) + \operatorname{rank} f_{f^{-1}(\Delta)} = \lambda(f) + \operatorname{rank} f ,$$

in contradiction with equality (1).

Moreover, if $V \neq \emptyset$, then equality (3) is true in the set $f(V)$, except for an F_σ–subset of $f(V)$ of dimension less than $\dim f(V)$.

Indeed, let $k > 0$, and suppose that the above is true when $\dim V < k$. (The case $\dim V = 0$ is trivial; see III. 4.3). Now let $\dim V = k$. Set

$$Z = \{z \in V^0 : \ \operatorname{rank}_z f < \operatorname{rank} f\} .$$

Then $\dim f(Z) < \dim f(V)$ (see §1, theorem 1; and n° 2, theorem 4). We have $V^* = \bigcup V_i \cup V'$, where the V_i are simple components of V^* for which $\dim f(V_i) = \dim f(V)$, while V' is the union of the remaining components, and hence $\dim f(V') < \dim f(V)$. Thus the equality (1) (combined with theorem 4 from n° 2) yields that $\lambda(f_{V_i}) < \lambda(f)$. Since $\dim V_i < k$ (see IV. 2.4, theorem 1), we have

$$\dim f_{V_i}^{-1}(w) = \lambda(f_{V_i}) \text{ in } f(V_i) \setminus H_i \ ,$$

where $H_i \subset f(V_i)$ is an F_σ-set of dimension less than $\dim f(V_i)$. The set

$$H = f(Z) \cup f(V') \cup \bigcup H_i$$

is an F_σ-set and $\dim H < \dim f(V)$. Now, if $w \in f(V) \setminus H$, then the fibre $f^{-1}(w)$ is of dimension $\leq \lambda(f)$ (and so $\dim f^{-1}(w) = \lambda(f)$). Indeed, it is disjoint from the sets V' and Z. Hence $f^{-1}(w) = f_{V_0}^{-1}(w) \cup \bigcup f_{V_i}(w)$ and $f_{V_0}^{-1}(w) \subset V \setminus C(f)$. Thus, by (2), we have $\dim f_{V_0}^{-1}(w) \leq \lambda(f)$, but also $\dim f_{V_i}^{-1}(w) \leq \lambda(f_{V_i}) \leq \lambda(f)$, and hence $\dim f^{-1}(w) \leq \lambda(f)$.

Without the assumption that V is irreducible we will prove

THEOREM 5 (CARTAN-REMMERT). *For any $k \in \mathbf{N}$, the set $\{z \in V : \dim l_z f \geq k\}$ is analytic.*

PROOF . Let $l = \dim V$. If $l \leq 0$, the theorem is trivial. Let $l > 0$, and assume that the result is true for varieties V of dimension $< l$. Set $Z_k(g) = \{z \in V' : \dim l_z g \geq k\}$ for any holomorphic mapping $g : V' \longrightarrow W'$ of analytic sets V', W'. One may assume that the set V is irreducible, as $Z_k(f) = \bigcup_i Z_k(f_{V_i})$, where $V = \bigcup V_i$ is the decomposition into simple components. (Indeed, $l_z f = \bigcup \{l_z f_{V_i} : V_i \ni z\}$; hence $\dim l_z f = \max\{\dim l_z f_{V_i} : V_i \ni z\}$; see II. 1.6.) Now, if $k \leq \lambda(f)$, then $Z_k(f) = V$, because of $(*)$. Assume that $k > \lambda(f)$ and set $C = C(f)$. Since $\dim C < \dim V$ (see IV. 2.5), it is enough to show that $Z_k(f) = Z_k(f_C)$. The inclusion $\supset$ is trivial. Now let $z \in Z_k(f)$, i.e., $\dim l_z f \geq k$. By (2), we must have $z \in C$. Consider the fibre $\Gamma = f^{-1}\big(f(z)\big)$. In view of the rank theorem (see C. 4.1) and (2), the set $\tilde{\Gamma} = \Gamma \setminus C$ is a $\lambda(f)$-dimensional submanifold, and so $\dim \tilde{\Gamma}_z \leq \lambda(f) < \dim l_z f$. As $\Gamma = (\Gamma \cap C) \cup \tilde{\Gamma}$ and $\Gamma \cap C = f_C^{-1}\big(f_C(z)\big)$, we have $l_z f = \Gamma_z = l_z f_C \cup \tilde{\Gamma}_z$, and so $\dim l_z f_C = \dim l_z f \geq k$ (see II. 1.6). Consequently, $z \in Z_k(f_C)$.

Using Remmert's proper mapping theorem (see 5.1 below) we obtain the following corollary ([19]) :

COROLLARY 1. *If the mapping f is proper, the set W is irreducible, and $\dim V \leq \dim W$, then the set $\{w \in W : \#f^1(w) < \infty\}$ is open and dense in W, and its complement in W is analytic (and nowhere dense).*

([19]) This corollary will not be used until Remmert's theorem is proved.

Indeed, since the fibres $f^{-1}(w)$ are compact, the complement of the above set in W is the set $Z = \{w \in W : \dim f^{-1}(w) \geq 1\}$ (see III. 4.3). The latter is analytic, because it is the image of the analytic set $\{z \in V : \dim l_z f \geq 1\}$. It is nowhere dense in W, for otherwise we would have $Z = W \neq \emptyset$ (see IV. 2.8, proposition 3), and consequently, $\dim V \geq 1 + \dim W$ (see n° 2, theorem 2), contrary to our assumption.

COROLLARY 2. *If M is an n–dimensional vector space and A is an analytic k–dimensional germ at $0 \in M$, then the set of coordinate systems $\varphi \in L_0(M, \mathbf{C}^n)$ in which the germ A is k–regular is the complement of a nowhere dense algebraic subset of $L(M, \mathbf{C}^n)$ (and hence it is open and dense in $L_0(M, \mathbf{C}^n)$; see II. 3.2).*

In fact, it is sufficient to show that the complement Z of our set in $L(M, \mathbf{C}^n)$ is algebraic (see III. 2.7). Let V be an analytic representative of the germ A in an open neighbourhood U of zero. One may assume that $k \geq 0$, i.e., that $0 \in V$. Let $N = \{z_1 = \ldots = z_k = 0\} \subset \mathbf{C}^n$. A coordinate system $\varphi \in L_0(M, \mathbf{C}^n)$ belongs to our set if and only if $\varphi(A) \cap N = 0$ (see III. 4.4) or, equivalently, if $\left(V \cap \varphi^{-1}(N)\right)_0 = 0$. It follows that the set Z is a cone. Now we have $l_{(0,\varphi)}\pi = \left(V \cap \varphi^{-1}(N)\right)_0 \times \varphi$ for $\varphi \in L(M, \mathbf{C}^n)$, where $\pi : \Lambda \longrightarrow L(M, \mathbf{C}^n)$ is the natural projection of the set

$$\Lambda = \{(z, \varphi) \in U \times L(M, \mathbf{C}^n) : z \in V, \varphi(z) \in N\}.$$

The latter is analytic in $U \times L(M, \mathbf{C}^n)$. Thus

$$Z = \{\varphi \in L(M, \mathbf{C}^n) : \dim l_{(0,\varphi)}\pi > 0\} \cup \Sigma,$$

where $\Sigma = L(M, \mathbf{C}^n) \setminus L_0(M, \mathbf{C}^n)$ is an algebraic set (see II. 3.2). Therefore, in view of theorem 5, the set Z is analytic, and so it is algebraic, according to the Cartan-Remmert-Stein lemma in II. 3.3.

Corollary 2 implies that the sets of coordinate systems in $L_0(\mathbf{C}^k, \mathbf{C}^k)$ and $L_0(M, \mathbf{C}^n)$, described in the proposition in III. 2.7 and its corollary (see footnotes ([10]) and ([10a]) in III.3), are the complements of nowhere dense algebraic sets in $L(\mathbf{C}^k, \mathbf{C}^k)$ and $L(M, \mathbf{C}^n)$, respectively. (Hence they are open and dense.)

4. Let V and W be locally analytic subsets of the manifolds M and N, respectively.

A mapping $f : V \longrightarrow W$ is said to be *biholomorphic* if it is bijective and the mappings f, f^{-1} are holomorphic. (In the case when V and W are submanifolds, the definition coincides with that given before; see C. 3.10).

Such a mapping is a homeomorphism. If there is a biholomorphic mapping $f : V \longrightarrow W$, we say that the subsets W and V are *biholomorphic*.

Obviously, the inverse of a biholomorphic mapping is biholomorphic and so is the composition of biholomorphic mappings. If $f : V \longrightarrow W$ is a biholomorphic mapping, then if $Z \subset V$ is a locally analytic set in M, the set $f(Z)$ is locally analytic in N and the mapping $f_Z : Z \longrightarrow f(Z)$ is biholomorphic. Note also that if $g : V \longrightarrow W$ is a holomorphic mapping, then the natural projection from the graph of g to V is biholomorphic.

Let $f : V \longrightarrow W$ be a biholomorphic mapping.

Biholomorphic mappings "preserve" regular points (and their dimensions) as well as singular points: $f(V^{(k)}) = W^{(k)}$, and hence $f(V^*) = W^*$. Indeed, let $a \in V^{(k)}$ and $b = f(a)$. The mapping f^{-1} is the restriction of a holomorphic mapping F of a neighbourhood of b in N in the manifold M. Therefore $F\big(f(z)\big) = z$ in a neighbourhood of a in V. Since $f_{V^{(k)}} : V^{(k)} \longrightarrow N$ is a holomorphic mapping of complex manifolds, we have $(d_b F)\big(d_a f_{V^{(k)}}(u)\big) = u$ in $T_a V$, and so the differential $d_a f_{V^{(k)}}$ is injective. Hence (see C. 3.14) the image under f of a neighbourhood in V of a, i.e., a neighbourhood in W of b, is a k–dimensional submanifold of the manifold N. This means that $b \in W^{(k)}$.

In particular, if V is a submanifold, then so is W.

It follows that every biholomorphic mapping preserves dimension:

$$\dim f(E) = \dim E \quad \text{for} \ \ E \subset V .$$

Now assume that V, W are analytic (in M and N, respectively).

Any biholomorphic mapping between analytic sets preserves analyticity and analytic constructibility. That is, if $E \subset V$ is an analytic (or analytically constructible) set in M, then $f(E)$ is analytic (or analytically constructible) in N. Any biholomorphic mapping (of analytic sets) preserves irreducibility of analytic subsets. This means that if a set $Z \subset V$ is an irreducible analytic subset of M, then its image $f(Z)$ is an irreducible analytic subset of N. Therefore any biholomorphic mappings between analytic sets preserves the decomposition of an analytic subset into simple components as well as the decomposition into components of constant dimension.

We say that a mapping $f : V \longrightarrow W$ is *biholomorphic at a point* $a \in V$, if there are open neighbourhoods Δ of a in V and Ω of $f(a)$ in W such that the mapping $f_\Delta : \Delta \longrightarrow \Omega$ is biholomorphic. If this is the case for each $a \in V$, we say that f is a *locally biholomorphic mapping* [20] .

---

[20] These definitions generalize those from C. 3.10.

(Of course, a locally biholomorphic mapping is holomorphic and is a local homeomorphism.) Note that if $f : V \longrightarrow W$ is biholomorphic at a point a, then $a \in V^0 \Longleftrightarrow f(a) \in W^0$.

PROPOSITION . *Suppose that* $f : V \longrightarrow W$ *is a holomorphic bijection and* W *is a submanifold of* N. *Then the following conditions are equivalent:*

(1) V *is a submanifold of* M;

(2) V *is of constant dimension*;

(3) f *is a homeomorphism*;

If any of the conditions is satisfied, then f *is biholomorphic* (^{21}) .

PROOF . One may assume that $W = N$. The implications $(1) \Longrightarrow (3)$ (with the last conclusion) and $(3) \Longrightarrow (2)$ follow from corollary 2 of theorem 2 in §1 and corollary 2 of theorem 2 in n° 2, respectively. Now suppose that (2) is satisfied. Then f is of constant dimension $n = \dim N$, because the natural projection $f \longrightarrow V$ is biholomorphic and the natural projection $f \longrightarrow N$ is bijective (see n° 2, corollary 1 from theorem 2). Take an arbitrary $(b, a) \in f^{-1} \subset N \times M$. It is enough to show that some neighbourhood of (b, a) in f^{-1} is the graph of a holomorphic mapping of an open set in N. Indeed, it would then follow that the mapping f^{-1} is holomorphic (since it would be biholomorphic at each $b \in N$); hence f would be biholomorphic and V would be a submanifold. Now, the germ $(f^{-1})_{(a,b)}$ is of constant dimension n. One may assume that $M = \mathbf{C}^m$, $N = \mathbf{C}^n$, and $(b, a) = 0$. As $(f^{-1})_0 \cap (0 \times M) = 0$, the germ $(f^{-1})_0$ is n–regular (see III. 4.4), and hence it has a normal triple of dimension n whose crown is a neighbourhood of the point 0 in f^{-1} (see IV. 1.4, proposition 1; and IV. 1.2). But the multiplicity of this triple must be 1, and so (see IV. 1.8) its crown is the graph of a holomorphic mapping (on an open neighbourhood of the origin).

The conditions (1)–(3) are essential. A counter-example is the holomorphic bijection $\mathbf{C}^2 \supset \{zw = 1\} \cup (0,0) \ni (z, w) \longrightarrow z \in \mathbf{C}$.

COROLLARY . *If a holomorphic mapping* $f : V \longrightarrow W$ *is a local homeomorphism and* W *is a submanifold of* N, *then* V *is a submanifold of* M *and* f *is locally biholomorphic.*

(^{21}) Note that if $f : V \longrightarrow W$ is a holomorphic bijection and V is a submanifold of M, then W need not to be a submanifold even if f is a homeomorphism and V, W are irreducible (see IV. 2.1, footnote (^{10})).

§4. Analytic spaces

Complex manifolds are locally modelled on open sets of the space $\mathbf{C}^n$. Similarly, analytic spaces are locally modelled on analytic subvarieties. Now we are going to define analytic spaces and state their elementary properties (in n° 1–3), which can be checked easily, usually in the same way as for manifolds.

1. By an *analytic space* we mean a topological Hausdorff space X with an *analytic atlas (in the restricted sense)*, i.e., with a family of homeomorphisms $\varphi_\iota : G_\iota \longrightarrow V_\iota$, where $\{G_\iota\}$ is an open cover of X and V_ι are locally analytic subsets of $\mathbf{C}^{n_\iota}$ such that the mappings

$$\varphi_\kappa \circ \varphi_\iota^{-1} : \varphi_\iota(G_\iota \cap G_\kappa) \longrightarrow \varphi_\kappa(G_\iota \cap G_\kappa)$$

are holomorphic ([22]) (and hence biholomorphic). Two such atlases (in the restricted sense) are said to be *equivalent* if their union is an analytic atlas. It is assumed that equivalent atlases introduce on X the same structure of an analytic space. Every atlas that is equivalent to the atlas $\{\varphi_\iota\}$ is called an *(analytic) atlas (in the restricted sense) of the analytic space X*. ([23])

In what follows, we will be assuming that there is a countable atlas on X or, equivalently, that there is a countable basis for the topology of X. Note that every analytic space is locally compact and hence is a Baire space. It is also a locally connected space ([24]) and hence its connected components are open.

Generally, an *analytic atlas* is defined by requiring in the above definition of an analytic atlas in the restricted sense that the V_ι are analytic subsets of manifolds M_ι, respectively (or, equivalently, that the V_ι are locally analytic subsets of some manifolds.) Two analytic atlases are said to be *equivalent* if their union is an analytic atlas. Any analytic atlas defines on X a structure of an analytic space. Namely, by taking for each of the manifolds M_ι a complex atlas $\{\psi_{\iota\kappa} : H_{\iota\kappa} \longrightarrow D_{\iota\kappa}\}$, we get the analytic atlas in the restricted sense

---

([22]) Each of the sets $\varphi_\iota(G_\iota \cap G_\kappa)$ is a locally analytic subset (in $\mathbf{C}^{n_\iota}$) since it is an open subset of the subset V_ι.

([23]) Exactly as in the case of the definition of a manifold (see C. 3.1, footnote ([26])) one can avoid the notion of equivalent atlases by furnishing X with a maximal analytic atlas (in the restricted sense).

Analytic spaces can also be defined in terms of sheaves. Then the above definition based on the notion of an atlas corresponds – according to the most frequently used terminology – to that of a *reduced analytic space* (see, e.g., [17a] or [16a]). See footnote ([21c]) in VI. 1.5 below.

([24]) See the theorem in IV. 1.6.

$\{\psi_{\iota\kappa} \circ \varphi_\iota : G_{\iota\kappa} \longrightarrow W_{\iota\kappa}\}$, where $G_{\iota\kappa} = \varphi_\iota^{-1}(H_{\iota\kappa})$ are open in X and $W_{\iota\kappa} = \psi_{\iota\kappa}(V_\iota \cap H_{\iota\kappa})$ are analytic in $D_{\iota\kappa}$, respectively. This atlas defines on X the structure of an analytic space which does not depend on either the choice of the complex atlases $\{\psi_{\iota\kappa}\}$ or the replacement of the analytic atlas $\{\varphi_\iota\}$ by an equivalent one. Every (analytic) atlas that is equivalent to the atlas $\{\varphi_\iota\}$ is called an *analytic atlas of the analytic space X*.

In particular, any locally analytic subset of a manifold is an analytic space (with the atlas consisting of the identity mapping of this subset).

A homeomorphism φ of an open subset of X onto an analytic subset of a manifold is said to be a *chart on the analytic space X* if $\{\varphi_\iota\} \cup \varphi$ is an atlas. (This definition is independent of the choice of the atlas $\{\varphi_\iota\}$ of the analytic space X.) In particular, the elements of any atlas of the analytic space X are charts on X. The atlases of the analytic space X are precisely the families of charts on X whose domains cover X.

In any open subset G of the analytic space X with the atlas $\{\varphi_\iota\}$, the induced structure of an analytic space is defined (in a unique fashion) by the atlas $\{(\varphi_\iota)_G\}$.

In the *Cartesian product of analytic spaces X and Y* with atlases $\{\varphi_\iota\}$, $\{\psi_\kappa\}$, respectively, the structure of an analytic space is defined (in a unique way) by the atlas $\{\varphi_\iota \times \psi_\kappa\}$. If φ and ψ are charts on the spaces X and Y, respectively, then $\varphi \times \psi$ is a chart on the space $X \times Y$.

2. Every complex manifold is an analytic space; it has the induced structure of an analytic space defined (in a unique fashion) by any complex atlas of the manifold. We say that an *analytic space X is an n–dimensional manifold* if its structure is induced by the structure of an n–dimensional manifold (the latter must then be unique; see below).

Note that an analytic atlas that consists of charts whose ranges are open subsets of $\mathbf{C}^n$ must be a complex atlas (see 3.4). It follows that:

The structure of an analytic space can be induced by at most one structure of an n–dimensional manifold. An analytic space X is an n–dimensional manifold if and only if it has an atlas that consists of charts onto open subsets of $\mathbf{C}^n$ (and this atlas is then the complex manifold atlas inducing the structure of an analytic space on X).

Let X be an analytic space. If it is an n–dimensional manifold, then so is each of its open subsets (with the induced structure of an analytic space). If $\{G_\iota\}$ is an open cover of the space X, then the latter is an n–dimensional manifold if and only if each of the analytic spaces G_ι is an n–dimensional manifold.

An analytic space X with an atlas $\{\varphi_\iota : G_\iota \longrightarrow V_\iota\}$ is an n–dimensional manifold if and only if all V_ι are n–dimensional submanifolds. (As for the sufficiency of the condition, it is enough to take the atlas $\{\psi_{\iota\kappa} \circ \varphi_\iota\}$, where $\{\psi_{\iota\kappa}\}$ are atlases of the manifolds V_ι, respectively. The condition is necessary, because every biholomorphic mapping of analytic sets preserves regular points and their dimension; see 3.4.)

In particular, a locally analytic subset of a manifold M, regarded as an analytic space, is an n–dimensional manifold precisely when it is an n–dimensional submanifold of the manifold M.

3. A subset Z of an analytic space X is said to be *analytic* or *locally analytic* if its image under every chart on an analytic subset of a manifold is analytic or locally analytic, respectively, in this manifold. It is enough if this is true for each chart of a given atlas $\{\varphi_\iota : G_\iota \longrightarrow V_\iota\}$. Then the set Z has the induced structure of an analytic space, defined (in a unique way) by the atlas of the restrictions $\{(\varphi_\iota)_Z : Z \cap G_\iota \longrightarrow \varphi_\iota(Z)\}$. The set Z furnished with this structure is called an *analytic subspace* of the space X. A subset of an analytic space is analytic if and only if it is locally analytic and closed. Clearly, every open subset G of an analytic space X is locally analytic ([25]) ; note that the charts of the analytic space G are precisely the charts of the space X whose domains are contained in G. The locally analytic subsets of the analytic space X are precisely the analytic sets in the open subsets of X (regarded as analytic spaces). If Z is an analytic subspace of the analytic space X, then a set $W \subset Z$ is locally analytic in Z if and only if it is locally analytic in X; then X and Z induce the same structure of an analytic space on W. In the case when Z is analytic (i.e., if, in addition, Z is closed), then a set $W \subset Z$ is analytic in Z if and only if it is analytic in X. If Z, W are (locally) analytic subsets of analytic spaces X, Y, respectively, then $Z \times W$ is a (locally) analytic subset of the space $X \times Y$. Then the Cartesian product $Z \times W$ of the analytic spaces Z and W coincides with the analytic subspace $Z \times W$ of the analytic space $X \times Y$. The union of a locally finite family and the intersection of a finite family of analytic subsets of an analytic space are analytic. If G is an open subset and Z is a (locally) analytic subset of the analytic space X, then $Z \cap G$ is a (locally) analytic subset of the analytic space G. If $\{G_\iota\}$ is an open cover of the analytic space X, then a set $Z \subset X$ is (locally) analytic if and only if each of the sets $Z \cap G_\iota$ is (locally) analytic in the analytic space G_ι, respectively.

If an analytic space X is a manifold, then the notions of an analytic and a locally analytic subset defined above coincide with those defined in II. 3.4.

---

[25] Obviously, the induced structure on G defined here coincides with that defined in n° 1.

If an analytic subspace Z of the analytic space X is an n–dimensional manifold, then the set (the analytic subspace) Z is called an n–*dimensional submanifold of the analytic space* X. Any k–dimensional submanifold (and in particular, any open subset) of an n–dimensional submanifold of an analytic space X, is also a k–dimensional (respectively, n–dimensional) submanifold of the space. If a set $Z \subset X$ is the union of a family of its open subsets, each of which is an n-dimensional submanifold, then the set Z is an n–dimensional submanifold. If $\{\varphi_\iota : G_\iota \longrightarrow V_\iota\}$ is an atlas of the analytic space X, where V_ι are analytic subsets of manifolds M_ι, respectively, then a set $Z \subset X$ is an n–dimensional submanifold if and only if all $\varphi_\iota(Z)$ are n–dimensional submanifolds of M_ι, respectively.

4. Let X be an analytic space, $E \subset X$, and $a \in X$. The *dimension of the set E at the point a* is well-defined by the formula $\dim_a E = \dim_{\varphi(a)} \varphi(E)$, where φ is any chart whose domain contains a. Next, we define $\dim E = \sup_{z \in X} \dim_z E$ (and we put $\dim \emptyset = -\infty$). Clearly, any non-empty countable set is of dimension zero. Note that the dimension of an analytic space can be infinite ([26]) , while each of its points has a neighbourhood of finite dimension. If X is a subspace of an analytic space Y, then the quantities $\dim_a E$, $\dim E$ defined above remain unchanged if E is regarded as a subset of Y. If X is a manifold, the above definitions of $\dim_a E$ and $\dim E$ coincide with those given in II. 1. If $Z \subset X$ is a non-empty submanifold, then $\dim Z$, as defined above, is also the dimension of the manifold Z. Moreover, we have the formula

$$\dim E = \sup\{\dim \Gamma : \ \Gamma \subset E \text{ is a submanifold of the space } X\} \ .$$

We say that the set E is *of constant dimension k* (where $k \in \mathbf{N}$) if $\dim_z E = k$ for $z \in E$. The dimension of the germ A of a set at a point a is well-defined by: $\dim A = \dim_a \tilde{A}$, where $\tilde{A}$ is a representative of the germ A. All the properties of the dimension listed in II. 1.3–6 ([27]) remain true in analytic spaces. They either follow directly from the definition or can be verified exactly as in the case of subsets of manifolds.

5. A *regular point (of dimension k) of an analytic space X* is a point of X a neighbourhood of which is a manifold (of dimension k). A point of X that is not regular is called a *singular point*. By X^*, X^0, and $X^{(k)}$ we denote respectively: the set of singular points of X, the set of regular points of X,

---

([26]) It is enough to take the disjoint union of the spaces $\mathbf{C}^1, \mathbf{C}^2, \mathbf{C}^3, \ldots$ (with the atlas $\mathbf{C}^n \ni z \longrightarrow z \in \mathbf{C}^n$, $n = 1, 2, \ldots$).

([27]) With regard to analytic germs, see n° 9 below.

and the set of regular points of dimension k of X. In particular, for each locally analytic subset V of an analytic space, we have thus defined the regular points (of dimension k), the singular points, and the sets $V^*, V^0, V^{(k)}$. In the case when X is a manifold, these definitions coincide with those given in IV. 2.1. For any chart, we have $\varphi(X^0) = \varphi(X)^0$, $\varphi(X^*) = \varphi(X)^*$, and $\varphi(X^{(k)} = \varphi(X)^{(k)}$. The space X is said to be *smooth* if $X^0 = X$. Therefore the n–dimensional manifolds are precisely the smooth spaces of constant dimension n.

Almost all properties from §2, Chapter IV, carry over to the case of (locally) analytic subsets of analytic spaces [28] . One proves them either in the same way as in §2, or directly from the definition and basic properties of (locally) analytic subsets of manifolds. Among others, the properties listed below remain true:

The elementary properties from n° 1 [29] , including proposition 1, theorem 1 from n° 4, the properties from n° 5 (in some of them one needs to assume that the dimension is finite or to replace "max" by "sup"), theorem 3 from n° 7. In particular, we have

THEOREM . *If X is an analytic space, then X^* is a nowhere dense analytic subset of X, whereas X^0 is an open dense subset of X. Moreover,* $\dim X^* < \dim X$, *provided that $X \neq \emptyset$ and $\dim X^* < \infty$.*

Irreducibility of a (locally) analytic subset (of an analytic space) is defined in the same way (see n° 8) and it is equivalent to irreducibility of the subset regarded as an analytic space. Propositions 2 and 3, together with their corollaries, remain true. Every irreducible analytic set and, in particular, every irreducible analytic space is of constant finite dimension (see corollary 1 of proposition 2). In particular, we have:

PROPOSITION α. *An analytic space $X \neq \emptyset$ is irreducible if and only if X^0 is a connected manifold. Then $\dim X = \dim X^0$.*

PROPOSITION β. *If X is an irreducible analytic space, then every proper analytic subset V is nowhere dense and $\dim V < \dim X$.*

Exactly as before, we can define the *decomposition of a (locally) analytic subset into simple components* (see n° 9) [30] . Theorem 4 and its corollaries remain true and we define in a similar fashion the *decomposition into components of constant dimension* (this decomposition can be infinite). In

---

[28] With regard to analytic germs, see n° 9 below.

[29] As far as invariance with respect to holomorphic mappings is concerned, see n° 7 below.

[30] It is at the same time the decomposition into irreducible components of the subset regarded as an analytic space.

particular:

Every analytic space X has a unique *decomposition into simple components*: it is the union of a unique locally finite family of irreducible analytic subsets X_i such that $X_i \not\subset X_j$ for $i \neq j$. These subsets are called the *simple components* of the space X. We have the formulae

$$(*) \qquad X^* = X_i^* \cup \bigcup_{i \neq j}(X_i \cap X_j) \quad \text{and} \quad X^0 = \bigcup_i (X_i^0 \setminus \bigcup_{\nu \neq i} X_\nu) \ .$$

Finally, proposition 5 from n° 9 and its corollary as well as theorem 5 from n° 10 remain true.

6. Let X and Y by analytic spaces.

We say that $f : X \longrightarrow Y$ is a *holomorphic mapping* if for each chart φ of the space X and for each chart ψ of the space Y the mapping $\psi \circ f \circ \varphi^{-1}$ is holomorphic ([30a]). It is enough to require that the above holds true for each pair of charts $\varphi_\iota, \psi_\kappa$ from a given pair of atlases $\{\varphi_\iota\}$, $\{\psi_\kappa\}$ of the spaces X and Y, respectively. If X and Y are locally analytic subsets of manifolds, the above definition of a holomorphic mapping coincides with that of 3.1. In the case $Y = \mathbf{C}$, we say that f is a *holomorphic function on X* ([31]) . The ring of holomorphic functions on X is denoted by $\mathcal{O}_X$. If X is irreducible, then $\mathcal{O}_X$ is an integral domain (by proposition α in n° 5; see C. 3.9). If X is irreducible, then $\mathcal{O}_X$ is an integral domain (by proposition α in n° 5; see C. 3.9).

The restriction of a holomorphic mapping $f : X \longrightarrow Y$ to any analytic subspace of X is holomorphic. If $Y' \subset Y$ is an analytic subspace, then the mapping $f : X \longrightarrow Y'$ is holomorphic precisely when the mapping $f : X \longrightarrow Y$ is holomorphic. If $\{G_\iota\}$ is an open cover of the space X, then the mapping $f : X \longrightarrow Y$ is holomorphic if and only if all the restrictions f_{G_ι} are holomorphic. The graph of the holomorphic mapping $f : X \longrightarrow Y$ is an analytic subset of the space $X \times Y$. The inverse image of a (locally) analytic subset of Y under a holomorphic mapping $f : X \longrightarrow Y$ is a (locally) analytic subset of X. The image of an irreducible locally analytic subset under a holomorphic mapping is irreducible, provided that it is locally analytic. The composition of holomorphic mappings is holomorphic. The diagonal product

---

([30a])   Naturally, for each of these mappings, the domain $\varphi(f^{-1}(\text{domain of }\psi))$ *should be open* in the range of φ. Clearly this is equivalent to the *continuity* of f.

([31])   In this case, it suffices if the composition $f \circ \varphi^{-1}$ is holomorphic for each chart φ from a given atlas on X.

as well as the Cartesian product of holomorphic mappings are holomorphic. The natural projections $X \times Y \longrightarrow X$ and $X \times Y \longrightarrow Y$ are holomorphic.

7. Let X and Y be analytic spaces. As in the case of locally analytic subsets of manifolds (see 3.4), a mapping $f : X \longrightarrow Y$ is said to be *biholomorphic* if it is bijective and the mappings f, f^{-1} are holomorphic. If there is a biholomorphic mapping $f : X \longrightarrow Y$, we say that the *spaces X and Y are isomorphic or biholomorphic*. Note that the charts of the space X are precisely the biholomorphic mappings of open subsets of X onto analytic subsets of manifolds.

Biholomorphic mappings of analytic spaces have all the properties listed in 3.4. In particular, they preserve (local) analyticity of subsets.

By an (*analytic*) *embedding of the space Y in the space X* we mean any biholomorphic mapping of the space Y onto an analytic subspace of the space X.

7a. Let $\{X_\iota\}$ be a family of analytic spaces. Just as in the case of manifolds (see C. 3.10a), the analytic space X with a family of biholomorphic mappings $\{f_\iota : X_\iota \longrightarrow X\}$ onto open subsets of X that constitute a cover of X is called a *gluing of the family* $\{X_\iota\}$. Each composition $f_{\iota\kappa} = f_\kappa^{-1} \circ f_\iota$ is a biholomorphic mapping of an open subset of the space X_ι onto an open subset of the space X_κ. We say that the space X, or – more precisely – the pair $X, \{f_\iota : X_\iota \longrightarrow X\}$, is a *gluing of the analytic spaces X_ι by the biholomorphic mappings $f_{\iota\kappa}$* [32] . The family $\{f_{\iota\kappa}\}$ satisfies the following conditions

$$(\#) \quad \begin{cases} f_{\iota\iota} \text{ is the identity on } X_\iota, \ f_{\iota\kappa} = f_{\kappa\iota}^{-1}, \ f_{\kappa\lambda} \circ f_{\iota\kappa} \subset f_{\iota\lambda}, \\ \text{the graph of } f_{\iota\kappa} \text{ is closed in } X_\iota \times X_\kappa; \ (\text{for all } \iota, \kappa, \lambda). \end{cases}$$

Conversely, if a family $\{f_{\iota\kappa}\}$ satisfies the conditions $(\#)$, where each $f_{\iota\kappa}$ is a biholomorphic mapping of an open subset of the space X_ι onto an open subset of the space X_κ, then there exists a gluing of the spaces X_ι by the biholomorphic mappings $f_{\iota\kappa}$ (and it is unique up to an isomorphism [33]). This can be verified in the same way as in the case of manifolds (see C. 3.10a).

If $\{f_{\iota\kappa}\}_{\iota \neq \kappa}$ is a family that satisfies the three last conditions in $(\#)$, then, after adding to the family the identities $f_{\iota\iota}$ of the spaces X_ι, respectively, the

---

[32] Obviously, if X_ι are manifolds, then so is X and the above definitions coincide with those given in C. 3.10a.

[33] See C. 3.10a, footnote [41].

family satisfies all the conditions in ($\#$). Hence these spaces comprise a gluing X. In such circumstances we say that X is a *gluing of the spaces* X_ι *by the biholomorphic mappings* $f_{\iota\kappa}$, $\iota \neq \kappa$.

8. Let $f : X \longrightarrow Y$ be a holomorphic mapping of analytic spaces.

Let $a \in X^0$. Then some open neighbourhood in Ω of the point a is a manifold. The rank of the mapping f at the point a is well-defined by the formula: $\mathrm{rank}_a f = \mathrm{rank}_a(\psi \circ f_\Omega)$, where ψ is a chart on Y whose domain contains $f(a)$ ([34]). Then $\mathrm{rank}_a f = \mathrm{rank}_{\varphi(a)} \psi \circ f \circ \varphi^{-1}$ for any charts φ, ψ of the spaces X, Y, whose domains contain a and $f(a)$, respectively. If X and Y are locally analytic subsets of manifolds, the above definition of $\mathrm{rank}_a f$ coincides with that from 3.2.

Next, we define (as in 3.2): $\mathrm{rank}\, f = \sup\{\mathrm{rank}_z f : z \in X^0\}$. Then $\mathrm{rank}\, f = \sup_{\iota\kappa} \mathrm{rank}(\psi_\kappa \circ f \circ \varphi^{-1})$ for any pair of atlases $\{\varphi_\iota\}$, $\{\psi_\kappa\}$ of the spaces X, Y, respectively.

We denote (as in 3.2) by $l_z f$ the germ at the point $z \in X$ of the fibre of this point. If the space X is irreducible, we define (as in 3.3) the generic dimension $\lambda(f)$ of the fibres of f, and also the set $C(f)$.

Theorems 2, 3, and 4 from 3.2, and their corollaries, as well as the properties in 3.3, together with the Cartan-Remmert theorem and corollary 1, remain true. They can be shown partly in the same way, partly by using properties of holomorphic mappings of (locally) analytic subsets of manifolds. We have the inequality

$$(\alpha) \qquad\qquad \dim X \geq k + \dim f(X) \;,$$

provided that $\dim f^{-1}(w) \geq k$ for each $w \in f(X)$.

Moreover, in the case when the space X is irreducible, we have the inequality

$$(\beta) \qquad\qquad \dim f^{-1}(w) \geq \dim X - \mathrm{rank}\, f \;\; \text{for} \;\; w \in f(X) \;.$$

9. Let X be an analytic space, and let $a \in X$.

The germs of analytic subsets of open neighbourhoods of a are called *analytic germs at the point a*. The dimension of the analytic germ is well-defined by the formula $\dim V_a = \dim_a V$ (see n° 4). Biholomorphic mappings preserve

---

([34]) The independence of ψ follows from the inequality $\mathrm{rank}_z(h \circ g) \leq \mathrm{rank}_z g$ for holomorphic mappings f, g of manifolds.

analyticity of germs, their irreducibility (as defined in II. 4.6), smoothness (as defined in II. 3.3), and dimension. In particular, the germ of a set at a is analytic (irreducible,...) if and only if its image via some (and hence each) chart whose domain contains a is analytic (irreducible,...).

This implies the existence and uniqueness of the decomposition into simple germs. Germs of constant dimension are defined in the same way. Propositions 1-4 from IV. 3.1 with the corollaries, and proposition 4 from IV. 2.8 with the remark and the corollary can now be deduced similarly.

Let $\mathcal{O}_a = \mathcal{O}_a(X)$ denote the ring of germs (at a) of holomorphic functions in some neighbourhood of the point a. The ring contains the field $\mathbf{C}$ [35] . For any analytic germ A at a we denote by $\mathcal{I}(A) = \mathcal{I}(A, X)$ the ideal $\{f \in \mathcal{O}_a : f_A = 0\}$. If X is an analytic subset of a manifold, the ring $\mathcal{O}_a(X)$ is identified with $\mathcal{O}_{X_a}$, and then the ideal $\mathcal{I}(A)$ is identified with $\mathcal{I}_{X_a}(A)$ (see IV. 4.2). If h is a biholomorphic mapping of a neighbourhood of a onto a neighbourhood of a point b in an analytic space Y such that $h(a) = b$, then the mapping $\mathcal{O}_b(Y) \ni \psi \longrightarrow \psi \circ h \in \mathcal{O}_a(X)$ is a $\mathbf{C}$–isomorphism; the ideal $\mathcal{I}(A)$ is the image of the ideal $\mathcal{I}(h(A))$. By using a chart h, one can carry over a number of properties from IV. 4.3–7.

In particular (see IV. 4.3), the ring $\mathcal{O}_a$ is always noetherian and local (with the maximal ideal $\mathfrak{m}_a = \{f \in \mathcal{O}_a : f(a) = 0\}$. It is an integral domain if and only if the germ X_a is irreducible. A germ from $\mathcal{O}_a$ is not a zero divisor precisely when it has a representative $\tilde{f} \not\equiv 0$ (see IV. 4.3).

The *Zariski dimension* of an analytic germ A at the point a is well-defined by the formula $\dim \operatorname{zar} A = \dim \operatorname{zar} \varphi(A)$ (see IV. 4.4) [36] , where φ is a chart whose domain contains a. We have the following properties:

$(\#\#)$ $\dim \operatorname{zar} X_a \geq \dim X_a$, and $\dim \operatorname{zar} X_a = \dim X_a \Longleftrightarrow a \in X^0$,

$(\#\#\#)$ $\dim \operatorname{zar} X_a \leq \dim \operatorname{zar} A + g(\mathcal{I}(A))$.

The locus of an ideal $I \subset \mathcal{O}_a$ is well-defined (as in II. 4.1 and 3) by the formula:

$$V(I) = V(f_1, \ldots, f_k) = \{\tilde{f}_1 = \ldots = \tilde{f}_k = 0\}_a ,$$

where $f_1, \ldots, f_k$ are generators of the ideal I, while $\tilde{f}_1, \ldots, \tilde{f}_k$ are their representatives. (The locus is clearly an analytic germ.)

---

[35] After having identified $\mathbf{C}$ with the subring of the germs of constants.

[36] Equivalently, it can be defined as in IV. 4.4, by introducing first the ring $\mathcal{O}_A$ as in IV. 4.2.

All the properties stated in II. 4 remain true and can be checked in the same way (except for the equivalence on generators of the ideal of a smooth germ). In particular:

$$I \subset J \Longrightarrow V(I) \supset V(J), \quad A \subset B \Longleftrightarrow \mathcal{I}(A) \supset \mathcal{I}(B),$$

$$V\big(\mathcal{I}(A)\big) = A \quad \text{and} \quad \mathcal{I}\big(V(I)\big) \supset I \ .$$

Furthermore: (A is irreducible)$\Longleftrightarrow$($\mathcal{I}(A)$ is prime).

If a germ $u \in \mathcal{O}_a$ is non-invertible and is not a zero divisor, then $\dim V(u) = \dim X_a - 1$. In fact, one may assume that X is not an analytic subset of a manifold M. Then $u = w_{X_a}$ for some $w \in \mathcal{O}_a(M)$, and we have $V(u) = V(w) \cap X_a$. It follows – by inequality $(*)$ in III. 4.6 – that $\dim V(u) \geq \dim X_a - 1$ (see II. 5.1). Now, if we had $\dim V(u) = \dim X_a$, then – in view of $(*)$ in II. 4.6 and proposition 2 in IV. 3.1 (by taking a simple component of $V(u)$ of largest dimension) – we would have $u = 0$ on a simple component of X_a. That contradicts our assumption (see IV. 4.3).

It follows that if, in addition, X_a is of constant dimension k, then $V(u)$ is of constant dimension $k-1$. (It is enough to apply the previous assertion to the germs $V(\check{u})$, $z \in V(\check{u})$, where $\check{u} \not\equiv 0$ is a holomorphic representative of u; compare this subsection with the corollary of proposition 4 in IV. 3.1).

Let us prove also that *Hilbert's Nullstellensatz* is true: $\mathcal{I}\big(V(I)\big) = \operatorname{rad} I$.

Indeed, we may assume that X is an analytic subset of a manifold M. Thus $\mathcal{O}_a(X) = \mathcal{O}_{X_a}$. Obviously, $\operatorname{rad} I \subset \mathcal{I}\big(V(I)\big)$ [37]. Now let $f \in \mathcal{I}\big(V(I)\big)$. We have $I = \sum_1^r \mathcal{O}_a(X)g_i$, where $g_i \in \mathcal{O}_a(X)$, and hence $f_{V(g_1,...,g_r)} = 0$. Moreover, $f = F_{X_a}$, $g_i = (G_i)_{X_a}$ $(i = 1,...,r)$, and $X_a = V(G_{r+1},...,G_s)$, where $F, G_1,...,G_s \in \mathcal{O}_a(M)$. Therefore $F = 0$ on $V(g_1,...,g_s) = V(G_1,...,G_s)$ and so, according to Hilbert's theorem from III. 4.1, we must have $F^m \in \sum_1^s \mathcal{O}_a(M)G_i$ for some m. Consequently, $f^m \in \sum_1^r \mathcal{O}_a(X)g_i$, which means that $f \in \operatorname{rad} I$.

In particular, $\mathcal{I}\big(V(I)\big) = I$ if the ideal I is prime. Thus we have the mutually inverse bijections $A \longrightarrow \mathcal{I}(A)$ and $I \longrightarrow V(I)$ between the set of simple analytic germs at a and the set of prime ideals of the ring $\mathcal{O}_a$.

Finally – as in III. 4.1 – if I is a proper ideal of $\mathcal{O}_a$ and $I_1,...,I_r$ are all its isolated ideals, then $V(I) = V(I_1) \cup ... \cup V(I_r)$ is the decomposition of $V(I)$ into simple germs.

10. The *Zariski dimension* of an analytic space X at a point $a \in X$ is defined by

$$\dim \operatorname{zar}_a X = \dim \operatorname{zar} X_a \ .$$

---

[37] For if $f^m \in I$, then $(f^m)_{V(I)} = 0$, and so $f_{V(I)} = 0$.

Then $\dim \operatorname{zar}_a X = \dim \operatorname{zar} \varphi(X)_{\varphi(a)}$, where φ is any chart whose domain contains a (see IV. 4.4) ([38]). It follows from proposition 3 in IV. 4.4 that $\dim \operatorname{zar}_a X$ is equal to the minimum of the dimensions of the manifolds containing an analytic set that is biholomorphic to an open neighbourhood of the point a. If $f : X \longrightarrow Y$ is a biholomorphic mapping of analytic spaces, then

$$\dim \operatorname{zar}_a X = \dim \operatorname{zar}_{f(a)} Y \quad \text{for each} \quad a \in X .$$

EXAMPLE (F.CATANESE) *of an irreducible analytic space which cannot be embedded in any manifold.*

Clearly, if X is an analytic subset of an n–dimensional manifold, then $\dim \operatorname{zar}_z X \leq n$ for $z \in X$ (see IV. 4.4). Thus it is enough to construct an irreducible analytic space X for which $\dim \operatorname{zar}_z X$ is not bounded for $z \in Z$.

Now, let M be a connected manifold of dimension $n > 0$, and let Z be its discrete subset which is the union of mutually disjoint subsets $Z_r = \{a_{r1}, \dots, a_{rr}\}$, $r = 1, 2, \dots$. For X we take the topological space obtained from M by identifying each of the sets Z_r with a single point a_r ([39]). (It is easy to check that such a space is a Hausdorff space with a countable basis for its topology.) Thus $X = G \cup \{a_1, a_2, \dots\}$, where $G = M \setminus Z$ is an open set also in X, while the set $\{a_1, a_2, \dots\}$ is discrete (and disjoint with G). Now let $\varphi_{r\nu} : U_{r\nu} \longrightarrow W_{r\nu}$ be charts of the manifold M at the points $a_{r\nu}$, respectively, where $U_{r\nu}$ are disjoint open neighbourhoods of these points and $W_{r\nu}$ are open neighbourhoods of zero in $\mathbf{C}^n$. Then the sets $U_r = a_r \cup \bigcup_{\nu=1}^r (U_{r\nu} \setminus a_{r\nu})$ are disjoint open neighbourhoods of the points a_r in X, whereas $W_r = \bigcup_{\nu=1}^r (0 \times \dots \times 0 \times W_r \times 0 \times \dots \times 0)$ are analytic in $W_{r1} \times \dots \times W_{r\nu}$, respectively. It is easy to check that the identity $G \longrightarrow G$, together with the homeomorphisms $\varphi_r : U_r \longrightarrow W_{r\nu}$ defined by

$$\varphi_r(a_r) = 0,$$
$$\varphi_r(z) = (0, \dots, 0, \varphi_{r\nu}(z), 0, \dots, 0) \quad \text{for} \quad z \in U_{r\nu} \setminus a_{r\nu} \quad (\nu = 1, \dots, r),$$

constitute an analytic atlas on X. The analytic space X so defined is irreducible, because $X^0 = G$ is a connected manifold (see n° 5, proposition α; and II. 3.6). Finally,

$$\dim \operatorname{zar}_{a_r} X = \dim \operatorname{zar} (W_r)_0 = nr, \quad \text{where} \quad r = 1, 2, \dots,$$

since $S_{(W_r)_0} = 0$ (see IV. 4.4, formula (**)).

11. For any mapping $f : Y \longrightarrow X$ of sets and any subset $E \subset X$, we denote by f^E the restriction $f_{f^{-1}(E)} : f^{-1}(E) \longrightarrow E$. We have $(f^E)^{-1}(A) =$

([38]) The ring $\mathcal{O}_a = \mathcal{O}_a(X)$ is isomorphic to the ring $\mathcal{O}_{\varphi(a)}(\varphi(X)) = \mathcal{O}_B$, where φ is any chart as above and $B = \varphi(X)_{\varphi(a)}$. It is also a complex vector space and its ideals are subspaces. The linear space $T_a^{\mathrm{zar}} X = (\mathfrak{m}_a / \mathfrak{m}_a^2)^*$ is called the *Zariski tangent space* at the point a to the analytic space X (see IV. 4.4). Therefore we have $T_a^{\mathrm{zar}} X \approx (\mathfrak{m}_B / \mathfrak{m}_B^2)^*$, and so
$$\dim \operatorname{zar}_a X = \dim T_a^{\mathrm{zar}} X .$$

([39]) See [16], Chapter 2, §4.

$f^{-1}(A)$ for $A \subset E$. In particular, the fibres of the mapping f^E are precisely the fibres $f^{-1}(z)$, where $z \in E$, because $(f^E)^{-1}(z) = f^{-1}(z)$. Obviously, $(f^E)^A = f^A$ if $A \subset E$, and $(f \circ g)^E = f^E \circ g^{f^{-1}(E)}$ for any $g : Z \longrightarrow Y$.

We say that *holomorphic mappings* of analytic spaces $f : Y \longrightarrow X$, $g : Z \longrightarrow X$ are *isomorphic*, and write $f \approx g$, if there exists a biholomorphic mapping $\varphi : Y \longrightarrow Z$ such that the following diagram commutes

(1)
$$
\begin{array}{ccc}
Y & \xrightarrow{\ \varphi\ } & Z \\
& f \searrow \quad \swarrow g & \\
& X &
\end{array}
$$

Clearly, it is an equivalence relation.

A property (α) of holomorphic mappings of analytic spaces (or manifolds) [40] is said to be *sublocal* if the following equivalence holds:

$$
(f : Y \longrightarrow X \text{ has the property } (\alpha)) \Longleftrightarrow (\text{all } f^{G_\iota} \text{ have the property } (\alpha))
$$

for holomorphic mappings f and open covers $\{G_\iota\}$ of X. Every property which is local [41] and stable [41a] is obviously sublocal. Furthermore, each of the following properties is sublocal: to be an injection, a surjection, a closed mapping, a proper mapping, a homeomorphism, a biholomorphic mapping, an embedding.

A property (α) of holomorphic mappings of analytic spaces (or manifolds) is said to be *rigid* if, in the diagram (1), for any $f \approx g$ that have the property (α) the biholomorphic mapping φ is unique. The following property of mappings $f : Y \longrightarrow X$

(m)   the set $\{w : \#f^{-1}(f(w)) = 1\}$ is dense in Y

───────────

[40] We admit here only the properties that are invariant with respect to the above equivalence relation, i.e., such that if f has the property (α) and $g \approx f$, then g has the property (α). It is certainly the case if the property (α) can be expressed solely in terms of the structure of an analytic space (or of a manifold).

[41] A property (α) of holomorphic mappings of analytic spaces (or manifolds) is called *local* if the following equivalence holds: (f has the property (α)) $\Longleftrightarrow$ (all f_{G_ι} have the property (α)) for holomorphic mappings $f : X \longrightarrow Y$ and open covers $\{G_\iota\}$ of X.

[41a] The property (α) of holomorphic mappings of analytic spaces (or manifolds) is called *stable* if the following equivalence holds:($f : X \longrightarrow H$ has property (α)) $\Longleftrightarrow$ ($f : X \longrightarrow Y$ has property (α)) for open $H \subset Y$ and holomorphic $f : X \longrightarrow H$.

Each of the following properties is local and stable: to be an open mapping, a local homeomorphism, an immersion, a submersion.

is rigid (^{42}) (and so is each property stronger then (m)). The property (m) is also sublocal.

12. Let (α) be a rigid and sublocal property.

Then
$$f \approx g \Longleftrightarrow (f^{H_\iota} \approx g^{H_\iota} \text{ for each } \iota)$$
for any holomorphic mappings $f : Y \longrightarrow X$, $g : Z \longrightarrow X$, and any open cover $\{H_\iota\}$ of X (^{43}).

We have the following proposition on gluing together mappings that have the property (α).

PROPOSITION . *Let* $\{G_\iota\}$ *be an open cover of an analytic space* X, *let* $f_\iota : Y_\iota \longrightarrow G_\iota$ *be holomorphic mappings (of analytic spaces* Y_ι) *that have the property* (α), *and suppose that* $f_\iota^{G_\iota \cap G_\kappa} = f_\kappa^{G_\iota \cap G_\kappa}$ *for each pair* ι, κ. *Then there is a unique (up to isomorphism) holomorphic mapping* $f : Y \longrightarrow X$ *(of an analytic space* Y) *such that* $f^{G_\iota} \approx f_\iota$ *for each* ι. *The mapping* f *also has the property* (α).

PROOF . Obviously, each such mapping must have the property (α). If $f' : Y' \longrightarrow X$ is also such a mapping, we must have $(f')^{G_\iota} \approx f^{G_\iota}$ for all ι, which implies $f' \approx f$. Therefore it is enough to show the existence of f.

Set $f_{\iota\kappa} = f_\iota^{G_\iota \cap G_\kappa}$, $f_{\iota\kappa\lambda} = f_\iota^{G_\iota \cap G_\kappa \cap G_\lambda}$, and $H_{\iota\kappa} = f_\iota^{-1}(G_\iota \cap G_\kappa)$, $H_{\iota\kappa\lambda} = f_\iota^{-1}(G_\iota \cap G_\kappa \cap G_\lambda)$ (the latter are open subsets of the spaces Y_ι). We have the commutative diagrams

(2)
$$
\begin{array}{ccc}
H_{\iota\kappa} & \xrightarrow{\;g_{\iota\kappa}\;} & H_{\kappa\iota} \\
& f_{\iota\kappa} \searrow \quad \swarrow f_{\kappa\iota} & \\
& G_\iota \cap G_\kappa &
\end{array}
$$

with some biholomorphic mappings $g_{\iota\kappa}$. Now there exists a gluing of the spaces Y_ι by the biholomorphic mappings $g_{\iota\kappa}$, because the conditions (#) in n° 7a are satisfied. Indeed, since the property (α) is rigid, the biholomorphic mapping $g_{\iota\iota}$ must be the identity of Y_ι, and $g_{\iota\kappa} = g_{\kappa\iota}^{-1}$. Next, since the diagrams above are commutative, so are the diagrams

$$
\begin{array}{ccc}
H_{\iota\kappa\lambda} & \xrightarrow{\;\alpha\;} & H_{\lambda\iota\kappa} \\
& f_{\iota\kappa\lambda} \searrow \quad \swarrow f_{\lambda\iota\kappa} & \\
& G_\iota \cap G_\kappa \cap G_\lambda &
\end{array}
$$

(^{42}) Indeed, for such mappings, the biholomorphic mapping φ in the diagram (1) is uniquely determined on a dense subset.

(^{43}) In the proof of "$\Longleftarrow$", gluing of the biholomorphic mappings is possible because of the rigidity of (α).

with both $\alpha = (g_{\iota\lambda})_{H_{\iota\kappa\lambda}}$ and $\alpha = g_{\kappa\lambda} \circ g_{\iota\kappa}$; hence, since the property (α) is rigid, we must have $g_{\kappa\lambda} \circ g_{\iota\kappa} \subset g_{\iota\lambda}$. Finally, observe that

$$g_{\iota\kappa} \subset \{f_{\iota}(x) = f_{\kappa}(y)\} \subset H_{\iota\kappa} \times H_{\kappa\iota} \subset Y_{\iota} \times Y_{\kappa} \; ;$$

since the graph $g_{\iota\kappa}$ is closed in $H_{\iota\kappa} \times H_{\kappa\iota}$, it is also closed in the set $\{f_{\iota}(x) = f_{\kappa}(y)\}$. But the latter is closed in $Y_{\iota} \times Y_{\kappa}$, which implies that $g_{\iota\kappa}$ is closed in $Y_{\iota} \times Y_{\kappa}$. Thus we have a gluing Y, $\{g_{\iota} : Y_{\iota} \longrightarrow Y\}$, and $g_{\kappa}^{-1} \circ g_{\iota} = g_{\iota\kappa}$. Now the holomorphic mappings $f_{\iota} \circ g_{\iota}^{-1} : g_{\iota}(Y_{\iota}) \longrightarrow X$ can be glued together into a holomorphic mapping

$$f = \bigcup f_{\iota} \circ g_{\iota}^{-1} : Y \longrightarrow X \; ,$$

as they are compatible. Indeed, we have $f_{\iota} \supset f_{\kappa} \circ g_{\iota\kappa}$ because the diagram (2) commutes; hence $f_{\iota} \circ g_{\iota}^{-1} \supset \left(f_{\kappa} \circ g_{\kappa}^{-1}\right)_{g_{\iota}(Y_{\iota})}$, and so $f_{\iota} \circ g_{\iota}^{-1} = f_{\kappa} \circ g_{\kappa}^{-1}$ on $g_{\iota}(Y_{\iota}) \cap g_{\kappa}(Y_{\kappa})$. Then $f^{-1}(G_{\iota}) = g_{\iota}(Y_{\iota})$, for $f^{-1}(G_{\iota}) = \bigcup_{\kappa} g_{\kappa}(H_{\kappa\iota})$, $H_{\iota\iota} = Y_{\iota}$, and $g_{\kappa}(H_{\kappa\iota}) \subset g_{\iota}(Y_{\iota})$ because $g_{\kappa} = g_{\iota} \circ g_{\kappa\iota}$ in $H_{\kappa\iota}$. Therefore we have the commutative diagram

$$
\begin{array}{ccc}
Y_{\iota} & \xrightarrow{\;g_{\iota}\;} & \\
 & {}_{f_{\iota}}\searrow \quad \nearrow_{f^{G_{\iota}}} & g_{\iota}(Y_{\iota}) = f^{-1}(G_{\iota}), \\
 & G_{\iota} &
\end{array}
$$

which means that the mapping f satisfies the required conditions.

13. Let X be an analytic space.

A *(proper) modification of the space* X is a proper holomorphic mapping $f : Y \longrightarrow X$ of analytic spaces such that there is a nowhere dense analytic set $S \subset X$ whose inverse image $f^{-1}(S) \subset Y$ is nowhere dense and such that the restriction $f^{X\backslash Y} : Y \setminus f^{-1}(S) \longrightarrow X \setminus S$ is biholomorphic [44] . Then we say that f is a *modification of the space* X *in the set* S.

Every modification is surjective [45] . Obviously, any modification in the set S is also a modification in each nowhere dense analytic set containing S. Both the image and the inverse image of a dense (respectively, nowhere dense) set via a modification is a dense (respectively, nowhere dense) set [46] .

---

[44] The condition "the restriction $f_{Y\backslash T} : Y \setminus T \longrightarrow X \setminus S$ is biholomorphic for some nowhere dense analytic sets $S \subset X$, $T \subset Y$" is not essentially more general, because then it must be $T = f^{-1}(S)$ (see footnote [7] in B. 2).

[45] Since $f(\overline{Y \setminus f^{-1}(S)}) = \overline{f(Y \setminus f^{-1}(S))} = \overline{X \setminus S}$.

[46] Due to the fact that $f^{X\backslash S}$ is a homeomorphism.

The modifications are precisely the proper holomorphic surjections satisfying the property (m) (see n° 11).

Indeed, if $f : Y \longrightarrow X$ is such a mapping, the sets $T = \{z \in X : \#f^{-1}(z) > 1\}$ and $f^{-1}(T)$ do not have interior points ([47]) . They are analytically constructible and nowhere dense by the versions of the lemma from 7.1 and lemma 2 from IV. 8.3 for analytic spaces; see the final remark in 7.3 ([48]) . The set $S = X^* \cup \bar{T}$ is analytic nowhere dense, and its inverse image is also nowhere dense ([49]) . Finally, $f^{X \setminus S}$ is a biholomorphic mapping (see 4.7 in reference to the proposition from 3.4). In conclusion, f is a modification.

As a consequence:

To be a modification is a sublocal and rigid property. In particular, the proposition on gluing from n° 12 holds for modifications.

Note also that the composition of modifications is a modification: if $f : Y \longrightarrow X$ is a modification in S and $g : Z \longrightarrow Y$ is a modification in T, then $f \circ g$ is a modification in $S \cup f(T)$ ([50]) .

§5. Remmert's proper mapping theorem

1. First, we will prove the following:

LEMMA . *Let $f : M \longrightarrow N$ be a holomorphic mapping of manifolds. If f is proper and has constant rank k, then $f(M)$ is an analytic subset of N of constant dimension k.*

To prove the lemma, take an arbitrary $c \in N$. The fibre $f^{-1}(c)$ is compact. Thus, by the rank theorem (see C. 4.1), there are open sets $G_1, \ldots, G_r$ in M that cover the fibre $f^{-1}(c)$ and are such that each of the sets $f(G_i)$ is a k–dimensional submanifold containing c. Therefore $f(G_i)$ is closed in an open neighbourhood U_i of the point c. By taking an open neighbourhood $U \subset \bigcap_1^r U_i$ of the point c such that $f^{-1}(U) \subset G_1 \cup \ldots \cup G_r$ (see B. 2.4), we have $U \cap f(M) = U \cap (g(G_1) \cup \ldots \cup f(G_r))$, and so $U \cap f(M)$ is an analytic subset of U of constant dimension k (see II. 3.4 and II. 1.5). Therefore the set $f(M)$ is an analytic subset of N of constant dimension k (see II. 3.4).

([47]) Because $f^{-1}(T)$ is the complement of the set $\{w : f^{-1}(f(w)) = 1\}$.

([48]) The equivalence we are proving will not be used in the present chapter.

([49]) Owing to the fact that $f^{-1}(\bar{T}) \subset \overline{f^{-1}(T)}$ and $f^{X \setminus \bar{T}}$ is a homeomorphism. (The former can be verified easily, because $\#f^{-1}(z) = 1$ for $z \in \bar{T} \setminus T$.)

([50]) The set $f(T)$ is analytic, in view of Remmert's theorem (see 5.1 below).

REMMERT'S PROPER MAPPING THEOREM . *If $f : X \longrightarrow Y$ is a proper holomorphic mapping of analytic spaces, then its range $f(X)$ is analytic in Y. Consequently, the image of each analytic subset of X is an analytic subset of Y.*

PROOF . One may assume that Y is a manifold. Indeed, assuming that this special case of the theorem has been proved, we can apply the result to the mapping $\varphi \circ f : f^{-1}(U) \longrightarrow N$, where φ is a chart of an open neighbourhood U of any point of the space Y onto an analytic subset of a manifold N (see B. 2.4); thus we conclude that the set $f(X) \cap U$ is analytic in the analytic space U. (See 4.3 and 4.7.)

Note that if $\{X_i\}$ is the family of simple components of the space X, then the family $\{f(X_i)\}$ is locally finite (see B. 2.4 and 4.5). Therefore one may assume that the dimension of the space X is finite (see 4. 3.5 and 6). Let n denote this dimension. The case $n = -\infty$ is trivial. Hence let $n \geq 0$ (i.e., $X \neq \emptyset$), and suppose that the theorem is true if the dimension of the space X is $< n$. We may assume (see above) that the space X is irreducible. Then X^0 is a connected n–dimensional manifold which is open and dense in X (see 4.5, proposition α and the theorem). Because X^* is an analytic subset of dimension $< n$ (see the theorem in 4.5), the set $Z = f(X^*)$ is analytic in Y $(^{51})$. Therefore the set $f^{-1}(Z)$ is analytic in X (see 4.6). If $f^{-1}(Z) = X$, then $f(X) = Z$. So, assume that

$$f^{-1}(Z) \subsetneq X .$$

Then the set $f^{-1}(Z)$ is nowhere dense in X (see 4.5, proposition β), and so $f^{-1}(Z) \cap X^0$ is a nowhere dense analytic subset of the manifold X^0 (see 4.3). Consequently, the set

$$X' = X \setminus f^{-1}(Z) = X^0 \setminus (f^{-1}(Z) \cap X^0)$$

is open and dense in X^0, and is a connected n–dimensional manifold (see II. 3.6). On the other hand, the set

$$G = \{z \in X' : \operatorname{rank}_z f = k\} ,$$

where

$$k = \operatorname{rank} f_{X'} = \operatorname{rank} f$$

$(^{51})$ We have applied the induction hypothesis to the mapping $f_{X^*} : X^* \longrightarrow Y$ (see 4.6 and B. 2.4).

(see C. 3.12 and 4.8), is open and dense in X' (see II. 3.7) and hence also in X^0. Now

$$\dim Z < k \;.$$

Indeed, all the fibres $\left(f_{g^{-1}(Z)}\right)^{-1}(w) = f^{-1}(w)$, where $w \in Z$, of the surjection $f_{f^{-1}(Z)} : f^{-1}(Z) \longrightarrow Z$ are of dimension $\geq n - k$ (see 4.8, inequality (β)). It follows (except for the trivial case $Z = \emptyset$) that $(n - k) + \dim Z \leq \dim f^{-1}(Z) < n$ (see 4.8, inequality (α); and 4.5, proposition β), that is, that $\dim Z < k$.

The mapping $f_{X'} : X' \longrightarrow Y \setminus Z$ is proper (since $X' = f^{-1}(Y \setminus Z)$; see B. 2.4) and $V = \{z \in X' : \text{rank}_z f < k\}$ is an analytic subset of X' of dimension $< n$ (see II. 3.7 and II. 1.2). Therefore $W = f(V)$ is an analytic set in $Y \setminus Z$ ([52]) and

$$\dim W < k$$

(see §1, theorem 1). As a consequence, $X' \not\subset f^{-1}(W)$, for otherwise we would have $W \supset f(X') \supset f(G)$. This is impossible, as $\dim f(G) \geq k$ ([53]) . Thus $f^{-1}(W) \cap X'$ is a nowhere dense analytic subset of X' (see II. 3.6), and so the set

$$X'' = X' \setminus f^{-1}(W) = f^{-1}\left((Y \setminus Z) \setminus W\right)$$

is open and dense in X'. Therefore it is open and dense in X and it is a submanifold. Since $X'' \subset X' \setminus V = G$, the mapping $f_{X''} : X'' \longrightarrow (Y \setminus Z) \setminus W$ has constant rank k. It is proper (see B. 2.4). Thus, in view of the lemma, $f(X'')$ is an analytic subset in $(Y \setminus Z) \setminus W$ and is of constant dimension k. Since $\dim W < k$, the Remmert-Stein theorem (see IV. 6.3) implies that $f(X'') \cup W$ is an analytic subset of $Y \setminus Z$. It is of constant dimension k, due to the fact that $W \subset f(X) = \overline{f(X'')}$. Therefore, as $\dim Z < k$, the Remmert-Stein theorem yields that $f(X'') \cup W \cup Z$ is an analytic subset in Y. But this set is equal to $f(X)$ (because $X'' \cup f^{-1}(W) \cup f^{-1}(Z) = X$), and so $f(X)$ is an analytic subset of Y.

THE CHEVALLEY-REMMERT THEOREM . *Let $f : M \longrightarrow N$ be a holomorphic mapping of manifolds. Let E be an analytically constructible set in M such that $f_{\bar{E}} : \bar{E} \longrightarrow N$ is proper. Then the image of E is analytically constructible in N. In particular, if the mapping f is proper, then the image of every analytically constructible set in M is an analytically constructible set in N.*

([52]) We have applied the induction hypothesis to the mapping $f_V : V \longrightarrow Y \setminus Z$ (see 4.6 and B. 2.4).

([53]) This inequality follows from the rank theorem (see C. 4.1) because $G \neq \emptyset$.

PROOF . According to the Remmert proper mapping theorem, if the restriction $f_F : F \longrightarrow N$ to a closed set $F \subset M$ is proper, then the image of every analytic set $W \subset F$ is analytic (in N). Furthermore, if W is irreducible, so is $f(W)$ (see IV. 2.8). We will need

LEMMA ([54]) . *If $W \subsetneq V \subset M$ are analytic sets, V is irreducible, and the restriction $f_V : V \longrightarrow N$ is proper, then there is an analytic set $Z \subsetneq f(V)$ such that $f(V) \setminus Z \subset f(V \setminus W)$.*

First of all, note that the lemma is true if $f(W) \subsetneq f(V)$ ([55]) . Set $s = \dim V - \dim f(V)$. If $s = 0$, then the lemma is true because $\dim f(W) \leq \dim W < \dim V = \dim f(V)$ (see §1, the corollary of theorem 1; and IV. 2.8, proposition 3), and so $f(W) \subsetneq f(V)$. Now, let $s > 0$, and suppose that the lemma is true if $\dim V - \dim f(V) < s$. One may assume that $f(W) = f(V)$. Then (in view of the irreducibility of $f(V)$) we must have $f(W_0) = f(V)$ for some simple component W_0 of W (see IV. 2.8, the corollary of proposition 4). We have $\dim W_0 < \dim V$. On the other hand, rank $f_{W_0} =$ rank f_V (see 3.2, theorem 4), and so the equality (1) from 3.3 shows that $\lambda(f_{W_0}) < \lambda(f_V)$. Thus

$$(\#) \qquad l_z f_{W_0} \subsetneq l_z f_V \quad \text{when} \quad z \in W_0 \setminus C(f_{W_0}) \,,$$

because $\dim l_z f_{W_0} = \lambda(f_{W_0}) < \lambda(f_V) \leq \dim l_z f_V$ (see 3.3, equalities (2) and (*)). This implies that

$$(\#\#) \qquad\qquad f(W_0 \setminus T) \subset f(V \setminus W) \,,$$

where $T = C(f_{W_0}) \cup (W_0 \cap W^*) \subsetneq W_0$ (see IV. 2.9, corollary 1 of theorem 4). Indeed, if $z \in W_0 \setminus C(f_{W_0}) \setminus W^*$, then (see IV. 2.9, corollary 4 of theorem 4) the set W_0 is the only simple component of W that contains z, and so, by $(\#)$, $f(z) = f(z')$ for some $z' \in V \setminus W$. Since $\dim W_0 - \dim f(W_0) < \dim V - \dim f(V)$, the induction hypothesis implies the existence of an analytic set $Z \subsetneq f(W_0) = f(V)$ such that $f(V) \setminus Z \subset f(W_0 \setminus T)$. In view of $(\#\#)$, this completes the proof of the lemma.

Now we can prove the theorem. By proposition 3 from IV. 8.3, it is enough to show that if $V, W \subset M$ are analytic sets and the restriction $f : V \longrightarrow N$ is proper, then the set $f(V \setminus W)$ is analytically constructible in N (see B. 2.4). Let $k = \dim f(V)$. The case $k \leq 0$ is trivial. Assume thus that $k > 0$ and the statement is true if $\dim f(V) < k$. One may assume that the

---

([54]) See Mumford [30], p. 37.

([55]) Since $f(V) \setminus f(W) \subset f(V \setminus W)$.

premises of the lemma are satisfied (see IV. 2.9 and B. 2.4), and so there is an analytic set $Z \subsetneqq f(V)$ such that $f(V) \setminus Z \subset f(V \setminus W)$. Then we have

$$f(V \setminus W) = \big(f(V \setminus W) \setminus Z\big) \cup \big(f(V \setminus W) \cap Z\big) = \big(f(V) \setminus Z\big) \cup f(V' \setminus W),$$

where $V' = V \cap f^{-1}(Z)$. But $f(V') = Z$ is of dimension $< k$, and so the induction hypothesis implies that the set $f(V' \setminus W)$ is analytically constructible. Consequently, the set $f(V \setminus W)$ is also analytically constructible.

REMARK . The Chevalley-Remmert theorem remains true (with the same proof) in the case of analytic spaces, provided that the definitions of analytically constructible subsets and their properties from IV. 8 are extended (via charts) to this case.

2. Let M and N be manifolds, and assume that N is compact.

PROPOSITION . *If $f : E \longrightarrow N$, where $E \subset M$, is a mapping whose graph is analytically constructible* ([56]) , *then there is an analytically constructible set $H \subset E$ which is smooth, open, and dense in E, and such that the restriction $f_H : H \longrightarrow N$ is holomorphic* ([57]) .

Indeed, there is an analytic set $W \subset N \times M$ which is nowhere dense in $\bar{f}$ and such that $\bar{f} \setminus W \subset f$ (see IV. 8.4, corollary 1 from proposition 7). Let $\pi : M \times N \longrightarrow M$ be the natural projection. By Remmert's proper mapping theorem (see n° 1), the set $V = \pi(W) \subset M$ is analytic. Now, the set $V \cap E$ is nowhere dense in E, because $\dim_z(V \cap E) < \dim_z E$ for $z \in V \cap E$ (see IV. 8.5). Indeed, if $z \in V \cap E$, then by taking a sufficiently small neighbourhood U of z we have

$$\dim_z(V \cap E) \leq \dim_z V \leq \dim(V \cap U) \leq \dim W_U < \dim \bar{f}_U =$$
$$= \dim f_U \leq \dim(E \cap U) = \dim_z E$$

(see V. 1, corollary 1 of theorem 1, IV. 2.5, IV. 8.5, and II. 1.4). Therefore the set $H = E^0 \setminus V$ is analytically constructible, smooth, and also open and dense in E. It is the union of the submanifolds $E^{(i)} \setminus V$ and these submanifolds are open in the set. As $f_{E^{(i)} \setminus V} = \bar{f}_{E^{(i)} \setminus V}$ (in view of $\bar{f} \setminus \pi^{-1}(V) \subset f \subset \bar{f}$), the set $f_{E^{(i)} \setminus V}$ is an analytic subset in $(E^{(i)} \setminus V) \times N$ (see IV. 8.3, proposition 5).

([56]) According to the Chevalley-Remmert theorem (see n° 1), the set E is also analytically constructible (since it is the image of the set f by the projection onto M).

([57]) Without the assumption of compactness of the manifold N, the proposition is no longer true. A counter-example is provided by the function $f : \mathbf{C}^2 \longrightarrow \mathbf{C}$ given by $f(z, w) = \frac{1}{z}$ on $\Sigma = \{(z, w) : z \neq 0, w = e^{1/z}\}$ and $f = 0$ on $\mathbf{C}^2 \setminus \Sigma$.

Consequently, by the analytic graph theorem (see §1, corollary 3 of theorem 2 and the remark), the mappings $f_{E^{(i)} \setminus V}$ are holomorphic and so is the mapping f_H.

§6. Remmert's open mapping theorem

LEMMA 1. *Let V be a locally analytic subset of the Cartesian product $M \times N$ of vector spaces. Suppose that $0 \in V$ and $r = \dim_0 V_0$ ([57a]). Then there exist a direct sum decomposition $N = L + T$ and arbitrarily small open convex neighbourhoods Δ, Ω, B of the zeros in the spaces M, L, T, respectively, such that $\dim L = r$, the set $V' = V \cap (\Delta \times (\Omega + B))$ is analytic in $\Delta \times (\Omega + B)$, the set $V'' = \pi'(V')$ is analytic in $\Delta \times \Omega$, where $\pi' : V' \longrightarrow \Delta \times \Omega$ is the natural projection, and finally, $\dim V'' = \dim V'$ and $\dim V''_z = \dim V'_z$ for $z \in \Delta$.*

PROOF . Since $(V_0)_0$ is an r–dimensional analytic germ in N, there is a direct sum decomposition $N = L + T$ such that $\dim L = r$ and $(V_0)_0 \cap T = 0$ (see, e.g., III. 4.5, formula (#)). Set $B_\rho = \{v \in T : |v| < \rho\}$ (where $|.|$ is a norm on T). There is an arbitrarily small $\varepsilon > 0$ and arbitrarily small open neighbourhoods Δ', Ω' of the origins in M, L, respectively, such that $\bar{B}_{2\varepsilon} \cap (\bar{V})_0 = 0$ and the set $V \cap (\Delta' \times (\Omega' + B_{2\varepsilon}))$ is closed in $\Delta' \times (\Omega' + B_{2\varepsilon})$. Then $V \cap (\Delta' \times (\Omega' + B_{2\varepsilon} \setminus B_\varepsilon)) \cap \bar{V} = \emptyset$. Therefore there are open convex neighbourhoods $\Delta \subset \Delta'$, $\Omega \subset \Omega'$ of the zeros in M, L, respectively, such that $\left(\Delta \times (\Omega + (\bar{B}_{2\varepsilon} \setminus B_\varepsilon))\right) \cap \bar{V} = \emptyset$. Hence the set $V' = V \cap (\Delta \times (\Omega + B_{2\varepsilon})) = V \cap (\Delta \times (\Omega + \bar{B}_\varepsilon))$ is analytic in the set $\Delta \times (\Omega + T)$, because it is locally analytic and closed in this set (see II. 3.4). Moreover, the natural projection $\pi' : V' \longrightarrow \Delta \times \Omega$ is proper (see B. 5.2) and has finite fibres (see the proposition in IV. 5). Therefore it follows from Remmert's proper mapping theorem (see 5.1) that $V'' = \pi'(V')$ is analytic in $\Delta \times \Omega$ and $\dim V'' = \dim V'$ (see 3.2, corollary 1 from theorem 2). Finally, for each $z \in \Delta$, we have $V'_z \subset \Omega + T$, the natural projection $\pi'_z : V'_z \longrightarrow \Omega$ has finite fibres, and $\pi'_z(V'_z) = V''_z$. This implies that $\dim V''_z = \dim V'_z$.

LEMMA 2. *If V is a locally analytic subset of the Cartesian product $M \times N$ of vector spaces such that*

$$\dim l_z \pi = r \quad for \quad z \in V ,$$

[57a] Here V_0, V'_z, V''_z denote fibres (see III. 1.2 footnote ([2])).

where $\pi : V \longrightarrow M$ is the natural projection, then each point $a \in V$ has an arbitrarily small open neighbourhood U in V such that $\pi(U)$ is locally analytic (in M) of dimension $\dim V_a - r$.

PROOF. One may assume that $a = 0$. Now, we can apply lemma 1 (using the same notation). Then we have $\dim V' = \dim V_a$. As the sets V_z, $z \in \Delta$, are of constant dimension r (because $\pi^{-1}(z) = z \times V_z$), so are the sets V'_z. Hence $V''_z = \Omega$ or $V''_z = \emptyset$, depending whether $V'_z \neq \emptyset$ or $V'_z = \emptyset$ (see II. 3.6). Therefore it follows that $V'' = \pi(V') \times \Omega$. Thus (see II. 3.4) $\pi(V')$ is an analytic subset of Δ of dimension $\dim V_a - r$ (since $\dim V'' = \dim V_a$).

THEOREM 1 (REMMERT'S RANK THEOREM). *If $f : X \longrightarrow Y$ is a holomorphic mapping of analytic spaces such that*

$$\dim l_z f = r \quad for \quad z \in X \ ,$$

then each point $a \in X$ has an arbitrarily small open neighbourhood whose image is locally analytic (in Y) of dimension $\dim X_a - r$.

Indeed, one may assume that X, Y are locally analytic subsets of vector spaces M, N, respectively. Then one can apply lemma 2 to the locally analytic set $f \subset M \times N$ (see 3.1) and to the natural projection $\pi : f \longrightarrow N$, using the fact that $l_{(z, f(z))} = (l_z f) \times f(z)$ for $z \in X$. This gives us the required result, because (see 3.4) the natural projection $\pi_0 : f \longrightarrow X$ is biholomorphic (and $f(U) = \pi(\pi_0^{-1}(U))$ for $U \subset X$).

For any holomorphic mapping $f : X \longrightarrow Y$ of analytic spaces and for any point $z \in X$, the number $\rho_z f = \dim X_z - \dim l_z f$ is called the *Remmert rank of the mapping f at the point z*. Therefore the dimension of the image (of any arbitrarily small neighbourhood of the point a) in theorem 1 is equal to the Remmert rank $\rho_a f$. In the case when X and Y are manifolds, the Remmert rank may be different from the ordinary rank. For instance, if $g : \mathbf{C} \ni z \longrightarrow z^2 \in \mathbf{C}$, then $\text{rank}_0 g = 0$, whereas $\rho_0 g = 1$. However, if f is a mapping with constant rank r, then $\rho_z f = r$ in X (see C. 4.1, the rank theorem). Consequently, in the general case, $\rho_z f = \text{rank}_z f$ in an open dense subset of X^0 (see 4.5 in ref. to IV. 2.1; see also, 4.8 and II. 3.7). If X and Y are manifolds, then we have

$$(*) \qquad\qquad\qquad \rho_z f \geq \text{rank}_z f \ .$$

Indeed, consider $a \in X$. By putting $k = \rho_a f$ and $n = \dim X$, we have $\dim \Gamma_a = n - k$, where $\Gamma = f^{-1}(f(a))$. Accordingly (see IV. 2.5), in any neighbourhood of a there is some $z \in \Gamma^{(n-k)}$. Since f is constant on Γ, we have $T_z \Gamma \subset \ker d_z f$ (see C. 3.11), and so $\text{rank}_z f = \text{codim}(\ker d_z f) \leq \text{codim } T_z \Gamma = k$. Hence, using lower semicontinuity of the function $z \longrightarrow \text{rank}_z f$ (see C. 3.12), we conclude that $\text{rank}_a f \leq k$.

Assume now that the space X is of constant dimension. Then, by theorem 3 from 3.2, the function $X \ni z \longrightarrow \rho_z f$ is lower semicontinuous and — by theorem 5 from 3.3 —

the set $\{z \in X : \rho_z \leq k\}$ is analytic for each $k \in \mathbf{N}$ [58].

In this case, *Remmert's rank theorem* can be reformulated as follows.

If the Remmert rank of the mapping f is constant and equal to k [59]*, then each point of the space X has an arbitrarily small open neighbourhood whose image is locally analytic (in Y) of constant dimension k.*

We will also show that in the general case (without the assumption that the space X is of constant dimension),

$$\operatorname{rank} f = \max\{\rho_z f; \ z \in X\} \ .$$

Indeed, the equality holds when X is irreducible, due to lower semicontinuity of the functions $X \ni z \longrightarrow \rho_z f$ and $X^0 \ni z \longrightarrow \operatorname{rank}_z f$. Owing to (*), it is enough to show that if $a \in X$, then $\rho_a f \leq \operatorname{rank} f$. Now, by taking the decomposition $X = \bigcup X_i$ into simple components, we have $\dim X_a = \dim(X_s)_a$ for some s (see II. 1.6), and so

$$\rho_a f = \dim X_a - \dim l_a f \leq \dim(X_s)_a - \dim l_a f_{X_s} = \rho_a f_{X_s} \leq \operatorname{rank} f_{X_s} \leq \operatorname{rank} f \ ,$$

(see 4.8 in ref. to corollary 1 from theorem 4 in 3.2).

LEMMA 3. *Let Δ, Ω be open convex neighbourhoods of zero in the vector spaces M, N, respectively, and let V be an analytic set in $\Delta \times \Omega$. If the natural projection $\pi : V \longrightarrow \Delta$ is open and $0 \times \Omega \subset V$, then $V = \Delta \times \Omega$.*

PROOF [60] . One may assume that $\dim M = 1$. Indeed, suppose the validity of the lemma in this case; then, taking any line $L \subset M$, since $0 \times \Omega \subset V_L$ and the projection $\pi_{V_L} : V_L \longrightarrow \Delta \cap L$ is open (see B. 2.1), we have $V_L = (\Delta \cap L) \times \Omega$. Now, suppose that $V \subsetneq \Delta \times \Omega$. Then $\dim V \leq n = \dim N$ (see II. 3.6). Hence the set $V_{\Delta \setminus 0} = V \setminus (0 \times \Omega)$ is of dimension $\leq n$, and so is its closure $\overline{V_{\Delta \setminus 0}}$ in $\Delta \times \Omega$, which is analytic in $\Delta \times \Omega$ (see IV. 8.3, proposition 5; and IV. 8.5). The set $\overline{V_{\Delta \setminus 0}} \cap (0 \times \Omega)$, which is nowhere dense in $\overline{V_{\Delta \setminus 0}}$, is of dimension $< n$ (see IV. 2.5). This implies that $\overline{V_{\Delta \setminus 0}} \cap (0 \times \Omega) \subsetneq 0 \times \Omega$, and so there is a point $a \in (0 \times \Omega) \setminus \overline{V_{\Delta \setminus 0}}$. Then some neighbourhood $U \subset V$ of a in U is contained in $0 \times \Omega$. Thus $\pi(U) = 0$, contrary to the fact that the mapping is open.

LEMMA 4. *Let M and N be vector spaces, and let $V \subset M \times N$ be a locally analytic set of constant dimension m. Let $W \subset M$ be a locally analytic set of constant dimension n that contains the image of the set V by the projection into M. If the natural projection $\pi : V \longrightarrow W$ is open, then*

$$\dim l_a \pi = m - n \quad \text{for each} \quad a \in V \ .$$

---

[58] Without the assumption that X is of constant dimension, these assertions are not true. As a counter-example take the restriction of the identity mapping on $\mathbf{C}^3$ to the set $X = \{z_1 = 0\} \cup \{z_2 = z_3 = 0\}$ and $k = 1$.

[59] i.e., $\rho_z f = k$ for $z \in X$.

[60] Simplified by P. Tworzewski and T. Winiarski.

PROOF . One may assume that $a = 0 \in V$. Also, it is enough to consider the case when $W = M$. Indeed, suppose that the lemma is true in this case. Since the germ W_0 is of constant dimension n, we may assume — in view of proposition 2 from IV. 1.5 — that $M = M_0 \times M_1$, where $\dim M_0 = n$, and that there exists an open neighbourhood U of 0 in W such that the natural projection $\tilde{\pi} : U \longrightarrow M_0$ is open and $\tilde{\pi}^{-1}(0) = 0$ (see IV. 1.1–3). Now, the set $\tilde{V} = \pi^{-1}(U)$ is open in V and is locally analytic (in $M_0 \times M_1 \times N$) of constant dimension m. The mapping $\pi_{\tilde{V}} : \tilde{V} \longrightarrow U$ is open and so is the composition $\tilde{\pi} \circ \pi_{\tilde{V}} : \tilde{V} \longrightarrow M_0$ (see B. 2.1). The latter is the natural projection of $\tilde{V}$ into M_0. Moreover, $l_0(\tilde{\pi} \circ \pi_{\tilde{V}}) = l_0 \pi$, and hence $\dim l_0 \pi = \dim l_0(\tilde{\pi} \circ \pi_{\tilde{V}}) = m - n$.

Set $r = \dim l_0 \pi = \dim_0(V_0)$. Apply lemma 1 (with the same notation). So we have $\dim V_0'' = \dim V_0' \geq r$, which implies that $V_0'' = \Omega$ (see II. 3.6), i.e., $0 \times \Omega \subset V''$. Moreover, the natural projection $\pi'' : V'' \longrightarrow \Delta$ is open, because $\pi_{V'} = \pi'' \circ \pi' : V' \longrightarrow \Delta$ is open and $\pi' : V' \longrightarrow V''$ is a continuous surjection (see B. 2.1). Therefore, by lemma 3, we have $V'' = \Delta \times \Omega$. Thus $\pi(V') = \Delta$ and for $z \in \Delta$ we have $\dim \pi_{V'}^{-1}(z) = \dim V_z' = \dim V_z'' = r$. Now theorem 2 from 3.2 implies that $m = \dim V' = r + \dim \Delta = r + n$. Consequently, $\dim l_0 \pi = r = m - n$.

THEOREM 2 (REMMERT'S OPEN MAPPING THEOREM). *Let $f : X \longrightarrow Y$ be a holomorphic mapping of analytic spaces. Assume that X is of constant dimension m, that Y is of constant dimension n, and Y is locally irreducible: that is, the germ Y_w is irreducible for each $w \in Y$. Then the mapping f is open if and only if one of the following equivalent conditions ([61]) is satisfied:*

(R) $$\dim l_z f = m - n \ for \ z \in X \ ,$$

(R') $$\dim f^{-1}(w) = m - n \ for \ w \in f(X) \ .$$

The assumption that Y is locally irreducible is essential only for the sufficiency of the condition (R) *([62])* .

([61]) The characterization (R') has been suggested by P. Tworzewski.

([62]) This assumption cannot be omitted if (R) is to be sufficient. As an example consider the mapping

$$\mathbf{C}^3 \supset X = \{z = t = 0\} \cup \{w = 0, t = 1\} \ni (z, w, t) \longrightarrow (z, w) \in Y = \{zw = 0\} \subset \mathbf{C}^2 \ .$$

The sets X, Y are of constant dimension 1, the fibres are finite, but nevertheless the mapping is not open: the image of the open subset $\{z = t = 0\}$ of X is the set $\{z = 0\}$ which is not open in Y. If we had assumed that Y was irreducible, it would not have sufficed

REMARK . The condition (R) means precisely that f has constant Remmert rank equal to n.

PROOF . The implication (R) $\Longrightarrow$ (R') is trivial. Conversely, suppose that the condition (R') is satisfied. Then obviously, $\dim l_z f \leq m - n$ for $z \in X$. On the other hand, for each simple component Z of the space X, we have $\dim l_z f_Z \geq \lambda(f_Z) = \dim Z - \dim f(Z)$ for $z \in Z$ (see 4.8 in ref. to 3.3 and 3.2, theorem 4). It follows (see IV. 2.9, corollary 3 of theorem 4) that $\dim l_z f \geq m - n$ for $z \in X$, which gives the equality (R).

Sufficiency of the condition (R). Let U be an arbitrary neighbourhood of any point $a \in X$. According to theorem 1, there is an arbitrarily small open neighbourhood Δ of a such that

$$(*) \qquad f(\Delta) \text{ is a locally analytic set in } Y \text{ of dimension } n \ .$$

By taking in turn: a neighbourhood $\Delta \subset U$ satisfying $(*)$, a sufficiently small irreducible open neighbourhood Ω of the point $f(a)$ in Y (see IV. 3.1, proposition 1) such that the set $\Omega \cap f(\Delta)$ is analytic in Ω, and a neighbourhood $\Delta_0 \subset \Delta$ satisfying $(*)$ and such that $f(\Delta_0) \subset \Omega$, we have $f(\Delta_0) \subset \Omega \cap f(\Delta) \subset Y$. Thus $\dim(\Omega \cap f(\Delta)) = n$, and so we must have $\Omega \cap f(\Delta) = \Omega$ (see 4.5, proposition β), and thus $\Omega \subset f(\Delta) \subset f(U)$. Therefore $f(U)$ is a neighbourhood of the point $f(a)$.

Necessity of the condition (R). One may assume that X and Y are locally analytic subsets of linear spaces M and N, respectively. Now, it is enough to apply lemma 4 to the locally analytic set $f \subset M \times N$ (see 3.1) and the natural projection $\pi : f \longrightarrow Y$. In fact, f is of constant dimension m, the projection π is open [63] , and $l_{(z,f(z))}\pi = (l_z f) \times f(z)$ for $z \in X$.

---

either. Indeed, consider the mapping

$$\mathbf{C}^3 \supset X = \{z = t^2 - 1, w = t^3 - t\} \ni (z, w, t) \longrightarrow (z, w) \in Y = \{w^2 = z^2(z + 1)\} \subset \mathbf{C}^2 \ .$$

The sets X, Y are of constant dimension 1, Y is irreducible (as the image of the set $\mathbf{C}$ by the mapping $\mathbf{C} \ni t \longrightarrow (t^2 - 1, t^3 - 1) \in \mathbf{C}^2$), the fibres are finite, but the mapping is not open: the image of some neighbourhood of the point $(0,0,1)$ in X is smooth, therefore it is not a neighbourhood of the point $(0,0)$ in Y, as the germ $Y_{(0,0)}$ is the union of two distinct smooth germs.

[63] Because the natural projection $\pi_0 : f \longrightarrow X$ is a biholomorphic mapping (see 3.4) and $\pi = f \circ \pi_0$.

§7. Finite holomorphic mappings

Let $f : V \longrightarrow W$ be a holomorphic mapping between non-empty analytic subsets V and W of manifolds M and N, respectively.

1. We have the following:

LEMMA . *If the mapping f is proper, then the sets*

$$Z_k = \{w \in W : \ \#f^{-1}(w) < k\}, \quad k = 1, 2, \ldots$$

are analytically constructible. If, in addition, the mapping f is open, then the sets Z_k are analytic.

PROOF . The set

$$\tilde{Z}_k = \{(z_1, \ldots, z_k, w) \in M^k \times N : (z_1, w), \ldots, (z_k, w) \in f,$$
$$z_1, \ldots, z_k \text{ are distinct}\}$$

is analytically constructible and contained in the closed set $F = \{(z_1, w), \ldots$ $\ldots, (z_k, w) \in f\}$. Since the restriction π_F of the natural projection $\pi : \ M^k \times N \longrightarrow N$ is proper (as $\pi_F^{-1}(E) \subset f^{-1}(E)^k \times E$ for $E \subset N$), the Chevalley-Remmert theorem (see 5.1) implies that the set $W \backslash Z_k = \pi(\tilde{Z}_k)$ is analytically constructible. Hence, so is the set Z_k. If, in addition, the mapping f is open, then the function $W \ni w \longrightarrow \#f^{-1}(w)$ is lower semicontinuous (see B. 2.1); hence the set Z_k is closed and thus analytic (see IV. 8.3, proposition 5).

According to Remmert's open mapping theorem (see §6):

The mapping $f : \ V \longrightarrow W$ with discrete fibres is open if and only if V and W are of constant dimension, W is locally irreducible, and $\dim \bar{V} = \dim W$.

In particular, we have the following proposition which we are going to prove directly.

PROPOSITION 1. *If the mapping f has only discrete fibres, V is of constant dimension, W is a submanifold, and $\dim V = \dim W$, then the mapping f is open.*

Indeed, one may assume that $W = N$. Then the analytic subset $f \subset M \times N$ is of constant dimension $n = \dim N$ [64] . It is enough to prove that the natural projection $\pi : \ f \longrightarrow N$ is open, i.e., that the image of any neighbourhood in f of any arbitrarily chosen point $a \in f$ is a neighbourhood

---

[64] Due to the fact that the natural projection $f \longrightarrow V$ is biholomorphic.

of $\pi(a)$ in N. We have $\dim f_a = n$. One may assume that M and N are vector spaces, and $a = 0$. Since $f_0 \cap (M \times 0) = 0$, the image under π of any neighbourhood of zero in f is a neighbourhood of zero in N (see III. 4.5).

Note that if the mapping f is proper, then its fibres are discrete precisely when they are finite.

We say that the mapping f is *light at a point* $a \in V$ if a is an isolated point of its fibre $f(f^{-1}(a))$ (see 2.1). The mapping f is said to be *finite* if it is proper and light at each point or, equivalently, if it is proper and its fibres are finite ([65]) . In the case when M is a vector space, f is finite if and only if it is proper (see the proposition in IV. 5) ([66]) .

Note that if f is light at a, then $f_\Omega : \Omega \longrightarrow \Delta$ is finite for some (arbitrarily small) open neighbourhoods Ω of a in V and Δ of $f(a)$ in W (such that $f(\Omega) \subset \Delta$). Indeed, one may assume that M is a vector space. Now, it is enough to take an open, relatively compact neighbourhood Ω_0 of the point a such that $f(a) \notin f(\partial \Omega_0)$, then to take Δ disjoint from $f(\partial \Omega_0)$, and set $\Omega = \Omega_0 \cap f^{-1}(\Delta)$. Then, if $E \subset \Delta$ is compact, so is $f_\Omega^{-1}(E) = \Omega_0 \cap f^{-1}(E) = \overline{\Omega_0} \cap f^{-1}(E)$ (as $\partial \Omega_0 \cap f^{-1}(E) = \emptyset$).

Note also that f is light at a if and only if the ring $\mathcal{O}_a(V)$ is finite over the subring $\mathcal{O}_{f(a)}(W) \circ f_a$.

Indeed, one may assume that $M = \mathbf{C}^r$, $N = \mathbf{C}^k$, $a = 0$, and $f(a) = 0$. Let $n = k + r$. Now, the fact that f is light at 0 is equivalent to the fact that the germ $A = \{(w, z) : w = f(z)\}_0$ is k–normal in $\mathbf{C}^n$ (see the proposition in III. 4.4), which, in turn, means that the ideal $I = \mathcal{I}(A)$ is k–normal (see III. 2.4). Equivalently, this says that $\hat{\mathcal{O}}_n$ is finite over $\hat{\mathcal{O}}_k$ (see III. 2.2). We have the epimorphism

$$\mathcal{O}_n \ni \varphi \longrightarrow \varphi \circ (f_0, e_0) \in \mathcal{O}_0(V), \quad \text{where} \quad e : V \hookrightarrow \mathbf{C}^r ,$$

under which $\mathcal{O}_k$ is mapped onto $\mathcal{O}_0(W) \circ f_0$. The kernel of the epimorphism is I, and hence it induces an isomorphism $\hat{\mathcal{O}}_n \longrightarrow \mathcal{O}_0(V)$, under which $\hat{\mathcal{O}}_k$ is mapped onto $\mathcal{O}_0(W) \circ f_0$. Therefore our condition is equivalent to finiteness of the ring $\mathcal{O}_0(V)$ over $\mathcal{O}_0(W) \circ f_0$.

PROPOSITION 2. *If the mapping f is finite and the set W is irreducible, then*

$$\dim V = \dim W \Longleftrightarrow (f \text{ is a surjection}) .$$

([65]) Or, also, if the mapping is closed and has only finite fibres (see B. 2.4).

([66]) The term "finite mapping" originates in algebraic geometry, where finiteness of a regular mapping $f : S \longrightarrow T$ between algebraic sets is defined as finiteness of the ring $R(S)$ over the subring $f^*(R(T))$. According to the proposition from VII. 12.1, this condition is equivalent to that of the finiteness defined above.

In fact, we then have $\dim f(V) = \dim V$ (see 3.2, corollary 1 of theorem 2) and, in view of Remmert's proper mapping theorem (see 5.1), the set $f(V)$ is analytic. Consequently (see IV. 2.8, proposition 3),

$$\dim f(V) = \dim W \Longleftrightarrow f(V) = W .$$

2. In general, the mapping f is said to be a *branched covering* if it is finite and there is a nowhere dense analytic set $\Sigma \subset W$ whose inverse image $f^{-1}(\Sigma) \subset V$ is nowhere dense, and for which the restriction

(#) $\qquad\qquad\qquad f^{W\backslash\Sigma} : V \setminus f^{-1}(\Sigma) \longrightarrow W \setminus \Sigma$

is a finite covering [67]. By taking a larger Σ one may require that $W \setminus \Sigma$ is smooth (see IV. 2.1, proposition 1; and B. 2.2) and then $V \setminus f^{-1}(\Sigma)$ is also smooth and the mapping (#) is locally biholomorphic (see the corollary of the proposition in 3.4).

We will consider here the situation when $W \setminus \Sigma$ is a *connected sub-manifold*. In such a case, f will be called a **-covering*. Then $V \setminus f^{-1}(\Sigma)$ is also a submanifold and has the same dimension, and the covering (#) is locally biholomorphic (see the proposition in 3.4). The set Σ is then called an *exceptional set* of the **-covering* f. The multiplicity p of the covering (#) is independent of the choice of the exceptional set Σ [68]; it is called the *multiplicity* of the **-covering*, and we say that f is a *p–sheeted *-covering*. [68a]

An example of a **-covering* is the projection $\pi : W \longrightarrow \Omega$ of a Weierstrass set (see III. 2.3), where the P_j's have non-zero discriminants [68b]; Z is an exceptional set. For π is finite (see B. 5.2 and 3) and $\pi^{\Omega\backslash Z}$ is a proper

---

[67] The condition "the restriction $f_{V\backslash Z} : V \setminus Z \longrightarrow W \setminus \Sigma$ is a finite covering for some nowhere dense analytic sets $Z \subset V$, $\Sigma \subset W$" is not necessarily more general, since then one must have $Z = f^{-1}(\Sigma)$ (see footnote [7] in B. 2.4).

[68] Since $f^{-1}_{V\backslash f^{-1}(\Sigma)}(w) = f^{-1}(w)$ for $w \in W \setminus \Sigma$.

[68a] Note that the assumption of finiteness of fibres is essential. For take a closure $\bar{f}$ in $\mathbf{C}^2 \times \bar{\mathbf{C}}$ of the mapping $f : \mathbf{C}^2 \backslash 0 \longrightarrow \mathbf{C}$ defined by $f(z, w) = w/z$ if $z \neq 0$ and $f(0, w) = \infty$ if $w \neq 0$ (see C. 3.19). By the Remmert-Stein theorem (see IV. 6.3), $\bar{f}$ is an analytic set. Now the natural projection $\pi : \bar{f} \longrightarrow \mathbf{C}^2$ is proper and its restriction $\pi_f : f \longrightarrow \mathbf{C}^2 \setminus 0$ is a one–sheeted covering, but the fibre $f^{-1}(0) = 0 \times \bar{\mathbf{C}}$ is not finite.

[68b] This assumption is not essential (because of the Andreotti-Stoll theorem below).

locally biholomorphic mapping (by the implicit function theorem; see C. 2.1), hence it is a finite covering (by proposition 1 in B. 3.2).

Observe that for a normal triple (Ω, Z, W), the natural projection $V \longrightarrow \Omega$ is a *–covering with an exceptional set Z (which has the same multiplicity). An important remark now follows (see IV. 1.5 and 2):

REMARK . For any point a of any locally analytic set X at which the germ X_a is of constant dimension, there is an arbitrarily small open neighbourhood V of a in X, and a *–covering $f : V \longrightarrow W$, where W is a (connected) manifold.

If a p–sheeted *–covering f is open, which is the case when the set W is a submanifold (proposition 1), or, more generally, when it is locally irreducible (theorem 2 from §6), then

$$\# f^{-1}(w) \le p \quad \text{for} \quad w \in W .$$

This means that $p = \sup\{\# f^{-1}(w) : w \in W\}$ [69] , for then the function $W \ni w \longrightarrow f^{-1}(w)$ is lower semicontinuous (see B. 2.1).

According to the Andreotti-Stoll theorem stated below, a finite mapping $f : V \longrightarrow W$ is a *–covering if and only if it satisfies the conditions

(∗) V is of constant dimension, W is irreducible, and $\dim V = \dim W$.

Consequently, if $\{V_i\}$ are the simple components of V, then the mapping f is finite if and only if all the restrictions $f_{V_i} : V_i \longrightarrow f(V_i)$ are *–coverings and the family $\{f(V_i)\}$ is locally finite. (See B. 2.4.)

THE ANDREOTTI-STOLL THEOREM . *A finite mapping* $f : V \longrightarrow W$ *is a* *–*covering if and only if the conditions* (∗) *are satisfied. If, in addition,* W *is a (connected) submanifold, then*

(1) $p = \sup\{\# f^{-1}(w) : w \in W\} < \infty$ *is the multiplicity of the* *–*covering* f,

(2) $\Sigma = \{w \in W : \# f^{-1}(w) < p\}$ *is its exceptional set* [70] ,

(3) $W \setminus \Sigma = \{w \in W : \# f^{-1}(w) = p\} =$

---

[69] The openness of f is essential, as the second example in footnote [62] shows.

[70] Clearly, it is the smallest exceptional set. It is called the *branching locus* of the *–covering f.

$$= \{w \in W : \; z \in f^{-1}(w) \Longrightarrow f \text{ is biholomorphic at } z\} \quad (^{70a}),$$

(4) *the set Z' of all points of V at which f is not holomorphic is analytic, and obviously, $f(Z') = \Sigma$.*

PROOF . The condition $(*)$ is necessary. Indeed, if f is a $*$–covering, it has the first two properties in $(*)$, since the submanifold $V \setminus f^{-1}(\Sigma)$ is open and dense in V (see IV. 2.5). However the submanifold $W \setminus \Sigma$ is open and dense in W, and is also connected (see IV. 2.8, corollary 2 from proposition 2). Hence the third property follows, since $\dim(V \setminus f^{-1}(\Sigma)) = \dim(W \setminus \Sigma)$ (see IV. 2.5). Suppose now that the condition $(*)$ is satisfied.

First consider the case when W is a (connected) submanifold.

Then, in view of proposition 1, the mapping f is open. We have $p = \sup\{\#f^{-1}(w) : \; w \in W\} < \infty$. Indeed, by lemma 1, the sets Z_k are analytic, but $W = \bigcup_1^\infty Z_k$; hence one of them must be equal to W (see IV. 2.8, the corollary from proposition 4). The set Σ defined by (2) is a nowhere dense analytic subset of W (by lemma 1) and its inverse image $f^{-1}(\Sigma)$ is a nowhere dense analytic subset of V (see B. 2.).

Now we are going to prove the equality (3). It implies that the mapping

$$f_{V \setminus f^{-1}(\Sigma)} : \; V \setminus f^{-1}(\Sigma) \longrightarrow W \setminus \Sigma,$$

which is proper and locally biholomorphic, is a p–sheeted covering. This means that f is a p–sheeted $*$–covering with the exceptional set Σ and, in this case, the condition $(*)$ is sufficient.

Now let $w \in W \setminus \Sigma$, that is, let $f^{-1}(w) = \{z_1, \ldots, z_p\}$. Take open disjoint neighbourhoods $U_1, \ldots, U_p$ of the points $z_1, \ldots, z_p$ in V. One may assume that $f(U_i) = \Delta$, $i = 1, \ldots, p$, where $\Delta \subset W \setminus \Sigma$ is an open neighbourhood of w, since the mapping f is open. Hence it is enough to take $\Delta = \bigcap_1^p f(U_i) \setminus \Sigma$ and replace U_i by $U_i \cap f^{-1}(\Delta)$. Then $f_{U_i} : \; U_i \longrightarrow \Delta$ must be bijective. Therefore, by the proposition from 3.4, the mapping $f_{U_i} : \; U_i \longrightarrow \Delta$ is biholomorphic $(i = 1, \ldots, p)$. Thus f is a biholomorphic mapping at each of the points z_i. Conversely, let $f^{-1}(w) = \{z_1, \ldots, z_q\}$, and suppose that f is biholomorphic at each of the points z_i. Consider open disjoint neighbourhoods U_i of these points in V such that the restrictions f_{U_i} are injective. Since the mapping f is proper, we have (see B. 2.4) $f^{-1}(\Delta) \subset U_1 \cup \ldots \cup U_q$ for some

---

$(^{70a})$ In the general case, since f^{W^0} is also a $*$–covering with the same multiplicity, the properties (1) and (3) hold with Σ defined by (2), where W is replaced by W^0, and $\Sigma \cup V^*$ is the smallest exceptional set.

neighbourhood Δ of w in W. There is a point $w' \in \Delta \setminus \Sigma$, and then we must have $q \geq \#f^{-1}(w') = p$. Thus $q = p$ (see (1)), and so $w \in W \setminus \Sigma$.

As far as (4) is concerned, it is enough to observe (see C. 3.13) that $Z' = V^* \cup Z_0$, where $Z_0 = \{z \in V^0 : \text{rank } d_z f < \dim V\}$ is an analytically constructible set (see 3.2, theorem 1). As the set Z' is closed, it must be analytic (see IV. 8.3, proposition 5).

In order to prove that the condition $(*)$ is sufficient in the general case, note first that if Λ is a nowhere dense analytic subset of W, then $f^{-1}(\Lambda)$ is a nowhere dense analytic subset of V. Indeed (see 3.2, corollary 2 of theorem 2; and IV. 2.8, proposition 3), we have $\dim f^{-1}(\Lambda) = \dim \Lambda < \dim W = \dim V$, which implies (see IV. 2.5) that the set $f^{-1}(\Lambda)$ is nowhere dense in V.

Now, the mapping $f_0 = f_{V_0} : V_0 \longrightarrow W^0$, where $V_0 = f^{-1}(W^0)$ is finite and satisfies the conditions $(*)$. Hence, by the first part of the proof, we have the finite covering

$$(\#\#) \qquad (f_0)_{V_0 \setminus f_0(\Sigma)} : V_0 \setminus f_0^{-1}(\Sigma_0) \longrightarrow W^0 \setminus \Sigma_0 ,$$

where $\Sigma_0 = \{w \in W^0 : \#f_0^{-1}(w) < p_0\} = W^0 \cap \{w \in W : \#f^{-1}(w) < p_0\}$ (with some $p_0 \in \mathbf{N}$). Furthermore, the set Σ_0 is closed and nowhere dense in W^0, and, in view of the lemma, it is analytically constructible. Therefore $\Sigma = \Sigma_0 \cup W^* = \bar{\Sigma}_0 \cup W^*$ is a nowhere dense analytic subset of W and its inverse image $f^{-1}(\Sigma)$ is a nowhere dense analytic subset of V. But $W \setminus \Sigma = W^0 \setminus \Sigma_0$, $V \setminus f^{-1}(\Sigma) = V_0 \setminus f_0^{-1}(\Sigma_0)$, and $f_{V \setminus f(\Sigma)} = (f_0)_{V_0 \setminus f_0^{-1}(\Sigma_0)}$. Hence $(\#\#)$ is the mapping $f_{V \setminus f(\Sigma)} : V \setminus f^{-1}(\Sigma) \longrightarrow W \setminus \Sigma$. Thus f is a $*$–covering.

COROLLARY . *Let M be a connected manifold, and let X be a vector space. For any subset $V \subset M \times X$, the following conditions are equivalent:*

(1) *V is analytic and the natural projection $V \longrightarrow M$ is a $*$-covering.*

(2) *V is analytic of constant dimension $\dim M$ and the natural projection $V \longrightarrow M$ is proper.*

(3) *There is a nowhere dense analytic subset $Z \subset M$ such that the pair $(Z, V_{M \setminus Z})$ is a quasi-cover in $M \times X$ with the adherence V.*

(See III. 1.2–3, the first lemma on quasi-covers.)

As an application of the Andreotti-Stoll theorem we will prove that *if a two-dimensional torus T contains only one one-dimensional subtorus R (such tori exist; see C. 3.21), then the translated subtori $c + R$, $c \in T$, are the only one-dimensional irreducible analytic subsets* ([71]).

([71]) See [41], VIII. 1.4.

Indeed, we have $T = X/\Lambda$ and $R = \pi(Z)$, where $\pi : X \longrightarrow T$ is the natural epimorphism and $Z \subset X$ is a Λ–distinguished line. Let $p : X \longrightarrow Y = X/Z$ be the natural mapping. Then $\Omega = p(\Lambda)$ is a net on Y. We have the one–dimensional complex torus $S = Y/\Omega$ and the holomorphic induced complex homomorphism $\tilde{p} : T \longrightarrow S$ whose kernel is $\pi(Z + \Lambda) = R$ (^{72}) and whose fibres are $c + R$, $c \in T$ (see C. 3.21). Suppose that there exists a one–dimensional irreducible analytic set $V \subset T$ which is different from each of those fibres. Then the holomorphic mapping $f = \tilde{p}_V : V \longrightarrow S$ is finite (see IV. 2.8, proposition 3; and III. 4.3) and, by the Andreotti-Stoll theorem, it is a $*$–covering. Moreover, there is a finite set $\Sigma \subset S$ and $r > 0$ such that $\#f^{-1}(w) \le r$ for $w \in S$, and $S \setminus \Sigma$ is the set of points $w \in S$ for which $\#f^{-1}(w) = r$. In addition, the restriction of f to $f^{-1}(S \setminus \Sigma)$ is locally biholomorphic. It follows that the mapping $\varphi_0 : S \setminus \Sigma \longrightarrow T$, defined by the formula

$$\varphi_0(w) = z_1 + \ldots + z_r, \text{ where } \{z_1, \ldots, z_r\} = f^{-1}(w) \text{ for } w \in S \setminus \Sigma,$$

is holomorphic. Now, the graph of φ_0 is contained in the compact set

$$\Psi = \{(w, z_1 + \ldots + z_r) \in S \times T : \tilde{p}(z_1) = \ldots = \tilde{p}(z_r) = w, \ z_1, \ldots, z_r \in V\} .$$

The sets $\Psi_w = \{z : (w, z) \in \Psi\}$ are finite. Therefore $\varphi = \bar{\varphi}_0 \subset \Psi$ and the sets φ_w are finite. They are also connected. In fact, taking a base of compact connected neighbourhoods $U_1 \supset U_2 \supset \ldots$ of the point w, we have $w \times \varphi_w = \bigcap \varphi_{U_\nu}$ (^{73}). Consequently, each of the sets φ_w consists of a single point. Thus we have the continuous mapping $\varphi : S \longrightarrow T$, which is holomorphic (see II. 2.5). Its range is (see the proposition in C. 3.21) a translated subtorus $\varphi(S) = c + R'$, where $c \in T$, and $R' \subset T$ is a subtorus of dimension ≤ 1. Now, since $f(\varphi_0(w)) = rw$ in $S \setminus \Sigma$, we have $\tilde{p} \circ \varphi = \tilde{r}e \ne 0$, where e is the identity mapping of the space Y (^{74}). Therefore $\tilde{p}(c) + \tilde{p}(R') = \tilde{r}e(S)$ consists of more then one point, which implies that R' is a one–dimensional subtorus different from R. This is in contradiction with our assumption.

3. The Andreotti-Stoll theorem implies the following result:

PROPOSITION 3. *If* $f : V \longrightarrow W$ *is a proper holomorphic mapping, V is of constant dimension, and W is a connected manifold of dimension* $\dim V$, *then the set*

$$(**) \qquad \{w \in W : z \in f^{-1}(w) \Longrightarrow f \text{ is biholomorphic at } z\} \quad (^{75})$$

is open and dense, and its complement is an analytic (and nowhere dense) set.

Indeed, the set $G = \{w \in W : \#f^{-1}(w) < \infty\}$ is open, and its complement Z is analytic and nowhere dense (see 3.3, corollary 1 of theorem 5).

(^{72}) Since $\tilde{p} \circ \pi = \pi' \circ p$, where $\pi' : Y \longrightarrow S$ is the natural epimorphism.

(^{73}) The intersection of a decreasing sequence of connected compact sets is connected.

(^{74}) See C. 3.21, footnote (^{69}).

(^{75}) Its inverse image is contained in V^0 (see footnote (^{66})).

Therefore the restriction $f_{f^{-1}(G)} : f^{-1}(G) \longrightarrow G$ is a p–sheeted *–covering. Moreover, (since the fibres are compact) the set $(**)$ is contained in G and its complement Σ in G is $\{w \in G : \#f^{-1}(w) < p\} = \{w \in W : \#f^{-1}(w) < p\}$. The latter is a nowhere dense analytic subset of G and, by the lemma from n° 1, it is analytically constructible in W. Thus the complement $Z \cup \Sigma = Z \cup \bar{\Sigma}$ of the set $(**)$ in W is analytic (see IV. 8.3, proposition 5) and nowhere dense.

4. Let $f : V \longrightarrow W$ be a p–sheeted *–covering, where W is a (connected) manifold. Let Σ be its exceptional set. Put $Z = f^{-1}(\Sigma)$.

Define $\beta^0 = \beta \circ f$ for $\beta \in \mathcal{O}_W$. The subring $\mathcal{O}_W^0 = \mathcal{O}_W \circ f$ of $\mathcal{O}_V$ is isomorphic to $\mathcal{O}_W$ (via the monomorphism $\mathcal{O}_W \ni \beta \longrightarrow \beta^0 \in \mathcal{O}_V$ $(^{75a})$). Hence it is integrally closed (see the corollary in II. 3.5).

Let $\tilde{\mathcal{O}}_V$ denote the ring of holomorphic functions on $V \setminus Z$ that are locally bounded on V. It is obviously an extension of $\mathcal{O}_V$ (via the identification $\mathcal{O}_V \ni \varphi \longrightarrow \varphi_{V \setminus Z} \in \tilde{\mathcal{O}}_V$). The ring $\tilde{\mathcal{O}}_V$ will be regarded also as a module over $\mathcal{O}_W$ (with the product defined by $\beta\varphi = \beta^0\varphi$ for $\beta \in \mathcal{O}_W$ and $\varphi \in \tilde{\mathcal{O}}_V$). Then $\mathcal{O}_V$ is its submodule.

For any $P = \sum \beta_\nu T^\nu \in \mathcal{O}_W[T]$, set $P^0 = \sum \beta_\nu^0 T^n u$ and $P(w, S) = \sum \beta_\nu(w) S^n u$ for $w \in W$ $(^{75b})$. Observe that $P(\beta)(w) = P(w, \beta(w))$ for $\beta \in \mathcal{O}_W$. If P is monic and Δ is its discriminant, then Δ^0 is the discriminant of P^0 and $\Delta(w)$ is the discriminant of $P(w, T)$ (see A. 4.3).

If $P \in \mathbf{C}[X_1, \ldots, X_p]$ is symmetric, then for every $\alpha \in \tilde{\mathcal{O}}_V$ there is a unique $P_\alpha \in \mathcal{O}_W$ such that

$$P_\alpha(w) = P\big(\alpha(\eta_1), \ldots, \alpha(\eta_p)\big) \quad \text{if } w \in W \setminus \Sigma \text{ and } f^{-1}(w) = \{\eta_1, \ldots, \eta_p\}.$$

Similarly, if $P \in \mathbf{C}[X_1, \ldots, X_p, T_1, \ldots, T_p]$ is symmetric in $(X_1, T_1), \ldots$ $\ldots, (X_p, T_p)$ $(^{75c})$, then for any $\alpha, \varphi \in \mathcal{O}_{\tilde{V}}$ there is a unique $P_{\alpha\varphi} \in \mathcal{O}_W$ such that

$$P_{\alpha\varphi}(w) = P\big(\alpha(\eta_1), \ldots, \alpha(\eta_p), \varphi(\eta_1), \ldots, \varphi(\eta_p)\big)$$
$$\text{if } w \in W \setminus \Sigma \text{ and } f^{-1}(w) = \{\eta_1, \ldots, \eta_p\}.$$

$(^{75a})$ It is injective because of the surjectivity of f.

$(^{75b})$ For any S from any ring containing $\mathbf{C}$.

$(^{75c})$ i.e., $P(X_{\alpha_1}, \ldots, X_{\alpha_p}, T_{\alpha_1}, \ldots, T_{\alpha_p}) = P$ for every permutation $(\alpha_1, \ldots, \alpha_p)$.

Indeed, each of the above formulae well-defines a holomorphic function on $W \setminus \Sigma$ which is locally bounded on W. Thus it extends holomorphically to W (see the proposition in II. 3.5).

Let α be any function from $\tilde{\mathcal{O}}_V$. We define its *characteristic polynomial* $Q_\alpha \in \mathcal{O}_W[T]$ by $Q_\alpha = T^p + (\sigma_1)_\alpha T^{p-1} + \ldots + (\sigma_p)_\alpha$, where $\sigma_1, \ldots, \sigma_p \in \mathbf{C}[X_1, \ldots, X_p]$ are the basic symmetric polynomials. Then D_α, where $D = \prod_{i<j}(X_i - X_j)^2$, is its discriminant, and we have

$$
\begin{aligned}
(1) \quad & Q_\alpha(w, T) = (T - \alpha(\eta_1)) \ldots (T - \alpha(\eta_p)) \\
(2) \quad & D_\alpha(w) = \prod_{i<j}(\alpha(\eta_i) - \alpha(\eta_j))^2
\end{aligned} \Bigg\} \quad \text{if } w \in W \setminus \Sigma \text{ and}
$$

$$
f^{-1}(w) = \{\eta_1, \ldots, \eta_p\}
$$

(see A. 4.1 and 3). It follows that $Q_\alpha^0(\alpha) = 0$ [75d], i.e., α is a root of $Q_\alpha^0 \in \mathcal{O}_V[T]$.

Thus the ring $\tilde{\mathcal{O}}_V$ is integral over $\mathcal{O}_W^0$, and each of its elements is a root of a monic polynomial (from $\mathcal{O}_W^0[T]$) of degree p.

Consider the Cramer identities from A. 2.2a with $n = p$. Substitute for X the Vandermonde matrix $X_{ij} = X_i^{j-1}$. Then $D = (\det X)^2$. Multiplying the identities by $\det X$, we get

$$
(3) \qquad\qquad DT_i = \sum_{j=0}^{p-1} C_j X_i^j, \quad i = 1, \ldots, p,
$$

where $C_j = (\det X)(\det X^{(j)}) \in \mathbf{C}[X_1, \ldots, X_p, T_1, \ldots, T_p]$ are symmetric in $(X_1, T_1), \ldots, (X_p, T_p)$ and linear in $T_1, \ldots, T_p$ [75e]. Let $\alpha, \varphi \in \tilde{\mathcal{O}}_V$. Take any $z \in V \setminus Z$ so that $f^{-1}(f(z)) = \{\eta_1, \ldots, \eta_p\} \subset V \setminus Z$. By substituting $X_\nu = \alpha(\eta_\nu)$ and $T_\nu = \varphi(\eta_\nu)$, $\nu = 1, \ldots, p$, in (3), we get $D\alpha(f(z))\varphi(\eta_i) = \sum_{j=0}^{p-1}(C_j)_{\alpha\varphi}(f(z))\alpha(\eta_i)^j$. Thus, as $z = \eta_s$ for some s and since $V \setminus Z$ is dense in V, we obtain

THE GRAUERT-REMMERT FORMULA [75f].

$$
(\text{GR}) \qquad\qquad D_\alpha^0 \varphi = \sum_{j=0}^{p-1} C_{j\alpha\varphi}^0 \alpha^j \quad for \quad \alpha, \varphi \in \tilde{\mathcal{O}}_V
$$

[75d]   Because for $z \in V \setminus Z$ we have $Q_\alpha^0(\alpha)(z) = Q_\alpha(f(z), \alpha(z))$.

[75e]   i.e., they are linear combinations of $T_1, \ldots, T_p$ with coefficients in $\mathbf{C}[X_1, \ldots, X_p]$.

[75f]   See [17a] p. 140, the theorem on the primitive element.

(here, $D_\alpha^0 = (D_\alpha)^0$, $C_{j\alpha\varphi}^0 = (C_{j\alpha\varphi})^0$, and $C_{j\alpha\varphi} = (C_j)_{\alpha\varphi}$).

We call α a *primitive element for f* if $D_\alpha \neq 0$ [75g] (in this case $D_\alpha^0 \not\equiv 0$; see B. 2.2). It means precisely (by (2)) that α is injective on some fibre $f^{-1}(w)$, $w \in W \setminus \Sigma$. *In such a case:*

(A) *The sequence $C_{0\alpha\varphi}, \ldots, C_{(p-1)\alpha\varphi}$ is the unique one in $\mathcal{O}_W^p$ for which (GR) holds.*

(B) D_α^0 *is the universal denominator of $\tilde{\mathcal{O}}_V$ over $\mathcal{O}_W^0[\alpha]$.*

(C) *The mapping $\tilde{\mathcal{O}}_V \ni \varphi \longrightarrow (C_{0\alpha\varphi}, \ldots, C_{(p-1)\alpha\varphi}) \in \mathcal{O}_W^p$ is a monomorphism of modules over $\mathcal{O}_W$. In particular, the image $L = L(f, \alpha)$ of $\mathcal{O}_V$ is a submodule of $\mathcal{O}_W^p$ and the restriction of the mapping is an isomorphism of $\mathcal{O}_V$ onto L (of modules over $\mathcal{O}_W$).*

To check (A), we take any $w \in \{D_\alpha \neq 0\} \setminus \Sigma$ so that $f^{-1}(w) = \{\eta_1, \ldots \ldots, \eta_p\} \subset V \setminus Z$. By evaluating (GR) at η_i, $i = 1, \ldots, p$, we get a system of linear equations for $C_{0\alpha\varphi}(w), \ldots, C_{(p-1)\alpha\varphi}(w)$, whose determinant is $\neq 0$, by (2). As for (C), the mapping is linear over $\mathcal{O}_W$, because the C_j's are linear in $T_1, \ldots, T_p$. The injectivity of that mapping follows from (GR).

REMARK 1. If $\alpha \in \mathcal{O}_V$ is a primitive element for f, then D_α^0 is a universal denominator of $\tilde{\mathcal{O}}_V$ over $\mathcal{O}_V$. (This follows from (B).)

REMARK 2. Recall that in the conclusion of the classic descriptive lemma (III. 3.2), the mapping $\pi : V \longrightarrow \Omega$ is p–sheeted *–covering with the exceptional set Z (see V. 7.2 and IV. 1.1), where p is the degree of $\tilde{p}_{k+1}$. Set $\zeta = (z_{k+1})_V$. Then $Q_\zeta(z_{k+1}) = \tilde{p}_{k+1}$ [75h], hence $D_\zeta = \tilde{\delta}$, and so ζ is a primitive element for π.

REMARK 3. If, e.g., V is a locally analytic subset of a linear space X, then there exists a primitive element $\alpha \in \mathcal{O}_V$ for f. Then, it is enough to take $\alpha = \lambda_V$, where $\lambda \in X^*$ is injective on some fibre $f^{-1}(w)$, $w \in W \setminus \Sigma$.

5. Assume now that V is irreducible. Then $\tilde{\mathcal{O}}_V$ is an integral domain (since $V \setminus Z$ is a connected manifold; see II. 3.6 and C. 3.9).

PROPOSITION (PŁOSKI-WINIARSKI [75i]). *Assume that there exists a prim-*

[75g]   See the proposition in n° 5 below.

[75h]   Since, for any $u \in \Omega \setminus Z$, the polynomials $t \longrightarrow O_\zeta(u, t)$ and $t \longrightarrow \tilde{p}_{k+1}(u, t)$ have p distinct roots in common.

[75i]   See [33a]. The authors give there a simple proof of the following (strengthened) version of a theorem of E. Platte: *Let h be a holomorphic function in an open connected*

itive element for f (e.g., by remark 3 in n° 4, that V is a locally analytic subset of $\mathbf{C}^n$) ([75j]). Then, for $\alpha \in \tilde{\mathcal{O}}_V$, the following three conditions are equivalent:

(a) α is a primitive element for f;

(b) α is a primitive element of $\tilde{\mathcal{O}}_V$ over $\mathcal{O}_W^0$ (see A. 8.3) or, equivalently, of $\mathcal{O}_V$ over $\mathcal{O}_W^0$ in the case when $\alpha \in \mathcal{O}_V$;

(c) Q_α^0 is the minimal polynomial of α over $\mathcal{O}_W^0$ (see A. 8.2).

PROOF . Let P be the minimal polynomial of α over $\mathcal{O}_W^0$. Then P must divide Q_α^0 (see A. 8.2). We show that (a)$\Longrightarrow$(c). If α is injective on a fibre $f^{-1}(w) = \{\eta_1, \ldots, \eta_p\} \subset V \setminus Z$, then, because of $P(\eta_i, \alpha(\eta_i)) = 0$ and since all $P(\eta_i, T)$ coincide (as the coefficients of P are constant on $f^{-1}(w)$), we must have $\deg P \geq p$, and so $P = Q_\alpha^0$. Let $K \subset L$ be the fields of fractions of $\mathcal{O}_W^0 \subset \tilde{\mathcal{O}}_V$, respectively. Then P is the minimal polynomial of α over K, and the condition (c) means precisely that α is of degree p over K. Next, the condition (b) means exactly that α is a primitive element of L over K, and the equivalence in (b) follows from remark 1 in n° 4 (see A. 1.15). Owing to the remark in A. 8.3, every element of L is algebraic of degree $\leq p$ over K and, by (a)$\Longrightarrow$(c), there is one of degree p. According to the corollary in A. 7, the field L is finite over K and (b)$\Longleftrightarrow$(c). Finally, (c)$\Longrightarrow$(a) by the proposition from A. 6.3: if $Q_\alpha^0 = P$, then Q_α^0 is irreducible in $K[T]$ (see A. 5.2); hence $D_\alpha^0 \neq 0$, and so $D_\alpha \neq 0$.

6. Let f be as in n° 4 (we no longer suppose that V is irreducible). Let α be a primitive element for f.

Take any open connected subset U of W, and set $\Omega = f^{-1}(U)$. Then $f^U : \Omega \longrightarrow U$ is a *-covering. It follows directly from the definitions that:

$$(4) \qquad D_{\alpha_\Omega} = (D_\alpha)_U \quad \text{and} \quad C_{j\alpha_\Omega\varphi_\Omega} = (C_{j\alpha\varphi})_U \quad \text{for} \quad \varphi \in \mathcal{O}_V.$$

Hence, putting $C_{j\alpha\varphi} = C_{j\alpha_\Omega\varphi}$ for $\varphi \in \mathcal{O}_\Omega$, we get by (GR) for f^U (and since $(\beta_U)^0 = (\beta^0)_\Omega$ for $\beta \in \mathcal{O}_W$) that:

$$(5) \qquad D_\alpha^0 \varphi = \sum_0^{p-1} (C_{j\alpha\varphi})^0 \alpha^j \quad \text{in} \quad \Omega \quad \text{for} \quad \varphi \in \mathcal{O}_\Omega.$$

---

*subset G of $\mathbf{C}^n$. Assume that $f = \operatorname{grad} h : G \longrightarrow H$, where H is an open subset of $\mathbf{C}^n$, f is a *-covering, and f has a unique zero in G. Then h is a primitive element for f.*

([75j]) One cannot drop this assumption. A counter-example is the mapping $F : \bar{\mathbf{C}} \longrightarrow \bar{\mathbf{C}}$ defined as a holomorphic extension of $\mathbf{C} \ni z \longrightarrow z^2 \in \mathbf{C}$.

Consider the submodule $L_U = L(f^U, \alpha_\Omega)$ of $\mathcal{O}_U^p$ (see (C) in n° 4). The second equality in (4) gives: $F \in L \implies F_U \in L_U$. Likewise, if $U_1 \subset U_2$ are open connected subsets of W, then $F \in L_{U_2} \implies F_{U_1} \in L_{U_1}$. This implies that for any $u \in W$ the set

$$L_u = \{F_u : \ f \in L_U, \ U \ni u\}$$

is a submodule of $\mathcal{O}_u^p$.

PROPOSITION . $L = \{F \in \mathcal{O}_W^p : \ F_u \in L_u \ for \ u \in W\}$.

PROOF . Since the inclusion $\subset$ is obvious, suppose now that $F = (F_0, \ldots \ldots, F_{p-1})$ belongs to the right hand side. Then, for every $a \in W$, there is an open connected neighbourhood U_a of a and some $\varphi_a \in \mathcal{O}_{\Omega_a}$, where $\Omega_a = f^{-1}(U_a)$, such that $(F_j)_U = C_{j\alpha\varphi_a}$. Hence, by (5), we have $D_\alpha^0 \varphi_a = \sum_0^{p-1} F_j^0 \alpha^j$ in Ω_a. Since $D_\alpha^0 \not\equiv 0$, the φ_a's coincide on the intersections of their domains, and therefore can be amalgamated into a function $\varphi \in \mathcal{O}_V$. Then $D_\alpha^0 \varphi_a = \sum_0^{p-1} F_j^0 \alpha^j$, and by the uniqueness in (GR) (see (A) in n° 4), we must have $F_j = C_{j\alpha\varphi}$. This means that $F \in L$.

The family of submodules $L_u \subset \mathcal{O}_u^p$, $u \in W$, is coherent (see VI. 1.4 below). This follows from the Grauert-Remmert finite mapping theorem ([17a] p. 64), via the isomorphism (C) (for the restrictions f^U). Here, a special case of $\mathcal{S} = \mathcal{O}$ is needed. It can be stated as follows.

Let $f : V \longrightarrow W$ be a finite mapping. For any finite $E \subset V$, define $\mathcal{O}_E = \mathcal{O}_E(V) = \prod_{z \in E} \mathcal{O}_z$ and $\psi_E = (\psi_z)_{z \in E} \in \mathcal{O}_E$ if ψ is a holomorphic function in a neighbourhood of E. If $z \in V$, then $\mathcal{O}_z$ is a module over $\mathcal{O}_{f(z)}$ (with the product defined by $\zeta\xi = (\zeta \circ f_z)\xi$ for $\zeta \in \mathcal{O}_{f(z)}$ and $\xi \in \mathcal{O}_z$) and so $\mathcal{O}_{f^{-1}(u)}$ is a module over $\mathcal{O}_u$ for any $u \in W$.

For every point of W, there is an open neighbourhood U and $\psi_1, \ldots, \psi_s \in \mathcal{O}_{f^{-1}(U)}$ such that if $u \in U$, then the module $\mathcal{O}_{f^{-1}(u)}$ is generated by $(\psi_1)_{f^{-1}(u)}, \ldots, (\psi_s)_{f^{-1}(u)}$.

As for the proof (see ibid.), one starts with the projection $\pi : \ W \longrightarrow \Omega$ of a Weierstrass set (see III. 2.3), where the P_j's have non-zero discriminants. Namely, the system $(v^s)_W$, $|s| < p$, where $v = (z_{k+1}, \ldots, z_n)$ and $p = \sum \deg P_j$, has the required property: the germs $((v^s)_W)_{\pi^{-1}(a)}$, $|s| < p$, generate $\mathcal{O}_{\pi^{-1}(a)}(W)$ for any $a \in V$. In the case of $n = k + 1$, this is shown by means of the Hensel lemma (see II. 3.5): if $\pi^{-1}(a) = \{b_1, \ldots, b_r\}$, $b_i = (a, c_i)$, then $P_{k+1} = Q_1, \ldots, Q_r$ in $\pi^{-1}(U_1)$, where U_1 is an open neighbourhood of a, $Q_i \in \mathcal{O}_{U_1}[z_{k+1}]$ are monic, and $Q_i(a, t) = (t - c_i)^{n_i}$. To see this one sets $S_i = \prod_{\nu \neq i} Q_\nu$. Then, for arbitrarily given $h = (h_1, \ldots, h_r) \in \mathcal{O}_{f^{-1}(a)}(W)$, i.e., $h_i = (H_i)_{W_{b_i}}$, where $H_i \in \mathcal{O}_{b_i}$, the preparation theorem (I. 1.4) gives $H_i(S_i)_{b_i}^{-1} = q_i(Q_i)_{b_i} + (R_i)_{b_i}$, with $R_i \in \mathcal{O}_U[z_{k+1}]$ and degree $< p$, and some open neighbourhood $U \subset U_1$ of a. By setting $R = \sum_1^r R_\nu S_\nu$ and taking restrictions to W_{b_i}, one obtains $h_i = (R_W)_{b_i}$ (since $((Q_i)_W)_{b_i} = 0$ and $((S_\nu)_W)_{b_i} = 0$ for $\nu \neq i$), i.e., $h = (R_W)_{f^{-1}(a)} \in \sum_0^{p-1} \mathcal{O}_a((z_{k+1})_W^\nu)_{f^{-1}(a)}$. The general case follows by a straightforward induction, using the previous case and two easy-to-check observations on finite mappings: (α) if the statement is true for $f : X \longrightarrow Y$, then it is also true for $f_S : S \longrightarrow T$ with any analytic $S \subset X$, $T \subset Y$ such that $f(S) \subset T$; (β) if the statement is true for $f : X \longrightarrow T$ and $g : Y \longrightarrow Z$, then it is also true for $g \circ f$. Next, by (α), the statement holds for $\pi_V : V \longrightarrow \pi(V)$ with any $V \subset W$ which is analytic in $\Omega \times \mathbf{C}^{n-k}$. Finally, for any finite mapping, since the statement is local on W ([75k]), we may assume that $V \subset \subset \mathbf{C}^1$ and $W \subset \mathbf{C}^k$ are locally analytic and contain 0. Then it is enough to prove the statement for the natural projection $f \longrightarrow W$ (since the natural projection $f \longrightarrow V$ is biholomorphic; see V. 3.4), i.e., for $F \longrightarrow W$, where $F = \{(w,z) : w = f(z)\}$. Since F_0 is k-normal (see the proposition in III. 4.4), we may assume (see III. 2.4) that F is contained in a Weierstrass set (see III. 2.3 with $n = k + 1$), where the P_j's have non-zero discriminants, and that F is analytic in $\Omega \times \mathbf{C}^1$. But this is the previous case, in which the statement has already been proved.

7. REMARK . The theorems and properties from §7 can be carried over, with the same proofs, to the case of analytic spaces. After extending the definition of analytically constructible sets to this case (in a natural way, via charts), it is easy to verify that the properties from IV. 8, theorem 1 from 3.2, and the Chevalley-Remmert theorem from 5.1 remain true.

§8. c–holomorphic mappings

Let X and Y be analytic spaces.

1. We say that a mapping $f : X \longrightarrow Y$ is c–*holomorphic* if it is continuous and its restriction $f_{X^0} : X^0 \longrightarrow Y$ is holomorphic.

This is an essential extension of the notion of a holomorphic mapping. For instance, the inverse of the holomorphic mapping $\mathbf{C} \ni t \longrightarrow (t^2, t^3) \in W = \{z^3 = w^2\} \subset \mathbf{C}^2$ is a c–holomorphic mapping (see IV. 2.1, footnote ([10])), as it is locally biholomorphic at any $t \neq 0$. However, it is not holomorphic, for otherwise it would be biholomorphic, contrary to the fact that 0 is a singular point of W.

THEOREM . *A mapping $f : X \longrightarrow Y$ is c–holomorphic if and only if it is continuous* ([76]) *and its graph is analytic in $X \times Y$.*

PROOF . One may assume that X is an analytic subset of a manifold M and that Y is a vector space. The condition is sufficient in view of the analytic graph theorem (see §1, corollary 3 of theorem 2) applied to the restrictions

---

([75k]) i.e., it suffices to prove it for f^U with an open neighbourhood U of any point of W.

([76]) Local boundedness suffices, as it yields continuity since the graph is closed. (See B.2.4.)

of the mapping f to the connected components of X^0. To prove that the condition is also necessary, it is enough to show that (if f is c–holomorphic) each point $a \in X$ has in M an open neighbourhood U such that the set $f_{X \cap U}$ is analytic in $U \times Y$.

Assume first that the germ X_a is irreducible. One may assume that X is the crown of a normal triple (Ω, Z, X) of dimension k in $\mathbf{C}^n$, that $a = 0$, and that $M = \Omega \times \mathbf{C}^{n-k}$ (see IV. 1.5). Now, since $X_{\Omega \setminus Z} \subset X^0$ (see IV. 1.1, property (3)), the set $f_{X_{\Omega \setminus Z}}$ is a closed submanifold in $(\Omega \setminus Z) \times \mathbf{C}^{n-k} \times Y$ and its closure in $\Omega \times \mathbf{C}^{n-k} \times Y = M \times Y$ is f (see IV. 1.1, property (4)). Furthermore, in view of property (1) from IV. 1.1, the natural projection $f \longrightarrow \Omega$ is proper. Thus the proposition from IV. 6.1 implies that f is analytic in $M \times Y$.

Now, consider the general case. According to proposition 3b from IV. 3, there is an open neighbourhood U' of the point a such that we have a decomposition into irreducible components $X \cap U' = X_1 \cup \ldots \cup X_r$ and the germs $(X_i)_a$ are irreducible. The mapping $f_{X_i^0 \setminus X^*}$ is holomorphic and the set $X^* \cap X_i^0$ is nowhere dense in X_i^0 (see IV. 2.9, theorem 4). Therefore the mapping $f_{X_i^0}$ is also holomorphic (see II. 3.5) and thus the mapping f_{X_i} is c–holomorphic ($i = 1, \ldots, r$). Hence there exists an open neighbourhood $U \subset U'$ of the point a such that the sets $f_{X_i \cap U}$ are analytic in $U \times Y$ and so is their union $f_{X \cap U}$.

2. Almost all properties and theorems from §§3–6 generalize easily to the case of c–holomorphic mappings of analytic spaces. For instance, we have the following two theorems on the c–holomorphic mapping

$$ f : X \longrightarrow Y $$

of the analytic spaces X and Y. In order to prove them, it is enough to note that the natural projection $\pi : f \longrightarrow Y$ is holomorphic and to apply to π the corresponding result for holomorphic mappings.

THE SEMICONTINUITY THEOREM . *The function* $X \ni z \longrightarrow \dim l_z f$ *is upper semicontinuous* [77] .

Indeed, since $\pi^{-1}(c) = f^{-1}(c) \times c$, we have $l_{(z, f(z))} \pi = (l_z f) \times f(z)$, and hence $\dim l_z f = \dim l_{(z, f(z))} \pi$.

REMMERT'S PROPER MAPPING THEOREM . *If the mapping* f *is proper, then* $f(X)$ *is an analytic subset of* Y.

---

[77] The germ $l_z f$ is defined in the same way as in 3.2.

For then the projection π is proper (because $\pi^{-1}(E) = f_{f^{-1}(E)}$) and $f(X) = \pi(f)$.

CHAPTER VI

NORMALIZATION

§1. The Cartan and Oka coherence theorems

1. First, we are going to prove two lemmas that form the core of the proof of Cartan's coherence theorem.

Let $G \subset \mathbf{C}^k$ be an open set, and let V be an analytic set in $G \times \mathbf{C}^{n-k}$. Suppose that there are monic polynomials $p_j \in \mathcal{O}_G[z_j]$, $j = k+1, \ldots, n$, (^{1}) that vanish on V and polynomials $Q_{k+2}, \ldots, Q_n \in \mathcal{O}_G[z_{k+1}]$ such that, if the discriminant of p_{k+1} is denoted by δ, then the Rückert formula holds:

$$(R) \quad V_{\{\delta \neq 0\}} = \{(u, v) \in G \times \mathbf{C}^{n-k} :$$

$$\delta(u) \neq 0, \ p_{k+1}(u, z_{k+1}) = 0, \ \delta(u)z_j = Q_j(u, z_{k+1}), \ j = k+2, \ldots, n\}$$

and $V_{\{\delta \neq 0\}}$ is dense in V. Let $N = \{z_1 = \ldots = z_{k+1} = 0\} \subset \mathbf{C}^k$. For $c \in G \times \mathbf{C}^{n-k}$, let J_c denote the ideal generated by the germs at c of the functions

$$p_j(u, z_j), \ j = k+1, \ldots, n, \quad \text{and} \quad \delta(u)z_j - Q_j(u, z_{k+1}), \ j = k+2, \ldots, n.$$

Let $m \geq$ the sum of the degrees of the polynomials p_j.

If an ideal I of the ring $\mathcal{O}_n$ satisfies the assumptions of Rückert's classic descriptive lemma (see II. 3.2), then there are: (i) an open neighbourhood $G \subset \mathbf{C}^k$ of zero, (ii) a representative V of the germ $V(I)$, and (iii) polynomials p_j, Q_j such that the above conditions are satisfied.

(^{1}) We have $\mathcal{O}_G[z_j] \subset \mathcal{O}_{G \times \mathbf{C}^{n-k}}$, after the natural identification $\mathcal{O}_G \hookrightarrow \mathcal{O}_{G \times \mathbf{C}^n}$.

LEMMA 1 (CARTAN). *Let* $c \in G \times \mathbf{C}^{n-k}$. *We have* $\mathcal{I}(V_c) \supset J_c : \delta_c^m$ *and*

$$\mathcal{I}(V_c) = J_c : \delta_c^m, \quad when \quad V \cap (c + N) \subset c.$$

PROOF. Let $f \in J_c : \delta_c^m$. Take a representative $\tilde{f}$ of the germ f. It follows from the formula (R) that $\tilde{f}\delta^m = V$ on $V_{\{\delta \neq 0\}}$ in a neighbourhood of c. Thus $\tilde{f} = 0$ on V in a neighbourhood of c, which means that $f \in \mathcal{I}(V_c)$. Therefore $J_c : \delta_c^m \subset \mathcal{I}(V_c)$.

Now, let $V \cap (c + N) \subset c$. The natural projection $V \longrightarrow G \times \mathbf{C}$ (parallel to N) is proper $(^2)$. Consequently, the point c has an arbitrarily small open neighbourhood $U = \Omega \times \Theta \times \Delta$, where $\Omega \subset G$, $\Theta \subset \mathbf{C}$, $\Delta \subset \mathbf{C}^{n-k-1}$ such that $V_{\Omega \times \Theta} \subset U$ (see B. 2.4). Moreover, in view of the preparation theorem, one may require that $p_{k+1} = gp$ in $\Omega \times \Theta$. Here g is a holomorphic function in $\Omega \times \Theta$ which is different from zero in $\Omega \times \Theta$, while $p \in \mathcal{O}_\Omega[z_{k+1}]$ is a monic polynomial of degree s such that $(u \in \Omega, \ p(u,t) = 0) \Longrightarrow t \in \Theta$ (see C. 2.2). Then

$$(1) \qquad u \in \Omega_0 \Longrightarrow \#\pi\big(V_u \cap (\Theta \times \Delta)\big) \geq s \ ,$$

where $\Omega_0 = \{\delta \neq 0\} \cap \Omega$ and $\pi : \ \Theta \times \Delta \longrightarrow \Theta$ is the natural projection. Indeed, if $u \in \Omega_0$, then all the roots of the polynomial $t \longrightarrow p(u,t)$ are simple, and so the polynomial $t \longrightarrow p_{k+1}(u,t)$ has s distinct roots in Θ. Therefore it follows from the formula (R) that the set $V_u \cap (\Theta \times \Delta) = V_u \cap (\Theta \times \mathbf{C}^{n-k-1})$ contains s points whose z_{k+1}–coordinates are distinct.

Let $\tilde{p}_j \in \mathcal{O}_k[z_j]$, $\tilde{\delta} \in \mathcal{O}_k$, $\tilde{Q}_j \in \mathcal{O}_k[z_{k+1}]$, and $\tilde{z}_j$ be the germs at 0 of the functions $p_j(c + z)$, $\delta(c + z)$, $Q_j(c + z)$, and $c_j + z_j$, respectively. Set $\tilde{v} = (\tilde{z}_{k+1}, \dots, \tilde{z}_n)$. By the preparation theorem (see I. 1.4), we have $\mathcal{O}_j = \mathcal{O}_j \tilde{p}_j + \sum_{\nu=0}^{s_j} \mathcal{O}_{j-1} z_j^\nu$, $j = n, \dots, k+1$, which implies easily that $\mathcal{O}_n = \sum_{j=k+1}^n \mathcal{O}_n \tilde{p}_j + \sum_{|q| \leq m} \mathcal{O}_k \tilde{v}^q$. But of course $(\tilde{\delta}\tilde{v})^q \in \mathcal{O}_k[z_{k+1}] + \sum_{j=k+2}^n \mathcal{O}_n(\tilde{\delta}\tilde{z}_j - \tilde{Q}_j)$ $(^3)$, and so we obtain

$$(2) \qquad \tilde{\delta}^m \mathcal{O}_n \subset \mathcal{O}_k[z_{k+1}] + \sum_{k+1}^n \mathcal{O}_n \tilde{p}_j + \sum_{k+2}^n \mathcal{O}_n(\tilde{\delta}\tilde{z}_j - \tilde{Q}_j).$$

Now assume that $f \in \mathcal{I}(V_c)$. Then the neighbourhood U can be chosen so small that the germ f has a holomorphic representative F in U which

---

$(^2)$ Since the natural projection $V \longrightarrow G$ is proper. (See B. 2.4.)

$(^3)$ Because $(\tilde{\delta}\tilde{v})^q$ is the product of $\tilde{\delta}\tilde{z}_j = \tilde{Q}_j + (\tilde{\delta}\tilde{z}_j - \tilde{Q}_j)$ and $\tilde{\delta}\tilde{z}_{k+1}$, $\tilde{Q}_j \in \mathcal{O}_k[z_{k+1}]$.

vanishes on $V \cap U$. Moreover, according to (2), one may require that

$$
(3) \qquad \delta^m F \in R(u, z_{k+1}) + \sum_{k+1}^{n} \mathcal{O}_U p_j(u, z_j) + \sum_{k+2}^{n} \mathcal{O}_U(\delta z_j - Q_j) ,
$$

where $R \in \mathcal{O}_\Omega[z_{k+1}]$. Replacing the polynomial R by the remainder from the division of R by p, we may also assume that the degree of R is $< s$. Now, if $u \in \Omega_0$, then, by (3) and (R), we have $R = 0$ on $u \times (V_u \cap (\Theta \times \Delta))$. Therefore it follows from (1) that the polynomial $t \longrightarrow R(u, t)$ vanishes at s distinct points and hence it is zero. Thus $R = 0$ since Ω_0 is dense in Ω. So, according to (3), we obtain $\delta_c^m f \in J_c$, which means that $f \in J_c : \delta_c^m$.

LEMMA 2 ([4]) . *Let V be an analytic set in an open neighbourhood of zero in $\mathbf{C}^n$. Assume that the germ V_0 is simple and regular. Then there exist holomorphic functions $h_1, \ldots, h_p$, $f_{11}, \ldots, f_{1l}, \ldots, f_{p1}, \ldots, f_{pl}$ in an open neighbourhood U of zero such that*

$$
\mathcal{I}(V_c) = \sum_{i=1}^{p} \mathcal{I}_{ic} : (h_i)_c \text{ for } c \in U ,
$$

where $\mathcal{I}_{ic}$ is the ideal generated by the germs $(f_{i1})_c, \ldots, (f_{il})_c$.

PROOF . The germ V_0 is simple and k–regular, where $0 \le k \le n$; hence its ideal $I = \mathcal{I}(V_0)$ is prime and k–regular (see II. 4.6 and III. 2.5). One may assume that V is analytic in $\Omega \times \mathbf{C}^{n-k}$, where Ω is an open neighbourhood of the origin in $\mathbf{C}^k$, and that for some $r \in \mathbf{N} \setminus 0$ we have

$$
\#V_u \le r \text{ for } u \in \Omega
$$

(see III. 2.3–4). Let $p = (n - k)r$. There are linear forms $\chi_i = \sum_{j=k+1}^{n} a_{ij} z_j \in (\mathbf{C}^{n-k})^*$, $j = 1, \ldots, p$, such that each set consisting of $n-k$ of them is linearly independent ([5]) . We can take neighbourhoods Δ_{ij} of the points a_{ij} in $\mathbf{C}$ such that if $c_{ij} \in \Delta_{ij}$, then each set which consists of $n - k$ of the forms $\zeta_i = \sum_{j=k+1}^{n} c_{ij} z_j$, $i = 1, \ldots, p$, is linearly independent; thus

$$
(*) \qquad
\begin{aligned}
&\text{for each } v_1, \ldots, v_r \in \mathbf{C}^{n-k} \setminus 0 \text{ there exists } s \text{ such that } \zeta_s(v_j) \ne 0, \\
&j = 1, \ldots, r.
\end{aligned}
$$

([4]) See Herve [24], IV. 1.

([5]) See, e.g., the proof of lemma 2 from III. 1.1.

Indeed, setting $\Lambda_j = \{i : \zeta_i(v_j) = 0\}$, we have $\#\Lambda_j < n - k$, and so $\#\bigcup\Lambda_j < p$. Thus there is an $s \leq p$ such that $s \notin \bigcup\Lambda_j$. Now, if $c_{ij} \in \Delta_{ij} \setminus 0$, then we have the linear automorphisms

$$\varphi_i = \left(z_1, \ldots, z_k, \zeta_i(z_{k+1}, \ldots, z_n), z_{k+2}, \ldots, z_n\right) : \mathbf{C}^n \longrightarrow \mathbf{C}^n .$$

In view of remark 1 following the classic descriptive lemma (see III. 3.2), we can choose $c_{ij} \in \Delta_{ij} \setminus 0$ so that the ideals $I \circ \varphi_i^{-1} = \mathcal{I}(\varphi_i(V)_0)$ satisfy the assumptions of that lemma. In view of lemma 1 (since $\varphi_i(V)$ is a representative of the germ $V(I \circ \varphi_i^{-1})$; see II. 4.5), there exist functions $\eta_1, \ldots, \eta_p, g_{11}, \ldots, g_{1l}, \ldots, g_{p1}, \ldots, g_{pl}$ that are holomorphic in an open neighbourhood $U' \subset \Omega \times \mathbf{C}^{n-k}$ of zero in $\mathbf{C}^n$, such that

$$\mathcal{I}\left(\varphi_i(V)_d\right) \supset J_{id} : (\eta_i)_d \quad \text{for} \quad d \in U' ,$$

where J_{id} is the ideal generated by the germs $(g_{i1})_d, \ldots, (g_{il})_d$, and the equality occurs when $\varphi_i(V) \cap (d + N) \subset d$. Let U be an open neighbourhood of zero in $\mathbf{C}^n$ such that $\varphi_i(U) \subset U'$ ($i = 1, \ldots, p$). Let $c \in U$. By taking $d = \varphi_i(c)$ and passing to the images under the isomorphism $\mathcal{O}_d \ni f \longrightarrow f \circ \varphi_i \in \mathcal{O}_c$, we obtain (see II. 4.4 and B. 4.1)

$$\mathcal{I}(V_c) \supset I_{ic} : (h_i)_c, \quad i = 1, \ldots, p ,$$

where $h_i = (\eta_i \circ \varphi_i)_U$ and I_{ic} is the ideal generated by the germs at c of the functions $f_{ij} = (g_{ij} \circ \varphi_i)_U$, $i = 1, \ldots, l$. Moreover, the equality occurs when $V \cap \left(c + \varphi_i^{-1}(N)\right) \subset c$. Hence, to have the equality stated in the lemma, it is enough to show that the last condition is satisfied for some i. Now, since $\varphi_i^{-1}(N) = 0 \times \ker\zeta_i$, that condition means precisely that $(V_u - v) \cap \ker \zeta_i \subset 0$, where $(u, v) = c$; in other words, $\zeta_i(w_\nu) \neq 0$ if $w_\nu \neq 0$, where $V_u - v = \{w_1, \ldots, w_s\}$, $s \leq r$. Thus it follows from $(*)$ that the condition is satisfied for some i.

2. Let X be an analytic space. Consider a family of submodules

$$M_z \subset \mathcal{O}_z^m, \quad z \in X .$$

A sequence of mappings $g_1, \ldots, g_k \in \mathcal{O}_X^m$ is called a *finite system of generators* of the family if the germs $(g_i)_z$ generate M_z for each $z \in X$ [6] .

---

[6] We identify here the sequence of functions $(f_1, \ldots, f_k) \in \mathcal{O}_X^k$ with their diagonal product $(f_1, \ldots, f_k) : X \longrightarrow \mathbf{C}^k$, and the germ $(f_1, \ldots, f_k)_z$ with the sequence of germs $((f_1)_z, \ldots, (f_k)_z)$.

It follows from the definition that if the families of submodules $M_z \subset \mathcal{O}_z^p$, $z \in X$, and $N_z \subset \mathcal{O}_z^q$, $z \in X$, have finite systems of generators, then so does the family

$$M_z \times N_z \subset \mathcal{O}_z^{p+q}, \quad z \in X ,$$

and in the case when $p = q$, so does the family

$$M_z + N_z \subset \mathcal{O}_z^p, \quad z \in X \quad (^7) .$$

The same holds for a finite number of such families.

A family of homomorphisms $h_z : \mathcal{O}_z^p \longrightarrow \mathcal{O}_z^q$, $z \in X$, is said to be *coherent* if it is of the form

$$h_z(\varphi_1, \ldots, \varphi_p) = \sum_1^p \varphi_i(g_i)_z, \quad \text{where} \quad g_1, \ldots, g_p \in \mathcal{O}_X^q \quad (^8) .$$

Then, if the family of submodules $M_z \subset \mathcal{O}_z^p$, $z \in M$, has a finite system of generators, so does the family of images

$$h_z(M_z) \subset \mathcal{O}_z^q, \quad z \in X .$$

(Indeed, if $f_\nu = (f_{\nu 1}, \ldots, f_{\nu p}) \in \mathcal{O}_X^p$ are generators of the family $\{M_z\}$, then, as $h_z((f_\nu)_z) = (\sum_1^p f_{\nu i} g_i)_z$, then the mappings $\sum_1^p f_{\nu i} g_i \in \mathcal{O}_X^q$ are generators of the family $\{h_z(M_z)\}$.

Let $F : T \longrightarrow X$ be an embedding of an analytic space T into the space X. In such a case, if a family of submodules $M_z \subset \mathcal{O}_z^m$, $z \in X$, has a finite system of generators, then so does the family

$$M_{F(t)} \circ F_t \subset \mathcal{O}_t^m, \quad t \in T .$$

(These are submodules, because the mappings $\mathcal{O}_{F(t)} \ni \varphi \longrightarrow \varphi \circ F_t \in \mathcal{O}_t$ are surjective. If $g_\nu \in \mathcal{O}_X^m$ are generators of the family $\{M_z\}$, then, since $(g_\nu)_{F(t)} \circ F_t = (g_\nu \circ F)_t$, it follows that the mappings $g_\nu \circ F \in \mathcal{O}_T^m$ are generators of the family $\{M_z \circ F_t\}$; see B. 4.2–3 and IV. 4.6.)

More generally, if $F : T \longrightarrow X$ is a holomorphic mapping of analytic spaces, then if a family M of submodules $M_z \subset \mathcal{O}_z^m$, $z \in X$, has a finite

---

$(^7)$ Indeed, if $f_i \in \mathcal{O}_X^p$, $g_j \in \mathcal{O}_X^q$ are generators of the families $\{M_z\}, \{N_z\}$, respectively, then (f_i, g_j) are generators of the family $\{M_z \times N_z\}$, whereas (if $p = q$) $f_1, \ldots, g_1, \ldots$ are generators of the family $\{M_z + N_z\}$.

$(^8)$ In other words, if the family of matrices of these homomorphisms is "induced" by a holomorphic mapping $g : X \longrightarrow \mathbf{C}_q^p$, that is, if $C_{h_z} = g_z$ for $z \in X$ (see A. 1.16).

system of generators, it follows that so does its *inverse image* F^*M *under the mapping* F. It is defined as the family of submodules $(F^*M)_t \subset \mathcal{O}_t^m$, $t \in T$, generated by $M_{F(t)} \circ F_t \subset \mathcal{O}_t^m$, respectively.

Note that if $G : S \longrightarrow T$ is a holomorphic mapping of analytic spaces, then $(F \circ G)^*M = G^*(F^*M)$.

For arbitrarily given $\mathbf{C}^p$–valued mappings $f_1, \ldots, f_m$ on X, we have the *family of modules of relations*

$$R_z(f_1, \ldots, f_m) \subset \mathcal{O}_z^m, \quad z \in X ,$$

defined by

$$R_z(f_1, \ldots, f_m) = \{ (\varphi_1, \ldots, \varphi_m) \in \mathcal{O}_z^m : \sum_1^m \varphi_i(f_i)_z = 0 \} .$$

2a. Let M be a manifold.

CARTAN'S CLOSEDNESS THEOREM . *Let* $a \in M$, *and let* $L \subset \mathcal{O}_a^r$ *be a submodule. Assume that a sequence* $f_\nu \in \mathcal{O}_M^r$ *converges almost uniformly to an* $f \in \mathcal{O}_M^r$. *If* $(f_\nu)_a \in L$ *for all* ν, *then* $f_a \in L$.

PROOF . We may assume that M is an open subset of $\mathbf{C}^n$ and that $a = 0$. For every k, consider the linear mapping of linear spaces

$$\Phi_k : \mathcal{O}_a^r \ni h \longrightarrow \left(\frac{\partial^{|p|} f}{\partial z^p}(a) \right)_{|p|<k} \in C_k = (\mathbf{C}^r)^{\{|p|<k\}} .$$

Then $\ker \Phi_k = \mathfrak{m}^k \mathcal{O}_a^r$ (see I. 1.7). Since the subspace $\Phi_k(L) \subset C_k$ is closed (because C_k is finite dimensional; see B. 5.1) and the sequence $\Phi_k((f_\nu)_a)$ converges to $\Phi_k(f_a)$ (see C. 1.9), we have $\Phi_k(f_a) \in \Phi_k(L)$, which means that $f_a \in L + \ker \Phi_k$. Therefore $f_a \in \bigcap_{k=1}^\infty (L + \mathfrak{m}^k \mathcal{O}_a^r) = L$, by the Krull intersection theorem (see A. 10.5, the remark).

Let $\mathcal{L} = \{ L_u \subset \mathcal{O}_u^r : u \in M \}$ be any family of submodules. An $F \in \mathcal{O}_M^r$ is said to be a section of $\mathcal{L}$ if $F_u \in L_u$ for each $u \in M$.

COROLLARY . *The set of sections of* $\mathcal{L}$ [8a] *is closed in* $\mathcal{O}_M^r$ *with respect to almost uniform convergence.*

Let α be a primitive element for f, as in V. 7.4. The corollary, combined with the proposition in V. 7.6, implies

---

[8a]   This is, of course, a submodule of $\mathcal{O}_W^p$.

REMARK . The submodule $L = L(f,\alpha) \subset \mathcal{O}_W^p$ is closed in $\mathcal{O}_W^p$ with respect to almost uniform convergence.

Let X be an analytic space. Using the Grauert-Remmert formula, one deduces ([8b]) the following

THEOREM (GRAUERT-REMMERT). *The limit of any almost uniformly convergent sequence of holomorphic functions on X is holomorphic.*

PROOF . We may assume that X is a locally analytic subset V of $\mathbf{C}^n$, and that it has a finite decomposition into simple components $V = V_1 \cup \ldots \cup V_r$ (see IV. 3.1, proposition 3b). Next, by a trick due to S. Cynk, it is enough to prove the theorem when V is of constant dimension. In fact, suppose that it is true in that case. Set $C_k = C^k \times 0 \subset \mathbf{C}^n$ and $k_i = \dim V_i$. Then the set $\tilde{V} = \bigcup_{i=1}^r v_i \times C_{n-k_i} \subset \mathbf{C}^{2n}$ is locally analytic of constant dimension n. Consider the natural projection $\pi : V \times \mathbf{C}^n \longrightarrow V$ and the mapping $\iota : V \ni z \longrightarrow (z,0) \in \tilde{V}$. Then $\pi_{\tilde{V}} \circ \iota = \mathrm{id}_V$. Now, if a sequence $\varphi_\nu \in \mathcal{O}_V$ converges to some φ almost uniformly, then $\varphi_\nu \circ \pi_{\tilde{V}}$ converges to $\varphi \circ \pi_{\tilde{V}}$ almost uniformly; hence $\varphi \circ \pi_{\tilde{V}}$ is holomorphic and so is $\varphi = \varphi \circ \pi_{\tilde{V}} \circ \iota$. Finally, by the remark in V. 7.2, we may assume that there exists a *-covering $f : V \longrightarrow W$, where W is a manifold. By remark 3 in V. 7.4, there exists a primitive element α for f. Suppose that a sequence $\varphi_\nu \mathcal{O}_V$ converges almost uniformly to φ. Then the sequences $C_{j\alpha\varphi_\nu}$ converge almost uniformly in $W \setminus \Sigma$, by the definition of $C_{j\alpha\varphi}$ (see V. 7.4 and B. 2.4). Therefore, by II. 3.9, $C_{j\alpha\varphi_\nu}$ converges to some $c_j \in \mathcal{O}_W$ almost uniformly in W. Now, by the remark, $(c_0, \ldots, c_{p-1}) \in L$, i.e., there is a $\psi \in \mathcal{O}_V$ such that $c_j = C_{j\alpha\psi}$. Also we have $D_\alpha^0 \psi = \sum_0^{p-1} c_j \alpha^j$, according to (GR). But, by passing to the limit in (GR) applied to φ_ν, we get $D_\alpha^0 \varphi = \sum_0^{p-1} c_j \alpha^j$. Thus, since φ is continuous, we obtain $\varphi = \psi$. ([8c])

3. We are going to prove Oka's theorem for $\mathbf{C}^n$, and then, using lemma 2 from n° 1, we will derive Cartan's theorem for $\mathbf{C}^n$.

OKA's THEOREM (the case of $\mathbf{C}^n$). *If $f_1, \ldots, f_m$ are $\mathbf{C}^p$-valued mappings, holomorphic in an open neighbourhood $U' \subset \mathbf{C}^n$ of zero, then there exists an open neighbourhood $U \subset U'$ of zero such that the family of modules of relations $R_z(f_1, \ldots, f_m)$, $z \in U$, has a finite system of generators.*

Following an idea due to Malgrange ([9]) , we will first formulate Oka's

---

([8b]) See [17a] pp. 145-146.

([8c]) For other proofs see [17c] p.171, or [22] V.B.5.

([9]) See [43], Chapter II, §6.

theorem in terms of flat rings. Let $\check{\mathcal{O}}_n$ denote the ring of germs at 0 of the mappings η defined on neighbourhoods of zero in $\mathbf{C}^n$ and with values $\eta(z) \in \mathcal{O}_z$ ([10]) . Then, after the identification

$$\mathcal{O}_n \ni f \longrightarrow (z \longrightarrow \tilde{f}_z)_0 \in \check{\mathcal{O}}_n ,$$

where $\tilde{f}$ is a representative of the germ f, the ring $\mathcal{O}_n$ becomes a subring of $\check{\mathcal{O}}_n$. Oka's theorem above is equivalent to the statement: *the ring $_n\check{\mathcal{O}}$ is flat over the ring $_n\mathcal{O}$* ([11]) .

In the proof of Oka's theorem we will need the following properties.

Let $u \in \mathbf{C}^{n-k}$, $v \in \mathbf{C}$. Then $\mathcal{O}_u \subset \mathcal{O}_{(u,v)}$ is a subring (after the identification). Obviously

(α) the germ $\zeta_{(u,v)} \in \mathcal{O}_{(u,v)}$ of the function $z \longrightarrow z_n$ is a transcendental element over $\mathcal{O}_u$.

Let $\check{\mathcal{Q}}_n \subset \mathcal{O}_n$ denote the subring of germs of mappings η such that $\eta(z)$ are germs of polynomials in z_n. Then

(β) $\check{\mathcal{Q}}_n$ is the set of germs at 0 of mappings of the form $z \longrightarrow \chi(z)(\zeta_z)$, where $\chi(u,v) \in \mathcal{O}_u[T]$.

If $\Omega \subset \mathbf{C}^{n-1}$ is an open set and $u \in \Omega$, then for $F = \sum a_\nu T^\nu \in \mathcal{O}_\Omega[T]$ we define $F_u = \sum (a_\nu)_u T^\nu \in \mathcal{O}_u[T]$ and $H_u = ((H_1)_u, \ldots, (H_p)_u) \in \mathcal{O}_u[T]^p$ for $H = (H_1, \ldots, H_p) \in \mathcal{O}_\Omega[T]^p$. For $F_1, \ldots, F_m \in \mathcal{O}_\Omega[T]^p$, we define the submodule $R_u(F_1, \ldots, F_m) = \{(\Psi_1, \ldots, \Psi_m) : \sum_1^m \Psi_i(F_i)_u = 0\} \subset \mathcal{O}_u[T]^p$. As before (see footnote ([11])), we can check that:

([10]) With the ring operations defined by: $\eta_0 + \chi_0 = (\eta + \chi)_0$, $\eta_0 \chi_0 = (\eta\chi)_0$, where $(\eta + \chi)(z) = \eta(z) + \chi(z)$ and $(\eta\chi)(z) = \eta(z)\chi(z)$.

([11]) The flatness of $\check{\mathcal{O}}_n$ over $\mathcal{O}_n$ (see A. 1.16a) means the following. If $f_1, \ldots, f_m$ are $\mathbf{C}^p$-valued holomorphic mappings in a neighbourhood of zero in $\mathbf{C}^n$, then there are $\mathbf{C}^m$-valued holomorphic mappings ψ_σ, $\sigma = 1, \ldots, r$ in a neighbourhood of zero (in $\mathbf{C}^n$) with germs $(\psi_\sigma)_0 \in R_0(f_1, \ldots, f_m)$, such that if $\eta(z) \in R_z(f_1, \ldots, f_m)$ for z from a neighbourhood of zero, then for each z from a neighbourhood of zero, $\eta(z)$ is a linear combination of $(\psi_1)_z, \ldots, (\psi_r)_z$ with coefficients from $\check{\mathcal{O}}_z$. Then, for z from some neighbourhood of zero, the germs $(\psi_\sigma)_z$ generate $R_z(f_1, \ldots, f_m)$; for if this were not true, there would exist a sequence $z_\nu \longrightarrow 0$ of points z_ν that do not have this property, and, by defining η in a neighbourhood of zero in such a way that $\eta(z_\nu) \in R_z(f_1, \ldots, f_m) \setminus \sum_{\sigma=1}^r \mathcal{O}_{z_\nu} (\psi_\sigma)_{z_\nu}$, we would get a contradiction.

(γ) $\breve{\mathcal{O}}_{n-1}[T]$ is flat over $\mathcal{O}_{n-1}[T]$ if and only if for any $F_1, \ldots, F_m \in \mathcal{O}_U[T]^p$, where U is an open neighbourhood of zero in $\mathbf{C}^{n-1}$, there is an open neighbourhood $U_0 \subset U$ of zero and $G_1, \ldots, G_r \in \mathcal{O}_{U_0}[T]^m$, such that $(G_\nu)_u$ generate $R_u(F_1, \ldots, F_m)$ for each $u \in U_0$ (^{12}).

If G is a biholomorphic mapping between two neighbourhoods of zero in $\mathbf{C}^n$ and $G(0) = 0$, then

(δ) the mapping $\breve{\mathcal{O}}_n \ni \varphi \longrightarrow \varphi \circ G \in \breve{\mathcal{O}}_n$ is an $\mathcal{O}_n$–automorphism (here, $\varphi \circ G = (z \longrightarrow \tilde{\varphi}_{G(z)} \circ G_z)_0$, where $\tilde{\varphi}$ is a representative of φ)   (^{13}).

PROOF of Oka's theorem (^{13a}). The case $n = 0$ is trivial. Let $n > 0$, and assume that the theorem is true for $n - 1$, i.e., that $\breve{\mathcal{O}}_{n-1}$ is flat over $\mathcal{O}_{n-1}$. First, we will show that then $\breve{\mathcal{Q}}_n$ is flat over $\mathcal{Q}_n$.

Let $\alpha_1, \ldots, \alpha_m \in \mathcal{Q}_n$ (^{14}). Then α_i, as an element of $\breve{\mathcal{O}}_n$, has a representative of the form $(u, v) \longrightarrow (\tilde{\alpha}_i(z_n))_{(u,v)} = (\tilde{\alpha}_i)_u(\zeta_{(u,v)})$, where $\tilde{\alpha}_i \in \mathcal{O}_U[T]$ and U is an open neighbourhood of zero in $\mathbf{C}^{n-1}$. We know that $\breve{\mathcal{O}}_{n-1}[T]$ is flat over $\mathcal{O}_{n-1}[T]$ (see A. 1.16a). Hence, by (γ), if U is chosen sufficiently small, there are $\tilde{\psi}_\nu = (\tilde{\psi}_{\nu 1}, \ldots, \tilde{\psi}_{\nu m}) \in \mathcal{O}[T]^m$, $\nu = 1, \ldots, r$, such that $(\tilde{\psi}_\nu)_u$ generate $R_u(\tilde{\alpha}_1, \ldots, \tilde{\alpha}_m)$ over $\mathcal{O}_u[T]$, for each $u \in U$. Then the germs $\psi_{\nu i}$ of the mappings $(u, v) \longrightarrow (\tilde{\psi}_{\nu i})_u(\zeta_{(u,v)}) = (\tilde{\psi}_{\nu i}(z_n))_{(u,v)}$ belong to $\mathcal{Q}_n$ and satisfy the equations $\sum_i \psi_{\nu i} \alpha_i = 0$, $\nu = 1, \ldots, r$ (^{15}). Now, let germs $\varphi_1, \ldots, \varphi_m \in \breve{\mathcal{Q}}_n$ satisfy the equations $\sum \varphi_i \alpha_i = 0$. Then, in view of ($\beta$), it follows, after taking a smaller U and choosing an open neighbourhood Δ of zero in $\mathbf{C}$, that the germs φ_i have representatives of the form $U \times \Delta \ni (u, v) \longrightarrow \tilde{\varphi}_i(u, v)(\zeta_{(u,v)})$, where $\tilde{\varphi}_i(u, v) \in \mathcal{O}_u[T]$, satisfying

---

(^{12}) We use the fact that the ring $\breve{\mathcal{O}}_{n-1}[T]$ is isomorphic to the ring of germs at 0 of mappings H from neighbourhoods of zero in $\mathbf{C}^{n-1}$, with values $H(u) \in \mathcal{O}_u[T]$, of uniformly bounded degrees. Then the subring $\mathcal{O}_{n-1}[T]$ corresponds to the subring of the germs of mappings of the form $u \longrightarrow F_u$, where $F \in \mathcal{O}_U[T]$ and $U \subset \mathbf{C}^{n-1}$ are open neighbourhoods of zero.

(^{13}) Because $(\eta \circ G)_z = \eta_{G(z)} \circ G_z$ for a holomorphic function η in a neighbourhood of zero.

(^{13a})   See [25] pp. 154-156.

(^{14}) To check flatness, according to the definition, it suffices to assume that $r = 1$ (see A. 9.6).

(^{15}) Since $\sum_i (\tilde{\psi}_{\nu i})_u (\tilde{\alpha}_i)_u = 0$.

$\sum \tilde{\varphi}_i(u,v)(\zeta_{(u,v)})(\tilde{\alpha}_i)_u(\zeta_{(u,v)}) = 0$ for $(u,v) \in U \times \Delta$. Consequently (α) implies that, for $(u,v) \in U \times \Delta$, we have $\sum \tilde{\varphi}_i(u,v)(\tilde{\alpha}_i)_u = 0$, which means that $(\tilde{\varphi}_1(u,v), \ldots, \tilde{\varphi}_m(u,v)) \in R_u(\tilde{\alpha}_1, \ldots, \tilde{\alpha}_m)$. Therefore $(\tilde{\varphi}_1(u,v), \ldots, \tilde{\varphi}_m(u,v))$ $= \sum_\nu \tilde{\lambda}_\nu(u,v)(\tilde{\psi}_\nu)_u$, where $\tilde{\lambda}_\nu(u,v) \in \mathcal{O}_u[T]$. Now, according to (β), the germs λ_ν of the mappings $(u,v) \longrightarrow \sum_\nu \tilde{\lambda}_\nu(u,v)(\zeta_{(u,v)})$ belong to $\check{\mathcal{Q}}_n$ and we have $(\varphi_1, \ldots, \varphi_m) = \sum \lambda_\nu(\psi_{\nu 1}, \ldots, \psi_{\nu m})$.

It is enough to show that if $F_1, \ldots, F_m \in \mathcal{O}_n$, then any solution $(\varphi_1, \ldots, \varphi_m) \in \check{\mathcal{O}}_n^m$ of the equation

$$(*) \qquad \sum_{i=1}^m \varphi_i F_i = 0$$

is a linear combination (with coefficients from $\check{\mathcal{O}}_n$) of solutions from $\mathcal{O}_n^m$ (see A. 9.6). Clearly, one may assume that $F_i \neq 0$. In view of the preparation theorem (see I. 2.1) and (δ), making a linear change of coordinates (see I. 1.4), one may assume that $F_i \in \mathcal{Q}_n$ are monic. It is enough to prove that $(\varphi_1, \ldots, \varphi_m)$ is a linear combination (with coefficients from $\check{\mathcal{O}}_n$) of the solutions from $\check{\mathcal{Q}}_n^m$, because each of the latter is a linear combination of solutions from $\mathcal{Q}_n^m$ (as $\check{\mathcal{Q}}_n$ is flat over $\mathcal{Q}_n$). According to the preparation theorem ([16]) (see I. 2.1), $F_m = \alpha\chi$, where $\chi \in \check{\mathcal{Q}}_n$ is the germ of a mapping whose values are distinguished ([17]) and α is an invertible germ in $\check{\mathcal{Q}}_n$. Furthermore, in view of the division version of the preparation theorem ([18]) (see I. 1.4), we have $\varphi_i = \gamma_i F_m + \varrho_i$, with some $\gamma_i \in \check{\mathcal{O}}_n$, $\varrho_i \in \check{\mathcal{Q}}_n$, $i = 1, \ldots, m-1$. Setting $\varrho_m = \varphi_m + \sum_1^{m-1} \gamma_i F_i$, we obtain the equality

$$(**) \quad (\varphi_1, \ldots, \varphi_m) = \gamma_1(F_m, 0, \ldots, 0, -F_1) + \ldots + \gamma_{m-1}(0, \ldots, F_m, -F_{m-1}) +$$
$$+ \alpha^{-1}(\alpha\varrho_1, \ldots, \alpha\varrho_m).$$

Now, $(F_m, 0, \ldots, 0, -F_1), \ldots, (0, \ldots, F_m, -F_{m-1})$ are solutions of the equation $(*)$ that belong to $\check{\mathcal{Q}}_n^m$. But also $(\alpha\varrho_1, \ldots, \alpha\varrho_m)$ is such a solution. Indeed, it follows from $(**)$ that $(\varrho_1, \ldots, \varrho_m)$ is a solution, and so $\sum_1^{m-1} F_i \varrho_i + \chi(\alpha\varrho_m) = 0$. Hence $\chi(\alpha\varrho_m) \in \check{\mathcal{Q}}_n$, which implies (see I. 2.1) that $\alpha\varrho_m \in \check{\mathcal{Q}}_n$. Therefore $(\alpha\varrho_1, \ldots, \alpha\varrho_m)$ is a solution which belongs to $\check{\mathcal{Q}}_n^m$.

---

[16] Applied to the germs of the representative at all points of a neighbourhood.

[17] If $c \in \mathbf{C}^n$, a germ $\eta \in \mathcal{O}_c$ is said to be *distinguished* if the germ $\eta \circ (c + z) \in \mathcal{O}_n$ is distinguished.

[18] As in footnote ([16]).

Let U be an open neighbourhood of zero in $\mathbf{C}^n$.

COROLLARY 1. *If a family of ideals $I_z \subset \mathcal{O}_z$, $z \in U$, has a finite system of generators and $g \in \mathcal{O}_U$, then there exists an open neighbourhood $U_0 \subset U$ of zero such that the family of the ideals $I_z : g_z$, $z \in U_0$, has a finite system of generators.*

Indeed, if $f_1, \ldots, f_k \in \mathcal{O}_U$ are generators of the family $\{I_z\}$, then $I_z :$ $g_z = h_z(R_z(g, f_1, \ldots, f_k))$, where $h_z : \mathcal{O}_z^{k+1} \in (\psi, \varphi_1, \ldots, \varphi_k) \longrightarrow \psi \in \mathcal{O}_z$, $z \in U$, is a coherent family of homomorphisms.

COROLLARY 2. *If families of ideals $I_z \subset \mathcal{O}_z$, $z \in U$, and $J_z \subset \mathcal{O}_z$, $z \in U$, have finite systems of generators, then there exists an open neighbourhood $U_0 \subset U$ of zero such that the family of the ideals $I_z \cap J_z$, $z \in U_0$, has a finite system of generators. The same is true for any finite number of families.*

Indeed, if $f_1, \ldots, f_k \in \mathcal{O}_U$ and $g_1, \ldots, g_l \in \mathcal{O}_U$ are systems of generators of those families, respectively, then $I_z \cap J_z = h_z(R_z(f_1, \ldots, f_k, g_1, \ldots, g_l))$, where

$$h_z : \mathcal{O}_z^{k+1} \in (\varphi_1, \ldots, \varphi_k, \psi_1, \ldots, \psi_l) \longrightarrow \sum \varphi_i(f_i)_z \in \mathcal{O}_z, \ z \in U \ ,$$

is a coherent family of homomorphisms.

CARTAN'S THEOREM (the case of $\mathbf{C}^n$). *If V is an analytic subset of an open set $G \subset \mathbf{C}^n$, then each point $c \in G$ has an open neighbourhood $U \subset G$ such that the family of ideals $\mathcal{I}(V_z) \subset \mathcal{O}_z$, $z \in U$, has a finite system of generators.*

PROOF . First suppose that the germ V_c is simple. Then, after an affine change of coordinates $L : \mathbf{C}^n \longrightarrow \mathbf{C}^n$ (as $\mathcal{I}(V) = \mathcal{I}(L(V)_{l(z)}) \circ L_z$; see II. 4.4), one may assume that $c = 0$ and that the assumptions of lemma 2 from n° 1 are satisfied; see n° 2, III. 2.7, and III. 3.2). Then, according to corollary 1 and the properties from n° 2, the theorem follows from that lemma. In the general case, one may assume (replacing G by a sufficiently small neighbourhood of the point c) that $V = \bigcup_1^k V_i$, where the V_i are analytic subsets of G and the germs $(V_i)_c$ are irreducible. Then $\mathcal{I}(V_z) = \bigcap_1^k \mathcal{I}((V_i)_z)$ for $z \in G$ (see II. 4.2) and, by the first part of the proof, the theorem follows from corollary 2.

4. Now we are going to prove the Oka and Cartan coherence theorems in the general case.

Let X be an analytic space. A family of submodules $M_z \subset \mathcal{O}_z^m$, $z \in X$, is said to be *coherent* if each point of the space X has an open neighbourhood

U such that the family M_z, $z \in U$, has a finite system of generators [19]. It follows from the definition that if $\{G_\iota\}$ is an open cover of the space X, then the family M_z, $z \in X$, is coherent if and only if each of the families M_z, $z \in G_\iota$, is coherent. Next, if $N_z \subset \mathcal{O}_z^m$, $z \in X$, is a coherent family of submodules and $a \in X$, then $(M_a \subset N_a) \implies (M_z \subset N_z$ in a neighbourhood of a). The properties from n° 2 imply the following:

If families of submodules $M_z \subset \mathcal{O}_z^p$, $z \in X$, and $N_z \subset \mathcal{O}_z^q$, $z \in X$, are coherent, then so is the family $M_z \times N_z \subset \mathcal{O}_z^{p+q}$, $z \in X$, and, in the case when $p = q$, so is the family $M_z + N_z \subset \mathcal{O}_z^p$, $z \in X$. The same holds for any finite number of families.

If $h_z : \mathcal{O}_z^p \longrightarrow \mathcal{O}_z^q$, $z \in X$, is a coherent family of homomorphisms and a family of submodules $M_z \subset \mathcal{O}_z^p$, $z \in X$, is coherent, then so is the family of the images $h_z(M_z) \subset \mathcal{O}_z^q$, $z \in X$.

If $F : T \longrightarrow X$ is an embedding of an analytic space T into the space X and if a family of submodules $M_z \subset \mathcal{O}_z^m$, $z \in X$, is coherent, then so is the family $M_{F(t)} \circ F_t \subset \mathcal{O}_t^m$, $t \in T$.

If $F : T \longrightarrow X$ is a holomorphic mapping of analytic spaces and if a family M of submodules $M_z \subset \mathcal{O}_z^m$, $z \in X$, is coherent, then so is its inverse image F^*M.

CARTAN'S THEOREM . *If $V \subset X$ is an analytic subset, then the family of ideals $\mathcal{I}(V_z)$, $z \in X$, is coherent.*

PROOF . First observe that if $g : X \longrightarrow Y$ is a biholomorphic mapping of analytic spaces, $W = g(V)$, and Cartan's theorem holds for W in Y, then it holds for V in X. (For $\mathcal{I}(V_z) = \mathcal{I}(W_{g(z)}) \circ g_z$; see V. 4.9 in ref. to II. 4.4.) Therefore, one may assume that X is an analytic subset of an open set $G \subset \mathbf{C}^n$. Then V is also an analytic subset of G. Cartan's theorem from n° 3 implies that the family of ideals $\mathcal{I}(V_z, G) \subset \mathcal{O}_z(G)$, $z \in G$, is coherent. Consequently, the theorem follows from the fact that $\mathcal{I}(V_z, X) = \mathcal{I}(V_z, G) \circ \iota_z$ for $z \in X$, where $\iota : X \hookrightarrow G$.

REMARK 1. If $c \in X$ and $f_1, \ldots, f_k$ are generators of the ideal $\mathcal{I}(V_c)$, then some of their representatives $\tilde{f}_1, \ldots, \tilde{f}_k$ in an open neighbourhood U of the point c are generators of the family $\mathcal{I}(V_z)$, $z \in U$ [20].

---

[19] According to Oka's theorem (see below), the coherent families of submodules $M_z \subset \mathcal{O}_z^m$, $z \in X$, are precisely the families of fibres of coherent subsheaves of the sheaf of the $\mathcal{O}$–modules $\mathcal{O}^n$ on X, where $\mathcal{O}$ is the sheaf of germs of holomorphic functions on X. (See, e.g., [17], Chapter IV; [25], Chapter VII; or [22], Chapters IV–V.)

[20] For, in some open neighbourhood U_0 of the point c, the family $\mathcal{I}(V_z)$, $z \in U_0$, has generators $g_1, \ldots, g_l$, and then the germs $(g_j)_c \in \mathcal{I}(V_c)$ are linear combinations of the germs f_i.

REMARK 2. If X is an n–dimensional manifold and $V \subset X$ is a closed submanifold of dimension $k < n$, then in a neighbourhood U of any point $z \in V$ we have $V \cap U = \{f = 0\}$, where $f = (f_1, \ldots, f_{n-k})$ is a submersion. Then $f_1, \ldots, f_{n-k}$ are generators of the family $\mathcal{I}(V_z)$, $z \in U$. (See II. 4.2. Obviously, 1 is a generator of the family $\mathcal{I}(V_z)$, $z \in X \setminus V$.)

OKA'S THEOREM . If $f_1, \ldots, f_m \in \mathcal{O}_X^p$, then the family of modules of relations $R_z(f_1, \ldots, f_m) \subset \mathcal{O}_z^m$, $z \in X$, is coherent.

PROOF . For any biholomorphic mapping $g : Z \longrightarrow Y$ of analytic spaces and any mappings $h_1, \ldots, h_k \in \mathcal{O}_Y^q$, we have the following equality:

$$(\#) \qquad R_{g(u)}(h_1, \ldots, h_k)) \circ g_u = R_u(h_1 \circ g, \ldots, h_k \circ g), \quad \text{for} \quad u \in Z \;.$$

It follows that if Oka's theorem holds in the space Y, then it holds in the space Z. Therefore one may assume that X is an analytic subset of an open set $G \subset \mathbf{C}^n$, and that $f_i = (F_i)_X$, where $F_i \in \mathcal{O}_G^p$ (see V. 3.1). Then, in view of Cartan's theorem (in n° 3), one may assume that the family of the submodules $\mathcal{I}(X_z) \times \ldots \times \mathcal{I}(X_z) \subset \mathcal{O}_z^p$, $z \in X$, has a finite system of generators $H_1, \ldots, H_k \in \mathcal{O}_G^p$. Now, it follows from Oka's theorem in n° 3 (which is true for any $c \in \mathbf{C}^n$ in place of 0, e.g., in view of the equality $(\#)$), that the theorem is true for G. Now, as it is easy to check, we have

$$R_z(f_1, \ldots, f_m) = h_z\big(R_z(F_1, \ldots, F_m, H_1, \ldots, H_k)\big) \circ \iota z \quad \text{for} \; z \in X \;,$$

where $\iota : X \hookrightarrow G$ and $h_z : \mathcal{O}_z(G)^{m+k} \ni (\varphi_1, \ldots, \varphi_m, \psi_1, \ldots, \psi_k) \longrightarrow (\varphi_1, \ldots, \varphi_m) \in \mathcal{O}_z(G)^m$ is a coherent family of homomorphisms. This implies the theorem.

COROLLARY 1. If $h_z : \mathcal{O}_z^p \longrightarrow \mathcal{O}_z^q$, $x \in X$, is a coherent family of homomorphisms, and a family of submodules $N_z \subset \mathcal{O}_z^q$, $z \in X$, is coherent, then so is the family of the inverse images $h_z^{-1}(N_z) \subset \mathcal{O}_z^p$, $z \in X$.

In fact, one may assume that the family $\{N_z\}$ has a finite system of generators $g_1, \ldots, g_k \in \mathcal{O}_X^q$. We have $h_z(\varphi_1, \ldots, \varphi_p) = \sum \varphi_i(f_i)_z$ with some $f_1, \ldots, f_p \in \mathcal{O}_X^q$. So, it is sufficient to observe that $h_z^{-1}(N_z) = \chi_z\big(R_z(f_1, \ldots \ldots, f_p, g_1, \ldots, g_k)\big)$, where

$$\chi_z : \mathcal{O}_z^{p+k} \in (\varphi_1, \ldots, \varphi_p, \psi_1, \ldots, \psi_k) \longrightarrow (\varphi_1, \ldots, \varphi_p) \in \mathcal{O}_z^p, \; z \in X$$

is a coherent family of homomorphisms.

COROLLARY 2. If families of submodules $M_z \subset \mathcal{O}_z^k$, $z \in X$, and $N_z \subset \mathcal{O}_z^k$, $z \in X$, are coherent, then so is the family $M_z \cap N_z$, $z \in X$. The same is true for any finite number of families.

For $M_z \cap N_z = \Delta_z^{-1}(M_z \times N_z)$, where $\Delta_z : \mathcal{O}_z^k \ni \varphi \longrightarrow (\varphi, \varphi) \in \mathcal{O}_z^{2k}$, $z \in X$, is a coherent family of homomorphisms.

COROLLARY 3. *If the families of ideals $I_z \subset \mathcal{O}_z$, $z \in X$, and $J_z \subset \mathcal{O}_z$, $z \in X$, are coherent, then so is the family $I_z : J_z$, $z \in X$.*

In fact, one may assume that these families have finite system of generators $f_1, \ldots, f_k$ and $g_1, \ldots, g_l$, respectively. Then $I_z : J_z = \bigcap_{i=1}^{l} h_z(R_z(g_i, f_1, \ldots, f_k))$, where $h_z : \mathcal{O}_z^{k+1} \ni (\varphi_0, \varphi_1, \ldots, \varphi_k) \longrightarrow \varphi_0 \in \mathcal{O}_z$ is a coherent family of homomorphisms.

LEMMA . *If a family of submodules $M_z \subset \mathcal{O}_z^m$, $z \in X$, is coherent, then the set $V = \{z \in X : M_z \not\subset \mathcal{O}_z^m\}$ is analytic.*

PROOF . One may assume that the family $\{M_z\}$ has a finite system of generators $g_i = (g_{i1}, \ldots, g_{im}) \in \mathcal{O}_X^m$, $i = 1, \ldots, r$. Then $z \notin V$ if and only if for any $\eta_1, \ldots, \eta_m \in \mathcal{O}_z$ the system of equations $\sum_{i=1}^{r} (g_{i\nu})_z \xi_i = \eta_\nu$, $\nu = 1, \ldots, m$, has a solution $\xi_1, \ldots, \xi_r \in \mathcal{O}_z$. Clearly, the latter condition implies the condition $\mathrm{rank}[g_{ij}(z)] \geq m$ (21) . Conversely, if the rank condition is satisfied, then some minor of order m is different from zero and Cramer's formulae imply that the above system of equations has a solution for any $\eta_\nu \in \mathcal{O}_z$. Thus $V = \{z \in X : \mathrm{rank}[g_{ij}] < m\}$ is an analytic set.

PROPOSITION . *If the families of submodules $M_z \subset \mathcal{O}_z^k$, $z \in X$, and $N_z \subset \mathcal{O}_z^k$, $z \in X$, are coherent, then the set $V = \{z \in X : M_z \not\subset N_z\}$ is analytic.*

PROOF . One may assume that the family $\{M_z\}$ has a finite system of generators $g_1, \ldots, g_r \in \mathcal{O}_X^k$. The family of homomorphisms $h_z : \mathcal{O}_z^r \ni (\varphi_1, \ldots, \varphi_r) \longrightarrow \sum_1^r \varphi_i(g_i)_z \in \mathcal{O}_z^k$, $z \in X$, is coherent. Therefore, by corollary 1 from Oka's theorem, the family of inverse images $h_z^{-1}(N_z) \subset \mathcal{O}_z^r$, $z \in X$, is coherent. In view of the lemma, it is enough to observe that $V = \{z \in X : h_z^{-1}(N_z) \not\subset \mathcal{O}_z^r\}$.

REMARK 3. In particular, if I is a coherent family of ideals, $I_z \subset \mathcal{O}_z$, $z \in X$, then the set $Z = \{z \in X : I_z \neq \mathcal{O}_z\}$, which is called the *zero set* of this family, is analytic. This follows also from the fact that if $f_1, \ldots, f_k$ are holomorphic in an open set $U \subset X$ and are generators of the family $\{I_z\}$, $z \in U$, then it is easy to verify that $Z \cap U = V(f_1, \ldots, f_k)$. If $F : T \longrightarrow X$ is a holomorphic mapping of analytic spaces, then $F^{-1}(Z)$ is the zero set of the family of ideals F^*I. Hence we have $V(I_z) = Z_z$ for $z \in X$.

COROLLARY 4. *If a family of ideals $I_z \subset \mathcal{O}_z$, $z \in X$, is coherent, then so*

---

(21) Because, by taking values at z, we can see that the system $\sum_{i=1}^{r} g_{i\nu}(z)u_i = v_\nu$, $\nu = 1, \ldots, m$ ($u_i, v_\nu \in \mathbb{C}$), always has a solution.

is the family rad I_z, $z \in X$.

This follows by Cartan's theorem, since rad $I_z = \mathcal{I}(V(I_z)) = \mathcal{I}(Z_z)$ (see Hilbert's theorem in III. 4.1).

5. Let X be an analytic space.

THE SCHUMACHER THEOREM ([21a]). *Let* $I_z \subset \mathcal{O}_z$, $z \in X$, *be a coherent family of ideals, and let* V *be its zero set. Assume that all the rings* $\mathcal{O}_z/I_z$, $z \in V$, *are Cohen-Macaulay. If the equality* $I_z = \mathcal{I}(V_z)$ ([21b]) *holds on a dense subset of* V, *then it must hold on the whole of* V. ([21c])

PROOF (Bonhorst ([21d])). Let $a \in V$. Let $I_a = \bigcap_1^r J_i$ be an irreducible primary decomposition. In view of corollary 1 in A. 14.3, all associated ideals $I_i = \text{rad } J_i$ are isolated. By replacing X with a sufficiently small neighbourhood of a, we may assume that holomorphic representatives on X of generators of the ideal J_i generate a coherent family of ideals $J_{iz} \subset \mathcal{O}_z$, $z \in X$ ($i = 1, \ldots, r$), and that (see n° 4):

$$(1) \qquad\qquad I_z = \bigcap_1^r J_{iz} \quad \text{for} \quad z \in X,$$

as, of course, $J_{ia} = J_i$. By corollary 4 in n° 4, the family $I_{iz} = \text{rad } J_{iz}$, $z \in X$, is coherent, and $I_{ia} = I_i$. Let V_i be its zero set. Then we have

$$(2) \qquad\qquad I_{iz} = \mathcal{I}((V_i)_z)$$

([21a]) See [36a] and [11a°] 1.4.

([21b]) Or, equivalently, rad $I_z = I_z$ (by Hilbert's theorem in III. 4.1).

([21c]) A *general analytic space* is defined (see [17a] Chapter 1 §1) as a sheaf of **C**–rings $(X, \mathcal{O})$ on a Hausdorff space X, which is locally isomorphic to sheaves of the form $\mathcal{S} = \bigcup\{\mathcal{O}_z/I_z : z \in V\}$ (with suitable topology and the natural projection $\mathcal{S} \longrightarrow V$), where $V = V(f_1, \ldots, f_k)$ is the analytic set defined by holomorphic functions $f_1, \ldots, f_k$ on a manifold L, and $I_z \subset \mathcal{O}_z$, $z \in X$, is the family of ideals with generators $f_1, \ldots, f_k$. Observe that if for a point $z \in L$ one of the following equivalent conditions is satisfied:

(a) $I_z = \mathcal{I}(V_z)$,

(b) rad $I_z = I_z$,

(c) $\mathcal{O}_z/I_z$ is without nilpotents,

then the ring $\mathcal{O}_z/I_z$ is naturally isomorphic to $\mathcal{O}_z(V)$ (see IV. 4.2). The analytic space $(\mathcal{O}, X)$ is said to be *reduced at a point* $a \in X$ if the ring $\mathcal{O}_a$ is without nilpotents. If this condition holds for each $a \in X$, the space is called *reduced*. This takes place if and only if there is on X the structure of an analytic space, as defined in V. 4, such that $(X, \mathcal{O})$ is isomorphic to the sheaf of germs of holomorphic functions.

Therefore *Schumacher's theorem* can be stated as follows (this is the original version): *Suppose that an analytic space* $(X, \mathcal{O})$ *is Cohen-Macaulay* (i.e., all $\mathcal{O}_a$ are Cohen-Macaulay). *If* $(X, \mathcal{O})$ *is reduced at each point of a dense subset of* X, *then* $(X, \mathcal{O})$ *is reduced.*

([21d]) See [11a°] 1.3 and 1.4.

(as $I_{iz} = \mathrm{rad}\, I_{iz} = \mathcal{J}(V(I_{iz}))$). Since $V_a = V(I_a) = \bigcup_1^r V(I_i)$ is the decomposition into simple germs (see V. 4.9) and $V(I_i) = (V_i)_a$, we may require – by suitable reduction of the size of X – that $V = \bigcup_1^r V_i$ is the decomposition into simple components (see V. 4.9 in ref. to proposition 3b in IV. 3.1). Fix any $s = 1, \ldots, r$. It is enough to show that $J_s = I_s$ (as this implies rad $I_a = I_a$). Now the set $S = \{z : J_{sz} \not\supset I_{sz}\}$ is obviously the zero set of the coherent family $J_{sz} : I_{sz}$, $z \in X$ (see corollary 3 in n° 4), hence it is analytic. Therefore, since $S_a = V(J_s : I_s)$ by Hilbert's Nullstellensatz (in V. 4.9), we must have

$$(3) \qquad\qquad \tilde{I} = \mathcal{I}(S_a)^k \subset J_s : I_s$$

for some k (see A. 9.2). According to the assumption (see V. 4.5 in ref. to theorem 4 in IV. 2.9), in a dense subset of $V_s \setminus V^*$, we have (by (1) and (2))

$$J_{sz} \supset I_z = \mathrm{rad}\, I_z = \mathcal{I}(V_z) = \mathcal{I}((V_s)_z) = I_{sz},$$

that is, $z \notin S$. This implies (since $a \in V_s$) that $(V_s)_a \not\subset S_a$, and so $I_s \not\supset \tilde{I}$ (by (2); see V. 4.9 in ref. to II. 4.5 and 3, and A. 1.5). Take $g \in \tilde{I} \setminus I_s$. If $f \in I_s$, then, by (3), $gf \in J_s$, which gives $f \in J_s$, since J_s is primary. Therefore $I_s \subset J_s$, and hence $I_s = J_s$.

6. Let M be a manifold of dimension n, and let $a \in M$.

Let $A \neq \emptyset$ be an analytic germ at a. We always have $g(\mathcal{I}(A)) \geq \mathrm{codim}\, A$ (since codim $A \leq p$ if $f_1, \ldots, f_p$ generate $\mathcal{I}(A)$; see III. 4.6 and II. 4.5). The germ A is said to be a *complete intersection* if $g(\mathcal{I}(A)) = \mathrm{codim}\, A$, i.e., if the ideal $\mathcal{I}(A)$ has a system $f_1, \ldots, f_p$ of $p = \mathrm{codim}\, A$ generators $(^{21e})$. In such a case, $A = V(f_1, \ldots, f_p)$ (see II. 4.5), and we say that $f_1, \ldots, f_p$ *realize A as a complete intersection*. Then these generators realize a set-theoretic complete intersection (see III. 4.7) $(^{21f})$, and so every complete intersection is a set-theoretic one $(^{21g})$.

Every smooth germ A is a complete intersection. Germs $f_1, \ldots, f_p \in \mathcal{O}_a$, where $p = \mathrm{codim}\, A$, realize A as a complete intersection if and only if rank $(df_1, \ldots, df_p) = p$ (see II. 42).

Observe that a non-empty principal germ is always a complete intersection (see A. 6.1 and Hilbert's theorem in III. 4.1).

$(^{21e})$ See Abhyankar [7aa].

$(^{21f})$ The converse is clearly not true: $z_1^2, z_2^2 \in \mathcal{O}_2$ is a counter-example.

$(^{21g})$ The converse is not true. As a counter-example consider a germ at 0 in $\mathbf{C}^4$ of the set

$$\{(u, v, x, y) : uv = 0, ux - vy = 0\} = \{u = v = 0\} \cup \{u = y = 0\} \cup \{v = x = 0\}.$$

It is of constant dimension 2. So it is a set-theoretic complete intersection. But it is not a complete intersection. Indeed, we have $I = \mathcal{I}(A) \subset \mathfrak{m}^2$. Consider the linear mapping $q : \mathfrak{m}^2 \ni \sum_2^\infty f_\nu \longrightarrow f_2 \in Q$, where f_ν is a form of degree ν and Q is the linear space of quadratic forms on $\mathbf{C}^4$. Now, if we had $I = \mathcal{O}_4 f + \mathcal{O}_4 g$, then $q(I) = \mathbf{C}q(f) + \mathbf{C}q(g)$ would be of dimension ≤ 2, but $uv, ux, vy \in q(I)$ are linearly independent, and we have a contradiction.

Let $f_1, \ldots, f_p$ be holomorphic functions, and let V be an analytic set in a neighbourhood of a. We say that these functions *realize V as a complete intersection at a* if their germs at a realize the germ V_a as a complete intersection. Then they form a complete intersection at any $z \in V$ from a neighbourhood of a. In fact, by Cartan's theorem (see remark 1 in n° 4), $(f_1)_z, \ldots, (f_p)_z$ generate $\mathcal{I}(V_z)$ for any $z \in V$ from a neighbourhood of a, hence codim $V_z \leq p$, and the equality follows (in a neighbourhood of a) by the lower semi-continuity of the left hand side (see II. 1.5). As corollaries, we have

If V_a is a complete intersection, then so is V_z for $z \in V$ from a neighbourhood of a.

If a germ is a complete intersection, then it must be of constant dimension ([21h]).

Let $f_1, \ldots, f_p$ be holomorphic functions in M, and set

$$V = V(f_1, \ldots, f_p) \quad \text{and} \quad S = \{z \in V : \text{rank}_z(f_1, \ldots, f_p) < p\}.$$

THEOREM (TSIKH ([21i])). *The functions $f_1, \ldots, f_p$ realize V as a complete intersection at each point of V if and only if S is nowhere dense in V. In this case $V^* = S$.*

PROOF. Clearly, the condition is implied by $V^0 \subset V \setminus S$. Thus it is necessary, and then $V^0 = V \setminus S$ and $V^* = S$. Suppose in turn that the condition is satisfied. Then V is of constant dimension $n - p$ (see IV. 2.5). This means that, for each $z \in V$, the germs $(f_1)_z, \ldots, (f_p)_z$ realize a set-theoretic complete intersection. Therefore, by the proposition in III. 4.7, for the family of ideals $I_z = \sum_1^p \mathcal{O}_z(f_i)_z$ the assumptions of the Schumacher theorem in n° 5 are satisfied. Consequently, since the equality $I_z = \mathcal{I}(V_z)$ holds for $z \in V \setminus S$ (see II. 4.2), it must hold for all $z \in V$, i.e., for each $z \in V$ the germs $(f_1)_z, \ldots, (f_p)_z$ realize V_z as a complete intersection.

7. Let M be a manifold of dimension n, and let $a \in M$.

ABHYANKAR'S LEMMA ([21j]). *Let $V \subset W$ be analytic germs at a. Assume that V is simple. If $V \not\subset W^*$ ([21k]), then the ring $\mathcal{L} = (\mathcal{O}_W)_{\mathcal{I}_W(V)}$ is regular.* ([21l])

PROOF. It suffices to prove that edim $\mathcal{L} \leq \dim \mathcal{L}$ (see A. 15.1). According to equality (∗) in A. 10.4, the lemma in IV. 4.3a applied to the rings $\mathcal{L}$ and $(\mathcal{O}_a)_{\mathcal{I}(V)}$, and by the regularity of the latter (see theorem 3 in A. 15 and proposition 2 in I. 1.8), we have edim $\mathcal{L} = n - \dim V - \text{rank } \overline{\mathcal{I}(W)}$ and $\dim \mathcal{L} = d_V W - \dim V$. Therefore it is enough to prove that

$$(1) \qquad \qquad \text{rank } \overline{\mathcal{I}(W)} \geq n - d_V W .$$

In order to do so, take representatives $\check{V} \subset \check{W}$ of V and W, both analytic in the same neighbourhood U of a (which can be arbitrarily small), and such that $\check{V}$ is irreducible (see proposition 1 in IV. 3.1). Let $z \in \check{V}^0$. Set $W' = \check{W}_z$ and $V' = \check{V}_z$.

([21h]) The converse is not true; see footnote ([21g]).

([21i]) See [43a⁰].

([21j]) See [7aa].

([21k]) For any analytic germ W at a, the germ $W^* = (\check{W}^*)_a$, where $\check{W}$ is a locally analytic representative of W, is well-defined.

([21l]) In fact, both conditions are equivalent; see [7aa] (9.3).

First, we prove that if z is sufficiently close to a, then for the extensions $\overline{\mathcal{I}(W')} \subset (\mathcal{O}_z)_{\mathcal{I}(V')}$ and $\overline{\mathcal{I}(W)} \subset (\mathcal{O}_a)_{\mathcal{I}(V)}$ (see A. 11.1), we have

(2)
$$\operatorname{rank} \overline{\mathcal{I}(W')} \leq \operatorname{rank} \overline{\mathcal{I}(W)} .$$

Now, if $f_1, \ldots, f_p$ and $g_1, \ldots, g_s$ are generators of $\mathcal{I}(W')$ and $\mathcal{I}(V')$, respectively, then $r' = \operatorname{rank} \overline{\mathcal{I}(W')}$ (since the least number of generators of $\overline{\mathcal{I}(W')}$ modulo $\overline{\mathcal{I}(V')}^2$; see A. 10.4 and A. 11.1) is the least number r' such that

$$q_s f_s \in \sum_1^{r'} \mathcal{O}_z h_i + \sum_{i,j} \mathcal{O}_z g_i g_j, \quad 1, \ldots, p ,$$

with some $h_i \in \mathcal{I}(W')$ and $q_s \in \mathcal{O}_z \setminus \mathcal{I}(V')$. The same holds for a instead of z, and so we have

(3)
$$\tilde{q}_s \tilde{f}_s = \sum_{i=1}^{r} \tilde{c}_{si} \tilde{h}_i + \sum_{i,j} \tilde{a}_{sij} \tilde{g}_i \tilde{g}_j, \quad 1, \ldots, p .$$

All the functions in (3) are holomorphic in the neighbourhood U (by making it smaller) and $r = \operatorname{rank} \overline{\mathcal{I}(W)}$. Moreover, $(\tilde{f}_s)_a$ and $(\tilde{g}_i)_a$ generate $\mathcal{I}(W)$ and $\mathcal{I}(V)$, respectively. Also $(\tilde{h}_i)_a \in \mathcal{I}(W)$ and $(\tilde{q}_s)_a \in \mathcal{O}_a \setminus \mathcal{I}(V)$. According to Cartan's theorem (see remark 1 in n° 4), if z is sufficiently close to a, then $(\tilde{f}_s)_z$ and $(\tilde{g}_i)_z$ generate $\mathcal{I}(W')$ and $\mathcal{I}(V')$, respectively, $(\tilde{h}_i)_z \in \mathcal{I}(W')$, and $(\tilde{q}_s)_z \in \mathcal{O}_a \setminus \mathcal{I}(V')$ (in view of the irreducibility of $\tilde{V}$). Therefore from (3) we conclude, by taking the germs at z, that r' must be $\leq r$. That is, we obtain (2).

To finish the proof, take a simple component $\tilde{W}_1 \supset V$ of $\tilde{W}$ of dimension $d_V W$ (such a component exists by the corollary of proposition 3b in IV. 3.1 and proposition 4 in IV. 2.8). Since $V \not\subset \tilde{W}_1^*$, we may assume that $z \in \tilde{W}^0 \cap \tilde{W}_1 \subset \tilde{W}_1^0 \setminus \tilde{W}^*$ (by $(*)$ in IV. 2.9), and then

(4)
$$d_{V'} W' = d_V W$$

(see IV. 2.9, theorem 4 with corollary 4). According to $(*)$ in A. 10.4 and the lemma in IV. 4.3a,

$$\operatorname{rank} \overline{\mathcal{I}(W')} = \dim(\mathcal{O}_z)_{\mathcal{I}(V')} - \dim(\mathcal{O}_{W'})_{\mathcal{I}_{W'}(V')} = n - d_{V'} W' ,$$

in view of the regularity of the rings on the right hand side (see theorem 3 in A. 15, and proposition 2 in I. 1.8). This, combined with (2) and (4), implies (1).

§2. Normal spaces. Universal denominators

Let X be an analytic space.

1. If the space X is smooth (see V. 4.5), then it has the following property (see II. 3.5):

(n) If $G \subset X$ is an open subset and $V \subset G$ is a nowhere dense analytic subset, then every holomorphic function on $G \setminus V$ which is locally bounded near V ([22]) has a holomorphic extension on G.

In general, an analytic space may not have this property even if it is locally irreducible. This is so, because – in general – c–holomorphic functions do not have to be holomorphic (see V. 8.1). Analytic spaces that have the above property (n) are called *normal*. Obviously, the space X is normal precisely in the case where for each open subset $\Omega \subset X$, every bounded holomorphic function in Ω^0 extends holomorphically to a function on Ω ([23]) .

A subset of an analytic space Y is said to be *l–bounded* if it is relatively compact in the domain of a chart whose range is a locally analytic set in a vector space (cf. II. 2.5). Obviously, any subset of an l–bounded set is l–bounded. Every finite subset of Y has an l–bounded neighbourhood. We say that a mapping $f : X \setminus Z \longrightarrow Y$, where $Z \subset X$ is a closed subset, is *l–bounded* near Z if every point of Z has a neighbourhood U in X such that the set $f(U \setminus Z)$ is l–bounded. It follows that:

If X is a normal space, $G \subset X$ is an open subset, and $V \subset G$ is a nowhere dense analytic subset, then every holomorphic mapping $f : G \setminus V \longrightarrow Y$ that is locally l–bounded near V has a holomorphic extension to G ([24]) . In particular, if a holomorphic mapping $f : G \setminus V \longrightarrow Y$ has a continuous extension to G, then this extension is holomorphic. ([25])

2. By a *universal denominator* on X we mean any holomorphic function $u \not\equiv 0$ on X that satisfies the condition: if $G \subset X$ is an open subset, $V \subset G$ is a nowhere dense analytic subset, and $f : G \setminus V \longrightarrow \mathbf{C}$ is a holomorphic

---

([22]) See C. 1.11.

([23]) Because every holomorphic function on $G \setminus V$ which is locally bounded near V extends holomorphically from $G^0 \setminus V$ to G^0, and the extension is locally bounded near G^*.

([24]) See II. 2.5, footnote ([6]).

([25]) In particular, *any c–holomorphic mapping of a normal space into an analytic space must be holomorphic.*

function locally bounded near V, then the function uf has a holomorphic extension to G. Obviously this is the case if and only if for each open subset $\Omega \subset X$ and for any bounded holomorphic function $f : \Omega^0 \longrightarrow \mathbf{C}$, the function uf has a holomorphic extension to Ω (as in the definition of normal space; see footnote $(^{23})$).

LEMMA . *Assume the conclusion of Rückert's descriptive lemma* (see III. 3.2). *Then* $\tilde{\delta}_V : V \longrightarrow \mathbf{C}$ *is a universal denominator on* V.

PROOF $(^{26})$. Let $G \subset \Omega \times \mathbf{C}^{n-k}$ be an open set, and let $f : V^0 \cap G \longrightarrow \mathbf{C}$ be a bounded holomorphic function. Since V' is dense in V^0, it is enough to show that $(\delta f)_{V'}$ extends holomorphically to a neighbourhood of any point $c \in V \cap G$. Now, there is a bounded open neighbourhood $U = \Omega_0 \times \Delta$ of the point c such that $\bar{U} \subset G$ and the natural projection $V' \cap U \longrightarrow \Omega_0 \setminus Z$ is an s–sheeted covering $(^{27})$. Put $\tilde{Q} = (\tilde{Q}_{k+2}/\tilde{\delta}, \ldots, \tilde{Q}_n/\tilde{\delta})$. If $u \in \Omega_0 \setminus Z$, then the Rückert formula implies that the set $V_u \cap \Delta$ consists of points $(\tau_i, \tilde{Q}(u, \tau_i))$, $i = 1, \ldots, s$, where the coordinates $\tau_i = \tau_i(u)$ are mutually distinct $(^{28})$. Therefore the function F defined on the set $(\Omega_0 \setminus Z) \times \Delta$ by the formula

$$F(u, t, w) = \sum_{i=1}^{s} \prod_{j \neq i} \frac{t - \tau_j}{\tau_i - \tau_j} \tilde{\delta}(u) f(u, \tau_i, \tilde{Q}(u, \tau_i))$$

is holomorphic. (It is well-defined due to symmetry of the right hand side with respect to $\tau_1, \ldots, \tau_s$.) It is also bounded (since τ_i are the roots of the polynomial $t \longrightarrow p_{k+1}(u, t)$, while $\tilde{\delta}(u)$ is its discriminant; see C. 2.1), and hence it extends holomorphically onto U. It is easy to check that $F = \tilde{\delta} f$ on $V' \cap U$.

OKA'S THEOREM . *Every point of the analytic space* X *has an open neighbourhood on which there is a universal denominator.*

PROOF . One may assume that X is an analytic subset V of a manifold M. Let $a \in V$. If the germ V_a is irreducible, then the theorem follows from the

---

$(^{26})$ See [44], Chapter IV, n° 4.

$(^{27})$ We have $c = (a, b)$, where $a \in \Omega$, $b \in \mathbf{C}^{n-k}$. Now, it is enough to take open bounded and disjoint sets $\Delta \ni b$, $\Delta' \supset V_a \setminus b$, and a bounded open connected neighbourhood Ω_0 of the point a such that $\overline{\Omega_0 \times \Delta} \subset G$ and $V_{\Omega_0} \subset (\Omega_0 \times \Delta) \cup (\Omega_0 \times \Delta')$ (see B. 2.4). For then the natural projection $V'_{\Omega_0} \longrightarrow \Omega_0 \setminus Z$ is a finite covering, and the set $V' \cap (\Omega_0 \times \Delta)$ is open and closed in V'_{Ω_0} (see B. 3.2).

$(^{28})$ Since V' is a locally topographic submanifold, it follows that in a neighbourhood of any point of $\Omega \setminus Z$ the functions $\tau_i(u)$ can be taken to be holomorphic (after a suitable ordering).

lemma (see III. 3.2; and III. 2.7 and 6); this follows also from the Grauert-Remmert formula (see remarks 1 and 2 in V. 7.4, and corollary 1 in III. 3.1).). In the general case, there exist locally analytic subsets $V_i \ni a$ such that $V_a = (V_1)_a \cup \ldots \cup (V_r)_a$ is the decomposition into simple germs. In view of proposition 3b from IV. 3.1, we may assume that $V = V_1 \cup \ldots \cup V_r$ is the decomposition into simple components, and moreover, that there are holomorphic functions u_i, φ_i, $i = 1, \ldots, r$, which are holomorphic on M, such that $(u_i)_{V_i}$ are universal denominators and $(\varphi_i)_a \in \mathcal{I}(\bigcup_{j \neq i}(V_j)_a) \setminus \mathcal{I}((V_i)_a)$ (see II. 4.5 and 6). Then $\varphi_i = 0$ on $\bigcup_{j \neq i} V_j$ and $\varphi_i \not\equiv 0$ on V_i (see IV. 2.8, corollary 3 of proposition 2). Now $u = \left(\sum_1^r \varphi_i u_i\right)_V$ is a universal denominator. Indeed, we clearly have $u \not\equiv 0$. Let $G \subset M$ be an open set, and let f be a bounded holomorphic function on $V^0 \cap G$. It is enough to show that uf extends holomorphically to a neighbourhood of any point $c \in V \cap G$. Since f_V is a bounded holomorphic function on $G \cap V_i \setminus V^*$, $u_i f_{V_i}$ extends holomorphically to the set $G \cap V_i$, because $G \cap V_i \cap V^*$ is a nowhere dense analytic subset (see IV. 2.9, theorem 4). Consequently, there is an open neighbourhood $U \subset G$ of the point c and there are holomorphic functions F_i on U, such that $u_i f_{U \cap V_i \setminus V^*} \subset F_i$. Then $uf = \varphi_i u_i f = \varphi_i F_i = \sum_1^r \varphi_i F_i$ in $U \cap V_i \setminus V^*$, and so we have $uf = \sum_1^r \varphi_i F_i$ in $V^0 \cap U$. [28a]

3. Let $a \in X$. The sets of the form U^0, where U are open neighbourhoods of the point a, are representatives of the germ $(X^0)_a$ [29] . Denote by $\tilde{\mathcal{O}}_a = \tilde{\mathcal{O}}_a(X)$ the ring of germs on $(X^0)_a$ of bounded holomorphic functions on the sets U^0. The ring $\tilde{\mathcal{O}}_a$ is an extension of the ring $\mathcal{O}_a$. For each such function, put $f_a = f_{(X^0)_a}$ [30] , according to the natural identification

$$\mathcal{O}_a \ni f \longrightarrow (\tilde{f}_{X^0})_a \in \tilde{\mathcal{O}}_a ,$$

where $\tilde{f}$ is a (bounded holomorphic) representative of the germ f.

A germ $u \in \mathcal{O}_a$ is said to be a *universal denominator* if it is a universal denominator of $\tilde{\mathcal{O}}_a$ over $\mathcal{O}_a$, that is, if it is not a zero divisor in $\mathcal{O}_a$ (i.e., it has a representative $\tilde{u} \not\equiv 0$; see V. 4.9) and $u\tilde{\mathcal{O}}_a \subset \mathcal{O}_a$ (see A. 1.15a). Obviously, a function $u \in \mathcal{O}_X$ is a universal denominator precisely when all of its germs u_z, $z \in X$, are universal denominators. Therefore, according to Oka's theorem from n° 2, *in the ring $\mathcal{O}_a$ there is always a universal denominator.*

[28a]   Instead of the above argument one can use the trick due to Cynk (see the proof of the Grauert–Remmert theorem in 1.2a.

[29]   See V. 4.5 in ref. to IV. 2.1.

[30]   The algebraic operations in $\tilde{\mathcal{O}}_a$ are given by $f_a + g_a = (f + g)_a$, $f_a g_a = (fg)_a$.

It follows that $\tilde{\mathcal{O}}_a$ is a subring of the *ring of fractions* $\mathcal{M}_a$ of $\mathcal{O}_a$: after the natural identification

$$\tilde{\mathcal{O}}_a \ni f \longrightarrow uf/u \in \mathcal{M}_a ,$$

where $u \in \mathcal{O}_a$ is a universal denominator (see A. 1.25a), we have

$$\mathcal{O}_a \subset \tilde{\mathcal{O}}_a \subset \mathcal{M}_a .$$

The ring $\tilde{\mathcal{O}}_a$ is always finite over $\mathcal{O}_a$. Indeed, taking a universal denominator $u \in \mathcal{O}_a$, we have the monomorphism $\tilde{\mathcal{O}}_a \ni f \longrightarrow uf \in \mathcal{O}_a$ of $\mathcal{O}_a$–modules; its range is an ideal of $\mathcal{O}_a$, and so it is finitely generated, since $\mathcal{O}_a$ is noetherian (see V. 4.9).

PROPOSITION (CARTAN). *The ring* $\tilde{\mathcal{O}}_a$ *is the integral closure of the ring* $\mathcal{O}_a$ *in the ring of fractions* $\mathcal{M}_a$.

Indeed, the ring $\tilde{\mathcal{O}}_a$ is finite over $\mathcal{O}_a$ and hence it is integral over $\mathcal{O}_a$ (see A. 8.1). Suppose now that $(f/g)^k + a_1(f/g)^{k-1} + \ldots + a_k = 0$, where $a_i, f, g \in \mathcal{O}_a$ and g is not a zero divisor. Therefore there is an open neighbourhood U of a, and there are bounded holomorphic representatives in U of those germs: $\tilde{a}_i, \tilde{f}$, and $\tilde{g} \not\equiv 0$ (see V. 4.9), such that $(\tilde{f}/\tilde{g})^k + \tilde{a}_1(\tilde{f}/\tilde{g})^{k-1} + \ldots + \tilde{a}_k = 0$ in $U \setminus V(\tilde{g})$ [31] . Hence the function $\tilde{f}/\tilde{g}$ must be bounded in $U \setminus V(\tilde{g})$ (see B. 5.3), and so $f/g \in \tilde{\mathcal{O}}_a$ [32] .

Let V be an analytic subset of an open subset G of an n–dimensional vector space M.

PROPOSITION (TSIKH) [33]). *Assume that* V *is of constant dimension* k *and* $f_1, \ldots$ $\ldots, f_{n-k} \in \mathcal{O}_G$ *vanish on* V. *Let*

$$D: G \ni z \longrightarrow d_z f_1 \wedge \ldots \wedge d_z f_{n-k} \in L = \Lambda^{n-k} X^* .$$

If W *is an open set in* V *and* h *is a bounded holomorphic function on* W^0, *then* hD *extends to a holomorphic function on* W. *Hence, if* $\lambda \in L^*$, *then the function* $u = (\lambda \circ D)_V$ [34] *is a universal denominator on* V, *provided that* $u \not\equiv 0$ [35].

---

[31] Before taking the representatives, we should multiply the previous equation by g^k.

[32] Because $\tilde{f}/\tilde{g}$ is equal in $U^0 \setminus V(\tilde{g})$ to a holomorphic function $\tilde{h} : U^0 \longrightarrow \mathbf{C}$ (see II. 3.5) whose germ $h \in \tilde{\mathcal{O}}_a$. Thus $f = gh$, and so $f/g = gh/g \in \tilde{\mathcal{O}}_a$ (since g is a denominator of h).

[33] See [43a°]; this is a more general version due to M. Jarnicki.

[34] In particular, if $M = \mathbf{C}^n$, then the restriction to V of any maximal minor of the matrix $[\partial f_i/\partial z_j]$ can be taken to be u.

[35] That is, if each simple component of V contains a point at which $u \neq 0$.

In fact, it is enough to prove that hD extends holomorphically to a neighbourhood of any point $a \in W$. One may assume that $a = 0$. Take a coordinate system in which the germ V_0 is k–regular. Let $F = \det(\partial f_i / \partial v_j)$, $(z = (u, v)$, $u \in \mathbf{C}^k$, $v \in \mathbf{C}^{n-k})$. We will prove that Fh extends to a holomorphic function in a neighbourhood of zero. To see this, note that by proposition 1 from IV. 1.4, the germ V_0 has a normal triple $(\Omega, Z, \tilde{V})$ with multiplicity p. It can be chosen so that $\tilde{V} = V \cap U$, where $U = \Omega \times \Delta$ is a neighbourhood of zero which can be taken to be arbitrarily small. Hence one can have $\tilde{V} \subset W$, and then h is holomorphic on $\tilde{V}_{\Omega \setminus Z}$. Next, in view of Hadamard's lemma (see C. 1.10 and footnote (^{18})) (^{36})

$$(1) \qquad f_i(u, v') - f_i(u, v) = \sum_{1}^{n-k} g_{ij}(u, v, v')(v_j' - v_j) \quad \text{for} \quad u \in \Omega, \ v, v' \in \Delta,$$

$$(2) \qquad g_{ij}(u, v, v) = (\partial f_i / \partial v_j)(u, v) \quad \text{in} \quad U,$$

where the g_{ij} are holomorphic and bounded in $\Omega \times \Delta^2$. Put $G = \det g_{ij}$. There exists a holomorphic function H in U such that

$$(3) \qquad H(u, v) = \sum_{\nu=1}^{p} h(u, v^\nu) G(u, v^\nu, v) \quad \text{if} \quad u \in \Omega \setminus Z, \ \{v^1, \ldots, v^p\} = \tilde{V}_u \text{ and } v \in \Delta,$$

because the right hand sides, due to symmetry with respect to v^ν, are holomorphic and bounded on $(\Omega \setminus Z) \times \Delta$. By (1), we have $G(u, v^\nu, v^\mu) = 0$ if $\nu \neq \mu$ (since then the left hand sides of the system (1) vanish) and, in view of (2), we have $G(u, v^\nu, v^\nu) = F(u, v^\nu)$. Therefore (3) implies that $H(u, v^\nu) = h(u, v^\nu) F(u, v^\nu)$. This means that $hF = H$ on $\tilde{V}_{\Omega \setminus Z}$ and hence on $\tilde{V}^0 = V^0 \cap U$.

Now, let $\pi_\alpha : (z_1, \ldots, z_n) \longrightarrow (z_{\alpha_1}, \ldots, z_{\alpha_n})$, where $\alpha = (\alpha_1, \ldots, \alpha_n)$ is any permutation. It follows from corollary 2 of theorem 5 in V. 3.3 that there exists a coordinate system $\varphi \in L(M, \mathbf{C}^n)$ such that the germ V_0 is k–regular in each coordinate system of the form $\pi_\alpha \circ \varphi$. (This can be achieved by ordering all the n–element permutations into a sequence $\alpha^1, \ldots, \alpha^{n!}$, and by choosing open sets $U_1 \supset \ldots \supset U_{n!} \neq \emptyset$ in $L(M, \mathbf{C}^n)$ so that the germ V_0 is k–regular in the coordinate system $\pi_{\alpha^i} \circ \varphi$ if $\varphi \in U_i$.) It follows from the first part of the proof that for each permutation α the functions $F_{\alpha_{k+1}, \ldots, \alpha_n} h$, where $F_\gamma = \det(\partial f_i / \partial z_{\gamma_j})$, extend to holomorphic functions in a neighbourhood of the origin. This completes the proof, since

$$D(z) = \sum_{\gamma_1 < \ldots < \gamma_{n-k}} E_\gamma(z) dz_{\gamma_1} \wedge \ldots \wedge dz_{\gamma_{n-k}} \ .$$

REMARK . It is easy to verify (see the previous footnote) that among the functions u described in the proposition, there is a universal denominator on V if and only if each simple component of the set V has a point at which $\mathrm{rank}(f_1, \ldots, f_{n-k})$ is equal to $(n - k)$. It can

---

(^{36}) Applied to the function $f_i(u, v + w) - f_i(u, v)$ in $\Omega \times \Delta \times (\Delta + \Delta)$, after taking a smaller U and a convex Δ such that $\Omega \times (\Delta + \Delta + \Delta)$ is a relatively compact subset of G.

be easily checked that the functions $\tilde{p}_{k+1}(u, z_{k+1})$, $\tilde{\delta}(u)z_j - Q(u, z_{k+1})$, $j = k+2, \ldots, n$, from Rückert's classic descriptive lemma (see III. 3.2) satisfy this condition. This, in turn, holds (see Tsikh's theorem in 1.6) if $f_1, \ldots, f_{n-k}$ realize the set V as complete intersection at each of its points. (Then there is a coordinate system such that each maximal minor of the matrix $[\partial f_i / \partial z_j]$ is a universal denominator.)

§3. Normal points of analytic spaces

Let X be an analytic space.

1. A point $a \in X$ is said to be *normal* and we say that the space X is *normal at the point* a if every bounded holomorphic function on the trace on X^0 of an open neighbourhood of a extends holomorphically across a, i.e., if $\tilde{\mathcal{O}}_a = \mathcal{O}_a$. As before (cf. footnote $(^{23})$). This is equivalent to the condition that every bounded holomorphic function on $U \setminus V$, where U is an open neighbourhood of a and $V \subset U$ is a nowhere dense analytic subset, extends holomorphically across a. It is clear that all regular points are normal.

It follows from the definition that *the space X is normal if and only if it is normal at each of its points.*

At any normal point a of the space X, its germ X_a must be simple. For otherwise (see V. 4.9 in ref. to proposition 3b from IV. 3.1, and V. 4.5 in ref. to theorem 4 from IV. 2.9), there would be an open neighbourhood U of the point a such that the set U^0 would have distinct connected components whose closures contain a. Then, to get a contradiction, it would suffice to take a function that is equal to different constants on different components of the set U^0. In conclusion, any normal space is locally irreducible.

Cartan's proposition from 2.3 yields the following algebraic characterization of the normality of an analytic space at a point (see A. 8.1 and V. 4.9).

PROPOSITION . *A point $a \in X$ is normal if and only if the ring $\mathcal{O}_a$ is integrally closed* $(^{37})$.

Let $u \in \mathcal{O}_a$ be a universal denominator. By the Zariski-Samuel lemma and the proposition in A. 11.1, we have the following

ZARISKI-SAMUEL CRITERION. *When u is non-invertible $(^{37a})$, a point $a \in X$ is normal if and only if all the localizations of $\mathcal{O}_a$ to the ideals associated with $\mathcal{O}_a u$ are discrete valuation rings.*

$(^{37})$ Obviously, the property that the ring $\mathcal{O}_a$ is integrally closed can be replaced by the condition that it is integrally closed in its ring of fractions (see the lemma in A.11.3).

$(^{37a})$ If u is invertible, then obviously a is normal.

2. Next, we have another characterization which will be used in the Grauert-Remmert proof of theorem 1 below.

THE GRAUERT-REMMERT LEMMA . *Let $a \in X$, and let $u \in \mathcal{O}_a$ be a universal denominator. The point a is normal if and only if every module endomorphism of the ideal $\mathcal{I}(V(u))$ is of the form $f \longrightarrow gf$, where $g \in \mathcal{O}_a$.*

PROOF . Set $I = \mathcal{I}(V(u))$. Suppose that the point a is normal, and let $\alpha : I \longrightarrow I$ be an endomorphism. Then $\alpha(f) = gf$ for $f \in I$, where $g = \alpha(u)/u \in \mathcal{M}_a$. Indeed, since $u, f \in I$, we have $\alpha(f) = u\alpha(f)/u = f\alpha(u)/u = gf$. Therefore $gI \subset I$, which proves (see A. 8.1) that g is integral over $\mathcal{O}_a$. It follows that, by the proposition from n° 1, we must have $g \in \mathcal{O}_a$. Assume, in turn, that the point a is not normal. Then $\mathcal{O}_a \subsetneq \tilde{\mathcal{O}}_a$. By Hilbert's Nullstellensatz, $I^m \subset \mathcal{O}_a u$ for some m (see A. 9.2), and hence $I^m \tilde{\mathcal{O}}_a \subset u\tilde{\mathcal{O}}_a \subset \mathcal{O}_a$. Therefore $I^{k+1}\tilde{\mathcal{O}}_a \subset \mathcal{O}_a$ and $I^k \tilde{\mathcal{O}}_a \not\subset \mathcal{O}_a$ for some k. Now, taking $h \in I^k \tilde{\mathcal{O}}_a \setminus \mathcal{O}_a$, we get $hI \subset \mathcal{O}_a$, which implies that $hI \subset I$ [38] . Thus the endomorphism $I \ni f \longrightarrow hf \in I$ is not of the form $f \longrightarrow gf$, where $g \in \mathcal{O}_a$, in contradiction with the choice of h.

THEOREM 1. *The set of non-normal points of the analytic space X is analytic.*

PROOF . By Oka's theorem from 2.2, each point of the space X has an open neighbourhood U on which there is a universal denominator u. In view of the Oka and Cartan coherence theorems (see 1.4), one can reduce U in size so that the family of ideals $I_z = \mathcal{I}(V(u_z)) \subset \mathcal{O}_z$, $z \in U$, has a finite system of generators $\varphi_1, \ldots, \varphi_r \in \mathcal{O}_U$ and the family of modules of relations $N_z = R_z(\varphi_1, \ldots, \varphi_r) \in \mathcal{O}_z^r$, $z \in U$, also has a finite system of generators $\psi_1, \ldots, \psi_s \in \mathcal{O}_U^r$. Now we fix an arbitrary $z \in U$. Then we have the epimorphism

$$\chi_z : \ \mathcal{O}_z^r \ni (f_1, \ldots, f_r) \longrightarrow \sum_1^r f_i(\varphi_i)_z \in I_z$$

whose kernel is N_z. Denote by $\tilde{\tau}$ the endomorphism of the module $\mathcal{O}_z^r$ whose matrix $\tau \in (\mathcal{O}_z)_r^r$. It is easy to check that for each matrix τ from the submodule $L_z = \{\tau : \ \tilde{\tau}(N_z) \subset N_z\} \subset (\mathcal{O}_z)_r^r$, there is a unique endomorphism $\alpha : I_z \longrightarrow I_z$ such that

(0) $\alpha \circ \chi_z = \chi_z \circ \tilde{\tau}$,

and the mapping $\tau \longrightarrow \alpha$ is a module epimorphism [39] with the kernel $K_z = \{\tau : \ \mathrm{im}\tilde{\tau} \subset N_z\}$. Moreover, the set of endomorphisms of the form

---

[38] Because h has a bounded representative.

[39] Because the equality (0) on the canonical basis of $\mathcal{O}_z^r$ means exactly that $\alpha((\varphi_i)_z)$ is

$I_z \ni f \longrightarrow gf \in I_z$, $g \in \mathcal{O}_z$, is the image of the submodule $\mathcal{O}_z E_z \subset L_z$, where $E \in (\mathcal{O}_U)_r^r$ is the identity matrix. Therefore, according to the Grauert-Remmert lemma, a point z is normal if and only if $L_z \subset \mathcal{O}_z E_z + K_z$. Now, denoting by $\varepsilon_1, \ldots, \varepsilon_r$ the canonical basis of $\mathcal{O}_U^r$, we have

$$K_z = \bigcap_1^r (\Phi_z^\nu)^{-1}(N_z) \quad \text{and} \quad L_z = \bigcap_1^s (\Psi_z^\mu)^{-1}(N_z) \,,$$

where

$$\Phi_z^\nu : (\mathcal{O}_z)_r^r \ni \tau \longrightarrow \tilde{\tau}\big((\varepsilon_\nu)_z\big) \in \mathcal{O}_z^r, \quad z \in U,$$

and

$$\Psi_z^\mu : (\mathcal{O}_z)_r^r \ni \tau \longrightarrow \tilde{\tau}\big((\psi_\mu)_z\big) \in \mathcal{O}_z^r, \quad z \in U \quad (\nu = 1, \ldots, r; \ \mu = 1, \ldots, s)$$

are coherent families of homomorphisms. By corollaries 1 and 2 of Oka's theorem (see 1.4), the families of submodules $K_z \subset (\mathcal{O}_z)_r^r$, $z \in U$, and $L_z \subset (\mathcal{O}_z)_r^r$, $z \in U$, are coherent. Since the set of points of U which are not normal coincides with $\{z : L_z \not\subset \mathcal{O}_z E_z + K_z\}$, it must be analytic in U [40] by the proposition from 1.4. [41]

COROLLARY 1. *The set of normal points of any analytic space is open.*

COROLLARY 2. *A point of an analytic space is normal if and only if an open neighbourhood of this point is a normal space.*

THEOREM 2. *If the space X is normal at a point $a \in X^*$, then $\dim_a X^* \leq \dim_a X - 2$.*

PROOF. The germ X_a is irreducible (see n° 1) and $k = \dim X_a > 0$ (see e.g., the theorem in V. 4.5). It is enough to prove that each locally analytic subset $V \ni a$ whose germ V_a is simple and $(k-1)$-dimensional contains some points from X^0. Let V be such a subset. Then the ideal $I = \mathcal{I}(V_a)$ is prime (see V. 4.9). By the proposition from n° 1, the ring $\mathcal{O} = \mathcal{O}_a$ is integrally closed. Now the ideal $\bar{I}$ of its localization $\mathcal{O}_I$ is principal. Indeed, the ring $\mathcal{O}_I$ is local, noetherian, and integrally closed (see V. 4.9 and A. 11.1), and so it is enough to check (see the proposition in A. 10.6) that $\bar{I}$ is its only non-zero prime ideal. The latter follows from the fact (see A.

---

the image of the i-th column of τ under the epimorphism χ_z, $i = 1, \ldots, r$. (So, for any α, there is a $\tau \in (\mathcal{O}_z)_r^r$ that satisfies (0). This implies that $\tau \in L_z$.)

[40] The family $\{\mathcal{O}_z E_z + K_z\}$ is coherent, because the family $\{\mathcal{O}_z E_z\}$ is, obviously, coherent. (See 1.4.)

[41] The original proof, in terms of the theory of sheaves, is shorter. (See [33], Chapter 6, n° 3.)

11.1 and V. 4.9) that there are no simple germs A such that $V(I) \subsetneqq A \subsetneqq X_a$ (for otherwise, since $V(I) = V_a$, one would have $\dim V_a < k - 1$; see V. 4.9 in ref. to proposition 2 from IV. 3.1). Consequently, the ideal $\bar{I}$ has a generator f/g, where $f \in I$ and $g \in \mathcal{O} \setminus I$. According to Cartan's coherence theorem (see 1.4), there is an open neighbourhood U of the point a such that $V \cap U$ is analytic in U and the family $\mathcal{I}(V_z)$, $z \in U$, has a finite system of generators $\varphi_1, \ldots, \varphi_r \in \mathcal{O}_U$. Since $(\varphi_i)_a \in I \subset \bar{I} = \mathcal{O}_I(f/g)$, we have $d_i(\varphi_i)_a = c_i f$ for some $c_i \in \mathcal{O}$, $d_i \in \mathcal{O} \setminus I$, $(i = 1, \ldots, r)$. One can choose a smaller U so that for some holomorphic representatives $\tilde{f}, \tilde{c}_i, \tilde{d}_i$ in U, we have $\tilde{d}_i \varphi_i = \tilde{c}_i \tilde{f}_i$ in U, and $\tilde{f} = 0$, $\tilde{d}_i \not\equiv 0$ on $V \cap U$ ([41a]). Also we may suppose that $\dim X_z = k$, $\dim V_z = k - 1$ for $z \in V \cap U$ (see V. 4.9 in ref. to the corollary of proposition 4 in IV. 3.1). Therefore, there exists a point $b \in V^0 \cap U$ such that $\tilde{d}_i(b) \neq 0$, $i = 1, \ldots, r$, (see the theorem in V. 4.5). Then $(\varphi_i)_b = (\tilde{c}_i/\tilde{d}_i)_b \tilde{f}_b$, and thus the germ $\tilde{f}_b$ generates the ideal $\mathcal{I}(V_b)$. Consequently – in view of the properties (###) and (##) from V. 4.9 – we have

$$\dim \operatorname{zar} X_b \leq \dim \operatorname{zar} V_b + 1 = \dim V_b + 1 = k = \dim X_b \, ,$$

which implies (because of (##)) that $b \in X^0$.

REMARK . The condition from theorem 2 does not characterize the normality of a point of an analytic space ([41b]). Even assuming local irreducibility and sufficiently high codimension $\dim_a X - \dim_a X^*$ of singularity, the point a does not have to be normal. This is shown by the following example ([42]) . The holomorphic mapping

$$F : \; \mathbf{C}^n \ni (t_1, \ldots, t_{n-1}, u) \longrightarrow (t_1, \ldots, t_{n-1}, u^2, u^3, t_1 u, \ldots, t_{n-1} u) \in \mathbf{C}^{2n}$$

is a proper injection, and hence a homeomorphism onto the range. Moreover, $0 \in \mathbf{C}^n$ is the only point at which the differential is not injective. By Remmert's theorem (see V. 5.1), the range $X \subset \mathbf{C}^{2n}$ of the mapping is an analytic set of constant dimension n (see, e.g., V. 3.2, corollary 1 of theorem 2). It has only one singular point $0 \in \mathbf{C}^{2n}$ (see C. III.14; and V. 1, corollary 2 of theorem 2) and is locally irreducible (see IV. 3.1, proposition 1; and IV. 2.8, proposition 2). Obviously, we have $\dim_a X - \dim_a X^* = n$. Nevertheless, X is not normal at 0. For if it were, it would be a normal space (see

---

([41a])   We choose U such that $(V \cap U)^0$ is connected (see V. 4.9 in ref. to proposition 1 from IV. 3.1, and V. 4.5 proposition α).

([41b])   See, however, the Oka-Abhyankar theorem in n° 3 below.

([42])   Due to K.Kurdyka.

n° 3), and then the mapping F^{-1} would be holomorphic, since it is continuous and the restriction $(F^{-1})_{X\setminus 0}$ is holomorphic (see 2.1). Thus F would be biholomorphic, which is impossible.

LEMMA . *If $a \in X$, the germ X_a is of constant dimension, and $\dim_a X^* \leq \dim_a X - 2$, then every holomorphic function $f : X^0 \longrightarrow \mathbf{C}$ is bounded on X^0 in a neighbourhood of the point a.* ([43])

PROOF . One may assume (see IV. 1.4 and 2) that $X = V$ is the crown of a normal triple (Ω, Z, V) in $\mathbf{C}^n$ of dimension k and multiplicity p, and that $\dim V^* \leq k - 2$ and $a = 0$. Since the natural projection $\pi : V \longrightarrow \Omega$ is proper, Remmert's theorem (see V. 5.1) implies that $S = \pi(V^*) \subset Z$ is analytic in Ω and of dimension $\leq k - 2$ (see V. 1, corollary 1 of theorem 1). The functions $a_1, \ldots, a_r$, well-defined on the set $\Omega \setminus Z$ by

$$ a_i(u) = \sigma\big(f(u, \eta_1), \ldots, f(u, \eta_r)\big), \text{ where } \{\eta_1, \ldots, \eta_r\} = V_u, \text{ for } u \in \Omega \setminus Z, $$

(where $\sigma_1, \ldots, \sigma_p$ are the basic symmetric polynomials) are holomorphic. They are locally bounded near $Z \setminus S$, because every point of this set has a compact neighbourhood $B \subset \Omega$ such that $V_B \cap V^* = \emptyset$ (see B. 2.4). Thus f is bounded in V_B. Hence the functions a_j extend holomorphically to $\Omega \setminus S$ (see II. 3.5), and so, by the Hartogs theorem (see III. 4.2), to Ω. Therefore they are bounded in some neighbourhood $\Omega_0 \subset \Omega$ of zero, which implies (see B. 5.3) that the function f is bounded on the set $(V^0)_{\Omega_0 \setminus Z}$. The latter is dense in $(V^0)_{\Omega_0}$ (see B. 2.1 and IV. 1.3), and so the function f is bounded in $(V^0)_{\Omega_0}$.

The lemma and theorem 2 imply

THEOREM 3. *The space X is normal if and only if for each open set $G \subset X$ every holomorphic function on G^0 has a holomorphic extension to G.*

Because of the Hartogs theorem (see III. 4.2) we obtain

COROLLARY . *If the space X is normal, then every holomorphic function on $X \setminus V$ extends holomorphically to X, for any V which is an analytic subset of X such that $\dim V_z \leq \dim X_z - 2$ for $z \in X$.*

3. Let M be a manifold of dimension n.

THE OKA-ABHYANKAR THEOREM ([43a]). *Let $V \subset M$ be an analytic set, and let $a \in V$. Assume that V_a is a complete intersection. Then V is normal at a if and only if $\dim_a V^* \leq \dim_a V - 2$.*

([43]) See [33], Chapter VI, n° 1, proposition 3.

([43a]) See [7aa].

PROOF . The condition is necessary by theorem 2 in n° 2. Suppose now that the condition is satisfied. The germ V_a is of constant dimension, say k; (see 1.6). There exists a universal denominator u of $\tilde{\mathcal{O}}_a(V)$ over $\mathcal{O}_a(V)$ (see 2.3). Omitting the trivial case, we may assume that u is non-invertible in $\mathcal{O}_a(V)$. According to the proposition in III. 4.7 (see 1.6), the ring $\mathcal{O}_a(V) = \mathcal{O}_{V_a} = \mathcal{O}_a/\mathcal{I}(V_a)$ is Cohen-Macaulay, hence so is $\mathcal{O}_a(V)/J$, where $J = u\mathcal{O}_a(V)$ (see corollary 2 in A. 14.3). Therefore, by corollary 1 in A. 14.3, all the ideals I_i associated with J are isolated for J. It follows that each $V(I_i)$ is maximal in the set of simple germs that are contained in $V(u)$ $(^{43b})$, hence it must be a simple component of $V(u)$ (by $(*)$ in II. 4.6). Consequently, it must be of dimension $k-1$, as $V(u)$ is of constant dimension $k-1$ (see V. 4.9); hence $V(I_i) \not\subset V_a^*$. Therefore, by the Abhyankar lemma (see 1.7), all the localizations $\mathcal{O}_a(V)_{I_i}$ are regular (we have $\mathcal{I}_{V_a}(V(I_i)) = I_i$ by Hilbert's theorem in V. 4.9). Finally (see A. 12.2), $h(\bar{I}_i) = h(I_i) \leq 1$ $(^{43c})$, which means that the ring $\mathcal{O}_a(V)_{I_i}$ is of dimension ≤ 1, and so its maximal ideal $\bar{I}_i$ is principal (as $g(\bar{I}_i) \leq 1$). Thus, by the proposition in A. 10.6, all $\mathcal{O}_a(V)_{I_i}$ are discrete valuation rings, and the Zariski-Samuel criterion gives the normality of V at a.

REMARK . It follows that in Kurdyka's example from the remark in n° 2, the germ X_0 is not a complete intersection.

Let $f_1, \ldots, f_p$ be holomorphic functions on M. Set

$$V = V(f_1, \ldots, f_p) \quad \text{and} \quad S = \{z \in V : \text{rank}_z(f_1, \ldots, f_p) < p\} \, .$$

The Oka-Abhyankar theorem and the Tsikh theorem in 1.6 imply the following result:

THE SERRE CRITERION. *The set V is normal and realized by $f_1, \ldots, f_p$ as a complete intersection at each of its points if and only if $\dim S \leq \dim V - 2$* $(^{43d})$.

REMARK . This condition is equivalent to each of the following ones:

(a) $\dim_z S \leq \dim_z V - 2$ for $z \in S$;

(b) $\dim S \leq n - p - 2$.

$(^{43b})$ If we had $V(I_i) \subsetneq V' \subset V(u) = V(J)$ with a simple germ V', then we would have $I_i \supsetneq \mathcal{I}_{V_a}(V') \supset \text{rad } J \supset J$. This is impossible, since I_i is isolated for J.

$(^{43c})$ If there were prime ideals $I' \subsetneq I'' \subsetneq I_i$, then we would have $V(I') \supsetneq V(I'') \supsetneq V(I_i)$, and hence $\dim V(I_i) \leq k - 2$.

$(^{43d})$ See [25b] App. I, 6.2. This criterion (its sufficiency) follows from the Serre criterion quoted in A. 11.3, footnote $(^{96})$: both implications hold for every ring $A = \mathcal{O}_a(V)$, $a \in V$. Indeed, A is Macaulay (by the Tsikh theorem and the proposition in III. 4.7), and so are A_I for all prime ideals I (see A. 14.3, footnote $(^{109})$), which gives the second implication. As for the first one, let $h(I) \leq 1$. Since $I = \mathcal{I}(C, V)$ with a simple germ $C \subset V_a$, we have (by the lemma in IV. 4.3a) $h(I) = \dim A_I = \dim V_a - \dim C$; hence $\dim C \geq \dim V_a - 1$, and so $C \not\subset V^*$. Therefore, by the Abhyankar lemma (in 1.7), the ring A_I is regular. Since it is of dimension ≤ 1, it is a discrete valuation ring (see the proposition in A. 10.6 condition (2)).

Observe that the above argument shows that the first implication holds, provided that V is of constant dimension and $\dim V^* \leq \dim V - 2$. Now, such a set does not have to be normal (e.g., the union of two planes in $\mathbf{C}^4$ whose intersection is 0). Thus the second implication is essential in applying the Serre criterion to the rings $\mathcal{O}_a(V)$.

PROOF of the criterion and remark. By the Tsikh and Oka-Abhyankar theorems, the condition (a) is necessary. It implies the condition in the criterion, which in turn implies (b) (since then a simple component of V of maximal dimension must contain some $z \notin S$; hence $\dim_z V = n - p$, and so $\dim V = n - p$). Finally, if (b) is satisfied, then, since $\dim_z V \geq n - p$ for $z \in V$ (by (*) in III. 4.6), S must be nowhere dense in V. Hence, by the Tsikh and Oka-Abhyankar theorems, condition (b) is sufficient.

§4. Normalization

Let X be an analytic space.

1. A *normalization* of the space X is defined as a finite holomorphic mapping $\pi : Y \longrightarrow X$ of a normal analytic space Y such that the set $\pi^{-1}(X^0)$ is dense in Y and the restriction $\pi^{X^0} : \pi^{-1}(X^0) \longrightarrow X^0$ is biholomorphic. The latter condition can be replaced by an equivalent one: $\#\pi^{-1}(z) = 1$ for $z \in X^0$ (see V. 4.7 in ref. to the proposition from V. 3.4). Therefore a normalization of X is precisely a modification in X^* with normal domain and finite fibres (see V. 4.13). Thus the property of being a normalization is a rigid property.

Every normalization is surjective (see V. 4.3). Obviously, we have $\pi^{-1}(X^0) \subset Y^0$. Irreducibility of the space Y is equivalent to irreducibility of the space X (see V. 4.5, proposition α). Of course the identity mapping of a normal space is a normalization.

Let $\pi : Y \longrightarrow X$ be a mapping between analytic spaces. If π is a normalization, then so is $\pi^G : \pi^{-1}(G) \longrightarrow G$ for any open set $G \subset X$. If $\{G_\iota\}$ is an open cover of the space X, then π is a normalization precisely when all $\pi^{G_\iota} : \pi^{-1}(G_\iota) \longrightarrow G_\iota$ are normalizations. In other words, the property of being a normalization is sublocal.

Of course, if π is a normalization, whereas $\iota : Z \longrightarrow Y$ and $\kappa : X \longrightarrow T$ are biholomorphic mappings of analytic spaces, then $\kappa \circ \pi \circ \iota : Z \longrightarrow T$ is also a normalization.

THEOREM (on uniqueness of normalization). *Any two normalizations of the space X are isomorphic: if $\pi_i : Y_i \longrightarrow X$, $i = 1, 2$, are normalizations, then there exists a biholomorphic mapping $\iota : Y_1 \longrightarrow Y_2$ such that the diagram*

$$
\begin{array}{ccc}
Y_1 & \xrightarrow{\ \iota\ } & Y_2 \\
{\scriptstyle \pi_1}\searrow & & \swarrow{\scriptstyle \pi_2} \\
& X &
\end{array}
$$

commutes. Moreover, ι is the unique homeomorphism with this property.

PROOF . Set $G_i = \pi_i^{-1}(X^0)$. Every mapping ι for which the above diagram commutes must contain the biholomorphic mapping $\iota_0 = (\pi_2)_{G_2}^{-1} \circ (\pi_1)_{G_1} : G_1 \longrightarrow G_2$. Therefore any such homeomorphism ι is unique. Now the mapping ι_0 has a holomorphic extension $\iota : Y_1 \longrightarrow Y_2$. Indeed, since the space Y_1 is normal and $Z = Y_1 \setminus G_1 = \pi_1^{-1}(X^*)$ is a nowhere dense analytic set, it is enough to check that the mapping $\iota_0 : G_1 \longrightarrow Y_1$ is l–bounded near Z (see 2.1). Let $a \in Z$. The set $\pi_2^{-1}(\pi_1(a))$ is finite, and so it has an l–bounded neighbourhood B (see 2.1). On the other hand, $\pi_2^{-1}(U) \subset B$ for some neighbourhood U of the point $\pi_1(a)$ (see B. 2.4). Finally, $\pi_1(W) \subset U$ for some neighbourhood W of the point a, and hence $\pi_1^{-1}(\pi_1(W)) \subset B$. Thus the set $\iota_0(W \setminus Z) \subset B$ is l–bounded (see 2.1). Similarly, the mapping $\iota_0^{-1} = (\pi_1)_{G_1}^{-1} \circ (\pi_2)_{G_2} : G_2 \longrightarrow G_1$ has a holomorphic extension $\kappa : Y_2 \longrightarrow Y_1$. Then the compositions $\kappa \circ \iota, \iota \circ \kappa$ are the identity mappings on the sets G_1 and G_2, respectively, and hence also on the sets Y_1 and Y_2. Consequently, $\iota : Y_1 \longrightarrow Y_2$ is a biholomorphic mapping. Clearly, $\pi_1 = \pi_2 \circ \iota$ in G_1, and hence also in Y_1.

Therefore the normal space Y of any normalization $Y \longrightarrow X$ is determined up to an isomorphism; it is called the *normalized space of the space* X.

Note that in the Puiseux theorem (the second version, see II. 6.2) we are dealing with a normalization of a representative of a one–dimensional simple analytic germ. Namely, it is easy to check that (using the same notation as in the theorem) the mapping

$$\{|t| < \delta\} \ni t \longrightarrow \left(t^p, h(t)\right) \in V$$

is a normalization, provided that p and h are chosen so that it is a homeomorphism.

Another example of normalization is the mapping $F : \mathbf{C}^n \longrightarrow X$ from the remark in 3.2 (the example of Kurdyka). Also, the mappings from footnote $(^{64})$ in V. 5 are normalizations.

2. We have the following property

PROPOSITION 1. *If $X = \bigcup X_\iota$ is the decomposition into irreducible components and $\pi_\iota : Y_\iota \longrightarrow X_\iota$ are normalizations such that the spaces Y_ι are disjoint* $(^{44})$, *then $\pi = \bigcup \pi_\iota : \bigcup Y_\iota \longrightarrow X$ is a normalization, where $\bigcup Y_\iota$*

$(^{44})$ Having the normalizations $\pi_\iota : Y_\iota \longrightarrow X_\iota$, one can achieve this requirement by replacing Y_ι by $\iota \times Y_\iota$ (with the structures induced via the natural bijections $y \longrightarrow (\iota, y)$).

has the structure of the sum of the analytic spaces Y_ι ([45]) .

Indeed, it is obvious that the mapping π is holomorphic. Since the family $\{X_\iota\}$ is locally finite, the mapping π is proper and has finite fibres. Furthermore, we have the equality $X^0 = \bigcup_\iota (X_\iota^0 \setminus \bigcup_{\nu \neq \iota} X_\nu)$, where the summands on the right hand side are dense in X_ι^0, respectively (see V. 4.5 in ref. to theorem 4 from IV. 2.9). It follows that the set $\pi^{-1}(X^0)$ is dense in $\bigcup Y_\iota$ (see V.4.13) and $\#\pi^{-1}(z) = 1$ for $z \in X^0$.

THE LOCAL NORMALIZATION LEMMA . *Let $a \in X$, and let the germ X_a be irreducible ([46]) . Assume that holomorphic functions $h_1, \ldots, h_r$ on X^0 in a neighbourhood of a are representatives of generators of $\tilde{\mathcal{O}}_a$ regarded as a module over $\mathcal{O}_a$ ([47]) . Then there is an open neighbourhood U of a such that the closure H in $U \times \mathbf{C}^r$ of the graph of the mapping $(h_1, \ldots, h_r)_U$ is analytic in $U \times \mathbf{C}^r$ and the natural projection $\pi : H \longrightarrow U$ is a normalization.*

PROOF . If U is a sufficiently small open neighbourhood of the point a, then $h = (h_1, \ldots, h_r)_U : U^0 \longrightarrow \mathbf{C}^r$ is a bounded holomorphic mapping, and (by taking a universal denominator in $\mathcal{O}_a$) we have $g h_i = f_i$ in U^0, where $g, f_1, \ldots, f_r$ are holomorphic functions in U and $g \not\equiv 0$ (see 2.3). The set $V = \{(z, w_1, \ldots, w_r) : g(z) w_i = f_i(z)\}$ is analytic in $U \times \mathbf{C}^r$. Now, the set

$$h \cap \{g \neq 0\} = V_{U^0} \cap \{g \neq 0\} = V \setminus (V_{U^*} \cup \{g = 0\})$$

is dense in h and its closure in $U \times \mathbf{C}^r$ is analytic (see V. 4.5 in ref. to theorem 5 from IV. 2.10). Therefore H is an analytic subset in $U \times \mathbf{C}^r$. Obviously, the natural projection $\pi : H \longrightarrow U$ is proper (see B. 5.2), and hence its fibres are finite (see the proposition in IV. 5). We have $\pi^{-1}(U^0) = h$ (as the graph of h is closed in $U \times \mathbf{C}^r$), and hence $\pi^{-1}(U^0)$ is dense in H and the restriction $\pi_{\pi^{-1}(U^0)} : \pi^{-1}(U^0) \longrightarrow U^0$ is biholomorphic (see V. 4.7 in ref. to V. 3.4).

It remains to prove that if the neighbourhood U is sufficiently small, then H is a normal space. Now, $H_a = b$, where $b \in \mathbf{C}^r$. Indeed, taking a base of open, relatively compact neighbourhoods $U \supset U_1 \supset U_2 \supset \ldots$ of the point a such that the U_i^0 are connected (see V. 4.9 in ref. to proposition 1 from IV. 3.1, and V. 4.5 proposition α), we have $a \times H_a = \bigcap \overline{h_{U_i}}$. But $\overline{h_{U_1}} \supset \overline{h_{U_2}} \supset \ldots$ are compact and connected, hence so is their intersection,

---

([45]) The *sum of a* (countable) *family of disjoint analytic spaces Y_ι* is defined as the space $\bigcup Y_\iota$ with the (uniquely determined) structure given by the union of the atlases of the spaces Y_ι. Then the spaces Y_ι are open subsets of the space $\bigcup Y_\iota$ and their structures coincide with the induced ones.

([46]) The lemma remains true without this assumption.

([47]) There are always such functions h_i, because $\tilde{\mathcal{O}}_a$ is finite over $\mathcal{O}_a$ (see 2.3).

and hence so is the set $H_a \subset \mathbf{C}^r$. Consequently, the latter consists of a single point (see the proposition in IV. 5). It follows that the family of sets $H_{U'}$, where U' are neighbourhoods of the point a, form a base of neighbourhoods of the point (a, b) in H (see B. 2.4). Therefore, according to corollary 2 from theorem 1 in 3.2, it is enough to prove that (a, b) is a normal point of the space H. Now, let f be a bounded holomorphic function on H^0 in a neighbourhood of (a, b). In view of $h \subset H^0$, if $W \subset U$ is a sufficiently small open neighbourhood of a, then the domain of f contains h_W, and so the function $f(z, h(z))$ is holomorphic and bounded in W^0 [48] . Thus we have $f(z, h(z)) = \sum_1^r a_i(z)h_i(z)$ in W^0 for some holomorphic functions a_i on W. Now the function $F(z, w_1, \ldots, w_r) = \sum_1^r a_i(z)w_i$ is holomorphic in $W \times \mathbf{C}^r$ and we have $f = F$ on h_W, and hence also on $(H^0)_W$. This means that the function f extends to a holomorphic function on H_W.

In view of proposition 1, we have (see V. 4.9 in ref. to proposition 3b from IV. 3.1):

COROLLARY . *Every sufficiently small open neighbourhood U of any point of an analytic space has a normalization $W \longrightarrow U$, and one may require that the space W is an analytic subset of a manifold* [49] .

The property of being a normalization is sublocal and rigid (see n° 1). Therefore the proposition on gluing from V. 4.12, combined with the uniqueness theorem from n° 1, implies:

THE NORMALIZATION THEOREM . *Every analytic space X has a normalization $\pi : Y \longrightarrow X$ which is unique up to an isomorphism.*

PROPOSITION 2. *Let $\pi : W \longrightarrow X$ be a normalization. For any $z \in X$, there are arbitrarily small open neighbourhoods $D_1, \ldots, D_r$ of the points of the fibre $\pi^{-1}(z)$, with locally analytic images $U_i = \pi(D_i)$, such that the restrictions $\pi_{D_i} : D_i \longrightarrow U_i$ are normalizations and $X_z = (U_1)_z \cup \ldots \cup (U_r)_z$ is the decomposition into simple germs.*

PROOF . Let $D_1', \ldots, D_r'$ be arbitrary disjoint open neighbourhoods of the points of the fibre $\pi^{-1}(z)$. There is an open neighbourhood U of the point z such that $\pi^{-1}(U) \subset D_1' \cup \ldots \cup D_r'$ (see B. 2.4). In view of the local normalization lemma, one may require that the neighbourhood U has a finite number of simple components $U_1, \ldots, U_s$ which have normalizations $\pi_i : H_i \longrightarrow U_i$, and that $X_z = (U_1)_z \cup \ldots \cup (U_s)_z$ is the decomposition into simple germs (see V. 4.9 in ref. to proposition 3b from IV. 3.1). Then the spaces H_i are connected (since they are irreducible; see n° 1) and one

---

[48] Therefore it is a representative of an element of $\tilde{\mathcal{O}}_a$.

[49] Even a locally analytic subset of a vector space.

may assume that they are pairwise disjoint (see footnote (^{44})). In accordance with proposition 1, we have the normalization $\tilde{\pi} : \bigcup_1^s H_i \longrightarrow U$, where $\bigcup_1^s H_i$ is endowed with the structure of the sum of the spaces H_i (see footnote (^{43})). But $\pi' = \pi^U : \pi^{-1}(U) \longrightarrow U$ is also a normalization, and hence, by the uniqueness theorem (see n° 1), there exists a biholomorphic mapping $\iota : \bigcup H_i \longrightarrow \pi^{-1}(U)$ such that the diagram

$$\bigcup H_i \quad \xrightarrow{\ \iota\ } \quad \pi^{-1}(U)$$
$$\tilde{\pi} \searrow \qquad \swarrow \pi'$$
$$U$$

commutes. Then the sets $D_i = \iota(H_i)$ are disjoint, open, and connected. Moreover, $\pi_{D_i} : D_i \longrightarrow U_i$ are normalizations (see n° 1). Since $\pi^{-1}(z) \subset D_1 \cup \ldots \cup D_s \subset D_1' \cup \ldots \cup D_r'$ and $D_i \cap \pi^{-1}(z) \neq \emptyset$, we have $s = r$ and the D_i are neighbourhoods of the points of the fibre $\pi^{-1}(z)$.

COROLLARY . If $\pi : W \longrightarrow X$ is a normalization, then for each $z \in X$ the number $\#\pi^{-1}(z)$ is equal to the number of simple components of the germ X_z.

3. In what follows, we will present Cartan's construction of a natural model of the normalization of the space X in which the normalized space is the set $\hat{X}$ of simple components of the germs of X furnished with a suitable structure of an analytic space. First we define a natural topology on $\hat{X}$.

For any analytic space Y we set $\hat{Y} = \bigcup\{[Y_z] : z \in Y\}$, where $[A]$ denotes the set of simple components of the analytic germ A.

If $G, H, G_\iota \subset X$ are open subsets, then obviously $\widehat{G \cap H} = \hat{G} \cap \hat{H}$ and $\widehat{\bigcup G_\iota} = \bigcup \hat{G}_\iota$. By Ritt's lemma (in IV. 2.8, the remark following proposition 4; see V. 4.9), if $U, V \subset X$ are analytic subsets and U is irreducible, then

$$(1) \qquad\qquad \hat{U} \cap \hat{V} \neq \emptyset \Longrightarrow U \subset V .$$

Furthermore,

$$(2) \quad \begin{cases} \text{if } X = \bigcup X_\iota \text{ is the decomposition into simple components,} \\ \text{then } \hat{X} = \bigcup \hat{X}_\iota \text{ and } \hat{X}_\iota \text{ are disjoint.} \end{cases}$$

Indeed, for each $z \in X$ we have the disjoint union $[X_z] = \bigcup[(X_\iota)_z]$ because, by (1), the sets $[(X_\iota)_z]$ are disjoint and their elements, taken together, form the decomposition of X_z into simple germs.

Let $\mathcal{B}(X)$ denote the class of locally analytic subsets $V \subset X$ such that $\hat{V} \subset \hat{X}$. Obviously, every open set in any set $V \in \mathcal{B}(X)$ — and in particular, every open subset of the space X — belongs to $\mathcal{B}(X)$. Because of (2), unions of irreducible components of any set $V \in \mathcal{B}(X)$ — and in particular, the irreducible components of the space X — belong to $\mathcal{B}(X)$. If $A \in \hat{X}$ and $V \in \mathcal{B}(X)$, then $A \subset V \Longrightarrow A \in \hat{V}$ (see V. 4.9 in ref. to II. 4.6). Thus, for any $U, V \in \mathcal{B}$, we have $U \subset V \Longleftrightarrow \hat{U} \subset \hat{V}$.

We can endow $\hat{X}$ with the topology whose base is $\{\hat{V} : V \in \mathcal{B}(X)\}$. (This is a base of a topology since if $U, V \in \mathcal{B}(X)$, then, by (1) and (2), we have $\hat{U} \cap \hat{V} = \hat{W}$. Here $W \in \mathcal{B}(X)$ is the union of all common irreducible components of the sets $U \cap G$ and $V \cap G$, while G is an open set in which the set $U \cap V$ is analytic.) Locally analytic irreducible representatives V of a germ $A \in \hat{X}$ belong to $\mathcal{B}(X)$, and the corresponding sets $\hat{V}$ form a base of neighbourhoods of A (because of (1) and (2); see V. 4.9 in reference to proposition 1 from IV. 3.1). The space $\hat{X}$ is a Hausdorff space (due to (1) and (2) (^{50})). If $V \in \mathcal{B}(X)$, then the topology of the space $\hat{V}$ coincides with that induced from $\hat{X}$ (^{51}) .

Define the natural mapping $\nu = \nu^X : \hat{X} \longrightarrow X$ by $\nu([X_z]) = z$ if $z \in X$. Clearly, $\hat{G} = \nu^{-1}(G)$ for every open set $G \subset X$, and $\nu^U = (\nu^X)_{\hat{U}}$ for each $U \in \mathcal{B}(X)$. Hence the mapping ν is continuous. Its restriction $\nu_{\widehat{X^0}} : \widehat{X^0} \longrightarrow X^0$ is a homeomorphism (since it is open and bijective (^{52})) that allows us to identify the set $\widehat{X^0}$ with the set X^0. The set $\widehat{X^0}$ is dense in $\hat{X}$ (see V. 4.9 in ref. to proposition 3a from IV. 3.1, and V. 3.5 in ref. to corollary 4 of theorem 4 from IV. 2.9, and (1)). Therefore, if the space X is irreducible, then the space $\hat{X}$ is connected (see V. 4.5, proposition α).

PROPOSITION 3. Let $\pi : W \longrightarrow X$ be a normalization. If $w \in W$, then, for a sufficiently small neighbourhood U of w, the germ $\vartheta(w) = \pi(U)_{\pi(w)}$ belongs to $[X_{\pi(w)}]$ and is independent of U. The mapping $\vartheta : W \longrightarrow \hat{X}$ is

---

(^{50}) For two distinct germs $A, B \in [X_a]$, it is enough to take their representatives from $\mathcal{B}(X)$ which are both analytic in one open neighbourhood of a and such that one of them is irreducible (see V. 4.9 in ref. to proposition 1 from IV. 3.1, and (1)).

(^{51}) Since the set $\hat{V}$ is open and $\mathcal{B}(V) = \{U \in \mathcal{B}(X) : U \subset V\}$, which means that $\{\hat{U} : U \in \mathcal{B}(V)\}$ is the family of all sets $\{\hat{U} : U \in \mathcal{B}(X)\}$ that are contained in $\hat{V}$.

(^{52}) Since the sets from $\mathcal{B}(X^0)$ are open in X^0.

the unique homeomorphism such that the diagram

$$(*) \qquad \begin{array}{ccc} W & \xrightarrow{\ \vartheta\ } & \hat{X} \\ & \pi \searrow \quad \swarrow \nu & \\ & X & \end{array}$$

is commutative.

PROOF . According to proposition 2, for an arbitrarily small neighbourhood U of the point w, the germ $\pi(U)_{\pi(w)}$ belongs to $[X_{\pi(w)}]$. Consequently, if the neighbourhood U is sufficiently small, the germ $\pi(U)_{\pi(w)}$ belongs to $[X_{\pi(w)}]$ and is independent of U ([53]) . Proposition 2 implies also that ϑ is bijective. Clearly, $\nu \circ \vartheta = \pi$. Now the mapping ϑ is continuous. In fact, let $A = \vartheta(w)$, $w \in W$, and let V be a locally analytic irreducible representative of A. There is an open neighbourhood U of the point w such that $\pi(U) \subset V$ ([54]) . Now if $w' \in U$, then $\vartheta(w') \subset V$, and hence $\vartheta(w') \in \hat{V}$ (since $\vartheta(w') \in \hat{X}$ and $V \in \mathcal{B}(X)$). Since the composition $\nu \circ \vartheta = \pi$ is proper, so is the mapping ϑ (see B. 2.4). Therefore ϑ is a homeomorphism. The uniqueness of the homeomorphism ϑ that satisfies $(*)$ follows from the fact that the mapping is uniquely determined on the dense set $\pi^{-1}(X^0)$.

The structure of a normal analytic space on the topological space $\hat{X}$ is said to be *natural* if (after the identification) it induces on X^0 the same structure of an analytic space as that induced by X, i.e., if $\nu_{\widehat{X^0}} : X^0 \longrightarrow \widehat{X^0}$ is a biholomorphic mapping. (The latter clearly holds when ν is a normalization.) There could be only one such structure, for the identity mapping of $\hat{X}$ onto the space $\hat{X}$ with another such structure must be biholomorphic (see 2.1; by proposition 3 and Remmert's theorem (V. 5.1) the set $\hat{X} \setminus X^0$ must be analytic). Proposition 3 and the normalization theorem imply:

THEOREM . *The topological space $\hat{X}$ has a natural structure of a normal analytic space. If $\hat{X}$ is endowed with this structure, then the natural mapping $\nu : \hat{X} \longrightarrow X$ is a normalization.*

We will give a proof of this theorem, and hence also a proof of the normalization theorem, without using the proposition on gluing from V. 4.12.

([53]) Indeed, let U' be a neighbourhood such that $\pi(U')_{\pi(w)} \in [X_{\pi(w)}]$. Then for any neighbourhood $U \subset U'$ there is a neighbourhood $U'' \subset U$ with the property $\pi(U'')_{\pi(w)} \in [X_{\pi(w)}]'$. Then $\pi(U'')_{\pi(w)} \subset \pi(U)_{\pi(w)} \subset \pi(U')_{\pi(w)}$, and hence $\pi(U'')_{\pi(w)} = \pi(U)_{\pi(w)} = \pi(U')_{\pi(w)}$.

([54]) Because, for some open neighbourhood U' of w, we have $\pi(U') \cap \Omega = V \cap \Omega$ for some open neighbourhood Ω of the point $\pi(w)$. Therefore it is enough to take $U = U' \cap \pi^{-1}(V)$.

Now, according to the corollary from the local normalization lemma (in n° 2), there exists a countable open cover $\{U_i\}$ of the space X such that each U_i has a normalization $\pi_i : W_i \longrightarrow U_i$, where W_i is an analytic subset of a manifold. Then $\{\hat{U}_i\}$ is an open cover of the space $\hat{X}$. The homeomorphisms $\vartheta_i : W_i \longrightarrow \hat{U}_i$ defined (as in proposition 3) by the commuting diagrams

$$
\begin{array}{ccc}
W_i & \xrightarrow{\ \vartheta_i\ } & \hat{U}_i \\
\pi_i \searrow & & \swarrow \nu_{\hat{U}_i} \\
& U_i &
\end{array}
$$

constitute an inverse analytic atlas on X. Indeed, by putting $W_{ij} = \pi_i^{-1}(U_i \cap U_j) = \vartheta_i^{-1}(\hat{U}_i \cap \hat{U}_j)$, we have the normalizations $\pi_{ij} = (\pi_i)_{W_{ij}} : W_{ij} \longrightarrow U_i \cap U_j$ (see n° 1), and the commuting diagrams:

$$
\begin{array}{ccc}
W_{ij} & \xrightarrow{\ \vartheta_j^{-1} \circ \vartheta_i\ } & W_{ji} \\
\pi_{ij} \searrow & & \swarrow \pi_{ji} \\
& U_i \cap U_j &
\end{array}
$$

Thus, in view of the uniqueness theorem (from n° 1), it follows that $\vartheta_j^{-1} \circ \vartheta_i$ are biholomorphic mappings. With the structure of an analytic space on $\hat{X}$ defined by the atlas $\{\vartheta_i\}$, the mappings $\nu_{\hat{U}_i} : \hat{U}_i \longrightarrow U_i$ are normalizations, since ϑ_i are biholomorphic (see n° 1). Therefore the mapping $\nu : \hat{X} \longrightarrow X$ is a normalization (see n° 1), and the structure we have just defined is the natural structure on $\hat{X}$ of a normal analytic space.

In view of the proposition from V. 3.4 (see V. 4.7), *the natural structure of a normal space on $\hat{X}$ is the only structure of a normal space* ([55]) *such that the natural mapping $\nu : \hat{X} \longrightarrow X$ is holomorphic.*

It follows that, *if $V \in \mathcal{B}(X)$, then the natural structure of the space $\hat{V}$ coincides with that induced by the natural structure of the space $\hat{X}$.* (For, since $\hat{V}$ is an open subset of the space $\hat{X}$, the induced structure makes it a normal space.) *Then the restriction $\nu_{\hat{V}} : \hat{V} \longrightarrow V$ is a normalization.*

If $X = \bigcup X_\iota$ is the decomposition into irreducible components, then $\hat{X} = \bigcup \hat{X}_\iota$ is the decomposition into connected components ([56]). (Indeed, the sets $\hat{X}_\iota$ are disjoint, open, and connected.)

([55]) On the topological space $\hat{X}$.

([56]) It is also the decomposition of the space $\hat{X}$ into simple components (see V. 4.5 in ref. to the corollary of proposition 5 from IV. 2.9, and 3.1).

Note that, among the holomorphic functions on X^0, the ones that extend holomorphically to $\hat{X}$ are precisely those locally bounded near X^* (^{57}). They are said to be the *weakly holomorphic functions* on X.

Finally, observe that if $a \in X$ and if V is a locally analytic irreducible representative of a germ $A \in [X_a]$, then every holomorphic function f which is bounded on $V \cap X^0$ has a limit at the point a (that is equal to the value at a of the holomorphic extension of the function f to $\hat{V}$) (^{58}).

Therefore, *if the space X is locally irreducible, there is no essential difference between weakly holomorphic and c–holomorphic functions. Namely, every weakly holomorphic function on X has a c–holomorphic extension to X.*

(^{57}) Due to the fact that the mapping ν is proper, the restrictions to X^0 of holomorphic functions on $\hat{X}$ are locally bounded near X^*. Conversely, the holomorphic functions on X^0 which are locally bounded near X^* are locally bounded near the set $\hat{X} \setminus X^0 = \nu^{-1}(X^*)$. Since the latter is analytic and nowhere dense, they extend to holomorphic functions on $\hat{X}$.

(^{58}) This fact can be derived directly, without the use of the normalization theorem. (Namely, Cartan's proposition from 2.3 implies that the set of limit points at a of the function f must be finite. See [33], Chapter IV, n° 1.)

CHAPTER VII

ANALYTICITY AND ALGEBRAICITY

§1. Algebraic sets and their ideals

Let M be a vector (or affine) space. Set $\mathcal{P} = \mathcal{P}(M)$.

1. For every algebraic set $V \subset M$ we define its *ideal*

$$\mathcal{I}(V) = \{f \in \mathcal{P} : \ f = 0 \text{ on } V\}$$

and for each ideal of the ring $\mathcal{P}$ we define its *zero set* or *locus*

$$V(I) = \bigcap_{f \in I} V(f) \ .$$

The ring $\mathcal{P}$ is noetherian (see A. 9.4), and thus

$$V(I) = V(f_1, \ldots, f_k) = V(f_1) \cap \ldots \cap V(f_k) \ ,$$

where $f_1, \ldots, f_k$ are generators of the ideal I. Therefore $V(I)$ is an algebraic set.

The properties listed below in §1 can be verified in the same way as those in II. 4, since the ring $\mathcal{P}$ is noetherian.

2. We have (for any algebraic set W)

$$V\big(\mathcal{I}(W)\big) = W \ .$$

Hence every algebraic set is the locus of an ideal. Moreover,

$$V \subset W \Longleftrightarrow \mathcal{I}(V) \supset \mathcal{I}(W) \text{ and } V = W \Longleftrightarrow \mathcal{I}(V) = \mathcal{I}(W)$$

(for any algebraic sets V and W).

It follows that every decreasing sequence of algebraic sets stabilizes after a finite number of steps.

Obviously,

$$\mathcal{I}(V_1 \cup \ldots \cup V_k) = \mathcal{I}(V_1) \cap \ldots \cap \mathcal{I}(V_k)$$

and

$$\operatorname{rad} \mathcal{I}(V) = \mathcal{I}(V) .$$

3. We have the following properties (for ideals I, J, I_i of the ring $\mathcal{P}$):

$$\mathcal{I}(V(I)) \subset I \quad (^1) ,$$
$$I \subset J \Longrightarrow V(I) \supset V(J) ,$$
$$V(\operatorname{rad} I) = V(I) ,$$
$$V(I_1 \cap \ldots \cap I_k) = V(I_1) \cup \ldots \cup V(I_k) ,$$

and

$$V(I_1 + \ldots + I_k) = V(I_1) \cap \ldots \cap V(I_k) .$$

More generally, $V\left(\sum I_\iota\right) = \bigcap V(I_\iota)$ for any family of ideals $\{I_\iota\}$ of the ring $\mathcal{P}$, where $\sum I_\iota$ denotes the ideal generated by $\bigcup I_\iota$. The intersection of any family $\{V_\iota\}$ of algebraic sets is algebraic, and $\bigcap V_\iota = V_{\iota_1} \cap \ldots \cap V_{\iota_r}$ for some $\iota_1, \ldots, \iota_r$.

Indeed, the ideal $\sum I_\iota$ is generated by a finite number of elements from $\bigcup I_\iota$ (see A. 9.2); hence for some $\iota_1, \ldots, \iota_r$, we have $\sum I_\iota \subset I_{\iota_1} + \ldots + I_{\iota_r}$, which implies

$$V\left(\sum I_\iota\right) \supset V(I_{\iota_1}) \cap \ldots \cap V(I_{\iota_r}) \supset \bigcap V(I_\iota) .$$

But $V\left(\sum I_\iota\right) \subset \bigcap V(I_\iota)$, and so

$$V\left(\sum I_\iota\right) = V(I_{\iota_1}) \cap \ldots \cap V(I_{\iota_r}) = \bigcap V(I_\iota) .$$

$(^1)$ By Hilbert's Nullstellensatz (see §10 below) we always have $\mathcal{I}(V(I)) = \operatorname{rad} I$. Thus, if the ideal I is prime, we have the equality $\mathcal{I}(V(I)) = I$ (see A. 1.11).

4. We have $V(f) = V(\mathcal{P}f)$ for $f \in \mathcal{P}$. The algebraic sets of this form are called *principal* (^{2}) . Obviously, we have $V(f) = V(g)$ if the polynomials $f, g \in \mathcal{P}$ are associated. We have

$$f \in I \Longrightarrow V(I) \subset V(f)$$

(for any ideal I of the ring $\mathcal{P}$). Also,

$$V(f_1 \ldots f_k) = V(f_1) \cup \ldots \cup V(f_k) \quad \text{for} \quad f_1, \ldots, f_k \in \mathcal{P} \ .$$

Consequently (in view of the factoriality of $\mathcal{P}$; see A. 6.2), for each principal algebraic set $V \subsetneq M$, there exists a polynomial $f \in \mathcal{P}$ with no multiple factors such that $V = V(f)$ (^{3}) .

5. If $\varphi : M \longrightarrow N$ is a vector (or affine) space isomorphism, then

$$V(f \circ \varphi^{-1}) = \varphi(V(f)) , \quad V(I \circ \varphi^{-1}) = \varphi(V(I)) ,$$
$$\mathcal{I}(\varphi(V)) = \mathcal{I}(V) \circ \varphi^{-1}$$

for any polynomial $f \in \mathcal{P}(M)$, any ideal $I \subset \mathcal{P}(M)$, and any algebraic set $V \subset M$.

6. A non-empty algebraic set $V \subset M$ is said to be *irreducible* or *simple* if it is not the union of two algebraic sets strictly contained in V (^{4}) (and then it is not the union of a finite number of algebraic sets strictly contained in V). Later on we will prove that this is the case if and only if the set V is irreducible when regarded as an analytic set.

An algebraic set $V \subset M$ is irreducible precisely when its ideal $\mathcal{I}(V)$ is prime.

If $V \subset M$ is an irreducible algebraic set, then

$$V \subset W_1 \cup \ldots \cup W_k \Longrightarrow (V \subset W_j \text{ for some } j)$$

(^{2}) So these are precisely the loci of the principal ideals.

(^{3}) In view of Hilbert's Nullstellensatz (see §10 below), it is unique up to a non-zero constant factor.

(^{4}) Hence an algebraic set V is not irreducible if and only if it is empty or is *reducible*, i.e., is the union of two algebraic sets strictly contained in V.

for any algebraic sets $W_i \subset M$.

Every algebraic set $V \subset M$ can be represented in a unique way as a finite union of irreducible algebraic sets V_i such that $V_i \not\subset V_j$ for $i \neq j$. This representation is called the *decomposition of V into simple components*, and the sets V_i are called the *simple components* of the set V. Later on (in 11.1), we will show that the above decomposition ([5]) coincides with the decomposition into simple components of the set V regarded as an analytic set.

If an algebraic set V is the union of irreducible algebraic sets $V_1, \ldots, V_k$, then its simple components are exactly the maximal elements of the set $\{V_1, \ldots, V_k\}$.

It will be shown below (in 6.2) that the simple components of an algebraic cone are also cones.

7. We will prove (in 15.3) that the regular points (of dimension k) of an algebraic set can be characterized in terms of polynomials, namely, by the condition (a) from III. 3.5, in which the mapping Φ is polynomial.

We will show (in 6.3) that the set of singular points of an algebraic set is also algebraic.

§2. The projective space as a manifold

1. Let X be an n–dimensional vector space. We have the natural mapping

$$\alpha = \alpha^X : X \setminus 0 \ni z \longrightarrow \mathbf{C}z \in \mathbf{P}(X) .$$

For every affine hyperplane $H \ni 0$ in the space X, the mappings

$$\alpha_H : H \not\ni z \longrightarrow \mathbf{C}z \in \mathbf{P}(X) \setminus \mathbf{P}(H_*)$$

and

$$\alpha_H^{-1} : \mathbf{P}(X) \setminus \mathbf{P}(H_*) \ni \mathbf{C}z \longrightarrow z/\lambda_H(z) \in H$$

are homeomorphisms. (See B. 6.10.) Now, the mappings α_H form an inverse complex atlas on $\mathbf{P}(X)$, because the compositions

$$\alpha_H^{-1} \circ \alpha_{H'} : H' \setminus H_* \ni z \longrightarrow z/\lambda_H(z) \in H \setminus H'_*$$

([5]) Its existence follows also from proposition 2 in 11.1.

are holomorphic. It defines the *natural structure of an* $(n-1)$–*dimensional manifold on the projective space* $\mathbf{P}(X)$.

Note that α_{H^ι} is an inverse atlas of the manifold $\mathbf{P}(X)$, provided that the family of affine hyperplanes $H^\iota \not\ni 0$ satisfies the condition $\bigcap H^\iota_* = 0$. For then the ranges of the inverse charts α_{H^ι} cover $\mathbf{P}(X)$, because $\bigcap \mathbf{P}(H^\iota_*) = \bigcap (H^\iota_*)\tilde{} = \emptyset$ (see B. 6.10).

2. The mapping $\alpha : X \setminus 0 \longrightarrow \mathbf{P}(X)$ is a surjective submersion with one–dimensional fibres. This is the case, since for any affine hyperplane $H \not\ni 0$ we have

$$\alpha_H^{-1} \circ \alpha \circ \gamma : H \times (\mathbf{C} \setminus 0) \ni (z,t) \longrightarrow z \in H ,$$

where $\gamma : H \times (\mathbf{C} \setminus 0) \ni (z,t) \longrightarrow tz \in X \setminus H_*$ is a biholomorphic mapping [6]. This implies that $\alpha_{X \setminus H_*}$ is a submersion (see C. 4.2). Therefore we have the following characterizations of submanifolds, of (locally) analytic subsets of $\mathbf{P}(X)$, and of holomorphic mappings defined on them with values in manifolds (see C. 4.2, II. 3.4 and V. 3.1):

A non-empty set $\Gamma \subset \mathbf{P}(X)$ is a submanifold or a (locally) analytic subset of constant dimension k if and only if the set $\Gamma\tilde{} \setminus 0 \subset X \setminus 0$ is, respectively, a submanifold or a (locally) analytic subset of (constant) dimension $k+1$. Then, if M is a manifold, a mapping $f : \Gamma \longrightarrow M$ is holomorphic precisely when the mapping

$$f \circ \alpha : \Gamma\tilde{} \setminus 0 \ni z \longrightarrow f(\mathbf{C}z) \in M$$

is holomorphic. In particular, this implies that:

Every projective subspace $\mathbf{P}(T)$ of dimension k (where $T \subset X$ is a linear subspace of dimension $k+1$) is a k–dimensional submanifold of $\mathbf{P}(X)$, and the induced manifold structure on $\mathbf{P}(T)$ coincides with its natural structure [7].

Every isomorphism of projective spaces is biholomorphic [8].

3. Consider the n–dimensional projective spaces $\mathbf{P}_n = \mathbf{P}(\mathbf{C}^{n+1})$. The mappings

$$\varphi_i : \mathbf{P}_n \setminus \mathbf{P}(H^i_*) \ni \mathbf{C}(z_0, \dots, z_{i-1}, 1, z_{i+1}, \dots, z_n) \longrightarrow$$

$$\longrightarrow (z_0, \dots, z_{i-1}, z_{i+1}, \dots, z_n) \in \mathbf{C}^n, \ i = 0, \dots, n ,$$

---

[6] Since $X \setminus H_* \ni \zeta \longrightarrow (\zeta/\lambda_H(\zeta), \ \lambda_H(\zeta)) \in H \times (\mathbf{C} \setminus 0)$ is the inverse of γ.

[7] Because $\mathbf{P}(T)\tilde{} = T$ and the mapping $\mathbf{P}(T) \hookrightarrow \mathbf{P}(X)$ is holomorphic, by the above criterion. (See C. 3.1, the remark)

[8] If $\varphi \in L_0(X,Y)$, then $\tilde{\varphi} \circ \alpha^X = \alpha^Y \circ \varphi$, and hence, by the above criterion, the mapping $\tilde{\varphi} : \mathbf{P}(X) \longrightarrow \mathbf{P}(Y)$ is holomorphic. Obviously, one can check this directly using charts.

where $H^i = \{(z_0, \ldots, z_n) : z_i = 1\}$, form an atlas of $\mathbf{P}$ [9] . They are called the *canonical charts* (or *Cartesian coordinate systems*) on the projective space $\mathbf{P}_n$. Note that if $i \neq j$, then the composition

$$\varphi_j \circ \varphi_i^{-1} : \{(t_0, \ldots, t_{i-1}, t_{i+1}, \ldots, t_n) : t_j \neq 0\} \longrightarrow$$

$$\longrightarrow \{(u_0, \ldots, u_{j-1}, u_{j+1}, \ldots, u_n) : u_i \neq 0\}$$

(i.e., the change from the i-th to the j-th Cartesian coordinate system) is given by

$$u_0 = t_0/t_j, \ldots, u_i = 1/t_j, \ldots, u_n = t_n/t_j .$$

For any $\lambda \in \mathbf{P}_n$, every $(z_0, \ldots, z_n) \in \lambda \setminus 0$ is called a sequence of *homogeneous coordinates* of λ. Thus, each $(z_0, \ldots, z_n) \in \mathbf{C}^{n+1} \setminus 0$ is a sequence of homogeneous coordinates of $\mathbf{C}(z_0, \ldots, z_n) \in \mathbf{P}_n$. Two sequences of homogeneous coordinates of one point differ by a factor from $\mathbf{C} \setminus 0$, and vice versa [10] .

Let W be a property of elements of $\mathbf{C}^{n+1} \setminus 0$ such that $(t \in \mathbf{C} \setminus 0,\ W(z)) \implies W(tz)$. A subset of the space $\mathbf{P}_n$ is said to be *given* (or *expressed*) *in homogeneous coordinates by the property W* if it is of the form $\{\mathbf{C}z : W(z)\}$. The images of such a set under the canonical charts are the sets

$$\{(z_0, \ldots, z_{i-1}, z_{i+1}, \ldots, z_n) : W(z_0, \ldots, z_{i-1}, 1, z_{i+1}, \ldots, z_n)\} \subset \mathbf{C}^n,$$

$$i = 0, \ldots, n .$$

Let $E \subset \mathbf{P}_n$, and suppose that a mapping $F : E^\sim \setminus 0 \longrightarrow Z$ satisfies the homogeneity condition $F(tz) = F(z)$ for $t \in \mathbf{C} \setminus 0$ and $z \in E^\sim \setminus 0$. Then we have the mapping $E \ni \mathbf{C}z \longrightarrow F(z) \in Z$, which is said to be *expressed* (or *given*) *in homogeneous coordinates* by the mapping F. For example, the canonical chart φ_i is given in homogeneous coordinates by

$$\{z_i \neq 0\} \ni (z_0, \ldots, z_n) \longrightarrow (z_0/z_i, \ldots, z_{i-1}/z_i, z_{i+1}/z_i, \ldots, z_n/z_i) \in \mathbf{C}^n .$$

According to the criterion from n° 2, a mapping of a locally analytic subset $\Gamma \subset \mathbf{P}_n$ into a manifold M is holomorphic if and only if it can be expressed in

---

[9] Because φ_i^{-1} is the composition of the chart α_{Hi} with the biholomorphic mapping $\mathbf{C}^n \ni (z_0, \ldots, z_{i-1}, z_{i+1}, \ldots, z_n) \longrightarrow (z_0, \ldots, z_{i-1}, 1, z_{i+1}, \ldots, z_n) \in H$ and $\{\alpha_{Hi}\}$ is an inverse atlas, (since $\bigcap H_*^i = 0$).

[10] Therefore $\mathbf{P}_n$ can be identified with the set of equivalence classes in $\mathbf{C}^{n+1} \setminus 0$, where two elements in $\mathbf{C}^{n+1} \setminus 0$ are considered equivalent if they differ by a factor from $\mathbf{C} \setminus 0$.

homogeneous coordinates by a holomorphic mapping (of the locally analytic subset $\Gamma^{\sim} \setminus 0$ into the manifold M).

4. Let $X_1, \ldots, X_k$ be vector spaces. The manifold $\mathbf{P}(X_1) \times \ldots \times \mathbf{P}(X_k)$ is called a *multiprojective space*. The mapping

$$\mathbf{s}: \ \mathbf{P}(X_1) \times \ldots \times \mathbf{P}(X_k) \ni (\mathbf{C}z_1, \ldots, \mathbf{C}z_k) \longrightarrow \mathbf{C}(z_1 \otimes \ldots \otimes z_k) \in$$
$$\mathbf{P}(X_1 \otimes \ldots \otimes X_k)$$

is an embedding of the multiprojective space $\mathbf{P}(X_1) \times \ldots \times \mathbf{P}(X_k)$ into the projective space $\mathbf{P}(X_1 \otimes \ldots \otimes X_k)$. It is called the *Segre embedding*.

Indeed, it is injective ([11]). Next, it is holomorphic, because $\alpha^{X_1} \times \ldots \times \alpha^{X_k}$ is a surjective submersion and the composition $\mathbf{s} \circ (\alpha^{X_1} \times \ldots \times \alpha^{X_k}) = \alpha^{X_1 \otimes \ldots \otimes X_k} \circ \tau$ is holomorphic, where $\tau: \ (z_1, \ldots, z_k) \longrightarrow z_1 \otimes \ldots \otimes z_k$ (see C. 4.2). Therefore, by Remmert's theorem (see V. 5.1), its range S is analytic. Now, for any $\alpha, \beta \in S$ there is a biholomorphic mapping $g: \ \mathbf{P}(X_1 \otimes \ldots \otimes X_k) \longrightarrow \mathbf{P}(X_1 \otimes \ldots \otimes X_k)$ such that $g(S) = S$ and $g(\alpha) = \beta$. In fact, let $\alpha = \mathbf{C}(a_1 \otimes \ldots \otimes a_k)$, $\beta = \mathbf{C}(b_1 \otimes \ldots \otimes b_k)$. Take automorphisms $f_i \in L_0(X_i, X_i)$ such that $f_i(a_i) = b_i$. Then the mapping $g = \tilde{f}$, where $f = f_1 \otimes \ldots \otimes f_k$ is biholomorphic, $g(\alpha) = \beta$, and $g(S) = S$, because $\mathbf{s} \circ (\tilde{f_1} \times \ldots \times \tilde{f_k}) = \tilde{f} \circ \mathbf{s}$. Consequently, all points of S are regular and of the same dimension (see IV. 2.1). This means that S is a submanifold. Thus (see V. 1, corollary 2 of theorem 2) $\mathbf{s}$ is a biholomorphic mapping onto the submanifold S.

An example of a compact manifold that cannot be embedded in a projective space is the two–dimensional torus that contains only one one–dimensional subtorus (see C. 3.21). For its one–dimensional irreducible analytic subsets are mutually disjoint (see V. 7.2), while any two–dimensional submanifold T of the projective space must contain two such sets that are distinct but intersect each other. In fact, for each hyperplane $H \not\supset T$, the intersection $T \cap H$ is of constant dimension 1 (because $\dim_z (T \cap H) \geq 1$ for $z \in T \cap H$; see III. 4.6, inequality (*)). Now, take a point $a \in T$, a hyperplane $H \ni a$, $H \not\supset T$, a simple component $V \ni a$ of the intersection $T \cap H$, a point $b \in V \setminus a$, a hyperplane $H' \ni a$, $H' \not\ni b$, and a simple component $V' \ni a$ of the intersection $T \cap H'$. Then we have $V \cap V' \neq \emptyset$ and $V \neq V'$.

The tori that can be embedded in projective spaces are called *abelian manifolds*. Later on (in 14.3) we will prove that every one–dimensional torus is an abelian manifold.

([11]) This follows directly from the theorem on bases of tensor products.

§3. The projective closure of a vector space

1. By the *projective closure* (or *compactification*) of an n–dimensional vector space X we mean the n–dimensional projective space

$$\bar{X} = \mathbf{P}(\mathbf{C} \times X) \,,$$

in which the complement of the hyperplane $X_\infty = \mathbf{P}(0 \times X)$, called the *hyperplane at infinity*, is identified with the space X via the biholomorphic mapping

$$\iota^X : X \ni z \longrightarrow \mathbf{C}(1, z) \in \bar{X} \setminus X_\infty \quad (^{12}) \,.$$

Thus the vector space X is an open dense subset of its projective closure.

In the one–dimensional case ($n = 1$), the set X_∞ consists of the single element $0 \times X$, which will be also denoted by X_∞.

In particular, we have $\overline{\mathbf{C}^n} = \mathbf{P}_n$ via the identification $\mathbf{C}^n \ni z \longrightarrow \mathbf{C}(1, z) \in \mathbf{P}_n \setminus \mathbf{C}_\infty^n$ $(^{13})$, i.e., the point $(z_1, \ldots, z_n) \in \mathbf{C}^n$ is identified with the point in $\mathbf{P}_n$ whose homogeneous coordinates are $(1, z_1, \ldots, z_n)$. The hyperplane at infinity $\mathbf{C}_\infty^n = (\mathbf{C}^n)_\infty$ is expressed in the homogeneous coordinates by the equation $x_0 = 0$.

The projective closure $\bar{\mathbf{C}} = \mathbf{P}_1$ is identified with the Riemann sphere $\bar{\mathbf{C}} = \mathbf{C} \cup \infty$ (see C. 3.19) via the natural biholomorphic mapping $\iota : \mathbf{C} \cup \infty \longrightarrow \mathbf{P}_1$ defined by

$$\iota(z) = \mathbf{C}(1, z) \text{ for } z \in \mathbf{C}, \text{ and } \iota(\infty) = \mathbf{C}_\infty = 0 \times \mathbf{C} \,.$$

So it is an extension of the identification $\iota^{\mathbf{C}}$. The mapping ι is biholomorphic, because (see C. 3.19) the compositions $\iota \circ \varphi^{-1} : \mathbf{C} \ni z \longrightarrow \mathbf{C}(1, z) \in \mathbf{P}_1$ and $\iota \circ \psi^{-1} : \mathbf{C} \ni z \longrightarrow \mathbf{C}(z, 1) \in \mathbf{P}_1$ are holomorphic (see, e.g., V. 1.1, corollary 2 from theorem 2).

Every projective space $\mathbf{P}$ is, up to an isomorphism, the projective closure of a vector space. Moreover, as the hyperplane at infinity and as the zero element one can take any hyperplane in the space $\mathbf{P}$ and any point that does not belong to this hyperplane. Namely, if $\mathbf{P} = \mathbf{P}(M)$ is an n–dimensional projective space (where M is a vector space of dimension $n + 1$), $H \subset \mathbf{P}$ is a hyperplane, $\lambda \in \mathbf{P} \setminus H$, and X is an n–dimensional vector space, then there is an isomorphism $\chi : \mathbf{P} \longrightarrow \bar{X}$ such that $\chi(H) = X_\infty$ and $\chi(\lambda) = 0$ (see B. 6.12).

$(^{12})$ It is the composition of the biholomorphic mapping $X \ni z \longrightarrow (1, z) \in 1 \times X$ with the chart $\alpha_{1 \times X}$.

$(^{13})$ It is the inverse of the 0-th canonical chart; see 2.3.

2. Let X and Y be vector spaces of the same dimension.

If $\varphi : \bar{X} \longrightarrow \bar{Y}$ is an isomorphism such that $\varphi(X) = Y$, then the restriction $\varphi_X : X \longrightarrow Y$ is an affine isomorphism. Conversely, every affine isomorphism of the space X onto the space Y can be (uniquely) extended to an isomorphism of the projective closures of these spaces.

Indeed, we have $\varphi = \tilde{\Phi}$, where $\Phi \in L_0(\mathbf{C} \times X, \mathbf{C} \times Y)$, and $\Phi(1, z) = (\gamma(z), \psi(z))$ in X, where $\gamma : X \longrightarrow \mathbf{C}$ and $\psi : X \longrightarrow Y$ are affine mappings. Now, if $z \in X$, then $\gamma(z) \neq 0$ [14] . Therefore γ must be a constant $c \neq 0$. Hence $\Phi(\mathbf{C}(1, z)) = \mathbf{C}(1, c^{-1}\psi(z))$ in X, and so (after identifications) we have $\varphi_X = c^{-1}\psi$. Thus the mapping $\varphi_X : X \longrightarrow Y$ is affine and, in view of its injectivity, it is an affine isomorphism. Conversely, if $\psi : X \longrightarrow Y$ is an affine isomorphism, then $\psi = b + \psi_0$, where $b \in Y$ and $\psi_0 \in L_0(X, Y)$. Then the mapping $\Phi : \mathbf{C} \times X \ni (t, z) \longrightarrow (t, bt + \psi_0(z)) \in \mathbf{C} \times Y$ is an isomorphism and $\Phi(\mathbf{C}(1, z)) = \mathbf{C}(1, \psi(z))$ in X. Thus we have $\psi(z) = \tilde{\Phi}(z)$ in X (after identifications).

3. In what follows, let X be an n–dimensional vector space.

If $N \subset X$ is a k–dimensional affine subspace, then its closure $\bar{N}$ in $\bar{X}$ is a k–dimensional projective subspace. Conversely, if $L \subset \bar{X}$ is a k–dimensional projective subspace which is not contained in X_∞, then $L \cap X \subset X$ is a k–dimensional affine subspace that is dense in L [15] .

Indeed, owing to the equivalence $L \not\subset X_\infty \Longleftrightarrow L^\sim \not\subset 0 \times X$ (see B. 6.10), the equality

$$(1) \qquad\qquad L^\sim \cap (1 \times X) = 1 \times N$$

establishes a bijection between the set of projective subspaces $L \subset \bar{X}$, $L \not\subset X_\infty$, of dimension k and the set of k–dimensional affine subspaces $N \subset X$. Moreover, the equality implies that $N = L \cap X$ (after identifications, since $\iota^X(N) = L \setminus X_\infty$) and $L = \bar{N}$ (because the set $L \cap X_\infty$ is nowhere dense in L; see IV. 2.8, propositions 3 and 2).

Let $T \subset X$ be a subspace. Then the projective closure $\bar{T}$ is equal to the closure of the set T in $\bar{X}$ [16] . Furthermore, the identifications are consistent:

---

[14] For otherwise $\Phi(1, z) \in 0 \times Y$, and thus (after identifications) we would get $\varphi(z) \in Y_\infty$, contrary to our assumption.

[15] Thus $N \longrightarrow \bar{N}$ and $L \longrightarrow L \cap X$ are mutually inverse bijections. We will prove in 4.5 that they are biholomorphic.

[16] Since $\bar{T}$ is a projective subspace of the space $\bar{X}$ that contains T. (see n° 1 and B. 6.1.)

$\iota^T \subset \iota^X$. Finally, we have the formula $T_\infty = \bar{T} \cap X_\infty$ ([17]).

We have the bijections $\mathbf{G}_k(X) \ni T \longrightarrow T_\infty \in \mathbf{G}_{k-1}(X_\infty)$, $k = 1, \ldots$
$\ldots, n$ ([18]), and in particular, the bijection $\mathbf{P}(X) \ni \lambda \longrightarrow \lambda_\infty \in X_\infty$ ([19]). So
$X_\infty = \{\lambda_\infty : \lambda \in \mathbf{P}(X)\} = \{z_\infty : z \in X \setminus 0\}$, where $z_\infty = (\mathbf{C}z)_\infty = 0 \times \mathbf{C}z$
for $z \in X \setminus 0$.

For every affine subspace $N \subset X$, set $N_\infty = \bar{N} \cap X_\infty$. By (1) we have
$0 \times N_* = \bar{N}\tilde{} \cap (0 \times X) = N\tilde{}_\infty$. This implies that:

If $M, N \subset X$ are affine subspaces of the same dimension, then
$$M \parallel N \Longleftrightarrow M_\infty = N_\infty .$$

4. If $T \subset X$ is a subspace, then we have a base of neighbourhoods
$\{\Omega_K\}_{K>0}$ of the set T_∞ in $\bar{X}$ given by
$$\Omega_K \cap X = \{t + u : |t| \geq K(1 + |u|)\},$$
$$\Omega_K \cap X_\infty = \{\{\lambda_\infty : \lambda \subset \{t + u : |t| \geq K|u|\}\}$$

($t \in T$, $u \in U$, $\lambda \in \mathbf{P}(X)$), where U is a linear complement of T (and
a norm on X is given). In particular, if T is a line, then the above gives
us a base of neighbourhoods in $\bar{X}$ of any point of ([20]), whereas the sets
$\{z \in X : |z| \geq K\} \cup X_\infty$, $K > 0$ form a base of neighbourhoods of the set
X_∞ in $\bar{X}$.

Indeed, let us introduce the norm $|(\tau, z)| = |\tau| + |z|$ on the space $\mathbf{C} \times X$.
Since $\mathbf{C} \times U$ is a linear complement of $0 \times T$, the sets
$$\Omega_K = \{\lambda \in \mathbf{P}(\mathbf{C} \times X) : \lambda \subset \{(\tau, t + u) : |\tau| + |u| \leq K^{-1}|t|\}\}, \quad K > 0 ,$$
(see B. 6.12) form a base of neighbourhoods of the set $T_\infty = \mathbf{P}(0 \times T)$ in
$\bar{X} = \mathbf{P}(\mathbf{C} \times X)$.

5. For any affine hyperplane $H \not\ni 0$ of the space X, we have the chart
$\beta_H = (\alpha_{\mathbf{C} \times H})^{-1} : \bar{X} \setminus \overline{H^*} \longrightarrow \mathbf{C} \times H$ on the projective closure $\bar{X}$. Observe
that for $\lambda \in \mathbf{P}(X)$,

(*)   (the domain of the chart β_H contains λ_∞) $\Longleftrightarrow \lambda \not\subset H_*$.

([17]) Because $(0 \times T)\tilde{} = (\mathbf{C} \times T)\tilde{} \cap (0 \times X)\tilde{}$ (see B. 6.10).

([18]) Due to the fact that the k–th mapping is the composition of the bijection $\mathbf{G}_k(X) \ni$
$T \longrightarrow (0 \times T) \in \mathbf{G}_k(0 \times X)$ and the bijection $\mathbf{G}_{k-1}(0 \times X) \ni 0 \times T \longrightarrow (0 \times T)\tilde{} \in$
$\mathbf{G}_{k-1}(\mathbf{P}(0 \times X))$ (see B. 6.9 and 12).

([19]) These mappings are even biholomorphic (see footnote ([18]) and 4.5 below).

([20]) This implies that a sequence $x_r \in X \setminus 0$ converges to an element $\lambda_\infty \in X_\infty$ (where
$\lambda \in \mathbf{P}(X)$ exactly when $|x_r| \longrightarrow \infty$ and $\mathbf{C}x_r \longrightarrow \lambda$).

It follows that the domains of the charts β_H (and even the domains of some charts $\beta_{H'}$, provided that $\bigcap H_*^\iota = 0$) cover the hyperplane at infinity X_∞. Therefore the charts β_H (or $\beta_{H'}$, as above), combined with the identification $(\iota^X)^{-1}$, form an atlas for the manifold $\bar{X}$. Furthermore, (after the identification through ι^X) we obtain the formulae (see 2.1)

$$(**) \quad \begin{cases} \beta_H(z) = \left(1/\lambda_H(z), z/\lambda_H(z)\right) \ \text{ for } \ z \in X \setminus \overline{H_*}, \\ \beta_H(z_\infty) = (0, z) \text{ if } z \in H, \ \text{ and } \ X_\infty \setminus \overline{H_*} = \{z_\infty : \ z \in H\}, \\ \beta_H^{-1}(t, z) = \begin{cases} z/t \ \text{ for } \ t \neq 0, \ z \in H, \\ z_\infty \ \text{ for } \ t = 0, \ z \in H. \end{cases} \end{cases}$$

They imply that, endowing X with a norm, and $\mathbf{C} \times H_8$ with the norm $|(t, z)| = |t| + |z|$, we have (see C. 3.2)

$$(***) \qquad \varrho_{\beta_H}(z, X_\infty \setminus \overline{H_*}) = |\lambda_H(z)|^{-1} \ \text{ for } \ z \in X \setminus H_* .$$

6. Let M be a manifold, and let Z be a closed subset of M. We say that a set $E \subset (M \setminus Z) \times X$ *satisfies the condition* (r) *near* Z if for each $a \in Z$ the following condition is satisfied: for some (and hence for each) chart φ of M, with domain $\Omega \ni a$ and a normed affine model, there is a neighbourhood $\Delta \subset \Omega$ of the point a such that (after fixing a norm on X)

$$(r) \qquad E_\Delta \subset \{(u, z) \in (\Delta \setminus Z) \times X : \ |z| \leq K\varrho_\varphi(u, Z \cap \Omega)^{-s}\}$$

for some $K, s > 0$. We say that the *mapping* $f : M \setminus Z \longrightarrow X$ *satisfies the condition* (r) *near* Z if its graph satisfies the condition (r) near Z; then the inclusion (r) means that

$$(r) \qquad |f(u)| \leq K\varrho_\varphi(u, Z \cap \Omega)^{-s} \ \text{ for } \ u \in \Delta \setminus Z .$$

LEMMA 1. *Let E be a closed subset of $(M \setminus Z) \times X$ such that the natural projection $E \longrightarrow M \setminus Z$ is proper. If the pair $\bar{E}, M \times X_\infty$ (where $\bar{E}$ is the closure of E in $M \times \bar{X}$) satisfies the condition of regular separation, then the set E satisfies the condition* (r) *near* Z.

PROOF . Let $a \in Z$. Let φ be a chart for the manifold M, with domain $\Omega \ni a$, modelled on a normed vector space L. Take any $b \in X_\infty$. Then $b = \lambda_\infty$, where $\lambda \in \mathbf{P}(X)$ (see n° 3). Fix a norm on X and take an affine hyperplane $H \not\ni 0$ of X such that $H_* \not\supset \lambda$. Then $|\lambda_H(z)| > \varepsilon|z|$ on $\lambda \setminus 0$ for some $\varepsilon > 0$.

The set $G = \{\mathbf{C}(t,z) : |\lambda_H(z)| > \varepsilon|z|\} \subset \bar{X}$ is open (see B. 6.1), it contains b, and we have

(#) $\qquad\qquad |\lambda_H(z)| > \varepsilon|z|$ for $z \in G \cap X$.

Now, $\gamma = \varphi \times \beta_H$ is a chart of $M \times \bar{X}$, modelled on the affine space $L \times \mathbf{C} \times H$ with the norm $L \times \mathbf{C} \times H_* \ni (v,t,z) \longrightarrow |v| + |t| + |z|$. Its domain $\Omega \times (\bar{X} \setminus \bar{H}_*)$ contains (a,b), in view of $(*)$. According to the condition of regular separation (see IV. 7.1), we have

$$\varrho_\gamma\big((u,z), \Omega \times (X_\infty \setminus \overline{H_*})\big) \geq c\varepsilon^{-1} \varrho_\gamma\Big((u,z), \bar{E} \cap \big(\Omega \times (X_\infty \setminus \overline{H_*})\big)\Big)^p$$
$$\text{if } (u,z) \in E \cap (\Delta \times U),$$

where $c, p > 0$, for some neighbourhoods $\Delta \subset \Omega$ and $U \subset G \setminus \overline{H_*}$ of the points a and b, respectively [21]. By the assumptions, $\bar{E} \cap (M \times X_\infty) \subset Z \times \bar{X}$ (see B. 5.2 and n° 4); hence it follows (in view of $\varrho_\gamma((u,z),(u',z')) = \varrho_\varphi(u,u') + \varrho_{\beta_H}(z,z')$) that $\varrho_{\beta_H}(z, X_\infty \setminus \overline{H_*}) \geq c\varepsilon^{-1} \varrho_\varphi(u, Z \cap \Omega)^p$ for $(u,z) \in E \cap (\Delta \times U)$. Therefore – according to $(***)$ and (#) – we get

(##) $\qquad 1/|z| \geq c\varrho_\varphi(u, Z \cap \Omega)^p$ for $(u,z) \in E \cap (\Delta \times U)$.

Now, a finite number of such neighbourhoods U covers X_∞ and one can find common Δ, c, p for them. These neighbourhoods, together with a ball $\{|z| < R\}$, cover the space $\bar{X}$ (see n° 4). Taking a smaller Δ, so that $c\varrho_\varphi(u, Z \cap \Omega) < 1/R$ in Δ, we obtain the inequality (##) for $(u,z) \in E \cap (\Delta \times \bar{X}) = E_\Delta$ (letting $1/0 = \infty$). This yields the conclusion of the lemma.

LEMMA 2. *Let L be a vector space. Then a closed set $E \subset L \times X$ satisfies the condition* (r) *near L_∞ if and only if (after fixing norms) we have*

(###) $\qquad E_{\{|u| \geq R\}} \subset \{(u,z) : |z| \leq C|u|^s\}$ *for some $R, C, s > 0$* .

PROOF . Suppose that the set E satisfies the condition (r) near L_∞. Let $a \in L_\infty$. Then the domain $\bar{L} \setminus \overline{H_*}$ of some chart β_H of the manifold $\bar{L}$ contains a (see n° 5). Thus, for some neighbourhood U of a, we have the inequality $|z| \leq K\varrho_{\beta_H}(u, L_\infty \setminus \overline{H_*})^{-s}$ in the set E_U for some $K, s > 0$. Therefore, in view of $(***)$, there is a $C > 0$ such that

$$|z| \leq C|u|^s \text{ for } (u,z) \in E_U .$$

---

[21] For, in the case when $(a,b) \notin \bar{E} \cap (M \times X_\infty)$, one must have $(a,b) \notin \bar{E}$. If one chooses a sufficiently small neighbourhood $\Delta \times U$, it is disjoint from $\bar{E}$.

Now a finite number of neighbourhoods U cover L_∞ and one can find common C, s for such neighbourhoods. They cover some set $\{|u| \geq R\}$ (see n° 4), and so we obtain the inclusion (###). Conversely, suppose that the inclusion (###) holds. Let $a \in L_\infty$. Then $a = \lambda_\infty$, where $\lambda \in \mathbf{P}(L)$ (see n° 3). Consider an affine hyperplane $H \not\ni 0$ of the space L such that $H_* \not\supset \lambda$. Then, according to (*), the domain $\bar{L} \setminus \bar{H}_*$ of the chart β_H contains the point a, and $|\lambda_H(u)| \geq 2\varepsilon|u|$ on λ for some $\varepsilon > 0$. The set $\Omega = \{\mathbf{C}(t, u) : |\lambda_H(u)| > \varepsilon|u|\} \subset \bar{L}$ is open (see B. 6.1), it contains a, and the inequality $|\lambda_H(u)| > \varepsilon|u|$ holds in $\Omega \cap L$. The set $\Delta = (\{|u| \geq R\} \cup L_\infty) \cap \Omega \setminus \bar{H}_*$ is a neighbourhood of the point a (see n° 4), and so, by (###) and (***), for $(u, z) \in E_\Delta$ we have the inequality $|z| \leq C|u|^s \leq C\varepsilon^{-s}|\lambda_H(u)|^s = C\varepsilon^{-s}\varrho_{\beta_H}(u, L_\infty \setminus \overline{H_*})^{-s}$. Thus, in conclusion, the set E satisfies the condition (r) near L_∞.

§4. Grassmann manifolds

Let X be an n–dimensional vector space.

1. Consider the Grassmann space $\mathbf{G}_k(X)$, where $0 \leq k \leq n$. The homeomorphisms introduced in B. 6.8

$$\varphi_{UV} : L(U, V) \ni f \longrightarrow \hat{f} \in \Omega(V), \quad \text{where} \quad U \in \mathbf{G}_k(X) \text{ and } V \in \Omega(U)$$

constitute an inverse atlas on $\mathbf{G}_k(X)$ which defines the *natural structure of a $k(n-k)$–dimensional complex manifold on the Grassmann space $\mathbf{G}_k(X)$.*

In fact, in order to show that any composition $(\varphi_{U',V'})^{-1} \circ \varphi_{UV}$ is holomorphic, it is enough to check that its graph is analytic in $L(U, V) \times L(U', V')$ (see V. 1, corollary 3 of theorem 2; and II. 3.4). Now, the graph is the set

$$\{(f, g) : \hat{f} = \hat{g}\} = \{(f, g) : \operatorname{rank}(f^\bullet \oplus g^\bullet) \leq k\} \quad (^{22}),$$

where $f^\bullet : U \ni u \longrightarrow u + f(u) \in X$ and $g^\bullet : U' \ni u \longrightarrow u + g(u) \in X$. This is so because $\operatorname{im} f^\bullet = \hat{f}$, $\operatorname{im} g^\bullet = \hat{g}$, and $\operatorname{im}(f^\bullet \oplus g^\bullet) = \hat{f} + \hat{g}$. Therefore the graph is analytic, being equal to the inverse image of an analytic subset under the affine mapping $L(U, V) \times L(U', V') \ni (f, g) \longrightarrow f^\bullet \oplus g^\bullet \in L(U \times U', X)$ (see II. 3.2).

$(^{22})$ For $F \in L(U, X)$, $G \in L(U', X)$, the mapping $F \oplus G \in L(U \times U', X)$ is defined by $(F \oplus G)(u, u') = F(u) + G(u')$. We have the isomorphism $L(U, X) \times L(U', X) \ni (F, G) \longrightarrow F \oplus G \in L(U \times U', X)$.

This can also be checked directly, without using the analytic graph theorem. In fact, the mapping $\Phi : B(U') \times (V')^k \ni (u,v) \longrightarrow f_{uv} \in L(U',V')$ is holomorphic by the lemma from n° 2 below. Take a basis $a_1, \ldots, a_k$ of U, and let $p : X \longrightarrow U'$, $q : X \longrightarrow V'$ be the projections corresponding to the direct sum $X = U' + V'$. We have the affine mappings

$$P : L(U,V) \ni f \longrightarrow (p(a_1 + f(a_1)), \ldots, p(a_k + f(a_k))) \in (U')^k,$$
$$Q : L(U,V) \ni f \longrightarrow (q(a_1 + f(a_1)), \ldots, q(a_k + f(a_k))) \in (V')^k.$$

Now, $(\varphi_{U'V'})^{-1} \circ \varphi_{UV} = \Phi \circ (P,Q)$. Indeed, the domains of both sides coincide $(^{23})$, and if $f \in P^{-1}(B(U'))$ and $g = \Phi(P(f), Q(f))$, then $g\big(p(a_i + f(a_i))\big) = q(a_i + f(a_i))$, which means that $a_i + f(a_i) \in \hat{g}$. Therefore $\hat{g} = \hat{f}$, and thus $g = (\varphi_{U'V'})^{-1}(\varphi_{UV}(f))$.

The natural structure of the manifold $\mathbf{G}_1(X)$ coincide with that defined earlier (see 2.1) for the projective space $\mathbf{P}(X) = \mathbf{G}_1(X)$. Indeed, for any affine hyperplane $H \not\ni 0$ and any line $\lambda \in \Omega(H_*)$, we have the affine isomorphism $\gamma_{\lambda H} : L(\lambda, H_*) \ni f \longrightarrow c + f(c) \in H$, where $c = \lambda \cap H$. Then $\varphi_{\lambda H_*} = \alpha_H \circ \gamma_{\lambda H}$.

2. We have the following

LEMMA . *Let Y be a vector space. The mapping $B_n(X) \times Y^n \ni (x,y) \longrightarrow f_{xy} \in L(X,Y)$, where f_{xy} is defined by $f_{xy}(x_i) = y_i$, $x = (x_1, \ldots, x_n)$, $y = (y_1, \ldots, y_n)$, is a (holomorphic) submersion.*

Indeed, it follows from the proof of the lemma in B. 6.3 that this mapping is holomorphic. (Since the mapping $B_n(X) \ni x \longrightarrow F_x^{-1} \in L(Y^n, L(X,Y))$ is holomorphic; see C. 1.12.) The differential of the mapping is surjective at each point $(x,y) \in B_n(X) \times Y^n$, because the linear mapping $Y^n \ni w \longrightarrow f_{xw} \in L(X,Y)$ is surjective.

The mapping defined in B. 6.1

$$\alpha = \alpha_k = \alpha^X = \alpha_k^X : B_k(X) \ni x \longrightarrow \sum_1^k \mathbb{C}x_i \in \mathbf{G}_k(X)$$

is a surjective submersion.

Indeed, for any chart $\varphi_{U,V}$, take the biholomorphic mapping

$$\vartheta : B_k(U) \times V^k \ni (u,v) \longrightarrow u + v \in \alpha^{-1}(\Omega(V)) .$$

$(^{23})$ Since $\hat{f} \in \Omega(V') \Longleftrightarrow P(f) \in B(U')$, because both conditions mean that $p_{\hat{f}} : \hat{f} \longrightarrow U'$ is an isomorphism.

The composition

$$\varphi_{UV}^{-1} \circ \alpha_{\alpha^{-1}(\Omega(V))} \circ \vartheta : \ B_k(U) \times V^k \ni (u,v) \longrightarrow f_{uv} \in L(U,V)$$

is a submersion, by the lemma. Therefore the restriction $\alpha_{\alpha^{-1}(\Omega(V))}$ is also a submersion. Since the sets $\alpha^{-1}(\Omega(V))$ cover $B_k(X)$ (see B. 6.8), the mapping α is a submersion.

As a consequence, we obtain the following characterizations of submanifolds and (locally) analytic subsets of the space $\mathbf{G}_k$, as well as holomorphic mappings of those objects into manifolds (see C. 4.2, and also II. 3.4 and V. 3.1):

A set $\Gamma \subset \mathbf{G}_k(X)$ is a submanifold or a (locally) analytic subset (of constant dimension) if and only if the set $\alpha^{-1}(\Gamma) \subset B_k(X)$ is a submanifold or a (locally) analytic subset (of constant dimension), respectively. Then, if M is a manifold, a mapping $f : \ \Gamma \longrightarrow M$ is holomorphic if and only if the mapping $f \circ \alpha : \ \alpha^{-1}(\Gamma) \longrightarrow M$ is holomorphic.

If $T \subset X$ is a subspace of dimension $\geq k$, then the space $\mathbf{G}_k(T) \subset \mathbf{G}_k(X)$ is a submanifold and the induced manifold structure on $\mathbf{G}_k(T)$ coincides with the natural one. Indeed, the set $B_k(T) = \alpha^{-1}(\mathbf{G}_k(T)) \subset B_k(X)$ is a submanifold and the mapping $\kappa : \ \mathbf{G}_k(T) \hookrightarrow \mathbf{G}_k(X)$ is holomorphic, since $\kappa \circ \alpha^T = (\alpha^X)_{B_k(T)}$. (See the remark in C. 3.10.)

Any isomorphism of Grassmann spaces is biholomorphic: if Y is a vector space and $\varphi : \ X \longrightarrow Y$ is an isomorphism, then $\tilde{\varphi} = \varphi_{(k)} : \mathbf{G}_k(X) \longrightarrow \mathbf{G}_k(Y)$ is a biholomorphic mapping. (It is enough to observe that the composition $\tilde{\varphi} \circ \alpha^X = \alpha^Y \circ (\varphi \times \ldots \times \varphi)_{B_k(X)}$ is holomorphic.)

The bijection defined in B. 6.4

$$\tau = \tau_k = \tau^X = \tau_k^X : \ \mathbf{G}_k(X) \ni V \longrightarrow V^\perp \in \mathbf{G}_{n-k}(X)$$

is biholomorphic.

Indeed, it follows from the proof in B. 6.4 of the continuity of the bijection τ that it is holomorphic (and hence also biholomorphic; see, e.g., V. 1, corollary 2 of theorem 2): namely, the composition $\tau \circ \alpha$ is holomorphic because, according to the lemma, the mappings $W \ni z \longrightarrow f_z \in B_{n-k}(X^*)$ are holomorphic and so are the restrictions $(\tau \circ \alpha)_W$.

3. If $U \subset X$ is a subspace of dimension $r \leq k$, then the Schubert cycle $\mathbf{S}^k(U) = \mathbf{S}^k(U,X) \subset \mathbf{G}_k(X)$ is a closed submanifold of dimension

$(n - k)(k - r)$. This is a consequence of the first formula in (#) from B. 6.5 (in view of the properties from n° 2).

If $0 \leq k \leq l \leq n$, then the subset of the manifold $\mathbf{G}_k(X) \times \mathbf{G}_l(X)$ given by

$$\mathbf{S}_k^l(X) = \{(U, V): \ U \subset V\}$$

is a closed submanifold of dimension $k(l - k) + l(n - l)$.

We are going to prove this together with the following

PROPOSITION . *Let* $0 \leq r, k, l \leq n$, *and* $k + l - n \leq r \leq k, l$. *Then the subset of the manifold* $\mathbf{G}_k(X) \times \mathbf{G}_l(X)$:

$$A = A_r^{kl} = \{(U, V): \ \dim U \cap V = r\}$$

is an analytically constructible submanifold of dimension $r(n-r)+(n-k)(k-r) + (n - l)(l - r)$ *and the mappings*

$$A \ni (U, V) \longrightarrow U \cap V \in \mathbf{G}_r(X) \ ,$$
$$A \ni (U, V) \longrightarrow U + V \in \mathbf{G}_{k+l-r}(X)$$

are holomorphic.

PROOF . First observe that the set $\mathbf{S}_p^q(X)$ $(0 \leq p \leq q \leq n)$ is analytic. Indeed (see II. 3.4), its inverse image under the surjective submersion $\alpha_p \times \alpha_q :$ $B_p(X) \times B_q(X) \longrightarrow \mathbf{G}_p(X) \times \mathbf{G}_q(X)$ (see C. 4.2 and n° 2), that is, the set

$$(\alpha_p \times \alpha_q)^{-1}\left(\mathbf{S}_p^q(X)\right) = \{(x_1, \ldots, x_p, y_1, \ldots, y_q) \in B_p(X) \times B_q(X) :$$
$$x_i \wedge y_1 \wedge \ldots \wedge y_q = 0, \ i = 1, \ldots, p\} \ ,$$

is analytic in $B_p(X) \times B_q(X)$. In particular, the set $\mathbf{S}_q^p(X)$ is closed. Since $\mathbf{S}_p^q(X) = A_p^{pq}$, the remaining properties of that set will be implied by the proposition.

It follows that the sets

$$\Sigma_i = \{(U, V, Z): \ Z \subset U \cap V\} \subset \mathbf{G}_k(X) \times \mathbf{G}_l(X) \times \mathbf{G}_i(X), \ \ i = r, r + 1 \ ,$$

and

$$\Delta = \{(U, V, W): \ U + V \subset W\} \subset \mathbf{G}_k(X) \times \mathbf{G}_l(X) \times \mathbf{G}_{k+l-r}(X)$$

are analytic. By Remmert's theorem (see V. 5.1), the projections of the sets Σ_i onto $\mathbf{G}_k(X) \times \mathbf{G}_l(X)$, i.e., the sets $\Sigma_i' = \{(U, V): \ \dim(U \cap V) \geq i\}$ are analytic. Hence the set $A = \Sigma_r' \setminus \Sigma_{r+1}'$ is locally analytic and analytically constructible. Now, for each pair $(U, V), (U', V') \in A$ there is an automorphism $\varphi \in L_0(X, X)$ such that $\varphi(U) = U'$ and $\varphi(V) = V'$. Therefore we have (see n° 2) the biholomorphic mapping $\Psi = \varphi_{(k)} \times \varphi_{(l)} : \mathbf{G}_k(X) \times \mathbf{G}_l(X) \longrightarrow \mathbf{G}_k(X) \times \mathbf{G}_l(X)$ such that $\Psi(U, V) = (U', V')$ and $\Psi(A) = A$. Consequently, it follows that each point of the set A is regular and of the same dimension (see IV. 2.1). This means that A is a submanifold. Both mappings described in the conclusion of our proposition are holomorphic. Indeed (see V. 1, the remark following corollary 3 of theorem 2), they are locally bounded and their graphs are analytic, because they are equal to the sets $\Sigma_r \cap (A \times \mathbf{G}_r(X))$, $\Delta \cap (A \times \mathbf{G}_{k+l-r}(X))$, respectively. Finally, observe that the first mapping is a surjection whose fibres $(\mathbf{S}^k(Z) \times \mathbf{S}^l(Z)) \cap \Omega$, $Z \in \mathbf{G}_r(X)$ are s–dimensional submanifolds, where $s = (n-k)(k-r) + (n-l)(l-r)$ and $\Omega = \{(U, V) : \dim U \cap V \leq r\}$ is open (see B. 6.6). Thus $\dim A = r(n-r) + s$ (see V. 3.2, theorem 2) $(^{24})$.

REMARK. It is easy to check that the sets A_r^{kl}, where $\max(0, k+l-n) \leq r \leq \min(k, l)$, are analytically constructible leaves $(^{25})$ and they form a complex stratification of the manifold $\mathbf{G}_k(X) \times \mathbf{G}_l(X)$.

4. We will now prove that every Grassmann manifold can be embedded in a projective space. Namely, the set

$$G_k(X) = \{\mathbf{C}(z_1 \wedge \ldots \wedge z_k) : (z_1, \ldots, z_k) \in B_k(X)\} \subset \mathbf{P}(\overset{k}{\bigwedge} X)$$

is a $k(n-k)$–dimensional compact submanifold and the bijection

$$\mathbf{p} = \mathbf{p}^X = \mathbf{p}_k^X : \mathbf{G}_k(X) \ni \sum_1^k \mathbf{C}z_i \longrightarrow \mathbf{C}(z_1 \wedge \ldots \wedge z_k) \in G_k(X)$$

$$((z_1, \ldots, z_k) \in B_k(X))$$

is biholomorphic. The embedding $\mathbf{p} : \mathbf{G}_k(X) \longrightarrow \mathbf{P}(\bigwedge^k X)$ is called the *Plücker embedding*, the submanifold $G_k(X)$ is called the *embedded Grassmannian*, and its cone

$$\mathcal{G}_k(X) = G_k(X)\tilde{} = \{z_1 \wedge \ldots \wedge z_k : (z_1, \ldots, z_k) \in B_k(X)\} \subset \overset{k}{\bigwedge} X$$

$(^{24})$ One can prove the proposition directly, using the implicit function theorem. However, the proof based on Remmert's theorem and properties of analytic sets is much shorter.

$(^{25})$ Their connectedness remains to be proved.

is called the *Grassmann cone* (^{26}).

Indeed, the mapping $\mathbf{p} :\ \mathbf{G}_k(X) \longrightarrow \mathbf{P}(\bigwedge^k X)$ is holomorphic, because (see n° 2) the composition

$$\mathbf{p} \circ \alpha :\ B_k(X) \ni (z_1, \ldots, z_k) \longrightarrow \mathbf{C}(z_1 \wedge \ldots \wedge z_k) \in \mathbf{P}(\overset{k}{\bigwedge} X)$$

is holomorphic (see 2.2). Hence it is enough to show that the set $G_k(X)$ is a submanifold (see 5.1, corollary 2 of theorem 2). According to Remmert's theorem (see V. 5.1), it is analytic. Next, for any pair $a, b \in G_k(X)$, we have $a = \mathbf{C}(a_1 \wedge \ldots \wedge a_k)$, $b = \mathbf{C}(b_1 \wedge \ldots \wedge b_k)$, where $(a_1, \ldots, a_k), (b_1, \ldots, b_k) \in B_k(X)$. There exists an automorphism $f \in L_0(X, X)$, such that $f(a_i) = b_i$, which induces an automorphism $g \in L_0(\bigwedge^k X, \bigwedge^k X)$, such that $g(a_1 \wedge \ldots \wedge a_k) = b_1 \wedge \ldots \wedge b_k$. Obviously, $g(\mathcal{G}_k(X)) = \mathcal{G}_k(X)$. So we have the biholomorphic mapping $\tilde{g} :\ \mathbf{P}(\bigwedge^k X) \longrightarrow \mathbf{P}(\bigwedge^k X)$ such that $\tilde{g}(a) = b$ and $\tilde{g}(G_k(X)) = G_k(X)$ (see B. 6.10). Consequently, each point of the set $G_k(X)$ is regular and of the same dimension (see IV. 2.1), and hence the set $G_k(X)$ is a submanifold.

Now consider the case when $X = \mathbf{C}^n$. In the space $\bigwedge^k \mathbf{C}^n$ we have the canonical basis $e_\alpha = e_{\alpha_1} \wedge \ldots \wedge e_{\alpha_k}$, $\alpha = (\alpha_1, \ldots, \alpha_s) \in \Lambda$, where $\Lambda = \{\alpha \in \mathbf{N}^k :\ 1 \le \alpha_1 < \ldots < \alpha_k \le n\}$ while $e_1, \ldots, e_n$ is the canonical basis for $\mathbf{C}^n$. We have the canonical isomorphism $\kappa :\ \mathbf{C}^\Lambda \ni \{\zeta_\alpha\} \longrightarrow \sum \zeta_\alpha e_\alpha \in \bigwedge^k \mathbf{C}^n$. The composition

$$\mathbf{p}_0 = \tilde{\kappa}^{-1} \circ \mathbf{p} :\ \mathbf{G}_k(\mathbf{C}^n) \longrightarrow \mathbf{P}(\mathbf{C}^\Lambda)$$

is an embedding, which is also called the *Plücker embedding*. By *Plücker coordinates of a k–dimensional subspace* $L \subset \mathbf{C}^n$ one means the homogeneous coordinates of its image $\mathbf{p}_0(L)$. They are determined up to a non-zero factor from $\mathbf{C}$. If

$$z_1 = (z_{11}, \ldots, z_{1n})$$
$$\cdots \cdots \cdots \cdots$$
$$z_k = (z_{k1}, \ldots, z_{kn})$$

is a basis of L, then the determinants $p_\alpha = \det z_{i\alpha_j}$, $\alpha \in \Lambda$, are the Plücker coordinates of L, since $z_1 \wedge \ldots \wedge z_k = \sum p_\alpha e_\alpha$ (see A. 1. 20). Therefore:

A system of Plücker coordinates of a k–dimensional subspace $L \subset \mathbf{C}^n$ is exactly the system of maximal minors of any matrix whose rows form a basis of L.

(^{26}) It is an algebraic cone (see 6.1 below).

In view of the characterization from 2.2, a set $\Gamma \subset \mathbf{G}_k(\mathbf{C}^n)$ is a k–dimensional submanifold if and only if the systems of the Plücker coordinates of the elements of Γ form a $(k+1)$–dimensional submanifold $\Gamma^{\#} \subset \mathbf{C}^{\Lambda}$. Then, if M is a manifold, a mapping $f : \Gamma \longrightarrow M$ is holomorphic if and only if it is holomorphic in Plücker coordinates, i.e., if $f(L) = g(\{p_{\alpha}\})$, where $\{p_{\alpha}\}$ is a system of Plücker coordinates of the subspace $L \in \Gamma$ and $g : \Gamma^{\#} \longrightarrow M$ is a holomorphic mapping (satisfying the homogeneity condition $g(tp) = p(p)$ for $t \in \mathbf{C} \setminus 0$, $p \in \Gamma^{\#}$).

By *Plücker dual coordinates of a k–dimensional subspace $L \subset \mathbf{C}^n$* one means Plücker coordinates of the subspace $L^{\perp}$ of $(\mathbf{C}^n)^*$ identified with $\mathbf{C}^n$ via the isomorphism $\mathbf{C}^n \ni c \longrightarrow (z \longrightarrow \sum c_i z_i) \in (\mathbf{C}^n)^*$. Thus:

A system of Plücker dual coordinates of a k–dimensional subspace $L \subset \mathbf{C}^n$ is exactly the system of the maximal minors of the coefficient matrix of any system of linear equations of the form

$$c_{11} z_1 + \cdots + c_{1n} z_n = 0,$$
$$\dotfill$$
$$c_{n-k,1} z_1 + \cdots + c_{n-k,n} z_z = 0$$

that describes this subspace. (Indeed, the left hand sides of the equations are linear forms that constitute a basis for the subspace $L^{\perp}$.)

In the same way one can characterize submanifolds of $\mathbf{G}_k(\mathbf{C}^n)$ and holomorphic mappings from such submanifolds into manifolds using Plücker dual coordinates. (It suffices to use the fact that $\tau : L \longrightarrow L^{\perp}$ is biholomorphic; see n° 2.)

5. In the space $\mathbf{G}'_k(X)$ one introduces the structure of a $(k+1)(n-k)$–dimensional manifold by transferring it from $\mathbf{G}_{k+1}(\mathbf{C} \times X) \setminus \mathbf{G}_{k+1}(0 \times X)$ through the bijection χ defined in B. 6.11.

Then the mapping

$$\beta : X \times B_k(X) \ni (z, x_1, \ldots, x_k) \longrightarrow z + \sum_1^k \mathbf{C} x_i \in \mathbf{G}'_k(X)$$

is a surjective submersion. Indeed, the mapping

$$\chi^{-1} \circ \beta : X \times B_k(X) \ni (z, x_1, \ldots, x_k) \longrightarrow \mathbf{C}(1, z) + \sum_1^k \mathbf{C}(0, x_i) \in \mathbf{G}_{k+1}(\mathbf{C} \times X)$$

is holomorphic (see n° 2) and so is the mapping β. Let $a = (c, a_1, \ldots, a_k) \in X \times B_k(X)$. Take a linear complement V of the subspace $U = \sum_1^k \mathbf{C} a_i$ and the mapping

$$\gamma : V^{k+1} \ni (w, v_1, \ldots, v_k) \longrightarrow (c + w, a_1 + v_1, \ldots, a_k + v_k) \in X \times B_k(X) .$$

Then the holomorphic mapping $\chi^{-1} \circ \beta \circ \gamma : V^{k+1} \longrightarrow \Omega(0 \times V)$ is bijective $(^{27})$ and hence biholomorphic (see V. 1, corollary 2 of theorem 2). Therefore the differential $d_0(\chi^{-1} \circ \beta \circ \gamma)$ is surjective and so is the differential $d_a(\chi^{-1} \circ \beta)$. Thus $\chi^{-1} \circ \beta$ is a submersion and so is β.

The mapping $\nu : \mathbf{G}'_k(X) \ni L \longrightarrow L_* \in \mathbf{G}_k(X)$ defined in B. 6.11 is a surjective submersion, since $\nu \circ \beta = \alpha \circ \pi$ and the natural projection $\pi : X \times B_k(X) \longrightarrow B_k(X)$ is obviously a submersion (see C. 4.2).

The mapping $\sigma : X \times \mathbf{G}_k(X) \ni (x, L) \longrightarrow x + L \in \mathbf{G}'_k(X)$ is also a surjective submersion, because $\sigma \circ (e \times \alpha) = \beta$, where e is the identity mapping of X (see C. 4.2).

The bijections

$$\psi_{UV} : L_1(U, V) \ni f \longrightarrow \hat{f} \in \nu^{-1}(\Omega(V)), \ U \in \mathbf{G}_k(X), \ V \in \Omega(U),$$

where $L_1(U, V)$ denotes the vector space of the affine mappings from U to V, and $\hat{f} = \{u + f(u) : u \in U\}$, form an inverse atlas on the manifold $\mathbf{G}'_k(X)$. Indeed, the sets $\nu^{-1}(\Omega(V))$ cover $\mathbf{G}'_k(X)$ and each of the mappings ψ_{UV} is biholomorphic. This is so because (see V. 1, corollary 2 of theorem 2) it is holomorphic, since $\psi_{UV}(f) = \beta(f(0), e_1 + f(e_1) - f(0), \dots, e_k + f(e_k) - f(0))$, where $e_1, \dots, e_k$ is a basis of U.

Note also that if $U \in \mathbf{G}_k(X)$ and $V \in \Omega(U)$, then $V \ni z \longrightarrow z + U \in \mathbf{G}'_k(X)$ is an immersion.

Let $\mathbf{P} = \mathbf{P}(X)$. In the space $\mathbf{G}_k(\mathbf{P})$ $(0 \leq k \leq n - 1)$, one introduces the structure of a manifold of dimension $(k + 1)(\dim \mathbf{P} - k)$, transferred from $\mathbf{G}_{k+1}(X)$ via the bijection $\omega = \omega^{\mathbf{P}}$ defined in B. 6.12.

Then the mapping $\mu : \mathbf{G}'_k(X) \setminus \mathbf{G}_k(X) \ni T \longrightarrow (CT)^{\check{}} \in \mathbf{G}_k(\mathbf{P})$ defined in B. 6.12 is a surjective submersion. For $\mu \circ \beta_{B_{k+1}(X)} = \omega \circ \alpha$ and $\beta(B_{k+1}(X)) = \mathbf{G}'_k(X) \setminus \mathbf{G}_k(X)$ (see C. 4.2).

Finally, observe that the bijection $\vartheta : \mathbf{G}_k(\bar{X}) \setminus \mathbf{G}_k(X_\infty) \ni L \longrightarrow L \cap X \in \mathbf{G}'_k(X)$ (see 3.3) is a biholomorphic mapping because $\vartheta = \chi \circ (\omega^{\bar{X}})^{-1}$.

§5. Blowings-up

Let X be an analytic space.

1. Let $f_1, \dots, f'_k$ be a sequence of holomorphic functions on X. Consider the mapping

$$\alpha \circ f : X \setminus S \ni z \longrightarrow \mathbf{C}f(z) \in \mathbf{P}_{k-1} ,$$

$(^{27})$ Because the sum $\mathbf{C} \times X = (\mathbf{C} \times U) + (0 \times V)$ is direct and $(1, c'), (0, a_1), \dots, (0, a_k)$ is a basis of the subspace $\mathbf{C} \times U$, where $c' \in U$ is the image of c under the projection parallel to V.

where $\alpha = \alpha^{\mathbf{C}^k}$ (see 2.1), $f = (f_1, \ldots, f_k)$, and $S = V(f_1, \ldots, f_k)$. The closure of its graph

$$Y = \overline{\alpha \circ f} \subset X \times \mathbf{P}_{k-1}$$

is an analytic set, because

$$\alpha \circ f = E(f) \setminus (S \times \mathbf{P}_{k-1}) \,,$$

where

$$E(f) = \{(z, \mathbf{C}w) : \ w \neq 0, \ w_i f_j(z) = w_j f_i(z), \ i,j = 1, \ldots, k\} \subset X \times \mathbf{P}_{k-1}$$

is an analytic subset (see V. 4.5 in ref. to theorem 5 from IV. 2.10). The natural projection

$$\pi : \ Y \longrightarrow X$$

(or the pair Y, π) is called the *(elementary) blowing-up of the space X by means of the functions $f_1, \ldots, f_k$*. It is holomorphic, proper, its range is $\overline{X \setminus S} = \overline{\{f \neq 0\}}$, and the restriction $\pi^{X \setminus S} : \ \alpha \circ f \longrightarrow X \setminus S$ is biholomorphic (see V. 4.7 in ref. to V. 3.4). Therefore it is a modification of the space X in the set S, provided that $f \not\equiv 0$, i.e., S is nowhere dense [28] . It always has the property (m) (see V. 4.11). The set Y is called the *blown-up space of X by means of the functions $f_1, \ldots, f_k$* or, shortly, the *blown-up space*. The analytic subset S is called the *centre of the blowing-up* and its inverse image $\pi^{-1}(S) \subset Y$ is called the *exceptional set* of the blowing-up.

If $G \subset X$ is an open subset, then π^G is obviously the *blowing-up of G by means of the restrictions $(f_i)_G$*.

If $V \subset X$ is an analytic subset, then the set

$$W = \overline{\pi^{-1}(V \setminus S)} = \overline{\alpha \circ f_V} \subset \pi^{-1}(V) \subset Y$$

(which is analytic [29]) is called the *proper inverse image* of the set V under the blowing-up π. The restriction $\pi_W : \ W \longrightarrow V$ is the blowing-up of V by means of the restrictions $(f_i)_V$. Clearly, $\pi^{-1}(V) = W \cup \pi^{-1}(V \cap S)$.

Any blowing-up by means of a single function $f \not\equiv 0$ is trivial, i.e., π is biholomorphic. (In such a case, $Y = X \times \mathbf{P}_0$ and $\mathbf{P}_0$ consists of a single point.)

---

[28] It can be always achieved by removing the irreducible components of X on which $f = 0$.

[29] See, e.g., V. 4.5 in ref. to the theorem from IV. 2.10.

2. Consider the blowing-up of $\mathbf{C}^n$ by means of the functions $z_1, \ldots, z_n$, that is, the *blowing-up of the space* $\mathbf{C}^n$ *at the point* 0 [30]. It consists of the set

$$\Pi_n = E(z) = \{(z, \mathbf{C}w) : \ w \neq 0, \ w_j z_i = w_i z_j\} \subset \mathbf{C}^n \times \mathbf{P}_{n-1}$$

($w = (w_1, \ldots, w_n)$) and the natural projection $\pi = \pi_n : \ \Pi_n \longrightarrow \mathbf{C}^n$ [31]. Now, the set Π_n is a closed n–dimensional submanifold, for its image under the chart $\psi_s = (\mathrm{id}_{\mathbf{C}^n}) \times \varphi_{s-1}$ (see 2.3) is the closed n–dimensional submanifold $\Pi^{(s)} = \{z_i = z_s w_i, \ i \neq s\} \subset \mathbf{C}^n \times \mathbf{C}^{n-1}$. The exceptional set $\Sigma = \pi^{-1}(0) = 0 \times \mathbf{P}_{n-1} \subset \Pi_n$ is an $(n-1)$–dimensional submanifold and the mapping $\pi^{\mathbf{C}^n \setminus 0} : \ \Pi_n \setminus (0 \times \mathbf{P}_{n-1}) \longrightarrow \mathbf{C}^n \setminus 0$ is biholomorphic. At each point of the exceptional set $(0, \eta)$, $\eta \in \mathbf{P}_{n-1}$, we have

$$T_{(0,\eta)} \Pi_n = \eta \times T_\eta \mathbf{P}_{n-1} \ .$$

(Because for some s we have $\eta = \mathbf{C}w$, $w_s = 1$; then $\psi_s(0, \eta) = (0, w_1, \ldots \ldots, w_{s-1}, w_{s+1}, \ldots, w_n)$, and so $\eta \times \mathbf{C}^{n-1}$ is the tangent space to $\Pi^{(s)}$ at the point $\psi_s(0, \eta)$.)

Note that if J is a family of ideals generated by $z_1, \ldots, z_n$ in $\mathbf{C}^n$, then its inverse image $\pi_n^* J$ is a family of principal ideals.

Indeed, consider $\zeta \in \Sigma$ (if $\pi(\zeta) \neq 0$, then $(\pi^* J)_\zeta = \mathcal{O}_\zeta(\Pi_n)$). Then $\zeta = (0, \eta)$, where $\eta \in \mathbf{P}_{n-1}$, and we have $\mathrm{im} \, d_\zeta \pi = \eta$. The germs $\tilde{z}_i = (z_i)_0 \circ \pi_\zeta$ that generate the ideal $(\pi^* J)_\zeta$ belong to the ideal $\mathcal{J}(\Sigma_\zeta)$, which is principal. But $d_\zeta \tilde{z}_i = z_i \circ d_\zeta \pi$, and so $\mathrm{im} \, d_\zeta \tilde{z}_i = z_i(\eta)$. Hence $d_\zeta \tilde{z}_i \neq 0$ for some i. Therefore (see II. 4.2) we must have $(\pi^* J)_\zeta = \mathcal{J}(\Sigma_\zeta)$.

Finally, note that if π is the blowing-up by means of the functions $f_1, \ldots, f_k \in \mathcal{O}_X$ (as in n° 1), then we have the commutative diagram

$$
\begin{array}{ccc}
Y & \xrightarrow{\ g\ } & \Pi_k \\
{\scriptstyle \pi}\downarrow & & \downarrow{\scriptstyle \pi_k} \\
X & \xrightarrow{\ f\ } & \mathbf{C}^k \ ,
\end{array}
$$

where $f = (f_1, \ldots, f_k)$, $g = (f \times e)_Y$, $e = \mathrm{id}_{\mathbf{P}_{k-1}}$.

3. Adopting the notation from n° 1, suppose now that X is an n–dimensional manifold and f is a submersion at each point of the set S. Thus S is an $(n-k)$–dimensional submanifold [32].

---

[30] See n° 6 below. Hence it is a modification of the space $\mathbf{C}^n$ at the point 0.

[31] We have $0 \times \mathbf{P}_{n-1} \subset \overline{E(z) \setminus (0 \times \mathbf{P}_n)}$, and so $\Pi_n = E(z)$.

[32] Therefore $\pi : \ Y \longrightarrow X$ is a modification in S.

Then $Y = E(f)$ is an n–dimensional manifold, the exceptional set $\pi^{-1}(S) = S \times \mathbf{P}_{k-1}$ is an $(n-1)$–dimensional submanifold, $\pi^S : S \times \mathbf{P}_{k-1} \longrightarrow S$ is the natural projection, and $\pi^{X \setminus S} : Y \setminus (S \times \mathbf{P}_{k-1}) \longrightarrow X \setminus S$ is biholomorphic.

Indeed, one may assume that X is connected and f is a submersion ([33]). It is enough to check that $E(f)$ is a connected manifold, because $S \times \mathbf{P}_{k-1} \not\subseteq E(f)$ (see n° 1). Now, $E(f)$ is the inverse image of the submanifold $\Pi_k \subset \mathbf{C}^k \times \mathbf{P}_{k-1}$ under the submersion $\chi = f \times (\mathrm{id}_{\mathbf{P}_{k-1}})$, and so it is a submanifold (see C. 4.2). It is connected, since $E(f) = \overline{\alpha \circ f} \cup (S \times \mathbf{P}_{k-1})$, the set $\overline{\alpha \circ f}$ is connected, and it intersects $z \times \mathbf{P}_{k-1}$ for each $z \in S$.

So $Y = \chi^{-1}(\Pi_k)$, hence we have (see n° 2 and C. 4.2)

$$T_{(z,\eta)}Y = (d_z f)^{-1}(\eta) \times T_\eta \mathbf{P}_{n-1} \quad \text{for} \quad (z, \eta) \in S \times \mathbf{P}_{k-1} ,$$

and then $\mathrm{im}\, d_{(z,\eta)}\pi = (d_z f)^{-1}(\eta)$. Therefore, if $z \in S$, then in view of $T_z S = \ker d_z f$ we have the biholomorphic bijection ([34]) :

$$\pi^{-1}(z) \ni \zeta \longrightarrow \mathrm{im}\, d_\zeta \pi \in \mathbf{S}^{n-k+1}(T_z S, T_z X) .$$

4. Let us adopt again the notation from n° 1. Let $g_1, \ldots, g_l$ be a sequence of holomorphic functions on X.

PROPOSITION 1. *If f_i and g_i generate the same ideal* ([35]) *, then the blowing-up by means of f_j is isomorphic to the blowing-up by means of g_j.*

PROOF. The blowing-up by means of g_j is the set $Z = \overline{\beta \circ g}$ with the natural projection $\pi' : Z \longrightarrow X$, where $\beta = \alpha^{\mathbf{C}^l}$ (see 2.1) and $g = (g_1, \ldots, g_l)$. We have $g_j = \sum_{i=1}^k a_{ij} f_i$ and $f_i = \sum_{j=1}^l b_{ij} g_j$, where a_{ij}, b_{ij} are holomorphic in X. Thus

$$(*) \qquad\qquad g(z) = a\big(z, f(z)\big) \quad \text{and} \quad f(z) = b\big(z, g(z)\big) ,$$

([33]) It suffices to prove the above properties for the connected components of a suitable neighbourhood of the set S.

([34]) If $\chi : L \longrightarrow \mathbf{C}^k$ is a linear mapping (of a vector space L) whose kernel N is r–dimensional, then the mapping $\mathbf{P}_k \ni \lambda \longrightarrow \chi^{-1}(\lambda) \in \mathbf{S}^{r+1}(N, L)$ is biholomorphic. Indeed, if $L = N + M$ is a direct sum, then this mapping is the composition of the biholomorphic mapping $(\chi_M^{-1})^{\tilde{}} : \mathbf{P}_k \longrightarrow \mathbf{P}(M)$ (see 4.2) and the biholomorphic bijection $\mathbf{P}(M) \ni \mu \longrightarrow N + \mu \in \mathbf{S}^{r+1}(N, L)$ (see 4.3).

([35]) In the ring $\mathcal{O}_X$, i.e., $\sum \mathcal{O}_X f_i = \sum \mathcal{O}_X g_j$.

where $a(z, u) = \left(\sum_1^k a_{1i}(z)u_i, \ldots, \sum_1^k a_{li}(z)u_i \right)$ and $b(z, v) = \left(\sum_1^l b_{1j}v_j, \ldots \right.$
$\left. \ldots, \sum_1^l b_{kj}(z)v_j \right)$. Consider the holomorphic mappings defined by

$$\tilde{a} : \ G \ni (z, \mathbf{C}u) \longrightarrow (z, \mathbf{C}a(z, u)) \in X \times \mathbf{P}_{l-1} \ ,$$

$$\tilde{b} : \ H \ni (z, \mathbf{C}v) \longrightarrow (z, \mathbf{C}b(z, v)) \in X \times \mathbf{P}_{k-1} \ ,$$

where $G = \{(z, \mathbf{C}u) : \ a(z, u) \neq 0\} \subset X \times \mathbf{P}_{k-1}$ and $H = \{(z, \mathbf{C}v) : b(z, v) \neq 0\} \subset X \times \mathbf{P}_{l-1}$ $(^{36})$. In view of $(*)$, we have $b(z, a(z, u)) = u$ when $(z, \mathbf{C}u) \in \alpha \circ f$, and hence also when $(z, \mathbf{C}u) \in Y$. This implies that $Y \subset G$. Similarly, $Z \subset H$. Now using $(*)$ we conclude that $\tilde{a}_{\alpha \circ f} : \ \alpha \circ f \longrightarrow \beta \circ g$ and $\tilde{b}_{\beta \circ g} : \ \tilde{\beta} \circ g \longrightarrow \alpha \circ f$ are mutually inverse biholomorphic mappings, and so are $\tilde{a}_Y : \ Y \longrightarrow Z$ and $\tilde{b}_Z : \ Z \longrightarrow Y$. Clearly, the diagram

$$
\begin{array}{ccc}
Y & \xrightarrow{\tilde{a}_Y} & Z \\
& \pi \searrow \quad \swarrow \pi' & \\
& X &
\end{array}
$$

commutes. Therefore $\pi \approx \pi'$.

COROLLARY. *If f_i and g_j are generators of the same family of ideals $I_z \subset \mathcal{O}_z$, $z \in X$ (see VI. 1.2), then the blowing-up by means of f_i is isomorphic to the blowing-up by means of g_j.*

In fact, let $\pi : \ Y \longrightarrow X$, $\pi' : \ Z \longrightarrow X$ be the blowings-up by means of f_i and g_j, respectively. Each point $z \in X$ has an open neighbourhood U_z such that $(f_i)_{U_z}$ and $(g_j)_{U_z}$ generate the same ideal, and so $\pi^{U_z} \approx (\pi')^{U_z}$. Since π and π' have the property (m) (see n° 1) which is sublocal and rigid, we have $\pi \approx \pi'$ (see V. 4.11–12).

5. A *blowing-up of the space X by means of the family of ideals $I_z \subset \mathcal{O}_z$, $z \in X$,* is defined to be a holomorphic mapping $\pi : \ Y \longrightarrow X$ of an analytic space Y (or, also, as a pair Y, π) such that each point of the space X has an open neighbourhood U for which π^U is isomorphic with the blowing-up by means of generators of the family I_z, $z \in U$. Therefore the family I_z, $z \in X$, must be coherent. The space Y is called the *blown-up space of X by means of the family $\{I_z\}$* or, shortly, the *blown-up space*.

In particular, the blowing-up by means of the holomorphic functions $f_1, \ldots, f_k$ is a blowing-up by means of the family of ideals generated by $f_1, \ldots, f_k$.

$(^{36})$ Since $e \times \alpha$, $e \times \beta$, where $e = \mathrm{id}_X$, are surjective submersions (see 2.2), the sets G, H are open and the mappings $\tilde{a}, \tilde{b}$ are holomorphic (see C. 4.2).

All blowings-up are proper mappings (^{37}) .

All blowings-up have the property (m) which is sublocal and rigid (see n° 1). Thus, according to the corollary of proposition 1 in n° 4, it follows (see V. 4.12) that:

All blowings-up by means of the same family of ideals are isomorphic. In particular, every blowing-up by means of the family of ideals generated by $f_1, \ldots, f_k$ is isomorphic to the blowing-up by means of $f_1, \ldots, f_k$.

If $\pi : Y \longrightarrow X$ is a mapping between analytic spaces and $X = \bigcup G_\iota$ is an open cover, then obviously:

(π is the blowing-up by means of the family $\{I_z\}$) $\Longleftrightarrow$

$\Longleftrightarrow$ (each π^{G_ι} is the blowing-up by means of the family $\{I_z\}_{z \in G_\iota}$.)

PROPOSITION 2. *For any coherent family of ideals $I_z \subset \mathcal{O}_z$, $z \in X$, there exists a unique (up to an isomorphism) blowing-up of the space X by means of this family.*

Indeed, there is an open cover $X = \bigcup U_\iota$ such that for each ι the family $\{I_z\}_{z \in U_\iota}$ has generators $f_1^\iota, \ldots, f_{k_\iota}^\iota$. Let $\pi_\iota : Y_\iota \longrightarrow U_\iota$ be the blowing-up by means of the functions $f_1^\iota, \ldots, f_{k_\iota}^\iota$. For any pair ι, κ, the restrictions $\pi_\iota^{U_\iota \cap U_\kappa}$ and $\pi_\kappa^{U_\iota \cap U_\kappa}$ are the blowings-up by means of $(f_i^\iota)_{U_\iota \cap U_\kappa}$ and $(f_j^\kappa)_{U_\iota \cap U_\kappa}$, respectively. Consequently, in view of the corollary of proposition 1 in n° 4, it follows that $\pi_\iota^{U_\iota \cap U_\kappa} \approx \pi_\kappa^{U_\iota \cap U_\kappa}$. According to the proposition on gluing from V. 4.12, there is a holomorphic mapping $\pi : Y \longrightarrow X$ of an analytic space Y such that $\pi^{U_\iota} \approx \pi_\iota$, and this clearly is a blowing-up of the space X by means of the family $\{I_z\}$.

Let $\pi : Y \longrightarrow X$ be the blowing-up by means of a coherent family I of ideals $I_z \subset \mathcal{O}_z$, $z \in X$.

The zero set $S \subset X$ of the family $\{I_z\}$ (see IV. 1.4, remark 3) is called the *centre of the blowing-up*. In the case when the family $\{I_z\}$ has generators $f_1, \ldots, f_k$, we have $S = V(f_1, \ldots, f_k)$, which is consistent with the case of the elementary blowing-up (see n° 1). The restriction $\pi^{X \setminus S} : Y \setminus \pi^{-1}(S) \longrightarrow X \setminus S$ is biholomorphic (^{38}) . The analytic subset $\pi^{-1}(S) \subset Y$ is said to be the

---

(^{37}) Since the property of being a proper mapping is sublocal.

(^{38}) Indeed, this is the case if the family $\{I_z\}$ has generators $f_1, \ldots, f_k$ (because then π is isomorphic to the elementary blowing-up by means of $f_1, \ldots, f_k$). Therefore, in the general case, $\pi^{U_\iota \setminus S}$ are biholomorphic for some open cover $X = \bigcup U_\iota$, and so is $\pi^{X \setminus S}$, since the property of being biholomorphic is sublocal.

exceptional set of the blowing-up. The blowing-up π is a modification of the space X in the set S, provided that S is nowhere dense or, equivalently, if $I_z \neq 0$ for $z \in X$.

If $V \subset X$ is an analytic subset, then the analytic (^{39}) subset

$$W = \overline{\pi^{-1}(V \setminus S)} \subset \pi^{-1}(V) \subset Y$$

is called the *proper inverse image of the set V under the blowing-up π*. Obviously we have $\pi^{-1}(V) = W \cup \pi^{-1}(V \cap S)$.

The restriction $\pi_W : W \longrightarrow V$ is the blowing-up of the subset V by means of the family $I_z \circ \iota_z \subset \mathcal{O}_z(V)$, $z \in V$, where $\iota : V \hookrightarrow X$.

Indeed, this is the case when the family $\{I_z\}$ has a finite system of generators (^{40}). In the general case, there is an open cover $X = \bigcup U_\kappa$ such that each of the families $\{I_z\}_{z \in U_\kappa}$ has a finite system of generators. The blowing-up of the set U_κ by means of the family $\{I_z\}_{z \in U_\kappa}$ is π^{U_κ}, and the proper inverse image of the set $V_\kappa = V \cap U_\kappa$ is, as can be easily checked, $W_\kappa = W \cap \pi^{-1}(U_\kappa)$. Therefore $(\pi_W)^{V_\kappa} = (\pi^{U_\kappa})_{W_\kappa}$ is the blowing-up of V_κ by means of the family $\{I_z \circ \iota_z\}_{z \in V_\kappa}$. But $V = \bigcup V_\kappa$ is an open cover, and hence π_W is the blowing-up of the subset V by means of the family $\{I_z \circ \iota_z\}_{z \in V}$.

The inverse image $\pi^* I$ is a family of principal ideals.

Indeed, one may assume that π is a blowing-up by means of $f_1, \ldots, f_k \in \mathcal{O}_X$. Then it is obvious that $I = f^* J$, where $f = (f_1, \ldots, f_k) : X \longrightarrow \mathbf{C}^k$ and J is a family of ideals generated by $z_1, \ldots, z_k$ in $\mathbf{C}^k$. In view of the commutative diagram from n° 2, we have $\pi^* I = g^*(\pi_k^* J)$ (see VI. 1.2). But $\pi_k^* J$ is a family of principal ideals (see n° 2) and hence so is $\pi^* I$.

Suppose that the zero set $S \subset X$ of the family I is nowhere dense.

THEOREM (HIRONAKA). *The blowing-up $\pi : Y \longrightarrow X$ of the space X by means of the family I is the unique (up to an isomorphism) holomorphic mapping of an analytic space into X that satisfies the conditions*

(1) $\pi^{-1}(S)$ *is nowhere dense and $\pi^* I$ is a family of principal ideals.*

(2) *If a holomorphic mapping $F : Z \longrightarrow X$ of an analytic space Z satisfies (in place of π) the condition (1), then $F = \pi \circ G$ with a unique holomorphic mapping $G : Z \longrightarrow Y$ (^{41}).*

(^{39}) See V. 4.5 in ref. to theorem 5 from IV. 2.10.

(^{40}) See n° 1. Then the restrictions of these generators to V are generators of the family $\{I_z \circ \iota_z\}_{z \in V}$.

(^{41}) In other words, a blowing-up π by means of a coherent family of ideals I is a universal holomorphic mapping with respect to the property (1). This implies that π is unique up to

REMARK . Assuming the second condition in (1), the first one is equivalent to the requirement that for each $w \in Y$ a generator of the ideal $(\pi^* I)_w$ is not a zero divisor.

PROOF . It remains to prove property (2) (see footnote (^{41})) and the remark. First assume that F satisfies the second condition in (1) and set $\Sigma = F^{-1}(S)$. If $z \in Z$, then the condition that a generator h of the ideal $(F^* I)_z$ is not a zero divisor means that a representative $\tilde{h}$ of h in a neighbourhood U of z is $\not\equiv 0$ (see IV. 4.3). But $\Sigma \cap U = \{\tilde{h} = 0\}$, provided that U is sufficiently small (see VI. 1.4, remarks 1 and 3). Therefore the condition is satisfied for each $z \in Z$ if and only if Σ is nowhere dense, and this proves the remark. Suppose now that F satisfies both conditions in (1). Since π is a modification in S, the mapping

$$G_0 = \pi_{X \setminus S}^{-1} \circ F : \ Z \setminus \Sigma \longrightarrow Y$$

is holomorphic. Uniqueness of the mapping G in the condition (2) follows from the fact that $G \supset G_0$. It is enough to show that G_0 has a holomorphic extension on Z, i.e., that each point $c \in Z$ has a neighbourhood Ω such that G_0 can be extended holomorphically from $\Omega \setminus \Sigma$ to Ω. Now, by taking a sufficiently small neighbourhood U of the point $F(c)$, we may assume that π^U is a blowing-up by means of $f_1, \ldots, f_k \in \mathcal{O}_U$. Then $S \cap U = V(f_1, \ldots, f_k)$, and we have the equality $G_0(z) = \left(F(z), \mathbf{C}\left(f_1\big(F(z)\big), \ldots, f_k\big(F(z)\big) \right) \right)$ for $z \in F^{-1}(U) \setminus \Sigma$. The ideal $(F^* I)_c$ has, according to the remark, a generator with representative $h \not\equiv 0$ in a neighbourhood $\Omega \subset \pi^{-1}(U)$ of the point c, and also $(f_i \circ F)_c$ are generators of this ideal. By making Ω smaller, we get $h = \sum a_i (f_i \circ F)$ and $f_i \circ F = b_i h$ in Ω with some $a_i, b_i \in \mathcal{O}_\Omega$. Therefore $h = \sum a_i b_i h$, and so $\sum a_i b_i = 1$. Thus $G_0(z) = \left(F(z), \mathbf{C}\big(b_1(z), \ldots, b_k(z) \big) \right)$ in $\Omega \setminus \Sigma$, and the right hand side is holomorphic in Ω (as $(b_1, \ldots, b_k) \neq 0$ in Ω).

6. By the *blowing-up of the space X in its analytic subset S* we mean the blowing-up of X by means of the family of ideals $\mathcal{I}(S_z)$, $z \in X$. This family is always coherent (due to Cartan's theorem; see VI. 1.4), and hence the blowing-up in any analytic subset always exists and is unique up to an isomorphism. The blown-up space Y is also referred to as the *space X blown-up in the subset S*.

---

an isomorphism. To be precise, if $\pi' : \ Y' \longrightarrow X$ is another such mapping, then $\pi' = \pi \circ \Phi$ and $\pi = \pi' \circ \Psi$ with some holomorphic mappings $\Phi : \ Y' \longrightarrow Y$, $\Psi : \ Y \longrightarrow Y'$. Then $\pi' = \pi' \circ \Psi \circ \Phi$ and $\pi = \pi \circ \Phi \circ \Psi$. Therefore (because of uniqueness in (2)) $\Psi \circ \Phi = \mathrm{id}_{Y'}$ and $\Phi \circ \Psi = \mathrm{id}_Y$, which means that Φ and Ψ are mutually inverse isomorphisms.

Accordingly, the blowing-up in the subset S is locally isomorphic to the elementary blowings-up by means of local generators of the family $\{\mathcal{I}(S_z)\}$.

In general, the blowing-up by means of functions $f_1, \ldots, f_k$ is not the same as a blowing-up in the set $V(f_1, \ldots, f_k)$. For example, the blowing-up of $\mathbf{C}^2$ by means of z_1, z_2^2 is not a blowing-up at 0, as the blown-up space is not a submanifold ([42]). In fact, the image of the latter under the chart ψ_1 (see n° 2) is the subset $\{z_2^2 = w_2 z_1\}$ of $\mathbf{C}^3$ for which 0 is a singular point.

Any blowing-up $\pi : Y \longrightarrow X$ in a set S is proper and, obviously, the centre of the blowing-up is the set S. If the latter is nowhere dense, the blowing-up is a modification of X in S. The restriction $\pi^{X \backslash S} : Y \backslash \pi^{-1}(S) \longrightarrow X \backslash S$ is biholomorphic.

Clearly, if $h : X \longrightarrow X'$ is a biholomorphic mapping of analytic spaces, then $h \circ \pi$ is the blowing-up of X' in $h(S)$.

If $\pi : Y \longrightarrow X$ is a mapping of analytic spaces, $S \subset X$ is an analytic subset, and $X = \bigcup G_\iota$ is an open cover, then we have the obvious equivalence

(π is the blowing-up of X in S) $\Longleftrightarrow$

$$\Longleftrightarrow \text{(each } \pi^{G_\iota} \text{ is the blowing-up of } G_\iota \text{ in } S \cap G_\iota).$$

If X is a manifold and the subset S is of constant codimension 1, then the blowing-up in S is trivial; it is a biholomorphic mapping. Indeed, in such a case (see II. 5.3; and IV. 3.1, corollary 1 of proposition 5), it is locally isomorphic to a blowing-up by means of a single function (see n° 1).

7. Now let $\pi : Y \longrightarrow X$ be a blowing-up of an n–dimensional manifold X in a closed submanifold S of dimension $k < n$ ([43]). It is, obviously, a modification in S.

Every point of the manifold X has an open neighbourhood U such that $S \cap U = \{f = 0\}$, where $f = (f_1, \ldots, f_{n-k}) : U \longrightarrow \mathbf{C}^{n-k}$ is a submersion (see C. 3.15). Then the blowing-up π^U of the neighbourhood U in the submanifold $S \cap U$ is isomorphic to the blowing-up of this neighbourhood by means of the functions $f_1, \ldots, f_{n-k}$. (These functions generate the family $\{\mathcal{I}(S_z)\}_{z \in U}$; see II. 4.2). By n° 3, we derive the following

PROPOSITION 3. *The blown-up space Y is an n–dimensional manifold, whereas the exceptional set $\pi^{-1}(S) \subset Y$ is an $(n-1)$–dimensional submanifold. The restriction $\pi^{X \backslash S} : Y \backslash \pi^{-1}(S) \longrightarrow X \backslash S$ is biholomorphic, while the*

([42]) See n° 2 or proposition 3 in n° 7 below.

([43]) In this case, the fact that the family $\{\mathcal{I}(S_z)\}$ is coherent can be verified easily, without use of Cartan's theorem (see VI. 1.4, remark 2).

restriction $\pi^S : \pi^{-1}(S) \longrightarrow S$ is a submersion whose fibres are biholomorphic to $\mathbf{P}_{n-k-1}$ [44] . *For each $z \in S$, we have the biholomorphic mapping*

$$\pi^{-1}(z) \ni \zeta \longrightarrow \operatorname{im} d_\zeta \pi \in \mathbf{S}^{k+1}(T_z S, T_z X) .$$

If $k = n-1$, then the blowing-up is trivial, i.e., π is a biholomorphic mapping.

We say that smooth germs A, B at a point a of the manifold X *intersect quasi-transversally* if the germ $A \cap B$ is smooth and of dimension $\dim(TA \cap TB)$. Then $T(A \cap B) = TA \cap TB$ (because we have the inclusion $\subset$ and equality of dimensions).

The fact that the germs A, B intersect quasi-transversally is equivalent to the property $\dim(A \cap B) = \dim(TA \cap TB)$. Indeed, suppose that the latter is satisfied. One may assume that X is a vector space and $a = 0$. Let $\pi : X \longrightarrow L = TA + TB$ be a projection. One can choose representatives $\tilde{A}, \tilde{B}$ of the germs A, B, respectively, so that they are submanifolds, the restrictions $\pi_{\tilde{A}} : \tilde{A} \longrightarrow \pi(\tilde{A})$, $\pi_{\tilde{B}} : \tilde{B} \longrightarrow \pi(\tilde{B})$ are biholomorphic , and $\tilde{A} \cap \tilde{B}$ is an analytic subset of both $\tilde{A}$ and $\tilde{B}$. Then the submanifolds $\pi(\tilde{A}), \pi(\tilde{B}) \subset L$ intersect transversally at 0 (owing to the fact that TA, TB are their tangent spaces). We have $\pi(\tilde{A} \cap \tilde{B}) \subset \pi(\tilde{A}) \cap \pi(\tilde{B})$, and by reducing the sizes of $\tilde{A}$ and $\tilde{B}$, one may require that the right hand side is a submanifold and the left hand side is its analytic subset. Because of the equality of the dimensions, one must have $\pi(\tilde{A} \cap \tilde{B}) = \pi(\tilde{A}) \cap \pi(\tilde{B})$, which implies that $\tilde{A} \cap \tilde{B}$ is a submanifold.

Using the above argument, one can easily check that there is a biholomorphic mapping of a neighbourhood of the point a in X onto a neighbourhood of the point 0 in $T_a X$ which takes a to 0 and maps A, B into TA and TB, respectively. (Cf. C. 3.16). Any quasi-transversal intersection is a transversal intersection on some submanifold.

We say that *submanifolds $V, W \subset X$ intersect quasi-transversally* if the germs V_z, W_z intersect quasi-transversally for each $z \in V \cap W$. Then $V \cap W$ is smooth [45] .

PROPOSITION 4. *Let $V \subset X$ be an l–dimensional closed submanifold that intersects the centre S quasi-transversally. Then its proper inverse image $W = \overline{\pi^{-1}(V \setminus S)}$ is an l–dimensional submanifold that intersects the exceptional submanifold $\pi^{-1}(S)$ in Y transversally. The restriction $\pi_W : W \longrightarrow V$ is the blowing-up of the manifold V in the smooth set $S \cap V$. If V intersects S transversally, then $W = \pi^{-1}(V)$.*

PROOF . The restriction $\pi_W : W \longrightarrow V$ is the blowing-up of the sub-manifold V by means of the family $\{\mathcal{I}(S_z) \circ \iota_z\}_{z \in V}$, where $\iota : V \hookrightarrow X$ (see n° 5). In order to prove that π_W is the blowing-up of V in $S \cap V$ (which – in view of proposition 3 – would imply that W is an l–dimensional submanifold), it is

---

[44] It is locally isomorphic with the natural projections $\Omega \times \mathbf{P}_{n-k-1} \longrightarrow \Omega$ (where the sets Ω are open in S).

[45] i.e., each of its connected components is a submanifold. (See IV. 8.5.)

enough to show that $\mathcal{I}(S_z) \circ \iota_z = \mathcal{I}((S \cap V)_z, V)$ when $z \in S \cap V$. Let f_i be generators of the ideal $\mathcal{I}(S_z)$; then the germs $(f_i)_{V_z}$ generate the ideal $\mathcal{I}(S_z) \circ \iota_z$, and so it suffices to prove that they also generate $\mathcal{I}((S \cap V)_z, V)$. Now (according to the equivalence from II. 4.2), the differentials $d_z f_i$ generate the subspace $(T_z S)^\perp \subset (T_z V)^*$, and hence – in view of $(T_z V) \cap (T_z S) = T_z(V \cap S)$ – the differentials $d_z((f_i)_{V_z}) = (d_z f_i)_{T_z V}$ generate the subspace $T_z(S \cap V)^\perp \subset (T_z V)^*$. Therefore it follows (by the same equivalence in II. 4.2) that $(f_i)_{V_z}$ are generators of the ideal $\mathcal{I}((S \cap V)_z, V)$. In order to prove transversality of the intersection of W and $\pi^{-1}(S)$, take $\zeta \in W \cap \pi^{-1}(S)$ and set $z = \pi(\zeta)$. As $\dim \pi^{-1}(S) = n - 1$, it is enough to check that $T_\zeta W \not\subset T_\zeta \pi^{-1}(S)$. Now, if this were not true, then in view of the fact that the sets S and V intersect transversally, we would have $(d_\zeta \pi)(T_\zeta W) \subset T_z S \cap T_z V = T_z(S \cap V)$. However, by proposition 3 applied to the blowing-up $\pi_W : W \longrightarrow V$ in $S \cap V$, the dimension of the left side is greater than that of the right one. Finally, if V and S intersect transversally, we have $\dim(S \cap V) = k + l - n$. Now, if $z \in S \cap V$, then $(\pi_W)^{-1}(z) \subset \pi^{-1}(z)$, and the dimensions of both sides are equal (because $l - (k + l - n) - 1 = n - k - 1$; see proposition 3). Thus $\pi^{-1}(z) = (\pi_W)^{-1}(z) \subset W$. Consequently, $\pi^{-1}(S \cap V) \subset W$, and so $W = \pi^{-1}(V)$ (see n° 5).

Without the assumption of quasi-transversality, $\pi_W : W \longrightarrow V$ may not be a blowing-up in the set $S \cap V$, even if the latter is a submanifold. For instance, if $X = \mathbf{C}^3$, $S = \{z_1 = 0, z_3 = z_2^2\}$, and $V = \{z_3 = 0\}$, then π is the blowing-up by means of the function $z_1, z_2^2 - z_3$, and then π_W is the blowing-up of the plane V by means of the functions z_1, z_2^2, but it is not a blowing-up at 0 (see the example in n° 6).

§6. Algebraic sets in projective spaces. Chow's theorem

1. Let X be a vector space.

A subset V of the projective space $\mathbf{P}(X)$ is said to be *algebraic* if its cone $V^\sim \subset X$ is algebraic or, equivalently, if its cone $V^\sim$ can be defined by homogeneous polynomials (see II. 3.3, the Cartan-Remmert-Stein lemma).

Therefore a subset V of the projective space $\mathbf{P}(X)$ is an algebraic subset if and only if it is of the form

$$V = \{\mathbf{C}z : z \neq 0, \ f_i(z) = 0, \ i = 1, \dots, r\} =$$
$$= \{\lambda \in \mathbf{P}(X) : \ f_i(\lambda) = 0, \ i = 1, \dots, r\} =$$
$$= (V(f_1, \dots, f_r) \cup 0)^\sim,$$

where $f_1, \ldots, f_r \in \mathcal{P}(X)$ are homogeneous polynomials. Then we say that V is *defined by the homogeneous polynomials* $f_1, \ldots, f_r$.

In the case of the space $\mathbf{P}_n$ (i.e., if $X = \mathbf{C}^{n+1}$), it means (see 2.3) that V is given in the homogeneous coordinates by the equations

$$f_i(z_0, \ldots, z_n) = 0, \quad i = 1, \ldots, r .$$

Then its images under the canonical charts are (see 2.3) the sets

$$\{f_i(z_0, \ldots, z_{s-1}, 1, z_{s+1}, \ldots, z_n) = 0, \ i = 1, \ldots, r\} \subset \mathbf{C}^n, \ s = 0, \ldots, n.$$

Obviously, every algebraic set in the space $\mathbf{P}(X)$ is analytic (see 2.2), but we also have the converse:

CHOW'S THEOREM . *All analytic subsets of the projective space* $\mathbf{P}(X)$ *are algebraic.*

Indeed, suppose that $V \subset \mathbf{P}(X)$ is an analytic set. One may assume that $V \neq \emptyset$. Then (see 2.2) the set $V^{\sim} \setminus 0$ is analytic in $X \setminus 0$ of dimension > 0. Thus, according to the Remmert-Stein theorem (see IV. 6.3), the cone $V^{\sim}$ is analytic in X. Hence it is algebraic by the Cartan-Remmert-Stein lemma (see II. 3.3).

2. Let V be an algebraic subset of the projective space $\mathbf{P} = \mathbf{P}(X)$.

If $V \neq \emptyset$, we have (see 2.2 and II. 1.6):

$$\dim V = \dim V^{\sim} - 1 = \dim(V^{\sim})_0 - 1 ,$$

and moreover, the set V is of constant dimension k if and only if the subset $V^{\sim}$ is of constant dimension $(k+1)$. Therefore

$$\operatorname{codim} V = \operatorname{codim} V^{\sim} = \operatorname{codim}(V^{\sim})_0 ,$$

which implies that for any algebraic subsets $V_1, \ldots, V_k \subset \mathbf{P}$ with non-empty intersection we have the inequality

$$\operatorname{codim}(V_1 \cap \ldots \cap V_k) \leq \sum_1^k \operatorname{codim} V_i$$

(see III. 4.6 and B. 6.10) ([46]) . Next,

$$\sum_1^k \operatorname{codim} V_i \leq \dim \mathbf{P} \implies V_1 \cap \ldots \cap V_k \neq \emptyset ,$$

([46]) Both relations are true for any (algebraic) V, V_i if we define $\dim \emptyset = -1$, and then the latter would imply the next statement.

since then $\dim(V_n \ldots \cap V_k)\tilde{}_0 > 0$ (by the same argument). In particular,

$$\dim V_1 + \dim V_z \geq \dim \mathbf{P} \implies V_1 \cap V_z \neq \emptyset .$$

Furthermore,

$$(V^0)\tilde{} \setminus 0 = (V\tilde{})^0 \setminus 0 \quad \text{and} \quad (V^*)\tilde{} = (V\tilde{})^* \cup 0 .$$

Indeed, we have $V\tilde{} \setminus 0 = \alpha^{-1}(V)$, $(V^0)\tilde{} \setminus 0 = \alpha^{-1}(V^0)$, and $(V^*)\tilde{} \setminus 0 = \alpha^{-1}(V^*)$ (see B. 6.10), which implies both equalities (see IV. 2.1).

Finally, if $V \neq \emptyset$, we have the equivalence

$$(V \text{ is irreducible}) \iff (V\tilde{} \text{ is irreducible})$$

(see 4 B. 6.10; and IV. 2.9, corollary 7 from theorem 4). Therefore $V = \bigcup V_i$ is the decomposition into simple components if and only if $V\tilde{} = \bigcup V_i\tilde{}$ is also of this type.

3. The algebraic subsets of a vector space X are exactly the traces (on X) of algebraic subsets of the projective closure $\bar{X}$.

Indeed (see A. 3.2), any polynomial $f \in \mathcal{P}(X)$ is of the form $f(x) = F(1, x)$, where $F \in \mathcal{P}(\mathbf{C} \times X)$ is a homogeneous polynomial (and vice versa). Thus (after using the identification ι^X; see 3.1), we have the equivalence $f(x) = 0 \iff F(x) = 0$ for each $x \in X$.

If $V \subset X$ is an algebraic subset, then its closure $\bar{V}$ in $\bar{X}$ is also algebraic (in $\bar{X}$).

Indeed, $V = W \setminus X_\infty$, where W is an algebraic subset of $\bar{X}$, which implies (see IV. 2.10, theorem 5) that the closure $\bar{V}$ in $\bar{X}$ is analytic, and hence it is algebraic by Chow's theorem.

Observe that if V is irreducible, then so is $\bar{V}$. If $V = V_1 \cup \ldots \cup V_r$ is the decomposition into simple components, then so is $\bar{V} = \bar{V}_1 \cup \ldots \cup \bar{V}_r$. If $W \subset X$ is algebraic, then the closure of any simple component of $W \cap X$ is a simple component of W.

4. A subset V of the multiprojective space $\mathbf{P}(X_1) \times \cdots \times \mathbf{P}(X_k)$ is said to be *algebraic* if it is of the form

$$V = \{(\lambda_1, \ldots, \lambda_k) \in \mathbf{P}(X_1) \times \cdots \times \mathbf{P}(X_k) : f_\nu(\lambda_1 \times \cdots \times \lambda_k) = 0,$$
$$\nu = 1, \ldots, r\},$$

where $f_\nu(x_1,\ldots,x_k)$ are polynomials on $X_1 \times \cdots \times X_k$ that are homogeneous with respect to each variable x_i $(i = 1,\ldots,k)$ [47]. Then we say that the set V is defined by the polynomials $f_1,\ldots,f_r$. Obviously, it is analytic, because

$$\beta^{-1}(V) = \{x \in (X_1 \setminus 0) \times \cdots \times (X_k) \setminus 0) : f_\nu(x) = 0, \ \nu = 1,\ldots,r\} \ ,$$

and $\beta = \alpha^{X_1} \times \cdots \times \alpha^{X_k}$ is a surjective submersion (see 2.2, C. 4.2, and also II. 3.4). The converse is also true.

CHOW'S THEOREM (for multiprojective space). *All analytic subsets of the multiprojective space* $\mathbf{P}(X_1) \times \cdots \times \mathbf{P}(X_k)$ *are algebraic.*

Indeed, assume that $V \subset \mathbf{P}(X_1) \times \cdots \times \mathbf{P}(X_k)$ is an analytic subset. Then its image $\mathbf{s}(V) \subset \mathbf{P}(X_1 \otimes \cdots \otimes X_k)$ under the Segre embedding (see 2.4) is analytic (see II. 3.4), and hence, by Chow's theorem from n° 1, it is algebraic. Thus

$$\mathbf{s}(V) = \{\lambda \in \mathbf{P}(X_1 \otimes \cdots \otimes X_k) : f_\nu(\lambda) = 0, \ \nu = 1,\ldots,r\} \ ,$$

where $f_\nu \in \mathcal{P}(X_1 \otimes \cdots \otimes X_k)$ are homogeneous polynomials. Consequently,

$$V = \mathbf{s}^{-1}\big(\mathbf{s}(V)\big) = \{(\lambda_1,\ldots,\lambda_k) \ni \mathbf{P}(X_1) \times \cdots \times \mathbf{P}(X_k) : F_\nu(\lambda_1 \times \cdots \times \lambda_k) = 0,$$
$$\nu = 1,\ldots,r\},$$

where $F_\nu(x_1,\ldots,x_k) = f(x_1 \otimes \cdots \otimes x_k)$ are polynomials on $X_1,\ldots,X_k$ that are homogeneous with respect to each of the variables x_i $(i = 1,\ldots,k)$.

The algebraic subsets of the space $X_1 \times \cdots \times X_k$ are precisely the traces (on the space) of algebraic subsets of the "multiprojective closure" $\bar{X}_1 \times \cdots \times \bar{X}_k$.

Indeed (see A. 3.2), every polynomial $f \in \mathcal{P}(X_1 \times \cdots \times X_k)$ is of the form $f(x_1,\ldots,x_k) = F(1,x_1,\ldots,1,x_k)$, where $F(t_1,x_1,\ldots,t_k,x_k)$ is a polynomial on $\mathbf{C} \times X_1 \times \ldots \times \mathbf{C} \times X_k$ which is homogeneous with respect to each pair (t_i,x_i). (Clearly, the converse is also true.) Then (after the identification $\iota^{X_1},\ldots,\iota^{X_k}$; see 3.1) we have the equivalence $f(x_1,\ldots,x_k) = 0 \iff F(x_1 \times \cdots \times x_k) = 0$ for $(x_1,\ldots,x_k) \in X_1 \times \cdots \times X_k$.

If $V \subset X_1 \times \cdots \times X_k$ is an algebraic set, then its closure $\bar{V}$ in $\bar{X}_1 \times \cdots \times \bar{X}_k$ is algebraic (in $\bar{X}_1 \times \cdots \times \bar{X}_k$).

---

[47] Thus an algebraic subset of the multiprojective space $\mathbf{P}_{n_1} \times \cdots \times \mathbf{P}_{n_k}$ is a set of sequences of k elements whose homogeneous coordinates $x_i = (x_i^{(0)},\ldots,x_i^{(n_i)})$, $i = 1,\ldots,k$, satisfy the equations $f_\nu(x_1,\ldots,x_k) = 0$. Here the f_ν are polynomials on $\mathbf{C}^{n_1+1} \times \cdots \times \mathbf{C}^{n_k+1}$, homogeneous with respect to each x_i $(i = 1,\ldots,k)$.

Indeed, in such a case, $V = W \setminus \Sigma$, where W is an algebraic subset of the multiprojective space $\bar{X}_1 \times \cdots \times \bar{X}_k$, while

$$\Sigma = (\bar{X}_1 \times \cdots \times \bar{X}_k) \setminus (X_1 \times \cdots \times X_k) = \bigcup_{i=1}^{k} \bar{X}_1 \times \cdots \times (X_i)_\infty \times \cdots \times \bar{X}_k$$

is an analytic set in $\bar{X}_1 \times \cdots \times \bar{X}_k$. Therefore (see IV. 2.10, theorem 5) the closure $\bar{V}$ in $\bar{X}_1 \times \cdots \times \bar{X}_k$ is analytic, and thus, by Chow's theorem, it is algebraic.

5. In what follows, let X be a vector space, and let M be a manifold. Let $\mathbf{P} = \mathbf{P}(X)$. Putting $e = \mathrm{id}_M$, we have the surjective submersion $\psi = e \times \alpha : M \times (X \setminus 0) \longrightarrow M \times \mathbf{P}$ whose fibres are one–dimensional (see 2.2 and C. 4.2).

We say that a set $V \subset M \times \mathbf{P}$ is defined by the holomorphic X–homogeneous functions $F_1, \ldots, F_r$ in $M \times X$ (see C. 3.18) if

$$V = \{(z, \lambda) \in M \times \mathbf{P} : F_i(z \times \lambda) = 0, \ i = 1, \ldots, r\} \ .$$

Then V is analytic, since

$$\psi^{-1}(V) = \{(z, x) \in M \times (X \setminus 0) : F_i(z, x) = 0, \ i = 1, \ldots, r\}$$

(see II. 3.4).

A set $V \subset M \times \mathbf{P}$ is said to be $\mathbf{P}$–*algebraic* if each point $a \in M$ has an open neighbourhood U such that the set $V_U \subset U \times \mathbf{P}$ is defined by X–homogeneous holomorphic functions in $U \times X$. Thus V is analytic. The converse is also true.

CHOW'S THEOREM (with a parameter). *All analytic subsets of the manifold $M \times \mathbf{P}$ are $\mathbf{P}$–algebraic.*

First, we will prove an analogue of the Cartan-Remmert-Stein lemma (IV. 3.3). We say that a set $E \subset M \times X$ is X–*conic* if it satisfies the condition: $(z, x) \in E, \ t \in \mathbf{C} \Longrightarrow (z, tx) \in E$ (i.e., for each $z \in M$ the set E_z is either a cone or the empty set).

LEMMA . *If $W \subset M \times X$ is a X–conic analytic subset, then each point $a \in M$ has an open neighbourhood U such that $W_U \subset U \times X$ is defined by X–homogeneous holomorphic functions on $U \times X$.*

PROOF of the lemma. There is an open relatively compact neighbourhood $\Omega \times \Delta$ of the point $(a, 0)$ in $M \times X$ such that $W \cap (\Omega \times \Delta) = \{f_1 = \ldots = f_p = 0\}$, where the f_i are holomorphic functions on $\Omega \times \Delta$. Taking smaller

Ω and Δ, one may require (see C. 3.18) that $f_i = \sum_{\nu=0}^{\infty} f_{i\nu}$ in $\Omega \times \Delta$ for $i = 1, \ldots, p$, and that the series is absolutely convergent, where the $f_{i\nu}$ are X–homogeneous holomorphic functions in $\Omega \times X$ of degree ν, respectively. Then it is easy to check (^{47a}) that one must have $W_\Omega = \bigcap_{i\nu}\{f_{i\nu} = 0\}$, i.e., $W_\Omega = \bigcap_0^\infty W_s$, where $W_s = \bigcap_{\nu=0}^s \{f_{1\nu} = \ldots = f_{p\nu} = 0\}$, $s = 0, 1, 2, \ldots$, is a decreasing sequence of analytic sets in $\Omega \times X$. By theorem 2 in IV. 3.2, taking an open relatively compact neighbourhood U of a in Ω, we have $W_\Omega \cap (U \times \Delta) = W_s \cap (U \times \Delta)$ for some s. This implies that $W_U = (W_s)_U$ (as W and W_s are X-conic), which means that the set W_U is defined by the functions $(f_{i\nu})_{U \times X}$ ($i = 1, \ldots, p; \ \nu = 0, \ldots, s$) in $U \times X$.

Before proving the theorem, note that the union of any locally finite family of **P**–algebraic sets $V_i \subset M \times \mathbf{P}$ is **P**–algebraic. Indeed, let $a \in M$. Take a compact neighbourhood Ω of the point a. Only finitely many sets from the family, say $V_{\iota_1}, \ldots, V_{\iota_k}$, have non-empty intersection with $\Omega \times \mathbf{P}$. Now take an open neighbourhood $U \subset \Omega$ of the point a such that each of the sets $(V_{\iota_\nu})_U$ is defined by X–homogeneous holomorphic functions on $U \times X$. Then $(\bigcup V_\iota)_U = \bigcup (V_{\iota_\nu})_U$ is also defined by such functions, by the argument used in II. 3.1.

PROOF of the theorem. Let $V \subset M \times \mathbf{P}$ be an analytic subset. According to the above remark, by replacing V by its irreducible components, one may assume that V is of constant dimension k. Therefore the set $V' = \psi^{-1}(V)$ is analytic in $M \times (X \setminus 0)$ and of constant dimension $k + 1$ (see II. 3.4). It is closed in $(M \times X) \setminus (W \times 0)$, where W is the image of V by the projection onto M (^{48}). Hence it is also analytic in $(M \times X) \setminus (W \times 0)$. Consequently, by Remmert's theorem (see V. 5.1), W is analytic of dimension $\le k$ (see V. 1, corollary 1 of theorem 1). Thus, according to the Remmert-Stein theorem, the set $V'' = V' \cup (W \times 0)$ is analytic in $M \times X$. But it is also X–conic, hence, by the lemma, each point $a \in M$ has an open neighbourhood U such that $(V'')_U = \{(z, x) : F_i(z, x) = 0, \ i = 1, \ldots, p\}$, where F_i are X–homogeneous holomorphic functions on $U \times X$. This implies that

$$V_U = \{(z, \lambda) : F_i(z \times \lambda) = 0, \ i = 1, \ldots, p\} \quad (^{49}).$$

PROPOSITION. *If $V \subset M \times \mathbf{P}$ is an analytic subset of constant dimension $n - 1$, where $n = \dim(M \times \mathbf{P})$, then each point $a \in M$ has an open neigh-*

(^{47a})   The inclusion $\supset$ follows in view of the fact that both sets are X-conic. As for $\subset$, see the end of the proof of the Cartan-Remmert-Stein lemma (IV. 3.3).

(^{48}) Since V' is a closed subset of the set $W \times (X \setminus 0)$, which is closed in $(M \times X) \setminus (W \times 0)$.

(^{49}) Indeed, $V = \{(z, \lambda) : \ z \times (\lambda \setminus 0) \subset V'\} = \{(z, \lambda) : \ z \times \lambda \subset V''\}$, and thus $V_U = \{(z, \lambda) : z \times \lambda \subset V''_U\}$.

bourhood U such that V_U is defined by a single X–homogeneous holomorphic function on $U \times X$.

PROOF . By Remmert's theorem (see V. 5.1), the image W of the set V by the projection onto M is analytic. One may assume that $a \in W$. According to Chow's theorem, V is **P**–algebraic, and hence there exists an open neighbourhood $\tilde{U}$ of the point a such that $V_{\tilde{U}}$ is defined by X–homogeneous holomorphic functions $G_1, \ldots, G_r$ on $\tilde{U} \times X$. It follows that the X–conic set

$$S = \bigcup \{z \times \lambda : (z, \lambda) \in V_{\tilde{U}}\} \subset \tilde{U} \times X$$

is analytic in $\tilde{U} \times X$ (^{50}) . In addition, it is of constant dimension n, since (see IV. 2.5) the set $\psi^{-1}(V_{\tilde{U}}) = S \setminus (\tilde{U} \times 0)$, which is open and dense in S, is of constant dimension n (see II. 3.4). We have $(a, 0) \in S$. Thus the germ $S_{(a,0)}$ is of constant dimension n, and so $S_{(a,0)} = V(f)$, where $f \in \mathcal{O}_{(a,0)} \setminus 0$ is without multiple factors (see II. 5.3). Let F be a holomorphic representative of f in an open neighbourhood $U' \times \Delta' \subset \tilde{U} \times X$ of the point a such that $S \cap (U' \times \Delta') = V(F)$ and $F = \sum_0^\infty F_\nu$ is uniformly convergent in $U' \times \Delta'$. Here the F_ν are holomorphic in $U' \times X$ and X–homogeneous of degree ν, $\nu = 0, 1, \ldots$ (see C. 3.18). One must have $S_{U'} \subset \bigcap_0^\infty V(F_\nu)$ (see footnote (^{47a})), and hence (see the theorem in II. 5.2) each of the germs $(F_\nu)_{(a,0)}$ is divisible by f. Therefore, by the corollary of the proposition from II. 3.8, there is a neighbourhood $\Omega \subset U' \times \Delta'$ of the point $(a, 0)$ such that $(F_\nu)_\Omega = G_\nu F_\Omega$, where G_ν are holomorphic functions on Ω ($\nu = 0, 1, \ldots$). Then $\sum G_\nu = 1$ and the series is almost uniformly convergent in $\Omega \setminus S$. Hence, the series is also uniformly convergent in Ω (see II. 3.9). It follows that $G_k(a, 0) \neq 0$ for some k. Taking an open neighbourhood $U \times \Delta \subset \Omega$ of $(a, 0)$ in which $G_k \neq 0$, we have $S \cap (U \times \Delta) = V((F_k)_{U \times \Delta})$. This implies that $S_U = V((F_k)_{U \times X})$ (since both sets are X-conic). Thus (in view of the equivalence $(z, \lambda) \in V_U \iff z \times \lambda \subset S_U$) the set V_U is defined by the function $(F_k)_{U \times X}$.

6. A set $V \subset M \times X$ is said to be *X–algebraic* if each point $a \in M$ has an open neighbourhood U such that the set V_U is defined by X–polynomials in $U \times X$ (see C. 3.18). Clearly, V is then analytic.

The algebraic X-subsets of the manifold $M \times X$ are precisely the traces (on $M \times X$) of the $\bar{X}$–algebraic subsets of the manifold $M \times \bar{X}$.

(^{50}) Because it is equal to $V(G_1, \ldots, G_r)$.

Indeed, if U is an open connected neighbourhood of a point $a \in M$, then (see C. 3.18) any X–polynomial f on $U \times X$ is of the form $F(z, 1, x)$, where F is a $(\mathbf{C} \times X)$–homogeneous holomorphic function on $U \times \mathbf{C} \times X$. (And also vice versa.) Then (after the identification ι^X — see 3.1) for $(z, x) \in U \times X$ we have the equivalence $f(z, x) = 0 \Longleftrightarrow F(z \times x) = 0$.

If $V \subset M \times X$ is an X–algebraic set, then its closure $\bar{V}$ in $M \times \bar{X}$ is $\bar{X}$–algebraic.

Indeed, $V = W \setminus (M \times X_\infty)$, where W is an analytic subset of the manifold $M \times \bar{X}$. Consequently (see IV. 2.10, theorem 5), the closure $\bar{V}$ in $M \times \bar{X}$ is analytic, and thus – according to Chow's theorem with a parameter (from n° 5) – it is $\bar{X}$–algebraic.

§7. The Rudin and Sadullaev theorems

1. Let M be an n–dimensional vector space, and let V be an algebraic subset of M.

Set $V_\infty = \bar{V} \cap M_\infty$. We have $\dim V_\infty = \dim V - 1$, provided $\dim V > 0$. Indeed, $\dim V_\infty < \dim V$ (because $V_\infty = \bar{V} \setminus V$; see IV. 8.5) and codim $V_\infty \le$ codim $V + 1$ (see 6.2 and IV. 8.5), which gives $\dim V_\infty \ge \dim V - 1$.

A linear subspace $Y \subset M$ is said to be a *Sadullaev subspace for V* if for some (and hence for each) linear complement X of Y (after endowing M with a norm) we have:

(Sd) $V \subset \{x + y : x \in X, y \in Y, |y| < C(1 + |x|)\}$ for some $C > 0$.

This holds exactly when $Y_\infty \cap V_\infty = \emptyset$. Indeed, both conditions are equivalent to $\bar{V} \cap Y_\infty = \emptyset$, which can be seen by taking the base of neighbourhoods of the set Y_∞ in $\bar{M}$ according to 3.4, and since (Sd) implies $V_\infty \subset \{\lambda_\infty : \lambda \subset \{x + y : |y| \le C|x|\}\}$ [50a].

Obviously, the condition (Sd) implies that the natural projection $v \longrightarrow X$ is proper (see V. 7.1 and B. 5.2). Moreover, it follows that if $\dim Y = \text{codim } V$, the projection is surjective (see V. 5.1, Remmert's theorem; and II. 1.4).

Let $1 \le l \le n$. Set $M_0 = 0 \times M$, and let $\iota : \mathbf{G}_l(M_0) \longrightarrow \mathbf{G}_l(M)$ denote the isomorphism induced by the natural isomorphism $M_0 \ni (0, x) \longrightarrow x \in M$.

[50a] See footnotes [20] in 3.4 and [26] in B. 6.1.

Let $p:\ \mathbf{S}_1^l(M_0) \longrightarrow \mathbf{G}_l(M_0)$, and let $\pi:\ \mathbf{S}_1^l(M_0) \longrightarrow \mathbf{P}(M_0) = M_\infty$ be the natural projections. Define $\Sigma = \iota\big(p(\pi^{-1}(V_\infty))\big)$. According to Remmert's theorem (see V. 5.1), it is an analytic subset of $\mathbf{G}_l(M)$.

The set of Sadullaev subspaces $Y \in \mathbf{G}_l(M)$ for V is equal to $\mathbf{G}_l(M) \setminus \Sigma$.

Indeed, the condition $Y \in \Sigma$ means that $0 \times Y \in p(\pi^{-1}(V_\infty))$, which is equivalent to $0 \times Y \supset \lambda \in V_\infty$ for some λ, i.e., to $Y_\infty \cap V_\infty \neq \emptyset$.

SADULLAEV'S THEOREM . *Let $0 \le k \le n$. For any algebraic subset $V \subset M$, a Sadullaev subspace $Y \in \mathbf{G}_{n-k}(M)$ exists if and only if $\dim V \le k$. Then the set of such subspaces is open and dense in $\mathbf{G}_{n-k}(M)$: it is the complement of a nowhere dense analytic ([51]) subset of $\mathbf{G}_{n-k}(M)$.*

PROOF ([52]) . One may assume that $k < n$. Suppose that $\dim V \le k$. Then $\dim V_\infty < k$. Now, the set of Sadullaev subspaces in $\mathbf{G}_{n-k}(M)$ is the complement of the set Σ (where $l = n - k$). Since the fibres of the projection π are the $k(n - k - 1)$–dimensional sets $\{\lambda\} \times \mathbf{S}^{n-k}(\lambda)$, $\lambda \in \mathbf{P}(M)$ (see 4.3), we have $\dim \Sigma = \dim p(\pi^{-1}(V_\infty)) \le \dim \pi^{-1}(V_\infty) \le k(n - k - 1) + \dim V_\infty$ (see II. 1.4). Therefore $\dim \Sigma < k(n - k) = \dim \mathbf{G}_{n-k}(M)$, which proves (see II. 1.2) that Σ is nowhere dense. On the other hand, if $\dim V > k$, then $\dim V_\infty \ge k$, and for each $Y \in \mathbf{G}_{n-k}(M)$ we have $\dim Y_\infty + \dim V_\infty \ge n - 1$. Thus (see 6.2) we must have $Y_\infty \cap V_\infty \neq \emptyset$, which means that Y is not a Sadullaev subspace for V.

COROLLARY . *For every algebraic subset $V \subset M$, there exists a projection $\pi:\ M \longrightarrow X$ onto a subspace $X \subset M$ whose restriction $\pi_V:\ V \longrightarrow X$ is finite and surjective. If, in addition, V is of constant dimension, then π_V is an open *–covering (see V. 7.2).*

2. Let X and Y be vector spaces.

PROPOSITION 1. *Let $V \subset X \times Y$ be an algebraic subset. If the natural projection $V \longrightarrow X$ is proper, then (after selecting norms on X and Y), we have*

$$V \subset \{(x,y):\ |y| \le M(1 + |x^k|)\} \text{ for some } M, k > 0 .$$

Indeed, the closure $\bar{V}$ of the set V in $\bar{X} \times \bar{Y}$ is algebraic (see 6.4), and hence the pair $\bar{V}, \bar{X} \times Y_\infty$ satisfies the condition of regular separation (see the theorem in IV. 7.1). Therefore, by lemma 1 from 3.6, the set V satisfies

---

([51]) It is even the complement of a nowhere dense algebraic set. See the proof, and 17.13–14 below (e.g., Chow's theorem).

([52]) See E. Fortuna [22a].

the condition (r) near X_∞, and so, by lemma 2 from 3.6, we get the required inclusion.

According to Liouville's theorem (see C. 1.8), we have the following corollary (a special case of Serre's theorem; see 16.3 below):

COROLLARY 1. *Every holomorphic mapping* $X \longrightarrow Y$ *with algebraic graph is a polynomial.*

COROLLARY 2. (^{53}) . *The inverse of any biholomorphic polynomial mapping is a polynomial.*

3. Let M be a manifold, and let $Z \subset M$ be a nowhere dense analytic subset. Let N be a vector space.

PROPOSITION 2. *Let* V *be an analytic subset of* $(M \setminus Z) \times N$ *of constant dimension* $m = \dim M$ *such that the natural projection* $V \longrightarrow M \setminus Z$ *is proper. Let* $\bar{V}$ *be the closure of* V *in* $\bar{M} \times \bar{N}$. *Then the following conditions are equivalent:*

(1) *the set* V *satisfies the condition* (r) *near* Z;

(2) *the pair* $\bar{V}$, $M \times N_\infty$ *satisfies the condition of regular separation;*

(3) *the set* $\bar{V}$ *is analytic in* $M \times \bar{N}$.

PROOF (^{54}) . We already have the implications (3) $\Longrightarrow$ (2) $\Longrightarrow$ (1) (see the theorem in IV. 7.1, and lemma 1 in 3.6). It remains to show the implication (1) $\Longrightarrow$ (3). Let $a \in M$, and assume the inclusion (r) from 3.6 with $E = V$ and $X = N$. It is enough to show that for some open neighbourhood U of a the closure of V_U in $U \times \bar{N}$ is analytic (see II. 3.4). Clearly, one may assume that M is an open subset of a normed vector space, and that $\varrho_\varphi = \varrho$ is the metric associated with the norm. Furthermore, one may assume that $Z = \{f = 0\}$, where $f \not\equiv 0$ is a holomorphic function on M, and that

$$V \subset \{(u,v) \in (M \setminus Z) \times N : |v| \le |f(u)|^{-s}\}, \quad \text{where } s > 0 \, .$$

Indeed, take a non-zero holomorphic function f on an open neighbourhood Δ of a which vanishes on Z. We may require that $|f(u)| \le \varepsilon \varrho(u, Z)$ in Δ with an arbitrarily small $\varepsilon > 0$. Then take a suitably smaller Δ and replace M by Δ, Z by $\{f = 0\}$, and V by $V_{\{f=0\}}$; thus one obtains the above inclusion, and it suffices to observe that the set $V_{\{f \neq 0\}}$ is dense in V (^{55}) . This is so (see B. 2.1), because the natural projection $V \longrightarrow \Delta$ is open (see V. 7.1, proposition 1), since its fibres are finite (see the proposition in IV. 5).

(^{53}) See T. Winiarski [45].

(^{54}) See E. Fortuna [22a].

(^{55}) Hence both sets have the same closure.

Consider the biholomorphic mapping $h : (M \setminus Z) \times N \ni (u,v) \longrightarrow$ $(u, f(u)^{s+1}v) \in (M \setminus Z) \times N$. The image $h(V)$, being contained in the set $F = \{(u,w) \in M \times N : |w| \le |f(u)|\}$, is analytic in $(M \times N) \setminus (0 \times Z)$. Since $\dim(0 \times Z) = \dim Z < \dim M = \dim h(V)$, then, by the Remmert-Stein theorem (see IV. 6.3), the closure $W = \overline{h(V)}$ in $M \times N$ is analytic. Since $W \subset F$, the set W is also analytic in $M \times \bar{N}$. By Chow's theorem with a parameter (see 6.5), it is $\bar{N}$–algebraic and hence also N–algebraic in $M \times N$ (see 6.6). Therefore there is an open neighbourhood U of a such that $W_U = \{p_i(u,w) = 0, \ i = 1, \dots, r\}$, where the p_i are N–polynomials on $U \times N$ (see C. 3.18). Since $W_{U \setminus Z} = h(V_U)$, it follows that $V_U = h^{-1}(W_{U \setminus Z}) = V'_{U \setminus Z}$, where $V' = \{p_i(u, f(u)^{s+1}v) = 0, \ i = 1, \dots, r\}$ is an N–algebraic subset of $U \times N$. Hence $V' = V'' \cap (U \times N)$, where V'' is an $\bar{N}$–algebraic subset of $U \times \bar{N}$ (see 6.6). Consequently, $V_U = V'_{U \setminus Z} = V'' \setminus ((U \times N_\infty) \cup (Z \times \bar{N}))$, which implies (see IV. 2.10, theorem 5) that the closure of V_U in $U \times \bar{N}$ is analytic.

4. Let M be a vector space.

RUDIN'S THEOREM . *An analytic set $V \subset M$ of constant dimension $k \ge 0$ is algebraic if and only if there is a k–dimensional vector subspace $X \subset M$ and its linear complement Y such that (after endowing M with a norm):*

(Rd) $V \subset \{x + y : x \in X, y \in Y, |y| \le C(1 + |x|^s)\}$ *for some $C, s > 0$* .

The condition (Rd) is necessary by Sadullaev's theorem (see n° 1). Suppose now that the condition (Rd) is satisfied. One may assume (after natural indentifications) that $M = X \times Y$. Then the natural projection $V \longrightarrow X$ is proper (see B. 5.2).

PROOF I ([56]) . According to lemma 2 from 3.6, the set V satisfies the condition (r) near X_∞. Therefore, by proposition 2, its closure $\bar{V}$ in $\bar{X} \times \bar{Y}$ is analytic. Hence it is algebraic (by Chow's theorem), and consequently, the set $V = \bar{V} \cap (X \times Y)$ is algebraic (see 6.4).

PROOF II. In view of the corollary of the Andreotti-Stoll theorem (see V. 7.2), there exists a nowhere dense analytic set $Z \subset X$ such that the pair $(Z, V_{X \setminus Z})$ is a quasi-cover with adherence V. Let p denote the multiplicity of the quasi-cover. According to the first lemma on quasi-covers (see III. 1.3), we have $V = F^{-1}(0)$, where $F : X \times Y \longrightarrow \mathbf{C}^l$ is a holomorphic mapping

---

([56]) See E. Fortuna [22a].

such that

$$(*) \qquad\qquad F(x, y) = P(\eta_1, \ldots, \eta_p; y)$$

when $\{\eta_1, \ldots, \eta_p\} = V_x$ and $(x, y) \in (X \setminus Z) \times Y$, while $P : Y^{p+1} \longrightarrow \mathbf{C}^l$ is a polynomial mapping. Since $|P(\eta_1, \ldots, \eta_p; y)| \leq K\big(1 + \sum_1^p |\eta_i| + |y|\big)^r$ in Y^{p+1}, where $K, r > 0$, the formula $(*)$ and the condition (Rd) imply that $|F(x, y)| \leq K'(1 + |x| + |y|)^{rs}$ in $(X \setminus Z) \times Y$ for some $K' > 0$, and hence the same holds in $X \times Y$. Therefore, by Liouville's theorem (see C. 1.8), the mapping F is polynomial, and hence V is algebraic.

COROLLARY (THE RUDIN-SADULLAEV THEOREM). *If $V \subset M$ is an analytic set of constant dimension $k \geq 0$, then the following conditions are equivalent:*

(1) *V is algebraic;*

(2) *there is a k–dimensional vector subspace $X \subset M$ and its linear complement Y such that (after endowing M with a norm)*

$$V \subset \{x + y : \ |y| < C(1 + |x|^s)\} \text{ for some } C, s > 0 \ ;$$

(3) *there is a k–dimensional vector subspace $X \subset M$ and its linear complement Y such that (after endowing M with a norm)*

$$V \subset \{x + y : \ |y| < C(1 + |x|)\} \text{ for some } C > 0 \ .$$

§8. Constructible sets. The Chevalley theorem

1. In the multiprojective space $\mathbf{P}(X_1) \times \cdots \times \mathbf{P}(X_k)$, where X_i are vector spaces, the *constructible sets* are defined as the analytically constructible sets. The class of those sets is, in view of Chow's theorem (see 6.4), the algebra of sets generated by the class of algebraic subsets (see IV. 8.4, corollary 3 of proposition 7).

Using the Chevalley-Remmert theorem (see V. 5.1), we derive the following properties.

The operations of composition, Cartesian product, diagonal product, and restriction to a constructible set, applied to mappings of subsets of multiprojective spaces, preserve the constructibility of graphs.

The image and the inverse image of any constructible set under a mapping with constructible graph are constructible sets.

2. Let X be a vector space (or – more generally – an affine space).

Its subsets of the form $V \setminus W$, where $V, W \subset X$ are algebraic, are called *quasi-algebraic*. Obviously, such sets are locally analytic. Note that a finite intersection of quasi-algebraic subsets is a quasi-algebraic subset (whereas the union of two quasi-algebraic subsets does not have to be quasi-algebraic). The Cartesian product of quasi-algebraic subsets is quasi-algebraic [57].

The *constructible sets* in the space X are defined to be the elements of the algebra of sets generated by the class of algebraic subsets of X. These sets are precisely finite unions of quasi-algebraic ones, i.e., the sets of the form

$$(K) \qquad \bigcup_1^r (V_i \setminus W_i), \quad \text{where} \quad V_i, W_i \subset X \quad \text{are algebraic} \quad (^{58}) .$$

Indeed, it is easy to check that the sets of this form constitute an algebra of sets.

Clearly, a finite union, a finite intersection, the complement, and the difference of constructible sets are constructible. If $Z \subset X$ is a vector subspace (or an affine subspace), then the trace on Z of any constructible set in X is constructible in Z, and for any $E \subset Z$ we have the equivalence:

$$(E \text{ is constructible in } Z) \Longleftrightarrow (E \text{ is constructible in } X) .$$

The Cartesian product of constructible sets is constructible.

Obviously, every constructible set is analytically constructible. Therefore we have the following properties (see IV. 8.3 and 5):

Every analytically constructible set is of calss F^σ and is a locally connected space.

If $E \subset X$ is a constructible set, then the set $\bar{E} \setminus E$ is nowhere dense in $\bar{E}$. If, in addition, $E \subset F \subset X$, then

$$(E \text{ is nowhere dense in } F) \Longleftrightarrow \text{int}_F E = \emptyset .$$

Let $E \subset X$ be a constructible set. Then

$$\dim \bar{E} = \dim E; \quad \dim_z \bar{E} = \dim_z E \quad \text{for} \quad z \in X;$$

$$\dim(\bar{E} \setminus E^{(k)}) < \dim E \quad \text{if} \quad k = \dim E \geq 0 .$$

---

[57] It is enough to observe that $(V \setminus W) \times (V' \setminus W') = (V \times V') \setminus ((V \times W') \cup (W \times V'))$.

[58] Moreover, one may require that $W_i \subsetneq V_i$ and V_i are irreducible.

Let $E \subset F \subset X$ be constructible sets. Then

$$(E \text{ is nowhere dense in } F \neq \emptyset) \Longleftrightarrow \dim E < \dim F;$$

$$(E \text{ is nowhere dense in } F) \Longleftrightarrow (\dim_z E < \dim_z F \quad \text{for} \quad z \in E).$$

3. Let $X_1, \ldots, X_k$ be vector spaces, and let $\bar{X}_1, \ldots, \bar{X}_k$ be their projective closures.

PROPOSITION 1. *A set $E \subset X_1 \times \cdots \times X_k$ is constructible in $X_1 \times \cdots \times X_k$ if and only if it is constructible in $\bar{X}_1 \times \cdots \times \bar{X}_k$. The trace on $X_1 \times \cdots \times X_k$ of any constructible set in $\bar{X}_1 \times \cdots \times \bar{X}_k$ is constructible in $X_1 \times \cdots \times X_k$.*

Indeed, the set $Y = X_1 \times \cdots \times X_k$ is constructible in $\tilde{Y} = \bar{X}_1 \times \cdots \times \bar{X}_k$. Now, if $E = \bigcup_1^r (V_i \setminus W_i)$, where $V_i, W_i \subset Y$ are algebraic, then $E = Y \cap \bigcup_1^r (\bar{V}_i \setminus \bar{W}_i)$ is constructible in $\tilde{Y}$, because $\bar{V}_i, \bar{W}_i$ are algebraic in $\tilde{Y}$ (see 5.4). On the other hand, if $F = \bigcup_1^r (V_i \setminus W_i)$, where $V_i, W_i \subset \tilde{Y}$ are algebraic in $\tilde{Y}$, then the trace $F \cap Y = \bigcup_1^r ((V_i \cap Y) \setminus (W_i \cap Y))$ is constructible in Y, as $V_i \cap Y$, $W_i \cap Y$ are algebraic subsets of Y (see 6.4).

It follows that for subsets of a vector space X we have (see 6.1, Chow's theorem):

PROPOSITION 2. *The closure of a constructible set is an algebraic set. A set is algebraic if and only if it is constructible and closed.*

For the closure of any set $E \subset X$ is the trace on X of the closure of E in the projective closure $\bar{X}$.

Therefore the interior of a constructible set is constructible (and moreover, it is the complement of an algebraic set).

COROLLARY 1. *If $E \subset X$ is a constructible set, then the following conditions are equivalent*

(1) *E is quasi-algebraic;*

(2) *$E = V \setminus Z$, where $V, Z \subset X$ are algebraic and Z is nowhere dense in V;*

(3) *E is locally closed.*

If any of the conditions is satisfied, then the representation (2) is unique. Namely, $V = \bar{E}$ and $Z = \partial E$.

Indeed, as far as the implication (3) $\Longrightarrow$ (2) is concerned, it is enough to set $V = \bar{E}$ and $Z = \partial E$ (see B. 1), whereas the implications (2) $\Longrightarrow$ (1) $\Longrightarrow$ (3) are trivial.

COROLLARY 2. *Any constructible set which is closed or open in a quasi-*

algebraic set is quasi-algebraic (^{59}) .

We are going to give a more direct proof of proposition 2. In view of (K), it suffices to show that if $V \subset X$ is an irreducible algebraic set, then every algebraic set $W \subsetneq V$ is nowhere dense in V. Therefore it is enough to prove the following lemma.

LEMMA . *Let $V \subset X$ be an irreducible algebraic set. Any polynomial $P \in \mathcal{P}(X)$ that vanishes on a neighbourhood in V of a point $a \in V$ must vanish on the whole set V.*

PROOF I. We have $V = V(F_1, \ldots, F_k)$, where $F_i \in \mathcal{P}(X)$. By Hilbert's Nullstellensatz applied to $\mathcal{R}_a$ (see 16.1 below (^{60})), we have $P^r = \Sigma(G_i/H)F_i$ in a neighbourhood of the point a for some $r \in \mathbf{N} \setminus 0$. Here $G_i, H \in \mathcal{P}(X)$ and $H(a) \neq 0$. Then $HP^r = \Sigma G_i F_i \in \mathcal{I}(V)$. Since the ideal $\mathcal{I}(V)$ is prime and $H \notin \mathcal{I}(V)$, one must have $P \in \mathcal{I}(V)$.

PROOF II. (^{61}). We have $P = 0$ in $V \cap U$, where U is a neighbourhood of the point a in X. Obviously, one may assume that $a = 0$. Let $I = \mathcal{I}(V)$. Then $V = V(I)$ (see 1.2). One may assume that $X = \mathbf{C}^n$, and that the assumptions of Rückert's lemma and the condition (1) from the proposition in 9.3 are satisfied (^{62}). Therefore (omitting the trivial case when $k = n$), since the projection $V \ni (u, v) \longrightarrow (u, z_{k+1}) \in \mathbf{C}^{k+1}$ is proper (see 9.3 (**)), we have $V_\Omega \subset U$ for some open neighbourhood Ω of zero in $\mathbf{C}^{k+1}$ (see B. 2.4). Hence Rückert's formula (R) and the theorem on continuity of roots (see B. 5.3) imply that

(#) $$V_{\{\delta \neq 0\}} \cap U \neq \emptyset .$$

Now, the ring $\hat{\mathcal{P}}_n$ is integral over the ring $\hat{\mathcal{P}}_k \approx \mathcal{P}_k$ (see 9.1), and so (see A. 8.2) the element $\hat{P} \in \hat{\mathcal{P}}_n$ has a minimal polynomial $\hat{Q} \in \hat{\mathcal{P}}_k[T]$, where $Q = T^m + a_1 T^{m-1} + \cdots + a_m \in \mathcal{P}_k[T]$. Thus $Q(P) \in I$, which means that $P^m + a_1 P^{m-1} + \cdots + a_m = 0$ on V. It follows by (#) (see the corollary from Rückert's lemma) that $a_m = 0$. Since the polynomial $\hat{Q}$ is irreducible (see A. 8.2), one must have $m = 1$. Therefore $P = 0$ on V.

Let X and Y be vector spaces. The Chevalley-Remmert theorem (see V. 5.1) yields

THE CHEVALLEY THEOREM . *The image under the natural projection $\pi : X \times Y \longrightarrow X$ of any constructible set in $X \times Y$ is a constructible set in X.*

Indeed, if the set $E \subset X \times Y$ is constructible, then, by proposition 1, it is also constructible in $\bar{X} \times \bar{Y}$. But $\pi(E) = \bar{\pi}(E)$, where $\bar{\pi} : \bar{X} \times \bar{Y} \longrightarrow \bar{X}$ is the natural projection. Thus $\pi(E)$ is a constructible set in $\bar{X}$, and hence, by proposition 1, it is also constructible in X.

COROLLARY . *The operations of composition, Cartesian product, diagonal product, or restriction to a constructible set, applied to mappings of subsets of vector spaces, preserve the constructibility of graphs* (^{63}) . *The image and*

(^{59}) Since it is then locally closed.

(^{60}) In the proof of that result, no properties of constructible sets are used.

(^{61}) Suggested by T. Winiarski.

(^{62}) In §9, we do not use any properties of constructible sets.

(^{63}) Thus the sum, the product, and the quotient (with a non-zero denominator) of functions (defined on a subset of a vector space) with constructible graphs is a function with constructible graph.

the inverse image of a constructible set under a mapping with constructible graph (in particular – under a polynomial mapping) is a constructible set.

We will also give a direct proof of Chevalley's theorem.

LEMMA 1. *If G is a finite subgroup of the group of automorphisms of the linear space X, then each G–invariant constructible subset of X is a finite union of sets of the form $A \setminus B$, where $A, B \subset X$ are G–invariant algebraic subsets.*

Indeed, we have $G = \{\varphi_1, \ldots, \varphi_l\}$ and every G–invariant constructible set $E \subset X$ is (according to the characterization (K) of its complement; see n° 2) a finite intersection of sets of the form $A \cup (\setminus B)$, where $A \subset B \subset X$ are algebraic. Since $E = \bigcap_1^l \varphi_i(E)$, the set E is a finite intersection of sets of the form $\bigcap_1^l (A_i \cup (\setminus B_i))$, where $A_i = \varphi_i(A)$ and $B_i = \varphi_i(B)$. Now it is enough to observe that, since $A_i \subset B_i$, we must have $\bigcup_1^l (A_i \cup (\setminus B_i)) = \bigcup_{s=0}^l (C_s \setminus D_s)$, with the sets $C_s = \bigcup_\alpha \bigcap_{\nu=1}^s A_{\alpha_\nu}$, $D_s = \bigcup_\alpha \bigcap_{\nu=s+1}^l B_{\alpha_\nu}$. Here $\alpha = (\alpha_1, \ldots, \alpha_l)$ are the permutations of $\{1, \ldots, l\}$ which are G–invariant and algebraic.

LEMMA 2. *Let $f_i, g_i \in \mathcal{P}_n$, $i = 1, \ldots, m$. Then the inverse image of any constructible set in $\mathbf{C}^m$ under the mapping*

$$\{f_1 \neq 0, \ldots, f_m \neq 0\} \ni z \longrightarrow (g_1(z)/f_1(z), \ldots, g_m(z)/f_m(z)) \in \mathbf{C}^m$$

is a constructible set in $\mathbf{C}^n$.

This follows from the characterization (K) of constructible sets (see n° 2).

Now let $n_1, \ldots, n_l \in \mathbf{N} \setminus 0$, and let $n = n_1 + \cdots + n_l$. The mappings $\pi_{\alpha_1} \times \cdots \times \pi_{\alpha_l} : \mathbf{C}^n \longrightarrow \mathbf{C}^n$, where $\alpha_i = (\alpha_{i1}, \ldots, \alpha_{in_i})$ are permutations of the sets $\{1, \ldots, n_i\}$, respectively, and $\pi_{\alpha_i} : \mathbf{C}^{n_i} \ni (\zeta_1, \ldots, \zeta_{n_i}) \longrightarrow (\zeta_{\alpha_1}, \ldots, \zeta_{\alpha_{n_i}}) \in \mathbf{C}^{n_i}$, form a finite subgroup S of the group of automorphisms of the vector space $\mathbf{C}^n$. The subsets of $\mathbf{C}^n$ and the mappings of $\mathbf{C}^n$ that are S–invariant will be called (temporarily) symmetric.

The mappings $\varphi \times e : \mathbf{C}^{n+1} \longrightarrow \mathbf{C}^{n+1}$, where $\varphi \in S$ and $e = \mathrm{id}_\mathbf{C}$, form a finite subgroup of automorphisms of $\mathbf{C}^{n+1}$. The subsets of $\mathbf{C}^{n+1}$ that are invariant with respect to this subgroup will be (temporarily) called symmetric.

Obviously, the image of a symmetric subset of $\mathbf{C}^{n+1}$ under the mapping $(z_1, \ldots, z_{n+1}) \longrightarrow (z_1, \ldots, z_n)$ is a symmetric subset of $\mathbf{C}^n$.

Let $\sigma_1^{(i)}, \ldots, \sigma_{n_i}^{(i)}$ be the basic symmetric polynomials from $\mathcal{P}_{n_i}$, and let $\sigma^{(i)} = (\sigma_1^{(i)}, \ldots, \sigma_{n_i}^{(i)}) : \mathbf{C}^{n_i} \longrightarrow \mathbf{C}^{n_i}$, $i = 1, \ldots, l$. Let $\sigma = \sigma^{(1)} \times \cdots \times \sigma^{(l)} : \mathbf{C}^n \longrightarrow \mathbf{C}^n$. This is a polynomial surjection (see B. 5.3).

The next lemma follows by induction on l from the theorem on symmetric polynomials (see A. 4.1).

LEMMA 3. *Every symmetric polynomial from $\mathcal{P}_n$ is of the form $F \circ \sigma$, where $F \in \mathcal{P}_n$.*

(To prove this statement, we apply the theorem on symmetric polynomials to the ring $\mathcal{P}_{n_1 + \cdots + n_{l-1}}[X_1, \ldots, X_{n_l}]$.)

Lemma 1, in view of lemma 0 from III. 1.1 and lemma 3, implies:

LEMMA 4. *Every symmetric constructible set in $\mathbf{C}^n$ is of the form $\sigma^{-1}(E)$, where $E \subset \mathbf{C}^n$ is a constructible set.*

(One uses the formula $V(F \circ \sigma) = \sigma^{-1}(V(F))$ for $F \in \mathcal{P}_n$.)

Set $P_i(c, t) = t^{n_i} + c_1^{n_i - 1} + \cdots + c_{n_i}$ for $c = (c_1, \ldots, c_{n_i}) \in \mathbf{C}^{n_i}$ and $t \in \mathbf{C}$, $i = 1, \ldots, l$. Let $1 \leq k \leq l$.

ELIMINATION LEMMA . *The image of the set*

$$\Theta = \{(a_1, \ldots, a_l, t) \in \mathbf{C}^{n+1} : P_1(a_1, t) = \ldots = P_k(a_k, t) = 0, \; P_{k+1}(a_{k+1}, t) \neq 0, \ldots$$
$$\ldots, P_l(a_l, t) \neq 0\}$$

under the projection $\mathbf{C}^{n+1} \ni (a_1, \ldots, a_l, t) \longrightarrow (a_1, \ldots, a_l) \in \mathbf{C}^n$ *is a constructible set.*

PROOF . Let $\pi : \mathbf{C}^{n+1} \ni (z_1, \ldots, z_n, t) \longrightarrow (z_1, \ldots, z_n) \in \mathbf{C}^n$. Set $Q_i(\eta, t) = (t - \eta_1) \ldots (t - \eta_{n_i})$ for $\eta = (\eta_1, \ldots, \eta_{n_i}) \in \mathbf{C}^{n_i}$ and $t \in \mathbf{C}$, $i = 1, \ldots, l$. Then $Q_i = P_i \circ (\sigma^{(i)} \times e)$ (see A. 4.1). Therefore

$$\Lambda = \{(\zeta_1, \ldots, \zeta_l, t) \in \mathbf{C}^{n+1} : Q_1(\zeta_1, t) = \ldots = Q_k(\zeta_k, t) = 0,$$
$$Q_{k+1}(\zeta_{k+1}, t) \neq 0, \ldots, Q_l(\zeta_l, t) \neq 0\} = (\sigma \times e)^{-1}(\Theta).$$

This is a symmetric set. Hence its projection $\pi(\Lambda)$ is symmetric. But $\pi(\Lambda)$ is constructible, because

$$\pi(\Lambda) = \bigcup_{\substack{1 \leq \nu_i \leq n_i \\ (i = 1, \ldots, k)}} \bigcup_{\substack{1 \leq \mu \leq n_j \\ (j = k+1, \ldots, l)}} \{\zeta_{j\mu} \neq \zeta_{1\nu_i} = \ldots = \zeta_{k\nu_k}\},$$

where $\zeta_i = (\zeta_{i1}, \ldots, \zeta_{in_i})$, $i = 1, \ldots, l$. By lemma 4, $\pi(\Lambda) = \sigma^{-1}(E)$, where $E \subset \mathbf{C}^n$ is a constructible set. Consequently, in view of the identities $\Theta = (\sigma \times e)(\Lambda)$ and $\pi \circ (\sigma \times e) = \sigma \circ \pi$, we get $\pi(\Theta) = \pi((\sigma \times e)(\Lambda)) = \sigma(\pi(\Lambda)) = E$.

PROOF of Chevalley's theorem. Naturally, it is enough to prove that the image of any constructible set in $\mathbf{C}^{m+1}$ under the projection $p: \mathbf{C}^{m+1} \ni (z, t) \longrightarrow z \in \mathbf{C}^m$ is a constructible set in $\mathbf{C}^m$. Since every constructible set in $\mathbf{C}^{m+1}$ is a finite union of sets of the form

$$E = \{F_1 = \ldots = F_k = 0, F_{k+1} \neq 0, \ldots, F_l \neq 0\}, \quad \text{where} \quad F_1, \ldots, F_l \in \mathcal{P}_{m+1},$$

it is sufficient to show that the image of any set of that form is constructible in $\mathbf{C}^n$.

We have $F_i(z, t) = f_{i0}(z)t^{n_i} + \cdots + f_{in_i}(z)$, and one may assume that $n_i \geq 1$, $i = 1, \ldots, l$. Set $n = n_1 + \cdots + n_k$. If $n = 1$, then $l = 1$, $n_1 = 1$, and the statement is easy to check. Let $n > 1$, and assume that the statement is true for $n - 1$. We have $E = (E \cap (B \times C)) \cup \bigcup_1^l E \cap \{f_{i0} = 0\}$, where $B = \{z : f_{10} \neq 0, \ldots, f_{l0} \neq 0\}$. By the induction hypothesis, the sets $\pi(E \cap \{f_{i0} = 0\})$ are constructible. Furthermore, lemma 2 and the elimination lemma imply that the set

$$\pi(E \cap (B \times C)) = \{z \in B : (f_{11}/f_{10}, \ldots, f_{1n_1}/f_{10}, \ldots, f_{l1}/f_{10}, \ldots, f_{ln_i}/f_{10})(z) \in \pi(\Theta)\}$$

is also constructible. This completes the proof.

Let X and Y be vector spaces.

LEMMA . *If $F \subset X \times Y$ is a constructible set, then the sets $\{z \in X;\ \#F_z \geq k\}$, $k = 1, 2, \ldots$, are constructible. In particular* [64] *, if $f :\ E \longrightarrow Y$, where $E \subset X$, is a mapping with constructible graph* [65] *, then the sets $\{w \in Y :\ \#f^{-1}(w) \geq k\}$, $k = 1, 2, \ldots$, are constructible.*

Indeed, the k-th of these sets is the image by the projection $X \times Y \longrightarrow X$ of the constructible set

$$\{(z, w_1, \ldots, w_k) \in X \times Y :\ (z, w_1), \ldots, (z, w_k) \in F,\ w_1, \ldots, w_k \text{ are distinct}\}.$$

4. In what follows, let X be an n–dimensional vector space.

In view of proposition 2 from n° 3, we have analogues of proposition 7 and corollary 2 from IV. 8.4 for constructible sets (in X).

A set $E \subset X$ is constructible if and only if the sets $V_i(E)$ are algebraic and $V_s = \emptyset$ for some s. Then $V_{i+1}(E)$ is nowhere dense in $V_i(E)$, $i = 0, 1, \ldots$, we have $V_i(E) = \emptyset$ for $i > n$, and if $2r > n$, then

$$E = \big(V_0(E) \setminus V_1(E)\big) \cup \ldots \cup \big(V_{2r-2}(E) \setminus V_{2r-1}(E)\big) .$$

The constructible sets are precisely the sets of the form $(V_0 \setminus V_1) \cup \ldots \cup (V_{2k} \setminus V_{2k+1})$, where $V_0 \supset \ldots \supset V_{2k+1}$ are algebraic subsets such that V_{i+1} is nowhere dense in V_0 $(i = 0, \ldots, 2k)$.

In other words, every constructible set $E \subset X$ has a decomposition into (disjoint) quasi-algebraic subsets S_i. Namely

(∗) $E = S_0 \cup \ldots \cup S_k$, where S_{i+1} is nowhere dense in ∂S_i, $i = 1, \ldots, k$.

The decomposition (∗) is unique (assuming that $S_k \neq \emptyset$ if $E \neq \emptyset$, and $k = 0$ if $E = \emptyset$). This is so because for a decomposition (∗) one must have $S_0 = \bar{E} \setminus \overline{(\bar{E} \setminus E)}$ [66] .

5. By a *constructible leaf* we mean a non-empty connected submanifold $\Gamma \subset X$ which is a constructible set or, equivalently, such that $\bar{\Gamma}$ and $\partial \Gamma$ are

---

[64] By interchanging X and Y.

[65] Then E, being the image of the graph of f by the projection $X \times Y \longrightarrow X$, is also constructible.

[66] Indeed, $\bar{E} = \bar{S}_0 \supset \ldots \supset \bar{S}_k$, so $S_1 \cup \ldots \cup S_k$ is nowhere dense in ∂S_0. Hence $\bar{E} \setminus E = \partial S_0 \setminus (S_1 \cup \ldots \cup S_k)$ is dense in ∂S_0, which means that $\partial S_0 = \overline{\bar{E} \setminus E}$.

algebraic. (Thus a constructible leaf is always quasi-algebraic.) By proposition 1 in n° 3, every analytically constructible leaf in X is a constructible leaf (see IV. 8.3, corollary 1 of proposition 5).

By a *constructible stratification* of an algebraic set V we mean a finite partition of V into constructible leaves Γ_ν^i such that $\dim \Gamma_\nu^i = i$ and each set $\partial \Gamma_\mu^k$ is the union of some Γ_ν^i, $i < k$ (⁶⁷).

For any finite family of constructible subsets of the space X, there is a constructible stratification of X that is compatible with this family.

Indeed, it is enough to take the complex stratification $\{\Gamma_\nu^i\}$ of the projective closure $\bar{X}$ which is compatible with this family and with the set X_∞ (see IV. 8.4, proposition 6). Then the leaves $\Gamma_\nu^i \subset X$ form the desired stratification.

In particular, every algebraic set has a constructible stratification.

The connected components of a constructible set are constructible sets; there is a finite number of them.

Finally (see n° 3 proposition 1, and IV. 8.5), if a set $E \subset X$ is constructible, then so are the sets E^0, E^*, and $E^{(k)}$ ($k = 0, \ldots, n$) (⁶⁸) . The connected components of the set E^0 are constructible leaves.

6. Let X and Y be vector spaces.

PROPOSITION 3. *If $f : E \longrightarrow Y$, where $E \subset X$, is a mapping with constructible graph (⁶⁹) , then there is a smooth quasi-algebraic set $H \subset E$ which is open and dense in E and such that the restriction $f_H : H \longrightarrow Y$ is holomorphic.*

In fact, it is sufficient to apply the proposition from V. 5.2 to the manifolds $\bar{X}, \bar{Y}$ and the mapping $f : E \longrightarrow \bar{Y}$ (see n° 3, proposition 1 and corollary 1 of proposition 2).

7. We will also prove that the graph of the mapping $\alpha = \alpha_k^X : B_k(X) \to \mathbf{G}_k(X)$ (see B. 6.1) is analytically constructible in both $(\bar{X})^k \times \mathbf{G}_k(X)$ and $\bar{X}^k \times \mathbf{G}_k(X)$.

First observe that in the case $k = 1$, the graph of the mapping $\alpha : X \backslash 0 \ni$

(⁶⁷) If the stratification is compatible with an algebraic set $W \subset V$, then the leaves $\Gamma_\nu^i \subset W$ form a *constructible stratification of W*.

(⁶⁸) Since $E^{(k)} \subset (\bar{E})^{(k)}$, the regular points (of dimension k) of a constructible set can be characterized in terms of polynomials. See 1.7.

(⁶⁹) Then, according to the Chevalley theorem, the set E is also constructible.

$z \longrightarrow \mathbf{C}z \in \mathbf{P}(X)$ is constructible in $\bar{X} \times \mathbf{P}(X)$. In fact, $\alpha = \{F(\lambda \times \mu) = 0\} \setminus \{G(\lambda \times \mu) = 0\}$, where $F(t, z, w) = z - tw$ and $G(t, z, w) = tz$.

In the general case, we have $\mathbf{p} \circ \alpha = \alpha_1^{\Lambda^k X} \circ \gamma$, where $\gamma : X^k \ni (z_1, \ldots, z_k) \longrightarrow z_1 \wedge \ldots \wedge z_k \in \Lambda^k X$ and $\mathbf{p} : \mathbf{G}_k(X) \longrightarrow \mathbf{G}_k(X)$ is the Plücker biholomorphic mapping. Thus (see n° 1) the graph of the mapping $\mathbf{p} \circ \alpha$, that is, $(e \times \mathbf{p})(\alpha)$ (where e is the identity mapping of $\overline{X^k}$), is constructible in $\overline{X^k} \times \mathbf{P}(\Lambda^k X)$, and hence (see IV. 8.3) it is analytically constructible in $\overline{X^k} \times \mathbf{G}_k(X)$. Therefore the graph of α is analytically constructible in $\overline{X^k} \times \mathbf{G}_k(X)$ (see IV. 8.3). Similarly, one verifies that the graph of α is analytically constructible in $(\bar{X})^k \times \mathbf{G}_k(X)$.

§9. Rückert's lemma for algebraic sets

Set $\mathcal{P}_n = \mathcal{P}(\mathbf{C}^n)$. After the appropriate identifications we have

$$\mathbf{C} = \mathcal{P}_0 \subset \mathcal{P}_1 \subset \ldots \subset \mathcal{P}_n$$

(see I. 1.1). By z_i we will also denote the polynomial $(z \longrightarrow z_i) \in \mathcal{P}_n$, $i = 1, \ldots, n$. Thus $\mathcal{P}_n = \mathbf{C}[z_1, \ldots, z_n]$.

1. Fix an ideal I of the ring $\mathcal{P}_n$.

We have the natural epimorphism $\mathcal{P}_n \ni f \longrightarrow \hat{f} \in \mathcal{P}_n/I$. Denote by $\hat{\mathcal{P}}_l$ the image of the subring $\mathcal{P}_l$, and set $\hat{P} = \sum \hat{a}_p T^p \in \hat{\mathcal{P}}_n[T_1, \ldots, T_r]$ for $P = \sum a_p T^p \in \mathcal{P}_n[T_1, \ldots, T_r]$. (We have the natural epimorphism $\mathcal{P}_n[T_1, \ldots, T_r] \ni P \longrightarrow \hat{P} \in \hat{\mathcal{P}}_n[T_1, \ldots, T_r]$.) Clearly, $P(g_1, \ldots, g_r)\hat{\ } = \hat{P}(\hat{g}_1, \ldots, \hat{g}_r)$ for $g_i \in \mathcal{P}_n$, and, more generally, $P(Q_1, \ldots, Q_r)\hat{\ } = \hat{P}(\hat{Q}_1, \ldots, \hat{Q}_r)$ for $Q_i \in \mathcal{P}_n[S_1, \ldots, S_q]$.

Let $0 \leq k \leq n$. The ideal I is said to be k–regular if it satisfies the following conditions

(1) I contains a polynomial from $\mathcal{P}_l$ which is monic in z_l, $l = k + 1, \ldots, n$;

(2) $I \cap \mathcal{P}_k = 0$.

This definition implies (see A. 3.3) that every proper ideal is k–regular for some k, after a suitable linear change of coordinates.

The condition (1) is equivalent to each of the following conditions:

(1') $\hat{z}_{k+1}, \ldots \hat{z}_n$ are integral over $\hat{\mathcal{P}}_k$,

(1") $\hat{\mathcal{P}}_n$ is finite (and hence integral $(^{70})$) over $\hat{\mathcal{P}}_k$.

Indeed, the condition (1) means that the element $\hat{z}_l$ is integral over $\hat{\mathcal{P}}_{l-1}$ $(l = k+1,\ldots,n)$. Hence (1") $\Longrightarrow$ (1) and also (1) $\Longrightarrow$ (1'), because $\hat{\mathcal{P}}_l = \hat{\mathcal{P}}_{l-1}[\hat{z}_l]$ (applied repeatedly) implies that the elements $\hat{z}_{k+1},\ldots,\hat{z}_n$ are integral over $\hat{\mathcal{P}}_k$ (see A. 8.1). Finally, since $\hat{\mathcal{P}}_n = \mathcal{P}[\hat{z}_{k+1},\ldots,\hat{z}_n]$, we obtain the implication (1') $\Longrightarrow$ (1") (see A. 8.1).

Note also that the condition (2) is equivalent to

(2') $\mathcal{P}_k \ni f \longrightarrow \hat{f} \in \hat{\mathcal{P}}_k$ is an isomorphism.

If the ideal is k-regular, then it has a finite system of generators from $\mathcal{P}_k[z_{k+1},\ldots,z_n]$. (One shows this as in III. 2.2.)

2. Any linear change of the coordinates $u = (z_1,\ldots,z_k)$ and any linear change of the coordinates $v = (z_{k+1},\ldots,z_n)$ does not change k–regularity of the ideal I. One can verify this statement as in III. 2.6, using the conditions (1") and (2).

3. Suppose now that the ideal I is prime and k–regular.

Then (see A. 8.2) the element $\hat{z}_j$, $j = k+1,\ldots,n$, has a (unique) minimal polynomial $\hat{p}_j \in \hat{\mathcal{P}}_k[T]$ over $\hat{\mathcal{P}}_k$, where $p_j \in \mathcal{P}_k[T]$. Obviously,

$$(*) \qquad\qquad p_j(z_j) \in I, \quad j = k+1,\ldots,n ,$$

and so

$$(**) \qquad\qquad V(I) \subset \{p_j(u,z_j) = 0, \ j = k+1,\ldots,n\} .$$

It follows that the projection $V(I) \ni (u,v) \longrightarrow u \in \mathbf{C}^k$ is proper.

As was done in III. 3.2, one can choose a linear change of the coordinates $z_{k+1},\ldots,z_n$ in such a way that $\hat{z}_{k+1}$ is a primitive element of the extension $\hat{\mathcal{P}}_n$ of the ring $\hat{\mathcal{P}}_k$. We have the following

PROPOSITION . *There is a linear change of the coordinates $z_{k+1},\ldots,z_n$ which makes $\hat{z}_{k+1}$ a primitive element of the extension $\hat{\mathcal{P}}_n$ of the ring $\hat{\mathcal{P}}_k$. Moreover, if $0 \in V(I)$, one may require that*

$$(1) \qquad\qquad V(I) \cap \{u = 0, \ z_{k+1} = 0\} = 0 .$$

$(^{70})$ See A. 8.1.

Furthermore, if $E \subset \mathbf{C}^n \setminus V(I)$ is a finite set, one may require that

(2) $$V(I) \cap \tilde{\pi}^{-1}\big(\tilde{\pi}(E)\big) = \emptyset, \quad \text{where} \quad \tilde{\pi} : z \longrightarrow (z, z_{k+1}) \, .$$

As for the condition (1), observe that the set $V(I) \cap \{u = 0\}$ is finite (because of the inclusion $(**)$). Thus $V(I) \cap \{u = 0\} \setminus 0 = \{(0, v_1), \ldots, (0, v_s)\}$. There is a linear form $\varphi(v) = \sum_{k+1}^n a_j z_j$ such that $\varphi(v_\nu) \neq 0, \nu = 1, \ldots, s$. Also, the linear form $\varphi(v) = \sum_{k+1}^n c_j z_j$ satisfies $\psi(v_\nu) \neq 0$, $\nu = 1, \ldots, s$, provided that the c_j belong to sufficiently small neighbourhoods U_j of the coefficients a_j, respectively. Then $V(I) \cap \{u = 0, \ \psi(v) = 0\} = 0$.

Now consider condition (2). Set $w' = (w_1, \ldots, w_k)$ and $w'' = (w_{k+1}, \ldots, w_n)$ for any $w = (w_1, \ldots, w_n) \in \mathbf{C}^n$. Hence we have $E = \{d_1, \ldots, d_r\}$. For each $i = 1, \ldots, r$, the set $V(I) \cap \{u = d_i'\}$ is finite (because of the inclusion $(**)$), i.e., it is of the form $\{\zeta_{i1}, \ldots, \zeta_{is_i}\}$. Of course, $\zeta_{i\nu}'' \neq d_i''$ (as $\zeta_{i\nu}' = d_i'$ and $V(I) \cap E = \emptyset$), and hence there exists a linear form $\varphi(v) = \sum_{k+1}^n a_j z_j$ such that $\varphi(\zeta_{i\nu}'') \neq \varphi(d_i'')$, $\nu = 1, \ldots, s_i$, $i = 1, \ldots, r$. We also have $\psi(\zeta_{i\nu}'') \neq \psi(d_i'')$ for the form $\psi(v) = \sum_{k+1}^n c_j z_j$, provided that the c_j belong to sufficiently small neighbourhoods U_j of the coefficients a_j, respectively. Then $V(I) \cap \bigcup_1^r \{u = d_i', \ \psi(v) = \psi(d_i'')\} = \emptyset$.

Since the sets $U_j \setminus 0$ are infinite, one can, as in III. 3.2, choose the coefficients $c_j \in U_j \setminus 0$ so that, after the change of coordinates $z \longrightarrow (u, \psi(v), z_{k+2}, \ldots, z_n)$, the element $\hat{z}_{k+1}$ becomes a primitive element of the extension $\hat{\mathcal{P}}_n$ of the ring $\hat{\mathcal{P}}_k$. Then it is obvious that the condition (1) (or the condition (2), respectively) will be satisfied.

RÜCKERT'S LEMMA . *Let the ideal I be prime and k–regular. If $\hat{z}_{k+1}$ is a primitive element of the extension $\hat{\mathcal{P}}_n$ of the ring $\hat{\mathcal{P}}_k$, then the following Rückert formula holds*

(R) $$V(I)_{\{\delta \neq 0\}} = \{(u, v): \ \delta(u) \neq 0, \ p_{k+1}(u, z_{k+1}) = 0,$$
$$\delta(u) z_j = Q_j(u, z_{k+1}), \ j = k + 2, \ldots, n\} \, ,$$

where $\delta \not\equiv 0$ is the discriminant of the polynomial p_{k+1} and the Q_j are polynomials ([71]) .

REMARK . The polynomial p_{k+1} is of positive degree (because $\hat{p}_{k+1}$ is the minimal polynomial of some element), and so the set (R) is non-empty.

([71]) See corollary 2 in §10 below.

Proof (^{72}) . According to the primitive element theorem (see A. 8.3), we have $\hat{\delta}\hat{z}_j = \hat{Q}_j(\hat{z}_{k+1})$, where $Q_j \in \mathcal{P}_k[T]$, and one may assume that

(#) $$Q_{k+1} = \delta T .$$

Thus $\delta z_j - Q_j(z_{k+1}) \in I$, $j = k+1, \ldots, n$. Now, we take a system of generators $f_i(z_{k=1}, \ldots, z_n)$ of the ideal I, where $f_i \in \mathcal{P}_k[X_{k+1}, \ldots, X_n]$, $i = 1, \ldots, r$ (see n° 1). Then

(α) $$\delta z_j - Q_j(z_{k+1}) = \sum_{i=1}^{r} a_{ij} f_i(z_{k+1}, \ldots, z_n), \quad j = k+1, \ldots, n ,$$

where $a_{ij} \in \mathcal{P}_n$. In view of ($*$), we have

(β) $$p_{k+1}(z_{k+1}) = \sum_{i=1}^{r} b_i f_i(z_{k+1}, \ldots, z_n), \quad \text{where} \quad b_i \in \mathcal{P}_n .$$

For some m and $F_i \in \mathcal{P}_k[X_{k+1}, \ldots, X_n]$, we have

(γ) $$\delta^m f_i(z_{k+1}, \ldots, z_n) = F_i(\delta z_{k+1}, \ldots, \delta z_n), \quad i = 1, \ldots, r.$$

Therefore

$$\hat{F}_i(\hat{Q}_{k+1}, \ldots, \hat{Q}_n)(\hat{z}_{k+1}) = \hat{F}_i(\hat{\delta}\hat{z}_{k+1}, \ldots, \hat{\delta}\hat{z}_n) = \left(\delta^m f_i(z_{k+1}, \ldots, z_n)\right)\hat{} = 0 ,$$

and hence (see A. 8.2) the polynomial $\hat{p}_{k+1}$ must be a divisor of the polynomial $\hat{F}_i(\hat{Q}_{k+1}, \ldots, \hat{Q}_n)$ in $\hat{\mathcal{P}}_k[T]$. That is, $\hat{F}_i(\hat{Q}_{k+1}, \ldots, \hat{Q}_n) = \hat{p}_{k+1}\hat{H}_i$, where $H_i \in \mathcal{P}_k[T]$, $i = 1, \ldots, r$. Thus (see 2' in n° 1)

(δ) $$F_i(Q_{k+1}, \ldots, Q_n) = p_{k+1}H_i, \quad i = 1, \ldots, r .$$

If Q_j, p_{k+1}, H_i are regarded as polynomials with respect to the variables (u, z_{k+1}), while f_i, F_i are considered as polynomials with respect to the variables $z = (u, v)$, then the identities (#) and (α) $-$ (δ) can be reformulated as follows:

(##) $$Q_{k+1}(u, z_{k+1}) = \delta(u)z_{k+1},$$

(a) $$\delta(u)z_j - Q_j(u, z_{k+1}) = \sum_{i=1}^{r} a_{ij}(z)f_i(z), \quad j = k+1, \ldots, n,$$

(b) $$p_{k+1}(u, z_{k+1}) = \sum_{i=1}^{r} b_i(z)f_i(z),$$

(c) $$\delta(u)^m f_i(z) = F_i\big(u, \delta(u)z_{k=1}, \ldots, \delta(u)z_n\big), \quad i = 1, \ldots, r,$$

(d) $$F_i\big(u, Q_{k+1}(u, z_{k+1}), \ldots, Q_n(u, z_{k+1})\big) = p_{k+1}(u, z_{k+1})H_i(u, z_{k+1}),$$
$$i = 1, \ldots, r.$$

_____

(^{72}) Cf. III. 3.1–2.

These equalities imply the formula (R). (The inclusion $\subset$ follows from (a) and (b), whereas the opposite inclusion is a consequence of (c), (d), and (##).)

Rückert's formula yields (see C. 2.1) the following

COROLLARY . *The set* (R) *is a* k-*dimensional locally topographic submanifold in* $\mathbf{C}^n$.

4. Let X be a vector space, and let $V \subset X$ be an algebraic set.

Rückert's lemma implies

PROPOSITION . *There is a universal denominator* $u \in R(V)$ [73] *on* V. *(Then, for each* $c \in V$, *the germ* $u_c \in \mathcal{O}_c(V)$ *is a universal denominator;* see VI. 2.3.)

Indeed, if V is irreducible, then one may assume that $X = \mathbf{C}^n$ and the assumptions of the Rückert lemma are satisfied for the ideal $I = \mathcal{I}(V)$ (see n° 1–2, the proposition from n° 3, and 1.6). Then $V = V(I)$ (see 1.2). Now, by taking $u = \delta_V$ and repeating the proof of the lemma from VI. 2.2, we arrive at the conclusion of the proposition. In the general case, we consider the decomposition into irreducible components $V = V_1 \cup \ldots \cup V_r$ (see 1.6) and we repeat the continuation of the proof of Oka's theorem from VI. 2.2 (taking $\varphi_i \in \mathcal{P}(X)$ such that $\varphi_i = 0$ on $\bigcup_{i \neq j} V_j$ and $\varphi_j \not\equiv 0$ on V_i).

Another proof by the Grauert-Remmert formula (see V. 7.4):
When V is of constant dimension, we have a direct sum $X = T + Y$ such that Y is a Sadullaev space and the natural projection $\pi : V \longrightarrow T$ is a *-covering (see 7.1). Then so are π^U, where $U \subset T$ is open and non-empty. There is a primitive element $\alpha = \lambda_V$, $\lambda \in X^*$, for π (see remark 3 in V. 7.4); then there is $\alpha_{\pi^{-1}(U)}$ for π^U (see V. 7.6). Now the conclusion follows with $u = D_\alpha^0$, in view of remark 1 in V. 7.4, since $D_\alpha \in \mathcal{P}(T)$ by Liouville's theorem in C. 1.8. The general case follows by the trick, due to Cynk, from the proof of the Grauert-Remmert theorem in VI. 1.2a.

§10. Hilbert's Nullstellensatz for polynomials

Let X be a vector (or – more generally – affine) space, and let $\mathcal{P} = \mathcal{P}(X)$.

HILBERT'S NULLSTELLENSATZ. *For any ideal* I *of the ring* $\mathcal{P}$, *we have* $\mathcal{I}(V(I)) = \text{rad } I$. *In other words* [74] , *if a polynomial* $g \in \mathcal{P}$ *vanishes on*

---

[73] By $R(V)$ we denote the ring of restrictions to V of the polynomials from $\mathcal{P}(X)$; see 12.1 below.

[74] The inclusion $\mathcal{I}(V(I)) \supset \text{rad } I$ is trivial (see 1.3).

$V(f_1, \ldots, f_k)$, *where* $f_i \in \mathcal{P}$, *then* $g^r \in \sum_1^r \mathcal{P} f_i$ *for some* $r \in \mathbf{N}$. *In particular,* $\mathcal{I}(V(I)) = I$ *if the ideal* I *is prime* (see A. 1.11).

PROOF I. First suppose that the ideal I is prime. One may assume that $X = \mathbf{C}^n$ and the ideal I is k–regular (see 1.5 and 9.1). In view of the proposition from 9.3, one may also suppose (see 9.2) that the assumptions of the Rückert lemma (see 9.3) are satisfied. Then (see the remark in 9.3) the image of the set (R) under the projection $(u, v) \longrightarrow u$ is the set $\{\delta \neq 0\}$ which is dense in $\mathbf{C}^n$, and so $\mathcal{P}_k \cap \mathcal{I}(V(I)) = 0$. It is enough to show that $\mathcal{I}(V(I)) \subset I$ (see 1.3). Let $f \in \mathcal{P}_n \setminus I$. Then (using the notation from 9.1) we have $\hat{f} \neq 0$, and hence $\hat{f}\hat{g} \in \hat{\mathcal{P}}_k \setminus 0$ for some $g \in \mathcal{P}_n$ (see the lemma in A. 8.3 and (1") in 9.1). Therefore $fg \in h + I \subset h + \mathcal{I}(V(I))$ for some $h \in \mathcal{P}_k \setminus I$. But we must have $h \notin \mathcal{I}(V(I))$, hence $fg \notin \mathcal{I}(V(I))$. Therefore $f \notin \mathcal{I}(V(I))$.

In the general case, we have $I = J_1 \cap \ldots \cap J_r$, where the J_i are primary ideals. Since the ideals rad J_i are prime (see A. 9.3), we obtain (see 1.2–3 and A. 1.5)

$$\mathcal{I}(V(I)) = \bigcap \mathcal{I}(V(J_i)) = \bigcap \mathcal{I}(V(\mathrm{rad}\ J_i)) = \bigcap \mathrm{rad}\ J_i = \mathrm{rad}\ I \ .$$

We are going to give a direct proof of the theorem $(^{75})$.

PROOF II. First we will prove the following

LEMMA . *If* K *is a countable field and* L *is its extension which is generated (over* K *) by a finite number of elements, then every monomorphism* $K \longrightarrow \mathbf{C}$ *can be extended to a monomorphism* $L \longrightarrow \mathbf{C}$.

In fact, it is enough to consider the case $L = K(\zeta)$, $\zeta \in L$. Let $\varphi : K \longrightarrow K'$ be an isomorphism onto a subfield $K' \subset \mathbf{C}$. When ζ is transcendental over K, let $\zeta' \in \mathbf{C}$ be a transcendental element over K' (such an element exists, because the set of algebraic elements over K' is countable). If ζ is algebraic over K, let $\zeta' \in \mathbf{C}$ be a root of the polynomial that corresponds via the induced isomorphism to the minimal polynomial of the element ζ. Then (see A. 5.3) the isomorphism φ extends to an isomorphism $K(\zeta) \longrightarrow K(\zeta')$.

Now we are going to prove the theorem. We may assume that $X = \mathbf{C}^n$, so $\mathcal{P} = \mathcal{P}_n$. As in the first proof, it is enough to prove the inclusion $\mathcal{I}(VI)) \subset I$ for any prime ideal I.

Let $f \in \mathcal{I}(V(I))$. The ideal I has generators $g_1, \ldots, g_r$. Then $V(I) = V(g_1, \ldots, g_r)$. Let $K \subset \mathbf{C}$ be a subfield generated by the coefficients of the

---

$(^{75})$ Following Lang [3].

polynomials $f, g_1, \ldots, g_r$. It is countable and $J = I \cap K[z_1, \ldots, z_n]$ is a prime ideal in $K[z_1, \ldots, z_n]$ (see A. 1.11). Obviously,

(1) $f \in K[z_1, \ldots, z_n]$ and $g_1, \ldots, g_r \in J$.

The ring $K[z_1, \ldots, z_n]/J$ is an integral domain. Let L be its field of fractions. We have the natural homomorphism

(2) $\varphi : K[z_1, \ldots, z_n] \longrightarrow L$ with kernel J .

Since $K \cap J = 0$ (because I cannot contain non-zero constants), $\varphi_K : K \longrightarrow K' = \varphi(K) \subset L$ is an isomorphism, and so is $\varphi_K^{-1} : K' \longrightarrow K \subset \mathbf{C}$. But L is generated over K' by $\varphi(z_1), \ldots, \varphi(z_n)$, hence, according to the lemma, the monomorphism $\varphi_K^{-1} : K' \longrightarrow \mathbf{C}$ extends to a monomorphism $\psi : L \longrightarrow \mathbf{C}$. Thus $\psi \circ \varphi : K[z_1, \ldots, z_n] \longrightarrow \mathbf{C}$ is a K–homomorphism, and so

$$\psi\big(\varphi(h)\big) = h(c) \quad \text{for} \quad h \in K[z_1, \ldots, z_n] ,$$

where $c = \Big(\psi(\varphi(z_1)), \ldots \psi(\varphi(z_n))\Big)$. So, by (1) and (2), we have $g_1(c) = \ldots = g_r(c) = 0$; that is, $c \in V(I)$, and so $\psi\big(\varphi(f)\big) = 0$, $f(c) = 0$, hence $\varphi(f) = 0$. Thus by (2) we obtain $f \in J \subset I$.

COROLLARY 1. *The mappings $V \longrightarrow \mathcal{I}(V)$ and $I \longrightarrow V(I)$ are mutually inverse bijections between the set of irreducible algebraic subsets of X and the set of prime ideals of $\mathcal{P}$. (See 1.6 and 1.2.)*

COROLLARY 2. *In Rückert's lemma (from 9.3), the set (R) is dense in $V(I)$. We have $\tilde{\pi}\big(V(I)\big) = \{(u, z_{k+1}) : p_{k+1}(u, z_{k+1}) = 0\}$, where $\tilde{\pi} : z \longrightarrow (u, z_{k+1})$.*

Indeed, the set $V(I)$ is irreducible. It follows from Rückert's formula that the set (R) is constructible, and hence its closure W is algebraic (see 8.3, proposition 2). We have $V(I) = W \cup W'$, where $W' = V(I)_{\{\delta = 0\}}$ is algebraic and, in view of the remark in 9.3, $W' \subsetneq V(I)$. Therefore we must have $W = V(I)$.

Now the natural projection

$$V' = \{(u, z_{k+1}) : p_{k+1}(u, z_{k+1} = 0\} \ni (u, z_{k+1}) \longrightarrow u \in \mathbf{C}^n$$

is an open surjection (see C. 2.1), and so (see B. 2.1) the set $V'_{\{\delta \neq 0\}}$ is dense in V'. Because of the inclusion (∗∗) from 9.3, the mapping $\tilde{\pi}_V$ is proper (see B. 5.2), and hence it is closed. Therefore, since $\tilde{\pi}\big(V(I)_{\{\delta \neq 0\}}\big) = V_{\{\delta \neq 0\}}$ (see (R)), we obtain $\tilde{\pi}\big(V(I)\big) = V'$.

We will also prove the following

HILBERT NULLSTELLENSATZ ("mixed" version). *Let M be a p-dimensional manifold, and let $a \in M$. Let $P_1, \ldots, P_k, Q$ be X-polynomials on $M \times X$. If $Q = 0$ on $V(P_1, \ldots, P_k)$, then there exist an exponent r, an open neighbourhood U of a, and X-polynomials $F_1, \ldots, F_k$ on $U \times X$, such that $Q^r = \sum_1^k F_i P_i$ in $U \times X$.*

PROOF (A. Ploski). One may assume that M is a neighbourhood of zero in $\mathbf{C}^p$, $a = 0$, $X = \mathbf{C}^n$, and $P_i(z, w)$, $Q(z, w)$ are polynomials whose coefficients are holomorphic in M. Then there exist forms $\tilde{P}_i(z, w, t)$, $\tilde{Q}(z, w, t)$ with respect to $(w, t) \in \mathbf{C}^{n+1}$, with holomorphic coefficients in M, such that $P_i(z, w) = \tilde{P}_i(z, w, 1)$ and $Q(z, w) = \tilde{Q}(z, w, 1)$ in $M \times \mathbf{C}^n$ (see C. 3.18). One may also require that $\tilde{Q}(z, w, 0) = 0$ in $M \times \mathbf{C}^n$. Then $\tilde{Q} = 0$ on $V(\tilde{P}_1, ..$
$.., \tilde{P}_k)$ [76]. According to the Nullstellensatz from III. 4.1, we have

$$(*) \qquad \tilde{Q}^r = \sum_1^k G_i P_i \quad \text{in a neighbourhood of zero in} \quad \mathbf{C}^{p+n+1}$$

for some r and some G_i that are holomorphic in a neighbourhood of zero. One may assume that the germs $\tilde{P}_i$, $\tilde{Q}$ at 0 are different from zero. We have a unique expansion $G_i = \sum_{\nu=0}^{\infty} G_i^{(\nu)}$ in an open neighbourhood $U \times \Delta$ of zero, where $U \subset M$ and $G_i^{(\nu)}(z, w, t)$ is a form of degree ν with respect to (w, t), with holomorphic coefficients in U (see C. 3.18). Let k_i, l denote the degrees of the forms $\tilde{P}_i, Q$, respectively. It follows from $(*)$ (by the uniqueness of expansions) that $\tilde{Q}^r = \sum_1^r G_i^{(lr-k_i)} \tilde{P}_i$ in $U \times \mathbf{C}^n$, where $G_i^{(\nu)} = 0$ for $\nu < 0$. Consequently, by putting $t = 1$, one gets the required equality with coefficients $F_i(z, w) = G_i^{(lr-k_i)}(z, w, 1)$.

REMARK . When $p = 0$, this is another proof of Hilbert's Nullstellensatz for polynomials, obtained by reducing the theorem to its germ version in III. 4.1.

§11. Further properties of algebraic sets.
Principal varieties. Degree

Let X be a vector space of dimension $n > 0$.

---

[76] Indeed, let $\tilde{P}_i(z, w, t) = 0$, $i = 1, \ldots, k$. If $t \neq 0$, then $P_i(z, w/t) = 0$, and hence $Q(z, w/t) = 0$. Therefore $\tilde{Q}(z, w, t) = 0$.

1. We have the following propositions.

PROPOSITION 1. *If $V \subset X$ is an algebraic set, then so is the set V^*.*

In fact, the closure $\bar{V}$ in $\bar{X}$ is algebraic, and hence so is the set $\bar{V}^* \subset \bar{X}$, which implies that $V^* = (\bar{V}^*) \cap X$ is algebraic in X. (See 6.3 and Chow's theorem from 6.1.)

PROPOSITION 2. *Every algebraic set $V \subset X$, regarded as an analytic set, has a finite number of simple components and each of them is algebraic.*

Indeed, the set V^0 is constructible (see 8.5), and so its decomposition into connected components is finite and each such component is constructible (see 8.5). This implies the proposition (see theorem 4 from IV. 2.9 and proposition 2 from 8.3).

The above yields (see IV. 2.9, proposition 5):

COROLLARY . *The connected components of any algebraic set $V \subset X$ are algebraic.*

PROPOSITION 3. *An algebraic set $V \subset X$ is irreducible (see 1.6) if and only if it is irreducible when regarded as an analytic set.*

For proposition 2 implies that an algebraic set is reducible if and only if it is reducible as an analytic set.

COROLLARY . *The decomposition of an algebraic set $V \subset X$ into simple components exists and coincides with the decomposition of V, regarded as an analytic set, into simple components* ([77]) .

2. Let $\mathcal{P} = \mathcal{P}(X)$.

If $f \in \mathcal{P}$, then it is obvious that:

(1) $V(f) = \emptyset \Longleftrightarrow (f$ is a non-zero constant$)$;

(2) $\emptyset \subsetneq V(f) \subsetneq X \Longleftrightarrow (f$ is of positive degree$)$.

Furthermore:

(3) $\big(V(f)$ is of constant dimension $n - 1\big) \Longleftrightarrow (f$ is non-zero$)$.

(Indeed, if $f \neq 0$, then $\dim V(f)_z = n - 1$ for $z \in V(f)$; see II. 5.1.)

PROPOSITION 4. *Let $f, g \in \mathcal{P}$, and suppose that f is irreducible. Then*

$$\big(g = 0 \text{ on } V(f)\big) \Longleftrightarrow (g \text{ is divisible by } f) .$$

([77]) See 1.6, footnote ([6]).

Indeed, the equivalence means that $\mathcal{I}(V(f)) = \mathcal{P}f$. Since f is a prime element of the ring $\mathcal{P}$ (see A. 6.1), this is a special case of Hilbert's Nullstellensatz from §10 (see A. 1.13).

We are going to give yet another proof of proposition 4 (based on Hilbert's Nullstellensatz for germs).

LEMMA . *If f is a holomorphic function on X and $P = fQ$, where $P \in \mathcal{P}$ and $Q \in \mathcal{P} \setminus 0$, then $f \in \mathcal{P}$.*

In fact, because of corollary 1 from 7.2, it is enough to observe that the graph of f is the closure of the constructible set $\{(z,t) : \ P(z) = tQ(z), \ Q(z) \neq 0\}$, and so (see 8.3, proposition 2) it is algebraic ([78]) .

Let us come back to the proof of proposition 4 (the implication $\Longrightarrow$). Choose one point from each of the simple components of $V(f)$. According to Hilbert's Nullstellensatz from III. 4.1, there exists $m \in \mathbf{N}$ such that the function g^m/f extends holomorphically across each of those points. Hence (see IV. 2.10, proposition 6) it also extends to the whole space X. In view of the lemma, this extension must be a polynomial. This means that g^m, and therefore the function g, too, is divisible (in $\mathcal{P}$) by the polynomial f because f is a prime element (see A. 6.1).

COROLLARY 1. *If $f \in \mathcal{P}$ is irreducible, then $V(f)$ is irreducible.* (See 1.6 and A. 6.1.)

COROLLARY 2. *If $f,g \in \mathcal{P}$ have no multiple factors, then*

$$V(f) = V(g) \Longleftrightarrow (f \text{ and } g \text{ are associated, i.e., they differ by a non-zero}$$
$$\text{constant factor}).$$

COROLLARY 3. *If a polynomial $g \in \mathcal{P}$ is of positive degree and $g = g_1^{k_r} \ldots g_r^{k_r}$ is its decomposition into irreducible factors (where $k_i > 0$ and there are no associated polynomials among the g_i's), then $V(g) = V(g_1) \cup \ldots \cup V(g_r)$ is the decomposition of $V(g)$ into simple components.*

COROLLARY 4. *If $f,g \in \mathcal{P}$ and f is irreducible, then*

$$V(g) = V(f) \Longleftrightarrow (g = af^r, \text{ where } a \in \mathbf{C} \setminus 0 \text{ and } r \in \mathbf{N} \setminus 0) .$$

([78]) The lemma can be also shown in a more elementary fashion. If $X = \mathbf{C}$, it is a trivial consequence of Liouville's theorem (see C. 1.8), and moreover, the degree of f is not greater than the degree of the polynomial P. In the general case, we introduce a linear system of coordinates of X so that the polynomial Q is monic with respect to each variable (see A. 3.3). Then $\frac{\partial^\nu f}{\partial z_i^\nu} = 0$ in X for $\nu > k$ $(i = 1, \ldots, n)$, where k is the degree of the polynomial P. Thus $D^p f = 0$ in X if $|p| > kn$.

According to the Hilbert theorem, $\mathcal{I}(V(f)) = \operatorname{rad} \mathcal{P}f$, and hence (see A. 6.1):

COROLLARY 5. *A non-zero polynomial has no multiple factors if and only if it generates the ideal of its locus.*

COROLLARY 6. *If $f \in \mathcal{P}$, then*

$$V(f) \text{ is a cone} \iff f \text{ is homogeneous and } f(0) = 0 .$$

Indeed, if $V(f)$ is a cone and the degree of f is positive, then $V(f) \subsetneq X$ and $f = f_1^{k_r} \ldots f_r^{k_r}$, where f_i are irreducible and $k_i > 0$. By the Cartan-Remmert-Stein lemma (see II. 3.3), there is a homogeneous polynomial $h \in \mathcal{P} \setminus 0$ that vanishes on $V(f) \supset V(f_i)$. Thus h is divisible by f_i, and hence (see A. 2.3) f_i is homogeneous, $i = 1, \ldots, r$.

3. We have the following proposition (cf. II. 5.3):

PROPOSITION 5. *If $f_1, \ldots, f_k \in \mathcal{P}$, then*

$$V(f_1, \ldots, f_k) = V(g) \cup B ,$$

where g is the greatest common divisor of the polynomials f_i, while B is an algebraic set of dimension $\leq n - 2$.

The proof is the same as that of the identity $(*)$ from II. 5.3 and is based on the following property: if $f, g \in \mathcal{P}$ are irreducible and relatively prime, then $\dim V(f) \cap V(g) \leq n - 2$. (It is a consequence of corollaries 1 and 2 of proposition 4; see IV. 2.8, proposition 3.)

As a result (cf. II. 5.2), we obtain

COROLLARY 1. *For any algebraic subset $V \subset X$, we have the equivalences*

$$(V \text{ is irreducible of dimension } n - 1) \Longleftrightarrow$$
$$\Longleftrightarrow (V = V(f), \text{ where } f \in \mathcal{P} \text{ is irreducible});$$
$$(V \text{ is of constant dimension } n - 1) \Longleftrightarrow (V = V(f), \text{ where } f \in \mathcal{P} \setminus 0) .$$

(Moreover, one may require that the polynomial f from the last condition has no multiple factors.)

Indeed, both implications $\Longleftarrow$ follow from corollary 1 of proposition 4 and property (3) from n° 2. The proof of both implications $\Longrightarrow$ is the same as in II. 5.3.

In view of the second of the above equivalences, we have (see IV. 2.5):

COROLLARY 2. *For any algebraic set* $V \subsetneq X$, *the following conditions are equivalent:*

(1) V *is a principal variety;*

(2) V *is of constant dimension* $n - 1$;

(3) V^0 *is a submanifold of dimension* $(n - 1)$ *(as a manifold).*

By proposition 5 we get (as in II. 5.3):

COROLLARY 3. *For any polynomials* $f_1, \ldots, f_k \in \mathcal{P}$, *we have the equivalence*

$$(f_1, \ldots, f_k \text{ are relatively prime}) \Longleftrightarrow \dim V(f_1, \ldots, f_k) \leq n - 2 .$$

COROLLARY 4. *For any polynomials* $f, g \in \mathcal{P} \setminus 0$ *we have :*

$$(f, g \text{ are relatively prime}) \Longleftrightarrow (V(f, g) \text{ is of constant dimension } n - 2) .$$

Indeed, by property (3) from n° 2 and the inequality (∗) from III. 4.6, we have $\dim_z (V(f) \cap V(g)) \geq n - 2$ for $z \in V(f) \cap V(g)$.

4. Recall that, by the proposition from A. 6.3, a polynomial $f \in \mathcal{P}_n$ which is monic in z_n has no multiple factors if and only if its discriminant is non-zero (f is regarded here as a polynomial from $\mathcal{P}_{n-1}[z_n]$).

PROPOSITION 6. *Let* $f \in \mathcal{P}$ *be a polynomial of positive degree, and let* $V = V(f)$. *Then the following conditions are equivalent:*

(1) f *has no multiple factors;*

(2) $d_z f \neq 0$ *for* $z \in V^0$;

(3) $V^* = \{z : f(z) = 0, \ d_z f = 0\}$.

PROOF . The equivalence (2)⟺(3) is trivial. To prove (1)⟹(2), suppose first that the polynomial f is irreducible. Then the set V is irreducible (see corollary 1 from proposition 4). The set V^0 contains a point a at which $d_a f \neq 0$, because in some linear coordinate system the polynomial f is monic with respect to z_n (see A. 3.3) and its discriminant is non-zero (see C. 2.1). Let $c \in V^0$. Then $c \in V^{(n-1)}$ (see property (3) from n° 1), and so $V(f_c) = V_c = V(\varphi_c)$, where φ is a holomorphic function in a neighbourhood of c and such that $\varphi(c) = 0$ and $d_c \varphi \neq 0$. Then the germ φ_c is irreducible (see I. 1.2) and therefore $f_c = h_c \varphi_c^k$, where h is a holomorphic function in a neighbourhood of c, such that $h(c) \neq 0$ and $k > 0$ (see II. 5.2, corollary 3). If one had $d_c f = 0$, one would have $k \geq 2$, and thus $d_z f = 0$ in a neighbourhood in V^0 of the point c. Hence, the same would hold on the whole set V^0 (as the latter is connected), and this is impossible. Therefore $d_z f \neq 0$ for $z \in V^0$. In the general case, we have $f = f_1 \ldots f_k$, where $f_i \in \mathcal{P}$ are irreducible and mutually non-associated. Thus $V = V_1 \cup \ldots \cup V_k$ with $V_i = V(f_i)$, is the decomposition into simple components (see corollary 3 of proposition 4). Now, if $z \in V^0$, then $z \in (V_s)^0 \setminus \bigcup_{i \neq 0} V_i$ for some s (see IV. 2.9, theorem 4), hence $d_z f_s \neq 0$, $f_s(z) = 0$ and $f_i(z) \neq 0$ for $i \neq s$. Therefore $d_z f \neq 0$.

It remains to show $(2) \Longrightarrow (1)$. In some linear coordinate system, the polynomial f is monic with respect to z_n (see A. 3.3). Hence the natural projection $V \longrightarrow \mathbf{C}^{n-1}$ is finite (see B. 5.2 and 3). By the Andreotti-Stoll theorem (see V. 7.2 and property (3) from n° 1), it is a *-covering. Consequently (in view of formula (3) from the Andreotti Stoll theorem), there is a point $a \in \mathbf{C}^{n-1}$ such that the line $a \times \mathbf{C}$ intersects V only at points $c_i = (a, \zeta_i) \in V^0$ $(i = 1, \ldots, s)$, and transversally at each of those points (see C. 3.17). This means that (because $d_{c_i} f \neq 0$) we have $\frac{\partial f}{\partial z_n}(a, \zeta_i) \neq 0$, and so (see C. 2.1) the discriminant of f at a is different from zero. Therefore f has no multiple factors.

5. Let $V \subsetneq X$ be a principal algebraic set.

Define the *degree* r of V as the minimum of the degrees of the non-zero polynomials from $\mathcal{I}(V)$. Clearly, $r = 0$ precisely when $V = \emptyset$.

If $V \neq \emptyset$, then r is equal to the degree of a polynomial $f \in \mathcal{P}$ without multiple factors such that $V = V(f)$. Indeed (see 1.4), such a polynomial f exists. Therefore we have $f = f_1, \ldots, f_k$, where $f_i \in \mathcal{P}$ are irreducible and relatively prime. Now, if $g \in \mathcal{I}(V)$, then $g \in \mathcal{I}(V(f_i))$. By proposition 4, the polynomial g is divisible by f_i, $(i = 1, \ldots, k)$, and thus it is also divisible by f.

Note that the principal sets of degree 1 are precisely the affine hyperplanes.

Obviously, each line $\lambda \in \mathbf{P}'(X)$ intersects V at at most r points, provided it is not contained in V.

PROPOSITION 7. *The set* $\{\lambda \in \mathbf{P}'(X) : \#(\lambda \cap V) = r\}$ *is open and dense in* $\mathbf{P}'(X)$.

PROOF . One may assume that $r > 0$. Then there exists a polynomial $f \in \mathcal{P}$ of degree r, without multiple factors, such that $V = V(f)$.

We are going to show that our set is open. Let $\lambda \in \mathbf{P}'(X)$ and $\#(\lambda \cap V) = r$. By taking a linear coordinate system in which λ_* is the z_n–axis, we may assume that f is a monic polynomial of degree r with respect to z_n. Then

$\lambda = a \times \mathbf{C}$, where $a \in \mathbf{C}^{n-1}$, which means that $\#\{z_n : f(a, z_n) = 0\} = r$, and so the discriminant of f is different from zero at a. Hence $\frac{\partial f}{\partial z_n} \neq 0$ on $\lambda \cap V$ (see C. 2.1). Therefore the line λ intersects the set V only at its regular points, and transversally at each of them (see C. 3.17). In view of the proposition from C. 3.16, we conclude that each line λ' from some neighbourhood in $\mathbf{P}'(X)$ of the line λ intersects the set V at at least r points. Consequently, $\#(\lambda' \cap V) = r$.

It remains to prove that our set is dense. Let $\lambda \in \mathbf{P}'(X)$. Take $c \in \lambda$. There is a line $\mu \in \mathbf{P}(X)$ arbitrarily close to the line λ_*, such that $\mu \not\subset V(f_r)$, where $f = f_0 + \ldots + f_r$, $f_r \neq 0$, is the decomposition into forms (for $V(f_r)\tilde{\ }$ is nowhere dense in $\mathbf{P}(X)$; see 6.2). Thus (in view of the continuity of the mapping σ; see B. 6.11) the line $\lambda' = c + \mu$ can be made arbitrarily close to the line λ (see B. 6.1). Now, taking a linear coordinate system in which μ is the z_n axis, one may assume that f is a monic polynomial of degree r with respect to z_n with non-zero discriminant, and we have $\lambda' = a \times \mathbf{C}$ for some $a \in \mathbf{C}^{n-1}$. Then the discriminant is different from zero at a point a' arbitrarily close to the point a, and so $\#(\lambda'' \cap V) = \#\{f(a', z_n) = 0\} = r$, where the line $\lambda'' = a' \times \mathbf{C}$ can be made arbitrarily close to the line λ' and hence also to the line λ.

REMARK. The condition from proposition 7 determines the number r in a unique fashion. Therefore it provides us with another, equivalent definition of the degree of a principal algebraic set $V \subsetneq X$: the degree of the set V is equal to the maximum of the number of points of intersection of this set with an affine line which is not contained in the set.

6. Let $\mathbf{P} = \mathbf{P}(X)$. An algebraic subset of the projective space $\mathbf{P}$ is said to be *principal* if it can be defined by a single homogeneous polynomial (on X).

If $V \subset \mathbf{P}$ is a non-empty algebraic set, then

$$(V \text{ is principal}) \iff (V\tilde{\ } \text{ is principal}) \quad (^{79}).$$

Indeed $(^{80})$, for homogeneous polynomials $f \in \mathcal{P}$, we have the equivalence:

$$(\#) \qquad (V \text{ is defined by } f) \iff V\tilde{\ } = V(f).$$

$(^{79})$ Obviously, the empty set $\emptyset \subset \mathbf{P}$ is principal, but its cone $0 = \emptyset\tilde{\ }$ is not principal – apart from the case $n = 1$ (see n° 2, property (3)).

$(^{80})$ See corollary 6 of proposition 4.

Therefore (by corollary 2 of proposition 5; see 6.2 and IV. 2.5), for any algebraic set $V \subsetneq \mathbf{P}$, the following conditions are equivalent:

(1) V is a principal set;

(2) V is of constant dimension $(m-1)$;

(3) V^0 is a submanifold of dimension $m-1$ (as a manifold),
     where $m = \dim \mathbf{P}$.

7. Let $V \subsetneq \mathbf{P} = \mathbf{P}(X)$ be a principal algebraic set.

We define the *degree* r of the set V as the minimum of degrees of homogeneous polynomials (on X) that have no multiple factors and define V. Of course, $r = 0$ exactly when $V = \emptyset$.

If $V \neq 0$, then the cone $V^{\sim}$ is a principal algebraic set of the same degree r, and r is the degree of a homogeneous polynomial (on X) that is without multiple factors and defines V (such a polynomial exists). This is a consequence of ($\#$) from n° 6 and corollary 6 of proposition 4.

Note that the principal sets of degree 1 are precisely the hyperplanes. (See B. 6.12.)

PROPOSITION 8. *The set* $\{L \in \mathbf{G}_1(\mathbf{P}) :\ \#(L \cap V) = r\}$ *is open and dense.*

PROOF . One may assume that $r > 0$. Then, in view of proposition 7, the set $\Omega = \{\lambda \in \mathbf{P}'(X) \setminus \mathbf{P}(X) :\ \#(\lambda \cap V^{\sim}) = r\}$ is open and dense in $\mathbf{P}'(X) \setminus \mathbf{P}(X)$. Hence so is $\Omega \setminus Z$, where $Z = \{\lambda \in \mathbf{P}'(X) :\ \lambda_* \in V\}$ is a nowhere dense set in $\mathbf{P}'(X)$ (since it is the inverse image of the nowhere dense set $V \subset \mathbf{P}$ under the open mapping $\lambda \longrightarrow \lambda_*$; see B. 6.11 and B. 2.1). Since the mapping $\mu :\ \mathbf{P}'(X) \setminus \mathbf{P}(X) \ni \lambda \longrightarrow (\mathbf{C}\lambda)^{\sim} \in \mathbf{G}_1(\mathbf{P})$ is an open and continuous surjection (see B. 6.12), it is enough to show that our set is the image under μ of the set $\Omega \setminus Z$.

Now, if $\lambda \in \big(\mathbf{P}'(X) \setminus \mathbf{P}(X)\big) \setminus Z$, then

$$(*) \qquad\qquad \#(\lambda \cap V^{\sim}) = \#\big((\mathbf{C}\lambda)^{\sim} \cap V\big) ,$$

because in this case the image of the set $\lambda \cap V^{\sim}$ under the injection $\lambda \ni z \longrightarrow \mathbf{C}z \in (\mathbf{C}\lambda)^{\sim}$ is the set $(\mathbf{C}\lambda)^{\sim} \cap V$. Furthermore, if $L \in \mathbf{G}_1(\mathbf{P})$ and $L \not\subset V$, then there is a line $\lambda \in \mathbf{P}'(X) \setminus \mathbf{P}(X)$ such that $\lambda_* \in L \setminus V$; then $L = (\mathbf{C}\lambda)^{\sim}$ and $\lambda \notin Z$ (hence $(*)$ holds). Therefore, if $\lambda \in \Omega \setminus Z$, then, by $(*)$, we have $\#\big((\mathbf{C}\lambda)^{\sim} \cap V\big) = r$. However if $L \in \mathbf{G}_1(\mathbf{P})$ and $\#(L \cap V) = r$, then $L \not\subset V$. Hence $L = (\mathbf{C}\lambda)^{\sim}$ for some $\lambda \in \big(\mathbf{P}'(X) \setminus \mathbf{P}(X)\big) \setminus Z$, and so $\lambda \in \Omega \setminus Z$ (by $(*)$).

REMARK 1. The condition from proposition 8 determines the number r in a unique fashion. Henceforth it gives an alternative equivalent definition of the degree of a principal algebraic set $V \subsetneq \mathbf{P}$.

REMARK 2. It follows from the proof of proposition 8 that either $\#(\lambda \cap V) \leq r$ or $\lambda \subset V$. Thus the degree of the set V is equal to the maximum of the number of points of intersection of this set with a line which is not contained in the set.

8. Let $V \subset X$ be an algebraic set of constant dimension k, where $0 \leq k \leq n$. We define the *degree* $\deg V$ of the set V as the unique number p such that $\#(L \cap V) = p$ for each L from some open dense subset of the space $\mathbf{G}'_{n-k}(X)$. This definition agrees with that given in n° 5 for principal sets (see proposition 7). The existence of such a number p follows from the lemma which we will prove below.

The set $G \subset \mathbf{G}_{n-k}(X)$ of Sadullaev's spaces for V (see 7.1) is the complement of a proper analytic subset and so is the set $G' = \{L : L_* \in G\} \subset \mathbf{G}'_{n-k}(X)$ (as the inverse image of G under ν; see 4.5) [81] . Note that $G' = \{L : L_\infty \cap V_\infty = \emptyset\}$ (see 7.1 and 3.3).

LEMMA . *There exists a number p such that the set*

$$\{L \in G' : \#(L \cap V) = p\} = \{L \in G' : L \text{ intersects } V \text{ transversally}\}$$

is open and dense in $\mathbf{G}'_{n-k}(X)$ [82] . *Moreover,* $\#(L \cap V) \leq p$ *if* $L \in G'$.

PROOF . Omitting trivial cases, one may assume that $0 < k < n$. The set $\Lambda = \{(L', L) : L' \parallel L\} \subset G' \times G$ is a connected submanifold of dimension $k(n - k + 1)$; it is the graph of the mapping $\nu_{G'} : G' \ni L \longrightarrow L_* \in G$, and the mapping $g : G' \ni L \longrightarrow (L, L_*) \in \Lambda$ is biholomorphic (see 4.5). Now, by the Andreotti-Stoll theorem (see V. 7.2), the mapping

$$f : V \times G \ni (z, L) \longrightarrow (z + L, L) \in \Lambda$$

is a p-sheeted *-covering. Indeed, $V \times G$ is of constant dimension $k + k(n - k)$. The fibres $f^{-1}((L', L)) = (L' \cap V) \times \{L\}$ are finite. Finally, the mapping f is proper, since (see B. 2.4) the mapping $\varphi : V \times G \ni (z, L) \longrightarrow z + L \in G'$ is proper. This is so because each point in G' has a neighbourhood of the form

---

[81] The complements of G and G' are even algebraic. See footnote [50] in 7.1, and 17.13 below.

[82] Its complement is an algebraic set (see proposition 10 below).

$\Omega' = \{z + L : z \in U,\ L \in \Omega\}$, where $U \subset X$, $\Omega \subset G$ are compact $(^{83})$. Then

$$\varphi^{-1}(\Omega') \subset (\bar{V} \cap \Omega^{\#}) \times \Omega \subset V \times G \quad (^{84}),$$

where $\Omega^{\#} = \bigcup\{\overline{z + L} : z \in U,\ L \in \Omega\} \subset \bar{X}$ is compact $(^{85})$.

Therefore we have equality (3) from the Andreotti-Stoll theorem, and hence that of the images under g^{-1}:

$$\{L \in G' : \#f^{-1}((L, L_*)) = p\} =$$
$$= \{L \in G' : z \in L \cap V \Longrightarrow f \text{ is biholomorphic at } (z, L_*)\}$$

(this set is open and dense in G'). We will check that this is the required identity from the lemma. Now, $f^{-1}((L, L_*)) = (L \cap V) \times \{L_*\}$, which implies that the left hand sides coincide, and that $\#(L \cap V) \le p$ $(^{86})$. In order to obtain equality of the right hand sides, it suffices to verify, for $z \in L \cap V^*$, the equivalence

$$(f \text{ is biholomorphic at } (z, L_*)) \Longleftrightarrow$$
$$\Longleftrightarrow (L \text{ intersects } V \text{ transversally at } z) \quad (^{87}).$$

The first condition is equivalent to the injectivity of the differential at z of the mapping $V \ni \zeta \longrightarrow \zeta + L_* \in G'$. The second one means that the differential at z of the projection, parallel to L_*, of V into a linear complement N of L_* $(^{88})$ is injective. The equivalence follows from the fact that the first of the two mappings is the composition of the second one with the immersion $N \ni \zeta \longrightarrow \zeta + L_* \in G'$ $(^{89})$.

Now, let $\mathbf{P} = \mathbf{P}(Y)$ be an n–dimensional projective space (i.e., Y is an $(n + 1)$–dimensional vector space), and let $V \subset \mathbf{P}$ be an algebraic set of constant dimension k, where $0 \le k \le n$. We define the *degree* $\deg V$ of the set V as the unique number p such that $\#(L \cap V) = p$ for each L from an open and dense subset of $\mathbf{G}_{n-k}(\mathbf{P})$. The definition agrees with that given in

---

$(^{83})$ Since σ is an open surjection (see B. 6.11).

$(^{84})$ Because if $z \in U$ and $L \in \Omega$, then $z + L \in G'$, and so $V^{\infty} \cap \overline{z + L} = \emptyset$.

$(^{85})$ See 3.3, footnote $(^{16})$.

$(^{86})$ See (1) in the Andreotti-Stoll theorem.

$(^{87})$ We have $(V \times G)^0 = V^0 \times G$ (see IV. 2.1).

$(^{88})$ See, e.g., C. 3.17.

$(^{89})$ See 4.5.

n° 6 for the principal sets (see proposition 8). Such a number p exists, owing to the following proposition.

PROPOSITION 9. *There exists p such that for $L \in \mathbf{G}_{n-k}(\mathbf{P})$ we have either* $\#(L \cap V) \leq p$ *or* $\dim(L \cap V) > 0$ [89a], *and the set*

$$\{L \in \mathbf{G}_{n-k} : \#(L \cap V) = p\} = \{L \in \mathbf{G}_{n-k}(\mathbf{P}) : L \text{ intersects } V$$
$$\text{transversally}\}$$

is open and dense in $\mathbf{G}_{n-k}(\mathbf{P})$, *its complement being analytic* [90].

PROOF . Omitting trivial cases, we may assume that $0 < k < n$ and $V \neq \emptyset$. Let Λ be the set of all hyperplanes $H \subset \mathbf{P}$ which do not contain any simple component of V. For any such hyperplane, we must have $\overline{V \setminus H} = V$ (see proposition 3 in IV. 2.8). We claim that if $H \in \Lambda$, then the set $G_H = \{L \in \mathbf{G}_{n-k}(\mathbf{P}) : L \cap V \cap H = \emptyset\}$ is open and dense in $\mathbf{G}_{n-k}(\mathbf{P})$, and that there is a p such that $L \in G_H \Longrightarrow \#(L \cap V) \leq p$ and the set

(*) $\{L \in G_H : \#(l \cap V) = p\} = \{L \in G_H : L \text{ intersects } V \text{ transversally}\}$

is open and dense in $\mathbf{G}_{n-k}(\mathbf{P})$. Indeed, one may assume that $\mathbf{P} = \bar{X}$ and $H = X_\infty$, where X is an n–dimensional vector space (see B. 6.12). Then the statement follows from the lemma applied to $V \cap X$, because the set G' is the image of the set G_H via the homeomorphism $\vartheta : \mathbf{G}_{n-k}(\mathbf{P}) \setminus \mathbf{G}_{n-k}(H) \ni L \longrightarrow L \cap X \in \mathbf{G}'_{n-k}(X)$ (see 4.5; recall that $G_H \cap \mathbf{G}_{n-k}(H) = \emptyset$ [91]) and $(V \cap X)_\infty = V \cap H$). As the set (*) is open and dense, it follows that p does not depend on H. Thus we have

(**) $\bigcup \{G_H : H \in \Lambda\} = \{L \in \mathbf{G}_{n-k}(\mathbf{P}) : \dim(L \cap V) = 0\}$.

Indeed, $\dim(L \cap V) > 0 \Longrightarrow L \cap H \neq \emptyset$ for $H \in \mathbf{G}_{n-k}(\mathbf{P})$ (see 6.2), and if $\dim(L \cap V) = 0$, then $L \cap V \cap H = \emptyset$ for some hyperplane $H \in \Lambda$, since the sets $\{H : H \ni \lambda\} \subset \mathbf{G}_{n-1}(\mathbf{P})$, $\lambda \in \mathbf{P}$, are nowhere dense [92]. The equality (**) implies that by taking the union, the equalities (*) give the equality in the proposition, and hence the set from the proposition is open and dense. It remains to check that it is analytically constructible (see IV. 8.3, proposition

---

[89a]   Even the number of connected components of $L \cap V$ is $\leq p$ (see below).

[90] It is even algebraic – see, e.g., 17.14, Chow's theorem.

[91] Since $L \cap V \neq \emptyset$ when $L \in \mathbf{G}_{n-k}(\mathbf{P})$; see 6.2.

[92] This is due to the fact that the Schubert cycles $\mathbf{S}^n(\lambda, Y)$, where $\lambda \in \mathbf{P}$, are nowhere dense (see B. 6.5 and 12).

5). But this follows from the fact that our set is the difference of the sets $\{L : \#(L \cap V) > r\} \subset \mathbf{G}_{n-k}(\mathbf{P})$, $r = p-1, p$, each of which is the image by the natural projection of the set

$$\{(L, z_1, \ldots, z_{r+1}) : z_i \in L \cap V \text{ are distinct}\} \subset \mathbf{G}_{n-k}(\mathbf{P}) \times \mathbf{P}^{r+1} ,$$

which is analytically constructible. This is so because the set $\{(L, z) : z \in L\} \subset \mathbf{G}_{n-k}(\mathbf{P}) \times \mathbf{P}$ is analytic, which follows from the analyticity of $\mathbf{S}_1^{n-k+1}(Y)$; see 4.3 and 4.5.)

COROLLARY . *If $V \subset \mathbf{P}$ is an algebraic set of dimension k (where $0 \le k \le n$), then the set $\{L : L \cap V = \emptyset\} \subset \mathbf{G}_{n-k}(\mathbf{P})$ is open and dense with analytic complement* ([93]) .

Indeed, one may assume that the set V is of constant dimension $l < k$ (by considering its simple components). Now, the set $\{L : L \cap V = \emptyset\}$ is non-empty. This is so, since $N \cap V$ is finite for some $N \in \mathbf{G}_{n-l}(\mathbf{P})$, and hence there is an $L \in \mathbf{G}_{n-k}(N)$ such that $L \cap V = L \cap (N \cap V) = \emptyset$ ([94]) . The complement of our set is analytic because it is the image under projection of the analytic set $\{(L, z) : z \in L \cap V\} \subset \mathbf{G}_{n-k}(\mathbf{P}) \times \mathbf{P}$.

If $V \subset X$ is an analytic set of constant dimension k, then so is $\bar{V} \subset \bar{X}$, and $\deg \bar{V} = \deg V$. (Indeed, $\#(\bar{L} \cap \bar{V}) = \#(L \cap V)$ for $L \in G'$, and the image of G' under the mapping $L \longrightarrow \bar{L}$ is open and dense in $\mathbf{G}_{n-k}(X)$; see 4.5 ([95]) .)

Let V, W be algebraic sets of constant dimension k in an n–dimensional projective or vector space. The definition of the degree (with the lemma and proposition 9) implies the following properties.

If $V \ne \emptyset$, then $\deg V > 0$. (See 6.2 and 7.1.)

If $V = V_\cup \ldots \cup V_r$ is the decomposition into simple components, then $\deg V = \deg V_1 + \ldots + \deg V_r$.

(Indeed, take an $(n - k)$-dimensional subspace L that intersects V transversally. Then it intersects each of the sets V_i transversally – see corollary 4 from theorem 4 in VI. 2.9 – and we have $\#(L \cap V) = \#(L \cap V_1) + \ldots + \#(L \cap V_r)$.)

Therefore: *If $V \subsetneq W$, then $\deg V < \deg W$.*

Finally, the lemma and proposition 9 imply the following

([93]) In fact, the complement is algebraic. See the proof below (and §17).

([94]) Due to the fact that the sets $\{L : L \ni z\} \subset \mathbf{G}_{n-k}(N)$ are nowhere dense; see footnote ([93]).

([95]) As $\mathbf{G}_{n-k}(X_\infty) \subset \mathbf{G}_{n-k}(\bar{X})$ are closed and nowhere dense (see B. 6.1).

PROPOSITION 10. *If $V \subset X$ is an algebraic set of constant dimension k and if $p = \deg V$, then the set*

$$\{L \in \mathbf{G}'_{n-k}(X) : \#(L \cap V) = p\}$$

is open and dense in $\mathbf{G}'_{n-k}(X)$ with analytic complement ([96]) .

Indeed, this set is contained in G' (because, if $\#(L \cap V) = p$, then $L_\infty \cap V_\infty \neq \emptyset$). Hence it coincides with the set from the lemma, and thus is open and dense in $\mathbf{G}'_{n-k}(X)$. One shows that it is constructible in the same way as in the proof of proposition 9, using the fact that $\{(L, z) : z \in L\} \subset \mathbf{G}'_{n-k}(X) \times X$ is analytic (which follows from the analyticity of $\mathbf{S}_1^{n-k+1}(\mathbf{C} \times X)$; see 4.3 and 4.5).

Following P. Tworzewski, we are going to prove some other properties of the degree. Let M be a linear or projective space of dimension n. For any algebraic $V \subset M$, define $\delta(V) = \sum \deg V_i$, where $V = \bigcup V_i$ is the decomposition into simple components. Thus $\delta(V) = \deg V$ when V is of constant dimension. Clearly, $\delta(V) > 0$ if $V \neq \emptyset$. Obviously, $\delta(V) = \delta(h(V))$ when $h : M \longrightarrow N$ is a (linear or projective) isomorphism. If M is linear, then $\delta(V) = \delta(\bar{V})$ (see 6.3).

It follows easily from the lemma and proposition 9 that if V is contained in a subspace of M, then $\delta(V)$ is the same when V is regarded as an algebraic subset of that subspace.

For any algebraic sets $W_1, \ldots, W_r \subset M$, we have

(1) $$\delta\left(\bigcup W_i\right) \leq \sum \delta(W_i).$$

(Since each simple component of $\bigcup W_i$ is a simple component of some W_i; see 1.6.)

For any algebraic $V \subset M$ and any subspace $L \subset M$, we have

(2) $$\delta(L \cap V) \leq \delta(V).$$

First consider the projective case. In view of (1), one may assume that V is irreducible. Since L is an intersection of a finite number of hyperplanes, it is enough to show (2) for a hyperplane L. Omitting trivial cases, we may assume that $V \not\subset L$ and $k = \dim V > 0$. Then $L \cap V$ is of constant dimension $k - 1$ (see $(*)$ in III. 4.6 and proposition 3 in IV. 2.8) and so, by proposition 9, there exists $N \in \mathbf{G}_{n-k+1}(M)$, $N \not\subset L$, such that $\deg(L \cap V) = \#(N \cap L \cap V) \leq \deg V$, since $N \cap L \in \mathbf{G}_{n-k}(M)$. Now the linear case follows, since we have $\delta(\overline{L \cap V}) \leq \delta(\bar{L} \cap \bar{V})$. This is so because each simple component of $\overline{L \cap V}$ is a simple component of $\bar{L} \cap \bar{V}$ (for $\bar{L} \cap \bar{V} \cap M = L \cap V$; see 6.3). In particular:

([96]) The complement is even algebraic. See the proof below (and §17).

The number of simple components of $L \cap V$ is $\leq \delta(V)$.

If V and W are algebraic subsets of linear spaces, then

$$(3) \qquad\qquad \delta(V \times W) = \delta(V)\delta(W).$$

It is enough to prove (3) when V and W are irreducible (see IV. 2.9, corollary 6 of theorem 4). According to the lemma, we take Sadullaev spaces L and N for V and W, respectively, that intersect V and W transversally. Then $L \times N$ is a Sadullaev space for $V \times W$, and it intersects $V \times W$ transversally. So, (3) follows by the lemma.

If $V, W \subset M$ are algebraic, then

$$(4) \qquad\qquad \delta(V \cap W) \leq \delta(V)\delta(W).$$

For, in the linear case, by (2) and (3), we have $\delta(V \cap W) = \delta(\Delta \cap (V \times W)) \leq \delta(V \times W) = \delta(V)\delta(W)$, where $\Delta \subset M^2$ is the diagonal. In the projective case, we may assume that $M = \bar{X}$, and that X_∞ does not contain any simple components of $V \cap W$ (see 3.1). In such a case, setting $S = V \cap X$ and $T = W \cap X$, we have $\bar{S} = V$, $\bar{T} = W$, and $\overline{S \cap T} = V \cap W$ (see proposition 3 in IV. 2.8). Therefore $\delta(V \cap W) = \delta(S \cap T) \leq \delta(S)\delta(T) = \delta(V)\delta(W)$.

9. Let $\mathbf{P}$ be an n–dimensional projective space, and let $V \subset \mathbf{P}$ be a locally analytic set of constant dimension m. Let $f : V \longrightarrow N$ be a proper holomorphic mapping into a connected manifold N of dimension $k \leq m$. The fibres of the mapping f are algebraic (see II. 3.4 and Chow's theorem in 6.1). The set

$$S = V^* \cup \{z \in V^0 : d_z f \text{ is not surjective}\}$$

is analytic in V (see V. 3.2, theorem 1; and IV. 8.3, proposition 5). The fibres $f^{-1}(w)$ which satisfy the condition

$$(\omega_f) \qquad\qquad S \cap f^{-1}(w) \text{ is nowhere dense in } f^{-1}(w)$$

are of constant dimension $p = m - k$ (because $f^{-1}(w) \setminus S$ is p–dimensional; see C. 4.2 and IV. 2.5).

PROPOSITION 11. *All fibres $f^{-1}(w)$ that satisfy the condition (ω_f) are of the same degree.*

First we will prove

LEMMA . *Let $g : X \longrightarrow Y$ be a holomorphic mapping of analytic spaces, and let $T = \{z \in X : \dim l_z g \geq p\}$, where $p \in \mathbf{N}$ [97] . Then $\dim l_z g_T \geq p$ for $z \in T$.*

_____

[97] According to the Cartan-Remmert theorem (see V. 4.8 in ref. to V. 3.3), it is analytic.

Indeed, if $z \in T$, there exists a simple component $F \ni z$ of dimension $\geq p$ of the fibre $g^{-1}(g(z))$; then obviously $F \subset T$, and so $\dim l_z g_T = \dim_z \left(g^{-1}(g(z)) \cap T \right) \geq \dim_z F \geq p$.

PROOF of the proposition. We have $\dim l_z f \geq p$ for $z \in V$ (see V. 3.2, the corollary of theorem 3). We will prove that the set

$$E = \{ w \in N : \ f^{-1}(w) \text{ satisfies } (\omega_f) \} = N \setminus f(\{ z \in V : \ l_z f \subset S \})$$

is connected. Indeed, we have $N \setminus E \subset f(\{ z \in S : \ \dim l_z f_S \geq p \})$ [98]. The right hand side is analytic in N [99]; hence it is enough to show that it is a proper subset of N (see II. 3.6). Since it is equal to $\bigcup f(T_i)$, where $T_i = \{ z \in S_i : \ \dim l_z f_S \geq p \}$ and S_i are the simple components of S, it is sufficient to prove that $\dim f(T_i) < k$. Now, if $\dim S_i = m$, then S_i must be a simple component of V (see IV. 2.8). Hence in the open and dense subset $S_i \setminus V^*$ of S_i (see IV. 2.9, theorem 4), the differential $d_z f_{S_i}$ is not surjective. This implies that $\dim f(S_i) < k$ (see V. 1, theorem 1; V. 5.1, the Chevalley-Remmert theorem; and IV. 8.5). If $\dim S_i < m$, then, according to the lemma, $\dim l_z f_{T_i} \geq p$ for $z \in T_i$. Therefore, by theorem 2 from V. 3.2, we have $\dim f(T_i) < m - p = k$.

Therefore it is enough to show that the function $E \ni w \to \deg f^{-1}(w)$ is locally constant. Let $c \in E$ and $r = \deg f^{-1}(c)$. According to proposition 9, there is a subspace $L \in \mathbf{G}_{n-p}(\mathbf{P})$ that intersects $f^{-1}(c)$ transversally, and then $L \cap f^{-1}(c) = \{ a_1, \ldots, a_r \}$. Moreover, one may also assume that $a_i \in V^0 \setminus S$, because $f^{-1}(c)$ satisfies (ω_f) (see the corollary of proposition 9). By the lemma from C. 4.2, there are disjoint open neighbourhoods U_i in V^0 of the points a_i, respectively, and a neighbourhood Ω of c in N such that for each i and $w \in \Omega$ the subspace L intersects $f^{-1}(w) \cap U_i$ at a single point, and transversally. Since $f^{-1}(c) \subset (V \setminus L) \cup \bigcup U_i$, there is a neighbourhood $\Omega_0 \subset \Omega$ of the point c such that if $w \in \Omega_0$, then $f^{-1}(w) \subset (V \setminus L) \cup \bigcup U_i$, and so $L \cap f^{-1}(w) \subset \bigcup U_i$ (see B. 2.4). Consequently, L intersects $f^{-1}(w)$ in precisely r points transversally, and hence (see proposition 9) $\deg f^{-1}(w) = r$.

REMARK 1. It follows from the proof of the proposition that the set of points $w \in N$ that satisfy the condition (ω_f) contains a connected subset which is open and dense in N.

10. We are going to present a construction, due to Hironaka, of some analytic – and, in particular, algebraic (see 17.11 below) – spaces which cannot

---

[98] Because, if $l_z f \subset S$, then $l_z f_s = l_z f$.

[99] By Remmert's theorem (see V. 5.1) and the Cartan-Remmert theorem (see V. 3.3).

be embedded in any projective space. The reason why such an embedding is impossible is given in remark 2 below, which follows from proposition 11 and the properties of the degree in n° 8.

First observe that the condition (ω_f) from n° 9 is well-defined for the fibres of a holomorphic mapping $f : V \longrightarrow N$ of any analytic space V into any manifold N. When V is smooth, it means exactly that $d_z f$ is surjective for any z in the fibre except on a nowhere dense subset.

REMARK 2. Let X be an analytic space, and let $\tilde{M}, \tilde{N} \subset X$ be analytic sets of the same constant dimension. Suppose that there are proper holomorphic mappings $f : \tilde{M} \longrightarrow M$, $g : \tilde{N} \longrightarrow N$ into connected manifolds M, N of the same dimension, such that $\alpha \subsetneqq \beta$ and $\alpha' \supset \beta'$ for some fibres α, α' of f that satisfy (ω_f) and some fibres β, β' of g that satisfy (ω_g). Then the set $\tilde{M} \cup \tilde{N}$ and, moreover, the space X cannot be embedded in any projective space.

For otherwise one might assume that $M, N \subset \mathbf{P}$, and then one would arrive at a contradiction: $\deg \beta' \leq \deg \alpha' = \deg \alpha < \deg \beta = \deg \beta'$.

LEMMA 1. *Let $g : S \longrightarrow M$ be a submersion of a two dimensional manifold S into a one-dimensional manifold M. Let $a \in M$, and let $p : \tilde{S} \longrightarrow S$ be the blowing-up at a point $b \in g^{-1}(a)$. Then the composition $f = g \circ p : \tilde{S} \longrightarrow M$ is a submersion except a single point of $\tilde{S}$, the fibre $f^{-1}(a)$ is of constant dimension one, and $f^{-1}(a) \supsetneqq p^{-1}(b)$.*

PROOF . The fibre $g^{-1}(a) \subset S$ is a one–dimensional submanifold (see C. 4.2). The exceptional set $p^{-1}(b) \subset \tilde{S}$ is a one–dimensional submanifold, and the restriction $p^{\tilde{S} \setminus b} : S \setminus p^{-1}(b) \longrightarrow \tilde{S} \setminus b$ is a biholomorphic mapping (see 5.7, proposition 3). So, the fibre $f^{-1}(a) = p^{-1}(b) \cup p^{-1}\left(g^{-1}(a) \setminus b\right)$ is of constant dimension one and $f^{-1}(a) \supsetneqq p^{-1}(b)$. The proper inverse image $L \subset \tilde{S}$ of the fibre $g^{-1}(a)$ is a one–dimensional submanifold, and the restriction $p_L : L \longrightarrow g^{-1}(a)$ is the blowing-up at b (see 5.7, proposition 4), and hence is biholomorphic (see 5.7, proposition 3). Let $c = p_L^{-1}(b)$. It remains to show that if $w \in \tilde{S} \setminus c$, then the differential $d_w f = d_{p(w)} g \circ d_w p$ is different from zero, which means that im $d_w p \not\subset$ ker $d_{p(w)} g$. This is obvious when $w \in \tilde{S} \setminus p^{-1}(b)$, because then dim im $d_w p = 2$. When $w \in p^{-1}(b) \setminus c$, then im $d_w p$, im $d_c p \subset T_b S$ are different one–dimensional subspaces (see 5.7, proposition 3). But ker $d_{p(w)} g =$ ker $d_b g = T_b g^{-1}(a) =$ im $d_c p_L \subset$ im $d_c p$, and so im $d_w p \not\subset$ ker $d_{p(w)} g$.

LEMMA 2. *Let X be a three-dimensional manifold, and let $M, N \subset X$ be closed connected one-dimensional submanifolds that intersect only at*

a single point a, quasi-transversally. Then there exists a modification f :
$\tilde{X} \longrightarrow X$ *with the following properties:* $\tilde{X}$ *is a three–dimensional mani-*
fold; the restriction $f^{X \setminus a}$ is the blowing-up in $(M \cup N) \setminus a$; the sets $\tilde{M} =$
$\overline{f^{-1}(M \setminus a)}$, $\tilde{N} = \overline{f^{-1}(N \setminus a)}$ *are two–dimensional submanifolds (that inter-*
sect each other transversally); the fibres $f_{\tilde{M}}^{-1}(a)$, $f_{\tilde{N}}^{-1}(a)$ satisfy the conditions
$(\omega_{f_{\tilde{M}}})$, $(\omega_{f_{\tilde{N}}})$, *respectively* $(^{99a})$, *and $f_{\tilde{M}}^{-1}(a) \not\supseteq f_{\tilde{N}}^{-1}(a)$.*

PROOF. Let $p: X' \longrightarrow X$ be the blowing-up in M. Then X' is a three–
dimensional submanifold, the proper inverse image of the submanifold N is a
one–dimensional manifold $N' \subset X'$, and $p_{N'}: N' \longrightarrow N$ is the blowing-up at
a. Hence it is biholomorphic (see 5.7, propositions 3 and 4). The submanifold
N' intersects the two–dimensional exceptional submanifold $M' = p^{-1}(M)$ at
a single point, namely, at $a' = p_{N'}^{-1}(a)$ $(^{100})$, and transversally at that point
(see 5.7, proposition 4). The restriction $p_{M'}: M' \longrightarrow M$ is a submersion
(see 5.7, proposition 3).

Let $q: \tilde{X} \longrightarrow X'$ be a blowing-up in N'. Then $\tilde{X}$ is a three–dimensional
manifold (see 5.7, proposition 3) and the composition $f = p \circ q: \tilde{X} \longrightarrow X$
is a modification (see V. 4.13). The proper inverse image of the submani-
fold $M' \subset X'$ is (see 5.7, proposition 4) the two–dimensional submanifold
$\tilde{M} = q^{-1}(M') = f^{-1}(M) = \overline{f^{-1}(M \setminus a)}$ $(^{101})$ which transversally intersects the two–dimensional exceptional submanifold $\tilde{N} = q^{-1}(N')$. The restriction $q_{\tilde{N}}: \tilde{N} \longrightarrow N'$ is a submersion (see 5.7, proposition 3). Thus
$\tilde{N} = q^{-1}(\overline{p^{-1}(N \setminus a)}) = \overline{f^{-1}(N \setminus a)}$ (see B. 2).

Since $q_{\tilde{M}}: \tilde{M} \longrightarrow M'$ is the blowing-up at $a' \in p_{M'}^{-1}(a)$ (see 5.7, proposition 4), lemma 1 applied to the composition $f_{\tilde{M}} = p_{M'} \circ q_{\tilde{M}}$ implies that the
fibre $f_{\tilde{M}}^{-1}(a)$ is of constant dimension one and contains as a proper subset the
fibre $f_{\tilde{N}}^{-1}(a) = q^{-1}(p^{-1}(a)) \cap \tilde{N} = q^{-1}(a') = q_{\tilde{M}}^{-1}(a')$. Furthermore, $f_{\tilde{M}}$ is a
submersion except at a single point, and hence the fibre $f_{\tilde{M}}^{-1}(a)$ satisfies the
condition $(\omega_{f_{\tilde{M}}})$. Also, the fibre $f_{\tilde{N}}^{-1}(a)$ satisfies the condition $(\omega_{f_{\tilde{N}}})$ because
$f_{\tilde{N}} = p_{N'} \circ q_{\tilde{N}}$ is a submersion.

Finally, $p^{X \setminus N}$ and $q^{X' \setminus M'}$ are the blowings-up in $M \setminus a$ and $N' \setminus a'$,

---

$(^{99a})$ $f_{\tilde{M}}$ and $f_{\tilde{N}}$ regarded as mappings into M and N, respectively.

$(^{100})$ Since $M' \cap N' = p^{-1}(M \cap N) \cap N' = a'$.

$(^{101})$ Because $\dim f^{-1}(a) = 1$, since $\dim p^{-1}(a) = 1$ and $f^{-1}(a) = q^{-1}(p^{-1}(a) \setminus N') \cup q^{-1}(a')$ (see 5.7, proposition 3).

respectively (see 5.6), while $q^{X' \setminus p^{-1}(N)}$ and $p^{X \setminus M}$ are biholomorphic. Thus $f^{X \setminus N} = p^{X \setminus N} \circ q^{X' \setminus p^{-1}(N)}$ is the blowing-up in $M \setminus a$ and $f^{X \setminus M} = p^{X \setminus M} \circ q^{X' \setminus M'}$ is the blowing-up in $N \setminus a$ (see 5.6). Therefore $f^{X \setminus a}$ is a blowing-up in $M \cup N) \setminus a$ (see 5.6).

HIRONAKA'S CONSTRUCTION. Let X be a three–dimensional manifold. Let $M, N \subset X$ be closed connected one–dimensional submanifolds that intersect at only two points $a \neq b$, quasi-transversally at each of these points.

In accordance with lemma 2 applied to $M \setminus b$, $N \setminus b \subset X \setminus b$, there is a modification $f : X_1 \longrightarrow X \setminus b$ such that X_1 is a three–dimensional manifold, the restriction $f^{X \setminus \{a,b\}}$ is a blowing-up in $(M \cup N) \setminus \{a,b\}$, the sets $M_1 = \overline{f^{-1}(M \setminus \{a,b\})}$, $N_1 = \overline{f^{-1}(N \setminus \{a,b\})}$ are two–dimensional submanifolds (that intersect each other transversally), and the fibres $f_{M_1}^{-1}(a) \supsetneqq f_{N_1}^{-1}(a)$ satisfy $(\omega_{f_{M_1}})$ and $(\omega_{f_{N_1}})$, respectively.

Again, by lemma 2 applied to $N \setminus a$, $M \setminus a \subset X \setminus a$ ([102]), there exists a modification $g : X_2 \longrightarrow X \setminus a$ such that X_2 is a three–dimensional manifold, the restriction $g^{X \setminus \{a,b\}}$ is the blowing-up in $(M \cup N) \setminus \{a,b\}$, the sets $N_2 = \overline{g^{-1}(N \setminus \{a,b\})}$, $M_2 = \overline{g^{-1}(M \setminus \{a,b\})}$ are two–dimensional submanifolds (that intersect each other transversally), and the fibres $g_{N_2}^{-1}(b) \supsetneqq g_{M_2}^{-1}(b)$ satisfy $(\omega_{g_{N_2}})$ and $(\omega_{g_{M_2}})$, respectively.

The restrictions $f^{X \setminus \{a,b\}}$, as the blowings-up in the same subset, are isomorphic (see 5.5–6). Hence the modifications f, g can be glued together (see V. 4.13 and the proposition in V. 4.12), and so there is a modification $h : \tilde{X} \longrightarrow X$ such that $h^{X \setminus b} \approx f$ and $h^{X \setminus a} \approx g$. We have the commuting diagrams

$$
\begin{array}{ccc}
X_1 & \xrightarrow{\ \varphi\ } & \tilde{X} \setminus h^{-1}(b) \\
 {\scriptstyle f}\searrow & & \nearrow{\scriptstyle h^{X \setminus b}} \\
 & X \setminus b &
\end{array}
$$

and

$$
\begin{array}{ccc}
X_2 & \xrightarrow{\ \psi\ } & \tilde{X} \setminus h^{-1}(a) \\
 {\scriptstyle g}\searrow & & \nearrow{\scriptstyle h^{X \setminus a}} \\
 & X \setminus a &
\end{array}
$$

with some biholomorphic mappings φ and ψ. Therefore $\tilde{X}$ is a three–dimensional manifold. Set $\tilde{M} = \overline{h^{-1}(M \setminus \{a,b\})}$ and $\tilde{N} = \overline{h^{-1}(N \setminus \{a,b\})}$. Due to the fact that $\varphi(f^{-1}(E)) = h^{-1}(E)$ for $E \subset X \setminus b$, the images of the

---

([102]) In the opposite order to that used previously!

sets M_1 and N_1 are $\tilde{M} \setminus h^{-1}(b)$ and $\tilde{N} \setminus h^{-1}(b)$, respectively. Thus they are two–dimensional submanifolds of $\tilde{X}$ (that intersect transversally) and the fibres $h_{\tilde{M}}^{-1}(a) \not\supsetneqq h_{\tilde{N}}^{-1}(a)$ satisfy $(\omega_{h_{\tilde{M}}})$ and $(\omega_{h_{\tilde{N}}})$, respectively $(^{102a})$. Similarly, $\tilde{N} \setminus h^{-1}(a)$ and $\tilde{M} \setminus h^{-1}(a)$ are two–dimensional submanifolds of the manifold $\tilde{X}$ (that intersect transversally) and the fibres $h_{\tilde{N}}^{-1}(b) \not\supsetneqq h_{\tilde{M}}^{-1}(b)$ satisfy $(\omega_{h_{\tilde{N}}})$ and $(\omega_{h_{\tilde{M}}})$, respectively. (Therefore $\tilde{M}$ and $\tilde{N}$ are two–dimensional submanifolds and they intersect transversally.) In view of remark 2 (applied to the mappings $h_{\tilde{M}} : \tilde{M} \longrightarrow M$ and $h_{\tilde{N}} : \tilde{N} \longrightarrow N$), we conclude that:

The set $\tilde{M} \cup \tilde{N}$, and hence the manifold $\tilde{X}$, cannot be embedded in any projective space.

§12. The ring of an algebraic subset of a vector space

Let X be a vector space of positive dimension. Set $\mathcal{P} = \mathcal{P}(X)$. Let $S \subset X$ be a non-empty algebraic subset.

1. *The ring of the algebraic set S* is defined as the ring of restrictions to S of the polynomials from $\mathcal{P}$:

$$R(S) = R(S, X) = \{F_S : F \in \mathcal{P}\} \,.$$

Obviously, if $Z \subset X$ is a subspace containing S, then $R(S, Z) = R(S, X)$. We have

$$R(S) \approx \mathcal{P}/\mathcal{I}(S) \,,$$

where the natural isomorphism is induced by the epimorphism

$$\varrho : \mathcal{P} \ni F \longrightarrow F_S \in R(S) \,.$$

Consequently, the ring $R(S)$ is always noetherian (see A. 9.4). We have the following equivalence (see A. 1.11 and 1.6):

$$(R(S) \text{ is an integral domain}) \Longleftrightarrow (S \text{ is irreducible}) \,.$$

Finally, note that $F \in R(V)$ is not a zero divisor if and only if $F \not\equiv 0$ $(^{103})$.

$(^{102a})$ Here $h_{\tilde{M}}$ and $h_{\tilde{N}}$ are regarded as mappings into M and N, respectively.

$(^{103})$ This can be checked in a manner similar to the case of analytic germs (see IV. 4.3) by using corollary 1 of proposition 2 in V. 2.8 (see 11.1 and C.3.9).

Let $S \subset X$ and $T \subset Y$ be algebraic subsets of vector spaces X and Y. A mapping $f : S \longrightarrow T$ is said to be *(algebraically) regular* if it is the restriction of a polynomial mapping $X \longrightarrow Y$ (cf. 16.2 below). If, in addition, f is bijective and f^{-1} is also regular, then the mapping f is called an *isomorphism of the algebraic sets* S and T ([104]).

For every regular mapping $f : S \longrightarrow T$, we have the C–homomorphism of rings $f^* : R(T) \ni \psi \longrightarrow \psi \circ f \in R(S)$. Obviously, $(g \circ f)^* = f^* \circ g^*$ (for any regular mapping g of T into another algebraic set), and if e_S denotes the identity mapping of S, then e_S^* is the identity mapping of $R(S)$. Thus, if f is an isomorphism, then f^* is a C–isomorphism of rings.

For any regular mappings $f : S \longrightarrow T$, $g : S \longrightarrow T$, we have $f^* = g^* \Longrightarrow f = g$. (Indeed, for any $z \in S$ and $\eta \in Y^*$, we have $\eta_T \circ f = \eta_T \circ g$, which means that $\eta(f(z) - g(z)) = 0$, and so $f(z) = g(z)$.

Conversely, each C–homomorphism $h : R(T) \longrightarrow R(S)$ is of the form $h = f^*$, where $f : S \longrightarrow T$ is a regular mapping. Moreover, if h is an isomorphism, then also f is an isomorphism. Thus

$$R(S) \overset{C}{\approx} R(T) \Longleftrightarrow (S \text{ and } T \text{ are isomorphic}) .$$

In fact, one may assume that $X = \mathbf{C}^n$ and $Y = \mathbf{C}^m$. Let $f_i = h((w_i)_T)$, where w_i denotes the form $\mathbf{C}^m \ni (w_1, \ldots, w_m) \longrightarrow w_i \in \mathbf{C}$, $i = 1, \ldots, m$, and let $f = (f_1, \ldots, f_m)$. For any $P = \sum c_p w^p \in \mathcal{P}(Y)$, we have

$$P \circ f = \sum c_p f^p = h\left(\sum c_p w_T^p\right) = h(P_T) .$$

Therefore $P \circ f = 0$ for $P \in \mathcal{I}(T)$, and hence $f(S) \subset V(\mathcal{I}(T)) = T$ (see 1.2). So, we have a regular mapping $f : S \longrightarrow T$. Therefore $\psi \circ f = h(\psi)$ for $\psi = P_T \in R(T)$, which means that $h = f^*$. If, in addition, h is an isomorphism, then $h^{-1} = g^*$, where $g : T \longrightarrow S$ is a regular mapping; hence $(g \circ f)^* = f^* \circ g^* = e_S^*$, and so $g \circ f = e_S$. Similarly, $f \circ g = e_T$. Therefore f is an isomorphism.

Note also that the *dimension of an algebraic set S is equal to the Krull dimension* (see A. 12.3) *of its ring $R(S)$:*

$$\dim S = \dim R(S) .$$

We prove this by the same argument as in proposition 1 from IV. 4.3 combined with the lemma, using proposition 3 in IV. 2.8, corollary 1 in §10, and A. 1.11.

PROPOSITION . *A regular mapping $f : S \longrightarrow T$ is finite* ([105]) *if and only if the ring $R(S)$ is finite over the ring $f^*(R(T)) = R(T) \circ f$.*

PROOF ([106]). The ring $R(S)$ is finitely generated over the subring $R(T) \circ f$, e.g., by the restrictions of forms $\zeta_1, \ldots, \zeta_n$ constituting a basis of the dual space X^* ([107]).

([104]) If such an f exists, then we say that the sets S and T are isomorphic. See 17.1 below.

([105]) I.e., proper; see V. 7.1.

([106]) Based on an idea due to T. Winiarski, P. Tworzewski, and Z. Jelonek.

([107]) Owing to the fact that the ring $R(T) \circ f$ contains $\mathbf{C}$ (as a ring of constants) and $R(S) = \mathbf{C}[\zeta_1, \ldots, \zeta_n]$.

Consequently, it is finite over $R(T) \circ f$ exactly when it is integral over $R(T) \circ f$ (see A. 8.1), i.e., when for each $\varphi \in R(S)$ there exists a polynomial

$$P = t^s + a_1 t^{s-1} + \cdots + a_s \in \mathcal{P}(Y \times \mathbf{C}), \quad \text{where} \quad a_i \in \mathcal{P}(Y),$$

such that

(#) $$P(f(z), \varphi(z)) = 0 \quad \text{for} \quad z \in S.$$

Suppose that this condition is satisfied, and let $E \subset T$ be a compact set. Then $a_i \circ f$ are bounded on $f^{-1}(E)$, and hence so is φ (see B. 5.3). In particular, the ζ_i are bounded on the set $f^{-1}(E)$, which implies compactness of the latter. Suppose in turn that f is proper. One may assume that S is irreducible, because by taking the decomposition $S = S_1 \cup \ldots \cup S_r$ into simple components and polynomials P_i for the restrictions φ_{S_i}, it is enough to consider $P = P_1 \ldots P_r$. Furthermore, one may assume that $f(S) = T = Y$. Indeed, by taking a proper surjection $\pi : f(S) \longrightarrow Y_0$ onto a subspace Y_0 (see the corollary in 7.1), we conclude then that $R(S)$ is finite over $R(Y_0) \circ \pi \circ f$. Hence it is also finite over $R(T) \circ f = R(f(S)) \circ f \supset R(Y_0) \circ \pi \circ f$. Then the set $W = (f, \varphi)(S) \subset Y \times \mathbf{C}$ is algebraic and irreducible ([108]), and the natural projection $W \longrightarrow Y$ is a proper surjection ([109]). Hence $\dim W = \dim Y$ ([110]). Therefore (see 11.3, corollary 1 of proposition 5), we have $W = V(P)$, where $P = a_0 t^s + \ldots + a_s \in \mathcal{P}(Y \times \mathbf{C})$, $a_i \in \mathcal{P}(Y)$, $a_0 \neq 0$. Then (#) holds and it is sufficient to show that a_0 is constant. Now, a_i/a_0 must be bounded on bounded subsets of the dense set $\{a_0 \neq 0\}$ (because the natural mapping $W \longrightarrow Y$ is proper). Consequently, $a_i = 0$ on $\{a_0 = 0\}$. So, if we had $a_0(w) = 0$, then it would follow that $w \times \mathbf{C} \subset W$, which is impossible. Therefore a_0 is constant.

In what follows, set $R = R(S)$.

2. Suppose that S is a cone.

Then every function $f \in R$ has a unique expansion into homogeneous elements, namely $f = \sum_0^\infty f_\nu$, where $f_\nu \in R$ is a homogeneous function of degree ν and $f_\nu = 0$ for a sufficiently large ν. (The uniqueness follows from the fact that for each such expansion we have $f(tz) = \sum_0^\infty f_\nu(z) t^\nu$ when $z \in S$ and $t \in \mathbf{C}$.) It follows that:

Every function from R which is homogeneous of degree k is the restriction of a form of degree k from $\mathcal{P}$.

In the case when the cone S is irreducible, if the product gh is homogeneous, where $g, h \in R \setminus 0$, then so are the factors g and h. (This can be verified in the same way as in A. 2.3, using the fact that R is an integral domain.)

Therefore (see A. 1.22):

---

[108] As $(f, \varphi) : S \longrightarrow Y \times \mathbf{C}$ is proper. See 8.3, 11.1, and V. 3.1.

[109] Because the inverse image of any set $E \subset Y$ is contained in $E \times \varphi(f^{-1}(E))$.

[110] By proposition 2 from V. 7.1.

The invertible elements of the ring $R(S)$ are precisely the non-zero constants ([111]) . *All non-zero homogeneous elements of degree 1 are irreducible.*

3. For any $f_1, \ldots, f_k \in R$, we set $V(f_1, \ldots, f_k) = \{z \in S : f_1(z) = \ldots = f_k(z) = 0\}$. Obviously,

$$V(f_1, \ldots, f_k) = V(f_1) \cap \ldots \cap V(f_k) ,$$
$$V(f_1 \ldots f_k) = V(f_1) \cup \ldots \cup V(f_k) .$$

For every algebraic set $V \subset S$ we define its *ideal in the ring* R

$$\mathcal{I}_S(V) = \{f \in R : f = 0 \text{ on } V\} ,$$

and for each ideal of the ring R we define its *locus* (or *zero set*):

$$V(I) = V(f_1, \ldots, f_k), \text{ where } f_1, \ldots, f_k \text{ are generators of the ideal } I.$$

The definition is independent of the choice of generators (cf. II. 4.3, footnote ([14])). The properties from 1.2 and 1.3 remain true in this case (and can be checked in the same way). Here are some of them (V, W, V_i are algebraic subsets of S and I, J, I_ν are ideals of R):

$$\mathcal{I}_S(V(I)) \supset I \quad (^{112}), \quad V(\mathcal{I}_S(W)) = W,$$
$$I \subset J \Longrightarrow V(I) \supset V(J),$$
$$V \subset W \Longleftrightarrow \mathcal{I}_S(V) \supset \mathcal{I}_S(W), \quad V = W \Longleftrightarrow \mathcal{I}_S(V) = \mathcal{I}_S(W),$$

$$V(I_1 \cap \ldots \cap I_k) = V(I_1) \cup \ldots \cup V(I_k),$$
$$\mathcal{I}_S(V_1 \cup \ldots \cup V_k) = \mathcal{I}_S(V_1) \cap \ldots \cap \mathcal{I}_S(V_k),$$
$$V(\text{rad } I) = V(I), \quad \text{rad } \mathcal{I}_S(V) = \mathcal{I}_S(V).$$

We also have the equivalence (for any algebraic subset V of S):

$$(V \text{ is irreducible}) \Longleftrightarrow (\mathcal{I}_S(V) \text{ is prime}) .$$

(The proof is the same as in II. 4.6.)

([111]) The assumption that S is a cone is essential. For instance, if $S = \{zw = 1\} \subset \mathbf{C}^2$, then the restrictions $z_S, w_S \in R$ are mutually inverse. On the other hand, the assumption that the cone is irreducible can be omitted, because the irreducible components of S are also cones (see 1.6 and 6.2).

([112]) According to Hilbert's Nullstellensatz (see n° 5 below), we have $\mathcal{I}_S(V(I)) = \text{rad } I$. Thus the equality occurs when the ideal I is prime (see A. 1.11).

4. For $f \in R$, we clearly have $V(f) = V(Rf)$. Thus $f \in I \implies V(I) \subset V(f)$ (for any ideal I of R).

If S is irreducible and of dimension m, then $V(f)$ is of constant dimension $(m-1)$ for every $f \in R \setminus 0$.

Indeed, $V(f) = V(F) \cap S$, where $f = F_S$, $F \in \mathcal{P} \setminus 0$. Hence, if $z \in V(f)$, then codim $V(f)_z \leq 1 + \text{codim } S_z$ (see III. 4.6, inequality $(*)$; and 11.2 property (3)), which gives $\dim V(f)_z \geq m-1$ (see IV. 2.8 and 11. 1). On the other hand, $V(f) \subsetneq S$, and so $\dim V(f)_z \leq m-1$ (see IV. 2.8, proposition 3).

5. *Hilbert's Nullstellensatz* remains true:

$\mathcal{I}(V(I)) = \text{rad } I$ *for every ideal I of the ring R. In particular, $\mathcal{I}(V(I)) = I$ if the ideal I is prime* (see A. 1.11).

Indeed, we have rad $I \subset \mathcal{I}_S(V(\text{rad } I)) = \mathcal{I}_S(V(I))$ (see n° 3). On the other hand, the ideal I has a system of generators $f_i = (F_i)_S$, where $F_i \in \mathcal{P}$, $i = 1, \ldots, k$, and we have $S = V(H_1, \ldots, H_r)$ with some $H_j \in \mathcal{P}$. Let I' be the ideal (in $\mathcal{P}$) generated by $F_1, \ldots, F_k, H_1, \ldots, H_r$. Then $V(I) = V(I')$ and $\varrho(I') \subset I$. Now, if $f = 0$ on $V(I)$, where $f = F_S$, $F \in \mathcal{P}$, then $F = 0$ on $V(I')$, and so Hilbert's Nullstellensatz (from §10) yields $F^r \in I'$ for some $r \in \mathbf{N}$, which gives $f^r \in I$. Thus $\mathcal{I}_S(V(I)) \subset \text{rad } I$.

Consequently, we have the equivalence (for any $f \in R$)

$$(f \text{ is an invertible element}) \iff V(f) = \emptyset \ .$$

For, if $V(f) = \emptyset$, then $1 \in Rf$.

6. An algebraic set S is said to be *factorial* if its ring $R = R(S)$ is factorial. Then S must be irreducible (see n° 1).

In the remaining part of §12 we will be assuming that S is factorial. Then it is of constant dimension $m = \dim S$.

It follows from Hilbert's Nullstellensatz (as in proposition 3 in 11. 2; see n° 4) that for $f, g \in R$, where f is irreducible, we have the equivalence

$$(g = 0 \text{ on } V(f)) \iff (g \text{ is divisible by } f) \ .$$

Consequently, one derives (as in 11.2) the following corollaries:

If $f \in R$ is irreducible, then $V(f)$ is irreducible.

If $f, g \in R$ are irreducible, then

$$V(f) = V(g) \iff (f \text{ and } g \text{ are associated}) \ .$$

If $g \in R \setminus 0$ is a non-invertible element and $g_1, \ldots, g_r$ are all its distinct factors (up to being associated), then $V(g) = V(g_1) \cup \ldots \cup V(g_r)$ is the decomposition of $V(g)$ into simple components.

7. In the same way we show that (as in 11. 3):

If $f_1, \ldots, f_k \in R$, then

$$V(f_1, \ldots, f_k) = V(g) \cup B ,$$

where g is the greatest common divisor of $f_1, \ldots, f_k$, and B is an algebraic set of dimension $\leq m - 2$. Hence (see n° 4 and 5)

$$(f_1, \ldots, f_k \text{ are relatively prime}) \Longleftrightarrow \dim V(f_1, \ldots, f_k) \leq m - 2 .$$

If V is an algebraic subset of S, then

$$\big(V \text{ is irreducible of dimension } (m-1) \big) \Longleftrightarrow$$
$$\Longleftrightarrow \big(V = V(f), \text{ where } f \in R \text{ is irreducible} \big).$$

8. Assume in addition that S is a cone. Then (see n° 2) in the decomposition of any homogeneous element into irreducible factors, each of the factors must be homogeneous.

As in 11.2 (corollary 6), we obtain the equivalences:

$$\big(V(f) \text{ is a cone} \big) \Longleftrightarrow \big(f \text{ is homogeneous and } f(0) = 0 \big) ,$$

for $f \in R$. Moreover (as in corollary 4 from 11.2; see n° 2), if $f, g \in R$ and f is an irreducible element, then

$$V(g) = V(f) \Longleftrightarrow (g = af^r, \text{ where } a \in \mathbf{C} \setminus 0 \text{ and } r \in \mathbf{N} \setminus 0) .$$

§13. Bézout's theorem. Biholomorphic mappings of projective spaces

1. Let X be an n–dimensional vector space.

LEMMA 1. *Let* $k_1, \ldots, k_n \in \mathbf{N} \setminus 0$, *and let* N *be the vector space of all the mappings* $f = (f_1, \ldots, f_n) : X \longrightarrow \mathbf{C}^n$, *where* f_i *is a form of degree* k_i $(i = 1, \ldots, n)$ [113]. *Then the set* $\Sigma = \{f : f^{-1}(0) \neq 0\} \subset N$ *is algebraic and nowhere dense.*

Indeed, the set Σ is the image under the natural projection $\pi : N \times X \longrightarrow N$ of each of the sets

$$A = \{(f, c) \in N \times X : f(c) = 0, \ c \neq 0\} \ ,$$
$$B = \{(f, c) \in N \times X : f(c) = 0, \ |c| = 1\}$$

(after setting a norm on X.) The set A is constructible, hence so is the set Σ, by Chevalley's theorem (see 8.3). Since the restriction π_B is proper (see B. 5.2), the set Σ is closed, and so (see 8.3, proposition 2), it is algebraic. It is nowhere dense, because $\Sigma \subsetneq N$, since (assuming that $X = \mathbf{C}^n$) $f = (z_1^{k_1}, \ldots, z_n^{k_n}) \in N \setminus \Sigma$.

LEMMA 2. *Let* $f = (f_1, \ldots, f_n) : X \longrightarrow \mathbf{C}^n$, *where* f_i *is a form of degree* k_i $(i = 1, \ldots, n)$, *and suppose that* $f^{-1}(0) = 0$. *Then* $\nu(f) = k_1 \ldots k_n$ (see V. 2.3).

In fact, one may assume that $X = \mathbf{C}^n$. Now, all k_i must be positive, for otherwise one would have $f_s = 0$ for some s, which would give codim $f^{-1}(0)_0 \leq \sum_{i \neq s} \text{codim } \{f_i = 0\}_0 \leq n - 1$ (see (*) in III. 4.6 and II. 5.1). Hence $\dim f^{-1}(0) \geq 1$, contradicting our assumption. By lemma 1, the set $G = \{g : g^{-1}(0) = 0\} \subset N$ is open and connected (see II. 3.6), so Rouché's lemma (see V. 2.3) applied to the holomorphic mapping $G \times X \ni (g, c) \longrightarrow g(c) \in \mathbf{C}^n$ implies that the function $G \ni g \longrightarrow \nu(g)$ is constant. Since $f \in G$, it is enough to note that $h = (z_1^{k_1}, \ldots, z_n^{k_n}) \in G$ and $\nu(h) = k_1 \ldots k_n$ (because $\#h^{-1}(w) = k_1 \ldots k_n$ for each $w = (w_1, \ldots, w_n)$ from the dense subset $\{w_1 \neq 0, \ldots, w_n \neq 0\}$ of $\mathbf{C}^n$; see V. 2.3, lemma 1).

2. Let Y be an $(n + 1)$–dimensional vector space, and let $\mathbf{P} = \mathbf{P}(Y)$. Consider the system of equations

(B) $$F_1(z) = \ldots = F_n(z) = 0 \ ,$$

where $F_1, \ldots, F_n \in \mathcal{P}(Y)$ are forms of degrees $k_1, \ldots, k_n$, respectively. A line $\lambda \in \mathbf{P}$ is said to be a *zero* (in $\mathbf{P}$) *of the system of equations* (B) if $F_1(\lambda) = \ldots = F_n(\lambda) = 0$.

---

[113] Obviously, it is finite dimensional. (See A. 3.2.)

If λ is an isolated point of the set of zeros in $\mathbf{P}$ of the system B, we say that λ is an *isolated zero* (in $\mathbf{P}$) of the system. We define the *multiplicity of the zero* λ as the multiplicity at any point $a \in \lambda \setminus 0$ of the restriction $F_H : H \longrightarrow \mathbf{C}^n$ of the mapping $F = (F_1, \ldots, F_n)$ to any affine hyperplane $H \subset Y$ such that $H \cap \lambda = a$ (see V. 2.1; such a point a is then an isolated point of the set $F_H^{-1}(0)$). This multiplicity does not depend on the choice of the point a and the hyperplane H.

Indeed, let H', H'' be affine hyperplanes such that $H' \cap \lambda = a' \neq 0$ and $H'' \cap \lambda = a'' \neq 0$. Then they do not contain the point 0. The mapping $G = \alpha_{H''}^{-1} \circ \alpha_{H'}$ maps the neighbourhood $U' = H' \setminus H''_*$ of the point a' in H' biholomorphically onto the neighbourhood $U'' = H'' \setminus H'_*$ of the point a'' in H'', and $G(a') = a''$ (because $\mathbf{C}a' = \lambda = \mathbf{C}a''$). Since $G(z) = z/\lambda_{H''}(z)$ in U' (see 2.1), we have $F_{U''} \circ G = \big(\gamma_1(F_1)_{U'}, \ldots, \gamma_n(F_n)_{U'}\big)$, where $\gamma_i(z) = \lambda_{H''}(z)^{-k_i} \neq 0$ in U' $(i = 1, \ldots, n)$. Therefore, in view of the corollary from V. 2.2 $(^{114})$, we obtain $m_{a''}F_{H''} = m_{a'}(F_{U''} \circ G) = m_{a'}F_{H'}$.

We have the following

BÉZOUT THEOREM . *If the number of zeros in* $\mathbf{P}$ *of the system* (B) *is finite, then the sum of their multiplicities is equal to the product* $k_1 \ldots k_n$ *of the degrees of the forms* $F_1, \ldots, F_n$.

PROOF . One may assume that $F(0) = 0$ $(^{115})$. Let $\lambda_1, \ldots, \lambda_s$ be all the zeros in $\mathbf{P}$ of the system (B), and let $p_1, \ldots, p_s$ denote their multiplicities. Then $F^{-1}(0) = 0 \cup \lambda_1 \cup \ldots \cup \lambda_s$. Take a hyperplane which does not contain any of the lines λ_i $(^{116})$. Then $F_H^{-1}(0) = 0$ and, by lemma 2,

$$\nu(F_H) = k_1 \ldots k_n .$$

Now, if $c \in Y \setminus H$, then $(c + H) \cap \lambda_i = a_i \neq 0$ and $F_{c+H}^{-1}(0) = \{a_1, \ldots, a_s\}$, and all a_i are distinct. Hence (see V. 2.3)

$$\nu(F_{c+H}) = \sum_1^s m_{a_i}F_{c+H} = \sum_1^s p_i .$$

For any point c from a fixed, open, bounded, and connected neighbourhood U of zero, define $f_c = F_{c+H} \circ \tau_c$, where $\tau_c : H \ni z \longrightarrow c + z \in c + H$, i.e., $f_c(z) = F(c + z)$ for $z \in H$. Then (see V. 2.3)

$$\nu(f_c) = \nu(F_{c+H}) .$$

$(^{114})$ We take $\psi : U' \times \mathbf{C}^n \ni (z, w) \longrightarrow (\gamma_1(z)w_1, \ldots, \gamma_n(z)w_n) \in \mathbf{C}^n$.

$(^{115})$ If $F(0) \neq 0$, the theorem is trivial, as then $k_s = 0$ for some s and the set of zeros is empty.

$(^{116})$ Such a hyperplane exists, because $G_n(Y) \setminus \bigcup_1^s \mathbf{S}^n(\lambda_i) \neq \emptyset$ (see B. 6.5).

Since $f_c \neq 0$ in $H \setminus E$ for $c \in U$ with some compact set $E \subset H$ ([117]), Rouché's lemma (see V. 2.3) applied to the holomorphic mapping $U \times H \ni (c, z) \longrightarrow F(c + z) \in \mathbf{C}^n$ implies that the function $U \ni c \longrightarrow \nu(f_c)$ is constant. By taking $c \in U \setminus H$, we get

$$\sum_1^s p_i = \nu(F_{c+H}) = \nu(f_c) = \nu(f_0) = \nu(F_H) = k_1 \ldots k_n \ .$$

REMARK . Consider the vector space M of the sequences $F = (F_1, \ldots, F_n)$, where $F_1, \ldots, F_n \in \mathcal{P}(Y)$ are forms of degrees $k_1, \ldots, k_n$, respectively ([118]). The sequences F for which the number of zeros of the system (B) in $\mathbf{P}$ is finite form an open and dense subset of M. In fact, the complement Z of this set is a nowhere dense algebraic subset of M.

Indeed, the set $W = \{(F, \lambda) : F(\lambda) = 0\} \subset M \times \mathbf{P}$ is analytic. For (see II. 3.4) its inverse image under the surjective submersion $\mathrm{id}_M \times \alpha : M \times (Y \setminus 0) \longrightarrow M \times \mathbf{P}$, where e denotes the identity mapping on the space M (see 2.2 and C. 4.2), is equal to the set $\{(F, z) \in M \times (Y \setminus 0) : F(z) = 0\}$. This set analytic in $M \times (Y \setminus 0)$. The natural projection $\pi : W \longrightarrow M$ is holomorphic and proper, the set $W_0 = \{(F, \lambda) \in W : \dim l_{(F,\lambda)} \pi > 0\}$ is closed in W (see V. 3.2, the semicontinuity theorem), and so the set $Z = \pi(W_0)$ is closed. In view of the Bézout theorem, it is the image by the projection onto M of the constructible set

$$\{(F, z_1, \ldots, z_k) \in M \times Y^k : F(z_i) = 0, \ z_i \neq 0, \ z_i \wedge z_j \neq 0 \text{ for } i \neq j\} \ ,$$

where $k = (k_1, \ldots, k_n) + 1$. By the Chevalley theorem (see 8.3), it is constructible and hence algebraic (see 8.3, proposition 2). Finally, $Z \subsetneq M$ because $(\zeta_1^{k_1}, \ldots, \zeta_n^{k_n}) \in M \setminus Z$ for any linearly independent forms $\zeta_1, \ldots, \zeta_n \in Y^*$.

3. Let X be an n–dimensional vector space.

LEMMA 3. *Consider the mapping* $F = (F_1, \ldots, F_n) : X \longrightarrow \mathbf{C}^n$, *where* $F_1, \ldots, F_n$ *are polynomials of degrees at most* $k_1, \ldots, k_n \geq 0$, *respectively. Let* f_ν *denote the homogeneous component of degree* k_ν *of the polynomial* F_ν ($\nu = 1, \ldots, n$). *If* $V(f_1, \ldots, f_n) = 0$, *then the set* $F^{-1}(0)$ *is finite and* $\nu(F) = k_1 \ldots k_n$.

PROOF . Fix a norm in X. Then $F_\nu(z)/|z|^{k_\nu} - f_\nu(z/|z|) \longrightarrow 0$ when $|z| \longrightarrow \infty$, and $\inf\{\sum_1^n |f_\nu(z)| : |z| = 1\} > 0$. It follows that $\sum_1^n |F_\nu(z)|/|z|^{k_\nu} > 0$ if $|z|$ is sufficiently large. Therefore the set $F^{-1}(0)$ is bounded and thus, being analytic, it is finite: $F^{-1}(0) = \{a_1, \ldots, a_r\}$.

([117]) It is enough to take the set $E = H \cap (F^{-1}(0) - \bar{U})$, since it is contained in the union of the compact sets $H \cap (-\bar{U})$, $H \cap (\lambda_i - \bar{U}) = \pi_i(-\bar{U})$, $i = 1, \ldots, n$. Here $\pi_i : Y \longrightarrow H$ denotes the projection that is parallel to λ_i.

([118]) Obviously, the space M is finite dimensional.

Let $G_\nu \in \mathcal{P}(\mathbf{C} \times X)$ be the form of degree k_ν such that

$$(*) \qquad\qquad F_\nu(z) = G_\nu(1, z) \quad \text{in } X \quad (\nu = 1, \ldots, n)$$

(see A. 3.2). Consider the system of equations

$$(\mathrm{B}') \qquad\qquad G_1(t, z) = \ldots = G_n(t, z) = 0 .$$

Every line in $\mathbf{P}(\mathbf{C} \times X)$ is either of the form $\mathbf{C}(0, a)$ where $a \in X \setminus 0$, or $\mathbf{C}(1, a)$, where $a \in X$. In the former case, it cannot be a zero of the system (B') (for otherwise one would have $a \in V(f_1, \ldots, f_k)$). Therefore – in view of $(*)$ – the only zeros of the system (B') in $\mathbf{P}(\mathbf{C} \times X)$ are the (distinct) lines $\mathbf{C}(1, a_i)$. Their multiplicities are equal to $m_{a_i} F$ (see n° 2), and so the Bézout theorem yields $\nu(F) = \sum_1^r m_{a_i} F = k_1 \ldots k_n$.

THE RUSEK-WINIARSKI INEQUALITY [119] . *Consider a mapping* $F = (F_1, \ldots, F_n) : X \longrightarrow \mathbf{C}^n$, *where* $F_1, \ldots, F_n \in \mathcal{P}(X)$ *are polynomials of degrees* $k_1, \ldots, k_n \geq 0$, *respectively. If the set* $F^{-1}(0)$ *is finite, then* $\nu(F) \leq k_1 \ldots k_n$.

PROOF . One may assume that $k_1, \ldots, k_n > 0$. Let N' be the vector space of all the mappings $H = (H_1, \ldots, H_n) : X \longrightarrow \mathbf{C}^n$, where $H_1, \ldots, H_n \in \mathcal{P}(X)$ are polynomials of degrees at most $k_1, \ldots, k_n$ [120] . Let us employ the notation from lemma 1. The mapping $\varphi : N' \ni (H_1, \ldots, H_n) \longrightarrow (h_1, \ldots, h_n) \in N$, where h_i denotes the homogeneous component of degree k_i of the polynomial H_i $(i = 1, \ldots, n)$, is an epimorphism. Therefore it is an open and continuous surjection (see B. 5.2). By lemma 1, the set $\Sigma' = \varphi^{-1}(\Sigma) \subset N'$ is nowhere dense (see B. 2.2). By lemma 3,

$$(\#) \qquad \text{the set } H^{-1}(0) \text{ is finite and } \nu(H) = k_1 \ldots k_n \quad \text{for } H \in N' \setminus \Sigma' .$$

Set $B_s = \{z \in X : |z| < s\}$ (after endowing X with a norm). We have $F^{-1}(0) \subset B_r$ for some $r > 0$. Put $B = B_{2r}$. The set $E = \bar{B} \setminus B_r$ is compact, its interior is non-empty, and so $|H| = \sup\{|H(z)| : z \in E\}$ is a norm on the space N'. Since $\varepsilon = \inf\{|F(z)| : z \in E\} > 0$, we have $|H(z)| \geq \frac{1}{2}\varepsilon$ in E, provided that $|H - F| < \frac{1}{2}\varepsilon$. Consequently, by taking the open ball $\Omega = \{H : |H - F| < \frac{1}{2}\varepsilon\} \subset N'$, we obtain

$$H_B(z) \neq 0 \quad \text{in } B \setminus B_r \quad \text{for } H \in \Omega .$$

---

[119] See [35a].

[120] This space is, clearly, finite dimensional.

In view of the Rouché lemma (see V. 2.3) applied to the holomorphic mapping $\Omega \times B \ni (H, z) \longrightarrow H_B(z) \in \mathbf{C}$ [121] , the set $H_B^{-1}(0)$ is finite for $H \in \Omega$ and the function $\Omega \ni H \longrightarrow \nu(H_B)$ is constant. Take $H \in \Omega \setminus \Sigma'$. Then, by (#), we obtain

$$\nu(F) = \nu(F_B) = \nu(H_B) \leq \nu(H) = k_1 \ldots k_n .$$

REMARK . In the space N' (see the above proof), the set of mappings F for which $F^{-1}(0)$ is finite contains an open and dense set; its complement Z is contained in a nowhere dense algebraic set [122].

Indeed, consider the vector space M' of the sequences $(G_1, \ldots, G_n)$, where $G_n \in \mathcal{P}(\mathbf{C} \times X)$ is a form of degree k_ν ($\nu = 1, \ldots, n$). We have the isomorphism $\psi : N' \ni (F_1, \ldots, F_n) \longrightarrow (G_1, \ldots, G_n) \in M'$, where the G_i are defined by the condition (∗) from the proof of lemma 3 (see A. 3.2). Now, if $Z' \subset M'$ is the complement of the set defined for M' according to the remark following the Bézout theorem, then $\psi(Z) \subset Z'$.

4. Let $\mathbf{P} = \mathbf{P}(Y)$, where Y is an $(n + 1)$–dimensional vector space.

We have the following version of Bézout's theorem in the particular case of n principal varieties that intersect transversally.

BÉZOUT'S THEOREM . *If the principal algebraic sets* $V_1, \ldots, V_n \subsetneq \mathbf{P}$ *of degrees* $r_1, \ldots, r_n$, *respectively, intersect transversally* [123] , *then* $\#(V_1 \cap \ldots \cap V_n) = r_1 \ldots r_n$.

PROOF . One may assume that the sets V_i are non-empty. Then V_i is defined by a homogeneous polynomial $h_i \in \mathcal{P}(Y)$ of degree r_i without multiple factors, and $V_i^\sim = V(h_i)$, $i = 1, \ldots, n$ (see 11.7 and 6). Thus $\bigcap_1^n V_i$ is the set of zeros in $\mathbf{P}$ of the system of equations

$$(\mathrm{B}_0) \qquad\qquad h_1(z) = \ldots = h_n(z) = 0 .$$

Therefore, by Bézout's theorem from n° 2, it suffices to prove that the multiplicity of each of the zeros in $\mathbf{P}$ of the system (B_0) is equal to 1.

---

[121] It is the restriction of a polynomial mapping.

[122] The set Z is constructible, because it is the image by the projection into N' of the constructible set

$$\{(H, z_1, \ldots, z_k) \in N' \times X^k : H(z_i) = 0,\ z_i \neq z_j \text{ for } i \neq j\} ,$$

where $k = (k_1 \ldots k_n) + 1$. On the other hand, it is not necessarily closed. For instance, if $X = \mathbf{C}^2$, the mappings $F_\nu(z, w) = (1 + \frac{1}{\nu}z, 1 + \frac{1}{\nu}z)$, $\nu = 1, 2, \ldots$, belong to Z, whereas their limit (1,1) does not belong to Z.

[123] See C. 3.11. Then the set $V_1 \cap \ldots \cap V_n$ is discrete (see C. 3.16) and compact, and thus it is finite.

In order to show this, let $\lambda \in \bigcap_1^n V_i$. Set $h = (h_1, \ldots, h_n)$. Take an affine hyperplane H such that $\lambda \cap H = a \neq 0$. It follows from the hypotheses that $\lambda \in V_i^0$, $i = 1, \ldots, n$, and $\bigcap_1^n T_\lambda V_i = 0$ (see C. 3.11). The images of the point a and the sets $V_i' = V_i^{\sim} \cap H$ under the biholomorphic mapping $\alpha_H : H \ni z \longrightarrow \mathbf{C}z \in \Omega = \mathbf{P} \setminus \mathbf{P}(H^*)$ are λ and the sets $V_i \cap \Omega$, respectively. Hence $a \in (V_i')^0$ ($i = 1, \ldots, n$), and $\bigcap_1^n T_a V_i' = 0$. Furthermore, $a \in (V_i^{\sim})^0$ (see 6.2), and so $d_a h_i \neq 0$ (see 11.4, proposition 6). Since $\lambda \subset \ker d_a h_i$, we must have $d_a(h_i)_H = (d_a h_i)_{H^*} \neq 0$. But $V_i' = \{(h_i)_H = 0\}$, and so $T_a V_i' = \ker d_a(h_i)_H$. It follows that $\ker d_a h_H = \bigcap \ker d_a(h_i)_H = 0$, and consequently, $d_a h_H$ is an isomorphism. Therefore (see V. 2.1) $m_a h_H = 1$, and thus the multiplicity of the zero λ of the system (B_0) is 1.

5. Let X and Y be $(n + 1)$-dimensional vector spaces. We will prove that the biholomorphic mappings of the space $\mathbf{P}(X)$ onto the space $\mathbf{P}(Y)$ are precisely the isomorphisms of projective spaces. This proof is due to P. Tworzewski.

LEMMA 4. *Let λ and μ be one–dimensional vector spaces. If $\psi : \bar{\lambda} \longrightarrow \bar{\mu}$ is a biholomorphic mapping such that $\psi(0) = 0$ and $\psi(\lambda_\infty) = \mu_\infty$, then the restriction $\psi_\lambda : \lambda \longrightarrow \mu$ is a linear mapping.*

In fact, one may assume that $\lambda = \mu = \mathbf{C}$, and interpret $\bar{\mathbf{C}}$ as the Riemann sphere (see 3.1–2). Then $\psi(\infty) = \infty$, and using the fact that ψ^{-1} is holomorphic at ∞, we get $|\psi(t^{-1})^{-1}| \geq \varepsilon|t|$ for a sufficiently small $t \in \mathbf{C} \setminus 0$ and for some $\varepsilon > 0$ (see C. 3.19). This implies that $|\psi(z)| \leq K(1 + |z|)$ in $\mathbf{C}$ for some $K > 0$. Thus, in view of Liouville's theorem (see C. 1.8), the restriction $\psi_{\mathbf{C}} : \mathbf{C} \longrightarrow \mathbf{C}$ is linear.

LEMMA 5. *Any biholomorphic mapping $\varphi : \mathbf{P}(X) \longrightarrow \mathbf{P}(Y)$ that maps every hyperplane onto a hyperplane must be an isomorphism of projective spaces.*

PROOF. One may assume that $\mathbf{P}(X) = \bar{M}$ and $\mathbf{P}(Y) = \bar{N}$, where M, N are n-dimensional vector spaces, and that $\varphi(M_\infty) = M_\infty$, $\varphi(0) = 0$ (see 3.1). Then $\varphi(M) = N$. If φ maps the $(k + 1)$–dimensional (projective) subspaces onto $(k + 1)$–dimensional subspaces, then the same holds for the subspaces of dimension k (see B. 6.12). Therefore φ maps lines onto lines. Consequently, if $\lambda \in \mathbf{P}(M)$, then $\bar{\lambda} \subset \bar{M}$ is a projective line and so is $\varphi(\bar{\lambda}) \subset \bar{N}$. Hence $\varphi(\bar{\lambda}) = \bar{\mu}$, where $\mu \in \mathbf{P}(N)$, and $\varphi(\lambda_\infty) = \mu_\infty$ (see 3.3). Thus, by lemma 4, the mapping $\varphi_\lambda : \lambda \longrightarrow \mu$ is linear. Consequently, the mapping $\varphi_M : M \longrightarrow N$ is homogeneous of degree 1, and hence (see C. 1.8) it must be linear. Therefore it is a linear isomorphism, and so (see 3.2) φ is an isomorphism of projective spaces.

LEMMA 6. *If hyperplanes* $H_1, \ldots, H_n \subset \mathbf{P}(X)$ *intersect at a singe point* (*i.e.*, $\#(H_1 \cap \ldots \cap H_n) = 1$), *then they intersect transversally.*

In fact, one may assume that $\mathbf{P}(X) = \bar{M}$ and $H_1 \cap \ldots \cap H_n = 0$, where M is an n–dimensional vector space (see 3.1). Now, $H_i \cap M \subset M$ are hyperplanes (see 3.3) and their intersection at 0 is transversal (see A. 1.18). This implies that the hyperplanes H_i intersect transversally.

THEOREM . *Every biholomorphic mapping of the space* $\mathbf{P}(X)$ *onto the space* $\mathbf{P}(Y)$ *is an isomorphism (of projective spaces)* [124] .

PROOF . Let $\varphi : \mathbf{P}(X) \longrightarrow \mathbf{P}(Y)$ be a biholomorphic mapping. In view of lemma 5, it is enough to prove that φ maps hyperplanes onto hyperplanes. Let $H \subset \mathbf{P}(X)$ be a hyperplane. There are hyperplanes $H = H_1, \ldots, H_n$ such that $\#(H_1 \cap \ldots \cap H_n) = 1$ [125] . By lemma 6, they intersect transversally. Their images $\varphi(H_1), \ldots, \varphi(H_n) \subset \mathbf{P}(Y)$ are closed submanifolds of dimension $n - 1$ which intersect at a single point, transversally at that point. In view of Chow's theorem (see 6.1), they are proper principal algebraic sets (see 11.6). Denote by $r_1, \ldots, r_n$, respectively, their degrees. By the Bézout theorem (from n° 4), we have $1 = r_1 \ldots r_n$, and so $r_1 = 1$. Consequently, $\varphi(H)$ is a hyperplane (see 11.7).

REMARK . In particular, the biholomorphic mappings of the Riemann sphere $\bar{\mathbf{C}} = \mathbf{P}_1$ onto itself are precisely the mappings $h : \mathbf{P}_1 \ni \mathbf{C}(t, z) \longrightarrow \mathbf{C}(at + bz, ct + dz)$, where $ad - bc \neq 0$. In other words (after the identification $\iota : \mathbf{C} \cup \infty \longrightarrow \mathbf{P}_1$, where $\iota(z) = \mathbf{C}(1, z)$, $\iota(\infty) = 0 \times \mathbf{C}$; see 3.1):

$$ h(z) = \frac{c + dz}{a + bz} \text{ for } z \neq -b/a, \ h(-b/a) = \infty, \ h(\infty) = d/b , $$

in the case when $b \neq 0$, and

$$ h(z) = c' + d'z, \ h(\infty) = \infty, \text{ where } d' \neq 0 $$

otherwise. Therefore they are the so-called *homographies* (see [5], Chapter IV, §8).

This can be also derived directly from lemma 4. Namely, if $f : \bar{\mathbf{C}} \longrightarrow \bar{\mathbf{C}}$ is biholomorphic, one may assume that $f(0) = 0$ and $f(\infty) = \infty$ [126] . Then the restriction $f_{\mathbf{C}} : \mathbf{C} \longrightarrow \mathbf{C}$ is linear, and so f is a homography.

---

[124] It is a special case of the theorem on biholomorphic mapping of factorial sets (see 18.2 below).

[125] It suffices to take $\lambda \in H$ and hyperplanes $H^\check{} = L_1, \ldots, L_n \subset X$ such that $\bigcap_1^n L_i = \lambda$, and set $H_i = L_i^\check{}$.

[126] Due to the fact that the homographies constitute a group, and if $\alpha, \beta, \gamma, \delta \in \bar{\mathbf{C}}$, $\alpha \neq \beta$, and $\gamma \neq \delta$, then there is a (unique) homography h such that $h(\alpha) = \gamma$ and $h(\beta) = \delta$.

§14. Meromorphic functions and rational functions

Let M be a complex manifold of dimension $m > 0$.

1. We say that a function f is *holomorphic nearly everywhere* on M if f is a holomorphic function on an open dense subset of the manifold M. The complement $Z \subset M$ of that set is closed and nowhere dense, and is called the *exceptional set* for the function f. A point $a \in Z$ is said to be a *removable singular point* if the function f extends holomorphically across this point (see II. 3.8, footnote (^{12})). Otherwise, it is called a *singular* point. The set of singular points of f is closed and nowhere dense. We denote it by S_f.

Let $\mathcal{O}'_M$ denote the set of functions that are holomorphic nearly everywhere on M. By the *complete elements* of the set $\mathcal{O}'_M$ we mean its maximal elements with respect to inclusion. A function $f \in \mathcal{O}'_M$ is complete exactly when S_f is its exceptional set.

We say that the functions $f, g \in \mathcal{O}'_M$ are equivalent, and write $f \simeq g$, if they coincide on the intersection of their domains. Obviously, it is an equivalence relation in the set $\mathcal{O}'_M$.

The equivalence class of a function $f \in \mathcal{O}'_M$ contains the greatest element $\hat{f}$ with respect to inclusion. It is the only complete element of the class. If $f \simeq g$, then $S_f = S_g$.

In fact, $\hat{f} = \bigcup \{g \in \mathcal{O}'_M : g \simeq f\}$ is a function from $\mathcal{O}'_M$ whose exceptional set is $S_{\hat{f}} = S_f$.

Let $f \in \mathcal{O}'_M$. If $G \subset M$ is an open set, then clearly $f_G \in \mathcal{O}'_G$ and $S_{f_G} = S_f \cap G$. If f is complete, so is f_G. If $\{G_\iota\}$ is an open cover of the manifold M, then: (f is complete) $\Longleftrightarrow$ (all $f_{G_\iota} \in \mathcal{O}'_{G_i}$ are complete).

2. If g and h are holomorphic functions on M and $h \not\equiv 0$, then by the *meromorphic fraction g/h on M* we mean the function $\{h \neq 0\} \ni z \longrightarrow g(z)/h(z) \in \mathbf{C}$. It is holomorphic nearly everywhere on M and its exceptional set is $V(h)$. If g/h and g'/h' are meromorphic fractions on M, then obviously

$$(*) \qquad\qquad g/h \simeq g'/h' \Longleftrightarrow gh' = g'h .$$

LEMMA 1. *Let g and h be holomorphic functions on a neighbourhood of a point $a \in M$. If the germs g_a, h_a are relatively prime and $h_a \neq 0$, then for a sufficiently small open neighbourhood U of the point a we have the following properties:*

(1) $f = g_U/h_U$ is a meromorphic fraction on U which is complete: $S_f = V(h_U)$,

(2) if $c \in S_f$, then the limit $\lim_{z \to c} f(z)$ is equal to ∞ or it does not exist, depending whether $g(c) \neq 0$ or $g(c) = 0$,

(3) $\dim\{z \in S_f : g(z) = 0\} < m - 1$.

Indeed, we have $\dim V(g_a, h_a) < m - 1$ (see II. 5.3), and hence, if U is a sufficiently small open neighbourhood of a, then the function $f = g_U/h_U$ is a meromorphic fraction on U whose exceptional set is $Z = V(h_U)$ and $\dim(Z \cap W) < m - 1$, where $W = V(g_U)$. Since $\dim_z Z \geq m - 1$ and $\dim_z W \geq m - 1$ for $z \in Z \cap W$ (see II. 5.1), the set $Z \cap W$ is nowhere dense in both Z and W (see IV. 2.5). Now, if $c \in Z \setminus W$, then obviously $\lim_{z \to c} f(z) = \infty$. If $c \in Z \cap W$, the limit does not exist, since in the latter case the function f takes the value zero in any neighbourhood of c (in $W \setminus Z$) and it attains arbitrarily large values (sufficiently close to points of the set $Z \setminus W$). This implies that $Z = S_f$.

A function $f \in \mathcal{O}'_M$ is said to be *meromorphic* on M (or a *meromorphic element of the set* $\mathcal{O}'_M$) if each point $a \in M$ has an open neighbourhood U such that the restriction f_U is equivalent to a meromorphic fraction g/h on U. (Therefore the equivalence class of any meromorphic element of $\mathcal{O}'_M$ consists of meromorphic elements only, and it contains the unique complete meromorphic function on M; see n° 1.)

Taking a smaller neighbourhood U and replacing the fraction g/h by an equivalent one, we can make the germs g_a, h_a relatively prime and the fraction g/h complete: this is achieved by (∗) (in view of the factoriality of $\mathcal{O}_a$) and by lemma 1. Then $f_U \subset g/h$ (see n° 1). Thus we conclude that:

A function $f \in \mathcal{O}'_M$, whose exceptional set is Z, is meromorphic if and only if each point $a \in Z$ has an open neighbourhood U such that

$$f(z) = g(z)/h(z) \quad \text{and} \quad h(z) \neq 0 \text{ in } U \setminus Z ,$$

where g and h are holomorphic functions on U. In addition, one may require that the germs g_a, h_a are relatively prime and the meromorphic fraction g/h on U is complete.

Next (see also lemma 1 and n° 1), for a function f defined on a subset of the manifold M, the following conditions are equivalent:

(1) f is a complete meromorphic function on M;

(2) each point $a \in M$ has an open neighbourhood U such that f_U is a meromorphic fraction of the form g/h on U, where the germs g_a, h_a are relatively prime;

(3) each point $a \in M$ has an open neighbourhood U such that f_U is a complete meromorphic fraction on U.

Suppose that $f, g \in \mathcal{O}'_M$ and $f \simeq g$. Then if f is meromorphic on M, so is g.

Let $f \in \mathcal{O}'_M$, and let $G \subset M$ be an open set. Let $\{G_\iota\}$ be an open cover of the manifold M. If f is (complete) meromorphic on M, then f_G is (complete) meromorphic on G. The function f is (complete) meromorphic on M if and only if f_{G_ι} is (complete) meromorphic on G_ι for each ι.

Now we are going to show that for any meromorphic function f on M the set of its singular points S_f is analytic of constant dimension $m - 1$. If $c \in S_f$, then the limit $\lim_{z \to c} f(z)$ is either equal to ∞ or it does not exist. In the former case, c is called a *pole* of the function f, while in the latter case, it is called an *indeterminate point* of f ([127]) . The set of indeterminate points of f is analytic of dimension $< m - 1$, and so it is nowhere dense in S_f. (Hence the set of poles of f is open and dense in S_f.) If $f \simeq g$, then f and g have the same poles and the same indeterminate points.

Indeed, for any point $a \in M$, there is an open neighbourhood U and a complete meromorphic fraction $f_0 = g/h$ on U containing f_U, and such that the germs g_a, h_a are relatively prime. Then $S_f \cap U = S_{f_U} = S_{f_0} = V(h)$ (see n° 1). Therefore the set S_f is analytic of constant dimension $m - 1$. In view of lemma 1, after taking a smaller U, if $c \in S_f \cap U$ then since the graph of f_U is dense in the graph of f_0, we have either that $\lim_{z \to c} f(z) = \lim_{z \to c} f_0(z) = \infty$ or neither of the two limits exists. Next (see (3) and (2)), the trace on U of the set of indeterminate points of f is analytic (in U) of dimension $< m - 1$. The last property follows from the fact that the graph of $f \cap g$ is dense in both f and g.

It is easy to check that the family $\mathcal{M}_M$ of all complete meromorphic functions on M, with addition and multiplication given by

$$(f, g) \longrightarrow \widehat{f + g} \quad \text{and} \quad (f, g) \longrightarrow \widehat{fg} \quad (^{128}) ,$$

is a ring (^{129}).

(^{127}) See the corollary from theorem 1' in n° 4 below.

(^{128}) The fractions $f + g$ and fg are defined in the intersection of the domains of the functions f, g and are meromorphic on M.

(^{129}) For any $a \in M$, we define the field $\mathcal{M}_a$ of *meromorphic germs* at a as the field of fractions of the ring $\mathcal{O}_a$. The *meromorphic germ* at a of a meromorphic function f on M is well-defined by the formula $f|_a = g_a/h_a$, where $g/h \approx h_U$ is a meromorphic fraction on an open neighbourhood of a (see (*)).

One defines the *sheaf of meromorphic germs* on M as the set $\mathfrak{M} = \bigcup \{\mathcal{M}_a : a \in M\}$ furnished with a suitable topology and the natural projection $\mathfrak{M} \longrightarrow M$ (see, e.g., [33],

If the manifold M is connected, then $\mathcal{M}_M$ is a field ([130]).

3. In the case when the manifold M is one–dimensional, a function $f \in \mathcal{O}'_M$ is complete meromorphic if and only if the set S_f is discrete and in a neighbourhood of any of its points a the function f is of the form

(m) $\qquad h(z)/z^k = b_k z^{-k} - \cdots + b_1 z^{-1} + h_1(z) \quad \text{for} \quad z \neq 0 \ .$

This condition is required to hold in some (and thus in each) coordinate system at a, with a holomorphic h in a neighbourhood of zero such that $h(0) \neq 0$ and $k > 0$ (then we have (m) with a holomorphic h_1 in a neighbourhood of zero, $b_i \in \mathbf{C}$, and $b_k \neq 0$). Therefore a is a pole and the exponent k (which is uniquely determined) is called the *multiplicity of the pole*. By putting $f(a) = \infty$, the function f becomes a holomorphic mapping of a neighbourhood of a into the Riemann sphere $\bar{\mathbf{C}}$ with multiplicity k at a. Conversely, any such mapping must be of the form (m). Consequently:

On any one–dimensional manifold M, the singular points of a meromorphic function are always (isolated) poles. A complete meromorphic function on M is the restriction of a holomorphic mapping $f : M \longrightarrow \bar{\mathbf{C}}$, $f \not\equiv \infty$, to the set $\{f \neq \infty\}$ and vice versa. Then $\{f = \infty\}$ is the set of its poles, and the multiplicity of any of these poles, say a, is equal to the multiplicity of the mapping f at the point a.

---

p. 88). The meromorphic functions on M are usually defined as the sections (on M) of this sheaf. They are the mappings $\varphi : M \longrightarrow \mathfrak{M}$ such that for each $a \in M$ there is a meromorphic fraction g/h on an open neighbourhood U of the point a, such that $\varphi(z) = g_z/h_z \in \mathcal{M}_z$ for $z \in U$. They constitute a ring $\mathfrak{M}_M$ containing $\mathcal{O}_M$ as a subring (after the identification $\mathcal{O}_M \ni h \longrightarrow (z \longrightarrow h_z) \in \mathfrak{M}_M$).

Now, the meromorphic functions defined as the sections of the sheaf $\mathfrak{M}$ correspond precisely to the complete meromorphic functions defined above. Indeed, it is easy to see that the mapping $\mathcal{M}_M \ni f \longrightarrow (z \longrightarrow f|_z) \in \mathfrak{M}_M$ is an $\mathcal{O}_M$–isomorphism of rings.

Finally, meromorphic functions can be regarded as the equivalence classes of the meromorphic elements of $\mathcal{O}'_M$.

([130]) It can happen that the field of fractions of the ring $\mathcal{O}_M$ is a proper subset of the field $\mathcal{M}_M$. Indeed, if M is compact, then $\mathcal{O}_M$ is the ring of constants (because of the maximum principle; see C. 3.9). On the other hand, on any multiprojective space of positive dimension there are non-constant meromorphic functions (see n° 7 below).

Consider a one–dimensional complex torus T. Then $T = \mathbf{C}/\Lambda$, where $\Lambda = \mathbf{Z}a + \mathbf{Z}b$ is a lattice on $\mathbf{C}$ (see C. 3.21). The natural homomorphism $\pi : \mathbf{C} \longrightarrow T$ is a doubly-periodic mapping with periods a and b. To every doubly-periodic mapping $f : \mathbf{C} \longrightarrow N$ (with periods a, b and values in a set N) corresponds in a one-to-one way a (unique) mapping $\tilde{f} : T \longrightarrow N$ such that $\tilde{f} \circ \pi = f$. Since π is a locally biholomorphic surjection, it follows (see C. 4.2) that if N is a manifold, then the above bijection establishes a one-to-one correspondence between the holomorphic doubly-periodic mappings of $\mathbf{C}$ into N (with periods a and b) and the holomorphic mappings of T into N. In particular, if $N = \bar{\mathbf{C}}$, then the meromorphic functions on the torus T correspond to the *elliptic functions* on $\mathbf{C}$ (with periods a and b) ([131]).

The residue theorem for the logarithmic derivative on a suitably chosen rectangle of periodicity $c + [0,1]a + [0,1]b$ implies (see [5], IX. 4) that every non-constant meromorphic function on the torus T (such a function must have a pole, by the maximum principle; see C. 3.9) attains each value exactly r times (counted with multiplicities), where r is the sum of the multiplicities of the poles of the function.

Consider the *Weierstrass elliptic function*

$$p(z) = z^{-2} + \sum_{c \in \Lambda \setminus 0} (z - c)^{-2} \quad ([132]) \ .$$

The function is even. Its derivative $q(z) = p'(z) = -2\sum_{\Lambda}(z-c)^{-3}$ is an odd elliptic function. Let $\tilde{p}$ and $\tilde{q}$ be the corresponding meromorphic functions on the torus T. Each has a unique pole at 0 (of multiplicities two and three, respectively). In particular, $\tilde{p}$ attains each value at two points. Moreover, $\tilde{p}(-\zeta) = \tilde{p}(\zeta)$ and $\tilde{q}(-\zeta) = -\tilde{q}(\zeta)$. The torus T can be embedded into the projective space $\mathbf{P}_2$ via the mapping $\tilde{g} : T \longrightarrow \mathbf{P}_2 = \overline{\mathbf{C}^2}$, defined by the formula

$$\tilde{g}(\zeta) = \big(\tilde{p}(\zeta), \tilde{q}(\zeta)\big) \in \mathbf{C}^2 \text{ for } \zeta \neq 0 \quad \text{and} \quad \tilde{g}(0) = \omega \in \overline{\mathbf{C}^2} \ ,$$

where the homogeneous coordinates of the point ω are $(0, 0, -2)$. Indeed, $\tilde{g}$ is an injection. To see this, let $\tilde{g}(\zeta) = \tilde{g}(\zeta')$. If $\zeta = 0$, then $\zeta' = 0$. Let $\zeta \neq 0$. Then $\tilde{p}(\zeta) = \tilde{g}(\zeta')$ and $\tilde{q}(\zeta) = \tilde{q}(\zeta')$; if, in addition, $\tilde{q}(\zeta) = 0$, then the multiplicity of $\tilde{p}$ at ζ is ≥ 2, and so $\zeta' = \zeta$. On the other hand, if $\tilde{q}(\zeta) \neq 0$, then $\tilde{q}(\zeta) \neq \tilde{q}(-\zeta)$, but $\tilde{p}(\zeta) = \tilde{p}(-\zeta)$, hence we must have $\zeta = \zeta'$. Now, $\tilde{g}$ corresponds to the mapping $g : \mathbf{C} \longrightarrow \mathbf{P}_2 = \overline{\mathbf{C}^2}$ defined by $g(z) = (p(z), q(z))$, for $z \in \mathbf{C} \setminus \Lambda$, and $g(\Lambda) = \omega$. It is enough to prove that g is an immersion, since then the mapping $\tilde{g}$ is also an immersion and, as it is simultaneously a homeomorphism onto its range, it must be an embedding (see C. 3.14). Now, if $z \notin \Lambda$, then one cannot have $p'(z) = q'(z) = 0$, for if this were true, the multiplicity of p at z would be ≥ 3, which is impossible. Furthermore, for $z \neq 0$ in a neighbourhood of zero, we have $p(z) = z^{-2} + a(z)$ and $q(z) = -2z^{-3} + a'(z)$, where a is a holomorphic function. Therefore, in a neighbourhood of zero, $(z^3, z + z^3 a(z), -2 + z^3 a'(z))$ are homogeneous coordinates of $g(z)$, and thus (by taking the 2nd canonical chart) we conclude that g is an immersion at 0. Thus, by double-periodicity, g is an immersion at each point of the lattice Λ.

([131]) i.e., to the doubly-periodic meromorphic functions. (See [5], IX. 4).

([132]) For every $\Lambda' \subset \Lambda$, the sum $\sum_{\Lambda'}(z - c)^{-2}$ is almost uniformly convergent in $\mathbf{C} \setminus \Lambda'$, because $\sum_{\Lambda \setminus 0} |c|^{-3} < \infty$. Double-periodicity of the function p is a direct consequence of the fact that the function is even and its derivative is doubly-periodic.

Thus the image $W = \tilde{g}(T) \subset \mathbf{P}_2$ is a submanifold that is biholomorphic to the one–dimensional (complex) torus. By Chow's theorem (see 6.1) it is algebraic, and hence $W_0 = W \cap \mathbf{C}^2 = \tilde{g}(T \backslash \omega) = g(\mathbf{C} \backslash \Lambda) = \{(p(t), q(t) : t \in \mathbf{C} \backslash \Lambda\} \subset \mathbf{C}^2$ is a principal algebraic set (see 6.3; and 11.3, corollary 1). Its degree is ≤ 3, because (see 11.5, proposition 7) the line $\{\alpha z + \beta w + \gamma = 0\}$ intersects the set in at most three points, since the meromorphic function $\alpha \tilde{p} + \beta \tilde{q} + \gamma$ has only one pole, its multiplicity is ≤ 3, and so the function vanishes at most at three points. Thus $W_0 = V(P)$, where P is a non-zero polynomial of degree ≤ 3, and we have the relation $P(p(z), p'(z)) = 0$ $(^{133})$.

4. We will now prove a theorem that characterizes meromorphic functions.

THEOREM 1. *Let* $f : M \backslash Z \longrightarrow \mathbf{C}$ *be a holomorphic function, where* $Z \subset M$ *is a nowhere dense analytic set. The following conditions are equivalent:*

(1) *the function* f *is meromorphic on* M;

(2) *the closure* $\bar{f}$ *in* $M \times \mathbf{C}$ *is analytic;*

(3) *the closure* $\bar{f}$ *in* $M \times \bar{\mathbf{C}}$ *is analytic;*

(4) *the graph of* f *is an analytically constructible set in* $M \times \mathbf{C}$;

(5) *the graph of* f *is an analytically constructible set in* $M \times \bar{\mathbf{C}}$;

(6) *the pair consisting of the closure* $\bar{f}$ *in* $M \times \bar{\mathbf{C}}$ *and the set* $M \times \infty$ *satisfies the condition of regular separation;*

(7) *the function* f *satisfies the condition* (r) *near* Z.

PROOF . The implication (1) $\Longrightarrow$ (7) follows from the proposition in IV. 7.2 $(^{134})$. Indeed, take $a \in Z$. There exists a coordinate system $\varphi :$ $G \longrightarrow U$ at a and a meromorphic fraction g/h on U containing $f \circ \varphi^{-1}$ (see n° 2). Then $\varphi(Z) \supset V(h) = h^{-1}(0)$, and (as $0 \in \varphi(Z)$) we have $|h(\zeta)| \geq c\varrho(\zeta, \varphi(Z))^s$ in the neighbourhood $\varphi(\Delta)$ of zero in $\mathbf{C}^m$. Here $\Delta \subset G$ is a compact neighbourhood of a, and $c > 0$, $s > 0$. Hence $|f(\varphi^{-1}(\zeta))| = |g(\zeta)/h(\zeta)| \leq K\varrho(\zeta, \varphi(Z))^{-s}$ for $\zeta \in \varphi(\Delta) \backslash \varphi(Z)$ with some $K > 0$, which means that $|f(z)| \leq K\varrho_\varphi(z, Z \cap G)^{-s}$ for $z \in \Delta \backslash Z$. (See 3.6.)

The conditions (3), (6), and (7) are equivalent, according to proposition 2 from 7.3, whereas the conditions (2), (4), as well as (3), (5) are equivalent

---

$(^{133})$ It can be shown (see [5], IX. 5) that $(p')^2 - 4p^3 + \alpha p + \beta = 0$ for some $\alpha, \beta \in \mathbf{C}$ that depend on Λ and are such that $\alpha^3 - 27\beta^2 \neq 0$. The converse is also true. For any such α, β, there is a lattice Λ for which the Weierstrass function satisfies the above equation. Since the polynomial $w^2 - 4z^3 + \alpha z + \beta$ is irreducible, it follows that the closure in $\overline{\mathbf{C}^2}$ of the algebraic set $\{w^2 - 4z^3 + \alpha z + \beta = 0\}$ is a submanifold which is biholomorphic to a one–dimensional torus, provided that $\alpha^3 - 27\beta^2 \neq 0$.

$(^{134})$ See also the remark following the proposition.

because of proposition 5 from IV. 8.3. The implication $(3) \implies (2)$ is trivial. If the condition (2) is satisfied, then the closure $\bar{f}$ in $M \times \mathbf{C}$ is analytic in $(M \times \bar{\mathbf{C}}) \setminus (Z \times \infty)$ [135] and of constant dimension m (see IV. 2.5). Therefore, since $\dim(Z \times \infty) < m$, condition (3) follows by the Remmert-Stein theorem (see IV. 6.3). Thus we have shown the equivalence of the conditions (2)–(7).

To finish the proof, it suffices to prove the implication $(3) \implies (1)$. Let $a \in Z$. The closure $\bar{f}$ in $M \times \bar{\mathbf{C}}$, where $\bar{\mathbf{C}} = \mathbf{P}_1 = \mathbf{P}(\mathbf{C}^2)$, is analytic of constant dimension m (see IV. 2.5). Therefore, in view of the proposition from 6.5, there is an open connected neighbourhood U of the point a such that $\bar{f}_U$ is defined by a holomorphic $\mathbf{C}^2$–homogeneous function in $U \times \mathbf{C}^2$. That is (see C. 3.18), it is defined by a function of the form $F(z, t, u) = \sum_0^r b_\nu(z) t^{k-\nu} u^\nu$, where $r \leq k$, the coefficients b_ν are holomorphic in U, and $b_r \not\equiv 0$. Thus, if $z \in U \setminus Z$ and $b_r(z) \neq 0$, then $f(z)$ is the only root of the equation $\sum_0^r b_\nu(z) u^\nu = 0$ (see 3.1, the identification $\mathbf{C} \hookrightarrow \mathbf{P}_1$), so we must have $r > 0$ and $b_{r-1}(z) = -r b_r(z) f(z)$. Therefore $f_U \simeq -b_{r-1}/r b_r$.

If $f \in \mathcal{O}'_M$, then also $(3) \implies (1)$, according to the last part of the proof. Furthermore $(1) \implies (3)$, because if the function f is meromorphic, then so is $\hat{f}$ and the graphs of both functions have the same closure in $M \times \bar{\mathbf{C}}$ (see n° 1). Theorem 1, combined with the analytic graph theorem (see V. 1, corollary 3 from theorem 2) and proposition 5 from IV. 8.3, implies the following

THEOREM 1'. *Let f be a continuous function on an open dense subset of the manifold M. Then the following conditions are equivalent:*

(1) *the function f is meromorphic on M;*

(2) *the closure $\bar{f}$ in $M \times \bar{\mathbf{C}}$ is analytic.*

Under the assumption that the domain of f is analytically constructible, the above conditions are equivalent to each of the following ones:

(3) *the closure $\bar{f}$ in $M \times \mathbf{C}$ is analytic;*

(4) *the graph of f is an analytically constructible set in $M \times \mathbf{C}$;*

(5) *the graph of f is an analytically constructible set in $M \times \bar{\mathbf{C}}$.*

COROLLARY . *If f is a meromorphic function on M and a is an indeterminate point of f, then $(\bar{f})_a = \bar{\mathbf{C}}$, where $\bar{f}$ is the closure in $M \times \bar{\mathbf{C}}$.*

Indeed, for each $z \in M$ we have $\#(\bar{f})_z < \infty$ or $(\bar{f})_z = \bar{\mathbf{C}}$, and the set $\{z \in M : (\bar{f})_z = \bar{\mathbf{C}}\}$ is closed. Therefore, if we had $\#(\bar{f})_a < \infty$, then for some open neighbourhood U of the point a the natural projection $\pi : f_U \longrightarrow U$ would have finite fibres. Since the analytic set $\bar{f}$ is of constant dimension m

---

[135] Since the set $\bar{f} \cap ((M \setminus Z) \times \bar{\mathbf{C}}) = f$ is analytic in $(M \setminus Z) \times \bar{\mathbf{C}}$.

(see IV. 2.5), the projection π would be open (see V. 7.1, proposition 1) and the function $U \ni z \longrightarrow \#(\bar{f})_z$ would be lower semicontinuous (see B. 2.1). But $\#(\bar{f})_a > 1$, and so we would have $\#(\bar{f})_z > 1$ in a neighbourhood of the point a, which is impossible. Consequently, $(\bar{f})_a = \bar{\mathbf{C}}$.

5. We say that the functions $f_1, \ldots, f_k \in \mathcal{O}'_M$ are *analytically dependent* if $\mathrm{rank}(f_1, \ldots, f_k) < k$. We say that they are *algebraically dependent* if $P(f_1, \ldots, f_k) = 0$ for some non-zero polynomial $P \in \mathcal{P}_k$ ([136]) .

THE SIEGEL-THIMM THEOREM . *Assume that the manifold M is compact. Then meromorphic functions $f_1, \ldots, f_k$ on M are algebraically dependent if and only if they are analytically dependent.*

PROOF ([137]) . One may assume that the functions f_i have a common domain G and that the complement of G is analytic ([138]) . Let $f = (f_1, \ldots, f_k) : G \longrightarrow \mathbf{C}^k$. By theorem 1 from n° 4, the sets $f_i \subset M \times \bar{\mathbf{C}}$ are analytically constructible and so is the set $f \subset M \times \bar{\mathbf{C}}^k$. According to the Chevalley-Remmert theorem (see V. 5.1), the set $f(G) = \pi(f)$, where $\pi : M \times \bar{\mathbf{C}}^k \longrightarrow \bar{\mathbf{C}}^k$ is the natural projection, is constructible in $\mathbf{C}^k$(see 8.3, proposition 1). Since the algebraic dependence of the functions $f_1, \ldots, f_k$ means that there exists a non-zero polynomial from $\mathcal{P}_k$ that vanishes on $f(G)$, it is equivalent to the condition $\mathrm{int} f(G) = \emptyset$ (see 8.1 and 8.3, proposition 2). Therefore it is equivalent to the analytic dependence of the functions $f_1, \ldots, f_k$ (see V. 1 theorem 1, and C. 4.2 ([139])).

COROLLARY . *Under the hypotheses of the theorem, if $k > m$, then the meromorphic functions $f_1, \ldots, f_k$ are algebraically dependent.*

Let V be an irreducible algebraic subset of a vector space M. Then the associated ring $R(V)$ is an integral domain (see 12.1) and its field of fractions $K(V)$ – called the *field of rational functions* on V – is an extension of the field $\mathbf{C}$. We will prove that:

dim V *is equal to the transcendence degree of the field $K(V)$ over $\mathbf{C}$, i.e., to the supremum of the number of elements of $K(V)$ which are algebraically independent over* $\mathbf{C}$ ([140]) .

([136]) Naturally, $(f_1, \ldots, f_k)$ denotes here the diagonal product of the restrictions to the intersection of the domains of $f_1, \ldots, f_k$. In each of the conditions, the functions f_i can be replaced by equivalent ones (see C. 3.12).

([137]) See Narasimhan [33], p. 135.

([138]) See footnote ([136]), and n° 1 and 2.

([139]) If $f_1, \ldots, f_k$ are not analytically dependent, then the mapping f is a submersion at some point of the set G.

([140]) Elements $x_1, \ldots, x_r$ of an extension L' of a field L are said to be *algebraically depen-*

Indeed, take m elements of the field $K(V)$. They are of the form $f_1/g, \ldots$
$\ldots, f_m/g$, where $f_i \in R(V)$, $g \in R(V) \setminus 0$. The graph of the mapping

$$\{g \neq 0\} \ni z \longrightarrow \big(f_1(z)/g(z), \ldots, f_m(z)/f(z)\big) \in \mathbf{C}^m$$

is constructible, so, by the Chevalley theorem (see 8.3), the range $H \subset \mathbf{C}^m$
of the mapping is constructible. Now, the algebraic independence of the elements $f_1/g, \ldots, f_m/g$ means, as can be easily checked $(^{141})$, that there is a
polynomial $P \in \mathcal{P}_m \setminus 0$ such that $P\big(f_1(z)/g(z), \ldots, f_m(z)/g(z)\big) = 0$ in the set
$\{g \neq 0\}$, i.e., it vanishes on H. This is equivalent to the condition $\dim H < m$
(because $\dim H = \dim \bar{H}$; see 8.1 and 8.3 proposition 2). Set $k = \dim V$. We
have $\dim H \leq k$ (see V. 1, corollary 1 of theorem 1), which means that any
m elements of $K(V)$, with $m > k$, must be algebraically dependent. On the
other hand, by taking $g = 1$ and $f_i = (F_i)_V$, where $F_1, \ldots, F_k \in M^*$ are
such that $V(F_1, \ldots, F_k)$ is a linear complement of the tangent space to the
k–dimensional manifold V^0 at one of its points, we get $\dim H = k$ (since
$(f_1, \ldots, f_k)_V$ is an immersion at that point). This means that $f_1, \ldots, f_k$ are
algebraically independent.

6. In what follows, let M be an m–dimensional vector space.

By a *rational function* on M we mean any function $\{h \neq 0\} \ni z \longrightarrow$
$g(z)/h(z) \in \mathbf{C}$, where $g, h \in \mathcal{P}(M)$ and $h \neq 0$. Such a function is a meromorphic fraction g/h on M, and hence it is a meromorphic function on M.

Obviously, the graph of a rational function on M is constructible in
$M \times \mathbf{C}$ $(^{142})$.

We will prove the following two propositions.

PROPOSITION 1. *The complete rational functions on M are precisely the
meromorphic fractions g/h on M, where $g, h \in \mathcal{P}(M)$ are relatively prime.*

PROPOSITION 2. *If f is a rational function on M, then so is $\hat{f}$.*

Indeed, if f is a meromorphic fraction of the form g/h, where $g, h \in$
$\mathcal{P}(M)$ are relatively prime, then the set of removable singular points of the
function f is open in $V(h)$, and hence it is of constant dimension $(m - 1)$;
it is contained in the algebraic set $V(g, h)$ of dimension $\leq m - 2$. (See 11.3,
corollaries 1 and 3 of proposition 5.) Therefore it must be empty, and thus

---

dent over L if $P(x_1, \ldots, x_r) = 0$ for some $P \in L[X_1, \ldots, X_r] \setminus 0$.

$(^{141})$ Using the fact that $T^s p(Z_1, \ldots, Z_m) = Q(TZ_1, \ldots, TZ_m)$ for some $s \in \mathbf{N}$ and $Q \in$
$\mathbf{C}[T, Y_1, \ldots, Y_m]$.

$(^{142})$ Since it is the set $\{g(z) - th(z) = 0, \ h(z) \neq 0\}$.

the function f is complete. Next, any rational function f on M is equivalent (owing to (*) from n° 2) to a meromorphic fraction g/h, where $g, h \in \mathcal{P}(M)$ are relatively prime; hence g/h must be complete, and so $\hat{f} = g/h$. If, in addition, the function f is complete, then $f = g/h$ (see n° 1).

It is easy to verify that the set of complete rational functions on M with addition and multiplication given by $(f, g) \longrightarrow \widehat{f + g}$ and $(f, g) \longrightarrow \widehat{fg}$, respectively, is a field ([143]) that is isomorphic to the field of fractions of the ring $\mathcal{P}(M)$. (The mapping that takes a meromorphic fraction g/h, where $g, h \in \mathcal{P}(M)$ are relatively prime, to the element g/h of the field of fractions of the ring $\mathcal{P}(M)$, is an isomorphism.)

7. Now, let $N = X_1 \times \cdots \times X_k$ and $\tilde{N} = \bar{X}_1 \times \cdots \times \bar{X}_k$, where X_i are vector spaces. Set $n = \dim N$.

THE HURWITZ THEOREM . *If $Z \subset N$ is an algebraic set of constant dimension $n - 1$, then for any function $f :\ N \setminus Z \longrightarrow \mathbf{C}$ we have the equivalence:*

$$(f \text{ is rational on } N) \Longleftrightarrow (f \text{ is meromorphic on } \tilde{N}) .$$

PROOF . If the function f is rational on N, then its graph is constructible in $\tilde{N} \times \bar{\mathbf{C}}$ (see n° 6; and 8.3, proposition 1). By theorem 1 (condition (5)), the function f is meromorphic on $\tilde{N}$ ([144]) . Conversely, suppose that the function f is meromorphic on $\tilde{N}$. According to theorem 1 (condition (3)) and Chow's theorem from 6.4, the closure $\bar{f}$ in $\tilde{N} \times \bar{\mathbf{C}}$ is algebraic of constant dimension n (see IV. 2.5), and so is its trace $\bar{f} \cap (N \times \mathbf{C})$ in $N \times \mathbf{C}$ (see 6.4). It follows (see 11.3, corollary 1 from proposition 5) that it is equal to $V(F)$, where $F(z, t) = \sum_0^r a_\nu(z) t^\nu$, $a_\nu \in \mathcal{P}(M)$, and $a_r \neq 0$. Hence, if $z \in N \setminus Z$ and $a_r(z) \neq 0$, then $f(z)$ is the only root of the equation $\sum_0^r a_\nu(z) t^\nu = 0$, and so we must have $r > 0$ and $a_{r-1}(z) = -r a_r(z) f(z)$. Thus $f \simeq -a_{r-1}/r a_r$ ([145]) and so $f \subset g/h$, where $g, h \in \mathcal{P}(M)$, $h \neq 0$ (see proposition 2 from n° 6). Since $Z = V(d)$, where $d \in \mathcal{P}(N) \setminus 0$ (see 11.3, corollary 1 of proposition 5), we have $f = gd/hd$. This means that the function f is rational on N.

THEOREM 2. *If $Z \subset M$ is an algebraic set of constant dimension $(m-1)$, then for a function $f :\ M \setminus Z \longrightarrow \mathbf{C}$ the following conditions are equivalent:*

(1) *the function f is rational on M;*

([143]) It is clearly a subring of the ring $\mathcal{M}_M$.

([144]) It can be verified directly by using the charts of the manifold $\tilde{N}$ (see (**) from 3.5).

([145]) The function f is also meromorphic on N (see n° 2).

(2) *the function f is continuous and its graph is constructible in $M \times \mathbf{C}$;*

(3) *the function f is holomorphic and*

$$(\#) \qquad |f(z)| \le K \left(\frac{1 + |z|}{\tilde{\varrho}(z, Z)} \right)^p \quad in\ M \setminus Z,\ for\ some\ K, p > 0\ ,$$

where $\tilde{\varrho} = \min(\varrho, 1)$ (after endowing M with a norm and putting $\tilde{\varrho}(z, \emptyset) = 1$.)

PROOF . The implication (1) $\Longrightarrow$ (2) is trivial. If the condition (2) is satisfied, then (see 8.3, proposition 1; and IV. 8.3, proposition 5) the closure $\bar{f}$ in $\bar{M} \times \bar{\mathbf{C}}$ is analytic. Therefore, in view of theorem 1' (condition (2)) combined with the Hurwitz theorem, the function f is rational on M. The equivalence (1) $\Longleftrightarrow$ (3) is implied by the following easy-to-check inequalities for polynomial $h \in \mathcal{P}(M)$ of degree $k \ge 0$:

$$\varepsilon \tilde{\varrho}(z, Z)^k \le |h(z)| \le C(1 + |z|)^k \tilde{\varrho}(z, Z) \quad in\ \ M\ ,$$

where $\varepsilon, C > 0$ and $Z = V(h)$ $(^{146})$. Indeed, if $f = g/h$, where $g, h \in \mathcal{P}(M)$ and $h \ne 0$, then $Z = V(h)$ and the condition $(\#)$ follows from the first inequality. On the other hand, assuming the condition $(\#)$, since $Z = V(f)$ for some $h \in \mathcal{P}(M) \setminus 0$ (see 11.3, corollary 1 of proposition 5), the second inequality yields $|f(z)h(z)^p| \le KC^p(1 + |z|)^{p+kp}$ in $M \setminus Z$. Therefore, by Liouville's theorem, fh^p is the restriction of a polynomial $g \in \mathcal{P}(M)$ (see C. 1.8 and II. 3.5), and so $f = g/h^p$.

One can prove that the condition $(\#)$ (for a locally bounded function f on $M \setminus Z$) is equivalent to the condition (r) on $\bar{M}$ near the set $Z \cup M_\infty = \bar{Z} \cup M_\infty$. This last set is algebraic in $\bar{M}$ (see 6.3) of constant dimension $m - 1$. Then the equivalence (1) $\Longleftrightarrow$ (3) is a consequence of Hurwitz's theorem and theorem 1 (condition (7)).

COROLLARY . *A continuous function $f : M \longrightarrow \mathbf{C}$ whose graph is constructible in $M \times \mathbf{C}$ is a polynomial.*

THEOREM 3. *Every continuous function whose graph $f \subset M \times \mathbf{C}$ is constructible, and whose domain is dense, is the restriction of a rational function on M $(^{147})$.*

$(^{146})$ As for the first inequality, one may assume that $M = \mathbf{C}^m$ and h is a monic polynomial in z_m (see A. 3.3). Then, if $z \in M$, we have $|h(z)| = |z - \zeta_1| \ldots |z - \zeta_k|$, where $\zeta_1, \ldots, \zeta_k \in Z$. Hence $|h(z)| \ge \varrho(z, Z)^k$. The second inequality is a simple consequence of the mean-value theorem.

$(^{147})$ Thus, it is enough to assume that the domain of f is open. Without any assumption on the domain of f, the theorem is no longer true (see n° 8 below, the remark following theorem 4).

PROOF . By the Chevalley theorem (see 8.3), the closure Z of the complement of the domain of f is algebraic and nowhere dense (see 8.3, proposition 2; and IV. 8.3, lemma 2). One may assume that $f : M \setminus Z \longrightarrow \mathbf{C}$ (because $f_{M \setminus Z}$ is dense in f ([148])). Now, $Z = Z_0 \cup Z_1$, where Z_0 is algebraic of constant dimension $(m - 1)$ while Z_1 is algebraic of dimension $< m - 1$. The graph of f is closed in $(M \setminus Z) \times \mathbf{C}$ and hence analytic (see IV. 8.3, proposition 5). Therefore, in view of the analytic graph theorem (see V. 1, corollary 3 of theorem 2), the function f is holomorphic in $M \setminus Z$. By the Hartogs theorem (see III. 4.2), it has a holomorphic extension $\tilde{f}$ on $M \setminus Z_0$. The graph of this extension is equal to $\bar{f} \cap ((M \setminus Z_0) \times \mathbf{C})$, where $\bar{f}$ is the closure in $M \times \mathbf{C}$, and so it is constructible in $M \times \mathbf{C}$ (see 8.3, proposition 2). Thus, by theorem 2, the function $\tilde{f}$ is rational on M.

8. Now we will prove the following

THEOREM 4 (ZARISKI'S CONSTRUCTIBLE GRAPH THEOREM). *For each function $f : S \longrightarrow \mathbf{C}$ with domain $S \subset M$ and graph constructible in $M \times \mathbf{C}$ ([149]) , there exists a quasi-algebraic set $T \subset S$ which is open and dense in S, and such that f_T is the restriction of a rational function on M, i.e.,*

$$f(z) = g(z)/h(z) \quad and \quad h(z) \neq 0 \quad for \quad z \in T ,$$

where $g, h \in \mathcal{P}(M)$.

REMARK . The function f itself, even if it is continuous, does not have to be the restriction of a rational function. This is the case, because there are continuous functions defined on an algebraic set and with algebraic graph which are not even holomorphic. (See the example from V. 8.1 ([150]) .)

In the proof of the theorem we will need the following four lemmas:

LEMMA 1. *Let V be an algebraic subset of the Cartesian product $X \times Y$ of vector spaces which is non-empty and of constant dimension equal to $\dim X$. Suppose that the natural projection $\pi : V \longrightarrow X$ is proper. Then there exist a nowhere dense algebraic set $Z \subset X$ and $p > 0$ such that $\#V_x = p$ for $x \in X \setminus Z$ and*

$$X \setminus Z = \{x : \ y \in V_x \Longrightarrow V \ is \ a \ topographic \ submanifold \ at \ (x, y)\} \ .$$

([148]) We take a complete rational extension of $f_{M \setminus Z}$. See n° 1, 2, and 6.

([149]) Then S is also constructible.

([150]) The function in that example is equal to w/z outside the point $(0, 0)$.

The lemma is contained in the Andreotti-Stoll theorem (see V. 7.2 and C 3.17; the projection π is then a p–sheeted $*$–covering whose exceptional set is $Z = \{x : \#V_x < p\}$). One only needs to check that Z is algebraic. But this is true, because Z is constructible (see 8.3, proposition 2): the complement of Z is the image by the projection onto X of the constructible set

$$\{(x, y_1, \ldots, y_p) \in X \times Y^p : (x, y_1), \ldots, (x, y_p) \in V, \ y_1, \ldots, y_n \text{ are distinct}\}.$$

LEMMA 2. *Let $\emptyset \neq V \subsetneq M$ be an irreducible algebraic set, and let Z be a nowhere dense algebraic subset of V. Let $f : V \setminus Z \longrightarrow \mathbf{C}$ be a continuous function with constructible graph in $M \times \mathbf{C}$, and let $E \subset M \setminus V$ be a finite set. Then there are: an irreducible algebraic set $\tilde{V} \supset V$ of codimension 1 and disjoint from E, its nowhere dense algebraic subset $\tilde{Z} \supset Z \cup \tilde{V}^*$, and a continuous function $\tilde{f} : \tilde{V} \setminus \tilde{Z} \longrightarrow \mathbf{C}$ with constructible graph in $M \times \mathbf{C}$, such that*

$$V \setminus \tilde{Z} \neq \emptyset \quad \text{and} \quad \tilde{f} = f \text{ in } V \setminus \tilde{Z} .$$

PROOF . The ideal $I = \mathcal{I}(V)$ is prime (see 1.6) and $V = V(I)$ (see 1.2). One may assume that $M = \mathbf{C}^n$ and that the ideal I is k–regular (see 9.1). Furthermore, (using the notation from §9) one may assume, in view of the proposition from 9.3 (and because of 9.2), that the hypotheses of Rückert's lemma from 9.3 are satisfied, and

$$(\#) \qquad V \cap \tilde{\pi}^{-1}\big(\tilde{\pi}(E)\big) = \emptyset, \quad \text{where } \tilde{\pi} : z \longrightarrow (u, z_{k+1}) .$$

Then (using the notation from Rückert's lemma) we have

$$(\mathrm{R}) \quad V_{\{\delta \neq 0\}} = \{(u, v) : \ \delta(u) \neq 0, \ p_{k+1}(u, z_{k+1}) = 0,$$
$$\delta(u) z_j = Q_j(u, z_{k+1}), \ j = k + 2, \ldots, n\} .$$

Therefore the set V is k–dimensional (see §10, corollary 2; and 8.2). The polynomial p_{k+1} is irreducible in $\mathcal{P}_{k+1}$, and hence also in $\mathcal{P}_n$ (see A. 2.3). Thus the algebraic set

$$(\#\#) \qquad\qquad \tilde{V} = \{z : \ p_{k+1}(u, z_{k+1}) = 0\} \subset M$$

is irreducible of codimension 1 (see 11.2, corollary 1 and (3)). As $p_{k+1} \in I$ (see 9.3, $(*)$), so $V \subset \tilde{V}$. Next, according to corollary 2 from §10, we have $\tilde{\pi}(V) = \tilde{\pi}(\tilde{V})$. Consequently, it follows from $(\#)$ that $\tilde{V} \cap E = \emptyset$. Set

$$\tilde{Z} = Z \cup \tilde{V}_{\{\delta = 0\}} .$$

By (##), we have $\tilde{V}_{\{\delta \neq 0\}} \subset \tilde{V}^0$, and so $\tilde{Z} \supset Z \cup \tilde{V}^*$. Since $\dim(\tilde{Z} \cap V) < k$ (see IV. 2.8, proposition 3; and then II. 1.4 and 9.3 (**)), we have $V \setminus \tilde{Z} \neq \emptyset$ (which implies that $\tilde{Z}$ is nowhere dense in $\tilde{V}$). Finally, the function

$$\tilde{f}: \tilde{V} \setminus \tilde{Z} \ni z \longrightarrow f\big(u, z_{k+1}, Q_{k+2}(u, z_{k+1})/\delta(u), \ldots, Q_n(u, z_{k+1}/\delta(u)\big)$$

is continuous with constructible graph (see 8.3, the corollary of Chevalley's theorem), and if $z \in V \setminus \tilde{Z}$, then, in view of (R), we have

$$z = \big(u, z_{k+1}, Q_{k+1}(u, z_{k+1})/\delta(u), \ldots, Q_n(u, z_{k+1})/\delta(u)\big) \ ,$$

and so $\tilde{f}(z) = f(z)$.

LEMMA 3. *If $Z \subset M$ is an algebraic set of codimension ≥ 2 and if $a \in M$, then the set $\{\lambda \in \mathbf{P}(M): (a + \lambda) \cap Z \subset a\}$ contains an open dense subset of $\mathbf{P}(M)$.*

Indeed, one may assume that $a = 0$ and then the complement of our set, being equal to the image of the set $Z \setminus 0$ under the mapping $\alpha = \alpha_1^M$, is constructible (see 8.1). That is because the graph of α is constructible in $\bar{M} \times \mathbf{P}(M)$ (see 8.7). The dimension of that complement is $\leq \dim Z < \dim \mathbf{P}(M)$ (see V. 1, corollary 1 of theorem 1), and thus its closure is nowhere dense in $\mathbf{P}(M)$ (see IV. 8.5 and II. 1.2).

LEMMA 4. *If $W \subset M$ is an algebraic set of constant dimension $(n - 1)$ and if $a \in W^0$, then the set*

$$\{\lambda \in \mathbf{P}(M): a + \lambda \ intersects \ W \ transversally\}$$

contains an open dense subset of $\mathbf{P}(M)$.

PROOF . One may assume that $a = 0$. The set $T = \alpha \cap (W \times \mathbf{P}(M))$ is constructible in $\bar{M} \times \mathbf{P}(M)$ (see 8.7; and 8.3, proposition 1) and it is analytic in $(M \setminus 0) \times \mathbf{P}(M)$. Being equal to the image of the set $W \setminus 0$ under the biholomorphic mapping $\beta: M \setminus 0 \ni z \longrightarrow (z, \mathbf{C}z) \in \alpha$, it is of constant dimension $n - 1$. The set $\bar{T} \subset \bar{M} \times \mathbf{P}(M)$ is algebraic and of constant dimension $n - 1$ (see IV. 8.3, proposition 5; and IV. 8.5). Furthermore, the natural projection $\pi: \bar{T} \longrightarrow \mathbf{P}(M)$ is holomorphic and proper. Therefore – in view of proposition 3 from V. 7.3 – there is an open dense subset $\Omega \subset \mathbf{P}(M)$ such that π is biholomorphic at the point $(z, \lambda) \in T$ if $\lambda \in \Omega$, and hence whenever $\lambda \in \Omega$ and $z \in \lambda \cap W \setminus 0$. Now, $\alpha_{W \setminus 0} = \pi \circ \beta'$, where $\beta' = \beta_{W \setminus 0}: W \setminus 0 \longrightarrow T$ is biholomorphic. Let $\lambda \in \Omega$ and $z \in \lambda \cap (W \setminus 0)$. Therefore $\alpha_{W \setminus 0}$ is biholomorphic at z, and hence $z \in W^0$ (see V. 3.4, the proposition) and the differential $d_z\alpha_{W\setminus 0} = (d_z\alpha)_{T_zW}$ is an isomorphism. Since $\alpha = \lambda$ on $\lambda \setminus 0$, so $\lambda \in \ker d_z\alpha$ (see C. 3.11), and hence $\lambda \not\subset T_zW$. Thus λ intersects W transversally at z. Finally, it is enough to observe that the set of lines $\lambda \in \mathbf{P}(M)$ that intersect W transversally at 0 (i.e., the set $\{\lambda : \lambda \not\subset T_0W\}$) is open and dense.

Let us prove the lemma in a more direct way.

We may assume that $M = \mathbf{C}^n$ and $a = 0$.

First suppose that W is irreducible. Then $W = V(P)$, where $P \in \mathcal{P}(M)$ is an irreducible polynomial (see 11.3, corollary 1 of proposition 5). Consider the polynomial

$$(\#\#\#) \qquad\qquad Q = \sum_1^n \frac{\partial P}{\partial z_i} z_i \ .$$

If $Q = 0$ on W, then Q is divisible by P (see 11.2, proposition 4), but the degree of Q is not greater than that of P, and so one must have $Q = aP$, where $a \in \mathbf{C}$; then P is a homogeneous polynomial ([151]), and hence W is a cone. Thus each line of the set $\mathbf{P}(M) \setminus W^{\check{}}$, which is open and dense (since $\dim W^{\check{}} = n-2$; see 6.2), can intersect W only at 0. On the other hand, if $V(Q) \not\supset W$, then the codimension of the set $W \cap V(Q)$ is ≥ 2 (see IV. 2.8, proposition 3). Therefore, by lemma 3, every line from an open dense subset of $\mathbf{P}(M)$ can intersect the set $W \cap V(Q)$ only at 0, i.e., it can intersect $W \setminus 0$ only at points z for which $Q(z) \neq 0$. Hence, in view of $(\#\#\#)$, it intersects only at regular points of W, transversally at each of those points.

In the general case, let $W = W_1 \cup \ldots \cup W_r$ be the decomposition into simple components. Then the codimension of the set $W' = \bigcup_{i\neq j} W_i \cap W_j$ is ≥ 2 (see IV. 2.8, proposition 3), and according to lemma 3, each line from an open dense subset of $\mathbf{P}(M)$ can intersect W' only at 0. Consequently, each line from an open dense subset of $\mathbf{P}(M)$ intersects the set $W \setminus 0$ transversally. Finally, it is enough to observe that the set of lines from $\mathbf{P}(M)$ that intersects W transversally at 0 is open and dense.

PROOF of theorem 4. One may assume that the function f is continuous, $S = V \setminus Z$, where $Z \subset V \subset M$ are algebraic sets with Z nowhere dense in V (see 8.6, proposition 3; and 8.3, corollary 1 of proposition 2), and, in view of theorem 3 from n° 7, that $\emptyset \neq V \subsetneq M$. So it is sufficient to prove that there exists a nowhere dense algebraic subset $Z' \supset Z$ of V such that $f_{V \setminus Z'}$ is the restriction of a rational function on M.

Let $V = V_1 \cup \ldots \cup V_r$ be the decomposition into simple components. Take $c_i \in V_i \setminus \bigcup_{j\neq i} V_i$, $i = 1,\ldots,r$. For each $i = 1,\ldots,r$, the set $Z \cap V_i$ is nowhere dense in V_i (see IV. 2.9, corollary 1 of theorem 4). Accordingly, lemma 2 implies the existence of: an irreducible algebraic subset $\tilde{V}_i \supset V$ of codimension 1 such that

$$(1) \qquad\qquad c_j \notin \tilde{V}_i \ \text{ for } \ j \neq i \ ,$$

---

[151] Because if $a \neq 0$ and $z \in M$, then the polynomial $f(t) = P(tz)$ satisfies the condition $tf'(t) = af(t)$, which implies (e.g., by comparing coefficients) that $f(t) = ct^a$ for some $c \in \mathbf{C}$. Therefore $P(tz) = t^aP(z)$ for $t \in \mathbf{C}$, $z \in M$.

its nowhere dense algebraic set

$$(2) \qquad\qquad \tilde{Z}_i \supset (Z \cap V_i) \cup \tilde{V}_i^* \ ,$$

and a continuous function $\tilde{f}_i : \tilde{V}_i \setminus \tilde{Z}_i \longrightarrow \mathbf{C}$ with constructible graph such that

$$(3) \qquad\qquad V_i \setminus \tilde{Z}_i \neq \emptyset \ \text{ and } \ \tilde{f}_i = f \ \text{ in } \ V_i \setminus \tilde{Z}_i \ .$$

The set

$$(4) \qquad\qquad \tilde{V} = \tilde{V}_1 \cup \ldots \cup \tilde{V}_r$$

is of constant dimension $n - 1$ and (4) is its decomposition into irreducible components (due to (1)). Therefore

$$(5) \ \ \tilde{Z} = \bigcup \tilde{Z}_i \cup \bigcup_{i \neq j} \tilde{V}_i \cap \tilde{V}_j \ \text{is nowhere dense in} \ \tilde{V} \ \text{and of codimension} \ \geq 2$$

(see B. 1; and IV. 2.8, proposition 3). Moreover, in view of (2),

$$(6) \qquad\qquad Z \supset \tilde{Z} \cup \tilde{V}^*$$

(see IV. 2.9, theorem 4).

Now take $a_i \in V_i \setminus \tilde{Z}$, $i = 1, \ldots, r$ (see (3)). Then, because of (6), we have $a_i \in \tilde{V}^0$, and so lemma 4 implies that for each line λ from an open dense subset of $\mathbf{P}(M)$ we have

$$(*) \qquad\qquad a_i + \lambda \ \text{intersect} \ \tilde{V} \ \text{transversally,} \ \ i = 1, \ldots, r \ .$$

By lemma 3 and (5), each line λ from an open dense subset of $\mathbf{P}(M)$ satisfies the conditions $(a_i + \lambda) \cap \tilde{Z} = \emptyset$, that is,

$$(**) \qquad\qquad a_i \in V_i \setminus (\tilde{Z} + \lambda), \ \ i = 1, \ldots, r \ .$$

Finally, for each line λ from an open dense set in $\mathbf{P}(M)$

$$(***) \qquad\qquad \lambda \ \text{is a Sadullaev line for} \ \tilde{V}$$

(see 7.1, Sadullaev's theorem).

Consequently, there exists a line $\lambda \in \mathbf{P}(M)$ satisfying the conditions $(*)$, $(**)$ and $(***)$. One may assume that $M = N \times \mathbf{C}$, where N is an $(n-1)$-dimensional vector space, and that $\lambda = 0 \times \mathbf{C}$. Let $\pi : M \longrightarrow N$ be the natural projection. Then (see 7.1) the projection $\pi_{\tilde{V}} : \tilde{V} \longrightarrow N$ is proper, and lemma 1 yields the existence of: a nowhere dense algebraic set $\Sigma_0 \subset N$ and $p > 0$ such that

$$(7) \qquad \#\tilde{V}_x = p \quad \text{for} \quad x \in N \setminus \Sigma_0,$$

$(8) N \setminus \Sigma_0 = \{x \in N : y \in \tilde{V}_x \Longrightarrow \tilde{V} \text{ is a topographic submanifold at } (x, y)\}.$

Therefore, owing to $(*)$ and the fact that $a_i + \lambda = \pi(a_i) \times \mathbf{C}$, we obtain (see C. 3.17)

$$(9) \qquad \pi(a_i) \in N \setminus \Sigma_0, \quad i = 1, \ldots, r .$$

Now, in view of (5), we have the disjoint union $\tilde{V} \setminus \tilde{Z} = \bigcup (\tilde{V}_i \setminus \tilde{Z})$, the terms of which are open in $\tilde{V} \setminus \tilde{Z}$ (in view of (6); see IV. 2.9, theorem 4). As a consequence, the function $\tilde{f} : \tilde{V} \setminus Z \longrightarrow \mathbf{C}$ defined by

$$\tilde{f} = \tilde{f}_i \quad \text{in} \quad \tilde{V}_i \setminus \tilde{Z}, \quad i = 1, \ldots, r ,$$

is continuous and its graph is constructible. Moreover,

$$(10) \qquad \tilde{f} = f \quad \text{in} \quad V \setminus \tilde{Z}$$

because, by (3), we have $\tilde{f} = \tilde{f}_i = f$ in $(\tilde{V}_i \setminus \tilde{Z}) \cap (V_i \setminus \tilde{Z}_i) = V_i \setminus \tilde{Z}$, $i = 1, \ldots, r$.

Consider now the Lagrange interpolation polynomial

$$L(y_1, \eta_1; \ldots; y_p, \eta_p; y) = \sum_{s=1}^{p} \prod_{i \neq s} \frac{y - y_i}{y_s - y_i} \eta_s ,$$

where $y_1, \eta_1, \ldots, y_p, \eta_p, y \in \mathbf{C}$ and $y_1, \ldots, y_p$ are distinct. Define

$$(11) \qquad \Sigma = \Sigma_0 \cup \pi(\tilde{Z}) .$$

(It is algebraic and nowhere dense, by (5), since the projection $\pi_{\tilde{V}}$ is closed.) Then the set

$$\tilde{H} = \{(x, y_1, \ldots, y_p, y, z) \in N \times \mathbf{C}^{p+2} : z = L(y_1, \tilde{f}(x, y_1); \ldots; y_p, \tilde{f}(x, y_p); y),$$
$$(x, y_1), \ldots, (x, y_p) \in \tilde{V}, \; x \in N \setminus \Sigma, \; y_1, \ldots, y_p \text{ are distinct}\}$$

is constructible $(^{152})$, and hence so is its image $H \subset M \times \mathbf{C}$ under the projection

$$N \times \mathbf{C}^{p+2} \ni (x, y_1, \ldots, y_p, y, z) \longrightarrow (x, y, z) \in M \times \mathbf{C} .$$

Now, according to (7), (8) , and (11), if $x \in N \setminus \Sigma$, then $\#V_x = p$ and $\tilde{V}$ is a topographic submanifold at each point $(x, y) \in \tilde{V}$. Hence (owing to the symmetry of L) the set H is the graph of a continuous function on the set $(N \setminus \Sigma) \times \mathbf{C} = M \setminus \pi^{-1}(\Sigma)$. By theorem 3 from n° 7, there are polynomials $g, h \in \mathcal{P}(M)$ such that

(12) $H(z) = g(z)/h(z)$ and $h(z) \neq 0$ in the set $M \setminus \pi^{-1}(\Sigma)$.

We claim that

(13) $\tilde{f} = H$ in $\tilde{V} \setminus \pi^{-1}(\Sigma)$.

Indeed, let $(x, y) \in \tilde{V} \setminus \pi^{-1}(\Sigma)$. Then $x \in N \setminus \Sigma$ and, by (7), we have $\tilde{V}_x = \{y_1, \ldots, y_p\}$. Therefore $y = y_s$ for some s. But $(x, y_1, \ldots, y_p, y, z) \in \tilde{H}$, where

$$z = L\big(y_1, \tilde{f}(x, y_1); \ldots; y_p, \tilde{f}(x, y_p); y\big) = \tilde{f}(x, y_s) = \tilde{f}(x, y) .$$

Thus $(x, y, z) \in H$, and so $\tilde{f}(x, y) = H(x, y)$.

Suppose now that $Z' = V \cap \pi^{-1}(\Sigma)$. Then $a_i \in V_i \setminus Z'$ because of (11). This is true in view of (9) and (**), $a_i \notin \pi^{-1}(\Sigma_0)$ and $a_i \notin \pi^{-1}(\pi(\tilde{Z})) = \tilde{Z} + \lambda$. Hence the set $Z' \cap V_i$ is nowhere dense in V_i, which implies (see IV. 2.9, corollary 1 of theorem 4) that the set Z' is nowhere dense in V. Next, by (11) and (6), we have $Z' \supset V \cap \tilde{Z} \supset Z$. Therefore $V \setminus Z' \subset V \setminus \tilde{Z}$, but also $V \setminus Z' \subset V \setminus \pi^{-1}(\Sigma)$, and so (10) and (13) yield that $f = H$ in $V \setminus Z'$. Thus, by (12), the function $f_{V \setminus Z'}$ is the restriction of a rational function on M.

Let us come back to the example from n° 3. Every elliptic function f is of the form $\varphi \circ \pi$, where φ is a meromorphic function on T, and so $f = \psi \circ \tilde{g} \circ \pi = \psi \circ g$, where ψ is a meromorphic function on W. Since the closure $\bar{\psi}$ in $\mathbf{P}_2 \times \bar{\mathbf{C}}$ is algebraic (see n° 4 theorem 1; and 6.4, Chow's theorem), therefore the graph of $\psi_{W_0} = \bar{\psi} \cap \mathbf{C}^3$ is algebraic in $\mathbf{C}^3$. By theorem 4, in the complement of a finite subset of W_0 we have $\psi = R$, where R is a rational function on $\mathbf{C}^2$. Thus every elliptic function f is (outside of a discrete subset of $\mathbf{C}$) of the form $f = R(p, p')$, where R is a rational function.

$(^{152})$ Since it is the image of the constructible set

$\{(x, y_1, \ldots, y_p, z, z_1, \ldots, z_p) \in N \times \mathbf{C}^{2p+2} : z = L(y_1, z_1; \ldots; y_p, z_p; y),$

$\qquad (x, y_i, z_i) \in \tilde{f}, (x, y_i) \in \tilde{V} (i = 1, \ldots, r), x \in N \setminus \Sigma, y_1, \ldots, y_p$ are distinct$\}$

under the mapping $(x, y_1, \ldots, y_p, y, z, z_1, \ldots, z_p) \longrightarrow (x, y_1, \ldots, y_p, y, z)$.

§15. Ideals of $\mathcal{O}_n$ with polynomial generators

1. We will identify the ring of polynomials $\mathcal{P}_n$ with a subring of the ring $\mathcal{O}_n$ via the monomorphism $\mathcal{P}_n \ni f \longrightarrow f_0 \in \mathcal{O}_n$. Then we have the inclusions $\mathcal{P}_n \subset \mathcal{Q}_n \subset \mathcal{O}_n$.

PROPOSITION ([153]) . *Let* $P_1, \ldots, P_q \in \mathcal{P}_n^p$. *Then the submodule*

$$M = \{(f_1, \ldots, f_q) \in \mathcal{O}_n^q : \sum_1^q f_i P_i = 0\}$$

of the module $\mathcal{O}_n^q$ *over* $\mathcal{O}_n$ *has a finite system of generators from* $\mathcal{P}_n^q$.

The proposition states exactly that $\mathcal{O}_n$ is flat over $\mathcal{P}_n$ (see A. 1.16a).

PROOF . The case $n = 0$ is trivial. Let $n > 0$, and suppose that the proposition is true for $\mathcal{O}_{n-1}$. Then $\mathcal{Q}_n = \mathcal{O}_{n-1}[z_n]$ is flat over $\mathcal{P}_n = \mathcal{P}_{n-1}[z_n]$ (see A. 1.16a and I. 1.5). Since $\mathcal{O}_n$ is noetherian (see I. 1.6), it is enough to show that if $P_1, \ldots, P_m \in \mathcal{P}_n$, then any solution $(f_1, \ldots, f_m) \in \mathcal{O}_n^m$ of the equation

$$(*) \qquad\qquad \sum_1^m f_i P_i = 0$$

is a linear combination of solutions from $\mathcal{P}_n^m$ (see A. 9.6). One may assume that the P_i are non-zero ([154]), and then that all the P_i are regular (by a linear change of coordinates); see I. 1.4. It is sufficient to prove that $(f_1, \ldots, f_m)$ is a linear combination of solutions of the equation $(*)$ from $\mathcal{Q}_n$, because every such solution must be a linear combination of solutions from $\mathcal{P}_n^m$. By the preparation theorem, $P_m = QR$, where $Q, R \in \mathcal{Q}_n$, R is distinguished, and Q is invertible in $\mathcal{O}_n$ (see I. 2.1). Again, by the preparation theorem (see I. 1.4), we have $f_i = g_i P_i + r_i$, where $g_i \in \mathcal{O}_n$, $r \in \mathcal{Q}_n$, $i = 1, \ldots, m-1$. By putting $r_m = f_m + \sum_1^{m-1} g_i P_i$, we obtain the identity

$$(f_1, \ldots, f_m) = g_1(P_m, 0, \ldots, 0, -P_1) + \cdots + g_{m-1}(0, \ldots, 0, P_m, -P_{m-1}) +$$
$$+ Q^{-1}(Qr_1, \ldots, Qr_m) \ .$$

Now, $(P_m, 0, \ldots, 0, -P_1), \ldots, (0, \ldots, 0, P_m, -P_{m-1})$ are solutions of the equation $(*)$ from $\mathcal{Q}_n$. But so is $(Qr_1, \ldots, Qr_m)$. Indeed, it follows from the identity that $(r_1, \ldots, r_m)$ is a solution, i.e., $\sum_1^{m-1} r_i P_i + (Qr_m)R = 0$, and so

---

([153]) This proposition is an analogue of Oka's theorem on modules of relations. The idea of the proof is the same. (See VI. 1.3 and [25] pp. 154-156.)

([154]) One reduces the statement to this case by using solutions of the form $(0, \ldots, 1, \ldots \\ \ldots, 0)$.

$(Qr_m)R \in \mathcal{Q}_n$. Hence $Qr_m \in \mathcal{Q}_n$ (see I. 2.1), and therefore $(Qr_1, \ldots, Qr_m)$ is a solution from $\mathcal{Q}_n^m$.

COROLLARY 1. *If $I_1, \ldots, I_k$ are ideals of the ring $\mathcal{O}_n$ and each of them has a finite system of generators from $\mathcal{P}_n$, then so does the ideal $I_1 \cap \ldots \cap I_k$.*

In fact, it is enough to consider the case of two ideals I, J of the ring $\mathcal{O}_n$. Suppose that $F_1, \ldots, F_k \in \mathcal{P}_n$ generate the ideal I and $G_1, \ldots, G_l \in \mathcal{P}_n$ generate the ideal J. The ideal $I \cap J$ is the image of the submodule

$$M = \{(f_1, \ldots, f_k, g_1, \ldots, g_l) \in \mathcal{O}_n^{k+l} : \sum_i f_i F_i - \sum_j g_j G_j = 0\}$$

under the module homomorphism $\mathcal{O}_n^{k+l} \ni (f_1, \ldots, f_k, g_1, \ldots, g_l) \longrightarrow \sum_i f_i F_i \in \mathcal{O}_n$. By the proposition, the module M has a finite system of generators belonging to $\mathcal{P}_n^{k+l}$. Their images belong to $\mathcal{P}_n$ and generate the ideal $I \cap J$.

COROLLARY 2. *If an ideal I of the ring $\mathcal{O}_n$ has a finite system of generators from $\mathcal{P}_n$ and $F \in \mathcal{P}_n$, then so does the ideal $I : F$.*

Indeed, let $F_1, \ldots, F_k \in \mathcal{P}_n$ be generators of the ideal I. The ideal $I : F$ is the image of the module $M = \{(f, f_1, \ldots, f_k) \in \mathcal{O}_n^{k+l} : fF - \sum_i f_i F_i = 0\}$ under the module homomorphism $\mathcal{O}_n^{k+1} \ni (f, f_1, \ldots, f_k) \longrightarrow f \in \mathcal{O}_n$. By the proposition, the module M has a finite system of generators from $\mathcal{P}_n^{k+1}$. Their images belong to $\mathcal{P}_n$ and generate the ideal $I : F$.

2. Using the notation from §9, assume the hypotheses of Rückert's lemma and condition (1) from the proposition in 9.3. As a special case of Cartan's lemma 1 in VI. 1.1, we have the following

CARTAN LEMMA . *Let $V = V(I)$. Then*

$$\mathcal{I}(V_0) = J : \delta_0^m$$

with some m, where J is the ideal of $\mathcal{O}_n$ generated by the germs at 0 of the polynomials

$$p_j(u, z_j), \quad j = k+1, \ldots, n, \quad and \quad \delta(u)z_j - Q_j(u, z_{k=1}), \quad j = k+2, \ldots, n \ .$$

Combining Cartan's lemma with corollary 2 from n° 1, one gets the following

COROLLARY . *The ideal $\mathcal{I}(V_0)$ has a finite system of generators from $\mathcal{P}_n$.*

3. Let V be an algebraic subset of an n–dimensional vector space X.

SERRE'S LEMMA . *For each $a \in V$, the ideal $\mathcal{I}(V_a)$ of the ring $\mathcal{O}_a$ has a finite system of generators that are germs of polynomials.*

PROOF . We may assume that $X = \mathbf{C}^n$ and $a = 0$. First assume that V is irreducible. Then $V = V(I)$, where $I = \mathcal{I}(V)$ is a prime ideal (see 1.2 and 1.6). One may assume (after a linear change of coordinates) that the hypotheses of Cartan's lemma are satisfied (see 9.1–3). By the corollary of Cartan's lemma, the ideal $\mathcal{I}(V_0)$ has a finite system of generators belonging to $\mathcal{P}_n$. In the general case, we have $V_0 = \bigcup (V_i)_0$, where the V_i are simple components of V that contain 0. According to the first part of the proof and corollary 1 from n° 1, the ideal $\mathcal{I}(V_0) = \bigcap \mathcal{I}((V_i)_0)$ has a finite system of generators from $\mathcal{P}_n$.

It follows from Serre's lemma that the space $S_{V_a} \subset X^*$ (defined in IV. 4.4) can be expressed in terms of polynomials, namely

$$S_{V_a} = \{d_a P : \ P \in \mathcal{I}(V)\}$$

(as both sides coincide with $\{d_a P : \ P \in \mathcal{P}(X), \ P_a \in \mathcal{I}(V_a)\}$ [155]). The *Zariski dimension of the set V at a point a* is defined by the formula

$$\dim \operatorname{zar}_a V = \operatorname{codim}\{d_a P : \ P \in \mathcal{I}(V)\} .$$

Therefore $\dim \operatorname{zar}_a V = \dim \operatorname{zar} V_a$.

Similarly, the *Zariski tangent space to the set V at the point a* is defined by

$$T_a^{\mathrm{zar}} V = \bigcap \{\ker d_a P : \ P \in \mathcal{I}(V)\} = S_{V_a}^\perp = T_{V_a}^{\mathrm{zar}}$$

(see IV. 4.4).

COROLLARY . *A point $a \in V$ is regular and of dimension k if and only if $V_a = \{\Phi = 0\}_a$ for some polynomial mapping $\Phi : X \longrightarrow \mathbf{C}^{n-k}$ whose differential $d_a \Phi$ is surjective. One may have $V \subset \{\Phi = 0\}$.*

Indeed, if V_a is a k–dimensional smooth germ, then, in view of corollary 2 and the equality (∗∗) from IV. 4.4, we have $\dim S_{V_a} = n - k$. This means that S_{V_a} has a basis $d_a P_1, \ldots, d_a P_{n-k}$, where $P_i \in \mathcal{I}(V)$. So, $V_a \subset \{P_1 = \ldots = P_{n-k} = 0\}$ and the differentials $d_a P_1, \ldots, d_a P_{n-k}$ are linearly independent. Therefore $V_a = \{P_1 = \ldots = P_{n-k} = 0\}_a$ [156] .

---

[155] Indeed, by taking the union V' of the simple components of V that do not contain a and $Q \in \mathcal{I}(V')$, $Q(a) = 1$, it is enough to observe that if $P_a \in \mathcal{I}(V_a)$, then $d_a P = d_a(PQ)$ and $PQ \in \mathcal{I}(V)$ (see IV. 2.8, proposition 2).

[156] One can also use the criterion from II. 4.2 to derive this property.

The above corollary can be derived in a more direct manner, following an idea of Z. Jelonek.

One may assume that V is irreducible and k–dimensional (see IV. 2.8, corollary 4 of theorem 4; and IV. 2.8, corollary 1 of proposition 2). First suppose that $k = n - 1$. Then $V = V(F)$, where $F \in \mathcal{P}(X)$ is irreducible (see 11.3, corollary 1). But $V_a = V(h)$, where $h \in \mathcal{O}_a$, $d_a h \neq 0$, and hence h is irreducible (see I. 1.2). Therefore $F_a = gh^k$, where $g \in \mathcal{O}_a$, $g(a) \neq 0$, and $k \geq 1$ (see II. 5.2, corollary 3). We must have $k = 1$, for otherwise we would have $\partial F/\partial z_i = 0$ on V, and so $\partial F/\partial z_i$ would be divisible by F (see 11.2, proposition 4), which is impossible. Consequently, $d_a F \neq 0$.

In the general case, in view of Sadullaev's theorem (see 7.1), one may assume that $X = \mathbf{C}^n$, $\pi : V \ni z \longrightarrow (z_1, \ldots, z_k) \in \mathbf{C}^k$ is proper, and $V \subset \mathbf{C}^k \times \mathbf{C}^{n-k}$ is topographic at a. Changing the coordinates $z_{k+1}, \ldots, z_n$, one can have $\pi_j^{-1}(\pi_j(a)) = a$, where $\pi_j :$ $V \ni z \longrightarrow (z_1, \ldots, z_k, z_j) \in \mathbf{C}^{k+1}$, $j = k + 1, \ldots, n$ ([157]). Moreover, the π_j are proper (see B. 2.4). Therefore $\pi_j(V)$ is algebraic, irreducible, and topographic at $\pi_j(a)$ ([158]). By the first part of the proof, there is a polynomial $\Psi_j(z_1, \ldots, z_k, z_j)$ that vanishes on $\pi_j(V)$ and satisfies $(\partial \Psi_j/\partial z_j)(\pi_j(a)) \neq 0$. Then the polynomial mapping $\Phi = (\Psi_{k+1} \circ \pi_{k+1}, \ldots, \Psi_n \circ \pi_n)$ vanishes on V and the differential $d_a \varphi$ is surjective. Hence it follows that $V_a = \{\Phi = 0\}_a$.

§16. Serre's algebraic graph theorem.
Zariski's analytic normality theorem

Let X be a linear space.

1a. Let $a \in X$. A germ in $\mathcal{O}_a$ is said to be *(algebraically) regular* if it is of the form $(F/G)_a$, where $F, G \in \mathcal{P}(X)$ and $G(a) \neq 0$. Such germs form a subring $\mathcal{R}_a$ of $\mathcal{O}_a$ isomorphic to the localization $\mathcal{P}(X)_{\mathcal{I}(a)}$ (see A. 11.1); obviously, the ideal $\mathcal{I}(a) = \{F \in \mathcal{P}(X) : F(a) = 0\}$ is prime via the isomorphism

$$\mathcal{P}(X)_{\mathcal{I}(a)} \ni F/G \longrightarrow (F/G)_a \in \mathcal{R}_a \quad (^{159}).$$

Therefore the ring $\mathcal{R}_a$ is noetherian and local (see A. 11.1). Its maximal ideal is $\mathfrak{n}_a = \{f \in \mathcal{R}_a : f(a) = 0\}$ and corresponds to the ideal $\overline{\mathcal{I}(a)}$ via the above isomorphism.

([157]) Indeed, the set $E = \pi^{-1}(\pi(a)) \setminus a$ is finite and $p(a) \notin p(E)$, where $p : z \longrightarrow (z_{k+1}, \ldots, z_n)$. Hence the set of linear forms $\zeta \in (\mathbf{C}^{n-k})^*$ such that $\zeta(p(a)) \notin \zeta(p(E))$ is dense and one can choose $n - k$ linearly independent forms from that set.

([158]) Indeed, some neighbourhood U of the point a in V is a topographic submanifold. There exists an open neighbourhood B_j of the point $\pi_j(a)$ such that $U_j = \pi_j^{-1}(B_j) \subset U$ (see B. 2.4), and then U_j is a topographic submanifold. Then the image $\pi_j(U_j) = B_j \cap \pi_j(V)$ is also a topographic submanifold and it is a neighbourhood of $\pi_j(a)$ in $\pi_j(V)$.

([159]) On the left hand side, F/G is a fraction, while on the right hand side it is a rational function.

Furthermore, we have the equalities

$$(*) \qquad\qquad \mathfrak{n}_a^\nu = \mathfrak{m}_a^\nu \cap \mathcal{R}_a, \quad \nu = 1, 2, \dots .$$

They follow from the fact that (after endowing X with a norm)

$$\mathfrak{m}_a^\nu = \{f_a \in \mathcal{O}_a : f(z) = o(|z - a|^{\nu-1})\} \quad (^{160})$$
$$\mathfrak{n}_a^\nu = \{h_a \in \mathcal{R}_a : h(z) = o(|z - a|^{\nu-1})\} ,$$

as the ideal $\mathfrak{n}_a^\nu$ corresponds to the ideal $(\overline{\mathcal{I}(a)})^\nu = \overline{\mathcal{I}(a)^\nu}$ (see A. 11.1). Also

$$\mathcal{I}(a)^\nu = \{F \in \mathcal{P}(X) : F(z) = o(|z - a|^{\nu-1})\} \quad (^{161}).$$

For any ideal I in the ring $\mathcal{R}_a$, we have the equality

$$(**) \qquad\qquad \mathcal{O}_a I \cap \mathcal{R}_a = I .$$

It means that if a germ from $\mathcal{R}_a$ is a linear combination of germs from $\mathcal{R}_a$ with coefficients from $\mathcal{O}_a$, then that germ is also a linear combination of those germs with coefficients from $\mathcal{R}_a$.

Indeed, if $F = \sum_1^s H_i F_i \in \mathcal{R}_a$, where $F_i \in \mathcal{R}_a$ and $H_i \in \mathcal{O}_a$, then, clearly, for each ν we have $H_i \in \mathcal{R}_a + \mathfrak{m}_a^\nu$. Therefore, in view of $(*)$, we get $F \in I + \mathfrak{n}_a^\nu$, where $I = \sum_1^s \mathcal{R}_a F_i$. Thus $F \in \bigcap_1^\infty (I + \mathfrak{n}_a^\nu) = I$ (see A. 10.5, corollary of Krull's theorem).

The above, combined with Hilbert's Nullstellensatz for $\mathcal{O}_a$ (see III. 4.1), implies *Hilbert's Nullstellensatz for $\mathcal{R}_a$*:

If $f_1, \dots, f_k \in \mathcal{R}_a$ and a germ $f \in \mathcal{R}_a$ vanishes on $V(f_1, \dots, f_k)$, then $f^r \in \sum_1^k \mathcal{R}_a f_i$ for some $r \in \mathbf{N}$.

1b. Let $V \subset X$ be an algebraic subset, and let $a \in V$. A germ from $\mathcal{O}_a(V)$ is said to be *(algebraically) regular* if it is of the form $(f/g)_a$, where $f, g \in R(V)$ and $g(a) \neq 0$. All such germs form a subring $\mathcal{R}_a(V)$ of the ring

---

(^{160}) Because this is so when $X = \mathbf{C}^n$ and $a = 0$ (see I. 1.7).

(^{161}) Indeed, one may assume that $X = \mathbf{C}^n$ and $a = 0$. The inclusion $\subset$ is obvious. On the other hand, if $F \in \mathcal{P}_n$ and $F(z) = o(|z|^{\nu-1})$, consider the decomposition into homogeneous polynomials $F = F_k + F_{k+1} + \cdots +$, where $F_k \neq 0$ (we omit here the trivial case $F = 0$). We have $k \geq \nu$ (for otherwise one would have $F_k(z) = o(|z|^k)$). Therefore $F \in \sum_{|p|=\nu} \mathcal{P}_n z^p = \mathcal{I}(0)^\nu$ (because $z_1, \dots, z_n$ generate $\mathcal{I}(0)$, and so z^p, $|p| = \nu$, generate $\mathcal{I}(0)^\nu$; see A. 1.7).

$\mathcal{O}_a(V)$. Recall that $\mathcal{O}_a \ni f \longrightarrow f_V \in \mathcal{O}_a(V)$ is an epimorphism with the kernel $\mathcal{I}(V_a)$ and the image under this mapping of the ideal $\mathfrak{m}_a$ is the ideal $\mathfrak{m}_a(V)$ (see IV. 4.2). Clearly, the image of the subring $\mathcal{R}_a$ is the subring $\mathcal{R}_a(V)$. Therefore the ring $\mathcal{R}_a(V)$ is noetherian and local with the maximal ideal $\mathfrak{n}_a(V) = \{h \in \mathcal{R}_a(V) : h(a) = 0\}$.

Set $\mathcal{O} = \mathcal{O}_a(V)$, $\mathcal{R} = \mathcal{R}_a(V)$, $\mathfrak{m} = \mathfrak{m}_a(V)$, $\mathfrak{n} = \mathfrak{n}_a(V)$. We have

$$\mathfrak{m}^k = \mathcal{O}\mathfrak{n}^k \quad \text{for} \quad k \in \mathbf{N} \ ,$$

since the ideals $\mathfrak{m}^k$ and $\mathfrak{n}^k$ have common generators ([162]) . Moreover,

$$\mathcal{O} = \mathcal{R} + \mathfrak{m}^k \quad \text{for} \quad k \in \mathbf{N} \ \ ([163]).$$

An element $(f/g)_a \in \mathcal{R}$ (where $f, g \in R(V)$, $g(a) \neq 0$) is a non-zero zero divisor if and only if $f_{V'} = 0$ and $f_{V''} \neq 0$ for some simple components V', V'' of V such that $a \in V' \cap V''$ ([164]) . Consequently,

($\mathcal{R}$ is an integral domain) $\Longleftrightarrow$

(only one of the simple components of V contains a ([165])).

Next, for $\varphi \in \mathcal{R}$,

(φ is a zero divisor in $\mathcal{R}$) $\Longleftrightarrow$ (φ is a zero divisor in $\mathcal{O}$)

(see IV. 4.3, the corollary of proposition 3b in IV. 3.1).

Let $\mathcal{M}$ and $\mathcal{N}$ denote the rings of fractions of the rings $\mathcal{O}$ and $\mathcal{R}$, respectively. We have $\mathcal{N} \subset \mathcal{M}$ after the identification through the well-defined monomorphism $\mathcal{N} \ni \varphi/\psi \longrightarrow \varphi/\psi \in \mathcal{R}$ (see A. 1.15a).

The ring $\mathcal{R}$ is isomorphic to the localization $R(V)_N$ to the prime ideal $N = \mathcal{I}_V(a) = \{f \in R(V) : f(a) = 0\}$ via the natural isomorphism

$$\mathcal{R} \ni (f/g)_a \longrightarrow \bar{f}\bar{g}^{-1} \in R(V)_N \ .$$

---

[162] Because (see A. 1.7) $\mathfrak{m}$ and $\mathfrak{n}$ have common generators: if $X = \mathbf{C}^n$ and $a = 0$, then $((z_i)_V)_0$, $i = 1, \dots, n$, are such generators.

[163] One takes the images by the epimorphism $\mathcal{O}_a \ni f \longrightarrow f_V \in \mathcal{O}_a(V)$ in the equality $\mathcal{O}_a = \mathcal{R}_a + \mathfrak{m}_a^k$.

[164] This can be obtained easily from the following obvious properties: $1°$ If $W \subset V$ is an algebraic set and $z \in V \setminus W$, then there is $f \in R(V)$ such that $f_V = 0$ and $f(z) \neq 0$. $2°$ If V is irreducible, then $f_W \neq 0 \Longrightarrow f_V \not\equiv 0$ for $f \in R(V)$; see IV. 2.8, proposition 2.

[165] Thus $\mathcal{R}$ is an integral domain and $\mathcal{O}$ is not (see IV. 4.3) if, e.g., V is irreducible and the germ V_a is irreducible. For instance, this is the case for the curve $V = \{w^2 - z^2 - z^3 = 0\} \subset \mathbf{C}^2$ with the "double" point $a = (0,0)$ (the curve V is the range of the immersion $\mathbf{C} \ni t \longrightarrow (t^2 - 1, t(t^2 - 1)) \in \mathbf{C}^2$.)

(For $\mathcal{I}_N = \{f \in R(V) : f_a = 0\}$; see A. 11.2.) The homomorphism $R(V) \ni f \longrightarrow f_a \in \mathcal{R}$ corresponds to the natural homomorphism $R(V) \ni f \longrightarrow \bar{f} \in R(V)_N$, whereas the maximal ideal $\mathfrak{n}$ corresponds to the maximal ideal $\bar{N}$.

Now, to any prime ideal I of the ring $\mathcal{R} \approx R(V)_N$ there corresponds the prime ideal $\{f \in R(V) : f_a \in I\} \subset N$ of the ring $R(V) \approx \mathcal{P}(X)/\mathcal{I}(V)$ (see A. 11.2). To this ideal there corresponds the prime ideal $\{F \in \mathcal{P}(X) : (F_V)_a \in I\}$ of the ring $\mathcal{P}(X)$ (see A. 1.10 and 11) which contains $\mathcal{I}(V)$ and is contained in $\mathcal{I}(a)$. Finally, to this ideal there corresponds the algebraic set

$$V'(I) = V(\{F \in \mathcal{P}(X) : (F_V)_a \in I\}) \subset X$$

contained in V and containing a (see §10, corollary 1). Conversely, to any irreducible algebraic set Z such that $a \in Z \subset V$, there corresponds the prime ideal $\mathcal{I}(Z)$ of the ring $\mathcal{P}(X)$ which contains $\mathcal{I}(V)$ and is contained in $\mathcal{I}(a)$. Then, to $\mathcal{I}(Z)$ there corresponds the prime ideal $\{f \in R(V) : f_Z = 0\} \subset N$ of the ring $R(V)$, and to this ideal there corresponds the prime ideal

$$\mathcal{I}'(Z) = \{\varphi \in \mathcal{R} : \varphi_{Z_a} = 0\}$$

of the ring $\mathcal{R}$. Therefore

$$\mathcal{I}'(V'(I)) = I \quad \text{and} \quad V'(\mathcal{I}'(Z)) = Z \ ,$$

and the mappings $I \longrightarrow V'(I)$ and $z \longrightarrow \mathcal{I}'(Z)$ are mutually inverse bijections between the set of prime ideals of the ring $\mathcal{R}$ and the set of irreducible algebraic subsets of V containing a.

In view of Serre's lemma (se 15.3), we have $\mathcal{I}(Z_a, V) = \mathcal{O}\mathcal{I}'(Z)$ (for any irreducible algebraic set Z such that $a \in Z \subset V$).

PROPOSITION 1. *For every ideal I of the ring $\mathcal{R}$, we have the equality*

$$(\mathcal{O}I) \cap \mathcal{R} = I \ ;$$

in particular, $\mathfrak{m}^n \cap \mathcal{R} = \mathfrak{n}^n$ for $n \in \mathbf{N}$. In other words, if a germ from $\mathcal{R}$ is a linear combination of germs from $\mathcal{R}$ with coefficients from $\mathcal{O}$, then it is also a combination of these germs with coefficients from $\mathcal{R}$.

Indeed, let $f = \sum h_i f_i \in \mathcal{R}$, where $f_i \in \mathcal{R}$, $h_i \in \mathcal{O}$. We have $f = F_{V_a}$, $f_i = (F_i)_{V_a}$, $h_i = (H_i)_V$, where $F, F_i \in \mathcal{R}_a$ and $H_i \in \mathcal{O}_a$. Therefore $(F - \sum H_i F_i)_{V_a} = 0$, and hence Serre's lemma (see 15.3) implies that $F - \sum H_i F_i = \sum H'_j F'_j$, where $F'_j \in \mathcal{I}(V_a) \cap \mathcal{R}_a$ and $H'_j \in \mathcal{O}_a$. Hence, according to $(**)$ (see n° 1a), we have $F = \sum G_i F_i + \sum G'_i F'_i$ for some $G_i, G'_j \in \mathcal{R}_a$, and so $f = \sum g_i f_i$, where $g_i = (G_i)_V \in \mathcal{R}$.

COROLLARY . *If a germ $h \in \mathcal{R}$ is not a zero divisor* ([166]) *, then for $f \in \mathcal{O}$ we have the implication $fh \in \mathcal{R} \Longrightarrow f \in \mathcal{R}$. In other words,*

$$\mathcal{N} \cap \mathcal{O} \subset \mathcal{R} .$$

This follows since $(\mathcal{O}h) \cap \mathcal{R} \subset \mathcal{R}h$.

2. Let $S \subset X$ be a quasi-algebraic set. A function $f : S \longrightarrow \mathbf{C}$ is said to be *(algebraically) regular* if each of its germs is regular, i.e., if $a \in S \Longrightarrow f_a \in \mathcal{R}_a(\bar{S})$. This means that each point of the set S has a neighbourhood U in S such that f_U is the restriction of a rational function on X. Clearly, the regular functions on S form a ring. The restriction of a regular function to a quasi-algebraic subset of the domain of this function is a regular function.

PROPOSITION 2. *If $V \subset X$ is an algebraic set, then every regular function $f : V \longrightarrow \mathbf{C}$ is the restriction of a polynomial from $\mathcal{P}(X)$. In other words, the ring $R(V)$ is equal to the ring of all regular functions on V.*

PROOF . For any $a \in V$, we have

$$(\#) \qquad Q_V f = P_V \quad \text{and} \quad Q(a) \neq 0 \quad \text{for some} \quad P, Q \in \mathcal{P}(X) .$$

Indeed, we have $f = P/Q$ and $Q \neq 0$, in a neighbourhood of a in V, where $P, Q \in \mathcal{P}(X)$. Then $Qf = P$ on each simple component $\tilde{V} \ni a$ of V, because V^0 is a connected submanifold which is dense in $\tilde{V}$ (see IV. 2.8, proposition 2; and IV. 2.1, proposition 1). Now, it is sufficient to multiply P and Q by a polynomial that is different from zero at a and vanishes on all the simple components which do not contain a.

Consider the ideal $I = \{Q \in \mathcal{P}(X) : \ Q_V f \in \mathcal{R}(V)\}$ of the ring $\mathcal{P}(X)$. Obviously, $\mathcal{I}(V) \subset I$, and so $V = V\big(\mathcal{I}(V)\big) \supset V(I)$ (see 1.2–3). On the other hand, because of $(\#)$, we have $V \cap V(I) = \emptyset$, and hence $V(I) = \emptyset$. Therefore Hilbert's Nullstellensatz (see §10) implies that $1 \in \text{rad } I$, and hence $1 \in I$. Consequently, $f \in \mathcal{R}(V)$.

An analogous argument shows that on any principal quasi-algebraic set, i.e., on a set of the form $S = V \setminus V(g)$, where $g \in \mathcal{P}(X)$, the regular functions are the restrictions of the rational functions of the form f/g^m, where $m \in \mathbf{N}$ and $f \in \mathcal{P}(X)$.

Therefore, for any quasi-algebraic set $S = V \setminus V(g_1, \ldots, g_k) = \bigcup_1^k (V \setminus V(g_i))$, any regular function on S restricted to any of the sets $V_i \setminus V(g_i)$ is the restriction of a rational function. In general, however, it is not the restriction of just one rational function to the

---

whole set S. As an example, take the regular function $\varphi = (z_1/z_2)_S \cup (z_3/z_4)_S$ on the set $S = V \setminus W$, where $V = \{z_1 z_4 = z_2 z_3\} \subset \mathbf{C}^4$ and $W = \{z_2 = z_4 = 0\} \subset \mathbf{C}^4$ $(^{167})$.

3. Let Y be a vector space.

In accordance with the definition from n° 2, a mapping $f : S \longrightarrow Y$ of a quasi-algebraic set $S \subset X$ is said to be *(algebraically) regular* if each point of the set S has a neighbourhood U in S such that $f = F/G$ and $G \neq 0$ in U, where $G \in \mathcal{P}(X)$ and $F : X \longrightarrow Y$ is a polynomial mapping. Obviously, when $Y = \mathbf{C}^m$, the mapping $f = (f_1, \ldots, f_m)$ is regular if and only if the functions $f_1, \ldots, f_m$ are regular.

It follows from proposition 2 in n° 2 that if $V \subset X$ is an algebraic set, then

$(f : V \longrightarrow Y$ is regular$) \Longleftrightarrow$

$\Longleftrightarrow (f$ is the restriction to V of a polynomial mapping $X \longrightarrow Y)$ $(^{168})$.

We have the following

SERRE ALGEBRAIC GRAPH THEOREM . *If $f : W \longrightarrow Y$ is a holomorphic mapping of a locally analytic set $W \subset X$ and the graph of f is constructible (in $X \times Y$), then the set W is quasi-algebraic and the mapping f is regular. In particular, if a holomorphic mapping $f : V \longrightarrow Y$ (of an analytic set V) has an algebraic graph, then it is the restriction of a polynomial mapping (and V is algebraic).*

PROOF . By Chevalley's theorem from 8.3, the set W is constructible and so it is quasi-algebraic (see 8.3, corollary 1 of proposition 2). It is enough to consider the case of a function (i.e., $Y = \mathbf{C}$). Indeed, one may assume that $Y = \mathbf{C}^m$ and $f = (f_1, \ldots, f_m)$, and then the functions f_i are holomorphic and their graphs are constructible (by Chevalley's theorem from 8.3).

According to the constructible graph theorem (theorem 4 in 14.8), we have $f = g/h$ and $h \neq 0$ in a quasi-algebraic set S which is dense in the

---

$(^{167})$ If we had $\varphi = f/g$ on S for some $f, g \in \mathcal{P}_4$, then we would have $V(g) \cap V \subset W$, and hence $g(tu, u, tv, v) = 0 \Longrightarrow u = v = 0$. The latter would imply that g_S is a non-zero constant (see 11.2), which is impossible, since φ is unbounded in a neighbourhood of zero (since $\varphi(t, u, t, u) = t/u$).

$(^{168})$ It is enough to observe that if $f : V \longrightarrow Y$ is a regular mapping and $\varphi : Y \longrightarrow Z$ is a linear mapping (into a linear space Z), then the composition $\varphi \circ f : V \longrightarrow Z$ is also a regular mapping. Consequently, we can assume that $Y = \mathbf{C}^m$, and then the regularity of the mapping $f = (f_1, \ldots, f_m) : X \longrightarrow \mathbf{C}^m$ implies the regularity of the functions $f_1, \ldots, f_m$.

set W, where $g, h \in \mathcal{P}(X)$. Now, if $a \in W$, then $f_a(h_W)_a = (g_W)_a$ (because $fh = g$ in W) and $(h_V)_a$ is not a zero divisor in $\mathcal{O}_a(W)$ (since S is dense in W). Therefore, by the corollary from proposition 1 in n° 16, the germ f_a is regular.

REMARK 1. In the case when the set V is algebraic, the assumption that the graph of f is algebraic in (the second part of) Serre's theorem is equivalent – by Rudin's theorem (see 7.4) – to the *condition of polynomial growth* $|f(z)| = 0(|z|^k)$ as $|z| \longrightarrow \infty$ for some k (with norms on X and Y) ([169]) . This implies a *theorem of Rusek and Winiarski* ([170]) : *Every holomorphic function on V which is of polynomial growth must be regular.*

One can also deduce the Rusek-Winiarski theorem from the Grauert-Remmert formula (see V. 7.4), combined with the corollary in n° 3 ([170a]). In fact, using the trick due to Cynk from the proof of the Grauert-Remmert theorem in VI. 1.2a, one may assume that V is of constant dimension, say k. By the Rudin-Sadullaev theorem from 7.4, we may suppose that $V \subset \{(x, y) \in X \times Y : |y| \leq 1 + |x|\}$, where X and Y are vector spaces with norms, and $\dim X = k$. Then the natural projection $\pi : V \longrightarrow X$ is a *-covering (e.g., by the Andreotti-Stoll theorem in V. 7.2). By remark 3 in V. 7.4, there is a primitive element $\alpha = \Lambda_V$, $\lambda \in (X \times T)^*$ for π. Let $\varphi \in \mathcal{O}_X$ be of polynomial growth. In view of the fact that $\pi^{-1}(x) = \{\eta_1, \ldots, \eta_p\} \Longrightarrow |\eta_i| < 1 + 2|x|$, it follows from the definition of the functions $D_\alpha, C_{j\alpha\varphi} \in \mathcal{O}_X$ (see V. 4.7) that they are of polynomial growth, and hence are polynomials (by Liouville's theorem in C. 1.8). Therefore, in (GR), the function D_α^0 and the right hand side are regular. Since $D_\alpha^0 \not\equiv 0$, the function φ is regular, by the corollary in n° 3.

REMARK 2. A germ $f \in \mathcal{O}_a$, the graph of a representative of which is a neighbourhood of $(a, f(a))$ in an algebraic subset of $X \times Y$, does not have to be regular. An example is the germ at $1 \in \mathbf{C}$ of a branch of the root $\sqrt{z}$ in $\{\operatorname{Re} z > 0\}$ ([171]) .

4. Let $V \subset X$ be an algebraic set, and let $a \in V$. Assume the notation from n° 1b: for the rings $\mathcal{O} = \mathcal{O}_a(V)$, $\mathcal{R} = \mathcal{R}_a(V)$, their maximal ideals $\mathfrak{m}, \mathfrak{n}$, and their rings of fractions $\mathcal{M}, \mathcal{N}$.

The normality of V at a ([172]) has been characterized (see the proposition in VI. 3.1) by the integral closedness of the ring $\mathcal{O}$. Now we are going to prove

---

([169]) Obviously, the latter condition is implied by algebraicity of the graph in view of Serre's theorem, but the implication can be also derived directly from the regular separation property (see 3.6 and the theorem in IV. 7.1).

([170]) See [35b].

([170a]) Observed by Winiarski.

([171]) If this germ were regular, then we would have $P^2 = Q^2 z$ for some non-zero polynomials P and Q, which is impossible (since the degree of the left hand side is even and that of the right hand side is odd).

([172]) V regarded as an analytic space.

the Zariski analytic normality theorem, which states that integral closedness of the ring $\mathcal{R}$ also characterizes the normality of V at a.

LEMMA 1. *If I is a prime ideal of the ring $\mathcal{R}$, then $\mathcal{O}I$ is a proper ideal and a finite intersection of prime ideals of the ring $\mathcal{O}$.*

Indeed, looking at the decomposition into simple components $V'(I)_a = A_1 \cup \ldots \cup A_s$, $s > 0$ [173] , we have (see n° 1b) $\mathcal{O}I = \mathcal{O}I'(V'(I)) = \mathcal{I}(V'(I)_a, V) = \mathcal{I}(A_1, V) \cap \ldots \cap \mathcal{I}(A_r, V)$, and moreover, $\mathcal{I}(A_i, V)$ are prime ideals in the ring $\mathcal{O}$ (see V. 4.9).

LEMMA 2. *The localization $\mathcal{O}_I$ to a prime ideal I of the ring $\mathcal{O}$ is always a ring without nilpotent elements.*

Indeed, it is enough to check that $f^n \in \mathcal{I}_I \Longrightarrow f \in \mathcal{I}_I$ (see A. 11.2). Let $f^n g = 0$, $f \in \mathcal{O}$, $g \in \mathcal{O} \setminus I$. Let $V_a = A_1 \cup \ldots \cup A_r$ be the decomposition into simple germs. Since the rings $\mathcal{O}_{A_i}$ are integral domains (see IV. 4.3), for each i we have $f_{A_i} = 0$ or $g_{A_i} = 0$, and thus $fg = 0$.

LEMMA 3. *If $I_1, \ldots, I_r$ are ideals of the ring $\mathcal{R}$, then*

$$\mathcal{O}(I_1 \cap \ldots \cap I_r) = \mathcal{O}I_1 \cap \ldots \cap \mathcal{O}I_r .$$

It is sufficient to show the inclusion $\supset$ for $r = 2$, i.e., that $\mathcal{O}(I \cap J) \supset \mathcal{O}I \cap \mathcal{O}J$ for ideals $I, J \subset \mathcal{R}$. Let $h \in \mathcal{O}I \cap \mathcal{O}J$. Since $\mathcal{O} = \mathcal{R} + \mathfrak{m}^n$ for each $n \in \mathbf{N}$ (see n° 1b), we have $h - \varphi_n$, $h - \psi_n \in \mathfrak{m}^n$ for some $\varphi_n \in I$, $\psi_n \in J$. Thus $\varphi_n - \psi_n \in \mathfrak{n}^n \cap (I - J)$, in view of proposition 1 from n° 1b. Therefore, by the Artin-Rees lemma (see A. 9.4a), there is k such that $\varphi_n - \psi_n \in \mathfrak{n}^{n-k}(I - J)$ for $n \geq k$. Then $\varphi_n - \psi_n = \alpha_n - \beta_n$, where $\alpha_n \in \mathfrak{n}^{n-k}I$, $\beta_n = \mathfrak{n}^{n-k}J$. But $\chi_n = \varphi_n - \alpha_n = \psi_n - \beta_n \in I \cap J$, and so $h = \chi_n + \alpha_n + (h - \varphi_n) \in \mathcal{O}(I \cap J) + \mathfrak{m}^{n-k}$ (for $n \geq k$). Hence, by the corollary from Krull's theorem (see A. 10.5), we conclude that $h \in \mathcal{O}(I \cap J)$.

LEMMA 4. *If I is a primary ideal of the ring $\mathcal{R}$ and $\chi \in \mathcal{R} \setminus \mathrm{rad}\, I$, then for $f \in \mathcal{O}$ we have the implication*

$$f\chi \in \mathcal{O}I \Longrightarrow f \in \mathcal{O}I .$$

Indeed, let $f\chi \in \mathcal{O}I$. Since $\mathcal{O} = \mathcal{R} + \mathfrak{m}^n$ for each n (see n° 1b), we have $f - \varphi_n \in \mathfrak{m}^n$ for some $\varphi_n \in \mathcal{R}$. Hence

$$\varphi_n \chi \in (\mathcal{O}I + \mathfrak{m}^n) \cap \mathcal{R} = \mathcal{O}(I + \mathfrak{n}^n) \cap \mathcal{R} = I + \mathfrak{n}^n ,$$

---

[173] Because $a \in V'(I)$ (see n° 1b).

by proposition 1 from n° 1b (in view of $\mathfrak{m}^n = \mathcal{O}\mathfrak{n}^n$; see n° 1b). Therefore $\varphi_n \chi \in (I + \mathfrak{n}^n) \cap \mathcal{R}\chi$. By the corollary from the Artin-Rees theorem (see A. 9.4a), there exists k such that $\varphi_n \chi \in I + \mathfrak{n}^{n-k}\chi$ for $n \geq k$. Then $\varphi_n \chi \in I + \alpha_n \chi$, where $\alpha_n \in \mathfrak{n}^{n-k}$, which means that $(\varphi_n - \alpha_n)\chi \in I$, and so $\varphi_n - \alpha_n \in I$ (because I is primary). Therefore $f = (\varphi_n - \alpha_n) + \alpha_n + (f - \varphi_n) \in \mathcal{O}I + \mathfrak{m}^{n-k}$ (for $n \geq k$), and so, by the corollary of Krull's theorem (see A. 10.5), we obtain $f \in \mathcal{O}I$.

LEMMA 5. *Suppose that $\mathcal{R}$ is an integral domain. Let I be its prime ideal and assume that the localization $\mathcal{R}_I$ is a discrete valuation ring. Let $\mathcal{O}I = J_1 \cap \ldots \cap J_r$, $r > 0$, be an irreducible intersection of prime ideals J_i of the ring $\mathcal{O}$ (see lemma 1). Then the localizations $\mathcal{O}_{J_i}$ are discrete valuation rings and $\mathcal{O}I^{(k)} = J_1^{(k)} \cap \ldots \cap J_r^{(k)}$ for $k \in \mathbf{N}$.*

PROOF. The ideal $\bar{I}$ of the localization $\mathcal{R}_I$ is principal: $\bar{I} = \mathcal{R}_I \varphi$, where $\varphi \in I$ (see A. 10.6 and A. 11.1). Consider an irreducible primary decomposition $\mathcal{R}\varphi = \bigcap I_i$. We have $\bar{I}_s = \bar{I}$ for some s, and thus $I_s = I$ (see A. 11.1). Take $\psi \in \bigcap_{i \neq s} I_i \setminus I$. So

$$(1) \qquad\qquad \psi I \subset \mathcal{R}\varphi .$$

Fix an arbitrary i. By taking $f \in \bigcap_{\nu \neq i} J_\nu \setminus J_i$, we have $f J_i \subset \mathcal{O}I$. It follows that

$$(2) \qquad\qquad J_i \cap \mathcal{R} \subset I ,$$

for if there were $\chi \in J_i \cap \mathcal{R} \setminus I$, then $f\chi \in \mathcal{O}I$, and by lemma 4 we would have $f \in \mathcal{O}I \subset J_i$. In view of (2),

$$(3) \qquad\qquad \psi \in \mathcal{O} \setminus J_i ,$$

and hence $g = \psi f \in \mathcal{O} \setminus J_i$. Because of (1), we have $g J_i \subset \psi \mathcal{O}I \subset \mathcal{O}\varphi$. Let us pass to the localization $\mathcal{O}_{J_i}$. Then $\bar{g}\bar{J}_i \subset \mathcal{O}_{J_i}\bar{\varphi}$, but $\bar{g}$ is invertible (see A. 11.2) and $\bar{\varphi} \in \bar{J}_i$, therefore

$$(4) \qquad\qquad \bar{J}_i = \mathcal{O}_{J_i}\bar{\varphi} .$$

Furthermore, by lemma 2, the localization $\mathcal{O}_{J_i}$ is without nilpotent elements, and thus it is a discrete valuation ring, according to the proposition from A. 10.6.

Now, let us prove the equality. As for the inclusion $\subset$, we have $I^{(k)} \subset J_i^{(k)}$, in view of $\mathcal{R} \setminus I \subset \mathcal{O} \setminus J_i$, which follows from (2) (see A. 11.2). It remains to show that

$$(5) \qquad\qquad J_1^{(k)} \cap \ldots \cap J_r^{(k)} \subset \mathcal{O}I^{(k)} \quad \text{for} \quad k \in \mathbf{N} .$$

Now, if $\bar{J}_i = 0$ for some i, then $J_i = \mathcal{I}_{J_i}$ consists of zero divisors (see A. 11.2), and since $I \setminus 0 \subset J_i$ does not contain zero divisors ([174]), we must have $I = 0$. Then $J_1 \cap \ldots \cap J_r = 0$ and both sides of the inclusion (5) are zeros. Suppose now that $\bar{J}_i \neq 0$ for each i. We will prove the inclusion (5) by induction with respect to k. Suppose that it holds for $k - 1$, where $k \geq 2$, and let $f \in J_1^{(k)} \cap \ldots \cap J_r^{(k)}$. Then $f \in \mathcal{O}I$, and so (1) implies that $\psi f \in \mathcal{O}\psi I \subset \mathcal{O}\varphi$, which means that $\psi f = h\varphi$, where $h \in \mathcal{O}$. We claim that $h \in J_i^{(k-1)}$ for each i. Indeed, passing to the localization $\mathcal{O}_{J_i}$, we have $\bar{\psi}\bar{f} = \bar{h}\bar{\varphi}$, and moreover, $\bar{\varphi} \neq 0$ in view of (4). If $\bar{h} = 0$, then $h \in \mathcal{I}_{J_i} \subset J_i^{(k-1)}$ (see A. 11.2). If $h \neq 0$, then taking the valuation ν on $\mathcal{O}_{J_i}$ (see A. 10.6), we have $\nu(\bar{\psi}) + \nu(\bar{f}) = \nu(\bar{h}) + \nu(\bar{\varphi})$. But according to (3), $\nu(\bar{\psi}) = 0$. Next, $\nu(\bar{f}) \geq k$ and $\nu(\bar{\varphi}) = 1$ (because $\bar{\varphi}$ generates $\bar{J}_i \neq 0$; see A. 10.6). Thus $\nu(h) \geq k - 1$, which means that $h \in J_i^{(k-1)}$. Consequently, $\psi f = h\varphi \in \mathcal{O}I^{(k-1)}I \subset \mathcal{O}I^{(k)}$ (see A. 1.11). Thus it follows from lemma 4 that $f \in \mathcal{O}I^{(k)}$.

ZARISKI'S ANALYTIC NORMALITY THEOREM . *Let $V \subset X$ be an algebraic set. A point $a \in V$ is a normal point of the analytic space V if and only if the ring $\mathcal{R}_a(V)$ is integrally closed* ([175]) .

PROOF . According to the proposition from VI. 3.1, it is enough to prove that the ring $\mathcal{O}$ is integrally closed if and only if the ring $\mathcal{R}$ is integrally closed. Let $\mathcal{O}'$ and $\mathcal{R}'$ denote the integral closures of $\mathcal{O}$ in $\mathcal{M}$ and of $\mathcal{R}$ in $\mathcal{N}$, respectively. (Hence $\mathcal{O}' = \tilde{\mathcal{O}}_a$; see VI. 2.3.) Obviously, $\mathcal{R}' \subset \mathcal{O}'$ (see n° 1b).

If the ring $\mathcal{O}$ is integrally closed, it is an integral domain and $\mathcal{O}' \subset \mathcal{O}$. Then $\mathcal{R}$ is an integral domain and $\mathcal{R}' \subset \mathcal{O} \cap \mathcal{N} = \mathcal{R}$ (see n° 1b, the corollary of proposition 1). Hence the ring $\mathcal{R}$ is integrally closed.

Now suppose that the ring $\mathcal{R}$ is integrally closed. It is sufficient to show that the ring $\mathcal{O}$ satisfies the assumptions of the Zariski-Samuel lemma (see A. 11.3). It is clearly a noetherian local ring without nilpotent elements. By the proposition from 9.4, the ring $\mathcal{O}'$ has a universal denominator $u \in \mathcal{R}$ over $\mathcal{O}$. Assume that u is non-invertible in $\mathcal{R}$. Then $\mathcal{R}u \subsetneq \mathcal{R}$. Let $I_1, \ldots, I_s$ be all the ideals associated with $\mathcal{R}u$. According to the proposition from A. 11.3, the localizations $\mathcal{R}_{I_i}$ are discrete valuation rings, and hence we have a primary decomposition of the form $\mathcal{R}u = \bigcap I_i^{(k_i)}$, where $k_i > 0$ (since u is

---

([174]) In $\mathcal{R}$ and hence in $\mathcal{O}$ (see n° 1b).

([175]) In the Zariski theorem, the condition that the ring $\mathcal{R}_a(V)$ is integrally closed can be replaced by the condition that it is integrally closed in its ring of fractions (see the lemma in A. 11.3).

not a zero divisor; see A. 11.2). By lemma 3, we have $\mathcal{O}u = \bigcap \mathcal{O}I_i^{(k_i)}$ and, by lemma 1, we get irreducible intersections $\mathcal{O}I_i = J_{i1} \cap \ldots \cap J_{ir_i}$ of prime ideals $J_{i\nu}$ of the ring $\mathcal{O}$. Therefore lemma 5 implies that the localizations $\mathcal{O}_{J_{i\nu}}$ are discrete valuation rings and $\mathcal{O}I_i^{(k_i)} = J_{i1}^{(k_i)} \cap \ldots \cap J_{ir_i}^{(k_i)}$. Consequently, $\mathcal{O}u = \bigcap J_{i\nu}^{(k_i)}$, and hence all ideals associated with $\mathcal{O}u$ are among the ideals $J_{i\nu}$. In conclusion, the ring $\mathcal{O}$ satisfies the assumptions of the Zariski-Samuel lemma and thus it is integrally closed.

PROPOSITION 3. *An irreducible algebraic set $V \subset X$ is a normal analytic space if and only if the ring $R(V)$ is integrally closed.*

Indeed, according to the Zariski theorem, it is enough to prove that the ring $R(V)$ is integrally closed if and only if all the rings $R_a(V) \approx R(V)_{\mathcal{I}_V(a)}$, $a \in V$, are integrally closed (see n° 1b). It is obvious that the former condition implies the latter (see A. 11.1). On the other hand, since the ring $R(V)$ is an integral domain, the equality

$$R(V) = \bigcap_{a \in V} R(V)_{\mathcal{I}_V(a)}$$

will yield the converse implication (se A. 8.1). In order to verify the above equality (note that the inclusion $\subset$ is trivial), take an element φ of the right hand side. For each $a \in V$, we have $\varphi = f^a/g^a$, where $f^a, g^a \in R(V)$, $g^a \neq 0$, hence there exists a regular function $h : V \longrightarrow \mathbf{C}$ such that $g^a h = f^a$ for each $a \in V$ [176]. But $h \in R(V)$ by proposition 2 from n° 2, hence $h \in R(V)$, and thus $\varphi = f^a/g^a$.

COROLLARY . *An algebraic set $V \subset X$ is a normal space if and only if its simple components V_i are disjoint and their rings $R(V_i)$ are integrally closed* [177] .

Indeed, if V is a normal space, then all of its germs V_a, $a \in V$, are simple (see VI. 3.1), and hence the simple components V_i must be disjoint (see IV. 3.1, the corollary from proposition 3b; and IV. 2.8, proposition 4).

5. Let $V \subset X$ be an algebraic set. Let $R'(V)$ be the integral closure of the ring $R(V)$ in its ring of fractions. We will prove that:

The set $R'(V)$ coincides with the set of fractions f/g ($g \neq 0$ [178]) which are locally bounded, i.e., such that the function $\{g \not\equiv 0\} \ni$

---

[176] Obtained by gluing together the functions $\{g^a \neq 0\} \ni z \longrightarrow f^a(z)/g^a(z)$.

[177] The integral closedness of the ring $R(V)$ is a sufficient condition for normality, but it is not necessary, as V can be a disconnected normal space. See also proposition 3a below.

[178] The denominators of those fractions are $\not\equiv 0$, because they are not zero divisors (see 12.1).

$z \longrightarrow f(z)/g(z)$ *is bounded on each bounded subset of the set $\{g \neq 0\}$ (this condition is independent of the pair f, g).*

The ring $R'(V)$ has a universal denominator over $R(V)$.

The ring $R'(V)$ is finite over $R(V)$.

Indeed, the locally bounded fractions constitute a ring, which we will denote by $R^\sim$. Now, the universal denominator $u \in R(V)$ from the proposition in 9.4 is a universal denominator for the ring $R^\sim$ over $R(V)$. To show this, let $f/g \in R^\sim$, $f, g \in R(V)$, $g \not\equiv 0$. For each $a \in V$, we have $g_a \eta^a = f_a$ with some $\eta^a \in \check{\mathcal{O}}_a$ (see VI. 2.3). Also $v^a = u_a \eta^a \in \mathcal{O}_a$, because the germ u_a is a universal denominator (see n° 1b and VI. 2.3). Therefore $g_a v^a = f_a u_a$ for $a \in V$, and since g_a are not zero divisors (see IV. 4.3), we have $v^a \in \mathcal{R}_a(V)$ (because of the corollary of proposition 1 in n° 1b). The last equalities imply that the v^a are the germs of a single regular function $v \in R(V)$ (see proposition 2, n° 2) $(^{179})$. We have $gv = fu$, and so $u(f/g) = v \in R(V)$. Therefore u is a universal denominator of $R^\sim$ over $R(V)$ $(^{180})$. It follows that $R^\sim$ is finite over $R(V)$: we have the monomorphism $R^\sim \ni \varphi \longrightarrow u\varphi \in R(V)$ of modules over $R(V)$ whose range is a finitely generated ideal $(^{181})$. Thus (see A. 8.1), the ring $R^\sim$ is integral over $R(V)$, and hence $R^\sim \subset R'(V)$. We also have the opposite inclusion, since it follows from the definition of integral elements $(^{182})$. Therefore $R = R'(V)$.

A similar argument shows that the integral closure $\mathcal{R}'$ of the ring $\mathcal{R} = \mathcal{R}_a(V)$ in its ring of fractions is equal to the set of fractions φ_a/ψ_a such that the function $z \longrightarrow \varphi(z)/\psi(z)$ is bounded on $\{\psi \neq 0\}$ in a neighbourhood of the point a.

PROPOSITION 3a. *An algebraic set $V \subset X$ is a normal analytic space if and only if the ring $R(V)$ is integrally closed in its ring of fractions, i.e., if $R'(V) \subset R(V)$.*

Indeed, let $V_1, \ldots, V_r$ be the simple components of V. If they are disjoint, then

$$(*) \qquad\qquad R(V) \approx R(V_1) \times \ldots \times R(V_r) \quad (^{183}).$$

Thus, if V is a normal space, then the corollary of proposition 3 in n° 4 implies that the ring $R(V)$ is integrally closed in its ring of fractions (see

---

$(^{179})$ Namely, v is obtained by gluing together representatives $\tilde{v}^a$ of the germs v^a which are holomorphic on neighbourhoods U_a, respectively, and such that $g\tilde{v}^a = fu$ in U_a.

$(^{180})$ Since u is obviously not a zero divisor (see 12.1).

$(^{181})$ Since the ring $R(V)$ is noetherian.

$(^{182})$ Owing to properties of the roots of a polynomial; see B. 5.3. See also the proof of Cartan's proposition from VI. 2.3.

$(^{183})$ Since in this case, in view of proposition 2 from n° 2, we have the isomorphism $R(V_1) \times \ldots \times R(V_r) \ni (f_1, \ldots, f_r) \longrightarrow f_1 \cup \ldots \cup f_r \in R(V)$.

A. 8.1). Conversely, if the latter condition is satisfied, i.e., if $R'(V) \subset R(V)$, then the components V_i must be disjoint. For if they were not, then we would have, e.g., $V_1 \cap W \neq \emptyset$, where $W = \bigcup_{i>1} V_i$. By taking $\varphi \in \mathcal{I}(V_1) \backslash \bigcup_{i>1} \mathcal{I}(V_i)$ and $\psi \in \bigcap_{i>1} \mathcal{I}(V_i) \backslash \mathcal{I}(V_1)$ (see A. 1.11 and 1.2-3), we would have $\varphi_{V_1} = 0$, $\varphi_W \not\equiv 0$, $\psi_W = 0$, $\psi_{V_1} \not\equiv 0$, and hence $(\varphi + \psi)_V \not\equiv 0$. Then the fraction $\varphi_V/(\varphi + \psi)_V$ would belong to $R'(V) \backslash R(V)$, owing to the fact that the function $z \longrightarrow \varphi(z)/(\varphi(z) + \psi(z))$ vanishes on a dense subset of V_1 and is equal to 1 on a dense subset of W ([184]) . Therefore, in view of $(*)$, the rings $R(V_i)$ are integrally closed (see A. 8.1), and so the corollary of proposition 3 in n° 4 implies that V is a normal space.

One can show in a similar way that if $\mathcal{R}' \subset \mathcal{R}_a(V)$, then only one simple component of V contains the point a, which means (see n° 1b) that the ring $\mathcal{R}_a(V)$ is an integral domain. Consequently, according to the Zariski theorem, *the point a is a normal point of the space V if and only if the ring $\mathcal{R}_a$ is integrally closed in its ring of fractions.*

PROPOSITION 4. *An algebraic set $V \subset X$, regarded as an analytic space, always has a regular normalization $\pi : W \longrightarrow V$, where W is an algebraic subset of a vector space. If $\Phi_1/\Psi, \ldots, \Phi_r/\Psi$ are generators of $R'(V)$ regarded as a module over $R(V)$, then the algebraic set*

$$W = \overline{\{(z, w) : \ \Psi(z) \neq 0, \ \Phi_i(z) = \Psi(z) w_i, \ i = 1, \ldots, r\}} \subset V \times \mathbf{C}^r$$

is a normal space and the natural projection $\pi : W \longrightarrow V$ is a normalization.

PROOF . The mapping π is proper (see B. 5.2) and its fibres are finite (see IV. 5, proposition 2). The set $\Omega_0 = \{\Psi \neq 0\}$ is open and dense in V, whereas the set $W_0 = \{\Psi(z) \neq 0, \ \Phi_i(z) = \Psi(z) w_i, \ i = 1, \ldots, r\}$ is open and dense in W. The restriction $\pi_{W_0} : W_0 \longrightarrow \Omega_0$ is a biholomorphic mapping with the inverse

$$\chi : \ \Omega_0 \ni z \longrightarrow \left(z, \Phi_1(z)/\Psi(z), \ldots, \Phi_r(z)/\Psi(z)\right) \in W_0 .$$

The mapping

$$V^0 \cap \Omega_0 \ni z \longrightarrow \left(\Phi_1/\Psi(z), \ldots, \Phi_r(z)/\Psi(z)\right) \in \mathbf{C}^r ,$$

whose graph H_0 is dense in W_0, extends to a holomorphic mapping on V^0 (see the proposition in II. 3.5) whose graph H is dense in W. Moreover, $\pi^{-1}(V^0) = H$ ([185]) , and thus $\pi^{V^0} : \ H \longrightarrow V^0$ is biholomorphic . Therefore π is a

---

([184]) The restriction of this function to $V \cap \{\varphi + \psi \neq 0\}$ cannot be the restriction of a continuous function on V, because for $a \in V_1 \cap W$ we have $(V_1)_a \not\subset W_a$ and $W_a \not\subset (V_1)_a$ (see IV. 2.8, proposition 4).

([185]) For H is the closure of H_0 in $\pi^{-1}(V^0)$, that is, $H = \bar{H}_0 \cap \pi^{-1}(V^0) = \pi^{-1}(V^0)$.

modification in V^* (see V. 4.13), and it suffices to prove that W is a normal space (see VI. 4.1) or, in view of proposition 3a, that $R'(W) \subset R(W)$. Take an arbitrary element from $R'(W)$. It is of the form F_W/G_W, where $F, G \in \mathcal{P}(X \times \mathbf{C}^r)$, $G_W \not\equiv 0$. Also the function $W_1 \ni \zeta \longrightarrow F(\zeta)/G(\zeta) \in \mathbf{C}$, where $W_1 = \{\zeta \in W_0 : G(\zeta) \neq 0\}$ is a dense subset of W_0, is bounded on bounded subsets of the set W_1. Then the function $\Omega_1 \ni z \longrightarrow F(\chi(z))/G(\chi(z)) \in \mathbf{C}$, where $\Omega_1 = \pi(W_1)$ is dense in Ω_0, is bounded on bounded subsets of the set Ω_1. We have $F(\chi(z))/G(\chi(z)) = P(z)/Q(z)$ in Ω_1, where $P, Q \in R(V)$, $Q \neq 0$ in Ω_1 [186]. It follows that $P/Q \in R'(V)$, and hence $P/Q = \sum_1^r H_i \Phi_i/\Psi$, where $H_i \in R(V)$. This implies

$$F(\chi(z))/G(\chi(z)) = \sum_1^r H_i(z)\Phi_i(z)/\Psi(z) = H(\chi(z)) \quad \text{in} \quad \Omega_1 \,,$$

where $H \in R(V \times \mathbf{C}^r)$ is defined by $H(z, w) = \sum_1^r H_i(z)w_r$. Hence $F = GH$ in the set W_1, and the equality persists in the set W as well. This gives $F_W = G_W H_W$, and so $F_W/G_W = H_W \in R(W)$. Consequently, $R'(W) \subset R(W)$, which completes the proof.

§17. Algebraic spaces

We are going to define (in n° 2) the notion of an algebraic space. It has been introduced by Weil [43a] and Serre [38a] under the name of an "abstract variety" or "variété algébrique" [187]. The definition given here is an adaptation of Serre's definition. It can be obtained from the definition of an analytic space (see V. 4.1) by replacing the category of holomorphic mappings of locally analytic sets by the category of regular mappings of quasi-algebraic sets, and adding the requirement that atlases are finite. (Or, according to Serre's algebraic graph theorem, by addition of the constructibility condition for the compositions $\varphi_\kappa \circ \varphi_\iota^{-1}$ of charts of a finite analytic atlas.) The elementary properties of algebraic spaces given below can be checked easily and, in general, in a fashion similar to the case of analytic spaces.

---

[186] It is sufficient to multiply both the numerator and the denominator by a suitable power of Ψ.

[187] See also J. Dieudonne [14], I.R. Šafarevič [41], D. Mumford [29a], and R. Hartshorne [23] under the names "variété algébrique" or "(abstract) variety". In [43a], [29a], and [23] the irreducibility of the space is assumed (see n° 5 below).

1. Let V be a quasi-algebraic subset of an affine space (see 8.2). By *algebraic sets* in V we mean traces on V of algebraic sets. They are precisely the closed constructible subsets of V. (Indeed, if E is such a set, then $E = E \cap \bar{V}$.) Their differences coincide with quasi-algebraic subsets of the set V; they are exactly the locally closed $(^{188})$ constructible subsets of the set V.

The complements of algebraic subsets of the set V, i.e., the constructible open subsets of the set V, are said to be *Z–open* sets in V. (Clearly, they are quasi-algebraic.)

Such sets form a topology (see 1.3) called the *Zariski topology* on V $(^{189})$. In this topology, closed sets are exactly algebraic subsets of V, while locally closed sets are precisely quasi-algebraic sets. If $Z \subset V$ is a quasi-algebraic set, then the Zariski topology on V induces the Zariski topology on Z. On the other hand, the Cartesian product of Zariski topologies, in general, does not coincide with the Zariski topology of the Cartesian product $(^{190})$.

According to Serre's theorem, the regular mappings of quasi-algebraic sets (see 16.4) are precisely the holomorphic mappings with constructible graphs. The following properties are obvious (see 8.3, the corollary from Chevalley's theorem):

The inverse image of a quasi-algebraic set under a regular mapping is quasi-algebraic. The restriction of a regular mapping to a quasi-algebraic set is regular. The mapping obtained by gluing together a finite number of regular mappings defined on Z–open sets is regular: if V, W are quasi-algebraic sets and $G_1, \ldots, G_s \subset V$ are Z–open subsets, then $f : G_1 \cup \ldots \cup G_s \longrightarrow W$ is regular if and only if all the f_{G_i} are regular. The composition of regular mappings is regular: if $f : V \longrightarrow W$ and $g : W_0 \longrightarrow Z$, where $W_0 \subset W$, are regular mappings of quasi-algebraic sets, then so is the composition $g \circ f : f^{-1}(W_0) \longrightarrow Z$. The Cartesian products, as well as the diagonal products of regular mappings, are regular mappings. If V and W are quasi-algebraic sets, then the projections $V \times W \longrightarrow V$ and $V \times W \longrightarrow W$ are regular.

By a *(regular) isomorphism of quasi-algebraic sets* we mean any regular bijection the inverse of which is also regular. By Serre's theorem, such mappings are precisely the biholomorphic mappings with constructible graphs. The inverse of an isomorphism is an isomorphism and the composition of isomorphisms is an isomorphism. If $f : V \longrightarrow W$ is an isomorphism of quasi-algebraic sets and $Z \subset V$ is a quasi-algebraic subset, then $f(Z)$ is quasi-algebraic and the restriction $f_Z : Z \longrightarrow f(Z)$ is an isomorphism.

LEMMA . *Every quasi-algebraic subset V of an affine space X is a finite*

$(^{188})$ In the whole space or, equivalently, in the set V (see B. 1).

$(^{189})$ This is not a Hausdorff topology, but it is a T_1–topology.

$(^{190})$ Consider the example $\mathbf{C} \times \mathbf{C}$.

union of Z-open subsets (of V) that are isomorphic to algebraic subsets of the space $X \times \mathbf{C}$.

Indeed, $V = Z \setminus V(g_1, \ldots, g_k) = \bigcup(Z \setminus V(g_i))$, where $Z \subset X$ is algebraic and $g_i \in \mathcal{P}(X)$. Then we have the isomorphisms

$$Z \setminus V(g_i) \ni z \longrightarrow (z, 1/g_i(z)) \in (Z \times \mathbf{C}) \cap \{(z,t) : t g_i(z) = 1\} .$$

2. A *(complex) algebraic space* is a Hausdorff topological space X endowed with an *algebraic atlas*. The latter is a finite family of homeomorphisms $\varphi : G_i \longrightarrow V_i$, where $\{G_i\}$ is a finite open cover of the space X while V_i are quasi-algebraic subsets of affine spaces such that the compositions

$$\varphi_j \circ \varphi_i^{-1} : \ \varphi_i(G_i \cap G_j) \longrightarrow \varphi_j(G_i \cap G_j)$$

are regular mappings of quasi-algebraic sets [191]. Then such mappings are isomorphisms or, equivalently, they are biholomorphic and have constructible graphs. Two algebraic atlases are said to be *equivalent* if their union is an algebraic atlas. (It follows from n° 1 that this is an equivalence relation.) It is assumed that equivalent atlases define on X the same structure of an algebraic space. Each atlas which is equivalent to the atlas $\{\varphi_i\}$ is called an *algebraic atlas of the algebraic space X* [192].

The condition that X is a Hausdorff space is equivalent to the requirement that the graphs of the mappings $\varphi_j \circ \varphi_i^{-1}$ are closed in $V_i \times V_j$, respectively, in the usual topology, or – equivalently – in the Zariski topology (i.e., that those graphs are algebraic in $V_i \times V_j$, respectively).

Any algebraic space is an analytic space: the structure of an algebraic space induces in a natural way a structure of an analytic space, because any algebraic atlas is an analytic atlas (and any two equivalent algebraic atlases are also equivalent as analytic atlases). (See V. 4.1.)

Every quasi-algebraic subset of an affine space is an algebraic space (with the atlas formed by the identity mapping).

A *chart on an algebraic space* X is a homeomorphism φ of an open subset onto a quasi-algebraic subset of an affine space such that $\{\varphi_i\} \cup \varphi$ is an atlas. Any such chart is biholomorphic. The composition of a chart with an isomorphism of quasi-algebraic sets is also a chart. Algebraic atlases of an algebraic space X are exactly families of charts whose domains cover X.

---

[191] Then each of the sets $\varphi_i(G_i \cap G_j)$ is Z-open in V_i.

[192] See Serre [38a], 31–32 and 34. Compare with the definition of an analytic space in V. 4.1.

In view of the lemma from n° 1, every algebraic space has an atlas $\varphi_i : G_i \longrightarrow V_i$ for which the V_i are algebraic subsets of vector spaces ([193]).

3. Let X be an algebraic space. A subset of X is said to be an *algebraic, quasi-algebraic, Z–open, or constructible set* if its image under any chart $\varphi : G \longrightarrow V$ is an algebraic, quasi-algebraic, Z–open, or constructible set, respectively. It suffices if this is true for each chart from a given atlas. Obviously, algebraic sets are analytic, and quasi-algebraic sets are locally analytic. The constructible sets form the algebra of sets generated by algebraic sets. The closure of any constructible set is algebraic. The algebraic sets are precisely the closed constructible sets. The union of a finite family and the intersection of any family of algebraic sets is algebraic (see 1.3). The quasi-algebraic sets are precisely the locally closed constructible sets (see n° 1) and, equivalently, the differences of algebraic sets (see B. 1). The Z–open sets, i.e., the complements of algebraic sets, are exactly the open constructible sets.

They constitute a topology called the Zariski topology on X.

The chart domains are Z–open (see n° 2, footnote ([191])).

Every cover of X by Z–open sets has a finite subcover (see 1.3).

This means that any algebraic space with the Zariski topology is quasi-compact.

A quasi-algebraic set $Z \subset X$ has the structure of an algebraic space defined (in a unique way) by the atlas consisting of the restrictions of the charts from any atlas of the space X. Then Z endowed with such a structure is called an *algebraic subspace* of the algebraic space X. A set $V \subset Z$ is constructible (or quasi-algebraic) in Z if and only if it is constructible (or quasi-algebraic) in X. If Z is algebraic (or Z–open), then a set $V \subset Z$ is algebraic (or Z–open) in Z if and only if it is algebraic (or Z–open) in X.

In particular, any Z–open set G is a subspace whose charts are exactly the charts of the space X the domains of which are contained in G. The quasi-algebraic sets of the space X are precisely the algebraic subsets of the Z–open subspaces of X. If $\{G_\iota\}$ is a Z–open cover of the space X, then a set V is algebraic or quasi-algebraic, Z–open or constructible if and only if each of the sets $V \cap G_\iota$ is algebraic or quasi-algebraic, Z–open or constructible in the subspace G_ι, respectively.

If a set is algebraic, quasi-algebraic, Z–open, or constructible, so are its connected components. There can be only a finite number of them (since the trace of a component is the union of the components of the trace; see 8.5).

([193]) See Dieudonné [14], vol.2, §2.2–3 and 6, and Mumford [29a], I. 4–6.

If a set V is algebraic (quasi-algebraic), then so is the set V^* ([194]) (see 6.3).

4. In the Cartesian product of two algebraic spaces X and Y with atlases $\{\varphi_i\}$ and $\{\psi_j\}$, respectively, the structure of an algebraic space is well-defined by the atlas $\{\varphi_i \times \psi_j\}$. The Cartesian product of algebraic, Z–open, quasi-algebraic, or constructible sets is algebraic, Z–open, quasi-algebraic, or constructible, respectively. If $Z \subset X, W \subset Y$ are algebraic subspaces, then the Cartesian product structure on $Z \times W \subset X \times Y$ coincides with the structure of the subspace.

5. A non-empty algebraic subset of an algebraic space is said to be *irreducible* or *simple* if it is not the union of two of its proper algebraic subsets. One shows (as in 11.1) that any algebraic subset, regarded as an analytic subset, has finitely many simple components and each of them is algebraic. It follows that:

Irreducibility of an algebraic set is equivalent to irreducibility of the same set regarded as an analytic set. The decomposition of an algebraic set into simple components always exists and coincides with the decomposition into simple components of this set regarded as an analytic set.

For any quasi-algebraic set, irreducibility and the decomposition into simple components are uniquely determined. This is done by regarding the set as an algebraic subset of a Z–open subspace (see IV. 2.8 and 9).

An algebraic space X is irreducible if and only if X^0 is connected (see V. 4.5, proposition α). If G is a non-empty Z–open subset of X, then

$$(X \text{ is irreducible}) \Longleftrightarrow (G \text{ is irreducible and dense}) \,.$$

Indeed, if X is irreducible, then $Z = X \setminus G$ is nowhere dense, and so $G^0 = X^0 \setminus Z$ is connected (see V. 4.5, proposition β; and II. 3.6), whereas if G is irreducible and dense, then $G^0 \subset X^0 \subset \bar{G}^0$, and hence X^0 is connected.

Therefore, if $\varphi : G \longrightarrow V$ is a (non-empty) chart, then the space X is irreducible if and only if G is dense and V is irreducible (see n° 3).

Thus (see n° 2 and n° 8 below) every irreducible algebraic space is a gluing of irreducible algebraic subsets $V_1, \ldots, V_s$ of linear spaces by bijections such that: their graphs $f_{ij} \subset V_i \times V_j$ are algebraic, each of them is an isomorphism of a Z–open dense subset of V_i onto a Z–open dense subset of V_j,

---

([194]) By the *regular (singular) points* of a quasi-algebraic set V we mean the regular (singular) points of the locally analytic set V. See 1.7 and the corollary in 15.3.

and the compatibility conditions: $f_{ii} = \mathrm{id}_{V_i}$, $f_{ij}^{-1} = f_{ij}$, and $f_{jk} \circ f_{ij} \subset f_{ik}$ are satisfied ([195]).

6. Let X and Y be algebraic spaces. A mapping $f : X \longrightarrow Y$ is said to be *regular* if the composition $\psi \circ f \circ \varphi^{-1}$ is regular ([196]) for each chart φ of X and each chart ψ of Y. It is enough that the condition is satisfied by each pair of charts φ, ψ from a given pair of atlases of the spaces X and Y. In the case when $Y = \mathbf{C}$, we say that f is a *regular function*. We also have (see n° 1, V. 4.6; and n° 4):

SERRE'S THEOREM . *A mapping $f : X \longrightarrow Y$ is regular if and only if it is holomorphic and its graph is constructible (in $X \times Y$). In this case the graph is algebraic.*

The restriction of a regular mapping to an (algebraic) subspace is regular. If $Y' \subset Y$ is a subspace, then the regularity of a mapping $f : X \longrightarrow Y'$ is equivalent to the regularity of the mapping $f : X \longrightarrow Y$. If $\{G_\iota\}$ is a Z–open cover of the space X, then the regularity of a mapping $f : X \longrightarrow Y$ is equivalent to the regularity of all the restrictions f_{G_ι}. The composition, the Cartesian product, and the diagonal product of regular mappings is a regular mapping. The projections $X \times Y \longrightarrow X$, $X \times Y \longrightarrow Y$ are regular.

The inverse image under a regular mapping of an algebraic, Z–open, quasi-algebraic, or constructible set is an algebraic, Z–open, quasi-algebraic, or constructible set, respectively. The image of a constructible set is constructible. If, in addition, the mapping is closed, then the image of any algebraic set is algebraic. The image of an irreducible quasi-algebraic set is irreducible, provided that it is quasi-algebraic.

A bijection $f : X \longrightarrow Y$ is said to be an *isomorphism* if both f and f^{-1} are regular. In view of Serre's theorem, the isomorphisms of algebraic spaces are precisely the biholomorphic mappings whose graphs are constructible. The inverse of an isomorphism is an isomorphism. The composition of isomorphisms is an isomorphism. The Cartesian product of isomorphisms is an isomorphism. If $f : X \longrightarrow Y$ is an isomorphism and $Z \subset X$ is a subspace, then $f(Z) \subset Y$ is a subspace and $f_Z : Z \longrightarrow f(Z)$ is an isomorphism. The isomorphisms preserve algebraicity, Z–openness, quasi-algebraicity, constructibility, and irreducibility. They also preserve the regular and singular points of quasi-algebraic subsets. The charts of X are exactly the isomorphisms of its Z–open subsets onto quasi-algebraic subsets of affine spaces. If $f : X \longrightarrow Y$ is a regular mapping, then the natural projection $f \longrightarrow X$ is an isomorphism.

([195]) See Weil [43a], VII. 3.

([196]) Its domain *is assumed to be Z-open* in the range of φ.

Two algebraic space structures on a set T coincide if and only if the identity mapping on T is an isomorphism between the space T endowed with the first structure and the same space with the second structure.

A mapping $f : X \longrightarrow Y$ is called a *(regular) embedding* if its range is a quasi-algebraic subset and the mapping is an isomorphism onto its range. The regular embeddings are precisely the analytic embeddings (see V. 4.7) of algebraic spaces that are regular ([197]).

7. An algebraic space is called an *algebraic manifold* if (as an analytic space) it is a complex manifold (see V. 4.2) or, equivalently, if it is smooth ([198]) and of constant dimension (see V. 4.5). If $\{\varphi_i : G_i \longrightarrow V_i\}$ is an atlas, then the space is an algebraic manifold if and only if the V_i are submanifolds of the same dimension (see V. 4.2). An algebraic submanifold of an algebraic space is just a smooth quasi-algebraic subset ([199]) of constant dimension.

8. Let X_i be a finite family of algebraic spaces. As in the case of analytic spaces (see V. 4.7a), an algebraic space X together with a family of isomorphisms $f : X_i \longrightarrow X$ onto Z–open subsets jointly covering X is said to be a *gluing* of the family $\{X_i\}$. Each composition $f_{ij} = f_j^{-1} \circ f_i$ is an isomorphism of a Z–open subset of X_i onto a Z–open subset of X_j. We say then that the space X, or, more precisely, the pair $X, \{f_i : X_i \longrightarrow X\}$, is a *gluing of the algebraic spaces X_i by the isomorphisms* f_{ij} ([200]). Then the family f_{ij} satisfies the following conditions

$$(\#) \qquad \begin{array}{l} f_{ii} \text{ is the identity on } X_i, \ f_{ij} = f_{ij}^{-1}, \ f_{jk} \circ f_{ij} \subset f_{ik}, \\ \text{the graph of } f_{ij} \text{ is closed in } X_i \times X_j \text{ (for all } i,j,k). \end{array}$$

Conversely, suppose we are given a family $\{f_{ij}\}$ such that f_{ij} is an isomorphism of a Z–open subset of the space X_i onto a Z–open subset of the space X_j and the conditions ($\#$) are satisfied. Then there exists a gluing of the spaces X_i by the isomorphisms f_{ij} (and it is unique up to an isomorphism ([201])).

---

[197] Because of the Serre theorem.

[198] i.e., smooth when regarded as an analytic space.

[199] i.e., smooth when regarded as a locally analytic set.

[200] Obviously, this pair is a gluing by f_{ij} of X_i regarded as analytic spaces, in accordance with the definition from V. 4.7a.

[201] See C. 3.10a, footnote ([41]).

If $\{f_{ij}\}_{i\neq j}$ is a family that satisfies the last three conditions (#), then by adding the identity mappings f_{ii} on X_i to the family, we get one that satisfies all the conditions (#). Therefore there is a gluing X and we also say that X is a gluing of the spaces X_i by the isomorphisms f_{ij}, $i \neq j$.

Clearly, when the X_i are algebraic manifolds, then their gluing X is also an algebraic manifold.

9. If $f : Y \longrightarrow X$ and $g : Z \longrightarrow X$ are regular mappings, with the property (m), between algebraic spaces and $f \approx g$, then the biholomorphic mapping φ from the diagram (1) in V. 4.11 must be an isomorphism. Indeed, $\varphi = \overline{(g^T)^{-1} \circ f^T}$, where $T = \{z : \#f^{-1}(z) = 1\} = \{z : \#g^{-1}(z) = 1\}$ is a constructible set (see the lemma in 8.3).

Now, let (α) be a sublocal and rigid property of holomorphic mappings of analytic spaces which is stronger than the property (m) (see V. 4.11) [202]. The proof of the following proposition on gluing is the same as that of the proposition in V. 4.12.

PROPOSITION . Let $X = \bigcup G_i$ be a finite cover by Z–open sets of the algebraic space X. Let $f_i : Y_i \longrightarrow G_i$ be regular mapping of algebraic spaces satisfying the property (α) and such that $f_i^{G_i \cap G_j} \approx f_j^{G_i \cap G_j}$ for each pair i,j. Then there is a unique, up to an isomorphism, regular mapping $f : Y \longrightarrow X$ of algebraic spaces such that $f^{G_i} \approx f_i$ for each i. It has the property (α).

Any regular modification of algebraic spaces $f : Y \longrightarrow X$ is always a modification of the space X in its algebraic subset. Indeed (see V. 4.13), the set $X^* \cup \bar{T}$ is algebraic, where $T = \{z; \#f^{-1}(z) > 1\}$ is constructible (see the lemma in 8.3).

10. Let X be an algebraic space. It follows from the Zariski analytic normality theorem (see 16.4) that $a \in X$ is a normal point of X if and only if the ring $\mathcal{R}_a \subset \mathcal{O}_a$ of germs of regular functions [203] is integrally closed.

Since the property of being a normalization is sublocal, rigid, and stronger than the property (m) (see VI. 4.1 and V. 4.13), we have the following theorem.

Every algebraic space X has a unique, up to an isomorphism, regular normalization $f : Y \longrightarrow X$ [204] , *where Y is a normal algebraic space.*

---

[202] It is enough to require that (α) is a sublocal and rigid property of regular mappings of algebraic spaces. Local and rigid properties are defined as in V. 4.11 (admitting only Z–open sets and finite covers).

[203] In (Z–open) neighbourhoods of the point a.

[204] See the uniqueness theorem from VI. 4.1 (where ι is an isomorphism of algebraic

In fact, take a finite cover $X = \bigcup G_i$ by Z–open sets which are isomorphic to algebraic subsets of vector spaces (see n° 2). By proposition 4 from 16.5, there are regular normalizations $f_i : Y_i \longrightarrow G_i$ (where the Y_i are normal algebraic spaces). Then $f_i^{G_i \cap G_j} \approx f_j^{G_i \cap G_j}$ (see VI. 4.1, the uniqueness theorem), and so by the proposition on gluing from n° 9, the required normalization exists.

11. Let X be an algebraic space.

Let $J_z \subset \mathcal{R}_z$, $z \in X$, be a family of ideals. We say that the regular functions $f_1, \ldots, f_r$ on X are *generators* of this family if for each $z \in X$ the germs $(f_i)_z$ generate the ideal J_z (cf. VI. 1.2). Then the f_i are also generators of the family of the ideals $\mathcal{O}_z J_z \subset \mathcal{O}_z$, $z \in X$. We say that the family $\{J_z\}$ is *coherent* if each point of the space X has a Z–open neighbourhood G such that the family J_z, $z \in G$, has a finite system of generators (cf. VI. 1.4). Then the family of the ideals $\mathcal{O}_z J_z \subset \mathcal{O}$, $z \in X$, is also coherent.

If A is a germ at $a \in X$ of an algebraic subset of X, we denote by $\mathcal{J}(A)$ the ideal of the ring $\mathcal{R}_a$ consisting of the germs that vanish on A (cf. II. 4.2). Serre's lemma (see 15.3) says that

$$\mathcal{I}(A) = \mathcal{O}_a \mathcal{J}(A) \ .$$

We have an analogue of Cartan's coherence theorem (see VI. 1.4):

If $S \subset X$ is an algebraic subset, then the family of the ideals $\mathcal{J}(S_z)$, $z \in X$, is coherent. We have $\mathcal{I}(S_z) = \mathcal{O}_z \mathcal{J}(S_z)$ for $z \in X$.

In fact, in the case when X is an algebraic subset of a vector space [205] , generators of the ideal $\mathcal{I}(S)$ of the ring $R(X)$ are generators of the family $\mathcal{J}(S_z)$, $z \in X$.

Indeed, let $\Phi_1, \ldots, \Phi_r \in R(X)$ be generators of the ideal $\mathcal{I}(S)$. Let $z \in X$. Then $S = S' \cup S''$, where S' and S'' are the unions of the simple components of S which contain or do not contain z, respectively. Thus $z \notin S''$, and there is an $H \in R(X)$ such that $H = 0$ on S'' and $H(z) \neq 0$. Let $f \in \mathcal{J}(S_z)$. We have $f = (F/G)_z$, where $F, G \in R(X)$, $G(z) \neq 0$, and $F = 0$ on S'. So $FH = 0$, and hence $FH \in \sum R(X) \Phi_i$. Therefore (as $H(z) \neq 0$) we conclude that $f \in \sum \mathcal{R}_z (\Phi_i)_z$.

12. Consider the projective space $\mathbf{P}(X)$, where X is a vector space. As

---

spaces; see n° 9).

[205] Recall that, in the general case, X is a finite union of Z–open sets that are isomorphic to algebraic subsets of vector spaces (see n° 2).

we already know (see 2.1), the homomorphisms

$$\alpha_{H_1}^{-1}, \ldots, \alpha_{H_s}^{-1}, \text{ where } H_i \subset X \setminus 0 \text{ are affine hyperplanes,}$$

constitute an atlas of the manifold $\mathbf{P}(X)$, provided that their domains cover $\mathbf{P}(X)$ (i.e., when $\bigcap(H_i)_* = 0$). They also constitute an algebraic atlas, since the compositions $\alpha_{H_i}^{-1} \circ \alpha_{H_j}$ are regular mappings of quasi-algebraic sets (see 2.1). All such atlases are mutually equivalent. Therefore they define the same *natural structure of an algebraic manifold* on $\mathbf{P}(X)$, which of course induces the natural structure of a complex manifold on $\mathbf{P}(X)$ (it is the only such structure; see n° 14, corollary 3 from Chow's theorem below).

The algebraic subsets of the algebraic manifold $\mathbf{P}(X)$ coincide with those defined in 6.1, and hence its constructible subsets coincide with those defined in 8.1. Indeed, by Chow's theorem, the former must be algebraic as in the definition from 5.1, whereas if $V \subset \mathbf{P}(X)$ is algebraic in the sense of that definition, then for each hyperplane $H \subset X \setminus 0$ the set $\alpha_H^{-1}(V) = V^{\sim} \cap H$ is algebraic in H.

The mapping $\alpha = \alpha^X$ (see 2.1) is regular, because so are the compositions $\alpha_H^{-1} \circ \alpha : X \setminus H_* \ni z \longrightarrow z/\lambda_H(z) \in H$.

The set $\Gamma \subset \mathbf{P}(X)$ is algebraic, Z–open, quasi-algebraic, or constructible if and only if the same holds for the set $\Gamma^{\sim} \setminus 0 \subset X \setminus 0$ [206]. If $\Gamma \subset \mathbf{P}(X)$ is a quasi-algebraic set and T is an algebraic space, then a mapping $f : \Gamma \longrightarrow T$ is regular if and only if the composition $f \circ \alpha : \Gamma^{\sim} \setminus 0 \longrightarrow T$ is regular [207].

If $T \subset X$ is a linear subspace, then the structure of an algebraic space induced on the projective subspace $\mathbf{P}(T) \subset \mathbf{P}(X)$ coincides with its natural structure [208].

Any isomorphism of projective spaces is also an isomorphism when the spaces are regarded as algebraic spaces [209].

If $X_1, \ldots, X_k$ are vector spaces, then the Segre embedding $s : \mathbf{P}(X_1) \times \ldots \times \mathbf{P}(X_k) \longrightarrow \mathbf{P}(X_1 \otimes \ldots \otimes X_k)$ (see 2.4) is a regular embedding. For its graph is constructible, being equal to the range of the regular mapping $(\alpha^{X_1} \times \ldots \times \alpha^{X_k}, \alpha^{X_1 \otimes \ldots \otimes X_k} \circ \tau)$; see 2.4.

---

[206] Because of the equalities $\alpha^{-1}(\Gamma) = \Gamma^{\sim} \setminus 0$ and $\alpha(\Gamma^{\sim} \setminus 0) = \Gamma$ from B. 10.6, and the criterion from 2.2.

[207] Indeed, if $f \circ \alpha$ is regular, then the graph of f is constructible, since it is the range of the regular mapping $(\alpha_{\Gamma^{\sim} \setminus 0}, f \circ \alpha)$.

[208] Because the embedding $\mathbf{P}(T) \hookrightarrow \mathbf{P}(X)$ has an algebraic graph and thus is a regular embedding – see 2.2, footnote [7]; and n° 6.

[209] If $\varphi \in L_0(X, Y)$, then $\tilde{\varphi} \circ \alpha^X = \alpha^Y \circ \varphi$ is regular and so is $\tilde{\varphi}$.

Note that algebraic and constructible sets in the Cartesian product $\mathbf{P}(X_1) \times \ldots \times \mathbf{P}(X_k)$ are the same as those defined in 6.4 and 8.1. (See Chow's theorem from 6.4 and corollary 1 from n° 14 below.)

13. Consider the Grassmann manifold $\mathbf{G}_k(X)$, where X is a vector space (see 4.1). The homeomorphisms $(\varphi_{U_i V_i})^{-1}$, $i = 1, \ldots, s$, constitute its atlas, provided that their domains $\Omega(V_j)$ jointly cover $\mathbf{G}_k(X)$ (such a finite sequence of homeomorphisms exists; see B. 6.8). They also constitute an algebraic atlas, because the argument used in 4.1 shows that the graphs of the compositions $(\varphi_{U_i V_i})^{-1} \circ \varphi_{U_j V_j}$ are algebraic (see II. 3.2). All such atlases are mutually equivalent. Hence they define the same *natural structure of an algebraic manifold* on $\mathbf{G}_k(X)$, which, clearly, induces the natural complex manifold structure on $\mathbf{G}_k(X)$ (in fact, it is the only such structure; see n° 14, corollary 3 from Chow's theorem below).

The natural structure of an algebraic manifold on $\mathbf{G}_1(X)$ coincides with the natural structure of an algebraic manifold on $\mathbf{P}(X) = \mathbf{G}_1(X)$, as defined in n° 12. Indeed, their atlases are equivalent because $\varphi_{\lambda H_*} = \alpha_H \circ \gamma_{\lambda H}$, where $\gamma_{\lambda H}$ are affine isomorphisms (see 4.1).

The mapping $\alpha = \alpha_k = \alpha_k^X$ (see 4.2) is regular, because each of the compositions $(\varphi_{UV})^{-1} \circ \alpha$ is regular. This is so since its graph

$$\{(x_1, \ldots, x_k, f) : \ q(x_i) = f(p(x_i))\} \, ,$$

where $p : \ X \longrightarrow U$ and $q : \ X \longrightarrow V$ are projections with respect to the direct sum $X = U + V$, is algebraic.

The mapping from the lemma in 4.2 is regular, because its graph

$$\{x \in B_n(X), \ y_i = f(x_i), \ i = 1, \ldots, n\} \subset X^n \times Y^n \times L(X, Y)$$

is constructible.

As in n° 12, we check the following criteria. A set $\Gamma \subset \mathbf{G}_k(X)$ is an algebraic, quasi-algebraic, Z–open, or constructible set if and only if the set $\alpha^{-1}(\Gamma) \subset B_k(X)$ is, respectively, an algebraic, quasi-algebraic, Z–open, or constructible set. If $\Gamma \subset \mathbf{G}_k(V)$ is a quasi-algebraic subset and T is an algebraic space, then $f : \ \Gamma \longrightarrow T$ is regular if and only if the composition $f \circ \alpha : \ \alpha^{-1}(\Gamma) \longrightarrow T$ is regular.

If T is a vector subspace, then $\mathbf{G}_k(T) \subset \mathbf{G}_k(X)$ is an algebraic submanifold and the structure of an algebraic space induced on $\mathbf{G}_k(T)$ coincides with the natural structure of the latter ([210]) .

([210]) Because the embedding $\mathbf{G}_k(T) \hookrightarrow \mathbf{G}_k(X)$ has an algebraic graph, and so it is a regular embedding; see 4.2 and n° 6.

Any isomorphism of Grassmann spaces is also an isomorphism when the spaces are regarded as algebraic spaces $(^{211})$.

The bijection $\tau : \mathbf{G}_k(X) \ni V \longrightarrow V^\perp \in \mathbf{G}_{n-k}(X^*)$ (see 4.2) is an isomorphism of algebraic spaces. Indeed, its graph is constructible because it is the image of the constructible set $\{x \in B_k(X), \varphi(x_1) = \ldots = \varphi(x_k) = 0\} \subset X^* \times X^k$ under the regular mapping $e \times \alpha$, where e denotes the identity mapping of X^*.

The Schubert cycles $\mathbf{S}_k(U) \subset \mathbf{G}_k(X)$, as well as the set $\mathbf{S}_k^l(X) \subset \mathbf{G}_k(X) \times \mathbf{G}_l(X)$ (see 4.3) are algebraic submanifolds. Because those sets are constructible, the set $\mathbf{S}_k^l(X)$ is constructible. For it is the image under the regular mapping $\alpha_k \times \alpha_l$ of the constructible set $(B_k(X) \times B_l(X)) \cap \{x_i \wedge y_1 \wedge \ldots \wedge y_l = 0, \ i = 1, \ldots, k\}$. Also, $\mathbf{S}_k(U)$ is constructible, being the inverse image of $\mathbf{S}_r^k(X)$ under the regular mapping $\mathbf{G}_k(X) \ni V \longrightarrow (U, V) \in \mathbf{G}_r(X) \times \mathbf{G}_k(X)$, where $r = \dim U$.

The set $A = \{\dim(U \cap V) = r\} \subset \mathbf{G}_k(X) \times \mathbf{G}_l(X)$ from the proposition in 4.3 is an algebraic submanifold, and the mappings $A \ni (U, V) \longrightarrow U \cap V \in \mathbf{G}_r(X)$ and $A \ni (U, V) \longrightarrow U + V \in \mathbf{G}_{k+l-r}(X)$ are regular. In fact (see the proof of that proposition), the set A and the graphs of those mappings are constructible, since they can be expressed by means of constructibility preserving operations applied to the sets $\Delta, \Sigma_r, \Sigma_{r+1}$, the constructibility of which follows from that of the sets $\mathbf{S}_k^l$.

Plücker's embedding $\mathbf{p} : \mathbf{G}_k(X) \longrightarrow \mathbf{P}(\Lambda^k X)$ (see 4.4) is a regular embedding and its range, the embedded Grassmannian $\mathbf{G}_k(X) \subset \mathbf{P}(\Lambda^k X)$, is an algebraic submanifold. Indeed (see n° 6), the mapping $\mathbf{p}$ is regular, because the composition $\mathbf{p} \circ \alpha_k^X = \alpha_1^{\Lambda^k X} \circ \varepsilon$, where $\varepsilon : B_k(X) \ni (z_1, \ldots, z_k) \longrightarrow z_1 \wedge \ldots \wedge z_k \in \Lambda^k X$ (see 4.4), is regular.

Similarly we check that, for the algebraic manifolds $\mathbf{G}_k'(X)$ and $\mathbf{G}_k(\mathbf{P})$ with the structures transferred via the bijections χ, ω defined in 4.5, the mappings β, ν, σ, μ are regular. Also ϑ is an isomorphism, whereas $(\psi_{UV})^{-1}$ are charts of the algebraic manifold $\mathbf{G}_k'(X)$.

14. We have the following

CHOW'S LEMMA . *For every compact irreducible algebraic space X, there exists an irreducible algebraic subset of a projective space and a regular surjection $Y \longrightarrow X$ a restriction of which is an isomorphism of dense Z–open subsets* $(^{212})$.

$(^{211})$ Indeed, if $\varphi \in L_0(X, Y)$, then $\tilde{\varphi} \circ \alpha^X$ is regular (see 4.2) and so is $\tilde{\varphi}$.

$(^{212})$ That is, using the terminology of algebraic geometry, it is a birational isomorphism between X and Y.

Proof ([213]). Let $\varphi_i : G_i \longrightarrow V_i$, $i = 1, \ldots, k$, be an atlas, where the V_i are algebraic subsets of vector spaces (see n° 2); then the G_i are Z-open and dense (see n° 5), and $\bigcup G_i = X$. Take the closures $W_i = \bar{V}_i$ in the projective closures of these vector spaces. The set $G = \bigcap G_i \subset X$ is Z–open and dense. The mapping

$$\varphi = ((\varphi_1)_G, \ldots, (\varphi_k)_G) : G \longrightarrow W, \quad \text{where} \quad W = W_1 \times \ldots \times W_k \,,$$

is regular, and thus the closure $\bar{\varphi}$ of its graph in $X \times W$ is algebraic (see n° 6, 5, and 3). The projections

$$p : X \times W \longrightarrow X \quad \text{and} \quad \pi : X \times W \longrightarrow W$$

are regular (see n° 6) and proper. We have $p(\bar{\varphi}) = \overline{p(\varphi)} = \bar{G} = X$. Therefore $p_{\bar{\varphi}} : \bar{\varphi} \longrightarrow X$ is a regular surjection and $p_\varphi : \varphi \longrightarrow G$ is an isomorphism (see n° 6). Finally, G is irreducible, and so are φ and $\bar{\varphi}$ (see n° 5). Consequently, it is enough to show that $\pi_{\bar{\varphi}} : \bar{\varphi} \longrightarrow Z = \pi(\bar{\varphi})$ is an isomorphism, since $Z \subset W$ is algebraic (see n° 6), and so it is isomorphic to its image Y under the Segre embedding (see n° 12).

Set $W^{(i)} = W_1 \times \ldots \times V_i \times \ldots \times W_k$, $i = 1, \ldots, k$. The mapping $W^{(i)} \ni w \longrightarrow \varphi_i^{-1}(w_i) \in X$, where $w = (w_1, \ldots, w_k)$, is regular (see n° 6). Hence the projection of its graph onto $W^{(i)}$ is an isomorphism (see n° 6). This means that, setting

$$\Gamma_i = \{(x, w) \in X \times W^{(i)} : x = \varphi_i^{-1}(w)\} = \{(x, w) \in G_i \times W : w_i = \varphi_i(x)\} \,,$$

we have the isomorphism

$$\pi_{\Gamma_i} : \Gamma_i \longrightarrow W^{(i)} \,.$$

Now, $\varphi \subset \Gamma_i$ and Γ_i is closed in both $X \times W^{(i)}$ and $G_i \times W$, which gives

$$\bar{\varphi} \cap (X \times W^{(i)}) \subset \Gamma_i \quad \text{and} \quad \bar{\varphi} \subset \bigcup (X \times W^{(i)}) \,,$$

because $\bar{\varphi} \cap (G_i \times W) \subset \Gamma_i \subset X \times W^{(i)}$, $i = 1, \ldots, k$. So, the restrictions of the projection π

$$\bar{\varphi} \cap (X \times W^{(i)}) \longrightarrow Z \cap W^{(i)}, \quad i = 1, \ldots, k,$$

([213]) See [41] p.282.

are isomorphisms (see n° 6). Since the $W^{(i)}$ are open, it follows that the restriction $\bar{\varphi} \longrightarrow Z$ is also an isomorphism (see n° 6).

REMARK . It follows from the proof of Chow's lemma that instead of compactness of the space X it is enough to assume the following condition:

(*) for every algebraic space U, the projection $U \times X \longrightarrow U$ maps algebraic subsets onto algebraic subsets [214] .

As a matter of fact, the property (*) is equivalent to compactness [215] . Indeed, if the space X has the property (*), then, by such a strengthened version of Chow's lemma, there is a regular surjection $V \longrightarrow X$ of a compact set V, and so X is compact.

Chow's lemma implies

CHOW'S THEOREM . *In a compact algebraic space X, every analytic set is algebraic.*

Indeed, one may assume that the space X is irreducible (one can take the simple components instead). According to Chow's lemma, there exists a regular surjection $f : V \longrightarrow X$ of an algebraic subset of a projective space. Now if $Z \subset X$ is an analytic subset, then, by Chow's theorem from 6.1, the set $f^{-1}(Z)$ is algebraic and so is the set $Z = f\big(f^{-1}(Z)\big)$ (see n° 6).

COROLLARY 1. *In a compact algebraic space, the analytically constructible sets are precisely the constructible sets.* (See n° 3; and IV. 8.4, corollary 3 of proposition 7.)

In view of Serre's theorem (from n° 6), we have:

COROLLARY 2. *Any holomorphic mapping between compact algebraic spaces is (algebraically) regular.*

In particular:

COROLLARY 3. *In a compact analytic space X, the analytic structure can be induced by at most one structure of an algebraic space.*

In fact, take two such algebraic space structures. The identity mapping from the space X with the first structure to the space X with the second one is biholomorphic, and hence is an isomorphism. Therefore both structures coincide (see n° 6).

A compact analytic space which admits an algebraic space structure that induces the analytic structure is said to be *algebraic*.

---

[214] i.e., the projection is closed in the Zariski topology. In algebraic geometry, the spaces with the property (*) are said to be complete.

[215] In other words, completeness of a (complex) algebraic space is equivalent to its compactness in the usual topology.

An example of a compact complex manifold which is not algebraic is the two–dimensional torus T with only one one–dimensional subtorus (see V. 7.3; its only one–dimensional irreducible analytic subsets are the translated one–dimensional subtori). If such a torus were algebraic, it would have a chart $\varphi : G \longrightarrow V$, where V would be a two–dimensional algebraic submanifold of a vector space (see n° 2 and 7). Then any given point $a \in G$ could be joined by a one–dimensional irreducible algebraic set with any point b in some neighbourhood of a; namely, by $\overline{\varphi^{-1}(Z)}$, where $Z \subset X$ is a one–dimensional irreducible algebraic subset joining $\varphi(a)$ with $\varphi(b)$. But that is not possible, because only one one–dimensional translated subtorus passes through a.

15. Let X be an algebraic space.

The elementary blowing-up of the space X by means of regular functions $f_1, \dots, f_k$ on X (see 5.1) is clearly a regular mapping, namely, the projection $\pi : Y \longrightarrow X$ of the algebraic subset $Y = \overline{\alpha \circ f}$ of the space $X \times \mathbf{P}_{k-1}$.

By a *blowing-up of the space X by means of a family of ideals* $J_z \subset \mathcal{R}_z$, $z \in X$, we mean a regular mapping $\pi : Y \longrightarrow X$ of algebraic spaces such that each point of the space X has a Z–open neighbourhood G for which π^G is isomorphic to the elementary blowing-up by means of the generators of the family J_z, $z \in G$ (see 5.5) $(^{216})$. Then π is also a blowing-up by means of the family of the ideals $\mathcal{O}_z J_z \subset \mathcal{O}_z$, $z \in X$. By applying the proposition about gluing from n° 9, we obtain $(^{217})$ the theorem:

For every coherent family of ideals $J_z \subset \mathcal{R}_z$, $z \in X$, *there exists a blowing-up, which is unique up to an isomorphism, of the space X by means of this family* $(^{218})$.

If $S \subset X$ is an algebraic subset, then the blowing-up by means of the family of the ideals $\{\mathcal{J}(S_z)\}$ is also the blowing-up by means of the family of the ideals $\{\mathcal{I}(S_z)\}$ $(^{219})$, i.e., it is the blowing-up in S (see 5.6). Therefore:

There exists a regular blowing-up $\pi : Y \longrightarrow X$ *in S (where the blown-up space Y is algebraic).*

Then the proper inverse image of any algebraic subset $V \subset X$ and the exceptional set $\pi^{-1}(S)$ are algebraic in Y (see 5.5–6).

If X is an algebraic manifold and S is its closed algebraic submanifold,

---

$(^{216})$ Then the family $\{J_z\}$ must be coherent.

$(^{217})$ In the same way as we obtained proposition 2 from 5.5, using the quasi-compactness of the space X in the Zariski topology (see n° 3).

$(^{218})$ The mapping φ from diagram (1) in V. 4.11 must be an isomorphism of algebraic spaces, because blowings-up have the property (m) (see 5.5 and n° 9).

$(^{219})$ Since $\mathcal{I}(S_z) = \mathcal{O}_z \mathcal{J}(S_z)$; see n° 11.

then the blowing-up Y is an algebraic manifold. The exceptional set is then an algebraic submanifold, and the proper inverse image of any algebraic submanifold that intersects S quasi-transversally is an algebraic submanifold. (See 5.7, propositions 3 and 4.)

16. We are going to check that in Hironaka's construction from 10.11 the manifolds $\tilde{X}, \tilde{M}, \tilde{N}$ can be made algebraic (and, in addition, $\tilde{X}$ can be made compact: it is enough to take a compact X). Therefore:

There exists a three–dimensional compact algebraic manifold that cannot be embedded in any projective space.

In order to prove this statement, note first that if in lemma 2 from 11.10 the manifold X and its submanifolds M, N are algebraic, then the modification $f : \tilde{X} \longrightarrow X$ can be made algebraic (with an algebraic manifold $\tilde{X}$). Indeed (see the proof of that lemma and n° 15), the blowing-up $X' \longrightarrow X$ can be made regular (with an algebraic manifold X'), the proper inverse image N' of the submanifold N is an algebraic submanifold, and the blowing-up $q : \tilde{X} \longrightarrow X'$ can be made regular (with the algebraic manifold $\tilde{X}$). Now, in the Hironaka construction (see 11.10), take a (compact) algebraic manifold X and its algebraic submanifolds M, N (e.g., $X = \mathbf{P}_3$ and M, N defined in the homogeneous coordinates by $y = z = 0$ and $yt - x^2 + t^2 = z = 0$, respectively). Then the modifications $f : X_1 \longrightarrow X \setminus b$ and $g : X_2 \longrightarrow X \setminus a$ can be made regular (with algebraic manifolds X_1, X_2), and hence the mapping $h : \tilde{X} \longrightarrow X$ obtained by gluing together those mappings is regular and the manifold $\tilde{X}$ is algebraic (see the proposition in n° 9). Then the manifolds $\tilde{M}$ and $\tilde{N}$ are also algebraic.

§18. Biholomorphic mappings of factorial subsets in projective spaces

1. A non-empty algebraic subset Σ of the projective space $\mathbf{P}(X)$ (where X is a vector space) is said to be *factorial* if its cone $\Sigma^\sim \subset X$ is factorial. Then the subset must be irreducible (see 6.2 and 12.6).

Let X and Y be vector spaces of positive dimensions. Let $\Sigma \subset \mathbf{P}(X)$ be a factorial set, and let $S = \Sigma^\sim \subset X$ be its cone. Then $m = \dim S \geq 1$ (see 6.2).

LEMMA . *For each holomorphic mapping $\varphi : \Sigma \longrightarrow \mathbf{P}(Y)$, there is a mapping $F : S \longrightarrow Y$ which is the restriction of a homogeneous polynomial* *mapping (from X to Y), such that*

$$\varphi(\mathbf{C}w) = \mathbf{C}F(w) \quad for \quad w \in (S \setminus 0) \setminus F^{-1}(0), \quad and \quad \dim F^{-1}(0) \le m - 2 .$$

Then $F(S)$ is a dense subset of $\varphi(\Sigma)\tilde{}$, provided that φ is not constant.

PROOF. We may assume that $\mathbf{P}(X)$ and $\mathbf{P}(Y)$ are the projective closures of the vector spaces M and $\mathbf{C}^l$ (i.e., $X = \mathbf{C} \times M$ and $Y = \mathbf{C}^{l+1}$), and

$$(\#) \qquad\qquad \Sigma \not\subset M_\infty \quad and \quad \varphi(\Sigma) \not\subset \mathbf{C}^l_\infty .$$

(Indeed, there exists a hyperplane in $\mathbf{P}(X)$ (resp., $\mathbf{P}(Y)$) that does not contain Σ (resp., $\varphi(\Sigma)$) (see 3.1, and also B. 6.9 and 12). Therefore $\mathbf{P}(X) = \bar{M}$, $\mathbf{P}(Y) = \overline{\mathbf{C}^l} = \mathbf{P}_l$, and we have the identifications (see 3.1):

$$(\#\#) \qquad M \ni z \longrightarrow \mathbf{C}(1,z) \in \bar{M} \quad and \quad \mathbf{C}^l \ni v \longrightarrow \mathbf{C}(1,v) \in \mathbf{P}_l .$$

The sets $\varphi \subset \bar{M} \times \mathbf{P}_l$ and $\varphi^{-1}(\mathbf{C}^l_\infty) \subset \bar{M}$ are analytic (see V. 3.1), and hence they have to be algebraic, by Chow's theorems (see 6.1 and 4). Hence (see 6.3) the set $\Sigma_0 = \Sigma \cap M \setminus \varphi^{-1}(\mathbf{C}^l_\infty)$ is quasi-algebraic in M and dense in Σ (because, in view of $(\#)$, we have $\Sigma \cap M_\infty \subsetneqq \Sigma$ and $\varphi^{-1}(\mathbf{C}^l_\infty) \subsetneqq \Sigma$; see IV. 2.8, proposition 3). The set $\varphi \cap (M \times \mathbf{C}^l) = \varphi_{\Sigma_0}$ is algebraic in $M \times \mathbf{C}^l$. By Zariski's constructible graph theorem (theorem 4 in 14.8), there is a dense subset Σ' of the set Σ_0 such that

$$\varphi(\zeta) = \big(Q_1(\zeta)/R(\zeta), \ldots, Q_l(\zeta)/R(\zeta)\big) \quad and \quad R(\zeta) \ne 0 \quad for \quad \zeta \in \Sigma' ,$$

where $R, Q_1, \ldots, Q_l \in \mathcal{P}(M)$. Take $F_i = (P_i)_S \in R(S)$, $i = 0, \ldots, l$, where $P \in \mathcal{P}(\mathbf{C} \times M)$ are forms of the same degree such that $P_0(1,z) = R(z)$ and $P_i(1,z) = Q_i(z)$ for $i > 0$ (see A. 3.2) $(^{220})$. In view of the identifications $(\#\#)$, we have

$$(\#\#\#) \qquad \varphi\big(\mathbf{C}(t,z)\big) = \mathbf{C}\big(F_0(t,z), \ldots, F_l(t,z)\big) \quad and \quad F_0(t,z) \ne 0$$
$$for \quad (t,z) \in S' \setminus 0 ,$$

where $S' = (\Sigma')\tilde{}$ is a dense cone in S (see B. 6.10). (Indeed, if $(t,z) \in S' \setminus 0$, then $\mathbf{C}(t,z) \in \Sigma' \subset \Sigma \setminus M_\infty$, and so $t \ne 0$ and $z/t \in \Sigma'$.) Dividing the elements $F_i \in R(S)$ by their greatest common divisor, we get the identity $(\#\#\#)$ in $S' \setminus 0$, with the elements $F_i \in R(S)$ being relatively prime and

---

$(^{220})$ In order to obtain the same degree, we multiply them by suitable powers of t.

homogeneous of the same degree (see 12.2 and A. 1.22). Therefore these elements are restrictions of some forms (on X) of the same degree (see 12.2). The mapping $F = (F_0, \ldots, F_l) :\ S \longrightarrow \mathbf{C}^{l+1}$ is then the restriction of a homogeneous polynomial mapping and the equality $\varphi\big(\mathbf{C}(t, z)\big) = \mathbf{C}F(t, z)$ holds in the set $(S\backslash 0)\backslash F^{-1}(0)$, since the set $S'\backslash 0$ is dense in $S\backslash 0$. Moreover, $\dim F^{-1}(0) \leq m - 2$ (see 12.7).

Finally, suppose that the mapping φ is not constant. Since the set $S'' = (S\backslash 0)\backslash F^{-1}(0)$ is dense in both S and $S\backslash 0$ [221] (and the mappings α^X, φ, F are continuous), it follows that: the set $\{\mathbf{C}w :\ w \in S''\}$ is dense in Σ, and the set $\{\mathbf{C}F(w) :\ w \in S''\}^{\sim} = F(S'')\cup 0$ [222] is dense both in $\varphi(\Sigma)^{\sim}$ (see B. 6.10) and in $F(S)$. Therefore $\varphi(\Sigma)^{\sim} = \overline{F(S)}$.

2. Let X and Y be vector spaces of the same dimension.

THEOREM . *If $\Sigma \subset \mathbf{P}(X)$ and $\Theta \subset \mathbf{P}(Y)$ are factorial sets, then each biholomorphic mapping of Σ onto Θ is the restriction of an isomorphism of the space $\mathbf{P}(X)$ onto the space $\mathbf{P}(Y)$.*

PROOF . Let $\varphi :\ \Sigma \longrightarrow \Theta$ be a biholomorphic mapping. The sets Σ, Θ, and their cones $S = \Sigma^{\sim} \subset X$, $T = \Theta^{\sim} \subset Y$ are irreducible (see 12.6 and 6.2). Omitting the trivial case, we may assume that $\dim \Sigma = \dim \Theta \geq 1$, and then (see 6.2)

$$(1) \qquad\qquad m = \dim S = \dim T \geq 2 \ .$$

We apply the lemma to φ and φ^{-1}. Thus we have mappings $F = P_S$ and $G = Q_S$, where $P :\ X \longrightarrow Y$ and $Q :\ Y \longrightarrow X$ are homogeneous polynomial mappings. Next, in view of (1),

$$(2) \qquad \overline{F(S)} = T, \quad \overline{G(T)} = S, \quad \text{and so} \quad \overline{G\big(F(S)\big)} = S \ ,$$

hence the mappings F and G must be homogeneous of positive degree. Hence $F(0) = 0$ and $G(0) = 0$. Obviously, $R(T) \circ F \subset R(S)$ and $R(S) \circ G \subset R(T)$. Finally,

$$(3) \quad \begin{aligned} &\varphi(\mathbf{C}z) = \mathbf{C}F(z) \text{ for } z \in S \setminus F^{-1}(0), \text{ and } \dim F^{-1}(0) \leq m - 2, \\ &\varphi^{-1}(\mathbf{C}w) = \mathbf{C}G(w) \text{ for } w \in T \setminus G^{-1}(0), \text{ and } \dim G^{-1}(0) \leq m - 2. \end{aligned}$$

---

[221] Since $\dim F^{-1}(0) < \dim S$ (see IV. 2.8, proposition 3).

[222] The equality follows from the fact that the set $S'' \cup 0$, being a cone, coincides with $\bigcup\{\mathbf{C}w :\ w \in S''\}$ (see A. 1.22), and that the mapping F must be homogeneous of positive degree. Hence $\mathbf{C}F(w) = F(\mathbf{C}w)$ for $w \in S$.

Now, we are going to prove that for some non-zero homogeneous elements $p, q \in R(S)$ we have the equality

(4) $p(z)G(F(z)) = q(z)z$ for $z \in S$.

Indeed, it follows from (3) (with $w = F(z)$) that

(5) $\mathbf{C}G(F(z)) = \mathbf{C}z$ for $z \in S \setminus S'$,

where $S' = F^{-1}(G^{-1}(0))$. Let $p = \psi_S$, where $\psi \in X^*$ is a linear form whose kernel $\psi^{-1}(0) \not\supset S$ (such a form exists by (1)). Let $q = p \circ G \circ F$. Thus $p, q \in R(S)$, $p \neq 0$, and so $q \neq 0$ (because of (2)). Now, if $z \in S \setminus S'$, then by (5) we have $G(F(z)) = \alpha z$ for some $\alpha \in \mathbf{C}$, which gives $p(z)G(F(z)) = p(\alpha z)z = p(G(F(z)))z = q(z)z$. Hence the equality (4) holds in $S \setminus S'$. But the set $S' = (G \circ F)^{-1}(0)$ is nowhere dense in S, for otherwise (see IV. 2.8, proposition 3) we would have $G \circ F = 0$, which is impossible in view of (2) and (1). Consequently, the identity (4) holds in S.

Next, we will prove that if irreducible homogeneous elements $f \in R(S)$, $g \in R(T)$ satisfy the condition

(6) $\varphi(V(f)^\tilde{}) = V(g)^\tilde{}$ ([223]),

then

(7) $g \circ F = af^r$ and $f \circ G = bg^s$,   where $a, b \in \mathbf{C} \setminus 0$ and $r, s \in \mathbf{N} \setminus 0$.

Indeed, we have the inclusions

(8) $V(f) \subset V(g \circ F) \cup F^{-1}(0)$ and $V(g \circ F) \subset V(f) \cup F^{-1}(0)$

because, for $z \in S \setminus F^{-1}(0)$, we have in view of (3) and (6) (see B. 6.10),

$$z \in V(f) \Longleftrightarrow \mathbf{C}z \in V(f)^\tilde{} \Longleftrightarrow \mathbf{C}F(z) \in V(g)^\tilde{} \Longleftrightarrow F(z) \in V(g) \Longleftrightarrow$$
$$\Longleftrightarrow z \in V(g \circ F) .$$

Since $f \neq 0$ and $g \circ F \neq 0$, in view of (2), the subsets $V(f)$ and $V(g \circ F)$ are of constant dimension $(m - 1)$ (see 12.4). Hence, because of (3), the set $F^{-1}(0)$ in the inclusions (8) can be deleted, that is, $V(f) = V(g \circ F)$. This

---

[223] Then $f(0) = 0$ and $g(0) = 0$, hence $V(f)$ and $V(g)$ are cones (see A. 1.22).

implies (see 12.8) the first equality in (7). The second one follows in the same way.

Finally, take an element $f \in R(S) \setminus 0$ which is homogeneous of degree 1 and hence irreducible (see 12.2), and which is not a divisor of the element q. (For instance, one can take the restriction $f = \chi_S$ of a linear form $\chi \in X^*$ such that $\chi^{-1}(0) \cap \{q \neq 0\} \neq \emptyset$ and $\chi^{-1}(0) \not\supset S$; such a form exists due to (1).) Next, there exists an irreducible homogeneous element $g \in R(S)$ for which (6) holds. This is so because, according to Chow's theorem (see 6.1), $\varphi(V(f)^\check{})^\check{}$ is an irreducible algebraic cone of dimension $m-1$ (see 6.1–2; 12.7–8; and 11.1, proposition 3). Thus we have the relations (7). Consequently, $f \circ G \circ F = cf^{rs}$, where $c = ba^s \neq 0$, but (4) yields that $q(z)f(z) = f(q(z)z) = p(z)f\big(G(F(z))\big)$ for $z \in S$. Hence $qf = cpf^{rs}$. Therefore we must have $rs = 1$, for otherwise f would be a divisor of q. Therefore we have $q = cp$, and the identity (4) implies that $\xi_s \circ F \circ G = c\xi_s$ for $\xi \in X^*$. This gives $G(F(z)) = cz$ for $z \in S$. So, the mappings F and G must be homogeneous of degree 1 (see A. 1.22), and thus the mappings P and Q must be linear (see A. 3.1). We have $Q(P(z)) = cz$ in the space $X_0 \subset X$ generated by the set S. Therefore the mapping $P_{X_0} : X_0 \longrightarrow Y$ is a monomorphism, and so it can be extended to an isomorphism $\Phi : X \longrightarrow Y$ [224]. In view of (3), we have $\varphi(\lambda) = \Phi(\lambda) = \tilde{\Phi}(\lambda)$ for $\lambda \in \Sigma \setminus F^{-1}(0)^\check{}$ (see B. 6.10). But the set $F^{-1}(0)^\check{}$ is nowhere dense in Σ, because (see IV. 2.8, proposition 3), by (3), $\dim F^{-1}(0)^\check{} < \dim \Sigma$ (see 6.2). Hence $\varphi = \tilde{\Phi}_\Sigma$, which completes the proof.

REMARK . In the special case $(\Sigma = \mathbf{P}(X),\ \theta = \mathbf{P}(Y))$, we obtain the theorem from 13.5: *Every biholomorphic mapping of projective spaces is an isomorphism (of projective spaces).*

From the theorem we can derive (see B. 6.12) the following

COROLLARY . *If two factorial subsets of projective spaces are biholomorphic, then they generate projective subspaces of the same dimension.*

It follows from the corollary that the set $\Sigma \subset \mathbf{P}_2$ given, in homogeneous coordinates, by the equation

$(*)$ $\qquad\qquad\qquad\qquad\qquad tw = z^2$ [225]

is not a factorial set, since it generates $\mathbf{P}_2$ and is biholomorphic to $\mathbf{P}_1$.

Indeed, the cone $\Sigma^\check{} = \{tw = z^2\}$ generates the space $\mathbf{C}^3$ (since it contains the points $(1,0,0), (0,0,1)$, and $(1,1,1)$), hence the set Σ generates the space $\mathbf{P}_2$ (see B.

---

[224] Because $\dim X = \dim Y$.

[225] Note that this set is the closure of the "parabola" $\{w = z^2\} \subset \mathbf{C}^2$ in $\mathbf{P}_2 = \overline{\mathbf{C}^2}$.

6.12). The set Σ is a connected submanifold because its images under the canonical charts $\{w = z^2\}$, $\{tw = 1\}$, $\{t = z^2\}$, $\{= z^2\}$ are connected submanifolds containing the images of the point $\mathbf{C}(1,1,1)$ (see 2.3). The holomorphic mapping $f : \mathbf{P}_1 \longrightarrow \Sigma$, defined in the homogeneous coordinates (see §2) by the mapping $\mathbf{C}^2 \setminus 0 \ni (t,z) \longrightarrow \mathbf{C}(t^2, tz, z^2) \in \Sigma$, is injective ([226]). Hence (see V. 1, corollary 1 of theorem 2) is a biholomorphic mapping onto the open subset $f(\mathbf{P}_1)$ of the manifold Σ. Since the mapping is proper, the set $f(\mathbf{P}_1)$ is also closed in Σ, and consequently, $f(\Sigma) = \mathbf{P}_1$. Therefore Σ is biholomorphic to $\mathbf{P}_1$.

Besides, it follows directly from the equation $(*)$ that the set is not factorial, because the equality $\tau = (t_{\Sigma^-})(w_{\Sigma^-}) = (z_{\Sigma^-})^2$ means that there is no uniqueness of decomposition of τ into irreducible factors.

§19. The Andreotti-Salmon theorem

Let $1 \leq k \leq n$. Let Λ denote the set of k–tuples $(\alpha_1, \ldots, \alpha_k)$ of natural numbers such that $1 \leq \alpha_1 < \ldots < \alpha_k \leq n$. We order the set by means of the relation

$$(\alpha_1, \ldots, \alpha_k) \leq (\beta_1, \ldots, \beta_k) \Longleftrightarrow \alpha_1 \leq \beta_1, \ldots, \alpha_k \leq \beta_k .$$

1. Consider the ring $\mathcal{P} = \mathcal{P}(\mathbf{C}^J)$, where $J = \{(i,j) : i = 1, \ldots, k; \ j = 1, \ldots, n\}$. This is the ring of polynomials in the complex variables

$$
\begin{matrix}
z_{11}, & \cdots, & z_{1n}, \\
\cdots & \cdots & \cdots \\
z_{k1}, & \cdots, & z_{kn}
\end{matrix}
$$

arranged in a $(k \times n)$–matrix. The maximal minors of this matrix (i.e., the minors of order k) are the polynomials:

$$P_\alpha = \det z_{i\alpha_j} = \det(z_{\alpha_1}, \ldots, z_{\alpha_k}), \quad \text{where} \quad \alpha = (\alpha_1, \ldots, \alpha_k) \in \Lambda ,$$

with $z_i = (z_{1i}, \ldots, z_{ki})$ denoting the i-th column of the matrix $(i = 1, \ldots, n)$.

The subring $\mathcal{H}$ of the ring $\mathcal{P}$ generated by the maximal minors P_α is called the *Hodge ring*. We have

$$\mathcal{H} = \{f \circ P : f \in \mathcal{P}(\mathbf{C}^\Lambda)\}, \quad \text{where} \quad P : \mathbf{C}^J \ni z \longrightarrow \{P_\alpha(z)\}_{\alpha \in \Lambda} \in \mathbf{C}^\Lambda .$$

Fix $p \in \mathbf{N}$. The polynomials from $\mathcal{H}$ that can be expressed as forms of degree p in the minors P_α form the vector space

$$\mathcal{H}_p = \{f \circ P : f \in \mathcal{P}(\mathbf{C}^\Lambda) \ \text{is a form of degree } p\} .$$

It is obviously generated by the polynomials

---

[226] Because $(\tilde{t}^2, \tilde{t}\tilde{z}, \tilde{z}^2) = c^2(t^2, tz, z^2) \Longrightarrow (\tilde{t}, \tilde{z}) = \pm c(t,z)$.

$$Q_{\lambda_1,\ldots,\lambda_p} = P_{\lambda_1} \ldots P_{\lambda_p}, \quad (\lambda_1,\ldots,\lambda_p) \in \Lambda^p \quad (^{227}) \, .$$

Moreover, we have

HODGE'S THEOREM . *The polynomials Q_λ, $\lambda \in \Theta_p$, where*

$$\Theta_p = \{(\lambda_1,\ldots,\lambda_p) \in \Lambda^p : \lambda_1 \leq \ldots \leq \lambda_p\} \, ,$$

form a basis of the vector space $\mathcal{H}_p$.

In the proof of Hodge's theorem we need the following

HODGE IDENTITY. *Let $1 \leq s \leq k$. For any triple of sequences $a_1,\ldots$ $\ldots,a_{k-s} \in \mathbf{C}^k$, $b_1,\ldots,b_{k+1} \in \mathbf{C}^k$, and $c_1,\ldots,c_{s-1} \in \mathbf{C}^k$, we have*

$$\sum_\varrho \varepsilon_\varrho \det(a_1,\ldots,a_{k-s},b_{\varrho_1},\ldots,b_{\varrho_s}) \det(b_{\varrho_{s+1}},\ldots,b_{\varrho_{k+1}},c_1,\ldots,c_{s-1}) = 0 \, .$$

Here $\varrho = (\varrho_1,\ldots,\varrho_{k+1})$ are the permutations of the set $\{1,\ldots,k+1\}$ such that $\varrho_1 < \ldots < \varrho_s$ and $\varrho_{s+1} < \ldots, \varrho_{k+1}$.

PROOF . We have the sum $\mathbf{C}^{2k} = \mathbf{C}' + \mathbf{C}''$, where $\mathbf{C}' = \mathbf{C}^k \times 0$, $\mathbf{C}'' = 0 \times \mathbf{C}^k$. We also have the isomorphisms $\mathbf{C}^k \ni x \longrightarrow x' \in \mathbf{C}'$, $\mathbf{C}^k \ni x \longrightarrow x'' \in \mathbf{C}''$, where $x' = (x,0)$ and $x'' = (0,x)$. Let $e_1,\ldots,e_k$ be the canonical basis for $\mathbf{C}^k$. For $x_1,\ldots,x_k \in \mathbf{C}^k$, we have $x_1 \wedge \ldots \wedge x_k = \det(x_1,\ldots,x_k)e_1 \wedge \ldots \wedge e_k$ (see A. 1.20), and hence

$$(*) \qquad \begin{aligned} x_1' \wedge \ldots \wedge x_k' &= \det(x_1,\ldots,x_k)e_1' \wedge \ldots \wedge e_k', \\ x_1'' \wedge \ldots \wedge x_k'' &= \det(x_1,\ldots,x_k)e_1'' \wedge \ldots \wedge e_k'' \, . \end{aligned}$$

Since the elements $b_1,\ldots,b_{k+1}$ are linearly dependent, so are the elements $b_1' + b_1'',\ldots,b_{k+1}' + b_{k+1}''$, and hence it follows that

$$a_1' \wedge \ldots \wedge a_{k-s}' \wedge (b_1' + b_1'') \wedge \ldots \wedge (b_{k+1}' + b_{k+1}'') \wedge c_1'' \wedge \ldots \wedge c_{s-1}'' = 0 \, .$$

Because the exterior product of more than k elements of $\mathbf{C}'$ or $\mathbf{C}''$ vanishes, the left hand side of the last equality is equal (see A. 1.20) to the sum

$$\sum_\varrho \varepsilon_\varrho a_1' \wedge \ldots \wedge a_{k-s}' \wedge b_{\varrho_1}' \wedge \ldots \wedge b_{\varrho_s}' \wedge b_{\varrho_{s+1}}'' \wedge \ldots \wedge b_{\varrho_{k+1}}'' \wedge c_1'' \wedge \ldots \wedge c_{s-1}'' \, .$$

$(^{227})$ If $p = 0$, then $Q_\lambda = 1$ $(\lambda = \emptyset)$.

This means, in view of $(*)$, that it is equal to the left hand side of the Hodge identity multiplied by $e_1' \wedge \ldots \wedge e_k' \wedge e_1'' \wedge \ldots \wedge e_k'' \neq 0$.

Now, let $\alpha = (\alpha_1, \ldots, \alpha_k) \in \Lambda$. Define $\Theta_p(\alpha) = \{(\lambda_1, \ldots, \lambda_p) \in \Theta_p : \lambda_1 \geq \alpha\}$ and take the linear mapping

$$\pi_\alpha : \mathbf{C}^J \ni [z_{ij}] \longrightarrow [z_{ij}^\alpha] \in \mathbf{C}^J, \quad \text{where} \quad z_{ij}^\alpha = \begin{cases} 0, & \text{if } j < \alpha_i, \\ z_{ij}, & \text{if } j \geq \alpha_i. \end{cases} \quad (^{228}).$$

We will need the following

MUSILI LEMMA $(^{229})$. *The polynomials $Q_\lambda \circ \pi_\alpha$, $\lambda \in \Theta_p(\alpha)$, are linearly independent.*

PROOF . The lemma is trivial when $p = 0$. Now, let $p > 0$. Suppose that it is true for $p-1$ and false for p. Then there is a maximal element $\alpha \in \Lambda$ for which the lemma (for p) is false. Consequently, the lemma is true for $\beta > \alpha$. In order to reach a contradiction, it is enough to show that the lemma is true for α. Let

$$\sum_{\Theta_p(\alpha)} c_\lambda Q_\lambda \circ \pi_\alpha = 0 \ .$$

Take any $\beta > \alpha$. Then, obviously, $\pi_\alpha \circ \pi_\beta = \pi_\beta$. Next, $P_\gamma \circ \pi_\beta = 0$ if $\gamma \not\geq \beta$. Indeed, for such a γ, we have $\gamma_s < \beta_s$ for some s. But $z_{i\gamma_j}^\beta = 0$ if $\gamma_j < \beta_i$, and hence also when $j \leq s \leq i$. Thus $P_\gamma \circ \pi_\beta = \det z_{i\gamma_j}^\beta = 0$. Therefore $Q_\lambda \circ \pi_\beta = 0$ when $\lambda \notin \Theta_p(\beta)$, and we obtain $\sum_{\Theta_p(\beta)} c_\lambda Q_\lambda \circ \pi_\eta = 0$. Hence $c_\lambda = 0$ for $\lambda \in \Theta_p(\beta)$.

Therefore $c_{\lambda_1, \ldots, \lambda_p} = 0$ when $\lambda_1 > \alpha$, and so

$$\sum_{\mu \in \Theta_{p-1}(\alpha)} c_{\alpha, \mu} (P_\alpha \circ \pi_\alpha)(Q_\mu \circ \pi_\alpha) = 0 \ .$$

But $P_\alpha \circ \pi_\alpha = z_{1\alpha_1} \ldots z_{k\alpha_k} \neq 0$. This implies that $c_{\alpha, \mu} = 0$ for $\mu \in \Theta_{p-1}(\alpha)$, which completes the proof.

$(^{228})$ That is,

$$\pi_\alpha([z_{ij}]) = \begin{bmatrix} 0 & \cdots & 0 & z_{1\alpha_1} & & \cdots & & z_{1n} \\ \vdots & & & & \ddots & \ddots & & \vdots \\ 0 & & \cdots & & 0 & z_{k\alpha_k} & \cdots & z_{kn} \end{bmatrix}$$

$(^{229})$ See [30a].

PROOF of the Hodge theorem. It follows from Musili's lemma (since $\Theta_p = \Theta_p(\varepsilon)$, where $\varepsilon = (1,\ldots,k)$) that the polynomials Q_λ, $\lambda \in \Theta_p$, are linearly independent. Therefore it is sufficient to prove that they generate the space $\mathcal{H}_p$.

Set $|\alpha| = \alpha_1 + \cdots + \alpha_k$ for $\alpha \in \Lambda$ and $\tilde{\lambda} = (|\lambda_1|,\ldots,|\lambda_p|) \in \mathbf{N}^p$ for $\lambda = (\lambda_1,\ldots,\lambda_p) \in \Lambda^p$. Endow $\mathbf{N}^p$ with the lexicographic order; then $\mathbf{N}^p$ is well-ordered. Now, if $\alpha,\beta \in \Lambda$ and $\alpha \not\leq \beta$, then the product $P_\alpha P_\beta$ is a linear combination of the products $P_{\alpha'} P_{\beta'}$, where $\alpha', \beta' \in \Lambda$ and $|\alpha'| < |\alpha|$. Indeed, $\alpha = (\alpha_1,\ldots,\alpha_k)$, $\beta = (\beta_1,\ldots,\beta_k)$, and $\alpha_s > \beta_s$ for some s. Hence $\beta_1 < \ldots < \beta_s < \alpha_s < \ldots \alpha_k$. Then it is enough to apply Hodge's identity to the triple $(z_{\beta_{s+1}},\ldots,z_{\beta_k})$, $(z_{\beta_1},\ldots,z_{\beta_s},z_{\alpha_s},\ldots,z_{\alpha_k})$, and $(z_{\alpha_1},\ldots,z_{\alpha_{s-1}})$. It follows that

(#) *if $\lambda \in \Lambda^p \setminus \Theta_p$, then Q_λ is a linear combination of the polynomials Q_μ, where $\mu \in \Lambda_p$ and $\tilde{\mu} < \tilde{\lambda}$*

Suppose that the set $\{Q_\lambda : \lambda \in \Theta_p\}$ does not generate the space $\mathcal{H}_p$. Then for any set $\Theta' \subset \Lambda^p$ such that $\{Q_\lambda : \lambda \in \Theta'\}$ generates $\mathcal{H}_p$, we would have $\Theta' \setminus \Theta_p \neq \emptyset$, and so we could define $\sigma(\Theta') = \max\{\tilde{\alpha} : \alpha \in \Theta' \setminus \Theta_p\}$. Now, it follows from the property (#) that for each such Θ' there exists a set $\Theta'' \subset \Lambda^\alpha$ such that $\{Q_\lambda : \lambda \in \Theta''\}$ generates $\mathcal{H}_p$ and $\sigma(\Theta'') < \sigma(\Theta')$. Therefore, starting with $\Theta' = \Lambda^p$, we would obtain a strictly decreasing infinite sequence $\sigma(\Theta') > \sigma(\Theta'') > \ldots$ of elements from the set $\mathbf{N}^p$, which is impossible.

We have the following

HODGE LEMMA. *The minors P_α are prime elements of the ring $\mathcal{H}$. If P_α is a divisor of a polynomial $F \in \mathcal{H}$ in the ring $\mathcal{P}$, then it is also in the ring $\mathcal{H}$.*

PROOF. Clearly, it is enough to consider just one minor P_ε, where $\varepsilon = (1,\ldots,k)$ [230]. Since P_ε is a prime element of the ring $\mathcal{P}$ (see A. 2.3 and A. 6.2), it is enough to prove the second part of the lemma. Moreover, one may assume that $F \in \mathcal{H}_p$, where $p \geq 1$. Indeed, we obviously have $F = \sum_0^s F_p$, where $F_p \in \mathcal{H}_p$, and this is the decomposition into forms (in $\mathcal{P}$; see A. 3.2), because F_p is a form of degrees pk for $p = 0,\ldots,s$. Hence, if P_ε divides F in $\mathcal{P}$, then P_ε also divides each of the polynomials F_p (see A. 2.1), and thus $F_0 = 0$.

---

[230] Indeed, extend a given sequence $\alpha = (\alpha_1,\ldots,\alpha_k) \in \Lambda$ to a permutation $(\alpha_1,\ldots,\alpha_n)$ of the set $\{1,\ldots,n\}$ and take the linear automorphism $\chi : \mathbf{C}^J \ni [z_{ij}] \longrightarrow [z_{i\alpha_j}] \in \mathbf{C}^J$. Then it is easy to check that $\mathcal{H} \ni F \longrightarrow F \circ \chi \in \mathcal{H}$ is a ring automorphism which maps the minor P_ε onto the minor P_α.

The case $k = n$ is trivial, since $\Lambda = \{\varepsilon\}$, and so $F = cP_\varepsilon^p$. Assume thus that $k < n$. By Hodge's theorem, $F = \sum_{\Theta_p} c_\lambda Q_\lambda = \sum_{\Theta_{p-1}} c_{\varepsilon,\mu} P_\varepsilon Q_\mu + \sum_{\Theta_p(\eta)} c_\lambda Q_\lambda$, where $\eta = (1, \ldots, k-1, k+1)$, because η is the smallest element in Λ that is greater than ε. Now, if the polynomial F is divisible by P_ε in the ring $\mathcal{P}$, then, since $P_\varepsilon \circ \pi = 0$, we have $0 = F \circ \pi_\eta = \sum_{\Theta_p(\eta)} c_\lambda Q_\lambda \circ \pi_\eta$. Therefore, according to Musili's lemma, $c_\eta = 0$ for $\lambda \in \Theta_p(\eta)$. Consequently, $F = P_\varepsilon \sum_{\Theta_{p-1}} c_{\varepsilon,\mu} Q_\mu$ and thus the polynomial F is divisible by P_ε in the ring $\mathcal{H}$.

Now we will prove

THE ANDREOTTI-SALMON THEOREM [231] . *The Hodge ring $\mathcal{H}$ is factorial.*

We will need the following lemmas (due to Andreotti and Salmon).

LEMMA 1. *A polynomial $f \in \mathcal{P}$ belongs to $\mathcal{H}_p$ if and only if it satisfies the condition*

$$(**)\qquad\qquad f(AZ) = (\det A)^p f(Z) \quad for \quad A \in \mathbf{C}_k^k \quad [232],$$

where $Z = (z_{ij})_{(i,j) \in J}$.

PROOF . For any matrix $H \in \mathcal{P}^J$ with columns $h_1, \ldots, h_n$, and any $\alpha = (\alpha_1, \ldots, \alpha_k) \in \Lambda$, denote by H_α the square matrix whose columns are $h_{\alpha_1}, \ldots, h_{\alpha_k}$. Thus $P_\alpha = \det Z_\alpha$. Now, each element of $\mathcal{H}_p$ is of the form $F \circ P$, where $F \in \mathcal{P}(\mathbf{C}^\Lambda)$ is a form of degree p. Such an element satisfies the condition $(**)$, because in view of $(AZ)_\alpha = AZ_\alpha$, we have $P(AZ) = (\det A)P(Z)$, and so $F(P(AZ)) = (\det A)^p F(P(Z))$. Conversely, it follows from Cramer's identities (see A. 2.2a) that $P_\varepsilon Z = Z_\varepsilon H$, where H is the $(k \times n)$–matrix whose non-zero elements are (up to a sign) the maximal minors of the matrix Z. Hence $H \in \mathcal{H}^J$. Therefore, if a polynomial $f \in \mathcal{P}$ satisfies the condition $(**)$, then it is a form of degree kp [233] and we have $P_\varepsilon^{kp} f = f(P_\varepsilon Z) = f(Z_\varepsilon H) = P_\varepsilon^p f(H)$. This gives $P_\varepsilon^{(k-1)p} f \in \mathcal{H}$, and hence, by the Hodge lemma, $f \in \mathcal{H}$. So, $f = f_0 + f_1 + \ldots$, where $f_\nu \in \mathcal{H}_\nu$, and this is the decomposition of f into forms (since f_ν is a form of degree $k\nu$). Therefore $f = f_p \in \mathcal{H}_p$.

---

[231] See [7b].

[232] It suffices that the condition $(**)$ is satisfied for invertible matrices A (since they form a dense subset of $\mathbf{C}_k^k$.

[233] We obtain homogeneity by taking $A = tI$, where I is the identity matrix.

For any polynomial mapping $Q:\ \mathbf{C}^m \longrightarrow \mathbf{C}_k^k$, let

$$\mathcal{H}_Q = \bigcup_{p \in \mathbf{N}} \{f \in \mathcal{P}:\ f(AZ) = (\det A)^p f(Z) \quad \text{for} \quad A \in Q(\mathbf{C}^m)\} \ .$$

LEMMA 2. *If the polynomial* $\delta = \det Q \in \mathcal{P}(\mathbf{C}^m)$ *is prime and if the range* $Q(\mathbf{C}^m)$ *contains the identity matrix* I, *then every divisor in* $\mathcal{P}$ *of any non-zero element of* $\mathcal{H}_Q$ *belongs to* $\mathcal{H}_Q$.

PROOF . Let g be a divisor in $\mathcal{P}$ of an element $f \in \mathcal{H}_Q \setminus 0$. One may assume that g is prime in $\mathcal{P}$ (for the set $\mathcal{H}_Q$ is closed under multiplication). Then $f = g^r h$ for some $h \in \mathcal{P}$ which is non-zero and non-divisible by g, where $r \geq 1$. It follows that for some $p \in \mathbf{N}$ we have

$$g\big(Q(t)Z\big)^r h\big(Q(t)Z\big) = \delta(t)^p g^r h \ .$$

Now, the polynomial g is also prime in the ring $\mathcal{P}' = \mathcal{P}(\mathbf{C}^m \times \mathbf{C}^J)$ (see A. 2.3) and cannot be a divisor of the polynomial $h\big(Q(t)Z\big)$ in $\mathcal{P}'$, because for some $c \in \mathbf{C}^m$, we have

$$Q(c) = I \ .$$

Hence g must be a divisor of the polynomial $g\big(Q(t)Z\big)$ in $\mathcal{P}'$. This means that

$(\#\#)$ $$g\big(Q(t)Z\big) = \varphi(t) g(Z)$$

for some $\varphi \in \mathcal{P}(\mathbf{C}^m)$, due to the fact that the degree with respect to Z of the polynomial $g\big(Q(t)Z\big)$ is not greater than that of g. Therefore $\varphi(t)^r h\big(Q(t)Z\big) = \delta(t)^p h(Z)$, and so (by substituting $Z = C$ so that $h(C) \neq 0$) we conclude that φ is a divisor of δ^p. But δ is prime and (owing to $(\#\#)$) $\varphi(c) = \delta(c) = 1$. Hence $\varphi = \delta^s$, where $s \in \mathbf{N}$. Thus, in view of $(\#\#)$, the proof is complete.

COROLLARY 1. *Every divisor in* $\mathcal{P}$ *of any non-zero element of* $\mathcal{H}_p$ *belongs to* $\mathcal{H}$.

Indeed, one can take the identity mapping of $\mathbf{C}_k^k$ as Q, and then, according to lemma 1, we have $\mathcal{H}_Q = \bigcup_{\mathbf{N}} \mathcal{H}_p$.

LEMMA 3. *For each matrix* $C \in \mathbf{C}_k^k$ *whose determinant is 1, there exists a polynomial mapping* $Q:\ \mathbf{C}^m \longrightarrow \mathbf{C}_k^k$ *whose determinant is 1, and whose range contains* I *and* $\mathbf{C}$.

PROOF . Let $Q_\nu(u_1, \ldots, u_{k-1})$ denote the matrix obtained from I by replacing the ν-th column by $(u_1, \ldots, u_{\nu-1}, 1, u_\nu, \ldots, u_{k-1})$ [234] . Then the

---

[234] It is the matrix of a *transvection* of the hyperplane $H = \{z_\nu = 0\}$, i.e., a linear automorphism of the space $\mathbf{C}^k$ whose restriction to any affine hyperplane H' that is parallel to H is a translation of H' (onto itself).

right hand side multiplication of a matrix by $Q_\nu(u_1, \ldots, u_{k-1})$ coincides with the operation of adding to the ν-th column the linear combination of the remaining ones with coefficients $u_1, \ldots, u_{k-1}$. Note that if the first $r - 1$ rows of a matrix A are those of I, where $1 \le r \le k$, then by applying the above operations one can obtain a matrix whose first r rows coincide with those of I. It follows that $CQ_{\nu_q}(-c_q) \ldots Q_{\nu_1}(-c_1) = I$, with some ν_i, c_i. Therefore, since $Q_\nu(t)^{-1} = Q_\nu(-t)$, we get $C = Q_{\nu_1}(c_1) \ldots Q_{\nu_q}(c_q)$ [235]. Hence it is enough to take $Q : \mathbf{C}^{q(k-1)} \ni (t_1, \ldots, t_q) \longrightarrow Q_{\nu_1}(t_1) \ldots Q_{\nu_q}(t_q) \in \mathbf{C}_k^k$.

Lemma 3 implies that the subring

$$\mathcal{H}' = \{f \in \mathcal{P} : \ f(CZ) = f(Z) \text{ when } C \in \mathbf{C}_k^k \text{ and } \det C = 1\}$$

is the intersection of all the sets $\mathcal{H}_Q$, where $Q : \mathbf{C}^m \longrightarrow \mathbf{C}_k^k$ is a polynomial mapping with the determinant 1 whose range contains I. Therefore, in view of lemma 2, we have

COROLLARY 2. *Every divisor in P of a non-zero element of $\mathcal{H}'$ belongs to $\mathcal{H}'$. Hence (see A. 6.1) the ring $\mathcal{H}'$ is factorial.*

PROOF of the Andreotti-Salmon theorem. According to corollary 2, it suffices to show that $\mathcal{H} = \mathcal{H}'$. The inclusion $\mathcal{H} \subset \mathcal{H}'$ follows from lemma 1. Now suppose that $f \in \mathcal{H}'$. Let $f = f_0 + f_1 + \ldots$ be the decomposition into forms. The uniqueness of such a decomposition implies that all f_p must belong to $\mathcal{H}'$. In view of corollary 1, the proof will be complete if we show that $f_p^k \in \mathcal{H}_p$. Now, if $A \in \mathbf{C}_k^k$ is invertible, then $A = aC$, where $a \in \mathbf{C}$, $C \in \mathbf{C}_k^k$, and $\det C = 1$. Then $\det A = a^k$ and we have $f_p(AZ) = a^p f_p(CZ) = a^p f_p(Z)$. Hence $f_p^k(AZ) = (\det A)^p f_p^k(Z)$. Thus $f_p^k \in \mathcal{H}_p$ by lemma 1 (see footnote [233]).

REMARK. We have just shown that the Hodge ring coincides with the set of all polynomials $f \in \mathcal{P}$ satisfying the condition

$$f(CZ) = f(Z) \text{ for each matrix } C \in \mathbf{C}_k^k \text{ with determinant 1.}$$

2. Let M be a vector space, and let $S \subset M$ be an irreducible algebraic cone.

LEMMA. *Suppose that there is a regular mapping Φ of the set S into a vector space L such that the restriction $\Psi : \ S \setminus \Phi^{-1}(\Sigma) \longrightarrow L \setminus \Sigma$ is bijective*

---

[235] We have just shown that every automorphism of a vector space with determinant equal to 1 is a composition of transvections.

for some $\Sigma = V(\lambda)$, *where* $\lambda \in \mathcal{P}(L)$ *is such that* $\lambda \circ \Phi$ *is a prime element of* $R(S)$. *Then the cone* S *is factorial.*

PROOF . The mapping $\mathcal{P}(L) \ni F \longrightarrow F_* \in R(S)$, where $F_* = F \circ \Phi$, is a **C**–monomorphism of rings because the set $\Phi(S) \supset L \setminus \Sigma$ is dense in L. By our hypothesis, λ_* is a prime element of the ring $R(S)$. Therefore λ is irreducible. (For otherwise λ_* would be the product of two non-constant elements of $R(S)$, and thus it would be reducible; see 12.2.) Note that

$$(1) \qquad \begin{cases} \text{for each } f \in R(S) \setminus 0 \text{ there is a polynomial } F \in \mathcal{P}(L) \text{ which is} \\ \text{not divisible by } \lambda \text{ and such that } F_* = f\lambda_*^r \text{ for some } r \in \mathbf{N}. \end{cases}$$

Indeed, according to the proposition from V. 3.4, the restriction Ψ is a homeomorphism, and so the function $f \circ \Psi^{-1} : L \setminus \Sigma \longrightarrow \mathbf{C}$ is continuous, has a constructible graph (see 8.3, the corollary of Chevalley's theorem), and is non-zero (since the set $S \setminus \Phi^{-1}(\Sigma)$ is dense in S; see IV. 2.8, proposition 3). Therefore, by Zariski's constructible graph theorem (theorem 4 in 14.8; see also 11.2, proposition 4 with corollary 4), we have $f \circ \Psi^{-1} = F/\lambda^r$ in $L \setminus \Sigma$ for some $r \in \mathbf{N}$ and some polynomial $F \in \mathcal{P}(L)$ that is not divisible by λ. Hence $F^* = f\lambda_*^r$ (because the set $S \setminus \Phi^{-1}(\Sigma)$ is dense in S).

Since the ring $R(S)$ is a noetherian integral domain (see 12.1), it is enough to prove that each of its irreducible elements is prime (see A. 9.5). Let $p \in R(S)$ be an irreducible element. Omitting the trivial case, one may assume that p is not divisible by λ_*. Suppose that p is a divisor of the product of elements $f, g \in R(S) \setminus 0$ ([236]) . For some $h \in R(S) \setminus 0$, we have

$$(2) \qquad\qquad ph = fg .$$

It follows from (1) that there are $P, H, F, G \in \mathcal{P}(L)$ and $r, s \in \mathbf{N}$ such that P is not divisible by λ and

$$(3) \qquad P_* = p\lambda_*^r, \quad H_* = h\lambda_*^{r+s}, \quad F_* = f\lambda_*^{r+s}, \quad G_* = g\lambda_*^{r+s} .$$

Now the polynomial P is irreducible. Indeed, if it were not, then (since P is non-constant) it would be equal to the product of non-invertible polynomials $P', P'' \in \mathcal{P}(L)$ that are not divisible by λ. Then $p\lambda_*^r = P'_* P''_*$, which would imply (since λ_* is prime) that $P'_* = p'\lambda_*^{r'}$, $P''_* = p''\lambda_*^{r''}$, and $p = p'p''$, where $p', p'' \in R(S)$. Moreover, the elements p', p'' would be non-invertible (for otherwise, see 12.2, one of them, e.g., p' would be equal to a constant $c \neq 0$, and then $P' = c\lambda^{r'}$). This would contradict the irreducibility of p.

([236]) We omit the trivial case $fg = 0$.

The identities (2) and (3) yield that $P_* H_* \lambda_*^s = F_* G_*$, hence $PH\lambda^s = FG$, and so, e.g., F is divisible by P, i.e., $PQ = F$ for some $Q \in \mathcal{P}(L)$. In view of (3), we have $pQ_* = f\lambda_*^s$, and therefore Q_* is divisible by λ_*^s (because p is not divisible by λ_*). Thus p is a divisor of f.

3. Let X be an n–dimensional vector space. Consider the embedded Grassmannian

$$G_k(X) \subset \mathbf{P}(\Lambda^k X)$$

(where $1 \le k \le n$) and its cone (i.e., the Grassmann cone)

$$\mathcal{G} = \mathcal{G}_k(X) = G_k(X)^{\tilde{}} = \{z_1 \wedge \ldots \wedge z_k : z_1, \ldots, z_k \in X\} \subset \Lambda^k X$$

(see 4.4).

Let $e_1, \ldots, e_n$ be a basis of the space X. Then $\{e_\alpha\}_{\alpha \in \Lambda}$, where $e_\alpha = e_{\alpha_1} \wedge \ldots \wedge e_{\alpha_k}$ is a basis of the space $\Lambda^k X$.

Take the polynomial mapping $\chi : \mathbf{C}^J \ni Z \longrightarrow \sum P_\alpha(Z) e_\alpha \in \Lambda^k X$. We have $\mathcal{H} = \{f \circ \chi : f \in \mathcal{P}(\Lambda^k X)\}$ [237] and $\mathcal{G} = \chi(\mathbf{C}^J)$, because

$$\left(\sum_1^n z_{1j} e_j\right) \wedge \ldots \wedge \left(\sum_1^n z_{kj} e_j\right) = \sum_\Lambda P_\alpha(Z) e_\alpha \quad \text{for} \quad Z = [z_{ij}] \in \mathbf{C}^J$$

(see A. 1.20). Now the ring homomorphism

$$\mathcal{P}(\Lambda^k X) \ni f \longrightarrow f \circ \chi \in \mathcal{P},$$

whose range is $\mathcal{H}$ and whose kernel is the ideal $\mathcal{I}(\mathcal{G})$ induces the isomorphism

$$(1) \qquad\qquad \mathcal{P}(\Lambda^k X)/\mathcal{I}(\mathcal{G}) \longrightarrow \mathcal{H} .$$

On the other hand, we have the natural isomorphism

$$(2) \qquad\qquad \mathcal{P}(\Lambda^k X)/\mathcal{I}(\mathcal{G}) \longrightarrow R(\mathcal{G})$$

(see 12.1). As a result, we conclude that:

The ring $R(\mathcal{G})$ of the Grassmann cone is isomorphic to the Hodge ring $\mathcal{H}$.

Consequently, we have the following version of

---

[237] Because $f \circ \chi = \hat{f} \circ P$, where $\hat{f}$ denotes the polynomial f in the coordinate system (in $\Lambda^k X$) corresponding to the basis $\{e_\alpha\}_{\alpha \in \Lambda}$.

THE ANDREOTTI-SALMON THEOREM. *The embedded Grassmannian* $G_k(X) \subset \mathbf{P}(\Lambda^k X)$ *is factorial.*

We are going to give another proof of this theorem. Namely, by using the lemma from n° 2 we will show that the Grassmann cone $\mathcal{G}$ is factorial.

Note first that $\mathcal{G} \subset \Lambda^k X$ is an irreducible algebraic cone. Indeed (see 6.2), the Grassmannian $G_k(X) \subset \mathbf{P}(\Lambda^k X)$ is a connected submanifold (see 4.4 and B. 6.8), and hence it is an irreducible algebraic set (see IV. 2.8, proposition 2; Chow's theorem in 6.1; and proposition 3 in 11.1). Next, in view of Hodge's lemma, the restriction $\mu_{\mathcal{G}}$ of the form

$$\mu: \; \Lambda^k X \ni \sum c_\alpha e_\alpha \longrightarrow c_\varepsilon \in \mathbf{C}, \quad \text{where} \quad \varepsilon = (1,\ldots,k)\,,$$

is a prime element of the ring $R(\mathcal{G})$. This is so because the images under the isomorphisms (1) and (2) of the equivalence class of $\mu \in \mathcal{P}(\Lambda^k X)$ are $\mu \circ \chi = P_\varepsilon$ and $\mu_{\mathcal{G}}$, respectively.

Set $H = \ker \mu$. Then $H = \sum_{\alpha \neq \varepsilon} \mathbf{C} e_\alpha$. We have the equality

$$(***) \quad \mathcal{G} \setminus H = \{a(e_1 + v_1) \wedge \ldots \wedge (e_k + v_k): \; a \in \mathbf{C} \setminus 0, \; v_1, \ldots, v_k \in \sum_{k+1}^n \mathbf{C} e_i\}\,.$$

Indeed, for any element $z \in X$ denote by z', z'' its components $z' \in \sum_1^k \mathbf{C} e_i$ and $z'' \in \sum_{k+1}^n \mathbf{C} e_i$. Now, the inclusion $\supset$ is obvious. Conversely, let $w_1 \wedge \ldots \wedge w_k \notin H$. Define $U = \sum_1^k \mathbf{C} w_i$. We have $w_1 \wedge \ldots \wedge w_k \in w_1' \wedge \ldots \wedge w_k' + H$, and hence $w_1' \wedge \ldots \wedge w_k' \neq 0$, which means that $w_1', \ldots, w_k'$ are linearly independent. It follows that the projection $U \ni z \longrightarrow z' \in \sum_1^k \mathbf{C} e_i$ is an isomorphism. Consequently, the space U has a basis of the form $e_1 + v_1, \ldots, e_k + v_k$, where $v_1, \ldots, v_k \in \sum_{k+1}^n \mathbf{C} e_s$ [238]. Thus $w_1 \wedge \ldots \wedge w_k = a(e - 1 + v_1) \wedge \ldots \wedge (e_k + v_k)$ for some $a \in \mathbf{C} \setminus 0$.

We have the direct sum $\Lambda^k X = L + N$, where L and N are the subspaces generated by $\{e_{\alpha_1,\ldots,\alpha_k} : \; \alpha_{k-1} \leq k\}$ and $\{e_{\alpha_1,\ldots,\alpha_k} : \; \alpha_{k-1} \geq k+1\}$, respectively. Accordingly,

$$(\#\#\#) \quad L \setminus H = \{a e_1 \wedge \ldots \wedge e_k + \Big(\sum_{k+1}^n b_{1j} e_j\Big) \wedge e_2 \wedge \ldots \wedge e_k + \ldots$$

$$\ldots + e_1 \wedge \ldots \wedge e_{k-1} \wedge \Big(\sum_{k+1}^n b_{kj} e_j\Big) : \; a \in \mathbf{C} \setminus 0, \; b_{ij} \in \mathbf{C}\}.$$

---

[238] It corresponds to the basis $e_1, \ldots, e_k$.

Let $\pi : \Lambda^k X \longrightarrow L$ be the natural projection (with kernel N). The proof would be complete if we could apply the lemma from n° 2 to the regular mapping $\pi_{\mathcal{G}} : \mathcal{G} \longrightarrow L$ and the polynomial μ_L. Therefore it is enough to check that the hypotheses of the lemma are satisfied. Now, because of the inclusion $N \subset H$, we have

$$\mu_L \circ \pi_{\mathcal{G}} = \mu_{\mathcal{G}} \quad \text{and} \quad \pi_{\mathcal{G}}^{-1}(H \cap L) = H \cap \mathcal{G} .$$

Hence $\mu_L \circ \pi_{\mathcal{G}} \in R(\mathcal{G})$ is a prime element. Since $V(\mu_L) = H \cap L$, it only remains to check that the mapping $\pi_{\mathcal{G} \setminus H} : \mathcal{G} \setminus H \longrightarrow L \setminus H$ is bijective. But because of (∗∗∗) and (###), this follows from the identity

$$\pi\left(a\left(e_1 + \sum_{k+1}^{n} c_{1j}e_j\right) \wedge \ldots \wedge \left(e_k + \sum_{k+1}^{n} c_{kj}e_j\right)\right) =$$

$$= ae_1 \wedge \ldots \wedge e_k + a\left(\sum_{k+1}^{n} c_{1j}e_j\right) \wedge e_2 \wedge \ldots \wedge e_k + \ldots + e_1 \wedge \ldots \wedge e_{k-1} \wedge a\left(\sum_{k+1}^{n} c_{kj}e_j\right)$$

for $a, c_{ij} \in \mathbf{C}$.

§20. Chow's theorem on biholomorphic mappings of Grassmann manifolds

1. In any Grassmann space $\mathbf{G}_k(X)$, where X is an n–dimensional vector space and $1 \leq k \leq n$, we define the relation of being adjacent (denoted by Υ) by the formula

$$U \Upsilon V \Longleftrightarrow \dim(U \cap V) = k - 1 \Longleftrightarrow \dim(U + V) = k + 1 .$$

Observe that

$$(*) \qquad\qquad U \Upsilon V \Longleftrightarrow U^{\perp} \Upsilon V^{\perp} .$$

Next,

$$(**) \qquad U \Upsilon V \Longleftrightarrow (U \neq V \text{ and } \mathbf{p}(U) + \mathbf{p}(V) \subset \mathcal{G}_k(X)) .$$

In fact, extend a basis $z_1, \ldots, z_{k-r}$ of the subspace $U \cap V$ to a basis $z_1, \ldots$ $\ldots, z_{k-r}, u_1, \ldots, u_r$ of the subspace U and to a basis $z_1, \ldots, z_{k-r}, v_1, \ldots, v_s$ of the subspace V. Then $\mathbf{p}(U) = \mathbf{C}u$ and $\mathbf{p}(V) = \mathbf{C}v$, where $u = z_1 \wedge \ldots \wedge$

$z_{k-r} \wedge u_1 \wedge \ldots \wedge u_r$ and $v = z_1 \wedge \ldots \wedge z_{k-r} \wedge v_1 \wedge \ldots \wedge v_r$. Now the condition $U \curlyvee V$ means exactly that $r = 1$. Thus the implication $\Longrightarrow$ is trivial. Conversely, suppose that the right hand side of $(**)$ holds. Then $r > 0$ and $u + v$ is simple. Therefore $u_1 \wedge \ldots \wedge u_r + v_1 \wedge \ldots \wedge v_r$ is simple and we must have $r = 1$ (see A. 1.21).

Let $\mathcal{Z}$ denote the class of sets $Z \subset \mathbf{G}_k(X)$ satisfying the condition: $(L, N \in \mathcal{Z}, L \neq N) \Longleftrightarrow L \curlyvee N$.

LEMMA 1. *If* $2 \leq k \leq n-2$, *then the only maximal elements (with respect to inclusion) of the class* $\mathcal{Z}$ *are the sets of the form*

$(\#)$ $\mathbf{S}^k(U)$, *where* $U \in \mathbf{G}_{k-1}(X)$,   *and* $\mathbf{G}_k(V)$, *where* $V \in \mathbf{G}_{k+1}(X)$.

PROOF. The sets in the first family in $(\#)$ are maximal. For if $N \in \mathbf{G}_k(X) \setminus \mathbf{S}^k(U)$, then $N \curlyvee L$ does not hold for $L = U + \mathbf{C}v \in \mathbf{S}^k(U)$, where $v \in (X \setminus U) \setminus (\mathbf{C}u + N)$ and $u \in U \setminus N$ (because $\mathbf{C}u + \mathbf{C}v \subset L$ and $(\mathbf{C}u + \mathbf{C}v) \cap N = 0$). This implies, in view of the property $(*)$ and the equality $(\#)$ from B. 6.5, that the sets in the second family in $(\#)$ are also maximal.

Now suppose that a set $Z \in \mathcal{Z}$ is maximal. First note that

$(\#\#)$ $(L_1, L_2, N \in Z, \ L_1 \neq L_2) \Longrightarrow (N \supset L_1 \cap L_2 \quad \text{or} \quad N \subset L_1 + L_2)$.

(Indeed, suppose that $N \not\supset L_1 \cap L_2$. Then $N \neq L_i$, and so $\dim(N \cap L_i) = k - 1$ $(i = 1, 2)$. Next, $N \cap L_1 \neq N \cap L_2$ because, since $N \cap L_1 \cap L_2 \subsetneq L_1 \cap L_2$, we have $\dim(N \cap L_1 \cap L_2) < k - 1$. Hence $N = N \cap L_1 + N \cap L_2 \subset L_1 + L_2$.) Now take $L_1, L_2 \in Z$, $L_1 \neq L_2$ and set $U = L_1 \cap L_2$, $V = L_1 + L_2$. Clearly, it suffices to prove that $Z \subset \mathbf{S}^k(U)$ or $Z \subset \mathbf{G}_k(V)$. Suppose this is not so. Then there would exist $L_3 \in Z \setminus \mathbf{G}_k(V)$. Then $L_3 \neq L_1$ and $L_3 \neq L_2$, but according to $(\#\#)$ one must have $L_3 \supset U$, and thus $L_1 \cap L_3 = L_2 \cap L_3 = U$ (because of equality of the dimensions). Furthermore, there would exist $N \in Z \setminus \mathbf{S}^k(U)$. Therefore, in view of $(\#\#)$, we would have $N \subset \bigcap_{i<j}(L_i + L_j) \subset U$ $(^{239})$. This is impossible, since $\dim N > \dim U$.

2. Let X and Y be n–dimensional vector spaces, and let $1 \leq k \leq n - 1$. We have the following

---

$(^{239})$ Indeed, by taking a complementary line λ_i to U in L_i $(i = 1, 2, 3)$, we get the direct sum $U + \lambda_1 + \lambda_2 + \lambda_3$, (since $L_2 \not\subset L_1$ and $L_3 \not\subset L_1 + L_2$), and so $\bigcap_{i<j}(L_i + L_j) = \bigcap_{i<j}(U + \lambda_i + \lambda_j) = U$.

CHOW THEOREM ([240]) . *Every biholomorphic mapping of Grassmann spaces* $f : \mathbf{G}_k(X) \longrightarrow \mathbf{G}_k(Y)$ *is an isomorphism of Grassmann spaces or, in the case when* $n = 2k$, *it is the composition of an isomorphism of* $\mathbf{G}_k(X)$ *onto* $\mathbf{G}_k(Y^*)$ *with the biholomorphic mapping* $\tau_k^{-1} : \mathbf{G}_k(Y^*) \longrightarrow \mathbf{G}_k(Y)$ ([241]) .

REMARK . If $n = 2k \geq 4$, then the mapping τ_k^Y is not an isomorphism ([242]) . Indeed, if we had $\tau_k = F_{(k)}$, where $F \in L_0(Y, Y^*)$, then $U^{\perp} = F(U)$ for $U \in \mathbf{G}_k(X)$. We would then have $F(u)(v) = 0$ for $u, v \in X$ (by taking $U \ni u, v$). This would mean that $F = 0$, which is impossible.

In proving Chow's theorem we will need the following lemmas.

LEMMA 2. *Biholomorphic mappings of Grassmann manifolds preserve adjacency: if* $f : \mathbf{G}_k(X) \longrightarrow \mathbf{G}_k(Y)$ *is a biholomorphic mapping, then for any* $U, V \in \mathbf{G}_k(X)$ *we have the equivalence*

$$U \Upsilon V \iff f(U) \Upsilon f(V) .$$

PROOF . Obviously, it is enough to show the implication $\Longrightarrow$. We have the biholomorphic mapping of the embedded Grassmannians $\mathbf{p}^Y \circ f \circ (\mathbf{p}^X)^{-1} :$ $G_k(X) \longrightarrow G_k(Y)$. According to the Andreotti-Salmon theorem (see 19.3), $G_k(X)$ and $G_k(Y)$ are factorial subsets of $\mathbf{P}(\Lambda^k X)$ and $\mathbf{P}(\Lambda^k Y)$, respectively. Hence, by the theorem from 18.2, we have $\mathbf{p}^Y \circ f \circ (\mathbf{p}^X)^{-1} \subset \tilde{\Phi}$ for some $\Phi \in L_0(\Lambda^k X, \Lambda^k Y)$. Thus $\tilde{\Phi}(G_k(X)) \subset G_k(Y)$, and so $\Phi(\mathcal{G}_k(X)) \subset \mathcal{G}_k(Y)$ (see B. 6.10). Now, if $U \Upsilon V$, then by (**) we have $U \neq V$ and $\mathbf{p}^X(U) + \mathbf{p}^X(V) \subset \mathcal{G}_k(X)$. Therefore $f(U) \neq f(V)$ and

$$\mathbf{p}^Y(f(U)) + \mathbf{p}^Y(f(V)) = \tilde{\Phi}(\mathbf{p}^X(U)) + \tilde{\Phi}(\mathbf{p}^X(V)) =$$
$$= \Phi(\mathbf{p}^X(U) + \mathbf{p}^X(V)) \subset \mathcal{G}_k(Y) .$$

So, by (**), we have $f(U) \Upsilon f(V)$.

LEMMA 3. *Suppose that* $2 \leq k \leq n - 2$. *If* $f : \mathbf{G}_k(X) \longrightarrow \mathbf{G}_k(Y)$ *is a biholomorphic mapping; then we have the disjunction*

([240]) See [12b].

([241]) It is an isomorphism of algebraic spaces (see 17.13 or 17.14, corollary 2 of Chow's theorem).

([242]) If $n = 2$, then the mapping τ_1 is an isomorphism of projective spaces, by the theorem from 13.5.

(A) *there exists a biholomorphic mapping* $g : \mathbf{G}_{k-1}(X) \longrightarrow \mathbf{G}_{k-1}(Y)$ *such that*

(1) $f\big(\mathbf{S}^k(U)\big) = \mathbf{S}^k\big(g(U)\big)$ *for* $U \in \mathbf{G}_{k-1}(X)$,

or

(B) *there exists a biholomorphic mapping* $h : \mathbf{G}_{k-1}(X) \longrightarrow \mathbf{G}_{k-1}(Y)$ *such that*

(2) $f\big(\mathbf{S}^k(U)\big) = \mathbf{G}_k\big(h(U)\big)$ *for* $U \in \mathbf{G}_{k+1}(X)$.

PROOF . According to lemmas 1 and 2, the image under f of a set of the form (#) is always a set of the form (#). Now, if for some $U_0 \in \mathbf{G}_{k-1}(X)$ we have $f(\mathbf{S}^k(U_0)) = \mathbf{S}^k(U_0')$, where $U_0' \in \mathbf{G}_{k-1}(Y)$, then for each $U \in \mathbf{G}_{k-1}(X)$ we have $f(\mathbf{S}^k(U)) = \mathbf{S}^k(U')$, where $U' \in \mathbf{G}_{k-1}(Y)$. Indeed, this is so when $U \curlyvee U_0$, because then $\mathbf{S}^k(U_0) \cap \mathbf{S}^k(U) = \{U_0 + U\}$. If we had $f(\mathbf{S}^k(U)) = \mathbf{G}_k(V')$ with $V' \in \mathbf{G}_{k+1}(X)$, then the set $f(\mathbf{S}^k(U_0) \cap \mathbf{S}^k(U)) = \mathbf{S}^k(U_0') \cap \mathbf{G}_k(V')$ would be infinite, which is impossible. Therefore it is enough to observe that for any $U \in \mathbf{G}_{k-1}(X)$ which is different from U_0, we have $U_0 \curlyvee U_1 \curlyvee \ldots \curlyvee U_s \curlyvee U$ for some $U_i \in \mathbf{G}_{k-1}(X)$. Clearly, in the remaining case, for each $U \in \mathbf{G}_{k-1}(X)$ we have $f(\mathbf{S}^k(U)) = \mathbf{G}_k(V')$, where $V' \in \mathbf{G}_{k+1}(X)$. Hence there exists a mapping g that satisfies the condition (1) or a mapping h that satisfies the condition (2). As both conditions imply injectivity (of the mapping g or h, respectively), it remains to show that the mapping is holomorphic (see V. 1, corollary 1 of theorem 2 ([243])).

Let $U_0 \in \mathbf{G}_{k-1}(X)$. Take lines $\lambda, \mu \subset X$ such that the sum $U_0 + \lambda + \mu$ is direct. Then (see B. 6.6) for each U from an open neighbourhood Ω of U_0 in $\mathbf{G}_k(X)$, the sum $U + \lambda + \mu$ is direct, hence $U + \lambda \neq U + \mu$ and $f(U + \lambda) \neq f(U + \mu)$. But $U + \lambda, U + \mu \in \mathbf{S}^k(U)$. Hence, in case (A), we have $f(U + \lambda), f(U + \mu) \in \mathbf{S}^k\big(g(U)\big)$, and so

$$g(U) = f(U + \lambda) \cap f(U + \mu) .$$

In case (B), we have $f(U + \lambda), f(U + \mu) \in \mathbf{G}_k\big(h(U)\big)$, and thus

$$h(U) = f(U + \lambda) + f(U + \mu) .$$

([243]) Then the mapping g (or h, respectively) is proper and its range is closed, which implies that the mapping is surjective, since the Grassmannian space is connected.

In view of the proposition from 4.3, the above equalities show that the mapping g (or h, respectively) is holomorphic in Ω.

REMARK . The case (B) can happen only when $n = 2k$. (This is so, because the dimensions of the manifolds $\mathbf{G}_{k-1}(X)$ and $\mathbf{G}_{k+1}(Y)$ are equal, which means that $(k-1)(n-k+1) = (k+1)(n-k-1)$.)

LEMMA 4. *Assume that* $k \geq 2$. *If* $F \in L_0(X,Y)$, $g = F_{(k-1)} : \mathbf{G}_{k-1}(X) \to \mathbf{G}_{k-1}(Y)$ *and* $f : \mathbf{G}_k(X) \longrightarrow \mathbf{G}_k(Y)$ *is a bijection such that* $f(\mathbf{S}^k(U)) = \mathbf{S}^k(g(U))$ *for* $U \in \mathbf{G}_{k-1}(X)$, *then* $f = F_{(k)}$.

Indeed, if $T \in \mathbf{G}_k(X)$, then $T = U + V$ for some $U, V \in \mathbf{G}_{k-1}(X)$, and so $F(T) = F(U) + F(V)$. Therefore $f(\{T\}) = f(\mathbf{S}^k(U) \cap \mathbf{S}^k(V)) = \mathbf{S}^k(F(U)) \cap \mathbf{S}^k(F(V)) = \{F(T)\}$.

PROOF of Chow's theorem. The case $k = 1$ is just the theorem from 13.5 (cf. also the remark in 18.2). Let $2 \leq k < \frac{1}{2}n$, and assume that the theorem is true for $k - 1$. By lemma 3 and the following remark, there exists a biholomorphic mapping $g : \mathbf{G}_{k-1}(X) \longrightarrow \mathbf{G}_{k-1}(Y)$ such that $f(\mathbf{S}^k(U)) = \mathbf{S}^k(g(U))$ for $U \in \mathbf{G}_{k-1}(X)$. Thus $g = F_{(k-1)}$, where $F \in L_0(X,Y)$. Hence, according to lemma 4, we have $f = F_{(k)}$. Therefore the theorem is true if $1 \leq k < \frac{1}{2}n$.

Now, let $\frac{1}{2}n < k \leq n-1$. We have the biholomorphic mapping

$$\tau_k^X \circ f^{-1} \circ (\tau_k^Y)^{-1} : \mathbf{G}_{n-k}(Y^*) \longrightarrow \mathbf{G}_{n-k}(X^*) .$$

Since $1 \leq n-k < \frac{1}{2}n$, the mapping is an isomorphism. Hence it is of the form $F_{(n-k)}^*$, where $F \in L_0(X,Y)$. That is, we have $\tau_k^X \circ f^{-1} \circ (\tau_k^Y)^{-1} = F_{(n-k)}^* = \tau_k^X \circ (F_{(k)})^{-1} \circ (\tau_k^Y)^{-1}$ (see B. 6.9), and hence $f = F_{(k)}$.

Finally, if $n = 2k$, we use lemma 3 again. In the case (A), we obtain $f = F_{(k)}$ with some $F \in L_0(X,Y)$, as in the first part of the proof. In the case (B), there exists a biholomorphic mapping $h : \mathbf{G}_{k-1}(X) \longrightarrow \mathbf{G}_{k+1}(Y)$ such that for $U \in \mathbf{G}_{k-1}(X)$ we have $f(\mathbf{S}^k(U)) = \mathbf{G}_k(h(U))$. This gives

$$(\tau_k^X \circ f)(\mathbf{S}^k(U)) = \mathbf{S}^k((\tau_{k+1}^Y \circ h)(U))$$

(see B. 6.5, the equalities (#)). But since the mapping $\tau_{k+1}^Y \circ h$ maps the space $\mathbf{G}_{k-1}(X)$ biholomorphically onto the space $\mathbf{G}_{k-1}(Y^*)$, it is equal to $F_{(k-1)}$, where $F \in L_0(X,Y^*)$. Consequently, in view of lemma 4, we have $\tau_k^Y \circ f = F_{(k)}$. This means that $f = (\tau_k^Y)^{-1} \circ F_{(k)}$, and the proof is complete.

REFERENCES

[1] BALCERZYK S., JÓZEFIAK T., *Commutative Rings* (Polish), BM vol. 40, Warszawa 1985.

[1a] BIAŁYNICKI-BIRULA A., *Algebra* (Polish), BM vol. 47, Warszawa 1976.

[2] BROWKIN J., *Field Theory* (Polish), BM vol. 49, Warszawa 1977.

[3] LANG S., *Algebra*, Addison-Wesley, Reading Mass. 1965.

[4] ATIYAH M.F., MACDONALD I.G., *Introduction to Commutative Algebra*, Addison-Wesley, Reading Mass. 1969.

[5] LEJA F., *Complex Functions* (Polish), BM vol. 29, Warszawa 1971.

[6] SICIAK J., *Introduction to the Theory of Analytic Functions of Several Variables* (Polish), an appendix in the book [5].

[7] ZARISKI O., Samuel P., *Commutative Algebra*, Van Nostrand, Princeton 1958-1960.

* * *

[7a] ABHYANKAR S.S., *Local Analytic Geometry*, Academic Press, New York 1964.

[7aa] ABHYANKAR S.S., *Concepts of order and rank on a complex space, and a condition for normality*, Math. Ann. 141 (1960), pp. 171-192.

[7b] ANDREOTTI A., SALMON P., *Anelli con unica decomponibilità in fattori primi*, Monatsh. f. Math. 61 (1957), pp. 97-142.

[8] ANDREOTTI A., STOLL W., *Analytic and Algebraic Dependence of Meromorphic Functions*, Lecture Notes in Math. 234, Springer, Heidelberg 1971.

[9] BALCERZYK S., *Introduction to Homological Algebra* (Polish), BM vol. 34, Warszawa 1970.

[10] BIAŁYNICKI-BIRULA A., *Linear Algebra with Geometry* (Polish), BM vol. 48, Warszawa 1976.

[10a] BIEBERBACH L., *Lehrbuch der Funktionentheorie* II, Teubner, Leipzig 1931.

[11] BOCHNER S., MARTIN W.T., *Several Complex Variables*, Princeton Math. Ser., Princeton 1948.

[11a] BOHNHORST G., *Einfache holomorphe Abbildungen*, Math. Annalen 275 (1986), pp. 513-520.

[11b] BOURBAKI N., *Algèbre Commutative*, Hermann, Paris 1961-1965

[12] CARTAN H., *Théorie élémentaire des fonctions analytiques d'une ou plusieurs variables complexes*, Hermann, Paris 1961.

[12a] CARTAN H., *Seminaire E.N.S., École Normale Supérieure*, Paris 1951-1952.

[12b] CHOW W.L., *On the Geometry of Algebraic Homogeneous Spaces*, Annals of Math. 50 (1949), pp. 32-67.

[13] DIEUDONNÉ J., *Eléments d'analyse*, Gauthier-Villars, Paris 1969-1971.

[14] DIEUDONNÉ J., *Cours de géométrie algébrique*, Presses Univ. France 1974.

[15] DIEUDONNÉ J., *Topics in Local Algebra*, Notre Dame Math. Lect. 10, Notre Dame 1967.

[16] ENGELKING R., *General Topology*, PWN, Warszawa 1977.

[16a] FISHER G., *Complex Analytic Geometry*, Lecture Notes in Math. 538, Springer, Heidelberg 1976.

[16b] FORTUNA E., LOJASIEWICZ S., *Sur l'algébricité des ensembles analitiques complexes*, J. Reine Angew. Math. 329 (1981), pp. 215-220.

[17] GRAUERT H., FRITSCHE K., *Several Complex Variables*, Springer, Heidelberg 1976.

[17a] GRAUERT H., REMMERT R., *Coherent Analytic Sheaves*, Springer, Heidelberg 1984.

[17b] GRAUERT H., REMMERT R., *Analytische Stellenalgebren*, Springer, Berlin 1971.

[18] GRIFFITHS P., ADAMS J., *Topics in Algebraic and Analytic Geometry*, Math. Notes Princeton Univ. Press, Princeton 1974.

[19] GRIFFITHS P., HARRIS J., *Principles of Algebraic Geometry*, Wiley Interscience, New York 1978.

[20] GUNNING R.C., *Lectures on Complex Analytic Varieties, The Local Parametrization Theorem*, Math. Notes, Princeton Univ. Press, Princeton 1970.

[21] GUNNING R.C., *Lectures on Complex Analytic Varieties, Finite Analytic Mappings*, Math. Notes, Princeton Univ. Press, Princeton 1974.

[22] GUNNING R.C., ROSSI H., *Analytic Functions of Several Complex Variables*, Prentice-Hall, Englewood Cliffs 1965.

[23] HARTSHORNE R., *Algebraic Geometry*, Springer, New York 1977.

[24] HERVÉ M., *Several Complex Variables*, Local Theory, Oxford Univ. Press, London 1963.

[24a] HODGE W.V.D., PEDOE D., *Methods of Algebraic Geometry*, vol. 2, Cambridge Univ. Press, Cambridge 1952.

[25] HÖRMANDER L., *An Introduction to Complex Analysis in Several Variables*, Van Nostrand, Princeton 1966.

[25a] KAUP L., KAUP B., *Holomorphic Functions of Several Variables*, Walter de Gruyter, Berlin 1983.

[25b] KRAFT H., *Geometrische Methoden in der Invariantentheorie*, Braunschweig 1985.

[26] KURATOWSKI K., *Introduction to Set Theory and Topology*, Pergamon Press, Oxford 1961.

[27] ŁOJASIEWICZ S., *Ensembles semi-analytiques*, I.H.E.S., Bures-sur-Yvette 1965.

[28] MARTINET J., *Singularités des fonctions et applications différentiables*, P.U.C., Rio de Janeiro 1974.

[29] MILNOR J., *Singular Points of Complex Hypersurfaces*, Annals of Math. Studies 61, Princeton Univ. Press, Princeton 1968.

[29a] MUMFORD D., *Introduction to Algebraic Geometry*, preprint, Harvard Univ.

[30] MUMFORD D., *Algebraic Geometry I, Complex Projective Varieties*, Springer, Heidelberg 1976.

[30a] MUSILI C., *Postulation formula for Schubert varieties*, J. of Indian Math. Soc. 36 (1972) pp. 143-171.

[31] NAGATA M., *Local Rings*, Wiley Interscience, New York 1962.

[32] NORTHCOTT D.C., *Ideal Theory*, Cambridge Univ. Press, Cambridge 1953.

[33] NARASIMHAN R., *Introduction to the Theory of Analytic Spaces*, Lecture Notes in Math. 25, Springer, Heidelberg 1966.

[34] RÜCKERT W., *Zum Eliminationsproblem der Potenzenreihenideale*, Math. Ann. 107 (1932), pp. 259-181.

[35] RUDIN W., *A geometric criterion for algebraic varieties*, J. Math. Mech. 17 (1968), pp. 671-683.

[35a] RUSEK K., WINIARSKI T., *Polynomial automorphisms of C^n*, Univ. Iagell. Acta Math. 24 (1984) pp. 143-150.

[36] SADULLAEV A., *Kriterii algebraičnosti analitičeskih množestv*, Sbornik o golomorfnih funkcjah mnogih kompleksnih pieremiennih, Krasnojarsk 1976, pp. 107-122.

[36a] SCHUMACHER G., *Ein topologisches Reduziertheitskriterium für holomorphe Abbildungen*, Math. Annalen 220 (1976) pp. 97-103.

[37] SERRE J.P., *Géométrie algébrique et géométrie analytique*, Ann. Inst. Fourier 6 (1955-56) pp. 1-42.

[38] SERRE J.P., *Faisceaux algébriques cohérents*, Ann. Math. 61 (1955), pp. 197-278.

[39] SPIVAK M., *Calculus on Manifolds*, W.A. Benjamin, Inc., New York 1965.

[40] ŠABAT B.W., *Introduction to Complex Analysis* (Polish), PWN, Warszawa 1974.

[41] SHAFAREVICH I.R., *Basic Algebraic Geometry*, Springer, Heidelberg 1974.

[42] THOM R., *Modèles mathématiques de la morphogenèse*, Union Générale d'Editions, Paris 1974.

[43] TOUGERON J.C., *Idéaux des fonctions différéntiables*, Springer, Heidelberg 1972.

[43a°] TSIKH A.K., *Weakly holomorphic functions on complete intersections, their holomorphic continuation*, Mat. Sb. (N.S) 133 (175) (1987) n° 4, pp. 429-445.

[43a] WEIL A., *Foundations of Algebraic Geometry*, Am. Math. Soc. Coll. Publ. XXIX, 1962.

[44] WHITNEY H., *Complex Analytic Varieties*, Addison-Wesley, Reading, Mass. 1972.

[45] WINIARSKI T., *Inverses of Polynomial Automorphisms of C^n*, Bull. Ac. Pol. Math., 27 (1979) pp. 673-674.

NOTATION INDEX

SUBJECT INDEX

(Prepared by the author)